Core-Shell Nano Constructs for Cancer Theragnostic

Jayvadan K. Patel • Namdev Dhas
Gaurav Kant Saraogi
Editors

Core-Shell Nano Constructs for Cancer Theragnostic

Current Scenario, Challenges and Regulatory Aspects

Springer

Editors
Jayvadan K. Patel
Viesain Pharma LLC, GA, USA

Gaurav Kant Saraogi
Aurobindo Institute of Pharmacy
Indore, Madhya Pradesh, India

Namdev Dhas
Manipal College of Pharmaceutical Sciences
Manipal Academy of Higher Education
Udupi, Karnataka, India

ISBN 978-981-96-3024-0 ISBN 978-981-96-3025-7 (eBook)
https://doi.org/10.1007/978-981-96-3025-7

This Springer imprint is published by the registered company Springer Nature Singapore Pte Ltd.
The registered company address is: 152 Beach Road, #21-01/04 Gateway East, Singapore 189721, Singapore

"Being a family means you are a part of something very wonderful. It means you will love and be loved for the rest of your life". This book is dedicated to my parents Laxmanrao and Parvati, my wife Ritu, my daughter Shivanya, my son Rudra, my grandfather Baburao and grandmother Laxmibai, my uncles Ramrao and Vitthalrao, my beloved sisters Sita, Neeta, and Mukta, and whole Dhas family, without whom this book would not have been possible.

—Dr. Namdev Laxmanrao Dhas

Family is not an important thing. It's everything. This book is dedicated to my parents Kantilal and Kamuben, my wife Sneha, and my son Shubh, as well as my nephews Sai and Yakin, lovely Hely, and Kiva, without whom this book would not have been possible.

—Dr. Jayvadan K. Patel

This book is dedicated to my parents, whose unwavering guidance and love have shaped every step of my journey. To my wife, whose constant support and belief in me provided the strength to see this project through to its completion. To my children, who inspire me daily with their curiosity and joy, and to my friends, whose encouragement and companionship have been invaluable. Without the contributions of each of you, this book would not have been possible.

—Dr. Gaurav Kant Saraogi

Foreword

It is my immense pleasure to write the foreword for this prompt and important book, *Core–Shell Nano Constructs for Cancer Theragnostics: Current Scenario, Challenges, and Regulatory Aspects*. Cancer remains a major global health challenge, and the development of innovative and effective diagnostic and therapeutic strategies is crucial for improving patient outcomes.

The emergence of nanotechnology has revolutionized the field of cancer research, and core–shell nanoconstructs have shown tremendous promise in theragnostic applications. These nanostructures offer unparalleled opportunities for targeted diagnosis, treatment, and monitoring of cancer, enabling personalized medicines and precision therapy.

This book provides a comprehensive and authoritative overview of the latest advances in core–shell nanoconstructs for cancer theragnostic, covering fundamental concepts, design and synthesis strategies, and preclinical applications. The contributors are renowned experts in their fields, and their insights and expertise make this book an invaluable resource for researchers, clinicians, and students.

I congratulate the editors and authors on their outstanding contribution to the field of cancer nanomedicine. My wife is inspired to read about cancer theragnostic, driven by a deep desire to understand the innovative advances in diagnostics and treatment. This book will undoubtedly inspire new research directions, foster collaborations, and accelerate the translation of core–shell nanoconstructs into clinical practice, ultimately improving the lives of cancer patients worldwide.

Captain, Cricket Team, United State of America
Atlanta, USA

Monank Patel

Preface

Cancer remains a leading cause of global mortality, with conventional treatments such as chemotherapy frequently resulting in toxic effects on healthy cells. This challenge has promoted the development of nanomaterials aimed at enhancing the precision of cancer theranostics. Nanoconstructs which integrate nanoparticles with specific ligands are engineered to selectively target cancer cells, thereby addressing issues related to toxicity and nonspecific drug delivery. These advanced materials can encapsulate substantial quantities of therapeutic agents and are increasingly fabricated using biologically derived methods to minimize toxicity and facilitate manufacturing.

Core–shell nanoconstructs represent a sophisticated class of material categorized by distinct inner core and outer shell; the core may consist of metal, semiconductors, or organic materials, while the shell is typically composed of materials that enhance stability through compatibility and functionality. This shall not only protect the core but also improve viability and provide a platform for targeting, making these constructs highly effective in medical applications. Core–shell nanoconstructs are widely utilized in biomedical imaging and drug delivery. These nanoconstructs typically consist of a two-component structure featuring a soft shell of a bimolecular ligand surrounded by a hard nanoparticle core. They are primarily utilized for drug delivery in cancer treatment, addressing challenges such as toxicity, nonspecific distribution, and uncontrolled release rates. These nanoparticles have an advantage due to their distinct characteristics, which amalgamate the benefits of both the core nanoparticle and the shell layer, while addressing their limitations, such as structural instability short circulation, and leakage of cargo materials. This design enhances the efficacy and selectivity of targeted therapies. Nanoparticles can be categorized by their shape and possess unique properties attributable to their small size and high surface-to-volume ratio. Various polymer-based constructs, such as poly(lactic-co-glycolic acid) (PLGA), polycaprolactone (PCL), polylactic acid, poly(vinyl alcohol) (PVA), and chitosan, are employed to improve the drug release profile and bioavailability. Additionally, lipid-based and inorganic constructs, including iron oxide and gold nanoparticles, are explored for their therapeutic potential.

The incorporation of various ligands into these constructs enhances their uptake and selectivity, thereby increasing the therapeutic efficacy of the drug moiety. The choice of ligand is critical as the ligands with high affinity for targeted site can significantly improve safety and efficacy. Factors such as binding affinity and immunogenicity play a pivotal role in determining the success of drug delivery. Stability issues can be addressed through modifications, such as PEGylation, which can improve pharmacokinetics.

Moreover, these nanoconstructs can be tailored through surface chemistry and targeting strategies such as ligand functionalization to enhance specificity against cancer cells while supporting both imaging and therapeutic functions in personalized cancer theranostics. The importance of biosensors in cancer diagnosis continues to grow due to their rapid sensitive detection of biomarkers. Core–shell nanoconstructs, with their unique properties, significantly enhance biosensor performance by improving sensitivity and stability.

The advancement in nanotechnology has propelled the development of nanomedicine, enhancing both efficacy and safety of cancer therapies. However, a key challenge remains the efficient delivery of these nanomedicines to solid tumours. Current strategies rely on passive targeting through the enhanced permeability and retention effect, as well as active targeting via surface modification for receptor-mediated uptake. Despite the promise of active targeting, its effectiveness is often limited by factors within the tumour microenvironment. Tumour microenvironment-responsive nanoconstructs integrate the therapeutic and diagnostic capabilities, responding to specific stimuli present in the tumour microenvironment, such as hypoxia acidity and elevator levels of reactive oxygen species. This responsiveness enhances the precise and effectiveness of cancer theranostics by facilitating targeted drug delivery and real-time monitoring of therapeutic outcomes. Such nanoparticles exploit common features of the tumours, enabling drug release that is triggered by specific tumour microenvironment conditions.

This book, *Core–Shell Nano Constructs for Cancer Theragnostics: Current Scenario, Challenges, and Regulatory Aspects*, brings 21 chapters with an overview of core–shell nanoconstructs, various types, their challenges and regulatory aspects. This book is divided into six sections. The *first section* contains eight chapters which deal with the introduction and basic considerations of core–shell nanoconstructs in cancer theragnostics. This section covers the introduction of core–shell nanoconstructs in cancer theragnostics, their physicochemical properties, surface modification and targeting strategies, tumour microenvironment-responsive nanoconstructs, core–multishell nanoconstructs, their characterization and evaluation techniques, their biomedical applications, and cancer biosensing. The *second section* deals with organic–organic core–shell nanoconstructs in cancer theragnostics, covering polymer–polymer core–shell nanoconstructs, polymer–lipid core–shell nanoconstructs, and lipid–polymer core–shell nanoconstructs. The *third section* contains inorganic core and organic shell nanoconstructs, which involve magnetic–organic core–shell nanoconstructs and non-magnetic–organic core–shell nanoconstructs. The *fourth section* involves organic–inorganic core–shell nanoconstructs, containing organic–magnetic core–shell nanoconstructs and organic–non-magnetic core–shell nanoconstructs. In the *fifth section*, inorganic–inorganic core–shell nanoconstructs are discussed, containing silica–non-silica-based core–shell nanoconstructs, semiconductor–non-semiconductor-based core–shell nanoconstructs, lanthanide-based core–shell nanoconstructs, and upconversion core–shell nanoconstructs, and finally, the *sixth section* deals with the toxicological and regulatory aspects, containing clinical applications and commercialization challenges.

We would like to express our heartfelt gratitude to our chapter contributors for their exceptional work to this rapidly evolving field of core–shell nanoconstructs. The publication of this book would not have been possible without the invaluable support of our chapter authors, and we extend our sincere appreciation to them. We also wish to acknowledge our Springer family for their unwavering support and diligence in ensuring the book's quality. The printing press and their support system have done an outstanding job in making this book both marketable and visually appealing. If you find any shortcomings in this work, we, as editors, take full responsibility and welcome your feedback, as it will help us learn and improve in future publications.

Visnagar, Gujarat, India — Jayvadan K. Patel
Manipal, Udupi, Karnataka, India — Namdev Dhas
Indore, Madhya Pradesh, India — Gaurav Kant Saraogi

Contents

Editors and Contributors

About the Editors

Namdev Dhas is currently working as an Assistant Professor in the Department of Pharmaceutics, Manipal College of Pharmaceutical Sciences, Manipal Academy of Higher Education (MAHE), Manipal, Udupi, Karnataka, India. He received his PhD from the Institute of Pharmacy, Nirma University, Ahmedabad, Gujarat, in 2021. His area of research includes materials science and engineering, multimodal therapeutic platforms, and inorganic nanocomposites for cancer therapy. His main focus is on synthesizing different metal-based nanoplatforms and investigating how their properties affect the efficiency of therapy and diagnosis. He has published 65 publications in the form of several research and review articles in international journals. Additionally, he has edited a book published by Taylor & Francis CRC Press and Springer Nature. As of now, he has achieved a cumulative impact factor of over 250 through the publication of articles in several national and international journals. His scientific articles have been cited more than 1411 times. Recently, he received a research grant from the SERB and the ICMR, Government of India, worth 3 crore rupees. He was recently awarded the "Excellent Researcher in Pharmaceutical Science" award by the Indian Drug Manufacturers' Association (IDMA) and Healthcopeia in July 2022. He has also received the "Young Researcher Award-2020" from the Institute of Scholars, Bangalore, India.

Jayvadan K. Patel is a Vice-President at Viesain Pharma, LLC, GA, USA and a Professor Emeritus, Faculty of Pharmacy, Sankalchand Patel University, India. He has more than 28 years of research, academic, and industry experience and has published more than 275 research and review papers in international and national Journals. He has co-authored 24 books and contributed 135 chapters to books published by well-reputed publishers. His papers have been cited more than 7000 times, with more than 40 papers receiving over 50 citations each. He has an h-index of 42 and i10-index of 162 for his credit. He has guided 106 M. Pharm students and mentored 50 Ph.D. scholars. He has already been decorated with over a dozen national and international awards and prizes. He is the recipient of the prestigious "AICTE-Visvesvaraya Best Teachers Award-2020" from the All India Council for Technical Education, Government of India, and the APTI Young Pharmacy Teacher Award (2014) from the Association of Pharmaceutical Teachers of India. He is a reviewer for more than 75 and an editorial board member of 20 reputable scientific journals. He has completed 12 industry- and government-sponsored research projects.

Gaurav Kant Saraogi earned his Ph.D. in Pharmaceutical Sciences from the Department of Pharmaceutical Sciences, Dr. H. S. Gour Central University, Sagar, India, in 2010. He is currently serving as Professor and Principal at the Sri Aurobindo Institute of Pharmacy, Indore, MP, with over 20 years of experience in research and academia. Dr. Saraogi has an impressive portfolio, with more than 40 publications, over 1500 citations, 8 book chapters, and more than 5 patents to his credit. He is also actively involved in scholarly activities, serving as a reviewer for many high-impact journals and as the Editor-in-Chief of the Scopus-indexed International Journal of Applied Pharmaceutics.

His research primarily focuses on nanotechnology and targeted tubercular therapy, with significant contributions in the fields of dendrimers, transdermal, and nasal delivery systems for the effective delivery of drugs and bio-actives. Dr. Saraogi is a life member of the Association of Pharmaceutical Teachers of India (APTI) and the Indian Pharmacy Graduates' Association (IPGA).

Contributors

Deshmukh Aaishwaryadevi School of Pharmaceutical Sciences, JSPM University, Pune, Maharashtra, India

Vaibhav Dagaji Aher Department of Pharmaceutical Medicine, Maharashtra University of Health Sciences, Nashik, Maharashtra, India

Mohit Angolkar Department of Pharmaceutics, JSS College of Pharmacy, JSS Academy of Higher Education & Research (JSS AHER), Mysuru, Karnataka, India

K. M. Asha Spandana Department of Pharmaceutics, JSS College of Pharmacy, JSS Academy of Higher Education & Research (JSS AHER), Mysuru, Karnataka, India

Dipak Bari Department of Pharmaceutics, R. C. Patel Institute of Pharmaceutical Education and Research, Shirpur, Maharashtra, India

Mahek Bhatt Arihant School of Pharmacy and Bio-Research Institute, Adalaj, Gandhinagar, Gujarat, India

Ankit Chaudhary Saraswati Institute of Pharmaceutical Sciences, Chiloda, Gandhinagar, Gujarat, India

Sunita Chaudhary Arihant School of Pharmacy and Bio-Research Institute, Adalaj, Gandhinagar, Gujarat, India

Viola Colaco Department of Pharmaceutics, Manipal College of Pharmaceutical Sciences, Manipal Academy of Higher Education, Manipal, Karnataka, India

Deepanjan Datta Department of Pharmaceutics, Manipal College of Pharmaceutical Sciences, Manipal Academy of Higher Education (MAHE), Manipal, Udupi, Karnataka, India

Namdev Dhas Department of Pharmaceutics, Manipal College of Pharmaceutical Sciences, Manipal Academy of Higher Education (MAHE), Manipal, Karnataka, India

Jeel Dobariya Pioneer Pharmacy College, Vadodara, Gujarat, India

Paniz Hedayat Department of Pharmaceutics, Manipal College of Pharmaceutical Sciences, Manipal Academy of Higher Education, Manipal, Karnataka, India

Sandesh Ramchandra Jadhav Department of Pharmaceutics, Manipal College of Pharmaceutical Sciences, Manipal Academy of Higher Education (MAHE), Manipal, Karnataka, India

Madhuchandra Kenchegowda Department of Pharmaceutics, JSS College of Pharmacy, JSS Academy of Higher Education & Research (JSS AHER), Mysuru, Karnataka, India

Madhusmruti Khandai Royal College of Pharmacy and Health Sciences, Berhampur, Ganjam, Odisha, India

Ritu Kudarha Department of Pharmaceutics, Manipal College of Pharmaceutical Sciences, Manipal Academy of Higher Education, Manipal, Karnataka, India

K. Kranthi Kumar College of Pharmaceutical Sciences, Dayananda Sagar University, Bengaluru, Karnataka, India

Shubhrat Maheshwari Faculty of Pharmaceutical Sciences, Rama University, Kanpur, India

Bioorganic and Medicinal Chemistry Research Laboratory, Department of Pharmaceutical Sciences, am Higginbottom University of Agriculture Technology and Sciences, Prayagraj, India

Snigdha Das Mandal Department of Pharmacology, Parul Institute of Pharmacy and Research, Parul University, Vadodara, Gujarat, India

Surjyanarayan Mandal Themis Medicare Ltd, Haridwar, Uttarakhand, India

Bharat Mishra Dr. Shakuntala Misra National Rehabilitation University, Lucknow, India

Shrishti Mishra Central Drug Research Institute, Lucknow, India

Alima Misiriya Department of Pharmaceutics, Yenepoya Pharmacy College and Research Centre, Yenepoya (Deemed to be University), Mangalore, Karnataka, India

Anoushka Mukharya Department of Pharmaceutics, Manipal College of Pharmaceutical Sciences, Manipal Academy of Higher Education, Manipal, India

Srinivas Mutalik Department of Pharmaceutics, Manipal College of Pharmaceutical Sciences, Manipal Academy of Higher Education, Manipal, India

Gaurisha Alias Resha Ramnath Naik Department of Pharmaceutics, Manipal College of Pharmaceutical Sciences, Manipal Academy of Higher Education, Manipal, Karnataka, India

Riyaz Ali M. Osmani Department of Pharmaceutics, College of Pharmacy, King Khalid University (KKU), Al Faraa, Abha, Saudi Arabia

Sharanya Paramshetti Department of Pharmaceutics, JSS College of Pharmacy, JSS Academy of Higher Education & Research (JSS AHER), Mysuru, Karnataka, India

Ashish Kumar Parashar Lloyd Institute of Management and Technology, Greater Noida, Uttar Pradesh, India

Chandrakantsing Pardeshi Department of Pharmaceutics, R. C. Patel Institute of Pharmaceutical Education and Research, Shirpur, Maharashtra, India

Komal Parmar ROFEL Shri G. M. Bilakhia College of Pharmacy, Vapi, Gujarat, India

Amitkumar K. Patel Saffron Health LLC, East Brunswick, NJ, USA

Dhara Patel Pioneer Pharmacy College, Vadodara, Gujarat, India

Grishma Patel Pioneer Pharmacy College, Vadodara, Gujarat, India

Jayvadan K. Patel Viesain Pharma LLC, GA, USA

Faculty of Pharmacy, Sankalchand Patel University, Visnagar, Gujarat, India

Nilam Patel SAL Institute of Pharmacy, Ahmedabad, Gujarat, India

Vivek Patel Sun Pharmaceutical Industries Ltd., Vadodara, Gujarat, India

Rahul Pokale Department of Pharmaceutics, Manipal College of Pharmaceutical Sciences, Manipal Academy of Higher Education, Manipal, Karnataka, India

Bhupendra Prajapati Shree S.K. Patel College of Pharmaceutical Education and Research, Ganpat University, Mehsana, Gujarat, India

Sneh Priya Department of Pharmaceutics, Nitte Gulabi Shetty Institute of Pharmaceutical Science, Nitte (Deemed to be University), Mangalore, Karnataka, India

S. P. Rachana Department of Pharmaceutics, Manipal College of Pharmaceutical Sciences, Manipal Academy of Higher Education, Manipal, Karnataka, India

Vivek Rajkule Department of Pharmaceutics, R. C. Patel Institute of Pharmaceutical Education and Research, Shirpur, Maharashtra, India

Amarjitsing Rajput Department of Pharmaceutics, Poona College of Pharmacy, Bharati Vidyapeeth Deemed University, Erandwane, Pune, Maharashtra, India

Amrita Arup Roy Department of Pharmaceutics, Manipal College of Pharmaceutical Sciences, Manipal Academy of Higher Education, Manipal, Karnataka, India

Shashi Ravi Suman Rudrangi Department of Product Development and Commercialisation, Kent Pharma UK Limited, Ashford, Kent, UK

Gaurav Kant Saraogi Sri Aurobindo Institute of Pharmacy, Indore, MP, India

Vandana Arora Sethi Lloyd Institute of Management and Technology, Greater Noida, Uttar Pradesh, India

Aditya Singh Department of Pharmacy, Integral University, Lucknow, Uttar Pradesh, India

Sandeep Kumar Singh Department of Pharmacy, University of Kota, Kota, Rajasthan, India

Fathima Thahsin Department of Pharmaceutics, Yenepoya Pharmacy College and Research Centre, Yenepoya (Deemed to be University), Mangalore, Karnataka, India

Archita Tiwari Khwaja Moinuddin Chishti Language University, Lucknow, India

Gaurav Tiwari PSIT-Pranveer Singh Institute of Technology (Pharmacy), Kanpur, Uttar Pradesh, India

Ruchi Tiwari PSIT-Pranveer Singh Institute of Technology (Pharmacy), Kanpur, Uttar Pradesh, India

Shradhha Tiwari School of Pharmacy, MIT Vishwaprayag University, Solapur-Pune highway, Kegaon, Solapur, India

K. Trideva Sastri Department of Pharmaceutics, JSS College of Pharmacy, JSS Academy of Higher Education & Research (JSS AHER), Mysuru, Karnataka, India

Sandhya Vasanth Department of Pharmaceutics, Yenepoya Pharmacy College and Research Centre, Yenepoya (Deemed To Be University), Mangalore, Karnataka, India

Vishal Kumar Vishwakarma Institute of Pharmacy, Deen Dayal Upadhyay Gorakhpur University, Gorakhpur, Uttar Pradesh, India

Part I

Introduction and Basic Considerations of Core Shell Nanoconstructs

Introduction to Core-Shell Nanoconstructs in Cancer Theragnostics

1

Dhara Patel, Grishma Patel, Jeel Dobariya, Vivek Patel, and Jayvadan K. Patel

Abstract

Worldwide, cancer is the second most common cause of death. Chemotherapy and other traditional cancer treatments have toxicities that affect normal cells in addition to their intended targets, necessitating the development of novel techniques for more effective cell-specific targeting. The use of nanomaterials as chemical biology tools in cancer theranostics has been thoroughly investigated and researched. Nanoparticles (NPs) and ligands combine to form nanoconstructs, which have the ability to transport loaded cargo to the intended site of action. For the creation of nanoconstructs that may be used for both therapeutic and diagnostic reasons, a variety of nanoparticulate platforms have been used. The main purpose of nanoconstructs is to get beyond the drawbacks of cancer treatments, which include toxicity, nonspecific medication distribution, and uncontrolled release rate. Nanoconstructs are made specifically to target the required spot and remove obstacles that prevent their proper deployment for the intended benefit. Because of their vast surface area and capacity to functionalize with a variety of biosubstrates, including aptamers, antibodies, DNA, and RNA, nanoparticles can encapsulate a huge number of molecules and aid in theragnostic action. In contrast to nanoparticles created using conventional methods, biologically derived nanomaterials are thought to offer advantages in terms of cost,

D. Patel (✉) · G. Patel · J. Dobariya
Pioneer Pharmacy College, Vadodara, Gujarat, India

V. Patel
Sun Pharmaceutical Industries Ltd., Vadodara, Gujarat, India
e-mail: vivek.patel@sunpharma.com

J. K. Patel
Viesain Pharma LLC, GA, USA

Faculty of Pharmacy, Sankalchand Patel University, Visnagar, Gujarat, India

J. K. Patel et al. (eds.), *Core-Shell Nano Constructs for Cancer Theragnostic*,
https://doi.org/10.1007/978-981-96-3025-7_1

simplicity of manufacture, and lower toxicity. In general, nanoconstructs have many advantages, but they also have many drawbacks. Therefore, computational modeling techniques and artificial intelligence/machine learning procedures are being investigated to address these issues. Consequently, distribution modes for nanoconstructs are more appropriately categorized as autonomous and nonautonomous kinds rather than as actively or passively targeted systems. This chapter highlights the uses of smart nanomaterials (such as organic nanoparticles (NPs), inorganic NPs, and carbon-based NPs) and provides an overview of several approaches in cancer theragnostic.

Keywords

Nanoconstructs · Nanoparticles · Cancer · Theragnostics · Drug delivery · Challenges · Opportunities

1 Introduction

In the last ten years, cancer has emerged as the prevailing life-threatening condition and a primary contributor to mortality (Fatiregun et al. 2020). Cancer stands as a significant health challenge in the twenty-first century, with projections indicating a potential 50% increase in its incidence over the next two decades. The Global Cancer Observatory (GCO) projects that 30 million cancer patients will pass away from the illness by 2030. Every nation on the planet needs to deal with the prevalence of one or more forms of cancer. According to GLOBOCAN 2020, there were a projected 10 million cancer-related deaths and 19.3 million cancer cases globally. Prostate cancer (1.41 million cases, 7.3%), lung cancer (2.21 million cases, 11.4%), and female breast cancer (2.26 million cases, 11.7%) accounted for the majority of these total cases worldwide. The total cancer-related mortality shows that lung (1.79 million deaths, or 18% of all cancer-related deaths), liver (830,000, or 8.3%), stomach (769,000, or about 7.7%), and breast cancer (680,000, or 6.9%) were the main causes of cancer-related deaths (Sung et al. 2021). Apart from its high death rate, cancer imposes a significant financial strain on both society and the families of cancer patients. Thus, it is crucial to make efforts in cancer detection, treatment, and prevention. This surge is primarily attributed to two major factors: environmental influences, accounting for 90–95% of cases, and genetic factors contributing to 5–10%. Environmental factors manifest in diverse ways, with approximately 25–30% linked to smoking, 30–35% to aspects of nutrition and obesity, and 15–20% associated with stress, ionizing radiation, infection, and exposure to environmental pollutants. Notably, cancer rates are on the rise in developing nations due to a combination of aging populations and lifestyle choices associated with cancer, including smoking, sedentary behavior, and the adoption of Western dietary patterns (Anand et al. 2008). Uncontrolled and aberrant cell proliferation, invasion, and metastasis are the hallmarks of this diverse disease (Pedraza-Fariña 2006). Cancer is difficult to treat because cancer cells can multiply and discriminate at startling rates, shift to different organs, and quickly develop resistance to standard medicines (Mansoori et al. 2017). They can also mutate to elude the usual mechanisms of cell death. Various imaging modalities, including ultrasound (US), computed tomography (CT), magnetic resonance imaging (MRI), positron emission tomography (PET), and single photon emission computed tomography (SPECT), have been employed in clinical practice. Although they significantly aid in the detection of disease, there are several drawbacks that should be noted, such as the radiation risks associated with CT, PET, and SPECT; the limited spatial resolution of PET, SPECT, and US; and the low temporal resolutions of MRI (Xiao et al. 2023). Conventional cancer therapies have been utilized extensively to treat various cancer types, including surgery, chemotherapy, and radiation therapy (Tannock 1998). Surgery can only be performed to the fullest extent possible without endangering the structure or function of the organ(s) involved or nearby, as well as the ability to accurately assess the extent of the disease both before and after surgery (Constantin et al. 2019). The capacity to precisely determine the extent of the disease, the tumor's susceptibility to ionizing radiation, and the tolerance of nearby normal tissues to the targeted tumor are the factors that limit the use of radiation treatment (Emami 2013). Chemotherapy is further hampered by its cytotoxicity, lack of selectivity, short half-life, inadequate solubility in physiological settings, appearance of stem-like cells, and multidrug resistance (Cheng et al. 2021). Furthermore, because these traditional treatments frequently fall short of eliminating the whole tumor mass and have the potential to harm neighboring healthy tissues and organs while in treatment, there is a danger of tumor recurrence (Abbas et al. 2018). Moreover, the inadequate pharmacokinetic properties of anticancer drugs, including poor solubility, stability, and metabolism, present additional challenges such as toxicity, reduced efficacy, and limited distribution within the body. Consequently, there is a pressing need to develop formulations capable of overcoming these obstacles by selectively targeting tumor sites while minimizing harm to healthy tissues. Thankfully, the fusion of nanoscience and biomedicine has given rise to the field of nanomedicine, presenting promising avenues for combating cancer through theragnostic. These approaches, integrating both diagnostic and therapeutic functions, leverage nanotechnology to provide concurrent and synergistic benefits in cancer treatment (Chatterjee et al. 2014).

Nanotechnology, an interdisciplinary domain, spans a wide array of scientific and technological disciplines, including biomedical, pharmaceutical, agricultural, environmental, materials science, chemistry, physics, electronics, and information technology. The synthesis, characteristics, and applications of materials and devices measuring below 100 nm have made significant contributions to various biomedical sectors, such as imaging agents, drug delivery systems, and diagnostic tools, thereby aiding in the preservation of human life and the advancement of other fields. Biomedical engineering has facilitated the integration of engineering principles into medical practices, encompassing surgical procedures, diagnosis, monitoring, treatment, and therapy. The compact size and heightened surface-to-volume ratio of nanoparticles (NPs) are fundamental attributes that render them invaluable in biomedical applications, allowing for the emergence of novel properties, facile functionalization, and biomolecule conjugation (Yan et al. 2016).

The word "theragnostic," which refers to a single biocompatible and biodegradable nanoconstruct that can contain both therapeutic and diagnostic chemicals, is currently beginning to gain traction in the medical and research fields. The basic objective of theragnostic is to expedite, enhance outcomes, and make treatment safer by combining crucial stages of medical care, such as diagnosis and therapy. Theragnostic for cancer combines multiple modalities into a single platform (Fig. 1.1). The primary goals of nanotechnology are to enable agents based on nanoparticles to distribute payload effectively and selectively without causing hazardous side effects, and to track the effectiveness of noninvasively delivered therapeutics over an extended period of time with reliability. Targeted, nontargeted, or stimuli-responsive theragnostic agents can be created. A targeting ligand that can attach to the receptors overexpressed in malignancies is included in targeted theragnostic carriers. Nowadays, the creation of nanoconstructs—which are made by fusing ligands with nanoparticles (NPs)—represents a promising strategy for the

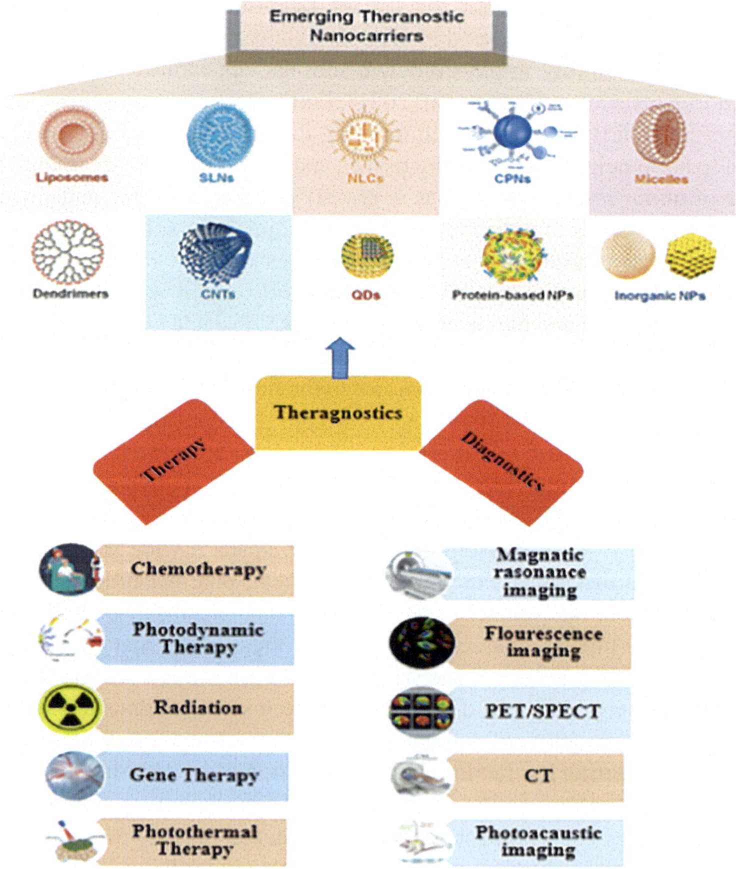

Fig. 1.1 Emerging theragnostic nanocarriers and few modalities of diagnosis and therapy in cancer theragnostic

therapy of cancer. Nanoconstructs are predominantly utilized to address the limitations associated with cancer therapies, including issues such as toxicity, nonspecific drug distribution, and uncontrolled release rates. The tactics used in nanoconstructs design are intended to improve the specificity and efficacy of theragnostic drugs, making them a viable cancer treatment strategy. These constructs are specifically tailored to target the intended site and circumvent barriers hindering their precise placement for optimal efficacy. As such, rather than categorizing nanoconstructs delivery modes as actively or passively targeted systems, it would be more reasonable to categorize them as autonomous and nonautonomous types. While nanoconstructs compromise manifold advantages, they are also beset by numerous challenges. Consequently, efforts are underway to address these challenges through computational modeling methods and the exploration of artificial intelligence (AI) and machine learning processes (Mishra et al. 2023). This chapter discusses the key components of the relevance, influence, and promise of nanoconstructs, with a focus on the most recent advancements in their use in cancer detection and therapy.

2 Core-Shell Nanoconstructs: Definition and Composition

The Greek word nanos, which means "a dwarf," is the foundation of the prefix nano. The prefix nano was formally accepted in 1947 at the 14th International Union of Pure and Applied Chemistry (IUPAC) conference to indicate the one-billionth part (10^{9}) of a unit. In various branches of current research, the prefix "nano" has become a common description in scientific writing to refer to small objects and processes. Nanoscience, nanotechnology, nanorobots, nanomagnets, nanoelectronics, nanoencapsulation, and other terminology are among them, although they are not the only ones (Joudeh and Linke 2022). Nanoconstructs consist of a dual-component architecture comprising a solid nanoparticle core and a flexible outer layer composed of biomolecular ligands, frequently employed for precise targeting in drug delivery applications. Different nanoparticulate platforms have been employed in the fabrication of nanoconstructs, serving dual roles in both diagnostics and therapeutics (Yan et al. 2016). Core-shell nanoparticles consist of a central core material enclosed within another material layer. In biological contexts, these nanoparticles offer significant advantages over conventional ones, resulting in enhanced properties such as:

1. Reduced cytotoxicity
2. Enhanced biocompatibility, dispersibility, and cellular environment compatibility
3. Enhanced conjugation capabilities with other bioactive molecules
4. Heightened thermal and chemical stability (Chatterjee et al. 2014)

Consequently, core-shell nanoconstructs hold more promise for biological applications compared to single nanoparticles (Chatterjee et al. 2014). After the nanoconstructs are administered, their fate will eventually depend on the surface chemistry and the vasculature as a whole, which includes pressure, velocities, and tissue heterogeneities. Therefore, to determine whether the nanoconstructs reached the artery walls or continued in blood, a computer modeling technique based on the 4S parameters (shape, surface, size, and stiffness) was used (King et al. 2019; Coclite et al. 2017). Consequently, computational techniques are applied to the study of complicated events and the creation of ideal nanostructures for use in biomedicine (Cervadoro et al. 2018).

2.1 Artificial Intelligence (AI) and Computational Modeling in Design of Nanoconstructs

AI would be crucial in maximizing the delivery of both unmodified and nanotechnology-modified medication combinations. At the moment, the following are the three primary areas in which large academic research laboratories, technology businesses, and biotechnology corporations apply AI methods:

1. Utilizing machine learning to forecast drug discovery targets and the therapeutic properties of chemical combinations.
2. Using sequence differentiation and recognition methods on medical images for quick disease detection and tracking (e.g., internal organs, bones, body surfaces and anomalies, retinal scans, and retinal images).
3. Creating unique predictive models by using artificial intelligence (AI) techniques to multimodal data sources, including genetic and clinical data (Hamilton and Kingston 2024).

Nanomedicine enters the picture by offering a variety of instruments for cancer diagnosis and treatment, helping to comprehend the molecular cause of cancer. This model was implemented using a number of techniques, including decision trees, support vector machine (SVM), and iterative stochastic elimination (ISE). Artificial intelligence algorithms are being refined to save time and improve medical staff's diagnostic abilities by detecting, characterizing, and continuously monitoring the reproducibility and accuracy of malignancies. If loading medications and imaging agents into the particle is necessary, one can use predictive AI systems to anticipate encapsulation efficiency (EE%). This was implemented using multiple algorithms. Because computation enhances information processing and modeling, it has the potential to have a major impact on nanomedicine (Zhang et al. 2023). Computation and informatics are essential tools for measuring nanoscale toxicity because of advancements in nanotechnology and increased computing capacity. Applications of nanoinformatics in nanomedicine include the structure-activity connection analysis of medications based on nanoparticles. Since nearly all physicochemical parameters, such as concentration, shape, size, surface area, or electrostatic properties, can affect their interaction with the adjacent media, computational and theoretical ab initio tools can handle biomaterial nanosafety at such a scale. Cancer is studied using a variety of computer models, including boundary-tracking models, lattice-based approaches, and off-lattice methods.

2.2 Building Blocks for Nanoconstructs

Scientific definitions define nanoparticles (NPs) as particles with a single dimension smaller than 100 nm and unique properties not seen in bulk samples of the same material (Cervadoro et al. 2018). The overall shape of the nanoparticle determines whether it is categorized as 0D (nanoparticles, quantum dots, and fullerenes), 1D (nanohorns, nanofibers, nanotubes, nanorods, and nanowires), 2D (nanofilms, nanolayers and nanosheets), or 3D (bulk powders, nanoparticle dispersions, arrays of nanowires and nanotubes, etc. are all included in this class) (Khan et al. 2022). These materials' exceptional qualities—such as their high surface-to-volume ratio, dissimilarity, submicron size, and enhanced targeting system—have made them more significant in transdisciplinary fields. NPs may contain one or more crystal solids and can be amorphous or crystalline. NPs come in two varieties: agglomerated and loose (Radu et al. 2023). NPs may be homogenous or have many layers. In the latter case, the layers are usually: (a) The surface layer, consisting of various small molecules, metal ions, polymers, and surfactants. (b) The shell layer is made of a material that differs chemically from the core layer. (c) The central region of the NP is known as the core layer. Because of their potential to effectively transport drugs and provide an enhanced permeability and retention (EPR) effect, nanoparticles (NPs) with a diameter range of 10–100 nm are frequently thought suitable for cancer therapy. Smaller particles (less than 1–2 nm) can easily leak from the normal vasculature, harming healthy cells, and can easily pass through the renal system (less than 10 nm in diameter). Particles larger than 100 nm are likely to be removed from circulation by phagocyte. Furthermore, the half-life and bioavailability of NPs can be impacted by the characteristics of their surfaces (Alven and Aderibigbe 2020). NPs coated in hydrophilic materials, such as polyethylene glycol (PEG), for instance, lessen opsonization and elude immune system clearance (Suk et al. 2016). The physical-chemical characteristics of core-shell nanoparticles enable their categorization into numerous groups. Core-shell metallic, core-shell magnetic, core-shell polymer, core-shell silica, core-shell upconversion, and carbon nanomaterial-based core-shell nanoparticles are among the basic categories into which core-shell nanoparticles can be divided (Tsamos et al. 2022). The other classifications based on core material are silica, iron oxide, gold, silver, platinum, palladium, superparamagnetic, and core-shell nanoparticles of these metals. Current nanosystems have the potential to enhance therapeutic response monitoring, diagnostics, and drug delivery (Nicolae-Maranciuc et al. 2022). In an effort to improve therapeutic outcomes, researchers tried to develop a theragnostic platform using multifunctional NPs, which have useful imaging properties. TNPs can therefore be made of metals, carbon, polymers, lipids, and ceramic materials.

Because of their excellent loading capacity for both hydrophilic and hydrophobic medicinal molecules, low toxicity, biodegradability, and biocompatibility, lipid nanoparticles are frequently used in the medical area. Additionally, by using a regulated release profile, they can enhance the pharmacodynamics and pharmacokinetics of medicinal drugs. Lipid NPs' capacity to be functionalized with aptamers, small molecules, peptides, or antibodies to carry out target therapy is another crucial feature (Begines et al. 2020).

Organic-based nanoparticles (NPs) having a diameter of less than 1 μm are often polymeric NPs. Depending on their composition, they can be referred to as nanospheres or nanocapsules. Since they can improve the solubility and bioavailability of hydrophobic drugs, these nanoparticles are frequently used as delivery systems (Begines et al. 2020).

Noble metals (Cu, Ag, and Au) are among the metal precursors used in the design of metallic nanoparticles. Gold nanoparticles (NPs) are the subject of most research in the biomedical field because of their distinct optical, electrical, and chemical inert properties. Additionally, they are particularly helpful for several medical applications, including bioimaging, photothermal therapy, and biosensing, due to their availability for surface functionalization. Enhanced optical properties, robust electrical conductivity, catalytic activity, chemical stability, and thermal conductivity are among the unique characteristics of silver nanoparticles. These nanoparticles (NPs) have several uses such as photonic, electronic, biosensors, medication transport, photothermal treatment, antibacterial and disinfection, and cellular imaging (Arvizo et al. 2012).

Well-known as quantum dots, semiconductor nanocrystals are another family of metallic nanoparticles. Numerous studies have documented their potential application in biomedical imaging, drug and gene delivery, and diagnostics because of their unique chemical and optical properties (Wagner et al. 2019).

Magnetic nanoparticles, or iron oxide NPs, are unique in their chemical, biological, and magnetic characteristics. These qualities include nontoxicity, chemical stability, biocompatibility, high magnetic susceptibility, and high saturation magnetization. Iron nanoparticles' primary flaw is their propensity to oxidize. Coating with a biocompatible shell—such as metals, polymers, or ceramics—is necessary to stop this undesirable process and avoid aggregation. Anticancer drugs, proteins, antibodies, and enzymes can also be used to functionalize iron oxide nanoparticles (NPs). These NPs are being researched for a variety of uses, such as targeted drug delivery, multimodal imaging, magnetic hyperthermia, magnetic resonance imaging (MRI) contrast agents, and gene therapy (Mittal et al. 2022).

Fullerenes and carbon nanotubes are two carbon-based NPs that show promise for biomedical uses. Fullerenes' unique globular network structure makes them appropriate for various functionalization stages. They are widely employed as photodynamic therapy photosensitizers, medication and gene delivery systems, antiviral drugs, and outstanding antioxidants. However, extended carbon nanotube designs have been used in drug administration, photothermal therapy, and diagnostic imaging techniques. The inorganic nonmetallic solids known as ceramic nanoparticles (NPs) are created through heating and subsequent cooling. As a result, these ceramic nanoparticles are widely employed in research as agents for photodegradation of dyes, imaging agents, photocatalysis, and catalysis (Gaur et al. 2021).

Developing and creating novel nanosystems has many difficulties. The novel tactic known as the "nanoparticle loaded nanoparticle" concept is made up of at least two distinct nanoparticles. To effectively penetrate tumors, silica-based porous nanoparticles have the potential to incorporate tiny gold nanoparticles conjugated with DNA within their pores. Interest in creating programmed nanoparticles using hybrid constructions has grown. DNA nanorobots, which are utilized for target therapy in cancer treatment, are constructed from a DNA robot and a DNA aptamer that provides molecular recognition of nucleolin (Rodríguez et al. 2022). Table 1.1 highlights the various approved and marketed nanopharmaceuticals for cancer therapy.

2.3 Therapeutic Applications for Theranostic Nanoparticles [33]

Theranostic NPs come in a variety of forms and can be developed for use in cancer treatment and diagnosis. Liposomes, whose unique properties make them widely used in clinical trials, serve as a prime example of their application. Table 1.2 lists a number of theranostic nanoparticles that have been employed in preclinical and clinical research for the diagnosis and treatment of cancer.

3 Advantages of Core-Shell Nanoconstructs

3.1 Enhanced Drug Delivery

Over the past six decades, advancements in healthcare have led to a significant increase in life expectancy in several developed nations, reaching between 80 and 90 years. Consequently, biomedical science and engineering have shifted their focus toward addressing noncommunicable diseases such as cancer and AIDS, with a predominant emphasis on cancer research. Recent data from the American Cancer Society indicates a decline in cancer-related deaths in the USA, likely attributed to advancements in early detection techniques and the development of more effective treatments, alongside a reduction in smoking rates. However, cancer mortality rates continue to rise in developing countries. Conventional cancer treatments like surgery, chemotherapy, radiation, and immunotherapy have limitations, particularly in their lack of selectivity in targeting cancer cells (Kumar et al. 2020). The core-shell structure allows for the encapsulation of therapeutic agents within the core, protecting them from degradation and facilitating controlled release at the target site. This enhances drug delivery efficiency and reduces off-target effects. Particle size, shape, drug-nanoparticle interaction, surface chemistry, hydrophilicity, hydrophobicity, surface functionalization, biodegradability, and responsiveness to temperature, pH, electric charge, light, sound, and magnetism are among the characteristics that are necessary for successful nanoparticle-based drug delivery systems. The choice of coating material, contingent on the intended therapeutic chemical, is also critical to the design of efficient drug delivery systems (Kumar et al. 2020).

Table 1.1 Approved and marketed nanopharmaceuticals for cancer therapy (Onaciu et al. 2020)

S. No.	Product	Drug	Nanotechnology platform	Cancer type	Approval	Advantages	Toxicity
1	SMANCS®	Zinostatin stimalamer	Conjugate polymer-protein	Primary hepatocellular carcinoma that is incurable	Japan, 1994	Increased accumulation and the impact of EPR	Liver dysfunction/ somewhat toxic
2	Doxil (Caelyx)	Doxorubicin hydrochloride	Pegylated liposome	AIDS-related Kaposi's sarcoma and ovarian cancer (FDA)	FDA, 1995	Extended drug half-lives, drug loading, and tumor targeting	Doxil® usage over time may put female patients at risk for oral squamous cell cancer
3	DaunoXome	Daunorubicin	Liposome	HIV-associated Kaposi sarcoma	FDA, 1996	Slow release into circulation and no polyethylene coating	Adverse cardiac effects
4	Lipo-Dox	Doxorubicin	Liposome	Ovarian, breast, and Kaposi's sarcoma	Taiwan 1998	Greater tolerance, longer half-life, decreased clearance, and longer duration of blood circulation	Gentle myelosuppression along with additional non-hematological toxins
5	Myocet	Doxorubicin	Liposome	Breast cancer	EMA 2000	Equivalent antitumor activity and less cardiotoxicity	Comparative uncertainty
6	Mepact	Muramyl phosphatidyl ethanolamine tripeptide	Liposome	Osteosarcoma that is not metastatic	EMA 2009	Greater half-life and reduced toxicity	Fever, headache, myalgias, chills, and exhaustion
7	Lipusu	Paclitaxel	Liposomes	Non-small-cell lung cancer and breast cancer	EMA 2013	Adjust the toxicity of paclitaxel without compromising its anticancer efficacy	Peripheral neuritis, vomiting, dyspnea, and nausea

8	NanoTherm	Fe_2O_3	Superparamagnetic iron oxide nanoparticles covered with amino silane	Prostate, pancreatic, and glioblastoma malignancies	2013 (EMA)	High tumor absorption and blood circulation time (EPR), heat generation when stimulated by electromagnetic fields (EMF), and teranostic qualities	Moderately negative outcome
9	Ameluz	5-Aminolevulinic acid	Gel containing 5-aminolevulinic acid, E211, SoyPC, and PG	Nodule or superficial basal cell carcinoma	2011 (EMA)	Durable release and little toxicity	Momentary discomfort and erythema
10	Depocyt	Cytarabine	Liposome	Meningitis with lymphomatous malignancy	1999 (FDA)	Enhance distribution while lowering	Neurotoxicity and arachnoiditis
11	Genexol-PM	Paclitaxel	Polymeric micelle	Lung cancer with non-small cells	(South Korea) 2006	Targeted medication delivery and controlled drug release	Myalgia, neutropenia, and neuropathy
12	Nanoxel	Docetaxel	Polymeric micelle	NSCLC, AIDS-related Kaposi's sarcoma, and malignancies of the breast and ovaries	(India) 2006	Reduced toxicity combined with good pharmacokinetic and efficacious qualities	Vomiting, diarrhea, alopecia, paresthesia, myalgia, nausea, and anemia
13	Marqibo	Vincristine	Liposome	Leukemia	2012 (FDA)	Slowly release into circulation; lack of polyethylene coating	Drug toxicity and unfavorable side effects
14	Onivyde pancreatic cancer	Irinotecan	Liposome	Pancreatic cancer	(FDA) 2015	Improved distribution and decreased systemic toxicity	Vomiting, diarrhea, neutropenia, and febrile neutropenia

(continued)

Table 1.1 (continued)

S. No.	Product	Drug	Nanotechnology platform	Cancer type	Approval	Advantages	Toxicity
15	Vyxeos	Daunorubicin with cytarabine	Liposome	Acute myeloid leukemia (AML)	(EMA) 2017	Enhanced general survivability	Tiredness, pneumonia, hypoxia, hypertension, bacteremia, sepsis, and febrile neutropenia
16	ONCASPAR	L-Asparaginase	PEGylated conjugate	Acute lymphoblastic leukemia	(FDA) 2006	Increased drug load stability	Hyperglycemia, pancreatitis, and venous thromboembolism
17	DPH107	Paclitaxel	Lipid nanoparticles	Advanced stomach cancer	(Korea) 2016	Lack of requirement for P-glycoprotein inhibitors or Cremophor polyoxyethylene glycerol triricinoleate 35 (EL) for absorption	Neutropenia and leukopenia
18	NBTXR3 (Hensify)	Hafnium oxide nanoparticles stimulated with external radiation to enhance tumor cell death via electron production	Nanoparticles of hafnium oxide	Locally advanced squamous cell carcinoma	(CE Mark) 2019	Enhancer for radiotherapy	Injection site discomfort, hypotension, and skin damage from radiation
19	Apealea	Paclitaxel	Polymeric micelles	Fallopian tube, peritoneal, and ovarian cancer	(EMA) 2018	Boost progression-free survival while taking carboplatin along	Peripheral neuropathy, nausea, vomiting, diarrhea, and neutropenia

20	Ontak	Denileukin diftitox	Cytotoxic protein produced from recombinant DNA	Cutaneous T-cell lymphoma	(FDA)1999	Targeted delivery	Asthenia, nausea/vomiting (which in some cases caused dehydration), acute hypersensitivity reactions, and infections
21	Eligard	Leuprolide acetate	Polymeric nanoparticles	Advanced prostate cancer	(FDA) 2002	Extended duration of circulation and controlled release	Hot flushes accompanied by gynecomastia, nausea, dizziness, and testicular shrinkage after malaise or exhaustion
22	Abraxane	Paclitaxel	Protein carrier	Various cancers including metastatic and pancreatic cancers	(FDA) 2013	Overcomes paclitaxel's extremely limited solubility	Paclitaxel's nonspecific binding to albumin
23	Kadcyla	DM1	Trastuzumab, covalently linked to DM1 via the stable thioether linker MCC	Breast cancer with HER2+ status	(FDA and EMA) 2013	Selectivity, tumor uptake (EPR), long blood circulation time, and low toxicity	Headache, constipation, diarrhea, thrombocytopenia, anorexia, increased liver enzymes, and epistaxis
24	Pazenir	Paclitaxel	Nanoparticles of paclitaxel bound to albumin were created. Powder for infusion and dispersion	Non-small cell lung cancer, metastatic pancreatic adenocarcinoma, and metastatic breast cancer	(EMA) 2019	Tumor uptake (EPR), blood circulation time, high solubility, and low toxicity	Reduced toxicity, tumor uptake (EPR), blood circulation time, and high solubility

Table 1.2 List of theranostic nanoparticles that have been employed in preclinical and clinical research for the diagnosis and treatment of cancer

Stage	Nanoparticle type	Therapeutic agent	Diagnostic agent	Pathology	Target
Preclinical	Liposomes (size: 100–200 nm)	Paclitaxel	Poly(ethylene oxide) (PEO)-modified poly (beta aminoester) nanoparticles that are pH-sensitive	Adenocarcinoma ovarian	Enhanced permeability and retention effect
	Silica (size range: 100–200 nm)	Paclitaxel and camptothecin	Superparamagnetic nanocrystals of iron oxide	Pancreatic cancer	Folic acid
	Oxide of iron (10–25 nm)	Anti-EGFRIgG	Nanoparticles of iron oxide	Glioblastoma	EGFR
	10 × 40 nm gold nanorod	Heat	Thermal/CT	Breast cancer	Enhanced permeability and retention effect
	Dots of quantum (30–50 nm)	Paclitaxel, doxorubicin, 5-fluorouracil	Quantum dots	Many cancers	CD44, folic acid
Clinical	Silica (6–7 nm)	cRGDY	Inorganic hybrid nanoparticles with ultrasmall size	Malignant brain tumors and melanoma	$\alpha_v\beta_3$ integrin
	Cyclodextrin (70 nm)	RNAi	Transferrin	Solid tumors	Transferrin receptor
	Silica-gold nanoshell	Photothermal ablation	Nanoshell (MR and optical)	Lung cancers, either primary or metastatic, and head/neck cancer	Enhanced permeability and retention effect
	Gold (at 27 nm)	Alpha tumor necrosis factor	Gold nanoparticles	Solid cancers	Enhanced permeability and retention effect (passive mechanism)
	Iron oxide	Endorem (superparamagnetic iron oxide particles)	Iron oxide	Volunteers in good health	None

EPR enhanced permeability and retention effect, *EGFR* epidermal growth factor receptor, *cRGDY* peptide cyclo-(Arg-Gly-Asp-Tyr), *rhTNF* recombinant human tumor necrosis factor alpha, *RNAi* ribonucleic acid interference, *MR* magnetic resolution

3.2 Targeted Therapy

Functionalization of the shell with targeting ligands enables specific recognition and binding to cancer cells, promoting selective uptake and accumulation of therapeutic payloads within tumor tissues while sparing healthy cells.

Targeted delivery of therapeutic cargo can be achieved through the following two mechanisms (Fig. 1.2):

(a) Active targeting
(b) Passive targeting

3.2.1 Active Targeting

When ligands like antibodies, peptides, or vitamins are applied to the surface of nanocarriers, they can attach to certain receptors on cell surfaces through a process known as receptor-mediated endocytosis. This is known as active targeting. The three stages of this process are as follows: (Fatiregun et al. 2020) ligand-receptor interaction; (Sung et al. 2021) endosome creation; and (Anand et al. 2008) endosome transport to the target region, where drug release occurs under physiological parameters specific to the area, such as pH or enzyme activity.

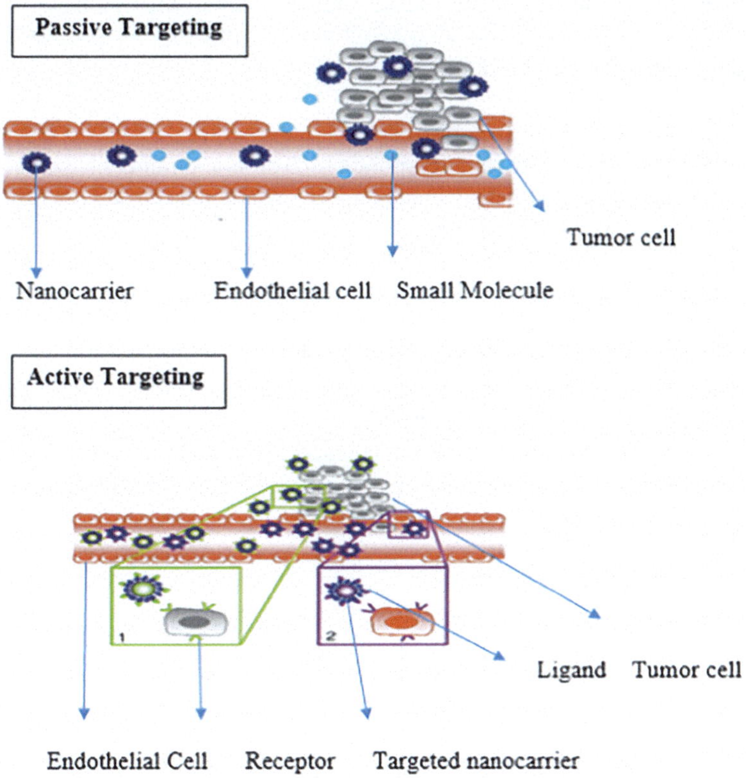

Fig. 1.2 Active and passive targeting methods in drug delivery

3.2.2 Passive Targeting

In contrast, passive targeting relies on natural conditions to guide the drug to the target tissue or organ, directing it to the site of action without the need for specific ligand-receptor interactions (Kumar et al. 2020).

3.3 Multimodal Imaging

Multimodal imaging refers to the use of multiple imaging techniques simultaneously or sequentially to acquire complementary information about a biological process or disease. Core-shell nanoconstructs play a crucial role in multimodal imaging by incorporating various imaging agents within their core or shell, allowing for the simultaneous or sequential visualization of tissues or organs using different imaging modalities. The following is an elaboration on how core-shell nanoconstructs facilitate multimodal imaging in cancer diagnosis and treatment monitoring:

3.3.1 Magnetic Resonance Imaging (MRI)

- Core-shell nanoconstructs can incorporate superparamagnetic iron oxide nanoparticles within their core, providing contrast enhancement for MRI.
- The magnetic properties of these nanoparticles alter the relaxation times of nearby protons, resulting in enhanced signal intensity in MRI images.
- MRI provides high-resolution anatomical images, allowing for precise localization and characterization of tumors.

3.3.2 Computed Tomography (CT)

- Core-shell nanoconstructs can incorporate high atomic number elements such as gold or iodine within their shell, enhancing X-ray attenuation and providing contrast enhancement for CT imaging.
- CT imaging offers excellent spatial resolution and tissue contrast, enabling the visualization of tumor morphology and surrounding vasculature.

3.3.3 Positron Emission Tomography (PET)

- Core-shell nanoconstructs can incorporate radioactive isotopes within their core or shell, serving as PET imaging probes.
- PET imaging provides functional information about metabolic activity and cellular processes within tumors, aiding in the assessment of tumor aggressiveness and response to therapy.

3.3.4 Fluorescence Imaging

- Core-shell nanoconstructs can incorporate fluorescent dyes or quantum dots within their shell, enabling fluorescence imaging.
- Fluorescence imaging offers high sensitivity and specificity, allowing for real-time visualization of tumor localization and monitoring of treatment response at the cellular level.

By integrating multiple imaging agents within a single core-shell nanoconstruct, multimodal imaging techniques provide complementary information about tumor morphology, physiology, and molecular characteristics. This comprehensive imaging approach enhances the accuracy of cancer diagnosis, enables precise monitoring of treatment response, and facilitates the real-time visualization of tumor localization during image-guided interventions. Overall, core-shell nanoconstructs play a pivotal role in advancing multimodal imaging techniques for improved cancer management and patient outcomes (Kumar et al. 2020).

3.4 Therapeutic Efficacy

The synergistic combination of diagnostic and therapeutic functionalities within a single nanoconstruct allows for personalized treatment strategies and optimization of therapeutic efficacy based on real-time feedback from imaging modalities.

3.5 Biocompatibility and Reduced Toxicity

Coating the core with biocompatible materials improves the overall biocompatibility of nanoconstructs, minimizing adverse effects on healthy tissues and organs. This is particularly advantageous in reducing systemic toxicity associated with conventional cancer treatments.

3.6 Stability and Prolonged Circulation

The shell provides protection against enzymatic degradation and immune recognition, resulting in prolonged circulation time in the bloodstream and enhanced stability of nanoconstructs, thereby improving their pharmacokinetic profile.

3.7 Customizable Design

Core-shell nanoconstructs offer versatility in design, allowing for precise control over size, shape, surface chemistry, and functionalization. This enables customization of nanoconstructs based on specific therapeutic and imaging requirements, leading to improved treatment outcomes.

3.8 Overcoming Biological Barriers

The unique properties of core-shell nanoconstructs, such as their small size, high surface area-to-volume ratio, and surface modifications, enable them to overcome biological barriers such as the blood–brain barrier and tumor microenvironment, enhancing their therapeutic efficacy in hard-to-reach locations.

Overall, core-shell nanoconstructs represent a promising platform for integrated cancer theranostics, offering significant advantages in terms of targeted drug delivery, multimodal imaging, biocompatibility, and therapeutic efficacy, ultimately contributing to improved patient outcomes in cancer diagnosis and treatment.

4 Disadvantages and Challenges

The enhanced permeability and retention (EPR) effect refers to the phenomenon where biomolecules enclosed within nanocarriers, such as lipoproteins, hormones, and albumins, exhibit increased permeation and prolonged retention in solid tumors due to characteristics such as leaky vasculature, heightened endothelial macromolecule transcytosis, and limited lymphatic drainage within the tumor interstitium. Achieving selective deposition of therapeutic agents based on the EPR effect requires minimal extravasation in normal tissues and avoidance of renal clearance. Nanoconstructs, with a reduced particle size typically ranging between 600 and 800 nm, are designed to exploit the EPR effect, allowing them to penetrate tumor vasculature while hindering extravasation through capillary pores and renal filtration. However, variations in the extracellular matrix (ECM) composition, tumor cell types, and physical properties of different tumors may impact the distribution and transport of nanoconstructs within the tumor microenvironment, thereby affecting their therapeutic efficacy. Certain tumors, such as hepatocellular carcinoma and Kaposi sarcoma, exhibit high levels of the EPR effect, whereas others like pancreatic ductal carcinoma and prostate cancer demonstrate low levels of this effect. Apart from the challenges associated with the EPR effect, nanoconstructs face biological and physical obstacles hindering their distribution and deposition in tumors, including protein adsorption, particle diffusion, aggregation, shear force destruction, and hydrolysis following intravenous administration.

Moreover, discrepancies between in vivo animal models and human cancers pose challenges in accurately assessing nanoconstructs' tumor accumulation and therapeutic efficacy, as human tumors typically have a lower body weight-to-tumor size ratio compared to animal models. Consequently, the actual amount of nanoconstructs reaching tumor sites in humans may fall short of the required therapeutic range (Chatterjee et al. 2014).

While core-shell nanoconstructs offer numerous advantages in cancer theragnostic, they also come with some disadvantages:

1. **Complex synthesis**: The fabrication of core-shell nanoconstructs often involves intricate synthesis processes, which can be time-consuming and require specialized equipment and expertise. This complexity may hinder their large-scale production and clinical translation.
2. **Limited loading capacity**: Despite their potential for targeted drug delivery, core-shell nanoconstructs may have limited loading capacities for therapeutic agents due to constraints imposed by the core-shell structure. This limitation could affect the efficacy of drug delivery, especially for payloads requiring high concentrations.

3. **Biocompatibility concerns**: While efforts are made to ensure the biocompatibility of core-shell nanoconstructs, there may still be concerns regarding potential toxicity or immunogenicity associated with the materials used in their synthesis. These issues need to be thoroughly addressed to ensure the safety of these nanoconstructs for clinical use.
4. **Stability challenges**: Core-shell nanoconstructs may face stability challenges, particularly in physiological environments, which could affect their performance and reliability over time. Factors such as aggregation, degradation, and premature drug release need to be carefully addressed to maintain the integrity and efficacy of these nanostructures.
5. **Tumor heterogeneity**: Tumors are known for their heterogeneity, which can pose challenges for targeted delivery using core-shell nanoconstructs. Variations in tumor vasculature, cellular composition, and microenvironment may affect the distribution and effectiveness of these nanoconstructs within different regions of the tumor.
6. **Biodegradation and clearance**: Core-shell nanoconstructs may undergo biodegradation or clearance from the body, potentially limiting their duration of action and necessitating repeated administrations for sustained therapeutic effects. Strategies to enhance their stability and prolong circulation time need to be explored.
7. **High cost**: The production and customization of core-shell nanoconstructs may involve high costs associated with materials, synthesis techniques, and quality control measures. This could pose economic challenges for widespread adoption and accessibility in clinical settings.

Overall, while core-shell nanoconstructs hold great promise for cancer theragnostic, addressing these disadvantages will be essential for optimizing their performance and advancing their clinical utility (Rasool et al. 2022).

5 Challenges and Opportunities of Core-Shell Nanoconstructs in Cancer Therapy

Despite the fact that cancer nanomedicine has been around for 30 years and has seen notable developments in the treatment of cancer, there are still significant drawbacks that need to be addressed. Since the discovery that fenestrated endothelium in the vasculature causes macromolecules to accumulate in tumor tissue (Danhier 2016), which is regarded as the "royal gate" in the drug delivery field (Nayak et al. 2021), more than sixty cancer nanoformulations have been authorized and made accessible as potential therapeutic interventions (Home-ClinicalTrials.Gov n.d.). Furthermore, clinical trials are being conducted on roughly 2000 developed nanoformulations with the goal of treating cancer (EU Clinical Trials Register-Update n.d.). However, the scientific and clinical community is encountering an increasing number of obstacles to the widespread development of Novel Drug Delivery System (NDD), such as high development costs, technological problems, and clinical translational failures. These factors explain why only a small percentage of the hundreds of produced NDDs find success in the marketplace. Due to the complexity of their formulations, nanomedicines can have significant effects on treatment efficacy and side effect profile from even slight changes in their technological process (Desai 2012). The European Medicines Agency (EMA) explicitly acknowledges the relevance of specific problems related to the development of nanomedicines. The EMA created several scientific recommendations for medicine makers addressing surface coatings and data requirements for intravenous colloidal nanoproducts (European Medicines Agency 2023). Despite the fact that nanodrugs are more specific and have higher clinical efficacy, nanomedicines are still less common than conventional chemotherapeutics, which are typically nonspecific and somewhat toxic (Narang et al. 2013). However, the creation of nanoformulations seeks to lessen the toxicity and improve the therapeutic efficacy of anticancer medications. By enhancing the pharmacokinetic features of anticancer drugs and making use of the targeted therapy approach (Gabizon et al. 2003) and the increased permeability and retention (EPR) effect, nanoformulations have the potential to improve their therapeutic efficacy. The negative effects are lessened by NDDs since they often have a lengthy systemic circulation and accumulate less in normal organ tissue than in tumor tissue. Nanomedicines have the potential to produce off-target effects in addition to reducing side effects, which are a hallmark of free anticancer drugs (Kreuter 2001; Park et al. 2017). Subsections that follow cover the effectiveness, side effects, biopharmaceutical characteristics, and off-target consequences of nanomedicines. The enhanced permeability and retention (EPR) effect is caused by the hyperpermeation and prolonged retention of biomolecules encapsulated in nanocarriers, such as lipoproteins, hormones, and albumins, retained in solid tumors with leaky vasculature, elevated endothelial macromolecule transcytosis, and a lack of functional lymphatic drainage inside its interstitium. To accomplish EPR effect-based selective tumor deposition, little extravasation of normal-type tissue and renal clearance are necessary. Because of the nanoconstructs' smaller particle size (600–800 nm), which is less than the permeability of tumor arteries, they were able to pass through with ease and create the EPR effect. Nevertheless, the larger particle size of the nanoconstructs than the blood capillary hole (6–12 nm) and renal filtration (5–6 nm) affects the NPs' targeting (Gabizon et al. 1994; Choi et al. 2009). Numerous malignancies may result in tumor interstitium with unique ECM (e.g., fibrin, fibronectin, hyaluronan collagen, and proteoglycans), tumor parenchyma, and stroma cell compositions due to variations in the kind and stage of the tumor. Variations in pressure ranges and physical stiffness may affect the transit and distribution of nanoconstructs inside the tumor, which may further affect these structures' therapeutic efficacy (Perry et al. 2017). Hepatocellular carcinoma, Kaposi sarcoma, renal cell carcinoma, and cancer of the neck and head are examples of high-level EPR tumors (Regev et al. 2005), whereas pancreatic ductal carcinoma and prostate cancer are examples of low-level EPR cancers (Maeda 2015). Beyond the limitations of the EPR effect, there are other variables that impede the dispersion and deposition of nanoconstructs in tumors, such as physical and biological barriers to the delivery of a sufficient dosage of a therapeutic medicine to the tumor's intended site (Bae and Park 2011). Following intravenous distribution, a variety of processes occur in nanoconstructs, including protein adsorption, particle transport, aggregation, shear force destruction, and hydrolysis (Lazarovits et al. 2015). The quantity of nanoconstructs that eventually make it to the tumor is influenced by events of this kind. To varying degrees, these events influence tumor accumulation in nanoconstructs, contingent upon the physicochemical features of the nanoconstructs (Gould et al. 2015). Inaccurate impressions resulting from the preclinical in vivo model selection present another difficulty. Studies conducted in vivo using cancer models developed from animals do not exhibit similarities with human tumors. When compared to tumors in animal models, the body weight-to-tumor size ratio in humans is significantly lower. It follows that the quantity of nanoconstructs that likely fall short of the necessary therapeutic range when they reach the tumor locations in humans is not surprising.

6 Conclusion and Future Prospective

The rapid growth of nanotechnology-based theranostics has significantly supported the advancement of cancer oncotherapy and detection techniques. The most recent developments in smart nanomaterials, including carbon-based, inorganic, and organic nanoparticles for cancer theranostic applications, are covered in this chapter. Theranostic uses for nanoconstructs in cancer enable enhanced detection, targeted medicine administration to tumors, and less toxic effects on healthy organs. Additionally, by utilizing specially designed probes, imaging techniques can assess drug efficacy in real time. Numerous platforms based on nanotechnology, including those derived from both organic and inorganic materials, have been investigated for theranostic applications. Both endogenous and exogenous stimuli are essential for regulating drug release and enhancing targetability's autonomous and nonautonomous mechanisms. For the creation of nanoconstructs for cancer theranosis, the full potential of DNA origami, nanorobots, artificial intelligence, and machine learning has not yet been investigated. Despite significant advancements in the creation of nanoconstructs for theranostic applications, relatively few of them were able to move on to the clinical stage of research. But cutting-edge nanomaterials are always the answer, and they will pave the way for the creation of multimodal NPs with improved therapeutic and diagnostic properties. The only way to achieve the intended results from the introduction of nanotechnology is to carry out multidisciplinary research on a sizable patient population.

Authorship Contribution Statement Dhara Patel: Writing—original draft, Grishma Patel: Writing, review & editing Jeel Dobariya: Writing, review & editing Vivek Patel: Editing & Review, Jayvadan K. Patel: Review.

References

Abbas M, Faggian A, Sintali DN, Khan GJ, Naeem S, Shi M, Dingding C. Current and future biomarkers in gastric cancer. Biomed Pharmacother. 2018;103:1688–700.

Alven S, Aderibigbe BA. Efficacy of polymer-based nanocarriers for co-delivery of curcumin and selected anticancer drugs. Nanomaterials. 2020;10(8):1556.

Anand P, Kunnumakara AB, Sundaram C, Harikumar KB, Tharakan ST, Lai OS, et al. Cancer is a preventable disease that requires major lifestyle changes. Pharm Res. 2008;25(9):2097–116.

Arvizo RR, Bhattacharyya S, Kudgus RA, Giri K, Bhattacharya R, Mukherjee P. Intrinsic therapeutic applications of noble metal nanoparticles: past, present and future. Chem Soc Rev. 2012;41(7):2943–70.

Bae YH, Park K. Targeted drug delivery to tumors: myths, reality and possibility. J Control Release. 2011;153(3):198–205.

Begines B, Ortiz T, Pérez-Aranda M, Martínez G, Merinero M, Argüelles-Arias F, Alcudia A. Polymeric nanoparticles for drug delivery: recent developments and future prospects. Nanomaterials. 2020;10(7):1403.

Cervadoro A, Coclite A, Di Mascolo D, Ferreira MA, Palange AL, Palomba R, et al. Smart nanoconstructs for theranostics in cancer and cardiovascular diseases. In: Challenges, strategies and prospects for novel carrier systems. Sawston: Woodhead Publishing; 2018.

Chatterjee K, Sarkar S, Jagajjanani Rao K, Paria S. Core/shell nanoparticles in biomedical applications. Adv Colloid Interf Sci. 2014;209:8–39.

Cheng S, Li Z, Gao R, Xing B, Gao Y, Yang Y, et al. A pan-cancer single-cell transcriptional atlas of tumor infiltrating myeloid cells. Cell. 2021;184(3):792–809.

Choi HS, Ipe BI, Misra P, Lee JH, Bawendi MG, Frangioni JV. Tissue- and organ-selective biodistribution of NIR fluorescent quantum dots. Nano Lett. 2009;9(6):2354–9.

Coclite A, Mollica H, Ranaldo S, Pascazio G, de Tullio MD, Decuzzi P. Predicting different adhesive regimens of circulating particles at blood capillary walls. Microfluid Nanofluid. 2017;21(11):168.

Constantin JM, Jabaudon M, Lefrant JY, Jaber S, Quenot JP, Langeron O, et al. Personalised mechanical ventilation tailored to lung morphology versus low positive end-expiratory pressure for patients with acute respiratory distress syndrome in France (the LIVE study): a multicentre, single-blind, randomised controlled trial. Lancet Respir Med. 2019;7(10):870–80.

Danhier F. To exploit the tumor microenvironment: since the EPR effect fails in the clinic, what is the future of nanomedicine? J Control Release. 2016;244(Pt A):108–21.

Desai N. Challenges in development of nanoparticle-based therapeutics. AAPS J. 2012;14(2):282–95.

Emami B. Tolerance of normal tissue to therapeutic radiation. Rep Radiother Oncol. 2013;1(1):123–7.

EU Clinical Trials Register-Update. n.d.. https://www.clinicaltrialsregister.eu/. Accessed 12 April 2023.

European Medicines Agency. Multidisciplinary: nanomedicines. 2023. https://www.ema.europa.eu/en/human-regulatory/researchdevelopment/scientific-guidelines/multidisciplinary/multidisciplinary-nanomedicines. Accessed 11 April 2023.

Fatiregun OA, Bakare O, Ayeni S, Oyerinde A, Sowunmi AC, Popoola A, et al. 10-year mortality pattern among cancer patients in Lagos State University Teaching Hospital, Ikeja, Lagos. Front Oncol. 2020;10:573036.

Gabizon A, Catane R, Uziely B, Kaufman B, Safra T, Cohen R, Martin F, Huang A, Barenholz Y. Prolonged circulation time and enhanced accumulation in malignant exudates of doxorubicin encapsulated in polyethylene-glycol coated liposomes. Cancer Res. 1994;54(4):987–92.

Gabizon A, Shmeeda H, Barenholz Y. Pharmacokinetics of pegylated liposomal Doxorubicin: review of animal and human studies. Clin Pharmacokinet. 2003;42(5):419–36.

Gaur M, Misra C, Yadav AB, Swaroop S, Maolmhuaidh FÓ, Bechelany M, Barhoum A. Biomedical applications of carbon nanomaterials: fullerenes, quantum dots, nanotubes, nanofibers, and graphene. Materials. 2021;14(20):5978.

Gould SE, Junttila MR, de Sauvage FJ. Translational value of mouse models in oncology drug development. Nat Med. 2015;21(5):431–9.

Hamilton S, Kingston BR. Applying artificial intelligence and computational modeling to nanomedicine. Curr Opin Biotechnol. 2024;85:103043.

Home-ClinicalTrials.Gov. n.d.. https://www.clinicaltrials.gov/. Accessed 12 April 2023.

Joudeh N, Linke D. Nanoparticle classification, physicochemical properties, characterization, and applications: a comprehensive review for biologists. J Nanobiotechnol. 2022;20(1):262.

Khan Y, Sadia H, Ali Shah SZ, Khan MN, Shah AA, Ullah N, Ullah MF, Bibi H, Bafakeeh OT, Khedher NB, et al. Classification, synthetic, and characterization approaches to nanoparticles, and their applications in various fields of nanotechnology: a review. Catalysts. 2022;12(11):1386.

King D, Başağaoğlu H, Nguyen H, Healy F, Whitman M, Succi S. Effects of advective-diffusive transport of multiple chemoattractants on motility of engineered chemosensory particles in fluidic environments. Entropy. 2019;21(5):465.

Kreuter J. Nanoparticulate systems for brain delivery of drugs. Adv Drug Deliv Rev. 2001;47(1):65–81.

Kumar R, Mondal K, Panda PK, Kaushik A, Abolhassani R, et al. Core-shell nanostructures: perspectives towards drug delivery applications. J Mater Chem B. 2020;8(39):8992–9027.

Lazarovits J, Chen YY, Sykes EA, Chan WCW. Nanoparticle–blood interactions: the implications on solid tumour targeting. Chem Commun. 2015;51(14):2756–67.

Maeda H. Toward a full understanding of the EPR effect in primary and metastatic tumors as well as issues related to its heterogeneity. Adv Drug Deliv Rev. 2015;91:3–6.

Mansoori B, Mohammadi A, Davudian S, Shirjang S, Baradaran B. The different mechanisms of cancer drug resistance: a brief review. Adv Pharm Bull. 2017;7(3):339–48.

Mishra S, Bhatt T, Kumar H, Jain R, Shilpi S, Jain V. Nanoconstructs for theranostic application in cancer: challenges and strategies to enhance the delivery. Front Pharmacol. 2023;14:1101320.

Mittal A, Roy I, Gandhi S. Magnetic nanoparticles: an overview for biomedical applications. Magnetochemistry. 2022;8(9):107.

Narang AS, Chang RK, Hussain MA. Pharmaceutical development and regulatory considerations for nanoparticles and nanoparticulate drug delivery systems. J Pharm Sci. 2013;102(11):3867–82.

Nayak PP, Narayanan A, Badekila AK, Kini S. Nanomedicine in cancer clinics: are we there yet? Curr Pathobiol Rep. 2021;9(2):43–55.

Nicolae-Maranciuc A, Chicea D, Chicea LM. Ag nanoparticles for biomedical applications-synthesis and characterization-a review. Int J Mol Sci. 2022;23(10):5778.

Onaciu A, Jurj A, Moldovan C, Berindan-Neagoe I. Theranostic nanoparticles and their spectrum in cancer. In: IntechOpen eBooks. London: IntechOpen; 2020.

Park IH, Sohn JH, Kim SB, Lee KS, Chung JS, Lee SH, et al. An open-label, randomized, parallel, phase III trial evaluating the efficacy and safety of polymeric micelle-formulated paclitaxel compared to conventional cremophor EL-based paclitaxel for recurrent or metastatic HER2-negative breast cancer. Cancer Res Treat. 2017;49(3):569–77.

Pedraza-Fariña LG. Mechanisms of oncogenic cooperation in cancer initiation and metastasis. Yale J Biol Med. 2006;79(3-4):95–103.

Perry JL, Reuter KG, Luft JC, Pecot CV, Zamboni W, DeSimone JM. Mediating passive tumor accumulation through particle size, tumor type, and location. Nano Lett. 2017;17(5):2879–86.

Radu ER, Voicu SI, Thakur VK. Polymeric membranes for biomedical applications. Polymers. 2023;15(3):619.

Rasool M, Malik A, Waquar S, Arooj M, Zahid S, Asif M, Shaheen S, Hussain A, Ullah H, Gan SH. New challenges in the use of nanomedicine in cancer therapy. Bioengineered. 2022;13(1):759–73.

Regev R, Yeheskely-Hayon D, Katzir H, Eytan GD. Transport of anthracyclines and mitoxantrone across membranes by a flip-flop mechanism. Biochem Pharmacol. 2005;70(1):161–9.

Rodríguez F, Caruana P, De la Fuente N, Español P, Gámez M, Balart J, et al. Nano-based approved pharmaceuticals for cancer treatment: present and future challenges. Biomolecules. 2022;12(6):784.

Suk JS, Xu Q, Kim N, Hanes J, Ensign LM. PEGylation as a strategy for improving nanoparticle-based drug and gene delivery. Adv Drug Deliv Rev. 2016;99(Pt A):28–51.

Sung H, Ferlay J, Siegel RL, Laversanne M, Soerjomataram I, Jemal A, et al. Global cancer statistics 2020: GLOBOCAN estimates of incidence and mortality worldwide for 36 cancers in 185 countries. CA Cancer J Clin. 2021;71(3):209–49.

Tannock IF. Conventional cancer therapy: promise broken or promise delayed? Lancet. 1998;351:SII9–16.

Tsamos D, Krestou A, Papagiannaki M, Maropoulos S. An overview of the production of magnetic core-shell nanoparticles and their biomedical applications. Metals. 2022;12(4):605.

Wagner AM, Knipe JM, Orive G, Peppas NA. Quantum dots in biomedical applications. Acta Biomater. 2019;94:44–63.

Xiao P, Liang M, Yang S, Sun Y, Li J, Gu Z, et al. A ratiometric near-infrared fluorescence/photoacoustic dual-modal probe with strong donor dithienopyrrole for in vivo nitric oxide detection. Biomaterials. 2023;294:121993.

Yan H, Zhao L, Shang W, Liu Z, Xie W, Qiang C, et al. General synthesis of high-performing magneto-conjugated polymer core–shell nanoparticles for multifunctional theranostics. Nano Res. 2016;10(2):704–17.

Zhang C, Xu J, Tang R, Yang J, Wang W, Yu X, Shi S. Novel research and future prospects of artificial intelligence in cancer diagnosis and treatment. J Hematol Oncol. 2023;16(1):114.

Physicochemical Properties of Core-Shell Nanoconstructs

2

Komal Parmar

Abstract

Core-shell nanoconstructs are highly functional material with modified characteristics. The nanoparticles have wide variety of applications in cancer diagnostics and therapy. The shell layer enhances the characteristics of the core, protects it from outside environment, provides a flexible surface for the target moiety, thereby enhancing the bioavailability of the active content. Various characteristics of core-shell nanosystems responsible for the working of such systems have been addressed in the following chapter.

Keywords

Upconversion properties · Molecular interaction · Catalytic behavior · Magnetic resonance imaging

1 Introduction

Ensuring a healthy life for every human is a major task for modern society. An estimation of 19.58 million new cancer cases and number of cancer-related deaths of about 6 million is projected only in the United States in 2023 (Siegel et al. 2023). According to projections, the number of deaths from cancer would rise to 13.1 million by 2030, reflecting an upsurge of cancer incidence in the next years (Chaturvedi et al. 2019). Nanotechnology has made a significant contribution to cancer treatment and offers a novel way to solve problems with current chemotherapy drugs (Jin et al. 2020). The primary advantages associated with nanoparticles used include small size and biocompatibility, because of which nanotechnology has gained

K. Parmar (✉)
ROFEL Shri G. M. Bilakhia College of Pharmacy, Vapi, Gujarat, India

J. K. Patel et al. (eds.), *Core-Shell Nano Constructs for Cancer Theragnostic*,
https://doi.org/10.1007/978-981-96-3025-7_2

distinction in cancer research, diagnosis, and therapy. In addition, it can be a noninvasive treatment that enhances the health of humans and can be applied as a molecular instrument for specific molecular medicinal operations. In the last decade, nanosystems have been used widely for targeted drug delivery in various cancerous conditions (Al-Obaidy et al. 2023; Yu et al. 2022; Helmi et al. 2021), tissue regeneration (Pooshidani et al. 2021; Yadid et al. 2019; Saderi et al. 2018), molecular imaging (Ghorbani et al. 2023; Kimura et al. 2022; Hu et al. 2019).

Theranosis in cancer is an essential instrument in the attempt to implement precision medicine in clinical settings. Precision medicine combines diagnostics and therapy by using imaging substances to target specific diseases at a deeper level. The primary goals of nanotechnology are to enable nanoparticles to distribute the content effectively and selectively without causing hazardous side effects and to consistently monitor the therapeutic efficiency of noninvasively delivered treatments over the course of time. Presently, nanoconstructs are a promising strategy toward cancer management. Table 2.1 demonstrates the recent reports on the application of nanoconstructs in cancer management. Nanoconstructs are obtained by combining nanoparticles with specific ligands that can transport the core substance at the chosen targeted spot, surmounting the obstacles that impede its appropriate deployment for the intended advantage. They are stable, have a straightforward geometry, and can be given actively or passively. Tailored nanoconstructs are considered essential for effective delivery and enhanced therapeutic efficacy in cancer treatment. The main goal of nanoconstructs is to get over the drawbacks of cancer treatments, which include toxicity, inefficient drug distribution, and unmanageable release rate.

Table 2.1 Nanoconstructs in cancer management

Materials	Drug	Application	References
Hyaluronic acid, zein	Curcumin and oxaliplatin	Colorectal cancer	Liu et al. (2022)
Gold, silver	Gold, silver	Cancer-associated fibroblasts	Kovács et al. (2020)
Gold, polyoxometalate	Zoledronate	Prostate adenocarcinoma	Tomane et al. (2021)
Polydopamine, silica	Gardiquimod	Photothermal therapy in cancer	Seth et al. (2020)
Polyacrylaminoester, poly-l-glutamic acid	Plasmids (pHR-pCas9)	Breast cancer	Duan et al. (2021)
Polydopamine, glucose oxidase, hyaluronic acid	–	Photothermal–chemodynamic synergistic therapy	Wu et al. (2021)
Egg yolk, silica	Doxorubicin	Prostate cancer	Tiburcius et al. (2022)
Silica	Bevacizumab	Non-small-cell lung cancer	Radhakrishnan et al. (2023)

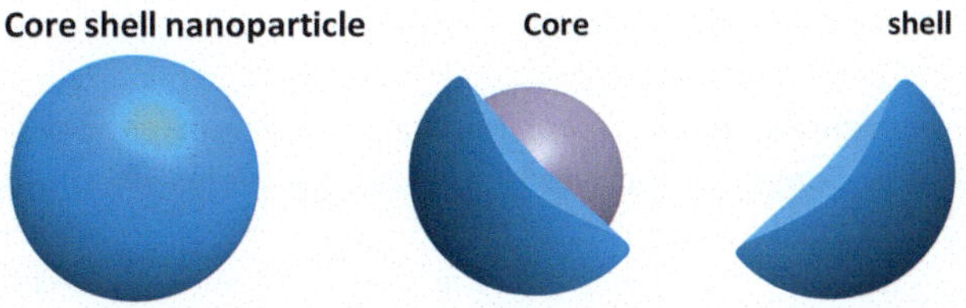

Fig. 2.1 Core-shell nanoparticle

2 Core-Shell Nanoconstructs

A "soft" shell of biomolecular ligands and a "hard" part of nanoparticle core make up nanoconstructs, which makes two constituent assemblies frequently utilized for targeted drug delivery (Dam et al. 2014) (Fig. 2.1). Drugs are stored, transported across biological barriers, and delivered as cargos at the required site using nanoconstructs consisting of active ingredients and carriers. The nanoconstructs should be engineered to hold a range of drugs within or on their surface, making them the perfect instrument for therapy. Following the administration of nanoconstructs, their ultimate destiny is determined by the vasculature as a whole, which includes pressure, rate, and tissue diversity, in addition to the surface chemistry. Hence, the presence of the nanoconstructs in blood or on vessel walls is determined by size, shape, surface and stiffness (Ye et al. 2018; Palange et al. 2017; Coclite et al. 2017). Various categories of core-shell structures include one metal core and another metal cover, metal core and nonmetal cover, metal core and polymer cover, nonmetal core and nonmetal cover, polymer core and nonmetal cover, and polymer core and polymer cover.

3 Properties of Core-Shell Nanostructures

Optimizing the size of core size and the thickness of shell is crucial for modifying the characteristics of core-shell nanomaterials. A perfect specimen of core-shell nanosystem should establish a stable assembly for the best control of properties. The core/shell nanostructure exhibit distinct physical, chemical, and biological properties that are reliant upon the composition's durability, surface morphology, and structural order.

4 Physical Properties

There are many advanced features associated with the core-shell nanosystems and surface modifications of core-shell nanomaterials. Below are the crucial physical properties of the core/shell nanostructures.

4.1 Upconversion Properties

Upconversion describes nonlinear optical processes that emit rays at a shorter wavelength than the wavelength of the pump after successively absorbing two or more than two pump photons via transitional longstanding energy conditions. The emission of photons with energies greater than the excitation photons is known as upconversion luminescence, where the extra energy for a single photon is obtained by absorbing two or more low-energy photons. Thus, materials capable of upconversion can transform near-infrared light into UV or visible light. This technique differs from the concurrent absorption of two or more photons through virtual states found in nonlinear multiple photon absorption in quantum dots and organic dyes (He et al. 2002). One of the distinctive luminous characteristics of lanthanide ions is their capacity to transform long-wavelength near-infrared excitation energy into shorter visible wavelengths. Since upconversion processes typically have efficiency levels of larger scale than those of nonlinear multiple photon absorption, upconversion can be created using an inexpensive continuous-wave diode laser rather than the ultra-short pulsed lasers required for excitation of nonlinear multiple photons. Figure 2.2 demonstrates a simple scheme of the formation of upconversion nanoparticles. Upconversion nanoparticles are well-suited for application in theranostics, which includes medication delivery, therapy, and imaging because of a variety of characteristics. They offer several unique features for medical diagnostics and

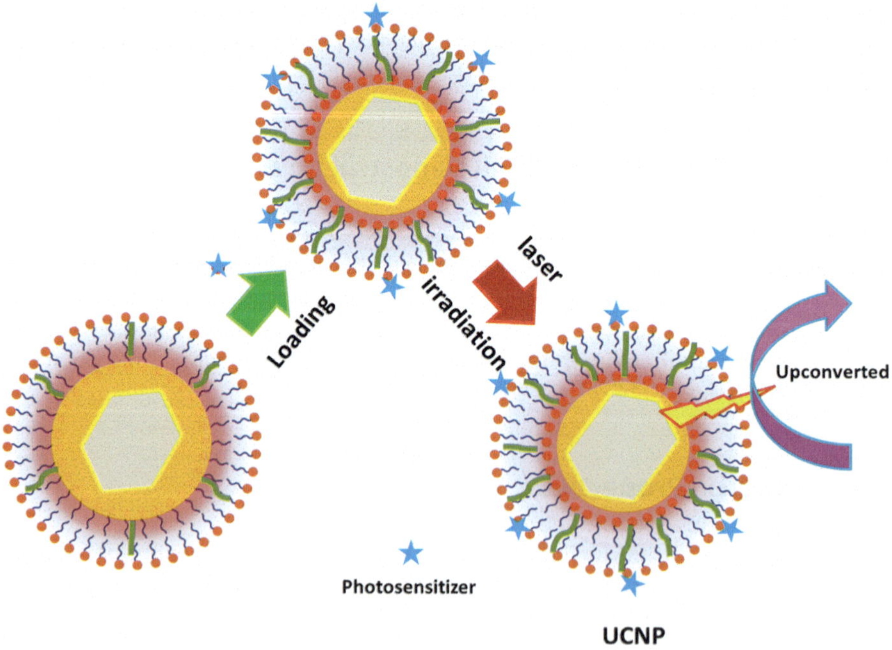

Fig. 2.2 Schematic diagram of the formation of upconversion nanoparticles (UCNP)

therapy because their special frequency conversion capacity is typically not available for prevailing endogenous and exogenous fluorophores. Comparing the upconversion strength of the nanoparticles in aqueous settings to their initial nanoparticles in organic solvents, it was discovered that the nanoparticles were severely quenched. This is caused due to vibrations of water molecules consisting of higher energy. The issue might be partially resolved by creating core-shell nanoparticles, which show better photophysical characteristics in water than the original core nanoparticles. Wang and coworkers synthesized biomimetic upconversion nanoparticles for new concurrent bi-modal imaging-guided photothermal treatment for cancer. The β-$NaYF_4$: Er^{3+}, Yb^{3+} was coated with Au nanoparticles. The cancer cell membrane-coated nanoparticles showed exceptional capacity to escape the immune system and target the site. The vastly precise and complex data obtained by cancer cell membrane-coated upconversion nanoparticles and high-resolution structural data obtained by the digital radiograph generated an accurate time and site for the photothermal therapy for cancer (Wang et al. 2020). In another work, a mesoporous heterostructure radiosensitizer containing Lu-based upconversion nanophosphor and Bi-based nanomaterial loaded with iron phthalocyanine for X-ray and near IR light dual-triggered tri-modal tumor therapy was reported. The upconversion nanoparticle was synthesized using the solvothermal method. It was layered with folic acid-modified amphiphilic poly ethylene glycol to generate a nanocomposite for bi-modal (upconversion luminescence/computed tomography) tumor imaging and tri-modal (Photothermal therapy/photodynamic therapy/radiotherapy) tumor therapy (Liu et al. 2020). A new nanohybrid based on Bi_2 Se_3-conjugated upconversion nanosystem has been effectively fabricated using an in situ growth approach. When subjected to 808 nm near-infrared laser irradiation, the upconversion nanoparticles can discharge bright visible light, whereas the Bi_2 Se_3 nanosystem displays effective photothermal conversion capability. Thus, nanohybrid shows effective cell upconversion glow, judicious computed tomography imaging, and estimable cancer cell removal ability (Zhao et al. 2020). A new upconversion luminescence-image-guided synergistic radiation and photodynamic therapy nanoplatform based on lanthanide ion-doped fluorides NaErF4:0.5%Tm@NaYF4@NaGdF4:15%Tb was reported for the photo-switchable near-infrared image-guided synergistic radiation and photodynamic therapy for deep cancer. The custom-made core/shell/shell structure was produced by a dry solvothermal method. Assisted with a cancer-targeting material folic acid, the upconversion luminescence-image-guided synergistic radiation and photodynamic therapy nanoplatform demonstrated an efficient buildup in tumor cells and enhanced anti-cancer activity when exposed to lower dose X-ray using synergistic radiation and photodynamic therapy (Feng et al. 2021).

4.2 Conductivity and Dielectric Properties

Certain polyelectrolytes can be electrostatically assembled layer by layer to cover nanoparticles in polymeric shells with ease. The ionic strength, polyelectrolyte concentration, and polyelectrolyte chain contour length can all improve the stability of colloidal dispersion. Core-shell nanocomposites exhibit strong conductivity at temperatures between 283 K and 423 K. It is observed that these characteristics are eminent for carbon and conductive polymer—polypyrrol-based core-shell nanostructures (Yang et al. 2011). Additionally, they reported a good discharge capacity in supercapacitors using these composites as the electrode materials in an electrolyte solution containing 1.0M $NaNO_3$. All of the samples exhibited semiconducting properties, according to the temperature dependency of conductivity. The high dielectric constant is another feature of core-shell nanocomposites across a broad frequency range. Methyl methacrylate (MMA) is subjected to in situ atom transmission radical polymerization to yield $BaTiO_3$@Poly(methyl methacrylate) (PMMA) core-shell nanoparticles, which exhibit a high dielectric constant across a frequency range of 0.1 Hz to 1 MHz. Elevated temperatures and low frequencies are the sole conditions that show the high dielectric loss. Additionally, they demonstrated low-base polymer MMA loss across a broad frequency range. By altering the PMMA shell thickness, one can customize the practical dielectric constant of core-shell nanoparticles hybrids (Xie et al. 2011). An Al@SiO_2 core-shell filler was effectively created by applying a SiO_2 shell layer via the sol–gel technique to the surface of metallic aluminum powder. The width of the Al@SiO_2 shell layer can be accurately regulated by altering the feeding TEOS concentration and reaction time. HTV/Al@SiO_2 composites were shown to have higher dielectric constants than HTV/Al composites with identical packing quantity. When compared to HTV/Al composites, the dielectric loss of HTV/Al@SiO_2 composites was higher at high frequencies and lower at low frequencies. Furthermore, compared to HTV/Al composites, the HTV/Al@SiO_2 composites had a higher electrical breakdown strength. Additionally, when the quantity of Al@SiO_2 core-shell filler grew, so did the composite's dielectric loss and dielectric constant (Huang et al. 2023). Core-shell nanoparticles with even amorphous Al_2O_3 shell coat containing $BaTiO_3$ (BT) particles were synthesized via heterogenous nucleation method. By adjusting the loadings of BT@Al_2O_3 nanoparticles, the dielectric behaviors of the polyvinylidene fluoride nanocomposite films were modified. The composite containing 20 vol% BT@ Al_2O_3 nanoparticles demonstrated a relatively low dielectric loss (0.02) and conductivity (2.3×10^{-8} S/m), which are approximately 48% and 78.8% less than those of the nanocomposite containing bare BT nanoparticles, respectively. Without compromising the dielectric constant or discharged energy storage density, the Al_2O_3 shell layer successfully lowers the interfacial polarization, electric conduction, and electric-field concentration in the composites, increasing their dielectric breakdown strength and reducing energy loss (He et al. 2017).

4.3 Molecular Interaction and/or Complexation

Nuclear magnetic resonance (NMR) studies strong host-guest interaction existing between beta-cyclodextrin (β-CD)-modified generation 5 (G5) poly(amidoamine) (PAMAM) (G5-CD) as core and adamantane-G3 PAMAM (G3-Ad) as shell components in the structure of core-shell tecto dendrimers (CSTDs). Also, the complexes demonstrated good antifouling property as confirmed by protein resistance assay (Song et al. 2021). In an assemblage of hard/soft nanosystem with cores containing magnetically hard cobalt ferrite enclosed by a magnetically soft nickel ferrite shell and inverse configuration, strong magnetic interaction between the core and the shell was evident from the preparation and magnetic characterization of core-shell nanosystems produced in both hard/soft and inverse soft/hard formations. Isothermal remanent magnetization and direct current demagnetization protocols have been used to analyze the interparticle interactions and the field dependency of the remanent magnetization at 5 K. The hard characteristics of cobalt ferrite seeds are affected by the generation of a soft nickel ferrite magnetic shell, resulting in a drop in $\mu_0 H_C$ from around ~1.3 to 0.8 T. In contrast, the coercivity of the soft seeds increases from approximately ~0.025 to 0.03 T due to the magnetically tougher cobalt ferrite shell (Omelyanchik et al. 2021). Cubic bi-magnetic hard/soft core/shell nanosystems were fabricated from cobalt ferrite nanoparticles and a shell of manganese ferrite. The fabrication of the heterostructures at the bulk and nanoscale levels was confirmed by combining the use of indirect (DC magnetometry) and direct (nanoscale chemical mapping by STEM-EDX) techniques. The outcomes demonstrated the production of thin-shell core-shell nanoparticles ($CoFe_2O_4$@$MnFe_2O_4$) through heterogeneous nucleation. Furthermore, homogeneous nucleation of manganese ferrite to generate a population of secondary nanoparticles was discovered (Sanna Angotzi et al. 2023). A hard ferrimagnetic (FiM) Co-ferrite shell or an antiferromagnetic (AFM) CoO shell at the surface of a soft FiM Fe_3-δO_4 core showed improved magnetic anisotropy energy when compared to Fe_3-δO_4 nanoparticles which might be due to magnetic coupling at the interface of core and shell. Compared to a CoO AFM shell, the Co-ferrite FiM shells improved magnetic anisotropy energy more. Moreover, the core size had an impact on the magnetic characteristics. Reducing the core increased the coercive field (HC) while increasing the blocking temperature (TB) with a bigger core. Co-ferrite and the fine coupling of Fe and Co cations were found to be consistent in each sample, as demonstrated by element-specific XMCD measurements. For example, the Fe_3-δO_4@CoO nanoparticles formed an interfacial layer doped with Co (Sartori et al. 2019).

4.4 Catalytic Behavior

Recently, graphene-based TiO_2@CeO_2 and CeO_2@TiO_2 core-shell heterostructures were fabricated for improved photocatalytic activity and cytotoxicity. The CeO_2 and TiO_2 NPs indicated 21.42 and 15.65% as well as CeO_2@TiO_2 and TiO_2@CeO_2 core-shell nanospheres showed 33.01 and 29% of Rhodamine B dye molecules adsorption in 180 min, respectively. The quantities of adsorption of CeO_2@TiO_2 and TiO_2@CeO_2 core-shell nanospheres were optimized after using rGO considerably. Furthermore, rGO-CeO_2@TiO_2 and rGO-TiO_2@CeO_2 ternary nanocomposites could absorb 80.79% and 76.28% of RhB dye during 30 min. The rGO-CeO_2@TiO_2 ternary nanocomposite exhibited more photocatalytic activity owing to effective photogenerated charge carrier separation, a decrease of charge recombination, and a lower band gap value as explained by the depicted mechanism (Malekkiani et al. 2022). Fe_3O_4 core-shell magnetic nanosphere by catechol-formaldehyde (CFR) resin by hydrothermal method and combined with graphene oxide and titanium dioxide to make their nanocomposites were demonstrated to exhibit catalytic activity for the adsorptive degradation of Evans blue dye. The enhanced adsorption could be because of the electrostatic interactions between graphene oxide π-electrons and the positively charged cationic dyes. It showed a maximum adsorption capacity of 0.1435 mg/g (Fe_3O_4@CFR@GO) and 9.345 mg/g (Fe_3O_4@CFR@TiO_2) nanocomposites. The kinetic study determined that Evans blue dye adsorption was in good analogy with the pseudo-first-order kinetic model (Jinendra et al. 2021). In another work, Fe_3O_4@TiO_2 core/shell connected to graphene was prepared by sol–gel method as a photocatalyst and was used for the solar degradation of cationic methylene blue in aqueous solution. The synthesized nanocomposites were connected to graphene oxide and reduced graphene oxide via implanting into 3-aminopropyltrimethoxysilane. The significance of the composite assembly in photocatalytic degradation was spectrophotometrically analyzed by mixing the formed nanosystems with wastewater containing methylene blue under solar irradiation. The appropriate dosage of 3-aminopropyltrimethoxysilane to link the Fe_3O_4@TiO_2 core/shell onto graphene oxide and reduced graphene oxide surfaces was determined to be 1 ml/g of Fe_3O_4@TiO_2 core/shell. It was observed that the interaction between the obtained core-shell nanostructure and graphene oxide layers was responsible for the degradation of the dye when subjected to solar radiation (Nazari and Salem 2019).

4.5 Magnetic Resonance Imaging

The Fe_3O_4 nanoparticles are known for their excellent magnetic characteristics. Recently, core-shell Fe_3O_4@C nano core-shell structure was developed. These nanostructures showed ideal relaxometric characteristics for T_2 magnetic resonance imaging contrast agents (Yue et al. 2024). In another work, $NaDyF_4$–$NaGdF_4$ core-shell nanoparticles were developed for targeting prostate cancer cells. The nanoparticles comprised of paramagnetic Dy^{3+} and Gd^{3+} (T_2- and T_1-contrast agents, respectively) demonstrate proton relaxivities of $r_1 = 20.2$ mM^{-1} s^{-1} and $r_2 = 32.3$ mM^{-1} s^{-1} at clinical 3 T and $r_1 = 9.4$ mM^{-1} s^{-1} and $r_2 = 144.7$ mM^{-1} s^{-1} at preclinical 9.4 T. The corresponding relaxivity values per NP are $r_1 = 19.4 \times 10^5$ $mM_{NP}{}^{-1}$ s^{-1} and $r_2 = 33.0 \times 10^5$ $mM_{NP}{}^{-1}$ s^{-1} at 3 T and $r_1 = 9.0 \times 10^5$ $mM_{NP}{}^{-1}$ s^{-1} and $r_2 = 147.0 \times 10^5$ $mM_{NP}{}^{-1}$ s^{-1} at 9.4 T (Dash et al. 2021). Bifunctional magnetic/fluorescent core-shell silica nanospheres consisting magnetic Fe_3O_4 core and N-(quinolin-8-yl)-2-(3-(triethoxysilyl) propylamino) acetamide) into the shell were fabricated. These functional nanoparticles were developed using a modified stöber method. The transverse relaxivity analysis demonstrated that the core-shell nanoparticles have T_2 relaxivity (r_2) of 155.05 mM^{-1} S^{-1} based on Fe amount on the 3.0 T scanner, signifying that the compound can be used as a negative contrast agent for magnetic resonance imaging (Qiu et al. 2021).

References

Al-Obaidy R, Haider AJ, Al-Musawi S, Arsad N. Targeted delivery of paclitaxel drug using polymer-coated magnetic nanoparticles for fibrosarcoma therapy: *in vitro* and *in vivo* studies. Sci Rep. 2023;13(1):3180. https://doi.org/10.1038/s41598-023-30221-x.

Chaturvedi VK, Singh A, Singh VK, Singh MP. Cancer nanotechnology: a new revolution for cancer diagnosis and therapy. Curr Drug Metab. 2019;20(6):416–29. https://doi.org/10.2174/1389200219666180918111528.

Coclite A, Mollica H, Ranaldo S, Pascazio G, de Tullio MD, Decuzzi P. Predicting different adhesive regimens of circulating particles at blood capillary walls. Microfluid Nanofluid. 2017;21(11):168. https://doi.org/10.1007/s10404-017-2003-7.

Dam DH, Lee RC, Odom TW. Improved in vitro efficacy of gold nanoconstructs by increased loading of G-quadruplex aptamer. Nano Lett. 2014;14(5):2843–8. https://doi.org/10.1021/nl500844m.

Dash A, Blasiak B, Tomanek B, Latta P, van Veggel FCJM. Target-specific magnetic resonance imaging of human prostate adenocarcinoma using NaDyF4-NaGdF4 core-shell nanoparticles. ACS Appl Mater Interfaces. 2021;13(21):24345–55. https://doi.org/10.1021/acsami.0c19273.

Duan J, Bao C, Xie Y, Guo H, Liu Y, Li J, Liu R, Li P, Bai J, Yan Y, Mu L, Li X, Wang G, Lu W. Targeted core-shell nanoparticles for precise CTCF gene insert in treatment of metastatic breast cancer. Bioact Mater. 2021;11:1–14. https://doi.org/10.1016/j.bioactmat.2021.10.007.

Feng Y, Liu X, Li Q, Mei S, Wu K, Yuan J, Tu L, Que I, Tamburini F, Baldazzi F, Chan A, Cruz LJ, Zuo J, Yao C, Zhang H. A scintillating nanoplatform with upconversion function for the synergy of radiation and photodynamic therapies for deep tumors. J Mater Chem C Mater. 2021;10(2):688–95. https://doi.org/10.1039/d1tc04930e.

Ghorbani F, Aminzadeh B, Borji N, Soudmand S, Montazerabadi A. Molecular MR imaging of prostate cancer by specified iron oxide nanoparticles with PSMA-11 peptides: a preclinical study. J Magn Reson Imaging. 2023;59(6):2204–14. https://doi.org/10.1002/jmri.28949.

He GS, Markowicz PP, Lin TC, Prasad PN. Observation of stimulated emission by direct three-photon excitation. Nature. 2002;415(6873):767–70. https://doi.org/10.1038/415767a.

He D, Wang Y, Chen X, Deng Y. Core–shell structured $BaTiO_3@Al_2O_3$ nanoparticles in polymer composites for dielectric loss suppression and breakdown strength enhancement. Compos A Appl Sci. 2017;93:137–43.

Helmi O, Elshishiny F, Mamdouh W. Targeted doxorubicin delivery and release within breast cancer environment using PEGylated chitosan nanoparticles labeled with monoclonal antibodies. Int J Biol Macromol. 2021;184:325–38. https://doi.org/10.1016/j.ijbiomac.2021.06.014.

Hu Z, Qin J, Li T, Guo J. Thyroid cancer MR molecular imaging via SHP2-targeted nanoparticles. Int J Nanomedicine. 2019;14:7365–73. https://doi.org/10.2147/IJN.S201358.

Huang B, Yu Y, Zhao Y, Zhao Y, Dai L, Zhang Z, Fei HF. Al@SiO2 core-shell fillers enhance dielectric properties of silicone composites. ACS Omega. 2023;8(38):35275–82. https://doi.org/10.1021/acsomega.3c05066.

Jin C, Wang K, Oppong-Gyebi A, Hu J. Application of nanotechnology in cancer diagnosis and therapy - a mini-review. Int J Med Sci. 2020;17(18):2964–73. https://doi.org/10.7150/ijms.49801.

Jinendra U, Bilehal D, Nagabhushana BM, Jithendra Kumara KS, Kollur SP. Nano-catalytic behavior of highly efficient and regenerable mussel-inspired Fe_3O_4@CFR@GO and Fe_3O_4@CFR@TiO_2 magnetic nanospheres in the reduction of Evans blue dye. Heliyon. 2021;7(1):e06070. https://doi.org/10.1016/j.heliyon.2021.e06070.

Kimura A, Utsumi S, Shimokawa A, Nishimori R, Hosoi R, Stewart NJ, Imai H, Fujiwara H. Targeted imaging of lung cancer with hyperpolarized 129Xe MRI using surface-modified iron oxide nanoparticles as molecular contrast agents. Cancer. 2022;14(24):6070. https://doi.org/10.3390/cancers14246070.

Kovács D, Igaz N, Marton A, Rónavári A, Bélteky P, Bodai L, Spengler G, Tiszlavicz L, Rázga Z, Hegyi P, Vizler C, Boros IM, Kónya Z, Kiricsi M. Core-shell nanoparticles suppress metastasis and modify the tumour-supportive activity of cancer-associated fibroblasts. J Nanobiotechnol. 2020;18(1):18. https://doi.org/10.1186/s12951-020-0576-x.

Liu J, Zhang J, Huang F, Deng Y, Li B, Ouyang R, Miao Y, Sun Y, Li Y. X-ray and NIR light dual-triggered mesoporous upconversion nanophosphor/Bi heterojunction radiosensitizer for highly efficient tumor ablation. Acta Biomater. 2020;113:570–83. https://doi.org/10.1016/j.actbio.2020.06.044.

Liu L, Yang S, Chen F, Cheng KW. Hyaluronic acid-zein core-shell nanoparticles improve the anticancer effect of curcumin alone or in combination with oxaliplatin against colorectal cancer via CD44-mediated cellular uptake. Molecules. 2022;27(5):1498. https://doi.org/10.3390/molecules27051498.

Malekkiani M, Ravari F, Heshmati Jannat Magham A, Dadmehr M, Groiss H, Hosseini HA, Sharif R. Fabrication of graphene-based TiO_2@CeO_2 and CeO_2@TiO_2 core-shell heterostructures for enhanced photocatalytic activity and cytotoxicity. ACS Omega. 2022;7(34):30601–21. https://doi.org/10.1021/acsomega.2c04338.

Nazari Y, Salem S. Efficient photocatalytic methylene blue degradation by Fe_3O_4@TiO_2 core/shell linked to graphene by aminopropyltrimethoxysilane. Environ Sci Pollut Res Int. 2019;26(24):25359–71. https://doi.org/10.1007/s11356-019-05740-8.

Omelyanchik A, Villa S, Vasilakaki M, Singh G, Ferretti AM, Ponti A, Canepa F, Margaris G, Trohidou KN, Peddis D. Interplay between inter- and intraparticle interactions in bi-magnetic core/shell nanoparticles. Nanoscale Adv. 2021;3(24):6912–24. https://doi.org/10.1039/d1na00312g.

Palange AL, Palomba R, Rizzuti IF, Ferreira M, Decuzzi P. Deformable discoidal polymeric nanoconstructs for the precise delivery of therapeutic and imaging agents. Mol Ther. 2017;25(7):1514–21. https://doi.org/10.1016/j.ymthe.2017.02.012.

Pooshidani Y, Zoghi N, Rajabi M, Haghbin Nazarpak M, Hassannejad Z. Fabrication and evaluation of porous and conductive nanofibrous scaffolds for nerve tissue engineering. J Mater Sci Mater Med. 2021;32(4):46. https://doi.org/10.1007/s10856-021-06519-5.

Qiu L, Zhou S, Li Y, Rui W, Cui P, Zhang C, Yu Y, Wang C, Wang X, Wang J, Jiang P. Silica-coated Fe_3O_4 nanoparticles as a bifunctional agent for magnetic resonance imaging and ZnII fluorescent sensing. Technol Cancer Res Treat. 2021;20:15330338211036539. https://doi.org/10.1177/15330338211036539.

Radhakrishnan D, Patel V, Mohanan S, Wong S, Netherton J, Karakoti A, Vinu A. Combinatorial treatment using bevacizumab/pemetrexed loaded core-shell silica nanoparticles for non-small cell lung cancer. Sci Technol Adv Mater. 2023;24(32):2274819.

Saderi N, Rajabi M, Akbari B, Firouzi M, Hassannejad Z. Fabrication and characterization of gold nanoparticle-doped electrospun PCL/chitosan nanofibrous scaffolds for nerve tissue engineering. J Mater Sci Mater Med. 2018;29(9):134. https://doi.org/10.1007/s10856-018-6144-3.

Sanna Angotzi M, Mameli V, Zákutná D, Secci F, Xin HL, Cannas C. Hard-soft core-shell architecture formation from cubic cobalt ferrite nanoparticles. Nanomaterials. 2023;13(10):1679. https://doi.org/10.3390/nano13101679.

Sartori K, Cotin G, Bouillet C, Halté V, Bégin-Colin S, Choueikani F, Pichon BP. Strong interfacial coupling through exchange interactions in soft/hard core-shell nanoparticles as a function of cationic distribution. Nanoscale. 2019;11(27):12946–58. https://doi.org/10.1039/c9nr02323b.

Seth A, Gholami Derami H, Gupta P, Wang Z, Rathi P, Gupta R, Cao T, Morrissey JJ, Singamaneni S. Polydopamine-mesoporous silica core-shell nanoparticles for combined photothermal immunotherapy. ACS Appl Mater Interfaces. 2020;12(38):42499–510. https://doi.org/10.1021/acsami.0c10781.

Siegel RL, Miller KD, Wagle NS, Jemal A. Cancer statistics, 2023. CA Cancer J Clin. 2023;73(1):17–48. https://doi.org/10.3322/caac.21763.

Song C, Gao Y, Chen J, Wang L, Bányai I, Shen M, Shi X. Physicochemical aspects of zwitterionic core-shell tecto dendrimers characterized by a thorough NMR investigation. Colloids Surf A Physicochem Eng Asp. 2021;618:126466.

Tiburcius S, Krishnan K, Jose L, Patel V, Ghosh A, Sathish CI, Weidenhofer J, Yang JH, Verrills NM, Karakoti A, Vinu A. Egg-yolk core-shell mesoporous silica nanoparticles for high doxorubicin loading and delivery to prostate cancer cells. Nanoscale. 2022;14(18):6830–45. https://doi.org/10.1039/d2nr00783e.

Tomane S, Wilhelm C, Boujday S, Fromain A, Miche A, Bourdreux F, Dolbecq A, Mialane P, Vallée A. Gold/polyoxometalate core/shell nanoparticles for combined chemotherapy–photothermal cancer therapy. ACS Appl Nano Mater. 2021;4(3):2339–44.

Wang R, Yang H, Fu R, Su Y, Lin X, Jin X, Du W, Shan X, Huang G. Biomimetic upconversion nanoparticles and gold nanoparticles for novel simultaneous dual-modal imaging-guided photothermal therapy of cancer. Cancer. 2020;12(11):3136. https://doi.org/10.3390/cancers12113136.

Wu H, Gu D, Xia S, Chen F, You C, Sun B. One-for-all intelligent core–shell nanoparticles for tumor-specific photothermal–chemodynamic synergistic therapy. Biomater Sci. 2021;9:1020–33.

Xie L, Huang X, Wu C, Jiang P. Core-shell structured poly(methyl methacrylate)/$BaTiO_3$ nanocomposites prepared by in situ atom transfer radical polymerization: a route to high dielectric constant materials with the inherent low loss of the base polymer. J Mater Chem. 2011;21:5897–906.

Yadid M, Feiner R, Dvir T. Gold nanoparticle-integrated scaffolds for tissue engineering and regenerative medicine. Nano Lett. 2019;19(4):2198–206. https://doi.org/10.1021/acs.nanolett.9b00472.

Yang C, Liu P, Wang T. Well-defined core-shell carbon black/polypyrrole nanocomposites for electrochemical energy storage. ACS Appl Mater Interfaces. 2011;3(4):1109–14. https://doi.org/10.1021/am1012529.

Ye H, Shen Z, Yu L, Wei M, Li Y. Manipulating nanoparticle transport within blood flow through external forces: an exemplar of mechanics in nanomedicine. Proc Math Phys Eng Sci. 2018;474(2211):20170845. https://doi.org/10.1098/rspa.2017.0845.

Yu Y, Wang A, Wang S, Sun Y, Chu L, Zhou L, Yang X, Liu X, Sha C, Sun K, Xu L. Efficacy of temozolomide-conjugated gold nanoparticle photothermal therapy of drug-resistant glioblastoma and its mechanism study. Mol Pharm. 2022;19(4):1219–29. https://doi.org/10.1021/acs.molpharmaceut.2c00083.

Yue H, Zhao D, Tegafaw T, Ahmad MY, Saidi AKAA, Liu Y, Cha H, Yang BW, Chae KS, Nam SW, Chang Y, Lee GH. Core-shell Fe_3O_4@C nanoparticles as highly effective T2 magnetic resonance imaging contrast agents: in vitro and in vivo studies. Nanomaterials. 2024;14(2):177. https://doi.org/10.3390/nano14020177.

Zhao S, Tian R, Shao B, Feng Y, Yuan S, Dong L, Zhang L, Wang Z, You H. UCNP-Bi2 Se3 upconverting nanohybrid for upconversion luminescence and CT imaging and photothermal therapy. Chemistry. 2020;26(5):1127–35. https://doi.org/10.1002/chem.201904586.

Surface Chemistries and Targeting Strategies of Core-Shell Nanoconstructs in Cancer Theragnostics

3

Ruchi Tiwari, Vaibhav Dagaji Aher, Sandeep Kumar Singh, Shashi Ravi Suman Rudrangi, and Namdev Dhas

Abstract

The unique properties and wide range of uses of core-shell nanoconstructs have attracted significant attention in the realm of cancer treatments. This study focuses on the surface chemistries and targeting strategies used in core-shell nanoconstructs for cancer thermodynamics. The stability, biocompatibility, and ability to target cancer cells of these nanoconstructs are significantly influenced by the surface chemistry of these materials. A range of surface modification methods have been employed to improve the specificity of core-shell nanoconstructs against cancer cells, such as functionalization with targeting ligands, polymers, and biomolecules. These targeting ligands, which attach to receptors or antigens overexpressed on cancer cells, can be antibodies, peptides, aptamers, or small molecules. This allows for targeted drug delivery and imaging. Furthermore, the surface characteristics and targeting effectiveness of nanoconstructs can be impacted by the selection of core and shell materials. The shell

R. Tiwari (✉)
PSIT-Pranveer Singh Institute of Technology (Pharmacy), Kanpur, Uttar Pradesh, India

V. D. Aher
Department of Pharmaceutical Medicine, Maharashtra University of Health Sciences, Nashik, Maharashtra, India

S. K. Singh
Department of Pharmacy, University of Kota, Kota, Rajasthan, India

S. R. S. Rudrangi
Department of Product Development and Commercialisation, Kent Pharma UK Limited, Kent, UK

N. Dhas
Department of Pharmaceutics, Manipal College of Pharmaceutical Sciences, Manipal Academy of Higher Education (MAHE), Manipal, Karnataka, India

J. K. Patel et al. (eds.), *Core-Shell Nano Constructs for Cancer Theragnostic*,
https://doi.org/10.1007/978-981-96-3025-7_3

materials, which are usually polymers or lipids, offer stability and biocompatibility, while the core materials, which include gold nanorods, quantum dots, and magnetic nanoparticles (NPs), can provide imaging and therapeutic functions. The application of core-shell nanoconstructs in cancer theragnostics is further enhanced by the incorporation of imaging modalities, including fluorescence imaging, positron emission tomography (PET), computed tomography (CT), and magnetic resonance imaging (MRI). Personalized cancer therapy is aided by these imaging modalities, which enable noninvasive monitoring of drug delivery and treatment response. The effectiveness of core-shell nanoconstructs in cancer theragnostics is largely dependent on their surface chemistries and targeting strategies. Subsequent investigations ought to concentrate on refining these approaches in order to enhance the therapeutic efficaciousness, biocompatibility, and specificity of core-shell nanoconstructs for the treatment of cancer.

Keywords

Core-shell nanoconstructs · Surface functionalization · Targeting strategies methods

1 Introduction

One of the leading causes of death worldwide, cancer may account for 70% of all deaths over the next 20 years if a novel and flexible anticancer platform is not developed. About 10 million cancer-related deaths and 19.3 million new cases of cancer were recorded in 2020, according to CA: A Cancer Journal for Clinicians. According to estimates, the number of deaths from cancer will rise to 13.1 million by 2030, reflecting a worsening of cancer incidence in the years to come. Presently, ligands and nanoparticles (NPs) are combined to create nanoconstructs, a potentially effective cancer treatment method. They can be supplied actively or passively and have straightforward geometry and stability. The toxicity and undesirable side effects of traditional cancer treatments are eliminated by nanoconstructs (Lee and Char 2009). The techniques used in the design of nanoconstructs contribute to the success of loaded theranostic agents as a cancer therapy by increasing their efficiency and specificity. Nanoconstructs are made specifically to target the required site and remove obstacles that prevent their proper placement for the intended benefit. Consequently, delivery modes for nanoconstructs are more appropriately categorized as autonomous and nonautonomous types rather than as actively or passively targeted systems. In general, nanoconstructs have many advantages, but they also have many drawbacks. Therefore, computational modeling techniques and artificial intelligence/machine learning procedures are being investigated to address these issues. The current review provides an overview of the attributes and applications offered by nanoconstructs as theranostic agent in cancer biodistribution (Tan et al. 2005).

The development of core-shell nanoconstructs for cancer thermometry is contingent upon surface chemistries and targeting strategies. These nanostructures, which usually comprise a core material encased in a shell, provide a versatile platform for the diagnosis and treatment of cancer (Mandal and Kruk 2012; Li et al. 2010). These nanoconstructs' surface chemistry is engineered to accomplish particular tasks like delivering therapeutic agents, imaging tumors, and focusing on cancer cells. This is accomplished by adding different molecules, like aptamers, peptides, or antibodies, to the surface of the nanoconstructs to increase their specificity by binding and recognizing particular cancer cell markers. These nanoconstructs use a variety of targeting techniques, such as passive targeting, in which they accumulate preferentially in cancerous tissues by taking advantage of the increased permeability and retention effect of tumors. On the other hand, active targeting entails using targeting ligands to bind to receptors that are overexpressed on cancer cells in an effort to increase the amount of nanoconstructs that accumulate at the tumor site (Mandal and Kruk 2012). Ultimately, the design of core-shell nanoconstructs for cancer theragnostics relies heavily on surface chemistries and targeting tactics to maximize on-target effects while improving tumor targeting, image processing, and treatment (Li et al. 2012).

2 Various Types of Chemistries Utilized for Surface Functionalization

In order to impart particular properties or functionalities, surface functionalization of nanoconstructs entails modifying the outer layer of nanoparticles with different chemical groups or molecules. Silanization, thiol chemistry, carboxylation, amine chemistry, click chemistry, electrostatic interactions, and biological conjugation are a few common chemistries used for surface functionalization. The surface of nanoparticles, especially those made of silica, can be functionalized by using silane molecules. The siloxane bond (Si-O-Si) found in silanes can create robust covalent bonds with the hydroxyl groups on the surface of silica nanoparticles, facilitating the attachment of different functional groups. Gold nanoparticles can be readily bonded to by thiol (-SH) groups, creating a strong gold-thiol bond. This chemistry is frequently used to functionalize gold nanoparticles with molecules—such as thiols, disulfides, and thioethers—that have thiol groups. There are a number of ways to add carboxylic acid (-COOH) groups to the surface of nanoparticles, including by attaching carboxylates or oxidizing alcohol groups (Tan et al. 2005; Blas et al. 2008). Subsequent functionalization, such as the conjugation of biomolecules, can be accomplished with these groups. Amines ($-NH_2$) can react with other functional groups or form stable amide bonds with carboxylic acid groups to functionalize nanoparticles. Biomolecules are frequently conjugated to nanoparticles using amines. Click chemistry is a class of fast, selective, high-yielding reactions that are perfect for functionalizing nanoparticle surfaces. The strain-promoted azide-alkyne cycloaddition (SPAAC) and the copper-catalyzed azide-alkyne cycloaddition (CuAAC) reaction are two examples. The noncovalent functionalization of nanoparticles with hydrophobic molecules can be accomplished through hydrophobic interactions. This can be accomplished by employing surfactants and amphiphilic molecules, or by altering the surface of nanoparticles to make them hydrophobic. Nanoparticles with charged molecules can be functionalized through electrostatic interactions. Different functional groups can attach to positively charged nanoparticles by interacting with negatively charged molecules and vice versa. This entails the binding of biomolecules to the surface of nanoparticles, including proteins, peptides, and antibodies (Li et al. 2010; Kim et al. 2008). A variety of chemistries, such as amine, thiol, and click chemistry, can be used to accomplish this, depending on the functional groups on the surface of the nanoparticle and the biomolecule (Fig. 3.1).

Some of the key chemistries utilized in surface functionalization include the following:

Silane chemistry: Silane chemistry is critical for the surface functionalization of core-shell nanoconstructs, especially those with silica or metal oxide cores. For this purpose, silanes like 3-mercaptopropyl trimethoxysilane (MPTMS) and aminopropyltriethoxysilane (APTES) are frequently used. They bind organic molecules to inorganic surfaces by acting as coupling agents. They include a hydrolysable group, like an alkoxy group (ethoxy or methoxy), and a reactive group, like an amino (-NH2) or thiol (-SH) group. When the silane is hydrolyzed in the presence of water, the hydrolysable group allows the formation of reactive silanol groups (Si-OH), which have the ability to attach to the surface. To reveal the silanol groups, silanes are first hydrolyzed in a solvent that contains water (Lee and Char 2009; Li et al. 2010). Usually, an acid or a base catalyzes this

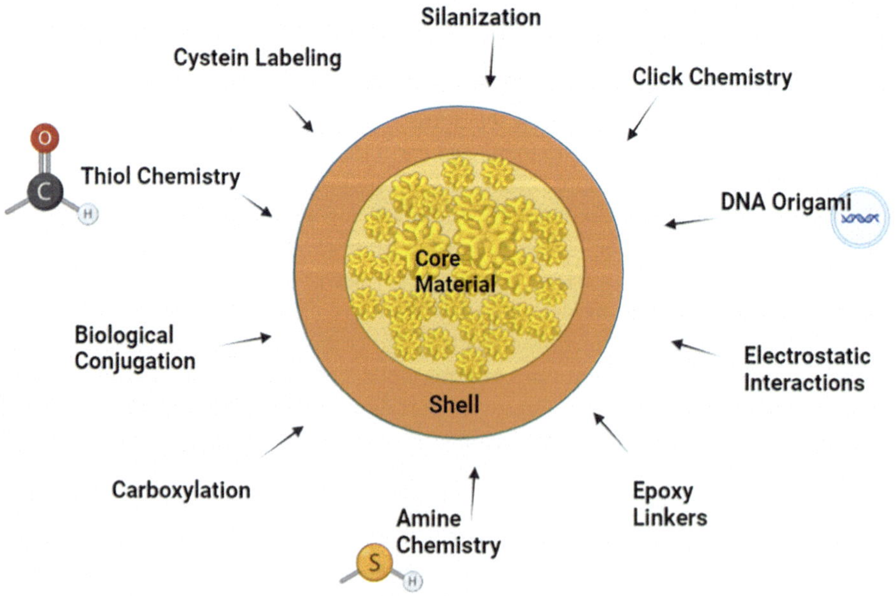

Fig. 3.1 Chemistries utilized to surface functionalization

hydrolysis. Following hydrolysis, the silanes go through condensation reactions in which the silanol groups condense with surface hydroxyl groups or with one another to form stable siloxane (Si-O-Si) bonds. The core-shell nanoconstructs' stability and biocompatibility are enhanced by their functionalization. Because the siloxane bonds that are created are stable in a physiological environment, the properties of the functionalized nanoconstructs are preserved throughout the body's circulation. Additionally, they enable exact control over the nanoconstructs' surface characteristics, such as charge, hydrophobicity, and reactivity. The creation of nanoconstructs with particular functions suited for cancer theranostics applications is made possible by this control (Li et al. 2010; Vasani et al. 2011).

Click chemistry: Click chemistry, particularly CuAAC and SPAAC, is an effective tool for surface functionalization of core-shell nanoconstructs. Click chemistry is the name given to a group of chemical reactions that are excellent for modifying sensitive biomolecules and nanomaterials because they are selective, high-yielding, and mildly conditionable. Functionalizing the surface of the nanoconstructs with either azide or alkyne groups is the first step in click chemistry. Using the proper silane coupling agents or other surface modification methods will help achieve this. In CuAAC, a copper catalyst facilitates the reaction between an azide group on one molecule and an alkyne group on a different one to create a triazole ring. Molecules with azide and alkyne groups covalently attach as a result of this extremely selective and efficient reaction. Under physiological conditions, SPAAC is a copper-free click reaction that proceeds quickly. It involves the formation of a stable triazole linkage through the reaction of an azide and cyclooctyne moiety (Li et al. 2012; Hu et al. 2000). When it comes to bioconjugation applications, where copper may be cytotoxic, SPAAC is especially helpful. Many biomolecules, including peptides, nucleic acids, and antibodies, can be conjugated to the surface of core-shell nanoconstructs via click chemistry. This makes it possible to create customized nanoconstructs for use in cancer theranostics applications. Stable covalent bonds are produced through click chemistry reactions, which improve the functionalized nanoconstructs' stability and biocompatibility. This guarantees that, in physiological settings, the molecules that are attached will stay attached. The precise control that click chemistry provides over the surface functionalization of core-shell nanoconstructs facilitates the controlled attachment of multiple molecules. This makes it possible to create intricately functional nanoconstructs for applications such as targeted drug delivery and imaging (Fu et al. 2003).

Thiol chemistry: Thiol chemistry, involving the reaction of thiol (-SH) groups with various functional groups, is widely utilized to functionalize the surfaces of core-shell nanoconstructs. There are several ways to introduce them onto the surface of the nanoconstructs, including thiolation of pre-existing functional groups on the surface or the use of silanes containing thiols. Maleimide groups and thiol groups on the surface of the nanoconstructs can react to form a stable thioether bond (Blas et al. 2008). Peptides, proteins, and polymers are examples of thiolated molecules that are frequently conjugated to the surface of nanoconstructs using this reaction. In disulfide exchange reactions, a thiol combines with a disulfide bond to create a new disulfide bond and a free thiol. This is another reaction in which they can take part. The reversible attachment of molecules made possible by this reaction enables the controlled release of payloads from the nanoconstructs. The stable covalent bonds produced by thiol chemistry improve the functionalized nanoconstructs' stability and biocompatibility. This guarantees that, in physiological settings, the molecules that are attached will stay attached. They provide fine control over the core-shell nanoconstructs' surface functionalization, enabling the controlled attachment of several molecules. This makes it possible to create intricately functional nanoconstructs for applications such as targeted drug delivery and imaging (Tan et al. 2005; Blas et al. 2008; Rao and Lopez 2000). Because of its mild reaction conditions and compatibility with biological molecules, thiol chemistry finds extensive application in bioconjugation applications. It permits biomolecules to adhere to the nanoconstructs' surface without changing their composition or functionality (Liu et al. 2014).

Carboxyl chemistry: Carboxyl chemistry is an additional significant technique for surface functionalization of core-shell nanoconstructs, especially when attaching molecules with amino groups or other nucleophiles. Carboxyl groups (-COOH) can be added to the nanoconstructs' surface using a variety of techniques, including the use of carboxylated silanes or polymers. Reactive intermediates that can react with amino groups or other nucleophiles can be formed when they are activated. Carbodiimide coupling reagents, such as NHS (N-hydroxysuccinimide) and EDC (1-ethyl-3-(3-dimethylaminopropyl) carbodiimide), are frequently used in activation procedures. Stable amide bonds can be created when activated carboxyl groups interact with amino groups on target molecules. Peptides, proteins, and other molecules containing amines are frequently conjugated to the surface of nanoconstructs via this reaction (Kim et al. 2008). Stable covalent bonds produced by carboxyl chemistry improve the functionalized nanoconstructs' stability and biocompatibility. This guarantees that, in physiological settings, the molecules that are attached will stay attached. They provide fine control over the core-shell nanoconstructs' surface functionalization, enabling the controlled attachment of several molecules. This makes it possible to create intricately functional nanoconstructs for applications such as targeted drug delivery and imaging. Because of their mild reaction conditions and compatibility with biological molecules, they are significantly utilized in biomedical applications. It permits biomolecules to adhere to the nanoconstructs' surface without changing their composition or functionality (Vasani et al. 2011; Saint-Cricq et al. 2015).

Hydroxyl chemistry: For the purpose of attaching molecules with complementary functional groups to the surface of core-shell nanoconstructs, hydroxyl chemistry is another crucial technique. There are a number of ways to add hydroxyl groups (-OH) to the surface of the nanoconstructs, including hydroxylated silanes or polymers. Reactive intermediates that can react with other functional groups can be formed by activating the hydroxyl groups on the surface of the nanoconstructs. Periodate oxidation is a common activation technique that transforms hydroxyl groups into aldehydes or other reactive species. Stable bonds can be formed between the activated hydroxyl groups and complementary functional groups on molecules of interest, such as amino groups or thiols. This enables different molecules to cling to the nanoconstructs' surfaces (Hu et al. 2000; Li et al. 2016). The stable covalent bonds produced by hydroxyl chemistry improve the functionalized nanoconstructs' stability and biocompatibility. This guarantees that, in physiological settings, the molecules that are attached will stay attached. This enables the controlled attachment of multiple molecules by providing exact control over the surface functionalization of core-shell nanoconstructs. This makes it possible to create intricately functional nanoconstructs for applications such as targeted drug delivery and imaging. Because of its mild reaction conditions and compatibility with biological molecules, it finds widespread use in biomedical applications (Fu et al. 2003). It permits biomolecules to adhere to the nanoconstructs' surface without changing their composition or functionality (Hu et al. 2018).

Phospholipid chemistry: Phospholipid chemistry is a specialized technique for the surface functionalization of core-shell nanoconstructs, with a focus on improving biocompatibility and stability. It is possible to create a lipid bilayer on the surface of the nanoconstructs by using phospholipids, which are essential parts of cell membranes. On the surface of the nanoconstructs, phospholipids, such as phosphatidylcholine (PC) or phosphatidylethanolamine (PE), can spontaneously self-assemble to form a lipid bilayer. This bilayer gives the nanoconstructs stability and biocompatibility by emulating the structure of cell membranes. Using targeting ligands like peptides or antibodies, they can be functionalized before they self-assemble on the nanoconstructs. This makes it possible to target cancerous tissues or cells specifically. The phospholipid-formed lipid bilayer improves the nanoconstructs' biocompatibility and lowers the possibility of an immunological reaction or toxicity. Pharmacological or imaging agents can be released from nanoconstructs under controlled conditions thanks to the ability of phospholipid bilayers to respond to external stimuli like temperature or pH changes (Rao and Lopez 2000). In biological settings, the lipid bilayer gives the nanoconstructs stability and guarantees that the attached molecules stay whole and functional. Liposomes, polymer nanoparticles, metallic nanoparticles, and other core-shell nanoconstructs can all be employed with phospholipid chemistry due to its versatility (Wang et al. 2017).

Polyethylene glycol (PEG)ylation: Polyethylene glycol (PEG)ylation is a popular method for surface functionalization of core-shell nanoconstructs, especially to improve biocompatibility and circulation time. A protective layer can be formed by conjugating PEG, a hydrophilic polymer, to the surface of the nanoconstructs. PEG molecules can be conjugated to the nanoconstructs' surface via a variety of techniques, including physical adsorption and covalent bonding. The hydrophilic barrier that the PEG layer creates around the nanoconstructs lessens their non-specific interactions with cells and proteins. By improving the nanoconstructs' biocompatibility, they lower the possibility of an immunological reaction or toxicity. Because PEGylation decreases the reticuloendothelial system's (RES) clearance of the nanoconstructs, it can lengthen their time in circulation in the bloodstream (Rao and Lopez 2000; Liu et al. 2014). The nanoconstructs have a "stealth" effect due to the PEG layer, which decreases their visibility to the immune system and lowers the possibility of recognition and clearance. By adjusting the PEG layer's length or density, they can also be used to regulate the release of medications or imaging agents from the nanoconstructs. It can attach targeting ligands or additional functions to the surface of the nanoconstructs and is suitable for a broad variety of biomolecules (Liu et al. 2017).

Biotin-streptavidin chemistry: Biotin-streptavidin chemistry is an extremely specific and strong noncovalent interaction that is commonly used to functionalize the surfaces of core-shell nanoconstructs. The vitamin biotin can be conjugated to the nanoconstructs' surface and subsequently complexed with the protein streptavidin, which binds to biotin exclusively. They are conjugated to the nanoconstructs' surface via a variety of techniques, including the use of polymers or molecules that have been derivatized with biotin. For streptavidin, the biotin molecules offer a particular binding site (Liu et al. 2014). Next, streptavidin—a tetrameric protein with four sites on which biotin can bind—is introduced to the mixture. A stable complex is formed when the streptavidin molecules specifically bind to the biotin molecules on the surface of the nanoconstructs. It is possible to conjugate molecules of interest to streptavidin, including targeting ligands, imaging agents, and therapeutic molecules. The biotin-streptavidin interaction then binds these molecules to the surface of the nanoconstructs. With a dissociation constant (Kd) in the femtomolar range, the biotin-streptavidin interaction is one of the strongest noncovalent interactions known to exist in biology. By doing this, the attachment of molecules to the surface of the nanoconstructs is guaranteed to have high specificity and affinity. It is adaptable and suitable for a large class of molecules and nanostructures. It permits the precise and controlled attachment of several molecules (Saint-Cricq et al. 2015; Chen et al. 2015).

3 Material and Chemistries for Targeting Strategies of Core-Shell Nanoconstructs in Cancer Theragnostics

In cancer theranostics, targeting strategies for core-shell nanoconstructs depend on particular materials and chemistries to guarantee precise delivery of therapeutic agents and imaging agents to cancer cells. As targeting ligands, a variety of substances, including aptamers, peptides, antibodies, and small molecules, bind to receptors or antigens that are overexpressed on cancer cells. The targeting ligands are affixed to the nanoconstructs' surface through various chemical techniques, including carboxyl activation, thiol chemistry, click chemistry, biotin-avidin interaction, and hydrazone chemistry. Because of their strong specificity and affinity for antigens expressed on cancer cells, antibodies are essential to the targeting strategies of core-shell nanoconstructs in cancer theranostics. Tumor-associated antigens (TAAs) are overexpressed on cancer cells, and monoclonal antibodies are designed or chosen to target these antigens (Li et al. 2016; Hu et al. 2018; Wang et al. 2017; Liu et al. 2017; Chen et al. 2015; Sun et al. 2018). Following that, these antibodies are conjugated to the surface of nanoconstructs via a variety of chemical processes, including NHS-ester chemistry and thiol-maleimide coupling. Once bonded, the antibodies help the nanoconstructs to selectively bind and internalize within the cancer cells. This targeted strategy takes advantage of the enhanced permeability and retention (EPR) effect to reduce off-target effects and improve the accumulation of nanoconstructs in tumors. Antibody-conjugated nanoconstructs are potent tools for tailored cancer treatment in cancer theranostics because they can carry therapeutic agents for targeted drug delivery and imaging agents for noninvasive tumor imaging. Targeting strategies of core-shell nanoconstructs in cancer theranostics also benefit from the use of peptides, aptamers, and small molecules, which provide low immunogenicity, high specificity for target molecules, and ease of synthesis. Peptides are short sequences of amino acids that can be engineered to attach to particular antigens or receptors on cancerous cells. Single-stranded DNA or RNA molecules known as aptamers have the ability to fold into distinctive three-dimensional structures and bind to target molecules extremely selectively. Organic compounds, known as small molecules, can be engineered to specifically target proteins or pathways implicated in the initiation and progression of cancer (Hu et al. 2018; Yu et al. 2013). It is possible to conjugate these targeting ligands to the surface of nanoconstructs through a variety of chemical processes, including biotin-avidin interaction, NHS-ester chemistry, and click chemistry. Peptides, aptamers, and other tiny molecules direct the nanoconstructs to cancer cells once they are attached, where they bind to their targets selectively. By precisely delivering therapeutic and imaging agents to cancer cells, this targeted approach maximizes the potency of cancer theranostics while reducing off-target effects (Rao and Lopez 2000; Hu et al. 2018). Furthermore, by customizing these ligands to target distinct molecular signatures of individual tumors, personalized cancer treatment is made possible (Fig. 3.2).

Targeting ligands can more easily be attached to the surface of core-shell nanoconstructs in cancer theranostics thanks to the crucial role chemistries play in designing targeting strategies. For this, a variety of chemistries are used, such as hydrazone chemistry, biotin-avidin interaction, NHS-ester chemistry, and thiol-maleimide coupling. The process of thiol groups on targeting ligands reacting with maleimide groups on nanoconstructs to form a stable bond is known as thiol-maleimide coupling (Wang et al. 2017). NHS-activated esters on the nanoconstructs that react with the amino groups on the targeting ligands are used in NHS-ester chemistry. The strong noncovalent interaction that exists between biotin and avidin or streptavidin is utilized by the biotin-avidin interaction to facilitate attachment. Under mild conditions, targeting ligands can be attached thanks to hydrazone chemistry, which forms hydrazone bonds between ligands functionalized with hydrazide and nanoconstructs functionalized with aldehyde or ketone. These chemistries enable precise and active targeting of cancer cells, increasing the efficacy of core-shell nanoconstructs in cancer theranostics (Liu et al. 2017; Chen et al. 2015; Sun et al. 2018; Yu et al. 2013; Ding et al. 2014; Kievit et al. 2009).

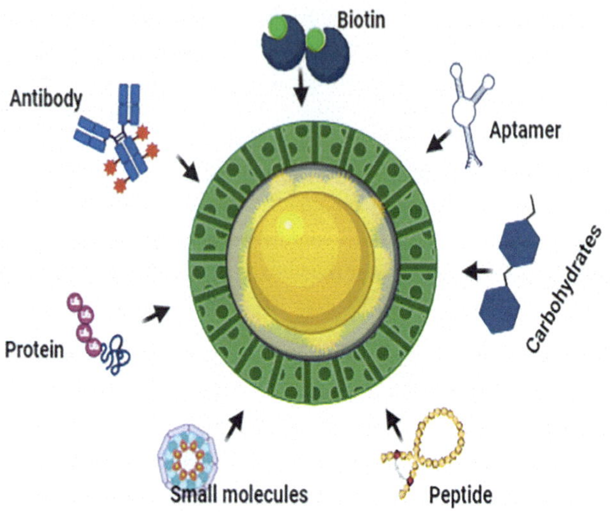

Fig. 3.2 Various targeting strategies of core-shell nanoconstructs

4 Methods Used for Core-Shell Nanoconstructs

4.1 Layer-by-Layer Assembly Method

Layer-by-layer (LbL) Assembly is a versatile and commonly employed technique for creating core-shell nanostructures with precise control over layer thickness and composition. It entails layer deposition onto a core material in a sequential manner. Numerous mechanisms, such as covalent bonds, hydrogen bonds, van der Waals forces, and electrostatic interactions, can accomplish this process (Liu et al. 2017; Chen et al. 2015). The final nanoconstruct's desired properties determine which assembly method is best. At the forefront of nanotechnology, layer-by-layer assembly provides an effective tool for creating core-shell nanoconstructs with specific properties. LbL-assembled nanoconstructs have a wide range of potential applications, including drug delivery, catalysis, and sensing (Kievit et al. 2010).

Steps involved: The layer-by-layer assembly method works best when the core and shell materials are carefully chosen. In addition to providing structural integrity, the core may have particular qualities like magnetic, optical, or electrical ones. In contrast, the shell serves as a barrier and can be designed to regulate interactions with the surroundings or accommodate extra features. This can be a nanoparticle, a microsphere, or another structure, and it serves as the focal point of the nanoconstruct. Choose the materials for the shell: These are the layers that the core will be covered with. Depending on the desired properties, they can be polymers, nanoparticles, or other materials. Usually, the core material is prepared as a stable solution or dispersion. This may entail the dispersion of pre-existing particles in an appropriate medium or the synthesis of nanoparticles. Prepare the materials' solutions that will go into the shell layers (Sun et al. 2018; Yu et al. 2013; Ding et al. 2014; Kievit et al. 2009, 2010; Stephen et al. 2016). Based on the desired shell properties, these solutions might include surfactants, polymers, or other substances. Soak the core material in one of the shell material solutions to start the layer-by-layer assembly process. The thickness of the deposited layer will depend on a number of factors, including the solution's concentration, immersion time, and other settings. Rinse the coated core material to get rid of extra solution and any unbound material after each immersion. This contributes to maintaining the deposited layer's purity. In order to prevent layer mixing, dry the coated core before adding the next layer. For every subsequent layer, repeat the immersion, rinsing, and drying procedures. To achieve the desired core-shell structure, alternate between various shell materials. Depending on the particular requirements of the application, the number of layers and the order of the materials can be changed. Verify the structure and properties of the resultant core-shell nanoconstructs by characterizing them with analytical techniques like spectroscopy and microscopy. Post-treatment procedures like cross-linking or extra coating may be used, depending on the application, to improve the stability or performance of the nanoconstructs. The layer-by-layer assembly method's capacity to produce nanoconstructs with customized properties is one of its main benefits. Researchers can create nanoconstructs with biocompatible shells in the field of drug delivery, for instance, to improve drug stability and regulate release kinetics (Yu et al. 2013; Wang et al. 2012). Similarly, the technique can be used to produce core-shell nanostructures with improved performance and optimized conductivity in electronics applying the core-shell nanoconstructs that were created for particular uses, like drug delivery, catalysis, sensing, or imaging. Furthermore, the technique makes it possible to engineer surfaces with antibacterial qualities, which makes it indispensable for the creation of medical implants (Lee et al. 2012).

4.2 Emulsion Polymerization Method

A common method for creating core-shell nanoconstructs, which consist of a polymer shell encircling a core material, is emulsion polymerization. This technique works especially well for creating polymer-coated nanoparticles. A common method for creating polymers in the form of colloidal particles scattered throughout a liquid medium is emulsion polymerization. This technique, when applied to core-shell nanoconstructs, entails creating a stable emulsion in which a polymer shell encases the core material, which is frequently a nanoparticle or droplet (Ding et al. 2014; Xing et al. 2013). Emulsion polymerization requires four essential ingredients: monomers, initiators, surfactants, and stabilizers. The monomers serve as the polymer chains' building blocks. Water-immiscible monomers are spread in an aqueous phase during emulsion polymerization with the aid of stabilizers and surfactants. Because they lessen the interfacial tension between the immiscible phases, surfactants are essential for stabilizing the emulsion. Stabilizers guarantee homogeneous shell formation around the core material by preventing the polymer particles from coalescing and clumping together. Polymer chains are formed when the polymerization reaction is triggered by initiators. Emulsion polymerization frequently uses water-soluble initiators to speed up the reaction in the aqueous phase. A variety of materials, such as polymers, biological entities, or even nanoparticles, can be used as the core material. The final core-shell nanoconstruct's characteristics and functionalities are dictated by the selection of the core material (Kievit et al. 2009; Alshaer et al. 2015).

Steps involved: Select the substance that will serve as the nanoconstruct's central component. This can include a range of materials, including organic and inorganic nanoparticles as well as other materials. For the polymer shell, choose monomers. Methacrylic and acrylic monomers are common; the choice relies on the desired characteristics of the polymer shell. Using stabilizers and surfactants, the water-immiscible monomers are dispersed in the aqueous phase to form the emulsion. Little droplets of monomers are consequently suspended in the ongoing aqueous phase. To initiate the polymerization of the monomers, incorporate initiators into the emulsion. The initiators exhibit solubility in either water or oil, contingent upon the specific system (Kievit et al. 2010). Polymer chains begin to grow from the surface of the core material upon initiation. To start the polymerization reaction inside the emulsion droplets, initiators are added. Little polymer particles are formed when polymer chains start to expand from the monomer droplets. Growing polymer chains surround the core material and form a shell around it as the polymerization process proceeds. The polymerization conditions can be changed to control the shell thickness. Following polymerization, the core-shell nanoconstructs are separated, usually by filtration or centrifugation. When you reach the desired shell thickness, stop the polymerization reaction. Reaction conditions can be changed or terminating agents can be added to achieve this. The nanoconstructs are usually cleaned to get rid of contaminants, surfactants, and unreacted monomers after polymerization. For purification, centrifugation or other separation methods could be used. Determine the size, shape, and shell thickness of the synthesized core-shell nanoconstructs using methods like transmission electron microscopy (TEM), dynamic light scattering, and other approaches (Stephen et al. 2016; Wang et al. 2012; Lee et al. 2012; Xing et al. 2013; Alshaer et al. 2015; Huang et al. 2009). Stabilize the nanoconstructs through the addition of stabilizing agents or by cross-linking the polymer shell. It is possible to apply post-treatment procedures to give the nanoconstructs' surfaces particular functions. Emulsion polymerization is used to create core-shell nanoconstructs, which are used in drug delivery systems. The shell offers stability and controlled release, while the core can contain therapeutic agents. In the creation of coatings, adhesives, and specialty materials, the controlled encapsulation of functional materials in emulsion-synthesized core-shell nanoconstructs is useful (Liu et al. 2012).

4.3 Sol-Gel Method

The sol-gel method is an easy and commonly employed approach for creating core-shell nanoconstructs, especially those containing inorganic or hybrid organic-inorganic materials. Using this process, a precursor sol (solution) is changed into a gel-like substance, which is then solidified to create the core-shell structure (Wang et al. 2012). By using the sol-gel method, a solution (sol) is changed into a network that resembles gel and eventually solidifies into a glass or ceramic substance. The sol-gel method has a special set of benefits when it comes to synthesizing core-shell nanoconstructs, one of which is the ability to precisely customize the shell's thickness and composition. This process represents a flexible and effective tool for the synthesis of core-shell nanoconstructs under control. Its uses in a variety of industries, including medicine and catalysis, highlight the breadth of possibilities offered by this innovative technique (Lee et al. 2012; Xing et al. 2013; Alshaer et al. 2015; Huang et al. 2009; Liu et al. 2012; Zhu et al. 2015).

Steps involved: Select the substance that will serve as the nanoconstruct's central component. This can apply to silica, metal oxides, or other inorganic substances. Decide on the shell material's precursors. This might be a hybrid organic-inorganic compound or just another inorganic substance. Assemble the precursor solutions for the materials in the shell and the core. Making a sol, or colloidal suspension of nanoparticles in a liquid medium (solvent), is the first step in the process. Usually, metal alkoxides or metal salts are dissolved in a solvent to create the sol, which is then converted into nanoparticles by hydrolysis and condensation reactions. The gelation process, which creates a three-dimensional network of nanoparticles, is allowed to occur in the sol. This can be accomplished in a number of ways, including heating the sol to encourage the formation of a gel or aging it at room temperature. After that, the gel is dried to get rid of the solvent and create a solid (Xing et al. 2013). Depending on the final material's desired properties, air drying, freeze drying, or supercritical drying can be used to achieve this. Sometimes, to further densify the material and get rid of any remaining organic components, the dried gel is put through a heat treatment procedure called calcination. Enhancing the material's thermal and structural qualities is a common use for calcination. The size, shape, composition, and other characteristics of the nanoconstructs are then ascertained by employing a variety of methods, including electron microscopy, X-ray diffraction (XRD), and spectroscopy. Finally, the nanoconstructs can be functionalized by incorporating additional materials, such as polymers, biomolecules, or nanoparticles, to impart specific properties or functionalities to the material (Xing et al. 2013; Alshaer et al. 2015). Catalytically active core-shell nanoconstructs, in which the shell increases stability and the core supplies catalytic sites, can be produced using the sol-gel technique. Applications like industrial processes and environmental remediation benefit greatly from this. Sol-gel synthesis yields core-shell nanoconstructs with applications in imaging, diagnostics, and drug delivery. Because the sol-gel process is tunable, nanostructures with regulated release profiles and improved biocompatibility can be designed (Liu et al. 2012; Zhu et al. 2015; Stecker et al. 2012).

4.4 Template-Assisted Synthesis Method

The technique known as "template-assisted synthesis" uses templates to direct the growth of materials around a central core in order to produce core-shell nanostructures. By using a template or scaffold to direct the deposition of materials, the template-assisted synthesis method produces nanostructures with distinct morphologies. When it comes to core-shell nanoconstructs, the template acts as a mold to ensure that a core material is precisely encapsulated inside a customized shell (Zhu et al. 2015). Because of this method's versatility, a wide range of nanostructures with simple to complex architectures can be created. This method gives you control over the final nanoconstructs' dimensions, composition, and form (Kang and Hah 2014).

Steps involved: Select a template for the nanostructure that has the appropriate size and shape. Templates can be made of a variety of materials, including metals, silica, and polymers. They can also be solid, like porous membranes and nanoparticles, or soft, like micelles and vesicles. In order to facilitate the adhesion of the material to be deposited, prepare the template by cleaning and functionalizing it as needed. Cover the template with the substance that needs to be synthesized. Depending on the material and template used, this can be accomplished using a variety of techniques, including chemical vapor deposition (CVD), electrodeposition, and physical vapor deposition (PVD). The synthesized nanostructure can be seen by removing the template (Stecker et al. 2012). This can be accomplished by removing the template material by etching or dissolving it, leaving the nanostructure in place. Using a variety of methods, including spectroscopy, X-ray diffraction, and electron microscopy, characterize the nanostructure to ascertain its dimensions, composition, and characteristics. In order to give the material particular qualities or functionalities, the nanostructure can also be functionalized by coating or doping it with other materials. In catalysis, the controlled structure that the template-assisted synthesis method offers is especially useful. In order to maximize reactivity and selectivity, core-shell nanoconstructs can be engineered with catalytically active cores and customized shells. For the purpose of producing core-shell nanoconstructs for biomedical applications, template-assisted synthesis is an excellent choice (Kang and Hah 2014; Bagalkot et al. 2007). Because of their controlled morphology and composition, these nanostructures can be customized for targeted therapies, drug delivery, or imaging contrast agents (Fig. 3.3).

4.5 Chemical Vapor Deposition (CVD) Method

Chemical vapor deposition (CVD) is a popular method for creating core-shell nanostructures, particularly for thin films and coatings. In the CVD process, vapor-phase precursors react chemically to deposit material onto a substrate in the form of a coating or thin film. Through the chemical reaction of gaseous precursors, thin films or nanostructures are grown on a substrate surface. When applied to core-shell nanoconstructs, CVD enables the selective deposition of various materials onto a central core, forming a distinct shell that encloses the core material. Because of the

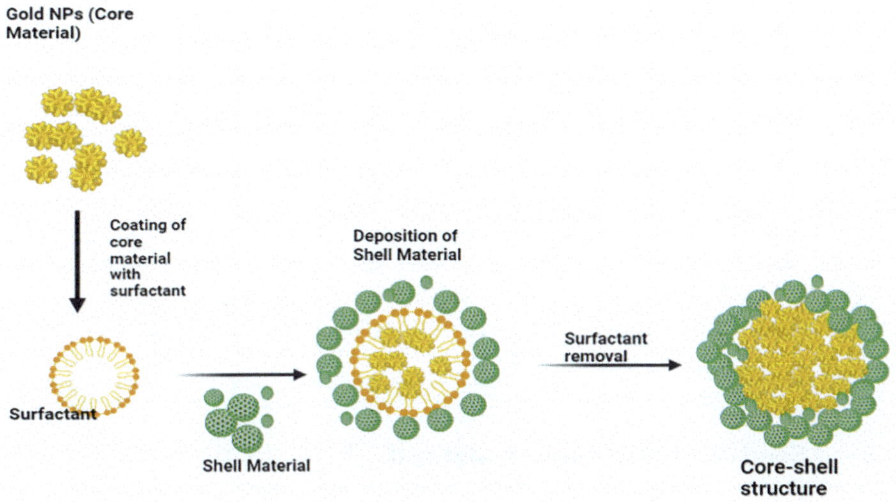

Fig. 3.3 Template-assisted synthesis method for nanoconstruct

method's great versatility, complex nanostructures with customized properties can be synthesized. With its unparalleled ability to control the growth of nanostructures, it has emerged as a key component in the production of sophisticated nanomaterials (Bagalkot et al. 2007; Wang et al. 2013). In the context of core-shell nanoconstructs, CVD provides a pathway to engineer materials with tailored properties for applications spanning from nanoelectronics to catalysis and energy storage.

Steps involved: Select precursor materials that, when exposed to the CVD process, will react to form the desired nanostructure. Gases, vapors, or volatile liquids that contain the components or compounds required for the synthesis can be considered precursors. Get the substrate ready for the nanostructure to be deposited on it. Usually made of a solid substance like silicon, glass, or metal foil, the substrate can also have a thin layer or film applied to it to aid in adhesion. Fill the CVD reactor chamber with the precursor materials. Depending on the desired deposition process (e.g., simultaneous or sequential CVD), the precursors are introduced into the chamber either simultaneously or sequentially. Reach the appropriate temperature by heating the precursor materials and substrate (Lin et al. 2014). Usually, the temperature is below the substrate and precursors' decomposition temperature, but high enough to thermally activate the precursor molecules and encourage the intended chemical reactions. To create the desired nanostructure, let the precursor molecules react on the substrate surface. Depending on the precursors and the intended product, the chemical reactions could entail pyrolysis, reduction, oxidation, or other processes. To regulate the nanostructure's growth, keep an eye on the deposition process. It is possible to regulate the growth rate, thickness, and morphology of the nanostructure by varying reaction time, temperature, pressure, and precursor flow rates. The sample can be cooled gradually or heated to a high temperature during annealing to enhance the adhesion, density, and crystallinity of the nanostructure.

Determine the size, shape, composition, and characteristics of the nanostructure by utilizing a variety of methods, including spectroscopy, transmission electron microscopy (TEM), X-ray diffraction (XRD), and scanning electron microscopy (SEM). Applications for CVD-synthesized core-shell nanostructures can be found in nanoelectronics, where sophisticated devices can be created through regulated material deposition. Encasing core materials, like semiconductor nanoparticles, in a shell can improve conductivity or add particular electronic features (Bagalkot et al. 2006). Improved electrode materials are available for energy storage devices like batteries and supercapacitors thanks to core-shell nanostructures produced by CVD. The thin shells that are created as a result of the controlled deposition improve the electrodes' stability and conductivity (Bagalkot et al. 2006; Liu et al. 2016; Boltz et al. 2011; Sierra and Tsao 2011).

4.6 Seed-Mediated Growth Method

Seed-mediated growth is a versatile method for synthesizing core-shell nanostructures, in which seeds (smaller nanoparticles) act as nucleation sites for the growth of a secondary material, resulting in a core-shell structure. This method makes use of nanoscale "seeds" as templates to allow for the regulated development of nanostructures. Within the framework of core-shell nanoconstructs, these seeds serve as centers around which extra materials are deposited, creating a distinct shell that encloses the core. The final nanostructures can be precisely controlled in terms of size, composition, and shape using this technique (Liu et al. 2016; Allen 2002). The resulting nanoconstructs' size, makeup, and morphology can all be controlled using this method.

Steps involved: Select the substance that will serve as the nanoconstruct's central component. Usually, this takes the form of nanoparticles. Typical seed materials consist of metals such as silver or gold. Decide on the shell material's precursors. These ought to be able to proliferate and nucleate on the seed nanoparticles' surface. Utilizing suitable techniques, such as chemical reduction, colloidal synthesis, or other methods, synthesize the seed nanoparticles. To attain homogeneity in the core, regulate the seed nanoparticles' size and form. To improve the adhesion of the shell material, it is optional to functionalize the surface of the seed nanoparticles. The process of surface functionalization may entail the ligand or surfactant attachment. Add the precursors of the shell materials to the reaction medium that contains the seed nanoparticles. In order to attain the required shell thickness, regulate reaction parameters like temperature, duration, and concentrations (Boltz et al. 2011). Assess the size, morphology, and structure of the synthesized core-shell nanoconstructs using analytical techniques like X-ray diffraction (XRD), microscopy (TEM and SEM), and other methods. To get rid of byproducts and unreacted precursors, wash the synthesized core-shell nanoconstructs. Use methods like centrifugation or dialysis to purify the nanoconstructs. Post-treatment procedures can be used to improve or change the core-shell nanoconstructs' characteristics, depending on the use case. Annealing, extra coating, or functionalization are a few examples of this. The design of core-shell nanoconstructs for use in nanomedicine is made easier by seed-mediated growth. Therapeutic agents can be incorporated into the core through controlled synthesis, and the shell can be designed for precise drug delivery and controlled release. The process works well for catalytic applications where effective catalysts can be made by adjusting the composition of the shell and core (Sierra and Tsao 2011). The ensuing core-shell nanostructures show increased stability and catalytic activity.

4.7 Hydrothermal Synthesis Method

Hydrothermal synthesis is a technique that uses high temperatures and pressure in an aqueous environment to promote the formation of nanomaterials. entails creating nanomaterials through the use of high-temperature, high-pressure aqueous environments to trigger chemical reactions. Hydrothermal conditions allow controlled growth of different materials in the context of core-shell nanoconstructs, resulting in a well-defined shell enclosing a central core. This approach offers benefits in terms of ease of use, economy, and the capacity to attain exact control over the morphology of the nanostructure. Core-shell nanostructures can be synthesized using this technique.

Steps involved: Select the substance that will serve as the nanoconstruct's central component. This may be a different kind of nanomaterial or a nanoparticle. Decide on the shell material's precursors (Allen 2002). It is necessary for these to dissolve or mix well in the hydrothermal reaction medium. The precursors of the core and shell materials should be dissolved or dispersed in the proper solvent or aqueous solution. Make sure the precursors can react in a hydrothermal environment and are compatible with one another. To create a uniform solution or dispersion, combine the precursor materials for the core and shell materials. Make sure the shell precursor surrounds and evenly distributes the core material. The reaction mixture should be placed in a sealed vessel and exposed to high pressure and temperature conditions that are usually associated with hydrothermal synthesis. One option is to autoclave the reaction vessel. Around the core material, the hydrothermal conditions encourage the nucleation and growth of the shell material. Cool the reaction vessel to room temperature or lower to stop the hydrothermal reaction (Peer et al. 2007). The produced core-shell nanoconstructs should be cleaned and purified to get rid of any impurities, byproducts, or unreacted precursors. Filtration or centrifugation are two possible methods for separation. Use analytical techniques like X-ray diffraction (XRD), microscopy (TEM and SEM), and other methods to characterize the size, morphology, and structure of the core-shell nanoconstructs. The characteristics of the core-shell nanoconstructs may be improved or modified by applying post-treatment procedures, depending on the application. This can entail functionalization, extra coating, or annealing. Nanoelectronic applications can benefit greatly from the creation of core-shell nanostructures through hydrothermal synthesis. The process makes it possible to create nanomaterials with specific electrical properties that are appropriate for use in electronic devices by regulating the growth conditions. Applications for core-shell nanostructures produced hydrothermally can be found in catalysis. The nanoconstructs' high surface area and carefully crafted morphology increase catalytic activity, which makes them useful in a variety of industrial processes (Leiro et al. 2015).

5 Conclusion and Future Perspective

The core-shell nanoconstructs provide a versatile platform for both diagnosis and treatment, and they have demonstrated encouraging potential in cancer theranostics. These nanoconstructs allow for targeted drug delivery, imaging, and therapeutic response monitoring all at the same time because they encapsulate diagnostic agents in the core and therapeutic agents in the shell. They are the perfect choice to overcome numerous obstacles in cancer treatment because of their special qualities, which include high surface area, tunable size, and biocompatibility. Cancer diagnosis, treatment, and monitoring could be revolutionized by incorporating core-shell nanoconstructs into cancer theragnostic approaches. In conclusion, the development of core-shell nanoconstructs for cancer theranostics depends critically on surface chemistries and targeting tactics. These nanoconstructs provide a flexible platform for combining therapeutic and diagnostic functions, allowing for more individualized and accurate cancer treatment plans. Because there are so many different surface chemistries to choose from, nanoconstructs can be tailored to have particular characteristics like stability, biocompatibility, and targeting efficiency. The potential for further advancements in cancer theranostics is great, and future perspectives on surface chemistries and targeting strategies of core-shell nanoconstructs are exciting. The development of new targeting ligands and techniques to increase the specificity of nanoconstructs for cancer cells will be the main goals of ongoing research. Aptamers, tiny compounds, or peptide ligands that target particular molecular markers on cancer cells with higher affinity and specificity. Further developments in stimuli-responsive nanoconstructs could allow for finer control of drug release in response to dynamic changes in the tumor microenvironment. Systems that react to various stimuli or those that enable on-demand drug release in response to outside stimuli may fall under this category. A primary goal will be to combine therapeutic and diagnostic capabilities into a single nanoconstruct. Advanced imaging modalities like magnetic resonance imaging (MRI), positron emission tomography (PET), or photoacoustic imaging may be incorporated into future nanoconstructs to allow for real-time drug delivery and treatment response monitoring. The transition of these technologies from the lab to the clinic will receive more attention as nanoconstructs get closer to being used in therapeutic settings. To guarantee the safe and efficient use of nanoconstructs in patients, this will entail addressing issues with scalability, reproducibility, and regulatory approval. Future developments in core-shell nanoconstruct targeting strategies and surface chemistries will, in general, continue to spur personalized cancer treatment innovation and provide patients with a variety of cancer types with new hope.

References

Allen TM. Ligand-targeted therapeutics in anticancer therapy. Nat Rev Cancer. 2002;2:750–63. https://doi.org/10.1038/nrc903.

Alshaer W, Hillaireau H, Vergnaud J, Ismail S, Fattal E. Functionalizing liposomes with anti-CD44 aptamer for selective targeting of cancer cells. Bioconjug Chem. 2015;26:1307–13. https://doi.org/10.1021/bc5004313.

Bagalkot V, Farokhzad OC, Langer R, Jon S. An aptamer-doxorubicin physical conjugate as a novel targeted drug-delivery platform. Angew Chem Int Ed. 2006;45:8149–52. https://doi.org/10.1002/anie.200602251.

Bagalkot V, et al. Quantum dot-aptamer conjugates for synchronous cancer imaging, therapy, and sensing of drug delivery based on bi-fluorescence resonance energy transfer. Nano Lett. 2007;10:3065–70. https://doi.org/10.1021/nl071546n.

Blas H, Save M, Pasetto P, et al. Elaboration of monodisperse spherical hollow particles with ordered mesoporous silica shells via dual latex/surfactant templating: radial orientation of mesopore channels. Langmuir. 2008;24:13132–7. https://doi.org/10.1021/la801897k.

Boltz A, et al. Bi-specific aptamers mediating tumor cell lysis. J Biol Chem. 2011;24:21896–905. https://doi.org/10.1074/jbc.m111.238261.

Chen Q, Wang X, Wang C, et al. Drug-induced self-assembly of modified albumins as nano-theranostics for tumor-targeted combination therapy. ACS Nano. 2015;9:5223–33. https://doi.org/10.1021/acsnano.5b00640.

Ding M, Zeng X, He X, et al. Cell internalizable and intracellularly degradable cationic polyurethane micelles as a potential platform for efficient imaging and drug delivery. Biomacromolecules. 2014;15:2896–906. https://doi.org/10.1021/bm500506v.

Fu Q, Rao GR, Ista LK, et al. Control of molecular transport through stimuli-responsive ordered mesoporous materials. Adv Mater. 2003;15:1262–6. https://doi.org/10.1002/adma.200305165.

Hu TJ, Gao J, Wu C. Swelling and shrinking of poly(N Isopropylacrylamide) chains adsorbed on the surface of polystyrene nanoparticles. J Macromol Sci Phys B. 2000;39:407–14. https://doi.org/10.1081/mb-100100395.

Hu DR, Chen LJ, Qu Y, et al. Oxygen-generating hybrid polymeric nanoparticles with encapsulated doxorubicin and chlorin e6 for trimodal imaging-guided combined chemo-photodynamic therapy. Theranostics. 2018;8:1558–74. https://doi.org/10.7150/thno.22989.

Huang YF, et al. Molecular assembly of an aptamer-drug conjugate for targeted drug delivery to tumor cells. ChemBioChem. 2009;10:862–8. https://doi.org/10.1002/cbic.200800805.

Kang S, Hah SS. Improved ligand binding by antibody-aptamer pincers. Bioconjug Chem. 2014;25:1421–7. https://doi.org/10.1021/bc500269y.

Kievit FM, Gunn JW, Lee D. PEI-PEG-chitosan copolymer coated iron oxide nanoparticles for safe gene delivery: synthesis, complexation, and transfection. Adv Funct Mater. 2009;19:2244–51. https://doi.org/10.1002/adfm.200801844.

Kievit FM, Veiseh O, Fang C, et al. Chlorotoxin labeled magnetic nanovectors for targeted gene delivery to glioma. ACS Nano. 2010;4:4587–94. https://doi.org/10.1021/nn1008512.

Kim J, Kim H, Lee N, et al. Multifunctional uniform nanoparticles composed of a magnetite nanocrystal core and a mesoporous silica shell for magnetic resonance and fluorescence imaging and for drug delivery. Angew Chem Int Ed Engl. 2008;47:8438–41. https://doi.org/10.1002/anie.200802469.

Lee H, Char K. Morphological changes from silica tubules to hollow spheres controlled by the intermolecular interactions within block copolymer micelle templates. ACS Appl Mater Interfaces. 2009;1:913–20. https://doi.org/10.1021/am900026s.

Lee SJ, Muthiah M, Lee HJ, et al. Synthesis and characterization of magnetic nanoparticle-embedded multi-functional polymeric micelles for MRI-guided gene delivery. Macromol Res. 2012;20:188–96. https://doi.org/10.1007/s13233-012-0023-4.

Leiro V, Garcia JP, Tomás H, et al. The present and the future of degradable dendrimers and derivatives in theranostics. Bioconjug Chem. 2015;26:1185–97. https://doi.org/10.1021/bc5006224.

Li J, Liu J, Wang D, et al. Interfacially controlled synthesis of hollow mesoporous silica spheres with radially oriented pore structures. Langmuir. 2010;26:12267–72. https://doi.org/10.1021/la101225j.

Li CY, Ma C, Wang F, et al. Preparation and biomedical applications of core-shell silica/magnetic nanoparticle composites. J Nanosci Nanotechnol. 2012;12:2964–72. https://doi.org/10.1166/jnn.2012.6428.

Li M, Li L, Zhan C, et al. Core-shell nanostars for multimodal therapy and imaging. Theranostics. 2016;6:2306–13. https://doi.org/10.7150/thno.15843.

Lin Z, Ma Q, Fei X, Zhang H, Su X. A novel aptamer functionalized CuInS2 quantum dots probe for daunorubicin sensing and near infrared imaging of prostate cancer cells. Anal Chim Acta. 2014;818:54–60. https://doi.org/10.1016/j.aca.2014.01.057.

Liu Z, et al. Novel HER2 aptamer selectively delivers cytotoxic drug to HER2-positive breast cancer cells in vitro. J Transl Med. 2012;10:1–10. https://doi.org/10.1186/1479-5876-10-148.

Liu J, Detrembleur C, Pauw-Gillet MD, Mornet S, et al. Heat-triggered drug release systems based on mesoporous silica nanoparticles filled with a maghemite core and phase-change molecules as gatekeepers. Mater Chem B. 2014;2:59–70. https://doi.org/10.1039/c3tb21229g.

Liu J, et al. Multifunctional nanoparticles for targeted drug delivery to circumvent cancer resistance. Biomaterials. 2016;91:44–56. https://doi.org/10.1016/j.biomaterials.2016.03.013.

Liu J, Chen Q, Zhu W, et al. Nanoscale-coordination-polymer-shelled manganese dioxide composite nanoparticles: a multistage redox/pH/H_2O_2-responsive cancer theranostic nanoplatform. Adv Funct Mater. 2017;27:1605926. https://doi.org/10.1002/adfm.201605926.

Mandal M, Kruk M. Family of single-micelle templated organosilica hol low nanospheres and nanotubes synthesized through adjustment of organo silica surfactant ratio. Chem Mater. 2012;24:123–32. https://doi.org/10.1021/cm202136r.

Peer D, et al. Nanocarriers as an emerging platform for cancer therapy. Nat Nanotechnol. 2007;2:751–60. https://doi.org/10.1038/nnano.2007.387.

Rao GV, Lopez GP. Encapsulation of poly(N-Isopropyl Acrylamide) in silica: a stimuli-responsive porous hybrid material that incorporates molecular nano-valves. Adv Mater. 2000;12:1692–5. https://doi.org/10.1002/1521-4095(200011)12:22<1692::aid-adma1692>3.0.co;2-c.

Saint-Cricq P, Deshayes S, Zink JI, et al. Magnetic field activated drug delivery using thermodegradable azo-functionalised PEG-coated core-shell mesoporous silica nanoparticles. Nanoscale. 2015;7:13168–72. https://doi.org/10.1039/c5nr03777h.

Sierra JR, Tsao MS. c-MET as a potential therapeutic target and biomarker in cancer. Ther Adv Med Oncol. 2011;3:S21–35. https://doi.org/10.1177/1758834011422557.

Stecker JR, Savage AA, Bruno JG, García DM, Koke JR. Dynamics and visualization of MCF7 adenocarcinoma cell death by aptamer-C1q-mediated membrane attack. Nucl Acid Ther. 2012;22:275–82. https://doi.org/10.1089/nat.2012.0355.

Stephen ZR, Dayringer CJ, Lim JJ, et al. Approach to rapid synthesis and functionalization of iron oxide nanoparticles for high gene transfection. ACS Appl Mater Interfaces. 2016;8:6320–8. https://doi.org/10.1021/acsami.5b10883.

Sun X, Zhang M, Du R, et al. A polyethyleneimine-driven self-assembled nanoplatform for fluorescence and MR dual-mode imaging guided cancer chemotherapy. Chem Eng J. 2018;350:69–78. https://doi.org/10.1016/j.cej.2018.05.157.

Tan B, Lehmler H, Vyas SM, et al. Fluorinated-surfactant-templated synthesis of hollow silica particles with a single layer of mesopores in their shells. Adv Mater. 2005;17:2368–71. https://doi.org/10.1002/adma.200500344.

Vasani RB, McInnes SJ, Cole MA, et al. Stimulus-responsiveness and drug release from porous silicon films ATRP-grafted with poly(N-isopropylacrylamide). Langmuir. 2011;27:7843–53. https://doi.org/10.1021/la200551g.

Wang C, Ravi S, Martinez GV, et al. Dual-purpose magnetic micelles for MRI and gene delivery. J Control Release. 2012;163:82–92. https://doi.org/10.1016/j.jconrel.2012.04.030.

Wang FB, et al. Recognition and capture of metastatic hepatocellular carcinoma cells using aptamer-conjugated quantum dots and magnetic particles. Biomaterials. 2013;34:3816–27. https://doi.org/10.1016/j.biomaterials.2013.02.018.

Wang L, Li D, Hao Y, et al. Gold nanorod - based poly (lactic-co-glycolic acid) with manganese dioxide core - shell structured multifunctional nanoplatform for cancer theranostic applications. Int J Nanomedicine. 2017;12:3059–75. https://doi.org/10.2147/ijn.s128844.

Xing H, et al. Selective delivery of an anticancer drug with aptamer-functionalized liposomes to breast cancer cells in vitro and in vivo. J Mater Chem B Mater Biol Med. 2013;1:5288–97. https://doi.org/10.1039/c3tb20412j.

Yu J, Li X, Luo Y, et al. Poly(ethylene glycol) shell-sheddable magnetic nanomicelle as the carrier of doxorubicin with enhanced cellular uptake. Colloids Surf B Biointerfaces. 2013;107:213–9. https://doi.org/10.1016/j.colsurfb.2013.02.009.

Zhu G, Niu G, Chen X. Aptamer-drug conjugates. Bioconjug Chem. 2015;26:2186–97. https://doi.org/10.1021/acs.bioconjchem.5b00291.

Tumor Microenvironment-Responsive Core-Shell Nanoconstructs in Cancer Theragnostics

4

Deshmukh Aaishwaryadevi and Jayvadan K. Patel

Abstract

Core shell nanoconstructs responsive to the tumor microenvironment (TME) have become a promising tool for advancing cancer theranostics, integrating both therapeutic and diagnostic capabilities to augment treatment accuracy and efficacy. These constructs, which consist of nanoparticles and targeted ligands, are specifically engineered to address challenges posed by traditional cancer therapies, such as non-specific drug distribution, systemic toxicity, and uncontrolled drug release. By leveraging the distinctive properties of the TME, including its acidic and hypoxic conditions, these nanoconstructs can precisely target tumors, improving the delivery and controlled release of therapeutic agents. Recent developments in the design of TME-responsive nanoconstructs focus on utilizing the unique attributes of the TME, such as pH fluctuations, low oxygen levels, and specific enzymes, to trigger the release of the therapeutic cargo. This enables a more localized and controlled drug delivery, minimizing side effects while enhancing treatment efficacy. These advances make TME-responsive nanoconstructs a promising approach to improving cancer treatment outcomes. Furthermore, the complex cellular and extracellular components of the TME, such as tumor cells, immune cells, and the surrounding matrix, play critical roles in cancer progression and response to treatments. Targeting these features with innovative nanoconstructs holds the potential to enhance the efficacy of traditional therapies and overcome resistance mechanisms that limit their effectiveness.

D. Aaishwaryadevi (✉)
School of Pharmaceutical Sciences, JSPM University, Pune, Maharashtra, India

J. K. Patel
Viesain Pharma LLC, GA, USA

Faculty of Pharmacy, Sankalchand Patel University, Visnagar, Gujarat, India

J. K. Patel et al. (eds.), *Core-Shell Nano Constructs for Cancer Theragnostic*,
https://doi.org/10.1007/978-981-96-3025-7_4

1 Introduction

Cancer remains a significant contributor to global mortality rates. Among the various treatments available, chemotherapy stands out as a widely adopted approach. However, the effectiveness of traditional small-molecule anticancer drugs against solid tumors is hindered by several challenges, including limited bioavailability, poor water solubility, and inadequate targeting (Xu et al. 2018a). The emergence of nanotechnology has transformed chemotherapy by enabling nanomedicine development that offers enhanced efficacy and safety compared to conventional treatments (Shi et al. 2017). Despite substantial progress in the field, many tumors still present challenges in terms of diagnosis and treatment, resulting in high mortality rates. A key barrier to cancer nanomedicines is their effective delivery to solid tumor targets. This delivery primarily relies on two mechanisms: active as well as passive targeting. The nanoparticles (NPs), which have prolonged circulation times, tend to accrue passively in the interstitial space of the tumor, leveraging heightened permeability and retention effect driven due to weakened lymphatics as well as leaky vasculature typical of rapidly-growing tumors. On the other hand, active targeting involves modifying the surfaces of nanoparticles with definite molecular ligands like peptides and/or antibodies, to enhance cellular uptake through receptor-mediated endocytosis (Du et al. 2015). While active targeting holds promise for increasing drug accumulation at tumor sites, its efficacy is often hindered by diverse factors like hypoxia, heterogeneity as well as limited endosomal escape like microenvironmental factors within tumors (Rosenblum et al. 2018).

There has been a surge in the stimuli-responsive polymers and nanoparticle development that have the capability of undergoing significant physicochemical changes in retort to changes in temperature, pH or enzymatic activity. These stimuli-responsive platforms offer the potential for targeted drug delivery by facilitating controlled drug release, uniform distribution within tumors, and augmented uptake in cells due to TME (Du et al. 2015). In comparison to healthy tissue, the tumor microenvironment (TME) exhibits distinctive features, including cell hypoxia, pH (acidic), and elevated concentrations of certain enzymes. While conventional nanoparticles bank on both passive as well as active targeting systems for tumor localization, nanoparticles that are TME-responsive offer numerous benefits. Active targeting hinges on the interface between target moieties or ligands besides receptors present on the surfaces of tumor cells. However, receptor allocation and concentration can vary among different tumor cell populations, limiting the pronounced applications of active targeting approaches. In contrast, TME-responsive nanoparticles exploit the typical biological characteristics shared by all types of cancer. This universal approach involves the liberation of anticancer drugs in a site-specific manner triggered by TME-associated abnormalities such as redox cellular environment, hypoxia, various enzymes, reactive oxygen species (ROS), and pH (Uthaman et al. 2018).

2 Tumor Microenvironment (TME)

Chemotherapy, surgery, and radiation therapy are conventional treatments for cancer, with the choice of modality depending on the stage of the disease. With the identification of various critical carcinogenic mutations, these treatment options have often been used as the foundation for the development of targeted cancer therapies (Huang 2021). Despite the effectiveness of these treatments, the increasing resistance to these drugs has resulted in frequent tumor recurrence, creating significant challenges in patient care. Cancerous tumor is not merely a collection of tumor cells, but a complex heterogeneous mixture of secreted factors, extracellular matrix, as well as resident infiltrating host cell. Tumor cells drive substantial changes at physical, cellular, and molecular levels in surrounding host tissues to foster the growth and progression of tumors. The tumor microenvironment (TME) is a dynamic and evolving structure that plays a crucial role in tumor development. Although the composition of the TME differs among tumor types, characteristic features comprise different types of cells and matrix-like ECM, blood vessels, immune as well as stromal cells. In advanced solid tumors, TME is quite complex and heterogeneous. Genetic variations among cancer cells primarily contribute to this complexity (Wu et al. 2021). The express proliferation of cancer cells often induces irreversible consequences, for instance hypoxia, which leads to metabolic changes and adaptations within TME. The interactions between cancer cells and neighboring stromal and immune cells further modulate the TME, often resulting in disorganized blood vessel formation and promoting metastasis. The evolving nature of tumor formation involves dynamic responses from both cancer cells and TME components to shifting environmental factors, which ultimately influence tumor growth (Lane et al. 2020). Promising therapeutic strategies influenced by TME factors include immunotherapies, antiangiogenic treatments, therapies targeting cancer-associated fibroblasts (CAFs), and the extracellular matrix (ECM). Other approaches under investigation or already approved for clinical use include cold atmospheric plasma, oncolytic viral therapies, bacterial therapies, nanovaccines, and repurposed drugs used in combination treatments.

3 Nanomedicines

Nanotechnology has revolutionized various scientific fields, with medicine being a major beneficiary, commonly referred to as nanomedicine (Leong and Ng 2014). Nanomedicines consist of a wide variety of particles that differ in size, shape, and material composition at the nanoscale, displaying exceptional physicochemical properties that offer vast potential for both diagnostics and therapeutic applications. These materials include organic substances like layer-by-layer assembled structures, lipid-based formulations, and even cell membrane-derived constructs along with inorganic counterparts alike silica, iron oxide, and/or gold (Tong and Langer 2015). The small size and biocompatibility of these nanomaterials make them effective at overcoming biological barriers encountered during systemic drug delivery.

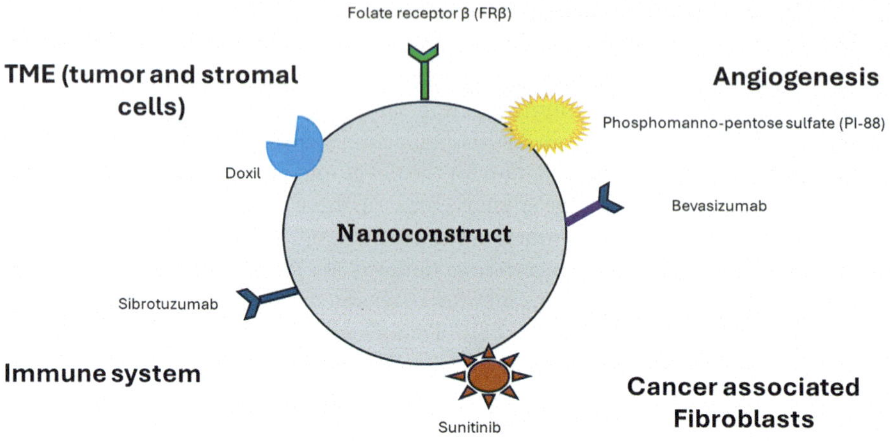

Fig. 4.1 Functionalization of nanomaterials with various ligands to distinct cell populations (target) within TME, including stromal and tumor cells, facilitating a collective approach for treatment

Additionally, their surface chemistry and high surface-to-volume ratio enable the incorporation of various biological moieties, allowing for customization of each nanomedicine for specific uses (Fig. 4.1) (Leong and Ng 2014). Furthermore, these nanocarriers support regulated sustained drug release, achieved owing to their design and external prompts such as changes in reduction potential or pH. This controlled release improves drug targeting, enhancing drug uptake and intracellular delivery while minimizing systemic toxicity. Depending on the specific biological modification, the aforementioned nanosystems can be applied in various applications like therapeutic, diagnostic as well as more recently as theranostics, providing a versatile platform that integrates both diagnostic and therapeutic functionalities into a single system (Overchuk and Zheng 2018).

Theranostics presents a unique ability to trigger drug release through the detection of specific biomarkers, while concurrently allowing real-time monitoring of disease progression. This capability is supported by distinct optical features, such as imaging technologies, fluorophores, or magnetic resonance (MR) contrast agents (Leong and Ng 2014). The integration of nanomedicine with innovative cancer therapeutic models has led to targeted strategies for the effective delivery of both single-agent as well as combination therapies (Fernandes et al. 2017).

A multifunctional nanocarrier can simultaneously target multiple cell types, benefiting from advances in understanding cancer pathophysiology and TME components, such as vascular irregularities, oxygen levels, blood flow, pH, and metabolic conditions. This knowledge supports the development of diverse nanosystems tailored to specific therapeutic needs (Leong and Ng 2014). By controlling the pharmacokinetics and dynamics of various drugs, nanomedicines have the potential to improve therapeutic outcomes, particularly in combination therapies, where drug co-delivery can prevent multidrug resistance. Additionally, nanomedicines can enhance conventional cancer treatments by aiding in tumor resection and boosting the effectiveness of radiotherapy. Nanomedicines accumulate in solid tumors through passive and/or active targeting mechanisms, with active targeting often relying on initial passive accumulation in the tumor. Passive targeting utilizes the enhanced permeability and retention (EPR) effect, taking advantage of the leakiness of tumor vasculature and inadequate lymphatic drainage (Chauhan and Jain 2013). Active targeting involves the modification of nanocarrier surfaces with ligands that recognize specific receptors expressed on the target cells.

Recent research has focused on using nanotechnology to address components of the TME. For example, Mardhian et al. conjugated human relaxin-2 (RLX) to superparamagnetic iron oxide nanoparticles (SPION) to prevent differentiation of stellate cells of the pancreas into cancer-associated fibroblast (CAF)-like myofibroblasts, leading to significant suppression of adenocarcinoma growth in pancreatic ducts (Mardhian et al. 2018). In another study, anti-angiogenic peptides-coated gold nanoparticles were effective in inhibiting blood vessel formation in ex ovo chick chorioallantoic membrane assay (Roma-Rodrigues et al. 2016). While the aforementioned nanomedicines show promise in overcoming some limitations of conventional and targeted therapies, FDA-approved nanoparticles such as Abraxane, DaunoXome, and Doxil/Caelyx have not shown substantial improvements in patient survival rates. This highlights the need for further optimization of nanomedicines, taking into account factors such as tumor type, stage, and TME characteristics. A one-size-fits-all approach seems insufficient (Chauhan and Jain 2013).

Wang and colleagues recently examined "tumor-on-a-chip" platforms, which are designed to aid the clinical application of nanoparticle-based therapies for cancer (Wang et al. 2018). Studies by Overchuck and Zheng proposed combining nanomedicines with pre-treatment approaches to enhance targeting efficiency. By adapting nanomedicine characteristics to the specific conditions of tumor microenvironment (TME) cells and considering the inherent variability within the TME, subtle modifications such as pre-treatments involving radiation, hyperthermia, and photodynamic therapy can promote the accumulation of nanoformulations in the TME. To optimize drug delivery, smart nanoformulations utilize the augmented expression of matrix metalloproteinases (MMPs) within the TME. Undoubtedly, nanomedicines are emerging as one of the most promising and effective cancer treatments, though there is still significant room for enhancement. Numerous nanoparticle-based therapies targeting the TME are currently in clinical trials, with one already approved in Europe: liposomal mifamurtide. This therapy targets the immune system by activating monocytes and macrophages (Siegler et al. 2016). A current observational study (NCT03737435) examined a patient's TME with mifamurtide-treated localized osteosarcoma, and a phase II trial reviewed its effectiveness to post-operative chemotherapy as an adjunct therapy in high-risk osteosarcoma. CRLX101 is another nanomedicine which is a polymeric nanoparticle consisting of Camptothecin directing HIF-1, and is presently involved in two active clinical trials. In the first trial CRLX101 in combination with Enzalutamide is used for pharmacotherapy of prostate cancer (NCT03531827), while the other trial incorporates CRLX101 with Olaparib for the treatment of relapsed/refractory small cell lung cancer (NCT02769962). Furthermore, Phase I trials (NCT00356980 and NCT00436410) are investigating CYT-6091, a gold nanoparticle formulation combined with TNFα, for targeted immune system modulation. All of these trial and studies suggest expanding interest and possibility of nanomedicine in transforming cancer treatment (Mishra et al. 2023).

4 Nanoconstructs

Nanoconstructs are dual-component systems consisting of a 'hard' nanoparticle core and a 'soft' biomolecular ligand shell, which are widely used in targeted drug delivery applications. The nanosystems play a role as transporters for functional therapeutic agents, permitting not just the conservation, but also the transportation amid biological barriers, and most importantly release of drugs in a controlled manner. The major advantages of chemotherapy which is nanoparticle-based is enhanced specificity therapy, greater accumulation of drug at target sites, and diminished systemic toxicity.

The successful design of nanoconstructs ensures their ability to retain a variety of agents, either on their surface or within their core, which makes them ideal for combination therapies. After administration, the behavior of nanoconstructs is influenced not only by their surface chemistry but also by factors such as vascular dynamics, including pressure, flow velocity, and tissue heterogeneity. Computational modeling approaches, based on parameters such as size, shape, surface properties, and stiffness (referred to as the 4S parameters), are utilized to predict whether nanoconstructs will interact with vessel walls or remain circulating in the bloodstream (Başağaoğlu et al. 2013; Coclite et al. 2017). As a result, computational techniques are vital for gaining insights into these complex phenomena and for designing optimal nanoconstructs for biomedical applications.

4.1 Nanoconstructs and Cancer-Targeting Approach

The advancement in drug or gene delivery systems having loading capacities higher for tumor targeting and curtailing damage of healthy cells are decisive for effective treatment. Understanding of different mechanisms of targeting systems and interaction processes concerning tumor cells, and nanoparticles (NPs) is crucial. There are various types of targeting systems including but not limited to active and passive targeting, stimuli-responsive targeting, as well as magnetic targeting. Passive targeting depends on diffusion-mediated drug delivery across complex drug-carrier systems. The complex is further transported through the systemic circulation to the tumor site. Leaky blood vessels and diminished lymphatic drainage, special features of solid tumors, are exploited by nanoconstructs which are designed for passive targeting permitting the particles to outflow from the tumor vasculature and collect in tumor tissues with the help of improved permeability and retention (EPR) effect (Black et al. 2014). There are many factors required for prominent passive drug targeting like particle size, surface characteristics, surface charge as well as

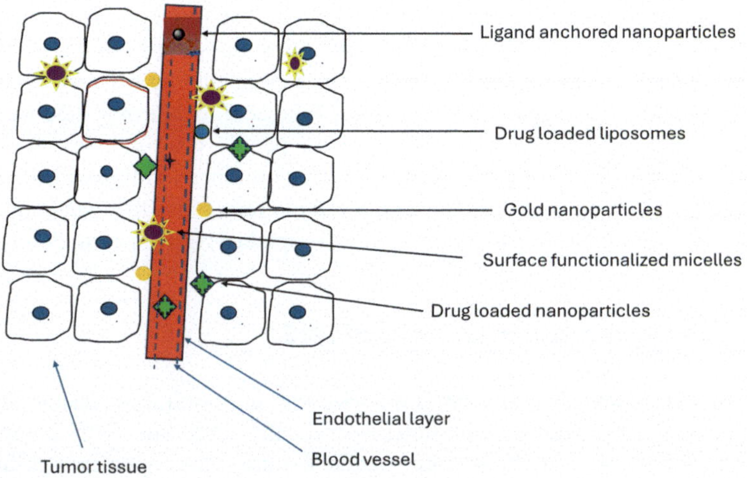

Fig. 4.2 Different targeting methods viz. passive and active

molecular weight. For instance, polyethylene glycol (PEG)-coated stealth liposomes demonstrate extensive circulation times in the bloodstream, wherein surface charge is a significant factor in their extended existence. Furthermore, nanoparticles based on poly(lactic-co-glycolic acid) (PLGA) have demonstrated potential in loading enormous quantities of drugs, while preserving superior stability, specificity along with controlled release pattern, and decreased degradation rates (Rezvantalab et al. 2018). Yet another material, polyvinyl alcohol (PVA) which is utilized for passive targeting, has assisted in encapsulating hydrophobic drugs and aiding their controlled release at peculiar sites on tumor (Wei et al. 2020).

As illustrated in Fig. 4.2, passive targeting remains a widely used approach for delivering drugs to cancerous cells.

Active targeting utilizes specific ligands or molecules that bind to receptors overexpressed on target cells. This method, known as ligand-mediated targeting, involves nanoparticles (NPs) that are functionalized with ligands designed to enhance uptake and retention, increasing their affinity for the target cells. Strong-binding ligands form ligand-receptor complexes, which initiate receptor-mediated endocytosis. Ligands like folate and transferrin on the cell surface capture specific target molecules, boosting NP attachment to cancer cells and facilitating improved drug penetration. Stimuli-responsive targeting capitalizes on the unique features of the tumor microenvironment (TME) and the abnormal cellular mechanisms in cancer. Nanoconstructs can be engineered to respond to external stimuli (e.g., light, magnetic fields) or internal stimuli (e.g., pH, enzymes, oxidative stress). Internal stimuli-responsive targeting uses compounds such as calcium carbonate and glutathione within nanocarriers to respond to pH changes. Transducers alter phototherapy or magnetic fields (external stimuli), into light radiation (physical stimuli), that in turn can be transformed into heat to successfully destroy tumor cells (Fang et al. 2021). Superparamagnetic nanoparticles coated with an anticancer drug are used by magnetic targeting and piloted to the tumor site with the virtue of a strong magnetic field, which has exhibited favorable conclusions, suggesting high specificity and fewer side effects. For developing nanoconstructs targeted drug delivery, ligands are necessary. A variety of ligands, like peptides, folic acid, hyaluronic acid as well as antibodies, are exploited to accomplish tumor-specific drug delivery. Moreover, ligands or peptides fostering cellular invasion are combined to improve the intracellular trafficking of antitumor drugs. Merging numerous types of ligands raises the target site selectivity besides increasing cellular uptake (Nguyen et al. 2021).

4.2 Aiming the Common Traits of TME

The tumor microenvironment (TME) possesses several unique characteristics that can be utilized to design nanoparticles specifically targeting TME components (Table 4.1) (Uthaman et al. 2018). One notable feature is the acidic extracellular pH

Table 4.1 TME-responsive nanoparticles (Uthaman et al. 2018)

Nanoparticle type	TME stimuli	Functionalities
Methoxy (polyethylene glycol) thioketal-poly(ε-caprolactone) (mPEG-TK-PCL) micelles	ROS	Drug release responsive to reactive oxygen species (ROS)
Dendrimer nanoparticles	pH	pH-responsive drug release
Human serum albumin nanoparticle	pH/H_2O_2	Degradation of nanoparticles under pH and H_2O_2 conditions, resulting in smaller polymer-drug conjugates
His-tagged fluorescent fusion protein chimera $NiFe_2O_4$-based magnetic nanoparticles	Enzyme	MMP-2 enzyme cleavable peptide linker
Polyethyleneimine (PEI) conjugated alkylated 2-nitroimidazole (NI) and hyaluronic acid (HA) conjugated chlorin e6 (Ce6)	Hypoxia	Drug release activated by hypoxia and light
Gold nanocluster	pH	pH-sensitive drug release
Gold nanoparticles	pH and redox	Drug release controlled by pH with disassembly mediated by glutathione (GSH)
Polymeric nanoparticles (PLGA)	Enzyme (MMP-2)	MMP-2 enzyme-sensitive peptide linker for drug release
Polymer-drug conjugates (PEG-b-PLA micelles)	pH and redox	pH-triggered drug release
Hollow mesoporous titanium dioxide nanoparticles	Hypoxia	Hypoxia-triggered drug release via ultrasound-induced microenvironment creation
Liposomal nanoparticles	Hypoxia	Drug release induced by hypoxic conditions
Heparosan- and deoxycholic acid-conjugated micelle	Redox	GSH-triggered drug release and degradation

within the TME, which typically ranges between pH 6.5 and pH 6.9, compared to pH of 7.2 to 7.5 (physiological range) in normal tissues, results from the raised glycolysis rate of tumor cells, producing lactic acid from glucose for ATP production. The disparity in pH offers significant potential for the development of cancer-targeting systems that are pH-responsive. A new defining trait of the TME is hypoxia, wherein cells located in the deeper regions of the tumor experience oxygen deprivation (Bogurcu et al. 2018). This condition arises due to irregular and poorly structured vasculature networks within solid tumors. Hypoxic cells tend to replicate gradually in comparison to well-oxygenated cells and exhibit resistance to standard anti-proliferative treatments. In addition to hypoxia and pH, the TME discloses amended expression levels of specific enzymes, which can be leveraged for targeted therapeutic release. Many enzymes overexpressed in the TME belong to protease families, such as lipases, MMPs, or phospholipase A2 (Radisky et al. 2017), the substrate specificity of which has driven the advancement in nanomaterials having enzyme-responsive approach for targeted drug delivery. Furthermore, cancer cells within the TME experience increased oxidative stress owing to high concentrations of free radicals like superoxide anion, hydroxyl radicals, and/or hydrogen peroxide (Wang et al. 2017). For alleviating the oxidative stress, cancer cells often increase the reduction potential through overexpression of redox molecules like glutathione (GSH) and superoxide dismutase (SOD). This adaptation leads to an overall imbalance in oxidative and reductive potentials, which can be exploited to design nanoparticles responsive to these redox changes. Cancer cells also generate elevated concentrations of ROS as compared to normal cells due to oncogenic transformations and aerobic metabolism (Mo and Gu 2016). These endogenous stimuli within the TME collectively provide a strong foundation for the development of nanoparticles that can be activated by the unique conditions of the TME.

Nanoparticles can respond to the tumor microenvironment (TME) due to their altered pH levels. Various pH-sensitive nanoparticles have been engineered with specific molecular structures and pKa values that align with the mildly acidic interstitial pH of tumors. These nanoparticles undergo structural transformations in response to the acidic TME. The acidic environment triggers the protonation of pH-sensitive components, disrupting the equilibrium between hydrophilic and hydrophobic domains. This process facilitates structural rearrangements and releases the encapsulated therapeutic agents. Typically, pH-sensitive nanoparticles are constructed using ionizable groups or acid-labile linkages (John et al. 2017). Poly(histidine) (pHis) is extensively employed in pH-sensitive drug delivery systems owing to its unique pH-dependent properties, stemming from the lone-pair electrons on the imidazole group's unsaturated nitrogen atom. Studies have shown the development of triblock copolypeptides, such as poly(ethylene glycol) methyl ether acrylate-block-poly(L-lysine)-block-poly(L-histidine), for targeted drug delivery. These nanoparticles remain stable under physiological conditions (pH 7.4) but destabilize in acidic environments due to the presence of pHis blocks (Johnson et al. 2015). This destabilization ensures controlled therapeutic agent release, resulting in dose-dependent cytotoxicity in murine cancer models.

Nanoparticles are also deliberated to exhibit pH-sensitive surface charge variations. Zwitterionic polymers, which contain both cationic and anionic groups, are particularly effective in this regard. Their surface charge transitions from positive in acidic conditions to negative in basic environments, while remaining neutral at physiological pH due to balanced hydrophobicity. This neutral charge shifts upon entry into tumor cells, enabling structural changes that promote drug release. Kang et al. (2018) introduced tumor microenvironment-sensitive theranostic nanoparticles with a fluorescence "on-off" mechanism. Encapsulating a photothermal dye (IR 825) in a carbonized zwitterionic polymer, these nanoparticles exhibited fluorescence quenching under neutral pH due to hydrophobic and π-π stacking interactions. However, pH shift in the TME restored fluorescence and released IR 825, enabling simultaneous diagnostic and therapeutic applications.

pH-sensitive nanoparticles have also been developed using acid-labile linkages, such as hydrazone, orthoester, imine, and phosphoramidate, to ensure drug release through hydrolysis. Liao et al. (2018) synthesized tumor-targeting nanoparticles for doxorubicin (DOX) delivery by conjugating DOX to a hyaluronic acid (HA) backbone via hydrazone linkage. In aqueous environments, these conjugates self-assembled into nanoparticles. HA facilitated receptor-mediated targeting by binding to CD44, a receptor highly expressed in cancer cells. These polymeric prodrugs selectively released DOX in response to pH changes. A major limitation of pH-sensitive nanoparticles is their inability to function effectively in the perivascular region, as the required acidic pH is typically present farther from blood vessels. Moreover, the slight pH difference between normal and tumor tissues may not always suffice to trigger the desired responsiveness.

4.2.1 Nanoparticles Activation by Hypoxia

Hypoxia significantly contributes to processes like tumor angiogenesis, metastasis, epithelial-to-mesenchymal transition, tumor invasiveness, and immune suppression. Consequently, substantial efforts have been made to design nanoparticles capable of targeting hypoxic regions within tumors. For instance, He et al. (2018) developed dual-sensitive nanoparticles that respond to both hypoxia and photoactivation for controlled anticancer drug release. These nanoparticles were created by self-assembling polyethyleneimine-nitroimidazole micelles (PEI-NI), which were further combined with Ce6-linked hyaluronic acid (HA). Nitroimidazole (NI), a hypoxia-sensitive electron acceptor, enabled activation under hypoxic conditions by converting its hydrophobic segments into hydrophilic 2-aminoimidazole, facilitating the release of doxorubicin (DOX) from the nanoparticles. There are few hypoxia-sensitive groups regularly exploited in nanoparticle design, azobenzene (AZO) being one of them which has been unified between polyethylene glycol (PEG) and polyethyleneimine (PEI) for producing nanocarriers for siRNA delivery (Perche et al. 2014). It was observed that in the hypoxic TME, the azobenzene bond gets cleaved, which in turn removes the PEG coating which further uncovers the positively charged PEI/siRNA nanoparticles, enhancing the cellular uptake.

Studies by Xie et al. (2018) showed that hypoxia-responsive nanoparticles could be developed for the co-delivery of DOX and siRNA. The nanoparticles were fabricated by coupling a polyethylene glycol (PEG) to polyamidoamine (PAMAM) dendrimer by an azobenzene linker which led to the formation of PAMAM-AZO-PEG (PAP). siRNA targeting hypoxia-inducible factor 1-alpha (HIF-1α) was attached electrostatically to its surface whereas DOX was encapsulated in the PAMAM dendrimer's hydrophobic core. The PEG coating which was present in the formulation diminished opsonization which in turn led to the prolongation of nanoparticle circulation time. Moreover, when exposed to hypoxic TME, PEG got cleaved because of activation of azobenzene, further exposing positively charged PAMAM improved cellular uptake coupled with endosomal escape through the proton sponge effect which ultimately permitted controlled release of DOX as well as siRNA. A new methodology purported by Yang et al. (2018a) signifies creating hollow silica nanoparticles by incorporating Ce6 and catalase (CAT) into the silica matrix. Enzyme like CAT, catalyzes the breakdown of hydrogen peroxide (H_2O_2) into molecules viz. water and oxygen, improving cellular hypoxic conditions. Further functionalization was conducted in these nanoparticles with (3-carboxypropyl) triphenyl phosphonium (CTPP) bromide which is a mitochondrial targeting molecule, as well as pH-responsive charge-convertible polymer through electrostatic interaction. It was shown that the polymer's charge shifted from negative to positive in the acidic TME, advancing cellular uptake. Photodynamic therapy (PDT)-induced cell death was also improved by the mitochondrial targeting moiety, whereas the decaying of H_2O_2 decreased the hypoxia, considerably enhancing PDT efficacy in solid tumors. Despite advancements in hypoxia-responsive nanoparticles, effectively delivering these systems to hypoxic regions within tumors remains challenging. These areas, often deep within tumors, lack sufficient vasculature, relying primarily on diffusion for transport. Inappropriately, however, suboptimal diffusion rates were exhibited by many nanoparticle systems in solid tumors. Hence, a realistic method for improving delivering hypoxia-activated prodrugs in the TME can be achieved by increasing the diffusion abilities of nanocarriers for small molecules.

4.2.2 Nanoparticles Activation by Enzymes

Matrix metalloproteinases (MMPs), especially MMP-2, play a critical part in not just tumor growth but also progression, especially in TME. Their levels are raised in the TME making them the top target for the development of enzyme-responsive nanoparticles. In a study by Sun et al. (2017a) MMP-2 activatable nanoprobes were developed by engineering self-assembling nickel ferrite nanoparticles and hexahistidine-tagged (His-Tagged) fluorescent proteins for accurate cancer imaging at cellular levels. The nickel ferrite nanoparticles worked as a binder for His-Tagged fluorescent proteins as well as the fluorescent quencher. The nanoprobes exhibited heightened cellular uptake and reinstated fluorescence, upon activation of MMP-2, permitting for effective visualization of the nanoparticles in the cancer tissues. Further, a polymeric conjugate was developed by Ma et al. (2018) specifically aimed at mitochondrial targeting plus paclitaxel (PTX) delivery. These conjugates were constructed by employing a PAMAM dendrimer core, to which PTX was coupled through the disulfide bonds and triphenylphosphine through amino bonds. For extension of its circulation time in the systemic circulation, polyethylene glycol (PEG) was connected via an MMP-2-sensitive peptide (GPLGIAGQ). These conjugates showed accumulation in tumor tissues due to improved permeability and retention (EPR) effect. Moreover, inside the tumor cells, MMP-2 cleavage of the peptide also lead to the separation of the PEG layer, revealing PAMAM dendrimer for mitochondrial targeting and the successive delivery of PTX into the cytoplasm.

In yet another study, Ansari et al. (2014) created theranostic nanoparticles with the dual function of enzyme-specific drug release and in vivo magnetic resonance imaging (MRI). These nanoparticles were prepared by conjugation of an FDA-approved iron oxide nanoparticle ferumoxytol with an azademethylcolchicine (ICT) and MMP-14-sensitive peptide forming CLIO-ICT. The study further showed that after achieving access to the tumor site, MMP-14 activity triggered the transition of the nanoparticle from a non-toxic to a toxic form, which in turn released the potent ICT. The overall model was assessed and a facilitation in the real-time monitoring of drug accumulation and localization within the tumor was found by MRI imaging. There are many enzymes overexpressed in different types of cancer, β-Galactosidase (β-gal) being one of them. This enzyme has been which has been exclusively utilized in enzyme-responsive prodrug design. Sharma et al. (2018) developed a theranostic prodrug for colon cancer treatment that utilized receptor-mediated targeting and enzyme-responsive activation. β-Gal was used to target asialoglycoprotein (ASGP) receptors and activate the prodrug. Upon entering colon cancer cells through receptor-mediated endocytosis, enzymatic activation triggered the release of the anticancer drug, enhancing the therapeutic effect. While enzyme-responsive therapies offer promising potential, they face significant challenges, particularly due to the heterogeneous expression of target enzymes across different cancer types and stages. To improve the efficacy and precision of these therapies, it is essential to gain a deeper understanding of the spatial and temporal distribution of enzyme expression at the target site (Fig. 4.3).

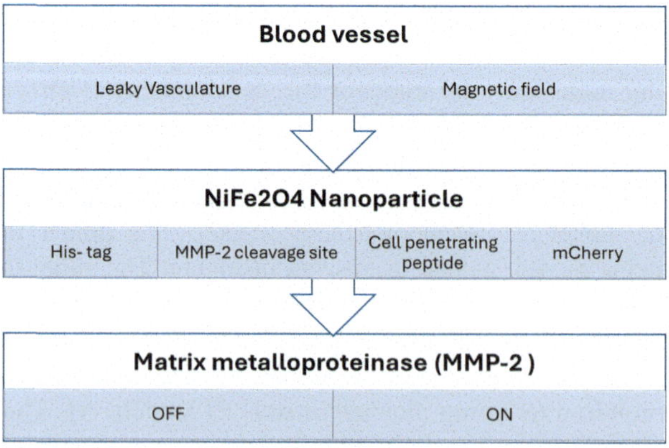

Fig. 4.3 Enzyme-responsive magnetic nanoprobe

4.2.3 Activation of Nanoparticles Through Redox Environment

The tumor microenvironment (TME) is classified according to the intracellular glutathione (GSH) concentrations, which are notably elevated ranging from 0.5 to 10 × 10^{-3}M as compared to those in normal tissues (concentrations are almost four times lower). Raised GSH concentrations are markedly evident in cell organelles like cytosol, mitochondria, and nucleus, as compared to the extracellular areas. These variations in GSH levels offer a new prospect for producing nanoparticles selectively releasing therapeutic agents upon their arrival in tumor cells (Luo et al. 2016). A conspicuous approach engages in deploying bio-reducible disulfide bonds, which further exhibits nanoparticles receptive to the reductive intracellular environments, facilitating the effective release of drugs. A redox-sensitive drug delivery system was developed by Sun et al. (2018) for treating laryngopharyngeal carcinoma. This delivery system elaborated on initiating an amphiphilic polymer by combining heparosan to deoxycholic acid utilizing disulfide bonds. This polymer formed nanoparticles then got disassembled on the cleavage of the disulfide bonds in a reductive environment, permitting drug release within the tumor cells. Similarly, the hybrid nanoparticles that combined zwitterionic polymers have also been persuaded to counter GSH as well as endosomal pH. There is yet another notable paradigm that is a redox-sensitive system developed by Zhou et al. (2017), which utilizes a dextran-indomethacin conjugate (DEX-SS-IND), which in turn employs a disulfide bridge (cystamine) for combining dextran with indomethacin, leading to the formation of micelles encapsulating doxorubicin (DOX). Under reducing conditions, the cleavage of the disulfide bond between dextran and indomethacin leads to the release of DOX. Certain in vivo studies have steered these micelles demonstrating potent antitumor activity rather than systems unresponsive to redox conditions.

Furthermore, Xia et al. (2018) expanded the research by creating polycarbonate-based core-crosslinked nanoparticles (CC-RRNs) with the aim of directed DOX delivery to tumor cells. The nanoparticles were generated by employing a click reaction involving α-lipoic acid, PEG-b-poly(MPC)n (PMPC) as well as 6-bromohexanoic acid, comprehended by crosslinking with a catalytic concentration of dithiothreitol (DTT). Further, the disulfide-linked core not only permitted the release of DOX under redox conditions in a controlled manner but also emphasized the perspective of multifunctional systems for nanomedicine targeting the TME. Nevertheless, the thorough intracellular mechanisms that control the conduct of redox-sensitive nanoparticles are yet to be comprehended entirely. Many studies have implied the role of surface thiols on cells in the uptake of disulfide-conjugated peptides (Aubry et al. 2009). Hence, more studies on the intracellular processes and nanoparticles activation in redox environments are crucial for expanding their therapeutic efficiency (Fig. 4.4).

4.2.4 ROS-Responsive Nanoparticles

An elevated level of ROS is found in cancer cells than normal cells, because of increased ROS production derived from aerobic metabolism piloted by oncogenic mutations. This increased ROS concentration offers an effective target for the development of nanoparticles responsive to ROS, which further aids in the regulated drug

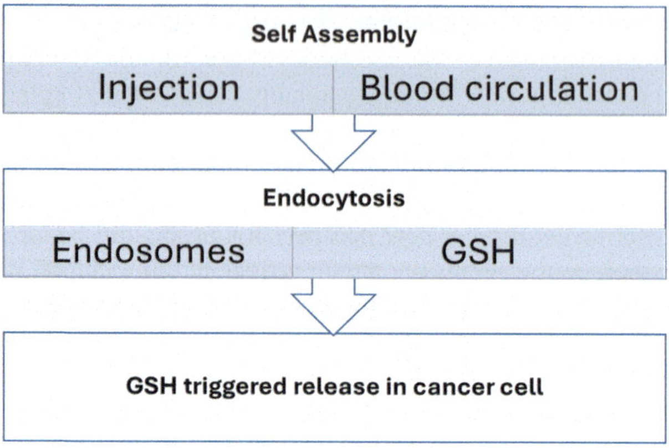

Fig. 4.4 Mode of release of DOX via self-assembled micelle and GSH-triggered mechanism

release explicitly at the tumor site. Functional groups like boronic esters, thioketals, and sulfides are generally employed in these. With all these developments in the current scenario, it is imperative that ROS-responsive nanoparticles are progressively being recognized as efficient tools for improving the delivery of chemotherapy drugs. Sun et al. (2017b) combined a ROS-sensitive thioketal linker having π-conjugated structure into methoxy (polyethylene glycol)-thioketal-poly(ε-caprolactone) (mPEG-TK-PCL) micelles. These micelles were responsive to ROS and had the anticipation of increasing the efficiency of drug delivery. These micelles were prepared by a self-assembly mechanism, encapsulating doxorubicin (DOX). The thioketal linker underwent cleavage when subjected to a prominent level of ROS in tumor cells, which further augmented the release of drug and improved cancer cell inhibition, both of which resulted in enhanced antitumor efficacy. Correspondingly, a ROS-sensitive prodrug was prepared by Xu et al. (2018b) by bringing together polyethylene glycol (PEG) to DOX via a thioketal bond. This prodrug showed self-assembly into micelles with DOX and demonstrated greater antitumor effects compared to non-responsive micelles made from poly(ethylene glycol)-block-polycaprolactone (PEG2k-PCL5k).

Yu et al. (2018) developed ROS-responsive polycarbonates for use in photodynamic therapy (PDT). These were synthesized through the ring-opening polymerization of cyclic carbonate monomers with ethyl selenide, phenyl selenide, or ethyl telluride groups. Amphiphilic block copolymers were prepared using PEG as a macro-initiator, forming nanoparticles less than 100 nm in size. These nanoparticles were stable in neutral conditions but completely dissociated when exposed to ROS. When loaded with DOX and Ce6, the nanoparticles exhibited enhanced drug release upon laser-induced generation of singlet oxygen (1O2), which triggered the breakdown of the nanoparticles and accelerated DOX liberation. Despite the advancements, there are several challenges associated with ROS-responsive nanoparticles. These include ensuring that the ROS-sensitive linkers are biocompatible, maintaining the stability of these systems during circulation and within healthy cells, and managing variability in ROS levels across individuals and conditions to enable personalized therapies. As a result, the careful selection of linkers and carriers is essential for developing efficient and tailored therapeutic solutions.

5 Cancer Theranostics and Nanoconstructs

Cancer theranostics represents a transformative approach that combines diagnostic and therapeutic functionalities, offering significant potential for both clinicians and patients. This strategy integrates real-time monitoring of treatment with targeted delivery systems, enhancing the precision and effectiveness of cancer therapies (Ryu et al. 2012). Theranostic applications leverage the detection of unique cellular phenotypes by specialized agents at tumor sites, facilitating simultaneous therapy and monitoring of these compounds (Table 4.2) (Mishra et al. 2023). Nanotechnology plays a pivotal role in advancing cancer diagnosis and treatment. A broad spectrum of organic nanoparticles—such as polymers, lipids, dendrimers, and liposomes—and inorganic nanoparticles—including metallic nanoparticles, carbon nanotubes,

Table 4.2 Theranostic applications of nanoconstructs in cancer

Carrier system	Material	Drug	Ligand	Indication
Polymeric nanoparticle	PLGA	Temozolomide	Cetuximab	Targeting EGFR-overexpressing cancers
Polymeric nanoparticle	PLGA	ETP	LF	Increased anticancer activity in glioblastoma cells
Polymeric nanoparticle	PLGA	Docetaxel	Transferrin	Improved targeting and reduced toxicity in breast cancer
Polymeric nanoparticle	PLGA	DTX	Anti-EGFR antibody	Enhanced cytotoxicity and targeting in non-small cell lung cancer
Bovine nanoparticles	BSA	Rg5	FA	Therapy for breast cancer
Bovine nanoparticles	BSA	Paclitaxel	HA	Ovarian cancer treatment
Silica or mesoporous silica nanoparticles	Silica	Paclitaxel	HA	Targeted therapy for breast cancer
Mesoporous silica	Mesoporous silica	DOX	HA	Improved targeting in HeLa cells
Mesoporous silica	Mesoporous silica	Zinc complexes	Chitosan-Biotin	Enhancing chemotherapy efficacy

(continued)

Table 4.2 (continued)

Carrier system	Material	Drug	Ligand	Indication
Mesoporous silica	Mesoporous silica	Epirubicin	GalNAc	Targeted therapy for hepatocellular carcinoma
Metallic nanoparticles	Monoolein	Copper acetylacetonate	HA	Targeting CD44-positive tumors
Metallic nanoparticles	Gold	siRNA	FA	Gene silencing for therapeutic purposes
Dendrimer	PAMAM	miRNA	Ferritin	Treating myeloid leukemia
Dendrimer	Selenium	pDNA	FA	Specific targeting of cancer cells
SLN	Stearic acid	Curcumin	Transferrin	Treatment for prostate cancer
SLN	Stearic acid	DOX	FA	Therapy for brain cancer
SLN	Stearic acid	DOX	Peptide	Targeting prostate cancer
SLN	Diacylglyceride	Tamoxifen citrate	Transferrin	Treatment of breast cancer
Carbon nanotube	Multiwalled carbon	DOX	FA	Breast cancer treatment
Carbon nanotube	Multiwalled carbon	DOX	Carbohydrate (galactose, mannose, lactose)	Targeting breast cancer
Carbon nanotube	Multiwalled carbon	GEM	HA	Colon cancer therapy
Liposomes	Phospholipids	DOX and SFB	Peptide	Therapy for breast cancer
Liposomes	Phospholipids	AChE	Transferrin	Liver cancer therapy
Liposomes	Phospholipids	Paclitaxel	PCSK9	Enhanced anticancer effects in various cell lines
Liposomes	Phospholipids	5-fluorouracil	FA	Improving drug delivery and cytotoxicity
Graphene oxide	Graphene oxide	Metformin	HA	Therapy for triple-negative breast cancer
Graphene oxide	Monoolein	Paclitaxel	EGFR antibody fragment	Ovarian cancer treatment
DNA nanorobot	DNA origami	Blood coagulation protease thrombin	DNA aptamer	Treatment for melanoma and ovarian cancer

and quantum dots—have been developed, offering imaging, therapeutic, and targeting capabilities. The application of theranostic nanoparticles has become a major area of research, with a significant focus on their use in chemotherapy, photothermal therapy, and the delivery of siRNA/miRNA, among other cancer treatments. These multifunctional nanoconstructs hold great promise for improving the efficacy and precision of cancer therapeutics.

5.1 Inorganic Material-Based Nanoconstructs

Inorganic nanoparticles (NPs), composed of materials such as elemental metals, metal oxides, and carbon-based substances, offer distinctive properties that make them highly suitable for medical applications, including imaging and drug delivery. Among these, black phosphorus (BP) has garnered attention for its two-dimensional nanomaterial properties. To enhance the stability of BP-based systems, heterogeneous doping with metals, polymers, or biomolecules like albumin and folic acid is often employed. These constructs offer targeted cancer treatments due to their function of reacting to particular stimuli, for instance, pH changes or near-infrared (NIR) irradiation (Pandey et al. 2021). One of the therapeutic approaches was the involvement of BP conjugated with poly-L-lysine (PLL) for delivering Cas13a/crRNA for Mcl-1 targeting, which is a protein linked to breast cancer cell proliferation. The results in in vitro studies uncovered a large decline in Mcl-1 expression and in turn the tumor cell activity (Schacter et al. 2014). Mesoporous silica nanoparticles (MSNs) can also be utilized to achieve pH-responsive drug-carrier phenomenon. Since these nanoparticles hitch the drug molecules by pH-sensitive linkages, a controlled delivery is achieved in acidic tumor environments. Equally, upconversion nanoparticles (UCNPs), linked to converting low-energy photons to high-energy photons, have also been exploited for theranostic application with both real-time imaging and therapy. By coating UCNPs with mesoporous silica and integrating metal-phenolic networks, researchers achieved simultaneous imaging and drug delivery, enhancing therapeutic outcomes (Hu et al. 2019). Hybrid nanoconstructs like copper sulfide (CuS) nanoparticles integrated with mesoporous silica nanoshells have also been explored. These systems, labeled with radiotracers, provide multimodal imaging capabilities and effective tumor eradication with minimal side effects (Goel et al. 2018). Another approach incorporates hyaluronic acid as a targeting moiety, enabling precise delivery to cancer cells overexpressing CD44 receptors. Trio-responsive systems combining photothermal agents, chemotherapeutics, and targeting moieties have shown enhanced therapeutic efficacy by reducing tumor volume and inducing apoptosis (Poudel et al. 2020). Efforts in targeted alpha therapy, a cutting-edge radiotherapy, involve conjugating radionuclides like 225Ac with silica nanoparticles. This technique demonstrated selective cytotoxicity and minimal off-target effects, showing potential for improving cancer treatment outcomes (Pallares et al. 2020; Dam et al. 2014). Research by Dam et al. (2014) displayed an improvement in the in vitro efficacy of nanoconstructs that consisted of gold, since gold has shown to enhance the loading of G-quadruplex aptamers. The data of this particular study demonstrated the avenue of loading the external surface of gold nanostars (AuNSs) through oligonucleotides and DNA aptamers by employing citrate-type buffers at low pH (Dam et al. 2014). Thus, these progresses keep an undertaking for conquering the confronts experienced in targeted alpha therapy and thus expanding its efficiency in cancer treatment.

5.2 Polymer-Based Nanoconstructs

Polymeric nanoconstructs are becoming increasingly valuable in cancer theranostics because of their ability to undergo surface modifications, respond to stimuli, and encapsulate both hydrophilic and lipophilic therapeutic or diagnostic agents. A specific example of these constructs is the deformable discoidal nanoconstructs, which represent a promising new platform for targeted drug delivery and imaging. These nanostructures are produced by polymerizing polyethylene glycol (PEG) and PLGA into a discoidal form, where the polymeric matrices feature both hydrophobic and hydrophilic domains that function as reservoirs for various compounds used in imaging and therapy. Their prolonged circulation time in the bloodstream helps avoid early clearance by the mononuclear phagocyte system (MPS), making them ideal for sustained delivery. Furthermore, these polymer matrices can incorporate agents like contrast materials, lipid-drug conjugates, and polymer-drug conjugates, thus enabling the creation of comprehensive theranostic agents. One novel technique to monitor these agents is Fluorescence Resonance Energy Transfer (FRET), which tracks drug release from the nanoparticles at tumor sites. For instance, DOX-loaded polyethylene glycol-block-peptide (FFKY)-block-tetraphenylethylene (PEG-PepTPE/DOX) NPs were tested in A549 cells, showing a cooperative effect between DOX and the self-assembled peptide, which allowed for real-time monitoring of the drug's release (Wang et al. 2021). In addition to imaging, PLGA-based nanoconstructs are being researched for their potential in radiodynamic therapy, a strategy that uses the generation of reactive oxygen species (ROS) at the tumor site. This approach is particularly useful in targeting tumors in hypoxic conditions, where oxygen scarcity leads to the production of ROS that induces tumor cell death. One innovative radiotherapy approach involves PLGA nanoparticles loaded with verteporfin and perfluorooctylbromide, which, under both normoxic and hypoxic conditions, significantly increase ROS production. This therapy has been shown to reduce the viability of pancreatic cancer cells by approximately 60% in human studies and to inhibit tumor growth over a period of 2 weeks. These findings emphasize the potential of radiodynamic therapy-based nanoconstructs as a non-invasive treatment for deeply located tumors in hypoxic regions (Clement et al. 2021). Another category of polymeric nanoconstructs with promising applications in cancer therapy is polyurethane (PU) nanoconstructs. As a part of the stimuli-responsive and biodegradable materials family, PU-based nanoconstructs serve as effective drug delivery systems. These constructs feature fast drug release rates, the ability to solubilize hydrophobic chemotherapy agents, and can be engineered to target specific cells. They are also sensitive to various stimuli such as pH, temperature, and external factors, which makes them versatile carriers for a wide range of applications (Gajbhiye et al. 2020).

Lipid-based nanoconstructs, such as PEGylated squalene (SQPEG), have been explored for their potential in cancer treatment as well. These nanostructures self-assemble with the lipophilic compound pyropheophorbide-a (Ppa) to create nanoparticles of approximately 200 nm in size, with a drug loading capacity of 18% (w/w). Notably, these constructs exhibit nearly complete fluorescence quenching, which allows researchers to assess their bioavailability and phototoxicity by light-induced eradication in vivo. In a study using a chick embryo model implanted with U87MG glioblastoma microtumors, these nanoconstructs showed excellent diagnostic capabilities (Adriouach et al. 2019). In addition to polymer- and lipid-based nanoconstructs, dendrimer-based nanostructures offer significant advantages as theranostic agents. Dendrimers, especially poly(amidoamine) (PAMAM) dendrimers, are commonly used in cancer nanomedicine due to their numerous functional groups and well-defined internal structures. However, early-generation dendrimers face challenges such as small size, which limits their drug loading capacity, and insufficient tumor targeting due to retention and permeability issues. To overcome these shortcomings, next-generation superstructured dendrimer nanoconstructs (SDNs) have been developed to enhance both loading capacity and targeting specificity.

5.3 Dendrimer-Based Nanoconstructs

Poly(amidoamine) (PAMAM) dendrimers are branched, having a spherical structure, which renders them greatly efficient at stabilizing metallic nanoparticles, while encapsulating the therapeutic agents. To validate this principle, a study verified that the development of the dendrimer-gold hybrid system in combination with fifth-generation PAMAM dendrimers with PEGylated amine-terminated $AuCl_4^-$ ions further loaded with MUC-1 aptamer linked curcumin demonstrated superior cytotoxic effects on HT29 and C26 cells compared to non-targeted systems, showcasing its potential in both cancer treatment and CT imaging applications (Alibolandi et al. 2018). Another application involved designing hyperbranched PAMAM micelles conjugated with the F3 peptide to target nucleolin, a protein overexpressed in MDA-MB-231 cells. These micelles, termed PAMAM-DOX-F3, were loaded with DOX and showed enhanced cellular uptake. PET imaging was performed using ^{64}Cu-chelated PAMAM-DOX-F3 micelles to track their behavior in vivo. Results revealed rapid and persistent accumulation in MDA-MB-231 tumors, outperforming non-targeted versions (Yang et al. 2018b).

5.4 Miscellaneous Nanoconstructs

Combining therapeutic agents or treatment modalities has become increasingly common to achieve synergistic effects in cancer therapy. While powerful drugs may work individually, their efficacy can often be improved when paired with complementary agents. For instance, an experiment in rabbits demonstrated that a single injection of αVβ3 integrin-targeted paramagnetic nanoparticles carrying fumagillin, combined with oral atorvastatin, produced prolonged antiangiogenic effects (Winter et al. 2008). Another innovative approach involved developing hybrid nanoparticles that encapsulated docetaxel (DTX) and gold nanorods modified with DOX. These nanostructures provided dual-chemotherapy and plasmonic photothermal therapy (PPTT). NIR radiation enabled controlled DOX release, while DTX diffused gradually. The system achieved synergistic cytotoxicity in MDA-MB231 cells after NIR exposure (Villar-Alvarez et al. 2019). Additionally, a multifunctional nanocarrier was created using superparamagnetic iron oxide nanoparticles (SPIONs) coated with folic acid-modified PAMAM dendrimers (FA-PAMAM). To enhance its therapeutic potential, the dendrimer was loaded with 3,4-difluorobenzylidene-curcumin (CDF), a potent hydrophobic anticancer drug. The resulting nanocarrier (SPIONs@FA-PAMAM-CDF) exhibited strong magnetic resonance imaging contrast and significant anticancer activity against SKOV3 and HeLa cancer cells (Luong et al. 2017).

6 Strategies to Enhance Nanoconstruct Delivery to Tumor Sites

Addressing the challenges related to nanoconstruct delivery can significantly improve their ability to penetrate and accumulate within tumors. A commonly used classification to facilitate effective drug release and improve nanoconstruct functionality includes autonomous and non-autonomous drug delivery systems. Non-autonomous systems can be further divided into actively and passively targeted nanoconstructs. One promising approach for cancer treatment involves altering the tumor microenvironment (TME) using nanoparticles. Tumor cells often reside in hypoxic conditions, where they switch to alternative energy pathways due to the lack of oxygen and ATP. Targeting this metabolic shift can potentially control cancer growth. For example, nanoparticles have been designed to deliver TRAP-1 inhibitors, a mitochondrial chaperone that could interfere with these energy production pathways in cancer cells. Iron oxide nanoparticles (IONs) conjugated with the Hsp90 inhibitor geldanamycin (GA) and mitochondria localization signal (MLS) peptides have been developed to selectively target tumors (Amash et al. 2020).

Additionally, studies have investigated the effect of nanoconstruct shape on endosomal trafficking. For instance, spiky gold nanoparticles were compared with spherical ones in terms of their impact on endosomal behavior, suggesting that surface curvature influences nanoparticle fate within the cell. This opens up avenues for optimizing nanoconstruct design for better intracellular processing (Lee et al. 2022). Autonomous delivery systems are often referred to as "therapeutic missiles" because they can direct nanoconstructs to target sites without relying on blood flow. Significant progress has been made with biohybrid bacteria used in cancer therapy. These bacteria, capable of self-propulsion, can transport nanomedicines to cancer sites and self-destruct once they reach the target, releasing the therapeutic agents. For example, motile *Salmonella* bacteria have been utilized to deliver doxorubicin (DOX)-loaded liposomes, achieving efficient tumor targeting and penetration, particularly in triple-negative cancer cells. The motility of these bacteria is influenced by glucose and pH levels within the TME, contributing to the therapeutic efficacy (Pontier-Bres et al. 2012). Additionally, the amalgamation of artificial intelligence (AI) in nanomedicine extends transformative possibilities, particularly in personalized treatment strategies. The fundamental confront in cancer therapy is drug synergy, further persuaded by several dynamics like dose, scheduling, and patient-specific traits. Machine learning and neural networks the mainstay AI technologies, can supply innovative resolutions for beating these confronts by yielding effective prediction and drug delivery system designs. For instance, artificial neural networks (ANNs) have been commissioned to simulate complicated arrangements, to name a few, controlled release drug delivery systems (CRDDS), where the correlation linking formulation and drug release is non-linear. These models can perfectly replicate the complex dynamics of drug release and assist in the optimization of treatment protocols (Cheung and Rubin 2021; Patel and Patel 2016).

Moreover, AI can enhance the competence of nanomedicine formulations by forecasting the encapsulation efficiency (EE%) in imaging agents as well as drugs, which is critical for boosting therapeutic outcomes. In recent times, a model employing Quantitative Structure-Property Relationships (QSPR) has attained an accuracy of more than 90% in predicting whether molecules can be effectively loaded into carriers on the basis of their chemical properties and processing conditions (Soltani et al. 2021). Likewise, AI can also improve the evaluation of key parameters specifically for personalized medicine i.e. nanoparticle cytotoxicity and biocompatibility. In medical imaging, AI also facilitates the analysis of therapeutic effectiveness and the biological distribution of nanomaterials. Nanorobotics, a cutting-edge technology that constructs robots at the nanoscale, holds exciting possibilities for the biomedical field. DNA-based nanomaterials, controlled by aptamer-encoded logic gates, represent a novel class of autonomous drug delivery systems. These DNA nanorobots, capable of executing complex tasks such as targeted drug delivery and real-time monitoring, are poised to revolutionize cancer therapy. DNA origami can be used to create nanorobots that deliver drugs to specific cancer cells, such as HER2-positive breast cancer, while simultaneously monitoring biomarkers and detecting cancer-associated miRNAs (Katsnelson 2012; Mirzaiebadizi et al. 2022).

7 Future Prospects

Nanoconstructs designed for theranostic purposes in cancer therapy hold significant potential for advancing detection methods, enhancing drug delivery precision to tumors, and minimizing side effects on healthy tissues. The use of tailored probes and imaging technologies allows real-time monitoring of treatment effectiveness, improving therapeutic outcomes. Computational modeling of the tumor microenvironment (TME) and cancer-related processes can guide the design of nanoconstructs with characteristics suited to the needs of individual patients. Both organic and inorganic nanomaterials are being explored for theranostic applications, with these platforms utilizing exogenous and endogenous stimuli to control drug release and enhance targeting, using both autonomous and non-autonomous mechanisms.

While technologies like DNA origami, nanorobots, AI, and machine learning hold enormous promise for advancing cancer theranostics, their full potential has yet to be realized. Although significant progress has been made, only a limited number of nanoconstructs have reached clinical investigation stages. The ongoing development of advanced nanomaterials presents the possibility of creating multimodal nanoparticles with enhanced diagnostic and therapeutic functions. To unlock the full potential of nanotechnology in cancer care, multidisciplinary research involving large patient populations will be essential. Collaboration across scientific fields will be crucial for translating these innovative nanoconstructs from the laboratory to clinical settings, ultimately improving cancer detection and treatment.

Therapeutic strategies targeting the TME show considerable promise as anticancer approaches. While some therapies targeting the TME have been successful, others have not delivered the expected results. Even successful treatments like immune checkpoint inhibitors (ICIs) are effective for specific tumor types and a small subset of patients, with an immunosuppressive TME often limiting their effectiveness. Ongoing research is focused on developing strategies to activate antitumor immunity, combat immunosuppressive processes, and improve the effectiveness of immune-targeted therapies. Despite the progress in immunotherapy, challenges such as low response rates, difficulties in determining clinical efficacy, and potential complications like autoimmune reactions or cytokine release syndromes continue to limit its broader clinical application.

The role of tumor-infiltrating immune cells, particularly T cells, is central to cancer immunotherapy, underscoring the importance of understanding immune cell behavior within the TME. Gaining deeper insights into the interactions between immune cells and cancer cells in the TME is crucial for identifying new therapeutic targets, developing predictive biomarkers, and refining immunotherapy approaches. A thorough understanding of the spatial organization of immune cells and cancer cells in the TME can offer valuable insights into cancer progression and may improve the efficacy of existing therapies. Future research on the TME will likely adopt a broader approach that considers the entire tumor ecosystem, rather than focusing solely on specific cell types. Additionally, comprehensive patient analysis, considering factors such as systemic conditions (e.g., gut microbiota, metabolism, lifestyle) and underlying health issues (e.g., aging, inflammation, obesity), will be vital, as these factors influence the TME and treatment responses. With increasing amounts of available data, the goal of targeting the TME to treat a larger population of cancer patients is becoming more achievable. Advances in TME research are expected to revolutionize cancer treatment and enhance outcomes for a wider group of patients affected by cancer.

References

Adriouach S, Vorobiev V, Trefalt G, Allémann E, Lange N, Babič A. Squalene-PEG: pyropheophorbide-a nanoconstructs for tumor theranostics. Nanomed Nanotechnol Biol Med. 2019;15:243–51. https://doi.org/10.1016/j.nano.2018.09.013.

Alibolandi M, Hoseini F, Mohammadi M, Ramezani P, Einafshar E, Taghdisi SM, et al. Curcumin-entrapped MUC-1 aptamer targeted dendrimer-gold hybrid nanostructure as a theranostic system for colon adenocarcinoma. Int J Pharm. 2018;549:67–75. https://doi.org/10.1016/j.ijpharm.2018.07.052.

Amash V, Paithankar K, Dharaskar SP, Arunachalam A, Amere SS. Development of nanocarrier-based mitochondrial chaperone, TRAP-1 inhibitor to combat cancer metabolism. ACS Appl Bio Mater. 2020;3:4188–97. https://doi.org/10.1021/acsabm.0c00268.

Ansari C, et al. Development of novel tumor-targeted theranostic nanoparticles activated by membrane-type matrix metalloproteinases for combined cancer magnetic resonance imaging and therapy. Small. 2014;10:566–75.

Aubry S, Burlina F, Dupont E, Delaroche D, Joliot A, Lavielle S, Chassaing G, Sagan S. Cell-surface thiols affect cell entry of disulfide-conjugated peptides. FASEB J. 2009;23:2956–67.

Başağaoğlu H, Allwein S, Succi S, Dixon H, Carrola JT, Stothoff S. Two- and three-dimensional lattice Boltzmann simulations of particle migration in microchannels. Microfluid Nanofluid. 2013;15:785–96. https://doi.org/10.1007/s10404-013-1191-z.

Black KCL, Wang Y, Luehmann HP, Cai X, Xing W, Pang B, et al. Radioactive 198Au-doped nanostructures with different shapes for in vivo analyses of their biodistribution, tumor uptake, and intratumoral distribution. ACS Nano. 2014;8:4385–94. https://doi.org/10.1021/nn406258m.

Bogurcu N, Seidel S, Garvalov BK, Acker T. Analysis of hypoxia and the hypoxic response in tumor Xenografts. Methods Mol Biol. 2018;1742:283–300.

Chauhan VP, Jain RK. Strategies for advancing cancer nanomedicine. Nat Mater. 2013;12:958–62.

Cheung HMC, Rubin D. Challenges and opportunities for artificial intelligence in oncological imaging. Clin Radiol. 2021;76:728–36. https://doi.org/10.1016/j.crad.2021.03.009.

Clement S, Guller A, Mahbub SB, Goldys EM. Oxygen-carrying polymer nanoconstructs for radiodynamic therapy of deep hypoxic malignant tumors. Biomedicine. 2021;9:322. https://doi.org/10.3390/biomedicines9030322.

Coclite A, Mollica H, Ranaldo S, Pascazio G, de Tullio MD, Decuzzi P. Predicting different adhesive regimens of circulating particles at blood capillary walls. Microfluid Nanofluid. 2017;21:168. https://doi.org/10.1007/s10404-017-2003-7.

Dam DHM, Lee RC, Odom TW. Improved in vitro efficacy of gold nanoconstructs by increased loading of G-quadruplex aptamer. Nano Lett. 2014;14:2843–8. https://doi.org/10.1021/nl500844m.

Du J, Lane LA, Nie S. Stimuli-responsive nanoparticles for targeting the tumor microenvironment. J Control Release. 2015;219:205–14.

Fang Z, Shen Y, Gao D. Stimulus-responsive nanocarriers for targeted drug delivery. New J Chem. 2021;45:4534–44. https://doi.org/10.1039/d0nj05169a.

Fernandes AR, Jesus J, Martins P, Figueiredo S, Rosa D, Martins LM, Corvo ML, Carvalheiro MC, Costa PM, Baptista PV. Multifunctional gold nanoparticles: a nanovectorization tool for the targeted delivery of novel chemotherapeutic agents. J Control Release. 2017;245:52–61.

Gajbhiye KR, Chaudhari BP, Pokharkar VB, Pawar A, Gajbhiye V. Stimuli-responsive biodegradable polyurethane nano-constructs as a potential triggered drug delivery vehicle for cancer therapy. Int J Pharm. 2020;588:119781. https://doi.org/10.1016/j.ijpharm.2020.119781.

Goel S, Ferreira CA, Chen F, Ellison PA, Siamof CM, Barnhart TE, et al. Activatable hybrid nanotheranostics for tetramodal imaging and synergistic photothermal/photodynamic therapy. Adv Mater. 2018;30:1704367. https://doi.org/10.1002/adma.201704367.

He H, Zhu R, Sun W, Cai K, Chen Y, Yin L. Selective cancer treatment via photodynamic sensitization of hypoxia-responsive drug delivery. Nanoscale. 2018;10(6):2856–65.

Hu F, Liu B, Chu H, Liu C, Li Z, Chen D, et al. Real-time monitoring of pH-responsive drug release using a metal-phenolic network-functionalized upconversion nanoconstruct. Nanoscale. 2019;11:9201–6. https://doi.org/10.1039/c9nr01892a.

Huang J. Current developments of targeting the p53 signaling pathway for cancer treatment. Pharmacol Ther. 2021;220:107720.

John JV, Uthaman S, Augustine R, Chen HY, Park IK, Kim I. pH/redox dual stimuli-responsive sheddable nanodaisies for efficient intracellular tumour triggered drug delivery. J Mater Chem B. 2017;5:5027–36.

Johnson RP, Zhang H, Schmidt T, Sun X, Young M, Demillo VG, et al. Poly(PEGA)-b-poly(L-lysine)-b-poly(L-histidine) hybrid vesicles for tumoral pH-triggered intracellular delivery of doxorubicin hydrochloride. ACS Appl Mater Interfaces. 2015;7:21770–9.

Kang EB, Lee JE, Mazrad ZAI, In I, Jeong JH, Park SY. pH-responsible fluorescent carbon nanoparticles for tumor selective theranostics via pH turn on/off fluorescence and photothermal effect in vivo and in vitro. Nanoscale. 2018;10:2512–23.

Katsnelson A. DNA robot could kill cancer cells. Nature. 2012;2012:10047. https://doi.org/10.1038/nature.2012.10047.

Lane AN, Higashi RM, Fan TW. Metabolic reprogramming in tumors: contributions of the tumor microenvironment. Genes Dis. 2020;7(2):185–98.

Lee K, Jung I, Odom TW. Delivery order of nanoconstructs affects intracellular trafficking by endosomes. J Am Chem Soc. 2022;144:5274–9. https://doi.org/10.1021/jacs.2c02276.

Leong DT, Ng KW. Probing the relevance of 3D cancer models in nanomedicine research. Adv Drug Deliv Rev. 2014;79-80:95–106.

Liao JH, Chen HY, Hsu YH, Chuang EY, Juang JH, Hsu WH, et al. Tumor-targeting and pH-responsive nanoparticles from hyaluronic acid for the enhanced delivery of doxorubicin. Int J Biol Macromol. 2018;113:737–47.

Luo C, et al. Facile fabrication of tumor redox-sensitive nanoassemblies of small-molecule oleate prodrug as potent chemotherapeutic nanomedicine. Small. 2016;12(46):6353–62.

Luong D, Sau S, Kesharwani P, Iyer AK. Polyvalent folate dendrimer-coated iron oxide theranostic nanoparticles for simultaneous magnetic resonance imaging and precise cancer cell targeting. Biomacromolecules. 2017;18:1197–209. https://doi.org/10.1021/acs.biomac.6b01885.

Ma P, et al. Overcoming multidrug resistance through the GLUT1-mediated and enzyme-triggered mitochondrial targeting conjugate with redox sensitive paclitaxel release. ACS Appl Mater Interfaces. 2018;15(10):12351–63.

Mardhian DF, Storm G, Bansal R, Prakash J. Nano-targeted relaxin impairs fibrosis and tumor growth in pancreatic cancer and improves the efficacy of gemcitabine in vivo. J Control Release. 2018;290:1–10.

Mirzaiebadizi A, Ravan H, Dabiri S, Mohammadi P, Shahba A, Ziasistani M, et al. An intelligent DNA nanorobot for detection of MiRNAs cancer biomarkers using molecular programming to fabricate a logic-responsive hybrid nanostructure. Bioprocess Biosyst Eng. 2022;45:1781–97.

Mishra S, Bhatt T, Kumar H, Jain R, Shilpi S, Jain V. Nanoconstructs for theranostic application in cancer: challenges and strategies to enhance the delivery. Front Pharmacol. 2023;14:1101320.

Mo R, Gu Z. Tumor microenvironment and intracellular signal-activated nanomaterials for anti-cancer drug delivery. Mater Today. 2016;19:274–83.

Nguyen PV, Allard-Vannier E, Chourpa I, Hervé-Aubert K. Nanomedicines functionalized with anti-EGFR ligands for active targeting in cancer therapy: biological strategy, design and quality control. Int J Pharm. 2021;605:120795.

Overchuk M, Zheng G. Overcoming obstacles in the tumor microenvironment: recent advancements in nanoparticle delivery for cancer theranostics. Biomaterials. 2018;156:217–37.

Pallares RM, Agbo P, Liu X, An DD, Gauny SS, Zeltmann SE, et al. Engineering mesoporous silica nanoparticles for targeted alpha therapy against breast cancer. ACS Appl Mater Interfaces. 2020;12:40078–84. https://doi.org/10.1021/acsami.0c11051.

Pandey A, Nikam AN, Padya BS, Kulkarni S, Fernandes G, Shreya AB, et al. Surface architectured black phosphorous nanoconstructs based smart and versatile platform for cancer theranostics. Coord Chem Rev. 2021;435:213826. https://doi.org/10.1016/j.ccr.2021.213826.

Patel J, Patel A. Artificial neural networking in controlled drug delivery. In: Patel J, Patel A, editors. Artificial neural network for drug design, delivery and disposition. Amsterdam: Elsevier; 2016. p. 195–218. https://doi.org/10.1016/B978-0-12-801559-9.00010-7.

Perche F, Biswas S, Wang T, Zhu L, Torchilin VP. Hypoxia-targeted siRNA delivery. Angew Chem Int Ed Engl. 2014;53:3362–6.

Pontier-Bres R, Prodon F, Munro P, Rampal P, Lemichez E, Peyron JF, et al. Modification of Salmonella typhimurium motility by the probiotic yeast strain Saccharomyces boulardii. PLoS One. 2012;7:e33796.

Poudel K, Banstola A, Tran TH, Thapa RK, Gautam M, Ou W, et al. Hyaluronic acid wreathed, trio-stimuli receptive and on-demand triggerable nanoconstruct for anchored combinatorial cancer therapy. Carbohydr Polym. 2020;249:116815.

Radisky ES, Raeeszadeh-Sarmazdeh M, Radisky DC. Therapeutic potential of matrix metalloproteinase inhibition in breast cancer. J Cell Biochem. 2017;118:3531–48.

Rezvantalab S, Drude NI, Moraveji MK, Güvener N, Koons EK, Shi Y, et al. PLGA-based nanoparticles in cancer treatment. Front Pharmacol. 2018;9:1260. https://doi.org/10.3389/fphar.2018.01260.

Roma-Rodrigues C, Heuer-Jungemann A, Fernandes AR, Kanaras AG, Baptista PV. Peptide-coated gold nanoparticles for modulation of angiogenesis in vivo. Int J Nanomedicine. 2016;11:2633–9.

Rosenblum D, Joshi N, Tao W, Karp JM, Peer D. Progress and challenges towards targeted delivery of cancer therapeutics. Nat Commun. 2018;9:1410.

Ryu JH, Koo H, Sun IC, Yuk SH, Choi K, Kim K, et al. Tumor targeting multi-functional nanoparticles for theragnosis: new paradigm for cancer therapy. Adv Drug Deliv Rev. 2012;64:1447–58. https://doi.org/10.1016/j.addr.2012.06.012.

Schacter JL, Henson ES, Gibson SB. Estrogen regulation of antiapoptotic Bcl-2 family member Mcl-1 expression in breast cancer cells. PLoS One. 2014;9:e100364.

Sharma A, Kim EJ, Shi H, Lee JY, Chung BG, Kim JS. Development of a theranostic prodrug for colon cancer therapy by combining ligand-targeted delivery and enzyme-stimulated activation. Biomaterials. 2018;155:145–51.

Shi JJ, Kantoff PW, Wooster R, Farokhzad OC. Cancer nanomedicine: progress, challenges and opportunities. Nat Rev Cancer. 2017;17:20–37.

Siegler EL, Kim YJ, Wang P. Nanomedicine targeting the tumor microenvironment: therapeutic strategies to inhibit angiogenesis, remodel matrix, and modulate immune responses. J Cell Immunother. 2016;2:69–78.

Soltani M, Moradi Kashkooli F, Souri M, Zare Harofte S, Harati T, Khadem A, et al. Enhancing clinical translation of cancer using nanoinformatics. Cancer. 2021;13:2481. https://doi.org/10.3390/cancers13102481.

Sun L, et al. Cell-permeable, MMP-2 activatable, nickel ferrite and his-tagged fusion protein self-assembled fluorescent nanoprobe for tumor magnetic targeting and imaging. ACS Appl Mater Interfaces. 2017a;9:39209–22.

Sun C, Liang Y, Hao N, Xu L, Cheng F, Su T, Cao J, Gao W, Pu Y, He B. A ROS-responsive polymeric micelle with a π-conjugated thioketal moiety for enhanced drug loading and efficient drug delivery. Org Biomol Chem. 2017b;15:9176–85.

Sun C, Li X, Du X, Wang T. Redox-responsive micelles for triggered drug delivery and effective laryngopharyngeal cancer therapy. Int J Biol Macromol. 2018;112:65–73.

Tong R, Langer R. Nanomedicines targeting the tumor microenvironment. Cancer J. 2015;21:314–21.

Uthaman S, Huh KM, Park I-K. Tumor microenvironment-responsive nanoparticles for cancer theragnostic applications. Biomater Res. 2018;22:22.

Villar-Alvarez E, Cambón A, Pardo A, Arellano L, Marcos AV, Pelaz B, et al. Combination of light-driven co-delivery of chemodrugs and plasmonic-induced heat for cancer therapeutics using hybrid protein nanocapsules. J Nanobiotechnol. 2019;17:106. https://doi.org/10.1186/s12951-019-0538-3.

Wang L, Huo M, Chen Y, Shi J. Tumor microenvironment-enabled nanotherapy. Adv Healthc Mater. 2017;7:1704490.

Wang Y, Cuzzucoli F, Escobar A, Lu S, Liang L, Wang SQ. Tumor-on-a-chip platforms for assessing nanoparticle-based cancer therapy. Nanotechnology. 2018;29:332001.

Wang K, Shang F, Chen D, Cao T, Wang X, Jiao J, et al. Protein liposomes-mediated targeted acetylcholinesterase gene delivery for effective liver cancer therapy. J Nanobiotechnol. 2021;19:31. https://doi.org/10.1186/s12951-021-00777-9.

Wei QY, Xu YM, Lau ATY. Recent progress of nanocarrier-based therapy for solid malignancies. Cancer. 2020;12:2783–837. https://doi.org/10.3390/cancers12102783.

Winter PM, Caruthers SD, Zhang H, Williams TA, Wickline SA, Lanza GM. Antiangiogenic synergism of integrin-targeted fumagillin nanoparticles and atorvastatin in atherosclerosis. JACC Cardiovasc Imaging. 2008;1:624–34. https://doi.org/10.1016/j.jcmg.2008.06.003.

Wu P, Gao Y, Zhang Z, et al. Adaptive mechanisms of tumor therapy resistance driven by tumor microenvironment. Front Cell Dev Biol. 2021;9:357.

Xia YC, Wang N, Qin Z, Wu J, Wang F, Zhang L, Xia X, Li J, Lu Y. Polycarbonate-based core-crosslinked redox-responsive nanoparticles for targeted delivery of anticancer drug. J Mater Chem B. 2018;6:3348–57.

Xie ZQ, et al. Targeting tumor hypoxia with stimulus-responsive nanocarriers in overcoming drug resistance and monitoring anticancer efficacy. Acta Biomater. 2018;71:351–62.

Xu L, Wang Y, Zhang Z, et al. A reactive oxygen species-responsive prodrug micelle with efficient cellular uptake and excellent bioavailability. J Mater Chem B. 2018a;6:1076–84.

Xu L, Yang Y, Zhao M, Gao W, Zhang H, Li S, et al. A reactive oxygen species-responsive prodrug micelle with efficient cellular uptake and excellent bioavailability. J Mater Chem B. 2018b;6:1076–84.

Yang GB, et al. Smart nanoreactors for pH-responsive tumor homing, mitochondria-targeting, and enhanced photodynamic-immunotherapy of cancer. Nano Lett. 2018a;18:2475–84.

Yang J, Lu W, Xiao J, Zong Q, Xu H, Yin Y, et al. A positron emission tomography image-guidable unimolecular micelle nanoplatform for cancer theranostic applications. Acta Biomater. 2018b;79:306–16. https://doi.org/10.1016/j.actbio.2018.08.036.

Yu L, Yang Y, Du FS, Li ZC. ROS-responsive chalcogen-containing polycarbonates for photodynamic therapy. Biomacromolecules. 2018;19:2182–93.

Zhou Y, Wang S, Ying X, Wang Y, Geng P, Deng A, Yu Z. Doxorubicin loaded redox-responsive micelles based on dextran and indomethacin for resistant breast cancer. Int J Nanomedicine. 2017;12:6153–68.

Core-Shell Nanoconstructs for Cancer-Based Biomedical Applications

5

Anoushka Mukharya, Rahul Pokale, Amrita Arup Roy, Viola Colaco, Gaurisha Alias Resha Ramnath Naik, Srinivas Mutalik, Namdev Dhas, and Ritu Kudarha

Abstract

In the growing field of cancer-associated biomedical applications, core-shell nanostructures have proven to be a paradigm shift, presenting improved accuracy and effectiveness in the treatment and screening domain. Here we have a summary of the diverse functions of these nanoparticles (NPs) throughout various cancer treatment techniques, which include combinational therapies that combine numerous treatments in a synergistic manner, chemotherapy, photothermal, photodynamic, gene therapy, and sonodynamic therapies. Core-shell nanoparticles dramatically boost the degree of sensitivity and selectivity of imaging tests like magnetic resonance imaging (MRI), positron emission tomography (PET) scans, and SPECTs in diagnostic evaluations, rendering tumor identification more precise and timely. The section emphasizes the need for further research centered on multimodal platforms, biocompatibility, targeted accuracy, and clinical translation while focusing on the future possibilities of these components in enhancing cancer treatment and diagnostics. These developments might usher in an era of innovation in cancer therapeutics by offering the prospect of more efficient, precise, and customized cancer treatments and diagnostic tools.

Keywords

Core-shell · Cancer · Sonodynamic · Photodynamic · Photothermal · Chemotherapy · Diagnostic techniques

A. Mukharya · R. Pokale · A. A. Roy · V. Colaco · G. A. R. R. Naik · S. Mutalik
N. Dhas · R. Kudarha (✉)
Department of Pharmaceutics, Manipal College of Pharmaceutical Sciences, Manipal Academy of Higher Education, Manipal, India

J. K. Patel et al. (eds.), *Core-Shell Nano Constructs for Cancer Theragnostic*,
https://doi.org/10.1007/978-981-96-3025-7_5

1 Introduction

A hollow or matrix core encased in a shell makes up core-shell nanoparticles (NPs). Benefits from this novel design may include enhanced bio-permeability, target-specific administration of drugs, configurable physicochemical characteristics, and multiple drug delivery at once (Dhas et al. 2018). Basically, core-shell nanoparticles consist of a center core material covered in a layer of another material above it. The attributes of these nanoparticles are superior when compared to the basic nanoparticles in various ways. These positive aspects include reduced cytotoxicity, increased rate of dispersion, better biocompatibility and cytocompatibility, stronger conjugation when combined with other bioactive molecules, greater chemical stability and thermal stability, etc. (Sounderya and Zhang 2024). When used for cancer-based biomedical applications, usually a benign substance is coated over the toxic anticancer drug core to make the overall system nontoxic and more biocompatible due to the harmless shell layer that simultaneously enhances the chemo toxic properties of the core material at the target site. For nanoparticles to be dispersed in biological systems (aqueous), their hydrophilicity is crucial. If the core material is hydrophobic, the issue of dispersibility and bio- and cytocompatibility can be resolved by coating the core surface with a hydrophilic substance in the structure of core-shell nanomaterial. Coating the outermost shell with an inert substance that will not change chemically or thermally when exposed to the surrounding environment usually prevents oxidation of the inner material and improves the durability and overall stability of the core particles. With regard to several bio-applications, the conjugation of biomolecules over particle surfaces is crucial. An appropriate biocompatible substance coated on the material of interest can help address the problem when the material is hard to conjugate with a particular kind of biomolecule (Chatterjee et al. 2013). As a result of their increased propensity for interacting with active pharmaceutical ingredients, ligands, receptors, and other molecules due to their surface chemistry, core/shell nanoparticles have been primarily developed for use in biomedical applications (Sahoo and Labhasetwar 2003). When a polymeric or inorganic layer, such as silica, is applied to a core nanoparticle, the hybrid structure gains a function or property on top of the core, allowing for the possibility of synergistically resulting operations (Kang et al. 2006; Naik et al. 2024). Despite the possibility that some particles have comparable compositions and materials, the characteristics of nanostructures are varied and cannot be generalized. As a result, we group them together according to the material that makes up the nanocomposite's shell and core (Colaco et al. 2024a). Hence core shells can be classified into (1) Inorganic core-shell nanoparticles such as metallic, semiconductor, and lanthanide nanoparticles. (2) Organic–inorganic hybrid core-shell nanoparticles such as inorganic core with organic shell nanoparticles and organic core with inorganic shell nanoparticles. Lastly, (3) polymeric core-shell nanoparticles.

Inorganic core-shell nanoparticles are usually comprised of semiconductors, metals, and lanthanides which make up the majority of the inorganic components of these economically available and extensively manufactured nanoparticles. Core-shell nanocomposites consisting of a silica shell over a gold or silver core are the most commonly employed metallic nanoparticles. The extent of the silica layer thickness affects the overall optical characteristics of the gold/silica nanoparticles, which are used in optical monitoring. The gold nanoparticles' silica coating renders them biocompatible (Botella et al. 2007). The most recent development in magnetic imaging for therapeutic purposes is the incorporation of iron oxide/silica and iron oxide/gold nanocomposite particles (Wu et al. 2007). Iron oxide core is usually microchip compatible due to its gold shell coating, and conjugation is made possible by the silica layer, which also inhibits aggregation. Similar core-shell nanostructures in this category include gold/palladium core-shell structures, which are employed as catalysts, and tin/tin oxide nanoparticles, which are utilized in the food industry and as humidity gauges (Pal and Chkaravorty 2005). Fluorophores engage with metals to enhance photostability, intensify fluorescence, and shorten longevity. Single particle imaging is facilitated by fluorescent core-shell nanoparticles, too. Copper/copper oxide nanoparticles are yet another kind of core-shell nanocomposite employed in optical bioimaging (Sargazi et al. 2022). Semiconductor-based core-shell nanoparticles are composed of an inorganic substance such as silica and an inorganic semiconductor component, metal oxide, or semiconductor alloy to form the shell (Cui et al. 2007). The frameworks may be either tertiary, comprising a core along with two shells, or simply binary, consisting of a core and the shell. Groups 3 and 5 or groups 4 and 6 metal alloys are the most prevalent binary structures, generally referred to as quantum dots. Nanoparticles CdSe/Cds, CdSe/ZnS, ZnSe/ZnS, and CdTe/CdS have been commonly employed for fluorescence bioimaging (Spanhel 2000; Patel et al. 2024). Compared to quantum dots in sols, nanoquantum dots exhibit superior luminosity and can be distributed throughout titania matrix. Any core-shell framework encompassing Co/Cd or Cd/Co in the form of quantum dots tends to exhibit magnetic properties (Petruska et al. 2004; Kim et al. 2005). Iron oxide, ZnSe, and CdSe are a few examples of tertiary frameworks with magnetic characteristics. These can be considered bifunctional, possessing both magnetic and fluorescent attributes. Consequently, it is feasible to perform further comprehensive in vivo imaging (Du et al. 2006). Similarly, lanthanide nanoparticles usually comprise a lanthanide group element core surrounded by an inorganic or lanthanide shell externally. Aqueous colloids containing rhabdophane, and lanthanide phosphate, fall under the same category and exhibit green luminescence. These particles have a $LnPO_4$-xH_2O shell and have been coated with Ce and Tb. To improve their luminescence, silica might be added to the aforementioned particles eventually. These have prospective applications in bioimaging and electronic devices as well (Thompson 2002; Rana et al. 2006).

Organic–inorganic hybrid core-shells have either cores made of organic substances, some of which may even be polymers or consist of a metal oxide, metal, silica, or any inorganic core along with an organic or inorganic shell based on the type of core used (Kumar et al. 2006). These are employed in the replacement of joints and may additionally be used to enhance the qualities of various materials because of their strong endurance against wear and corrosion. A large number of these synthesized nanostructures are classified as polymers or silica. A few examples of these are polycaprolactum/silica and polystyrenemethylmethacrylate/silica

(Liu et al. 2007; Avella et al. 2006; Allouche et al. 2006; Menaa et al. 2007). In a study by Xia et al., poly(methyl methacrylate) (PMMA) was stabilized by adopting core-shell organic–inorganic hybrid polymer nanoparticles (Si-ASA HPNs), which have a styrene-acrylonitrile copolymer (SAN) shell and a silicone-modified butyl acrylate copolymer (PBA) core. Besides enhancing the interfacial bond between the PBA components and the SAN copolymer material, silicone additionally serves as a chain supplement and compatibilizer, enhancing the chain linking of poly(acrylonitrile-styrene-acrylate) (ASA) (Xia et al. 2021). Recently, a flexible, rubberized core and a glassy shell make up core-shell composite particulates, which have drawn a lot of attention for their role as a strengthening agent for brittle and unstable polymers (Si et al. 2007). Inorganic metals, including Au, Ag, and Fe, are easily accessible, nontoxic, and have special qualities that can be utilized to transform for any required application (Chiozzi and Rossi 2020). Such metals have a characteristic surface plasmonic resonance (SPR) frequency, which is also a crucial component of Targeted Photodynamic therapy given that it enables the synthesis of a desired size and shape in a visible to near-infrared spectral area or wavelength in focus (Ghosh Chaudhuri and Paria 2012; Soman et al. 2024). For suspension media to remain colloidally stable and to provide activation sites where therapeutic medicines or ligands specific to cancer have the potential to be attached, enabling active targeting, an organic shell or simply a polymer coating must be added (Debele et al. 2015; Datta et al. 2025). The efficiency of inorganic NPs can be enhanced by organic shells, which are primarily composed of proteins, complex polysaccharides, or natural or synthetic polymers. These elements also serve as host platforms for covalent connections and can shield inorganic NPs from oxidation. A well-designed shell thickness can enhance the nanoparticles' thermal and chemical endurance and offer the opportunity to regulate and tune the expulsion of biomolecules from the core. New findings in the years to come will ensure improvements in the synthesis, uses, and characteristics of this ever-evolving class of nanomaterials (Chiozzi and Rossi 2020). Figure 5.1 below gives a summary of the various applications of core-shell nanoparticles in the field of theranostics.

2 Therapeutic Approaches to Treat Cancer

2.1 Chemotherapy

Heterogeneous nanoparticles that have numerous elemental components are called core-shell nanoparticles (López-Lorente et al. 2011). The core and shell components of core-shell nanoparticles (NP) may be formed using a broad variety comprising organic and inorganic nanomaterials (Nomoev et al. 2015). One of the most promising composite nanostructures for a variety of potential functions consists of the noble metal/semiconductor core-shell composite.

According to studies carried out by Nguyen-Tri et al., dumbbell-shaped Fe_3O_4-Ag nanoparticles were synthesized in hybrid form (FeAg NPs, 8–16 nm) and uniform nanosphere-like silver nanoparticles (Ag NPs, 5–10 nm) employing a seeded

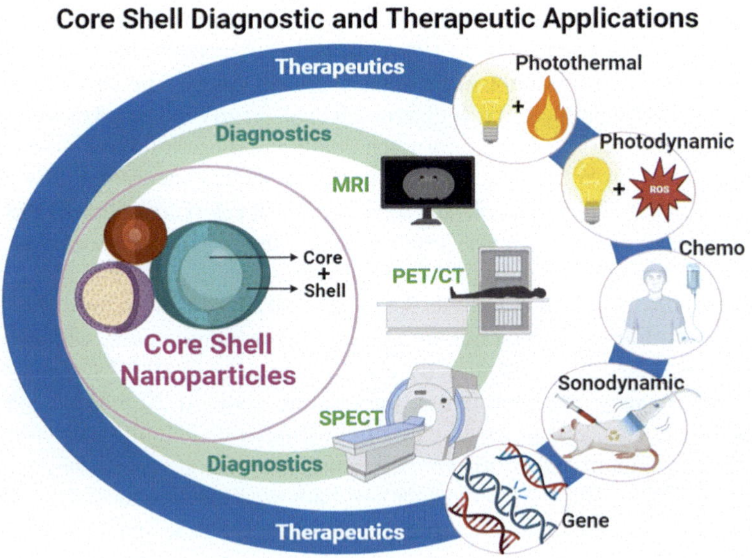

Fig. 5.1 Core-shell in diagnostic and therapeutic applications

development method using surfactants like oleic acid (OA) and oleylamine (OLA). Owing to its significant absorbance band on the visible and UV spectrum and the presence of surface plasmon resonance (SPR), which is crucial in influencing the optical response of nanoparticles, silver nanoparticles were chosen as the outermost shell on the magnetite nanosphere core. Nanoscale Raman spectroscopy has demonstrated that the biological activity of silver nanoparticles is enhanced by the transport of electrons from Ag NPs to Fe_3O_4 NPs. Nanoscale AFM-IR also indicates that the lamellae morphology that exists within spherulite of FeAg NPs/high-density polyethylene (HDPE) nano constructs appears to be crucial for its antibacterial activity (Nguyen-Tri et al. 2019). Hence Fe_3O_4–Ag nanocomposites have a self-sterilizing characteristic that prevents the growth of biofilms, the most perilous way for hazardous bacteria to propagate throughout the environment, enhancing the contrasting effect in magnetic resonance imaging (MRI) to detect cancer. Noble metal nanoparticles (NPs) such as Ag and Au display surface plasmonic resonance (SPR) (Liang et al. 2012). Around the plasmon resonances, plasmonic phenomena amplify scattering and absorption. Moreover, the size, shape, and embedded media all affect the plasmon resonance's strength and location in absorption spectra. Various applications of SPR include photothermal, therapeutic, and biosensing activities too (Doria et al. 2012; Brullot et al. 2012). Therefore the consequent formulation finds intriguing possibilities including chemotherapeutics over the integration of magnetic semiconductors within a core-shell nanostructure that incorporates the plasmonic effects produced by noble metals (Yeneayehu et al. 2021). A multitude of polymers such as chitosan, polyethylene glycol (PEG), poly-N-vinylpyrrolidone (PVP), nanocrystalline cellulose (NCC), hydroxyl ethylene cellulose (HEC), heparin-poloxamer (HP), polyethyleneimine (PEI), poly(N-isopropyl acrylamide) (PNIPAAm), and polyacrylic acid (PAA) have been successfully coated onto the Fe_3O_4 outer surface for tumor-targeted delivery of drugs. The investigation and results of the study conducted by Muhammadi et al. clearly indicated that the plasma surface alteration can result in a top-notch coating and cross-linking whether used independently or in conjunction with other components incorporated within the polymeric coating. They used plasma treatment to develop cross-linked PEG coating on magnetic nanoparticles. A cross-linked framework of PEG chains is produced via plasma-induced graft polymerization, leading to a stiff and dense surface that prevents the drug's burst release. A range of different grain sizes can be generated via the traditional co-precipitation process of magnetite core, which involves adding chitosan or PEG shell directly and then heating at 80 °C temperature for 30 min continuously, and then letting the mixture cool. At 20 mg/mL concentrations, the Fe_3O_4 surface augmented with PEG formulation exhibits noteworthy antibacterial activity against *Bacillus cereus*, *Escherichia coli*, and *Fusarium avenaceum* (Munir et al. 2023). HepG2 liver cancer cell lines were used to investigate the chemotherapeutic activity of magnetite-polymer nanoparticles, and this formulation proved to be appropriate for hyperthermia therapy in the treatment of carcinoma (Mohammadi et al. 2021). Similarly, polymer-coated magnetite core shells are more specifically used for tumor targeting based on the type of polymer used. It has vast potential in the chemotherapeutic or anticancer drug delivery systems. Ferromagnetic cores are specifically utilized in biosensors designed for in vivo cellular labeling because such invasive tissue diagnostic techniques demand the elimination of particles, or sometimes even the particles adhering to certain cells. Using particular biomarker molecules that bind to the targeted ligand equivalents articulated on the exterior of relevant cell types, cell labeling takes place effectively. Employing ferromagnetic cores in the form of amperometric nanosensors, has lately been used to recognize multiple types of malignant or dysfunctional cells based on a comprehensive collection of marker chemicals. A study conducted by Gupta et al. demonstrated how to derivatize superparamagnetic iron oxide nanoparticles (SPION) along with selectively targeting ligands of interest such as lactoferrin (Lf) or ceruloplasmin (Cp), which function as specific markers for cells, and the manner in which nanoparticles like these target surface receptors expressed across the surface of human fibroblasts without the cells internalizing them (Gupta and Gupta 2005). In yet another example, Heparin-poloxamer (HP)-coated SPIONs demonstrated a controlled release of doxorubicin (DOX) for a maximum of 120 h without any apparent initial burst release during evaluation of the core-shell system for chemotherapeutic drug administration (Hoang Thi et al. 2019).

2.2 Photothermal Therapy

In recent years, photothermal therapy (PTT) has garnered significant attention for being a potentially effective chemotherapeutic approach (Sarkar and Levi-Polyachenko 2020; Liu et al. 2019). It makes use of photo-absorbing nano-sized biomaterials to produce hyperthermia in the presence of near-infrared (NIR) light exposure, which kills tumor cells indefinitely (Wang et al. 2016). When viewed alongside popular anticancer treatment techniques like surgery, radiation, and chemotherapy, PTT provides the added benefit associated with being noninvasive, extremely spatially and temporally precise, accomplishing targeted therapy, and minimizing adverse effects on surrounding healthy tissues (Shi et al. 2019; Cheng et al. 2014). The vast majority of solid malignancies are not easily cured once they have metastasized. To increase the recovery rate of cancer patients, new treatment approaches are therefore essentially needed. Clinics have shown significant success employing checkpoint-blocking antibodies, as well as vaccines against cancer, and also adoptive T cell translocation for cancer immunotherapy (He et al. 2016; Chao et al. 2019). PTT has the potential to not only directly kill cancer cells but also trigger antitumor immune responses, hence boosting the effectiveness of systemic immunotherapy by generating tumor-linked agents derived from the remnants of ablated tumor cells (Fu et al. 2020; Guo et al. 2019). Guo et al. established a theranostic strategy recently that shows tremendous potential for setting up robust cancer photoimmunotherapy by virtue of the mutual advancement across magnetic-targeted accumulation of Magnetic-responsive immunostimulatory nanoagents (MINPs) in the tumor and MINP-based PTT, which leads to photothermal/immuno collaborative therapeutic outcomes that work against distant and primary tumors. Multitasking nanoagents—MINPs that encompass both the immunoadjuvant (CpG ODN) and the photoabsorber (SPIO)—were employed to demonstrate notable magnetic resonance (MR)/PA bimodal imaging operations in addition to accurately direct the PTT of the primary tumors. Hence this research supports the advancement of MINP-mediated photoimmunotherapy as a cutting-edge, complementary therapeutic approach for the efficient eradication of both primary and metastatic cancers (Guo et al. 2019). Similarly, Li et al. have studied tumor multimode ultrasonic (US), computed tomography (CT), photoacoustic (PA), thermal, and gold- coated nanostar (NS), further encapsulated with perfluorohexane (PFH) inside a hollowed-out mesoporous silica nanocapsule (HMS) enhanced with poly (ethylene glycol) (PEG) (PTT). The resulting formulation exhibits colloidal stability, appears noncytotoxic within the investigated concentration spectrum, and may be used for intravenous or intratumoral injection-based multiple modes operated via US/CT/PA/thermal surveillance of tumors in vivo. Therefore, the multifunctional HAPP that has been produced may be employed for PTT and tumor multimode imaging as a unique multifaceted theranostic nano platform (Li et al. 2017). A multipurpose Fe_3O_4@Au core-shell nanostar (NSs) enabling targeted multimodal magnetic resonance imaging (MR), PA, CT, and PTT of malignancies has also been designed by Hu et al. The NSs may be utilized for both MR imaging and CT examinations of malignancies because of the coexistence of the Fe_3O_4 core and the star-shaped Au shell. Similarly, the NSs may be employed for PA imaging and PTT of tumors owing to their near IR plasmonic absorption attribute. Consequently, this research presents a novel theranostic nano platform for efficient multimode-guided imaging-PTT addressing tumors, which might be expanded for translational medicine theranostics of other medical conditions as well (Hu et al. 2016). The temperature generated by photothermal treatment is dependent upon both the concentration of nanoparticles and the laser power density, according to the research conducted by Pandesh et al. Within C57BL/6 mice, the implanted melanoma tumors were eradicated by employing Fe_3O_4@Au core-shell nanoparticles. The tumor volume was considerably lower in the nanoparticles along with magnet and the laser condition group compared to the control group, according to the results. Fe_3O_4@Au core-shell NP photothermal therapy is thereby a successful cancer treatment strategy that can be employed widely (Pandesh et al. 2021). Incorporating core-shell magnetic gold (Fe_3O_4@Au) nanoparticles along with radio-photothermal treatment may be applied to combat cervical cancer with synergistic effectiveness. Cervical cancer cells in vitro have been demonstrated to perish in a time-dependent fashion with exposure to a brief dose of near-infrared light with a relatively small amount of Fe_3O_4@Au NPs. Moreover, RT and PTT together exhibited synergistic anticancer actions in vitro. Additionally, by facilitating better internalization, a magnetic field from the exterior may dramatically increase the synergistic effectiveness of Fe_3O_4@Au NPs as was observed by Hu et al. (2017). NIR photothermal treatment (PTT) guided by the image has tremendous possibilities by virtue of nanostructured materials with minimal tissue toxicity, high photothermal conversion efficiency, and multifaceted imaging capabilities. bifunctional nanoparticles (BFNP), approximately 16 nm) were developed implementing a one-pot solvothermal reaction process with ferrocene as the only source. It was composed of a carbon shell that was fluorescent in nature (roughly 3.4 nm) and a Fe_3O_4 magnetic core (about 9.1 nm). When combining T2-weighted magnetic resonance imaging ($r2 = 264.76\ mM^{-1}\ s^{-1}$) and fluorescence imaging prompted under multiple wavelengths from 405 nm to 820 nm, FNPs exhibit a dual-mode imaging power both in vitro and in vivo. According to Wang et al.'s research on mice with C6 glioblastoma, these BFNP have the ability to absorb and transform NIR light to heat, facilitating the overall photothermal intervention (Wang et al. 2018).

2.3 Sonodynamic Therapy

Sonodynamic Therapy (SDT) is a novel approach to treating malignancies that uses ultrasonography (US) as an external stimulus element for triggering sonosensitizers and producing a high concentration of reactive oxygen species (ROS), which kills malignant tumors (Cox and Beard 2015). The deep penetration depth, cheap cost, and minimum invasiveness of SDT are only a few of its numerous benefits (You et al. 2018). According to Sazgarnia et al., gold-based nanoparticles (Au NPs) are an effective catalyst of cavitation primarily because of their abrasive surface, which functions as a nucleation site and minimizes the cavitation threshold (Sazgarnia et al. 2013; Qian et al. 2016). Au NPs also address the drawbacks of traditional chemical sonosensitizers, such as their limited stability, limited solubility in water, and photosensitivity toward the skin. There is an unmet demand for novel US-activated sonosensitizers that produce ROS effectively, exhibit significant sensitivity along with penetration depth, and are nontoxic. Lin et al., have recently made use of this technique where Janus Au-MnO nanoparticles (JNPs) with dual-responsive vesicles that respond to both glutathione (GSH) and ultrasound (US) were coated with a ROS-sensitive polymer and PEG. The vesicles would split down into tiny Janus Au-MnO nanoparticles (NPs) exhibiting enhanced penetrating capacity when exposed to US radiation. The initial step in the synthesis of an asymmetric Janus Au-MnO NPs was to allow MnO to grow heteroepitaxially upon one of the edges of the Au NPs. US along with GSH dual-responsive hybridized vesicles made from amphiphilic Janus Au-MnO NPs can function as effective sonosensitizers by self-assembling. For particular tumor imaging, the vesicles showed tumor-induced NIT-II PAI and T1-MRI signals. This JNP Ve is a potential theranostic facilitator for image-guided invasive tumor-targeted treatment considering the vesicles initially start by deconstructing into tiny Au-MnO NPs for increased tumor penetration and then subsequently disassembled into smaller Au NPs. Then, smaller Au NPs in addition to several cavitation sites for nucleation and Mn^{2+} facilitating chemodynamic therapy (CDT) were concurrently released by GSH-triggered MnO degradation, leading to increased reactive oxygen species (ROS) formation. Due to the liberated Mn^{2+}, this also permitted T1-MR imaging and dual-modality photoacoustic imaging within the second near-infrared (NIR) window. Moreover, it prevented the development of orthotopic liver tumors by synergistic SDT/CDT (Lin et al. 2020). Shen et al. have also synthesized Fe_3O_4 nanoparticles (Fe_3O_4@TiO_2 NPs) encased in titanium dioxide in an attempt to enable targeted medication administration and combination treatment. Doxorubicin (DOX) can be loaded in a pH-dependent manner using titanium dioxide (TiO_2), which is also utilized in the form of a sonosensitizer in sonodynamic Therapy (SDT). Enhanced cytotoxicity and greater therapeutic effectiveness were the outcomes of the combination therapy in contrast to chemotherapeutic or sonodynamic treatment independently. As a result, the built-in NPs have many activities that enable them to transport the multimodal therapeutic dose to tumors selectively, improving treatment efficacy while minimizing adverse effects (Shen et al. 2015). Enhancing ultrasonic energy using nanoparticles to convert it into heat for tumor elimination is a potential way to increase the effectiveness of thermal-based cancer therapy. To assess the performance of the core-shell form of nanoparticles utilized for hyperthermia-based therapy of prostate cancer, as well as the heat loss resulting from blood flow in major arteries under various ultrasonication patterns, a research investigation was conducted by Mohammadi et al. The tumor's size, shape, and acoustic amplitude all had an impact on its thermal responses. When it came to producing heat in the medium and attenuating ultrasonic waves, titania nanoparticles outperformed magnetite nanoparticles. This study could contribute to understanding the variables influencing the hyperthermia-based therapy of prostate cancer using ultrasound and nanoparticles (Mohammadi and Rafii-Tabar 2024). Dynamic treatments are becoming more and more appealing as a means of treating tumors because they produce reactive oxygen species (ROS) in situ through both endogenous and external activation. Nonetheless, the effectiveness of customized stimulus-triggered dynamic treatment is significantly restricted by the intricacy of the malignancy. Bimetallic core-shell nanoparticles of copper along with ruthenium (Cu@Ru) have been employed presently for dynamic treatment driven by endogenous stimulation. ROS formation can be enhanced by the core-shell heterojunction interface's ability to swiftly distinguish electron-hole pairs produced by ultrasonic as well as light stimulation. These pairs then start interactions with adsorbed smaller molecules. This works in tandem alongside ROS from endogenous activation to enhance tumor therapy. Cu@Ru nanoparticles have been experimentally demonstrated in vitro and in vivo to trigger ferroptosis and apoptosis in tumor cells via producing ROS (Zheng et al. 2024).

2.4 Gene Therapy

A promising approach for improved cancer therapy is the simultaneous administration of gene therapy alongside chemotherapy. While RNA-cleaving DNAzyme was recently identified as a potentially useful tool for gene silencing, the absence of an efficient co-delivery mechanism to enable adequate intracellular DNAzyme activation that necessitates the presence of certain metal ions as a cofactor has compromised its ability to be used with chemotherapeutic medications. In a study conducted by Liu et al., Rapamycin (RAP), a hydrophobic chemotherapeutic, self-assembles into an absolute drug nanocore. An autophagy-inhibiting DNAzyme (DZ) is coloaded onto the surface of the metal-organic framework (MOF) shell, which is based on the coordination associated with Mn^{2+} and tannic acid (TA). The payloads are efficiently delivered into tumor cells via the nanosystem, and during endocytosis, the MOF shell disintegrates to release the interventions triggered due to its reaction with intracellular glutathione (GSH) and an acidic endo/lysosome milieu. Pure RAP core and DZ-encapsulating MOF shell were implemented to efficiently put together a smart nanosystem for drug-gene arrangements to combat triple-negative breast cancer (TNBC). Additionally, both in vitro and in vivo investigations demonstrated that activating the DZ could effectively suppress Beclin 1 protein expression, which subsequently prevented RAP-induced cell autophagy and increased RAP's antitumor activity. Furthermore, in vivo complementary therapy with gene–drug combinations showed improved tumor growth apprehension in comparison to single-drug therapy (Liu et al. 2020a). In yet another study conducted by Miao et al., A zinc oxide (ZnO) nanocore along with a polydopamine (PDA) shell constitute the core-shell nanosystem, which was designed to combine photothermal (PDA layer), gene (DNAzyme, DZ), and chemotherapeutic (doxorubicin, DOX) therapies in a single system. Due to its greater stability, biocompatibility, and consistent action, DZ was utilized for tumor-related gene (survivin) modulation rather than short-interfering RNAs. Physisorption and also covalent conjugation were the techniques employed to load DOX as well as amino-modified DZ upon the PDA shell, respectively. In this regard, the ZnO nanocore was engineered to function as a reservoir for metal cofactors, releasing Zn^{2+} in the presence of intracellular cues. This, in turn, activated DZ to silence genes subsequent to cell endocytosis. The optimized antitumor performance produced by these multimodal nanocomposites was established in both in vitro and in vivo examinations, which further emphasized the positive outcomes concerning these nano-drug delivery mechanisms in minimizing the adverse effects of DOX (Liu et al. 2020b).

Although messenger RNA (mRNA) and DNA-based treatments are revolutionizing the biomedical profession, the primary obstacle to the widespread adoption of these techniques in clinical settings is still the transmission of genetic components. The establishment of an innovative approach was carried out by Tarakanchikova et al. for synthesizing 50–150 nm vaterite calcium carbonate particulates and then employing them to create a prototype developing polymeric core-shell nanoparticles. Researchers have used these core-shell nanoparticles as effective and nontoxic carriers for mRNA and pDNA, to the extent that 99% of the primary T-lymphocytes proactively assimilate these nanocarriers, which have a viability of >90% and negligible cytotoxicity. As a representation of the hard-to-transfect kind that is frequently employed in gene and cell therapy operations, these nanocarriers prompt more effective transfection within primary human T-lymphocytes with respect to the usual electroporation technique (90% versus 51% for mRNA and 62% versus 39% for plasmid DNA). Thereby, these nanoscale core-shell enclosures offer a potential non-viral generic foundation for efficient and safe deployment of the gene for various cancer and non-cancer therapeutic approaches (Tarakanchikova et al. 2020). A hollow, mesoporous silica nanoparticle encased with polyamidoamine-aptamer for the simultaneous administration of sorafenib and CRISPR/Cas9 was conducted by Zhang et al. The gene–drug combo could be delivered specifically, with controlled release, at ultrahigh drug loading, and with good stability attributed to the core-shell nanoparticles. In all nine comparable sites, the nanocomplex demonstrated >60% Epidermal Growth Factor Receptor (EGFR)-editing effectiveness without off-target effects, modulated the EGFR-PI3K-Akt pathway to prevent angiogenesis, and had an additive effect on cell proliferation. Notably, the co-delivery nanosystem produced 85% tumor suppression in a mouse model and successfully influenced EGFR gene therapy to combat Hepatocellular carcinoma. Moreover, the nanocomplex demonstrated good safety with no severe organ damage and a substantial aggregation at the malignant site in vivo. These characteristics enable accurate gene editing and synergistic suppression by offering a flexible delivery method for effective co-loading of gene–drug pairings (Zhang et al. 2020).

3 Diagnostic Applications

3.1 Core Cell Nanoconstruct for MRI

Magnetic resonance imaging (MRI) is a crucial tool for cancer diagnosis, providing high-resolution images of soft tissues. Core-shell nano constructs have significantly improved MRI capabilities by acting as contrast agents, enhancing activity and specificity. Tailored supra-magnetic iron oxide nanoparticles for detecting specific cancer types are feasible given the current rate of progress in the field. The development of next-generation SPION-based MRI contrast agents involves conjugating tumor-specific targeting moieties with the SPION core after covering it with appropriate materials. These developments have made it easier to use MRI for tracking the effectiveness of treatment and early cancer detection. In one of the studies, one of the main areas where MRI is essential is the detection of breast cancer. Iron oxide nanoparticles provide an alternative to conventional MRI contrast agents, such as compounds based on gadolinium, which have better magnetic and biocompatibility qualities but have the drawbacks of potential toxicity and suboptimal targeting. Iron oxide core-shell nanoparticles with a silica or dextran shell were created by researchers studying breast cancer. The magnetic qualities required for MRI contrast were supplied by the core, and the shell guaranteed biocompatibility and functionalization for the purpose of targeting breast cancer cells. These core-shell nanoparticles were administered to patients suspected of having breast cancer as part of a clinical investigation. Tumor delineation was made easier by the nanoparticles' improved ability to contrast healthy and malignant tissues, as demonstrated by MRI scans. Targeting HER2 receptors that are overexpressed in some breast malignancies, the functionalized shell allowed for targeted accumulation at tumor locations. According to the study, iron oxide core-shell nanoparticles may increase breast cancer detection precision by lowering false positives and promoting early diagnosis. Research showed with superior resolution and specificity, superparamagnetic iron oxide nanoparticles (SPIONs) can improve the MRI diagnosis of prostate cancer. Prostate-specific membrane antigen (PSMA) ligands were used to functionalize and coat synthetic SPIONs, which were then used to target prostate cancer cells. Clinical trials showed enhanced MRI contrast in prostate cancer patients, allowing for better tumor visualization and differentiation from healthy tissue. Core-shell upconversion nanoparticles (UCNPs) with a NaGdF4 shell and an Er/Yb-doped NaYF4 core were created by Jianlin Shi et al. These nanoparticles showed enhanced T1-weighted MRI sensitivity and fluorescence imaging. They also encapsulated doxorubicin (DOX) and functionalized the surface of the nanoparticles with TAT peptides to create a nanotheranostic system that is directed toward cell nuclei (Liu et al. 2012). The remarkable ability of hybrid mesoporous silica nanospheres (MSN-Gd) to improve MR images was created and studied by Taylor et al. Their efficacy as optical and magnetic resonance imaging contrast agents has been amply proven in vitro. When employed at larger dosages, MSN-Gd has been demonstrated to be a superb T2 contrast agent for MR imaging of soft tissues and a highly effective T1 contrast agent for intravascular MR imaging. These hybrid nanomaterials are presently being tested by the team as target-specific contrast agents for multimodal imaging of cancer and inflammatory arthritis in animal models. This discovery emphasizes how useful and adaptable MSN-Gd nanospheres are as MR contrast agents (Taylor et al. 2008). Aldo Isaac Martínez-Banderas et al. created nanowires with an iron core and an iron oxide shell. These nanowires are biocompatible and have tunable magnetic characteristics, which makes them great materials for applications involving MRI contrast enhancement. When these core-shell nanowires' longitudinal and transverse magnetic relaxivities were measured at 1.5 T, they demonstrated excellent performance as T2 contrast agents. At 7 T, several surface coatings and oxidation levels were examined, and the customized transverse relaxivities showed how they affected T2 contrast. The successful identification of nanowire-labeled breast cancer cells was shown in T2-weighted pictures, both in vivo in the mouse brain and in vitro in tissue-mimicking phantoms. These iron-iron oxide core-shell nanowires, which served as T2 contrast agents, allowed for effective and long-term cellular identification using PT. These nanowires hold promise as a theranostic platform for exploring new areas in cell therapy due to their ability to be guided by magnetism and to initiate cellular responses, as demonstrated by the significant photothermal response that was recently seen (Martínez-Banderas et al. 2020). To improve MR imaging performance, a strong multilayered silica-Gd-silica core-shell nanoparticle system was created by Kobayashi et al. These nanoparticles are made up of two layers of silica: one coated with a gadolinium (Gd) shell and the other with another silica shell. The enhanced efficacy of the silica-Gd-silica core-shell nanoparticles as MR contrast agents was demonstrated by the much increased MR signal intensity observed in T1-weighted images when compared to water (Kobayashi et al. 2007).

3.2 Core Cell Nanoconstruct for PET/CT

A nuclear medicine technique called positron emission tomography (PET) scans produces three-dimensional (3D) pictures of the body using a radioactive tracer to assess organs and tissues and identify diseases. Using radiolabeled tracers that attach to particular chemicals expressed by cancer cells is the method used in this procedure. Early cancer identification can be aided by PET scans, which can reveal details about the metabolic activities of cancerous cells. PET is becoming a powerful tool for cancer detection and functional imaging of cancer cells as well as other illnesses, despite its exceptional sensitivity, unlimited depth of penetration, and measurable capabilities. Dual imaging methods refer to the use of MRI and CT or PET and CT together. In the end, these tools will aid in obtaining accurate knowledge of the molecular makeup of the human body (Gupta et al. 2024). Tantalum(V) ethoxide (TaOx), as a contrast agent, was developed by Lee et al. Tantalum(V) ethoxide was dissolved in a microemulsion containing Fe_3O_4 nanoparticles to create Fe_3O_4/TaOx core-shell nanoparticles. These multipurpose Fe_3O_4/TaOx core-shell nanoparticles provide supplementary data for MRI and CT scans. The injection of core-shell nanoparticles into tumor-bearing mice resulted in a definite preferential amplification of blood arteries in CT scans (Lee et al. 2012). A more effective and promising technique for creating multiple coated nanoparticles with cerium oxide and an effective internal core labeling of (CONPs) with the therapeutic PET radioisotope, Zr-89, has been reported by McDonagh et al. This allows for comprehensive PET scanning because it increases the rate of renal clarity and therapeutic efficacy (McDonagh et al. 2018).

Zyuzin et al. demonstrated the methods for radiolabeling calcium carbonate core-shell particles of varying sizes ($CaCO_3$ submicron-sized core-shell particles or SubCSPs) in order to accurately ascertain their in vivo biodistribution following intravenous administration in rats. To address this, a number of radiolabeling techniques have been devised, in which the positron emitter (^{68}Ga) is either deposited onto the surface of the shell (either layer coating or adsorption approaches) or into the core of the particle (co-precipitation approach). Based on the collected data, the co-precipitation method has demonstrated the best radiochemical stability in human serum (96–98.5% for both types of core-shell particles), and radiochemical binding and stability of ^{68}Ga greatly depend on the employed radiolabeling methodology. They demonstrated the size-dependent impact of the distribution of core-shell particles on the particular organ uptake by combining radiometry of individual organs with imaging methods such as PET, CT, and MRI (Zyuzin et al. 2020).

Xue et al. used mesoporous silica nanoparticles to create a multimodal imaging platform 128. They created mesoporous silica doped with fluorescent (NIR dye Cy5.5) dye, put it into a shell of mesoporous silica, and demonstrated its use in MRI, CT, and fluorescence imaging on superparamagnetic iron oxide nanoparticles. Increased fluorescence signal intensity of the nanocomposite was seen in in vitro investigations, and the accumulation of contrast agent in the liver was investigated by CT, and fluorescence imaging in an in vivo study (Xue et al. 2014). With advanced surface modification techniques, Chen et al. created hollow mesoporous silica nanoparticles (HMSN) for effective in vivo tumor targeting, multimodal imaging, biodistribution, and improved antitumor drug delivery. They coupled the targeting moiety (anti-CD105 antibody), NIR dye (ZW800), and PET isotope (^{64}Cu) to the nanostructure in order to deliver DOX to the tumor and perform in vivo tumor- targeted PET/NIRF imaging. (f)11) To avoid the self-quenching effect, a 1:2 ideal ratio between HMSN and ZW800 was kept. Mice injected with targeted nanoparticles showed increased tumor optical intensity according to in vivo NIRF imaging using ZW800 (Chen et al. 2014). In one more study, Cai's group designed multifunctional hollow mesoporous silica nanoparticles (HMSNs) for targeted tumor vasculature drug delivery and PET imaging. Amine functionalized HMSNs were conjugated to NOTA, further PEGylated and loaded with Sunitinib (anti-angiogenesis drug). Cyclo(Arg-Gly-Asp-D-Tyr-Lys) (cRGDyK) peptide was used as a targeting ligand and radiolabeled with ^{64}Cu for PET imaging. The targeted nanoprobe (^{64}Cu-NOTA-HMSN-PEG-cRGDyK) displayed integrin- specific uptake in both in vitro and in vivo studies. Enhanced uptake of targeted nanoprobe (~8% ID/g) was observed in the PET results, signifying the specificity of active targeting by nanoprobe in U87MG tumors and by passive targeting due to EPR effect (Chakravarty et al. 2015). Hayashi et al. created clusters of NIR fluorescent gold nanoparticles coated in silica and demonstrated the use of dual-modal imaging (CT and NIR fluorescence) to visualize lymph nodes and lymph vessels (Colaco et al. 2024b). In cancer patients, lymph nodes and lymph arteries are tiny, difficult to detect, and frequently removed in order to avoid cancer metastases. The development of high X-ray absorption coefficient NIR fluorescent silica-coated gold nanoparticle clusters ($Au@SiO_2$) revealed the accumulation of nanoparticles in lymph nodes via lymph veins. Without requiring dissection, CT imaging accurately identified the tumor and measured the size of the lymph nodes and lymph arteries (Hayashi et al. 2013).

3.3 Core Cell Nanoconstruct for SPECT

The nuclear medicine tomographic imaging method known as single photon emission computed tomography (SPECT) imaging uses gamma rays to offer three-dimensional information on the distribution of radioactive tracers within the body. A tiny quantity of radioactive material is injected into certain tissues, organs, or cellular receptors, where it builds up. A gamma camera detects the released gamma rays, and a computer interprets the signals to create cross-sectional pictures. Conditions impacting the heart, brain, bones, and other organs can be diagnosed and tracked with the use of SPECT imaging. It differs from conventional anatomical imaging methods in that it can see functional processes. Oncology, neurology, and cardiology all make extensive use of SPECT. In one of the research works by Li et al. the development of metal-drug coordination polymer (CP)-shell core-shell nanoconstructs, which are composed of a gold nanostar (AuNS) core. The CP shell, which included gadolinium and the chemotherapy medication gemcitabine, allowed for chemotherapy and MRI imaging, while the AuNS core allowed for the plasmonic photothermal effect and two-photon photoluminescence (TPL) imaging. In addition to combining photothermal therapy and chemotherapy to stop tumor growth, this multimodal nanoconstruct was able to track the location of nanoparticles using MRI and analyze their behavior in tumors using TPL imaging (Li et al. 2016). By chelating Ca^{2+} ions in hydroxyapatite, which are abundant in bone malignancies, the alendronic acid (ADA) unit in the vesicle demonstrates a strong affinity toward bone tissues. Furthermore, ADA-99mTc chelation makes it an excellent bone scintigraphic agent for SPECT imaging targeted at bone tumors (Zhou et al. 2021).

4 Conclusions and Future Perspectives

The development of core-shell nano structures, which provide notable advancements in therapeutic and diagnostic methods, signifies a pivotal moment in the realm of carcinoma- centered biomedical applications. These emerging nanoparticles have demonstrated remarkable promise in augmenting the effectiveness and selectivity of several cancer therapies, such as photothermal therapy, photodynamic therapy, sonodynamic therapy, chemotherapy, gene therapy, and even a combination of various techniques to combat cancer specifically. The capacity of core-shell nanoconstructs for administering theranostic drugs with decreased side effects and enhanced selectivity has rendered them extremely desirable for implementation in cancer-related biomedical applications. Core-shell designs improve patient outcomes by minimizing adverse effects on healthy biological tissues while simultaneously decreasing systemic toxicity by tailored dosage administration and localized stimulation of chemotherapeutic components. Core-shell nano constructs have transformed imaging techniques for diagnosis like MRI, CT/PET, and SPECT. Having the capacity to easily be programmed with numerous layers of functional components and targeting ligands improves visualization contrast and accuracy, which is crucial for timely intervention alongside effective cancer management. The potential to detect tumors accurately and promptly is made attainable via these nano constructs. To be able to navigate current obstacles and further develop these cutting-edge nanotechnologies into clinical practice, more multidisciplinary innovation and research within this domain remain crucial for enhancing the quality of life of those diagnosed with cancer globally in the decades to come. Numerous platforms centered around nanotechnology, including those derived from both inorganic and organic compounds, are still being investigated for theranostic applications. Both endogenous and exogenous stimuli are essential for modulating the release of drugs and enhancing targetability's autonomous and nonautonomous operations. For the fabrication of nanoconstructs for carcinoma theranosis, the full scope of DNA origami, nanorobots, artificial intelligence, and machine learning has not yet been investigated extensively. Despite significant advancements in the production of nanoconstructs for theranostic applications, only a handful of them were able to proceed ahead to the clinical stage of research. The only way to achieve the intended results from the introduction of nanotechnology is to carry out interdisciplinary research on a sizable patient population (Mishra et al. 2023).

References

Allouche J, Boissière M, Hélary C, Livage J, Coradin T. Biomimetic core–shell gelatine/silica nanoparticles: a new example of biopolymer-based nanocomposites. J Mater Chem. 2006;16(30):3120–5.

Avella M, Bondioli F, Cannillo V, Pace E, Errico M, Ferrari A, et al. Poly(ε-caprolactone)-based nanocomposites: influence of compatibilization on properties of poly(ε-caprolactone)–silica nanocomposites. Compos Sci Technol. 2006;66:886–94.

Botella P, Corma, Navarro M. Single gold nanoparticles encapsulated in monodispersed regular spheres of mesostructured silica produced by pseudomorphic transformation. Chem Mater. 2007;19:1979–83.

Brullot W, Valev VK, Verbiest T. Magnetic-plasmonic nanoparticles for the life sciences: calculated optical properties of hybrid structures. Nanomedicine. 2012;8(5):559–68.

Chakravarty R, Goel S, Hong H, Chen F, Valdovinos HF, Hernandez R, et al. Hollow mesoporous silica nanoparticles for tumor vasculature targeting and PET image-guided drug delivery. Nanomedicine. 2015;10(8):1233–46.

Chao Y, Chen G, Liang C, Xu J, Dong Z, Han X, et al. Iron nanoparticles for low-power local magnetic hyperthermia in combination with immune checkpoint blockade for systemic antitumor therapy. Nano Lett. 2019;19(7):4287–96.

Chatterjee K, Sarkar S, Korukonda JR, Paria S. Core/shell nanoparticles in biomedical applications. Adv Colloid Interf Sci. 2013;209:8–39.

Chen F, Hong H, Shi S, Goel S, Valdovinos HF, Hernandez R, et al. Engineering of hollow mesoporous silica nanoparticles for remarkably enhanced tumor active targeting efficacy. Sci Rep. 2014;4:5080.

Cheng L, Wang C, Feng L, Yang K, Liu Z. Functional nanomaterials for phototherapies of cancer. Chem Rev. 2014;114(21):10869–939.

Chiozzi V, Rossi F. Inorganic–organic core/shell nanoparticles: progress and applications. Nanoscale Adv. 2020;2(11):5090–105.

Colaco V, Roy AA, Naik GARR, Mondal A, Mutalik S, Dhas N. Advancement in lipid-based nanocomposites for theranostic applications in lung carcinoma treatment. OpenNano. 2024a;15:100199.

Colaco V, Pokale R, Mukharya A, Patel JK, Pathak YV, Mutalik S, et al. Lymphatic system: history, anatomy, physiology, challenges, and opportunities. In: Dhas N, Patel JK, Pathak YV, editors. Advanced targeting of the lymphatic system. Cham: Springer; 2024b. p. 1–21. https://doi.org/10.1007/978-3-031-64828-1_1.

Cox B, Beard P. Super-resolution ultrasound. Nature. 2015;527(7579):451–2.

Cui X, Li CM, Zang J, Yu S. Highly sensitive lactate biosensor by engineering chitosan/PVI-Os/CNT/LOD network nanocomposite. Biosens Bioelectron. 2007;22(12):3288–92.

Datta D, Colaco V, Bandi SP, Sharma H, Dhas N, Giram PS. 7 - Classes/types of polymers used in oral delivery (natural, semisynthetic, synthetic), their chemical structure and general functionalities. In: Parambath A, editor. Polymers for oral drug delivery technologies. Elsevier: Amsterdam; 2025. p. 263–333. https://www.sciencedirect.com/science/article/pii/B9780443137747000074.

Debele TA, Peng S, Tsai HC. Drug carrier for photodynamic cancer therapy. Int J Mol Sci. 2015;16(9):22094–136.

Dhas NL, Raval NJ, Kudarha RR, Acharya NS, Acharya SR. Chapter 9 - Core–shell nanoparticles as a drug delivery platform for tumor targeting. In: Grumezescu AM, editor. Inorganic frameworks as smart nanomedicines. William Andrew Publishing: Norwich; 2018. p. 387–448. https://www.sciencedirect.com/science/article/pii/B9780128136614000092.

Doria G, Conde J, Veigas B, Giestas L, Almeida C, Assunção M, et al. Noble metal nanoparticles for biosensing applications. Sensors. 2012;12(2):1657–87.

Du GH, Liu Z, Lu Q, Xia X, Jia LH, Yao K, et al. Fe_3O_4/CdSe/ZnS magnetic fluorescent bifunctional nanocomposites. Nanotechnology. 2006;17:2850.

Fu Q, Li Z, Ye J, Li Z, Fu F, Lin SL, et al. Magnetic targeted near-infrared II PA/MR imaging guided photothermal therapy to trigger cancer immunotherapy. Theranostics. 2020;10(11):4997–5010.

Ghosh Chaudhuri R, Paria S. Core/shell nanoparticles: classes, properties, synthesis mechanisms, characterization, and applications. Chem Rev. 2012;112(4):2373–433.

Guo Y, Ran Y, Wang Z, Cheng J, Cao Y, Yang C, et al. Magnetic-responsive and targeted cancer nanotheranostics by PA/MR bimodal imaging-guided photothermally triggered immunotherapy. Biomaterials. 2019;219:119370.

Gupta AK, Gupta M. Synthesis and surface engineering of iron oxide nanoparticles for biomedical applications. Biomaterials. 2005;26(18):3995–4021.

Gupta D, Roy P, Sharma R, Kasana R, Rathore P, Gupta TK. Recent nanotheranostic approaches in cancer research. Clin Exp Med. 2024;24(1):8.

Hayashi K, Nakamura M, Ishimura K. Near-infrared fluorescent silica-coated gold nanoparticle clusters for X-ray computed tomography/optical dual modal imaging of the lymphatic system. Adv Healthc Mater. 2013;2(5):756–63.

He C, Duan X, Guo N, Chan C, Poon C, Weichselbaum RR, et al. Core-shell nanoscale coordination polymers combine chemotherapy and photodynamic therapy to potentiate checkpoint blockade cancer immunotherapy. Nat Commun. 2016;7(1):12499.

Hoang Thi TT, Nguyen Tran DH, Bach LG, Vu-Quang H, Nguyen DC, Park KD, et al. Functional magnetic core-shell system-based iron oxide nanoparticle coated with biocompatible copolymer for anticancer drug delivery. Pharmaceutics. 2019;11(3):120.

Hu Y, Wang R, Wang S, Ding L, Li J, Luo Y, et al. Multifunctional Fe_3O_4@Au core/shell nanostars: a unique platform for multimode imaging and photothermal therapy of tumors. Sci Rep. 2016;6(1):28325.

Hu R, Zheng M, Wu J, Li C, Shen D, Yang D, et al. Core-shell magnetic gold nanoparticles for magnetic field-enhanced radio-photothermal therapy in cervical cancer. Nanomaterials. 2017;7(5):111.

Kang SM, Lee KB, Kim DJ, Choi IS. Biomimetic approach to the formation of gold nanoparticle/silica core/shell structures and subsequent bioconjugation. Nanotechnology. 2006;17(18):4719–25.

Kim H, Achermann M, Balet LP, Hollingsworth JA, Klimov VI. Synthesis and characterization of Co/CdSe Core/shell nanocomposites: bifunctional magnetic-optical nanocrystals. J Am Chem Soc. 2005;127(2):544–6.

Kobayashi Y, Imai J, Nagao D, Takeda M, Ohuchi N, Kasuya A, et al. Preparation of multilayered silica–Gd–silica core-shell particles and their magnetic resonance images. Colloids Surf A Physicochem Eng Asp. 2007;308(1):14–9.

Kumar A, Pushparaj VL, Murugesan S, Viswanathan G, Nalamasu R, Linhardt RJ, et al. Synthesis of silica-gold nanocomposites and their porous nanoparticles by an in-situ approach. Langmuir. 2006;22(21):8631–4.

Lee N, Cho HR, Oh MH, Lee SH, Kim K, Kim BH, et al. Multifunctional Fe_3O_4/TaO(x) core/shell nanoparticles for simultaneous magnetic resonance imaging and X-ray computed tomography. J Am Chem Soc. 2012;134(25):10309–12.

Li M, Li L, Zhan C, Kohane DS. Core-shell nanostars for multimodal therapy and imaging. Theranostics. 2016;6(13):2306–13.

Li X, Xing L, Zheng K, Wei P, Du L, Shen M, et al. Formation of gold nanostar-coated hollow mesoporous silica for tumor multimodality imaging and photothermal therapy. ACS Appl Mater Interfaces. 2017;9(7):5817–27.

Liang A, Liu Q, Wen G, Jiang Z. The surface-plasmon-resonance effect of nanogold/silver and its analytical applications. TrAC Trends Anal Chem. 2012;37:32–47.

Lin X, Liu S, Zhang X, Zhu R, Chen S, Chen X, et al. An ultrasound activated vesicle of janus Au-MnO nanoparticles for promoted tumor penetration and sono-chemodynamic therapy of orthotopic liver cancer. Angew Chem. 2020;132(4):1699–705.

Liu YR, Huang YD, Liu L. Thermal stability of POSS/methylsilicone nanocomposites. Compos Sci Technol. 2007;67:2864–76.

Liu J, Nan, Bu W, Min PL, Zhang S, Chen F, Zhou L, et al. Simultaneous nuclear imaging and intranuclear drug delivery by nuclear-targeted multifunctional upconversion nanoprobes. Biomaterials. 2012;33(29):7282–90.

Liu Y, Bhattarai P, Dai Z, Chen X. Photothermal therapy and photoacoustic imaging via nanotheranostics in fighting cancer. Chem Soc Rev. 2019;48(7):2053–108.

Liu P, Liu X, Cheng Y, Zhong S, Shi X, Wang S, et al. Core–shell nanosystems for self-activated drug–gene combinations against triple-negative breast cancer. ACS Appl Mater Interfaces. 2020a;12(48):53654–64.

Liu M, Peng Y, Nie Y, Liu P, Hu S, Ding J, et al. Co-delivery of doxorubicin and DNAzyme using ZnO@polydopamine core-shell nanocomposites for chemo/gene/photothermal therapy. Acta Biomater. 2020b;110:242–53.

López-Lorente AI, Simonet BM, Valcárcel M. Analytical potential of hybrid nanoparticles. Anal Bioanal Chem. 2011;399:43–54.

Martínez-Banderas AI, Aires A, Plaza-García S, Colás L, Moreno JA, Ravasi T, et al. Magnetic core–shell nanowires as MRI contrast agents for cell tracking. J Nanobiotechnol. 2020;18:42.

McDonagh PR, Sundaresan G, Yang L, Sun M, Mikkelsen R, Zweit J. Biodistribution and PET imaging of 89-zirconium labeled cerium oxide nanoparticles synthesized with several surface coatings. Nanomedicine. 2018;14(4):1429–40.

Menaa B, Takahashi M, Innocenzi P, Yoko T. Crystallization in hybrid organic–inorganic materials induced by self-organization in basic conditions. Chem Mater. 2007;19:1946–53.

Mishra S, Bhatt T, Kumar H, Jain R, Shilpi S, Jain V. Nanoconstructs for theranostic application in cancer: challenges and strategies to enhance the delivery. Front Pharmacol. 2023;14:1101320.

Mohammadi M, Rafii-Tabar H. Modeling hyperthermia-based prostate cancer treatment in the presence of core-shell nanoparticles and large blood vessels under different ultrasound sonication patterns. Int J Therm Sci. 2024;203:109123.

Mohammadi MA, Asghari S, Aslibeiki B. Surface modified Fe_3O_4 nanoparticles: a cross-linked polyethylene glycol coating using plasma treatment. Surf Interfaces. 2021;25:101271.

Munir T, Mahmood A, Rasul A, Imran M, Fakhar-e-Alam M. Biocompatible polymer functionalized magnetic nanoparticles for antimicrobial and anticancer activities. Mater Chem Phys. 2023;301:127677.

Naik G, Alias RR, Roy AA, Mutalik S, Dhas N. Unleashing the power of polymeric nanoparticles — creative triumph against antibiotic resistance: a review. Int J Biol Macromol. 2024;278:134977.

Nguyen-Tri P, Nguyen VT, Nguyen TA. Biological activity and nanostructuration of Fe_3O_4-Ag/high density polyethylene nanocomposites. J Composites Sci. 2019;3(2):34.

Nomoev AV, Bardakhanov SP, Schreiber M, Bazarova DG, Romanov NA, Baldanov BB, et al. Structure and mechanism of the formation of core–shell nanoparticles obtained through a one-step gas-phase synthesis by electron beam evaporation. Beilstein J Nanotechnol. 2015;6(1):874–80.

Pal B, Chkaravorty D. Electrical properties of composites with tin tin oxide core shell nanostructure and their sensing behaviour. J Phys D Appl Phys. 2005;38:3537–42.

Pandesh S, Haghjooy Javanmard S, Shakeri-Zadeh A, Shokrani P. Targeted photothermal therapy of melanoma in C57BL/6 mice using Fe_3O_4@Au core-shell nanoparticles and near-infrared laser. J Biomed Phys Eng. 2021;11(1):29–38.

Patel V, Arde G, Kashid G, Patel JK, Dhas N. Metal-based carriers for targeting lymphatic system. In: Dhas N, Patel JK, Pathak YV, editors. Advanced targeting of the lymphatic system. Cham: Springer; 2024. p. 271–93. https://doi.org/10.1007/978-3-031-64828-1_14.

Petruska MA, Bartko AP, Klimov VI. An amphiphilic approach to nanocrystal quantum dot-titania nanocomposites. J Am Chem Soc. 2004;126(3):714–5.

Qian X, Zheng Y, Chen Y. Micro/nanoparticle-augmented sonodynamic therapy (SDT): breaking the depth shallow of photoactivation. Adv Mater. 2016;28(37):8097–129.

Rana S, Rawat J, Sorensson MM, Misra RDK. Antimicrobial function of Nd3+-doped anatase titania-coated nickel ferrite composite nanoparticles: a biomaterial system. Acta Biomater. 2006;2(4):421–32.

Sahoo SK, Labhasetwar V. Nanotech approaches to drug delivery and imaging. Drug Discov Today. 2003;8(24):1112–20.

Sargazi S, Fatima I, Hassan Kiani M, Mohammadzadeh V, Arshad R, Bilal M, et al. Fluorescent-based nanosensors for selective detection of a wide range of biological macromolecules: a comprehensive review. Int J Biol Macromol. 2022;206:115–47.

Sarkar S, Levi-Polyachenko N. Conjugated polymer nano-systems for hyperthermia, imaging and drug delivery. Adv Drug Deliv Rev. 2020;163:40–64.

Sazgarnia A, Shanei A, Taheri AR, Meibodi NT, Eshghi H, Attaran N, et al. Therapeutic effects of acoustic cavitation in the presence of gold nanoparticles on a colon tumor model. J Ultrasound Med. 2013;32(3):475–83.

Shen S, Wu L, Liu J, Xie M, Shen H, Qi X, et al. Core–shell structured Fe_3O_4@TiO_2-doxorubicin nanoparticles for targeted chemo-sonodynamic therapy of cancer. Int J Pharm. 2015;486(1):380–8.

Shi H, Sun Y, Yan R, Liu S, Zhu L, Liu S, et al. Magnetic semiconductor Gd-doping CuS nanoparticles as activatable nanoprobes for bimodal imaging and targeted photothermal therapy of gastric tumors. Nano Lett. 2019;19(2):937–47.

Si QB, Zhou C, Yang HD, Zhang HX. Toughening of polyvinylchloride by core–shell rubber particles: influence of the internal structure of core–shell particles. Eur Polym J. 2007;43(7):3060–7.

Soman S, Kulkarni S, John J, Vineeth P, Ahmad SF, George SD, et al. Transferrin-conjugated UiO-66 metal organic frameworks loaded with doxorubicin and indocyanine green: a multimodal nanoplatform for chemo-photothermal-photodynamic approach in cancer management. Int J Pharm. 2024;665:124665.

Sounderya N, Zhang Y. Use of core/shell structured nanoparticles for biomedical applications. Recent Patents Biomed Eng. 2024;1(1):34–42.

Spanhel L. Raman characterization of CdTe/CdS-"core-shell"-clusters in colloids and films. J Crystal Growth. 2000. https://www.academia.edu/16291800/Raman_characterization_of_CdTe_CdS_core_shell_clusters_in_colloids_and_films.

Tarakanchikova Y, Muslimov A, Sergeev I, Lepik K, Yolshin N, Goncharenko A, et al. A highly efficient and safe gene delivery platform based on polyelectrolyte core–shell nanoparticles for hard-to-transfect clinically relevant cell types. J Mater Chem B. 2020;8(41):9576–88.

Taylor KML, Kim JS, Rieter WJ, An H, Lin W, Lin W. Mesoporous silica nanospheres as highly efficient MRI contrast agents. J Am Chem Soc. 2008;130(7):2154–5.

Thompson S. Oxo-metal-polyimide nanocomposites. 2. Enhancement of thermal, mechanical, and chemical properties in soluble hexafluoroisopropylidine-based polyimides via the in situ formation of oxo-lanthanide(III)-polyimide nanocomposites. Chem Mater. 2002. https://www.academia.edu/73369345/Oxo_Metal_Polyimide_Nanocomposites_2_Enhancement_of_Thermal_Mechanical_and_Chemical_Properties_in_Soluble_Hexafluoroisopropylidine_Based_Polyimides_via_the_in_Situ_Formation_of_Oxo_Lanthanide_III_Polyimide_Nanocomposites.

Wang D, Zhou J, Chen R, Shi R, Zhao G, Xia G, et al. Controllable synthesis of dual-MOFs nanostructures for pH-responsive artemisinin delivery, magnetic resonance and optical dual-model imaging-guided chemo/photothermal combinational cancer therapy. Biomaterials. 2016;100:27–40.

Wang H, Mu Q, Revia R, Wang K, Tian B, Lin G, et al. Iron oxide-carbon core-shell nanoparticles for dual-modal imaging-guided photothermal therapy. J Control Release. 2018;289:70–8.

Wu W, He Q, Chen H, Tang J, Nie L. Sonochemical synthesis, structure and magnetic properties of air-stable Fe_3O_4/Au nanoparticles. Nanotechnology. 2007;18(14):145609.

Xia J, Luo X, Huang J, Ma J, Yang J. Preparation of core/shell organic–inorganic hybrid polymer nanoparticles and their application to toughening poly(methyl methacrylate). RSC Adv. 2021;11(54):34036–47.

Xue S, Wang Y, Wang M, Zhang L, Du X, Gu H, et al. Iodinated oil-loaded, fluorescent mesoporous silica-coated iron oxide nanoparticles for magnetic resonance imaging/computed tomography/fluorescence trimodal imaging. Int J Nanomedicine. 2014;9:2527–38.

Yeneayehu K, Senbeta T, Mesfin B. Enhancement of the optical response of Fe_3O_4@ Ag core-shell nanoparticles. Physica E. 2021;134:114822.

You Y, Liang X, Yin T, Chen M, Qiu C, Gao C, et al. Porphyrin-grafted lipid microbubbles for the enhanced efficacy of photodynamic therapy in prostate cancer through ultrasound-controlled in situ accumulation. Theranostics. 2018;8(6):1665–77.

Zhang BC, Luo BY, Zou JJ, Wu PY, Jiang JL, Le JQ, et al. Co-delivery of Sorafenib and CRISPR/Cas9 based on targeted core–shell hollow mesoporous organosilica nanoparticles for synergistic HCC therapy. ACS Appl Mater Interfaces. 2020;12(51):57362–72.

Zheng N, Hu X, Yan L, Yun DL, Feng J, Li D, et al. Bimetallic Cu@Ru core–shell structures with ligand effects for endo–exogenous stimulation-mediated dynamic oncotherapy. Nano Lett. 2024;24(20):6165–73.

Zhou X, Yan N, Cornel EJ, Cai H, Xue S, Xi H, et al. Bone-targeting polymer vesicles for simultaneous imaging and effective malignant bone tumor treatment. Biomaterials. 2021;269:120345.

Zyuzin MV, Antuganov D, Tarakanchikova YV, Karpov TE, Mashel TV, Gerasimova EN, et al. Radiolabeling strategies of micron- and submicron-sized core-shell carriers for in vivo studies. ACS Appl Mater Interfaces. 2020;12(28):31137–47.

Core-/Multi-Shell Type of Core-Shell Nanoconstruct for Cancer Theragnostics

6

Bharat Mishra, Archita Tiwari, Shrishti Mishra, Aaishwaryadevi B. Deshmukh, and Jayvadan K. Patel

Abstract

Nanoconstructs are nanoformulations that are composed of two-component structures of soft shell and hard nanoparticle core. Nanoconstructs, a combination of ligand and nanoparticles, have the ability to reach the intended site of action with loaded cargo. Nanosystems incorporate both the active cargo (the drug) and the carrier (the delivery vehicle) to facilitate targeted drug delivery. The medications are maintained, transported across biological barriers, and delivered as payloads. Poly(propylene), poly-l-lysine (PLL), poly(ε-caprolactone), and poly (lactic-co-glycolic acid) (PLGA) are some polymers used in nanoconstruct for the treatment of cancer. Core-shell-type nanoparticles are biphasic materials characterized by an interior core structure surrounded by an exterior shell composed of multiple components. The distinctive properties of these particles, arising from the interaction of material, shape, and design within the core and shell, have garnered significant attention. Some have also been designed to improve reactivity and thermal stability. They are therefore widely applicable in fields including catalysts, semiconducting materials, electrical, and

B. Mishra (✉)
Dr. Shakuntla Misra National Rehabilitation University, Lucknow, India

A. Tiwari
Khwaja Moinuddin Chishti Language University, Lucknow, India

S. Mishra
Central Drug Research Institute, Lucknow, India

A. B. Deshmukh
School of Pharmaceutical Sciences, JSPM University, Pune, Maharashtra, India

J. K. Patel
Viesain Pharma LLC, GA, USA

Faculty of Pharmacy, Sankalchand Patel University, Visnagar, Gujarat, India

J. K. Patel et al. (eds.), *Core-Shell Nano Constructs for Cancer Theragnostic*,
https://doi.org/10.1007/978-981-96-3025-7_6

biomedicine. The particles of core-shell are synthesized in two stages: the core structure is synthesized first, and the shell material is subsequently coated over the core structure. Solution synthesis makes use of these techniques. Gas-phase synthesis is most commonly achieved via pulsed laser deposition (PLD) or chemical vapor deposition (CVD). These processes involve the use of substrates, where the shell material is deposited onto a pre-formed core structure, often through a multi-step procedure. Nanoscale materials employed for the treatment of various neurological disorders, including Parkinson's disease, Alzheimer's disease, Huntington's disease, frontotemporal dementia, amyotrophic lateral sclerosis, Prion disease, spinal cord injury, brain tumor, and stroke. Researchers are developing strategies to bypass the blood–brain barrier (BBB) and enhance the effectiveness of therapeutic treatments.

Keywords

Nanomedicines · Nanoconstruct · Core-shell nanoconstruct · Cancer

1 Introduction

Nanoconstruct drug delivery is one of the common uses for "nanoconstructs," which are two-component structures consisting of a "soft" shell of biomolecular ligands and a "hard" nanoparticle core (Mishra et al. 2023).

Nanoconstructs, a combination of ligand and nanoparticles, have the ability to reach the intended site of action with loaded cargo. A range of nanoparticulate matters have been employed to create nanoconstructs that have the potential to fulfill two purposes: for both therapeutic and diagnostic purposes. The main purpose of nanoconstructs is to get beyond the drawbacks of cancer treatments, which include toxicity, nonspecific medication distribution, and uncontrolled release rate. The tactics used in the creation of nanoconstructs contribute to the effectiveness and selectivity of loaded theragnostic drugs, making them a viable cancer treatment method. Nanoconstructs are made specifically to target the required location and get past obstacles that prevent their proper deployment for the intended advantage (Allen and Cullis 2004).

Medications are delivered using nanosystems, which also house the carrier and active cargo. The medications are preserved, transported across biological barriers, and subsequently released as therapeutic payloads (Murthy 2007). According to Schmidt, the main advantages of nanoparticle-based chemotherapy include decreased peripheral toxicity, increased drug deposition at the targeted location, and increased treatment specificity. The nanoconstructs ought to be made in a way that allows them to hold onto a range of agents either inside their core or on their surface, forming a perfect instrument for combo treatment (Yao et al. 2020). Following the administration of nanoconstructs, their final outcome is influenced not only by the surface chemistry but also by the characteristics of the entire vasculature, including factors such as pressure, flow velocities, and tissue heterogeneities (Zhao et al. 2019).

Consequently, distribution modes for nanoconstructs are appropriately characterized as autonomous and nonautonomous kinds rather than as passively or actively targeted systems (Saw et al. 2021).

1.1 Building Blocks for Nanostructures

Nanoparticles are those that have diameters less than 1000 nm with unique properties. Depending on the overall shape of the nanoparticle, they are classified as 0D, 1D, 2D, and 3D (Nguyen et al. 2021a). Nanoparticles are comprised of three main parts, the surface layer, shell, and core (Srinivasarao and Low 2017). Their uniqueness, submicron size, high surface-to-volume ratio, and improved targeting capabilities have made them extremely attractive across a wide range of fields (Jose et al. 2019). Polymer-based nanoconstructs, Poly(ε-caprolactone), poly(propylene) poly (lactic-co-glycolic acid) (PLGA) and poly-l-lysine (PLL) are some polymers used in the construction of polymer-based nanoconstruct for the treatment of cancer. The use of polymers offers several advantages, including enhanced drug release, improved bioavailability and solubility, controlled degradation, and reduced toxicity. Target cancer cells, tumor microenvironment, and toxicity status are taken into consideration while choosing polymers for nanoconstructs. The formation of nanoconstructs is also being investigated on a number of lipid-based nanoplatforms (Duwa et al. 2020). For their possible therapeutic advantages, inorganic nanoconstructs depend on gold, silver, iron oxide, and other non-metallic materials have also been investigated in addition to organic material-based nanoconstructs. Among them, black phosphorus has gained a lot of popularity in two-dimensional nanomaterials. Particular characteristics of it include thermal, optical, electrical, and biocompatibility due to its unique properties and structure which is responsible for its increased demand in comparison to nanomaterials containing graphene (Banstola et al. 2020).

1.2 Ligands for Forming Nanoconstruct

Target selectivity is typically lacking in conventional chemotherapy, which results in notable side effects that diminish the efficacy of a given medication. Therefore, drug delivery strategies involves that the medications are released in a way that preserves the patient's quality of life by releasing them in very site-selective and effective way with low-toxicity into the intracellular level. A key strategy for tumor-specific drug delivery involves choosing the ligands that bind to specific receptors present on cancer cells. For the tumor-targeted drug delivery systems various ligands, such as hyaluronic acid, peptides, folic acid, and antibodies, have been extensively used. Ligands or peptides that facilitate cell entry through tight junctions are also employed to elevate the intracellular movement of anticancer medicines. The use of different ligand combinations further improves cellular uptake and target selectivity (Nguyen et al. 2021a; Srinivasarao and Low 2017).

1.3 Process for Ligand Selection

One highly effective strategy to improve a drug's safety and efficacy is delivering the therapeutic agent using a specific ligand with high affinity for the target pathological site. These systems provide benefits like providing flexibility for drug optimization and avoiding toxicity to healthy cells. For this, a variety of ligands have been produced, including small compounds, aptamers, and antibody fragments. For the diagnosis ligand-targeted delivery systems also plays an important role. The size of the ligand, which influences the drug's pharmacokinetics, and its ability to attach to specific receptor surfaces are critical factors in selecting a targeting ligand. The concentration of the drug reaching the target site depends on the ligand's binding affinity, while its chemical properties can impact its binding capacity with receptors. Furthermore, factors such as immunogenicity, commonly observed with antibody–drug conjugates, can influence the drug's concentration and half-life. Other important considerations in ligand selection include the target selectivity, as well as challenges such as time and production cost. While techniques like PEGylation can improve stability, these aspects must be carefully evaluated to ensure effective receptor-targeted drug delivery (Table 6.1).

1.4 Core-Shell Nanoconstruct

Core-shell nanoparticles are comprised of an inner core material and an outer shell, which can be made from various combinations, such as inorganic/organic, inorganic/inorganic, organic/inorganic, or organic/organic materials. The shell material selection depends on the specific application. Multiple core-shell particles are created when a single shell material coats many small core particles. This design helps mitigate the hazards of bare nanoparticles, which may harm or irritate host tissues (Ghosh Chaudhuri and Paria 2012). Core-shell nanoparticles showed better characteristics than bare nanoparticles, due to reduced cytotoxicity, biocompatibility, high dispersibility, and improved conjugation with biomolecules and medicines because of better surface qualities as well as increased chemical and thermal stability. It is also commonly recognized that the hydrophilicity of nanoparticles is important in biological applications (Chatterjee et al. 2014). Dispersibility can therefore be improved by coating hydrophobic nanoparticles with hydrophilic elements, making them ideal for biological applications. Core-shell nanoparticles can be further functionalized using materials such as silane, polymers, dendrimers, and gold. This is a result of the core material's prominence over shell composition. For effective formulation delivery and optimal therapeutic outcomes in biomedical and drug delivery applications, biomolecules are conjugated on the surface of nanoparticles as a crucial step. While a variety of methods are available for surface engineering inorganic nanoparticles, conjugating biomolecules with the desired material surface often remains a challenge. This can be solved by covering the selected core material with a shell made of appropriate materials, which permits the conjugation of biomolecules. This idea is fundamental to biomedical engineering and targeted and

Table 6.1 Building blocks for nanostructures in cancer

Carrier	Material	Drug	Ligand	Indication	References
Polymeric nanoparticle	PLGA	Docetaxel	Transferrin	Improved target selectivity in breast cancer cells along with decreased toxicity	Jose et al. (2019)
	PLGA	Temozolomide	Cetuximab	For treatment of Epidermal Growth Factor Receptor (EGFR) overexpressing cancers	Duwa et al. (2020), Banstola et al. (2020)
	PLGA	DTX	anti-EGFR antibody	Improved cytotoxicity and site specificity in non-small cell lung cancer	Patel et al. (2018), Viswanadh et al. (2020)
Bovine nanoparticles	BSA	Paclitaxel	HA	Ovarian cancer therapy	Edelman et al. (2019), Yang et al. (2023)
Silica or mesoporous silica nanoparticles	Silica	Paclitaxel	HA	Breast cancer therapy	Li et al. (2018)
	Mesoporous silica	Zinc complexes chitosan	Chitosan-biotin	Enhanced chemotherapy	Kundu et al. (2021), Sabio et al. (2019)
Metallic nanoparticles	Monoolein	Copper acetylacetonate	HA	Selective targeting of CD44-expressing tumors	Pramanik et al. (2022), Palma et al. (2023)
	Gold	siRNA	FA	For gene silencing	Mbatha et al. (2019)
SLN	Stearic acid	Curcumin	Transferrin	Prostate cancer therapy	Akanda et al. (2021), Akanda (2015)

controlled drug distribution. Certain materials react differently to different external stimuli, such as heat and pH. These substances can be applied as coatings on appropriate core materials and utilized as stimuli-responsive nanocarrier materials (Sperling and Parak 1915).

There are two types of nanomaterials: core and shell core materials that are frequently utilized include silica, metal, magnetic, gene, drug, and protein, and the second is nanoparticles with amino acid. Polymers, silica, polysaccharides, proteins, metals, and metal oxides are some examples of materials used for shells of nanoparticles. Several categories can be used to group core-shell nanoparticles based on their physical characteristics. The fundamental varieties include core-shell silica nanoparticles, carbon nanomaterial-based core-shell nanoparticles, core-shell metallic nanoparticles, core-shell upconversion nanoparticles, and core-shell polymer nanoparticles (Nam et al. 2013). Core-shell nanoparticles of silver, gold, platinum, iron oxide, palladium, superparamagnetic, and silica nanoparticles are among the other classes that are based on core material. We provide a schematic discussion of the many kinds of core-shell nanoparticles in the sections that follow. Nanoparticles become less poisonous and more biocompatible when biocompatible materials are coated as the shell over the core ingredients. The benefits of shell materials go beyond only lowering toxicity; they may also alter, induce, or improve the qualities of the core material. For instance, coating or doping semiconductor with other materials improves their optical characteristics (Erathodiyil and Ying 2011).

Core-shell nanostructures are highly sought after for their distinctive geometry and innovative design. They are made up of an internal core made of one material (biomolecules or metal) encased in a shell made of another material. Core-shell nanoparticles are very important because of their superior solubility, reduced toxicity, and good thermal stability. Core-shell nanoparticles are biphasic materials featuring an exterior shell made of multiple components and an interior core structure (Gonçalves and Martins 2014). These particles have gained attention for their unique properties resulting from the interaction between the geometry, materials, and design of the core-shell nanoconstruct. Core-shell nanoparticles are often engineered with a cost-efficient core material that supports a thin, high-value shell or improves the core's reactivity, thermal stability, and resistance to oxidation. Consequently, core-shell nanoparticles are extensively utilized in diverse fields, including biomedicine, electronics, semiconductors, and catalysis (Gong and Winnik 2012).

The synthesis of core-shell particles typically involves a two-step solution-based process. In the first step, the core structure is synthesized and second step followed by the coating of the core structure with the shell material. There are two major steps involved in synthesis of core-shell nanoparticles: In the first step, core structure is synthesized and second step followed by the coating of the core structure with the shell material. These approaches are employed in solution synthesis. Chemical vapor deposition and pulsed laser deposition are the two common methods used in gas-phase synthesis. Substrates are used in these procedures, which often need several steps and the shell material to be deposited onto an already-formed core structure (Ma et al. 2020). The different criteria influencing the production of particles are known, and the methods of formation for these well-established procedures have been studied. By employing electron beam evaporation, the scientists have created core-shell Cu–Si and Ag–Si type particles in a novel manner. This

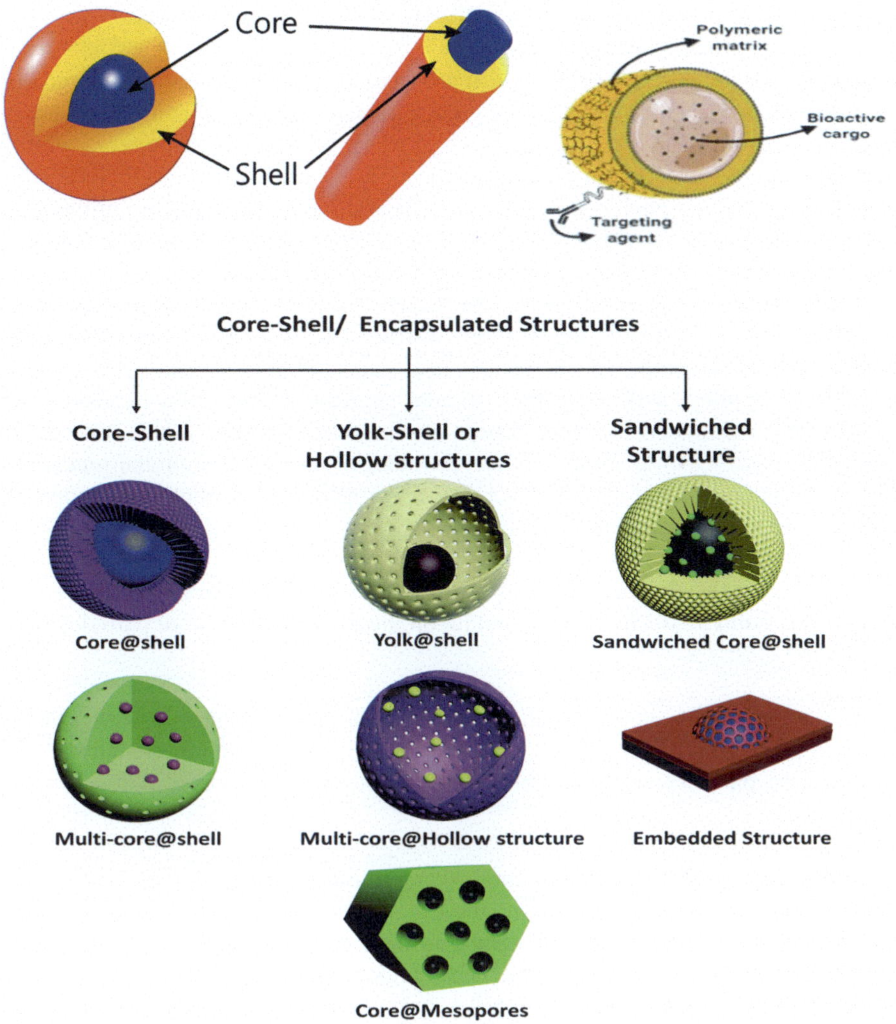

Fig. 6.1 Structures of core-shell nanoconstruct

method involves synthesizing the core-shell particles directly from the gas phase in a single step (Menath et al. 2021).

Using no substrates, the particles of core-shell are produced in a single step straight from the gas phase in this approach. Rather than using chemical compounds that break down into the final core or shell materials, elemental materials are employed as precursors. Therefore, this approach could potentially be a more cost-effective way to synthesize specific kinds of core-shell particles. Using this technique, nanopowders have been created at rates of kg/h (Fig. 6.1).

1.4.1 General Procedure for Synthesis of Multidimensional Core-Shell Nanoparticle

Nanomaterials are synthesized by two approaches; the first one is top-down and the other is bottom-up. The "top-down" approach involves traditional manufacturing or microfabrication techniques, where external tools are used to shape, cut, or mill materials into desired structures and configurations. Common methods include various forms of lithography (e.g., UV, electron beam, ion beam, scanning probe, and optical near-field lithography), laser-beam processing, and mechanical techniques such as machining, grinding, and polishing (Chen et al. 2022).

The "bottom-up" approach relies on the chemical properties of molecules to facilitate their self-assembly into functional structures. Techniques commonly used include chemical synthesis, chemical vapor deposition, laser-induced assembly (e.g., laser tapping), self-assembly, colloidal aggregation, and film deposition or growth. This method enables the creation of ultra-small particles and is considered potentially more cost-effective. It offers advantages such as exceptional precision, full process control, and reduced energy loss compared to the top-down approach (Abid et al. 2022; Tripathy et al. 2023).

1.4.2 Two-step Bottom-up Process for Synthesis of Core and Shell

Bottom-up synthesis technique depending on the availability of core particles may be of two types:

1. The core particles are first synthesized and then integrated into the system with appropriate surface modifications to enable the coating of the shell material.
2. The core particles are first synthesized in situ and then coated with the shell material.

Step 1: Mixing: It is the first method where a suspension of sol or colloidal powder is created by mechanically mixing the colloidal particles in water at an appropriate pH, ensuring no precipitation occurs. In the other two methods, metal salts or metal alkoxides serve as precursors. These undergo hydrolysis in the presence of water, forming hydrated metal hydroxides that subsequently condense to create oxo bridges (M–O–M). Once a sufficient number of M–O–M bonds form in a localized region, they collectively result in the formation of colloidal particles or sol. The pH of the medium and the water-to-precursor ratio influenced the particle size.

Step 2: Casting: Due to decreased viscosity of sol liquid, it is easily cast into a mold for shaping.

Step 3: Gel formation: Over time, colloidal particles and condensed species link together to create a three-dimensional network, progressively increasing the viscosity until the material transitions into a solid state.

Step 4: Aging: Aging the gel involves keeping the cast object immersed in a liquid for a specific period, allowing processes like syneresis, polycondensation, coarsening, and phase transformation to occur.

Step 5: Drying: This process involves the removal of the liquid incorporated in the interconnected pore network.

Table 6.2 Applications and synthesis methods of metallic core-shell nanoparticles

S. No.	Types of core-shell nanoparticle	Method		Applications	References
		Core synthesis	Shell synthesis		
1.	Fe_3O_4–Ag	Fe_3O_4 nanoparticles were synthesized, using neem leaf extract and $FeSO_4*7H_2O$, and $Fe(NO_3)_3*9H_2O$	Hydrothermal method by adding $AgNO_3$ in Fe_3O_4 solution used for the preparation of Fe_3O_4-Ag	Promising anticancer agents	Rajkumar et al. (2019); Cursaru (2011)
2.	Fe_3O_4–Au	Fe_3O_4 was prepared by coprecipitation method using $FeCl_3*6H_2O$ and $FeCl_2*4H_2O$	Fe_3O_4 mixed with $HAuCl_4$ at 90 °C for 5 min followed by the addition of sodium citrate	In cancer therapy	Yeamsuksawat et al. (2021), Li et al. (2011), Cui et al. (2004)
3.	Fe_3O_4–Ag	An eco-friendly method involves Precursors: V. vinifera stem extract, $FeCl_3*6H_2O$, and sodium acetate for the preparation of Fe_3O_4 nanoparticles	$AgNO_3$ was added to the Fe_3O_4 solution, with continuous stirring for 2 h. The core-shell nanoparticles is nearly spherical with ~32 nm diameter and the Particle size of nanoparticles is controlled by the addition of Vitis Vinifera stem extract	Extensive antibacterial potential	Cursaru (2011), Madhubala et al. (2023)
4.	Fe_3O_4–chitosan–Ag	Commercial magnetite and chitosan powders are used in suspension technique for Fe_3O_4–chitosan synthesis	Fe_3O_4–chitosan was immersed in dimethylformamide and then mixed with $AgNO_3$ and glucose	Potential antimicrobial agents (additives in medical, biological, food packaging, and textile applications)	Moradi et al. (2022), Yildiz et al. (2020)
5.	Fe_3O_4–Au	Coprecipitation method using $FeCl_3*6H_2O$ and $FeCl_2*4H_2O$	Green synthesis method: Fe_3O_4 nanoparticles were suspended in $HAuCl_4$ solution using J. regia (walnut) as a reducing agent	In treatment of cancer	Li et al. (2011), Izadiyan et al. (2019))

(continued)

Table 6.2 (continued)

S. No.	Types of core-shell nanoparticle	Method		Applications	References
		Core synthesis	Shell synthesis		
6.	Fe_3O_4–heparine–poloxamer (HP)	By coprecipitation method using $FeCl_3$*$6H_2O$ and $FeCl_2$*$4H_2O$, Fe_3O_4 was prepared	Ultrasonication is used for preparation of Fe_3O_4–HP core-shell system	Cancer treatment	Cursaru (2011), Hoang Thi et al. (2019)
7.	Fe_3O_4–PNIPAAm	Coprecipitation in Ar atmosphere by using $FeCl_2$ and $FeCl_3$	In the presence of Fe_3O_4 at 90 °C for 5 days polymerization of NIPAAm is performed	Targeted drug delivery against lung and breast cancer	Cui et al. (2015), Sharma (2021), Regmi (2011)

Step 6: Chemical stabilization or dehydration: This step involves the removal of surface metal-hydroxide bonds, transforming the material into an ultra-porous stable solid with interconnected optical transparency, porosity, and enhanced mechanical strength (Table 6.2).

Step 7: Densification: Following dehydration, the solid material is subjected to high temperatures to reduce internal porosity, thereby increasing its density (Gawande et al. 2015; El-Toni et al. 2016).

1.5 Role of Nanoconstruct in Various Diseases

Nanoscale materials have been extensively used in the treatment of various neurological disorders, including Parkinson's disease, Alzheimer's disease, frontotemporal dementia, Huntington's disease, amyotrophic lateral sclerosis, spinal cord injury, Prion disease, stroke, and brain tumor. Researchers are actively developing innovative strategies to enhance the effectiveness of therapeutics and address challenges posed by the blood–brain barrier (BBB) in cancer treatment (Waris et al. 2022). The nano-based approach is central to these advancements. Nanotechnology involves the use of various engineered nanoparticles and nanomaterials, their size typically ranging from 1 to 100 nm in at least one dimension. These nanoscale materials exhibit unique properties, including small size, a high surface-to-volume ratio, excellent stability, and strong interactions with biological systems. A distinct type of nanoparticles, quantum dots (QDs) are utilized in the treatment of Alzheimer's disease. They have the ability to cross the BBB (Nguyen et al. 2021a; Sriramoju et al. 2014; Barchet and Amiji 2009), vascular diseases and malignancies are easily diagnosed and treated through impactful nanomedicine-based approaches. Worldwide, the leading causes of illness and mortality are vascular disorders. Due to their lack of site-specific action, quick drug clearance, and low therapeutic index, current vascular treatment techniques that include the direct systemic delivery of medicines perform suboptimally and frequently cause severe systemic side effects such as coagulopathy and bleeding. Vascular therapies could be completely transformed by targeted nanomedicine strategies that use bioengineered nanoconstructs to deliver drugs selectively to specific vascular sites and release them under regulated conditions through the use of disease-specific molecular and cellular mechanisms (Gupta 2011; Gupta et al. 2019). Nanoconstructs should be capable of effectively carrying multiple agents on their surface and within their core, making them ideal tools for combination therapies. Additionally, a drug can be combined with a contrast agent, enabling both therapy and diagnosis in a single approach, a strategy increasingly referred to as "theragnostic" (Cervadoro et al. 2018). Silica nanoparticles have been utilized for targeted drug delivery to the heart through both passive and active mechanisms. Adenosine-loaded silica nanoparticles effectively restored blood pressure and significantly reduced infarct size compared to free adenosine administration. Additionally, multifunctional silica nanoparticles carrying cardioprotective compounds demonstrated a preference for interacting with cardiomyocytes and non-myocytes in the infarcted region. These nanoparticles have shown promising efficacy in actively targeting the endocardial layer of the injured heart, enhancing therapeutic outcomes (Skourtis et al. 2020). To detect early onset and progression of Alzheimer's disease by non-invasive methods a multifunctional theragnostic nanoconstruct system has been developed. The system enhances magnetic resonance contrast signals in disease-affected areas and protects primary cortical neurons from inflammation and oxidative stress (He et al. 2020). Polymeric nanoparticles can be directed to the target area through endothelial cell transcytosis or receptor-mediated endocytosis, and their delivery can be enhanced by coating them with polyethylene glycol (PEG) or antibodies. In Alzheimer's disease, senile plaques are marked by elevated concentrations of Aβ-42, changes in terminal amino acids, and an increase in their overall concentration (Bilal et al. 2020).

Significant interest has emerged in exploring various nanoparticles for their ability to inhibit synuclein amyloid formation. Among them, phosphorus-doped carbon nanotubes have demonstrated strong interactions with synuclein, suggesting their potential as a treatment for Parkinson's disease (Moni et al. 2021). Lactoferrin conjugated polyethylene glycol-lpolylactide-polyglycolide nanoparticle (Lf-NP), was introduced to evaluate the in vitro and in vivo delivery properties of a novel biodegradable brain drug delivery system. Autopsy samples taken from the brains of deceased Parkinson's disease (PD) patients showed dense protein aggregates known as amyloid fibrils and Lewy bodies within the brain tissue (Hu et al. 2011).

Nanoparticles with PLGA (poly [lactic-co-glycolic acid]) offer numerous merits over conventional drugs, including resistance to enzymatic degradation, enhanced drug solubility, improved targeting efficiency, and better cellular internalization. These nanoparticles have also shown promising effects in the treatment of Huntington's disease (Cardoso et al. 2015).

A nanotechnology-based drug delivery is also used for frontotemporal dementia, which is a very common type of dementia that affects people under the age of 65 years (Nguyen et al. 2021b).

Perfluorocarbon nanoparticles (PFC-NPs) demonstrate significant potential in revolutionizing stroke diagnosis and treatment by enabling early and precise detection of therapeutic responses through enhanced MRI imaging. In a detailed experimental setup, magnetic nanoparticles were functionalized by cross-linking with thrombin to induce clot formation. These nanoparticles were administered via tail vein injection and directed to the common carotid artery, where they facilitated rapid thrombus formation within minutes. The clots, now complex structures, were subsequently dislodged and transported into the cerebral artery, ultimately resulting in arterial blockages that mimic stroke conditions. This approach highlights the utility of PFC-NPs in advancing both diagnostic and therapeutic strategies for cerebrovascular diseases (Yan et al. 2005).

1.6 Role of Multi-Shell/Core-Shell Nanoconstruct in Various Diseases

Core-shell nanoparticles are made up of a coating of one substance covering the core of another. By improving characteristics including decreased cytotoxicity, enhanced dispersibility, enhanced bio- and cytocompatibility, enhanced conjugation with other bioactive compounds, and greater thermal and chemical durability, these core-shell architectures provide a number of advantages over plain nanoparticles. For a variety of bioapplications, conjugating biomolecules onto particle surfaces is essential. One way to overcome this difficulty is to coat the particles with an appropriate biocompatible substance. In recent years, research on the application of core-shell nanomaterials in the biomedical branch has grown. These tiny particles can address numerous issues related to simple nanoparticles and have demonstrated superior performance over simple nanoparticles (Chatterjee et al. 2014). Because of these unique advantages, core-shell nanoparticles may be more easily used to combine two or more different functional capabilities in a single nanosystem, thereby obtaining the physicochemical properties required for effective targeted drug administration. Numerous core-shell nanoparticle varieties, such as metallic, magnetic, silica-based, upconversion, and carbon-based core-shell nanoparticles, have been developed for use in drug delivery application. Numerous core-shell nanoparticle varieties, such as metallic, magnetic, silica-based, upconversion, and carbon-based core-shell nanoparticles, have been developed for use in drug delivery applications (Kashkooli et al. 2020).

Combining treatments and diagnostics, theragnostics is a new therapeutic approach. For theragnostic applications, core-shell mesoporous silica nanoparticles have emerged as a promising multipurpose nanoplatform. Researchers have created core-shell particles with better qualities than bulk materials, and these particles are now used in fields like drug delivery, treatment of tumors, biosensor imaging, and microfluidic devices. Because nanoparticles can reach therapeutic sites at the appropriate timing and concentrations, they are frequently used as drug-delivery vehicles. This feature lessens negative effects while increasing the efficacy of cancer treatments (Bae et al. 2011).

1.7 Disease-Specific Targeting Through Nanoconstruct

1.7.1 Cancer Targeting Through Nanoconstruct

Successful cancer treatment requires the development or engineering of a drug release or gene technology with a remarkable carrying capacity that is capable of targeting tumor cells while endangering healthy ones. Understanding the targeting mechanisms and the pattern of interactions among NPs, tumors, and carcinoma cells is essential. Active, passive, stimuli-responsive, and magnetic targeting are the few common types into which targeting systems can be divided (Kashkooli et al. 2020).

1.7.1.1 Passive Targeting

Passive targeting is a diffusion-mediated drug delivery method that includes complex formation of drug carriers (Mishra et al. 2023). To reach the intended location, the medication and carrier complex is delivered via blood flow. Nanoconstructs that target solid tumors passively take advantage of their distinct characteristics, such as their leaking vasculature and decreased lymphatic clearance, to enable nanoconstructs to escape from the tumor's vasculature and concentrate inside the tumor's tissues through enhanced photoreduction (EPR) Molecular weight, surface charge, surface type, and surface area are all critical factors for passive medication targeting that works (Mishra et al. 2023; Bae et al. 2011; Black et al. 2014).

Drugs can be loaded into PLGA-based nanoparticles (NPs) with great potential for specificity, stability, controlled release, and reduced degradation. In order to encapsulate hydrophobic medications and enable their controlled release at certain tumor areas. In passive targeting Polyvinyl alcohol (PVA) is also utilized. Passive targeting is the most widely utilized technique for medication delivery in cancerous cells (Wei et al. 2020).

1.7.1.2 Active Targeting

In order to achieve effective targeting, particular ligands or substances must attach to receptors or molecules that are markedly overexpressed on the target cells. Another name for this targeting technique is ligand-mediated targeting. To boost affinity in this case NPs must be near the target if they include ligands with certain properties, such as absorption and longevity. There is less damage to normal cells when active targeting is used to introduce nanostructures into the tumor microenvironment (TME). This approach starts the creation of complexes of ligands and receptors by using potent binding ligands that interact with specific receptors (Kuo and Chen 2015). Receptors such as transferrin and folate attach to specific molecules of interest or agonists, such as proteins, carbohydrates, and amino acids, on the cell surface. This process involves the movement of macromolecules from the extracellular fluid. This technique improves drug penetration by facilitating changes in NP binding to cancer cells (Chaturvedi et al. 2019).

1.7.1.3 Stimuli-Responsive Targeting

The distinct TME and aberrant cells, together with their intricate process, are characteristics of cancer. The nonspecific targeting of traditional cancer therapy poses a challenge. Both internal and environmental factors, including light, oxidative stress, pH, and enzymes, can affect how well nanostructures work. Two compounds that are used with nanocarriers for pH responsiveness during internal stimuli-responsive targeting are glutathione and calcium carbonate. By using protein-based biomineralization of Mn^{2+}, manganese dioxide (MnO_2) nanoparticles (NPs) encapsulated using human plasma albumin (HSA) were designed to respond to pH and H_2O_2. Transducers frequently convert environmental cues like electromagnetic waves or photo therapy into tangible amounts like light radiation that can subsequently be converted into heat in order to effectively kill cells (Gazeau et al. 2008).

1.8 Nanoconstruct in Cancer Therapy

A new intervention with potential benefits for doctors and patients is cancer therapy. This system has real-time diagnostic capabilities that are in line with the desired delivery and tracking of cancer treatment. Cancer theranostics relies on the identification of various cellular phenotypes at the tumor site, which are targeted by theranostic agents for both therapy and observation. Nanoconstructs are being developed and used in contemporary therapies to overcome the drawbacks of conventional chemotherapy for cancer, including toxicity, unpredictable release rates, and imprecise distribution of medications. The techniques used in the design of nanoconstructs contribute to the success of loaded theranostic drugs as a cancer therapy by increasing their efficiency and specificity. Nanoconstructs are made specifically to target the required spot and remove obstacles that prevent their proper deployment for the intended benefit. Consequently, distribution modes for nanoconstructs are more appropriately categorized as autonomous and nonautonomous kinds rather than as actively or passively targeted systems. In general, nanoconstructs have many advantages, but they also have many drawbacks. It has been determined that nanotechnology may be useful for both cancer diagnosis and treatment. The majority of the inorganic NPs (metal NPs, quantum dots, nanotubes made of carbon, etc.) and organic NPs (lipids, dendrimers, polymers, liposomes, etc.) provide examination. treatment, and targeting abilities. The use of theranostic NPs in photothermal treatment, chemotherapy, siRNA/miRNA transportation, and other treatment options for carcinomas is currently the subject of extensive investigation (Behranvand et al. 2022).

2 Therapeutic Approaches for Cancer Treatment

Anticancer treatments have advanced significantly over the past several years, but cancers continue to be resistant, mostly because of medication intransigence or recurrence. In the therapy of cancers, cancer stem cells (CSCs) have become possible targets or drug receptors. CSCs are a tiny subset of cancer cells that have high tumorigenic potential, self-renewal capabilities, and resistance to both radiation and chemotherapy. Researchers from all over the world are currently attempting to comprehend CSCs and create treatment strategies that specifically target them. Although CSCs had the ability to self-renew, their hyperproliferation and metastasis were strictly governed by internal and external stimuli. Additionally, they could develop into cancerous cells with markers such as alpha-fetoprotein, albumin, CK7, CK8, CK18, and others. According to the best available data, signaling pathways like Wnt/-catenin and Notch, Hh, are crucial for the development of cancer. Numerous elements of the tumor microenvironment, including inflammatory cells, cancer-associated fibroblasts, angiogenic vascular cells, endothelial cells, and cancer cells, affect the characteristics of CSCs. Anti-angiogenic therapies and anti-inflammatory drugs can be used in cancer prevention and treatment (Pan et al. 2018).

Enhancing chemotherapeutic transport to cancer cells is part of cancer therapy, which aims to lower systemic adverse effects and enhance the standard of life. Many therapeutic procedures are employed for the successful management of malignancies, but the main tactics rely on surgically removing the tumor mass. Other therapeutic modalities, such as radiotherapies, chemotherapy, and biotherapies, among others, also effectively contributed to the fast proliferation of cancerous cells (Amjad et al. 2020). Traditional methods of care have been used for a while, but there have also been significant recent advancements in treatment, including specific treatment, stem cell treatment, surgical removal treatment, naturally occurring antioxidant nanocarriers, radionics, sonodynamic management, chemodynamic medication, and ferroptosis-based medical care (Debela et al. 2021).

2.1 Approaches

2.1.1 Chemotherapy

Chemotherapy is used to ease cancer symptoms and treat cancer by slowing or stopping the cancer cell growth, which is divided and grows in an uncontrolled manner. Many types of cancers can be treatable by chemotherapy. In order to affect the macromolecular production and function of neoplastic cells, conventional chemotherapy medications usually interfere with the formation of RNA, DNA, or amino acids, or change the usual functioning of the preformed molecule. This inhibits cell proliferation and tumor expansion. The chemotherapeutic agent directly kills cells or triggers cell death when there is sufficient disruption in the creation or function of macromolecules (Ravindran Girija and Balasubramanian 2019). Drugs for chemotherapy can activate multiple signaling pathways and enhance the release of inflammatory mediators (Behranvand et al. 2022). Certain genetic alterations in particular types of cancers also modify some of these pathways, which vary even between people with the same tumor. The bulk of these pathways are associated with cell death and the AKT, ERK, and nuclear factor-κB pathways. These pathways are mostly triggered by upregulating the JNK/p38 MAP-kinases or by inhibiting these survival networks. Consequently, the balance between a tumor's cell death and survival pathways determines how effective a particular chemotherapy is. Changing these survival pathways could contribute to improving the effectiveness of chemotherapy (Perona et al. 2007). Chemotherapy has transformed cancer from a terminal and catastrophic disease to a treatable and, in some cases, curable illness. It can be used as a neoadjuvant, adjuvant, combination, or malignant treatment, among other scenarios. Chemotherapy is frequently used to achieve curative purposes in tumors such as head and neck, lung, and anal cancer, or to reduce tumors before operations. The first medications were derived from a very potent class of alkylating chemicals known as nitrogen mustards. The crucial damage that causes cancer cells to die is the ability of these extremely electrophilic chemicals to attach alkyl groups onto DNA bases, despite the fact that they can react with a vast array of biological nucleophiles. Concurrently, a second category of cancer chemotherapy drugs emerged, known as antimetabolites (e.g., aminopterin and amethopterin, which obstruct the production of folate). The antimetabolites, as opposed to nitrogen mustards, either mimic or disrupt the synthesis of DNA precursors, which stops or causes errors during DNA replication, eventually leading to the disintegration of tumor cells (Abraham and Staffurth 2016). Contraindications are frequently linked to chemotherapy drugs. Mucositis, myelitis, bloating, constipation, baldness, exhaustion, vomiting, sterility, infusion responses, infertility, and an elevated risk of infections as a result of immunosuppression are some of the adverse reactions that are linked to their mode of action. A number of processes, such as drug inactivation, efflux, target modification, and cell death inhibition, can result in chemotherapy resistance (Amjad et al. 2020). The adverse effects of chemotherapeutic agents have led to scientific advancements in a variety of delivery mechanisms for drugs, including aptamers, antibody–drug conjugates, peptides, organic nanocarriers (albumins, polymers, liposomes, and micelles), and inorganic nanocarriers (gold nanocarriers, mesoporous silica nanoparticles, platinum nanocarrier, and carbon nanotubes). These targeted drug delivery techniques offer several advantages, such as increased overall chemotherapeutic efficacy, reduced toxicity, enhanced bioavailability, and site-specific drug administration (Yadav et al. 2022).

2.1.2 Photothermal Therapy

In order to generate heat for thermal ablation of malignant tumors, photothermal therapy (PTT), a minimally invasive and possibly successful therapeutic treatment, uses pulsed laser irradiation at near-infrared (NIR) to activate photosensitizing chemicals. The method known as photothermal treatment (PTT) uses light radiation to kill cancer cells. A photosensitive probe (PS) is excited by a near-infrared (NIR) laser in photothermal therapy (PTT)-based theranostics, a sophisticated type of photodynamic therapy (PDT). PTT eliminates cancer cells by using the heat produced by this excitation. Radiative decay (fluorescence emission) and nonradiative decay (heat) must be balanced, though. In living cells, the heat generated can cause denaturation, cytosol evaporation, protein aggregation, and cell lysis. Because of their effective photothermal conversion capabilities and high optical absorption in the near-infrared spectrum, graphene-based nanomaterials make excellent PTT candidates (Shuel 2022). When exposed to light at their resonance wavelength, gold-silica nanoshells created by Hirsch et al. produce synchronized oscillations that either scatter or absorb light. It is possible to tune the light scattering to absorption ratio for particular uses (Wu et al. 2006).

2.1.3 Photodynamic Therapy

In order to cause irreversible photodamage to tumor tissues, photodynamic treatment (PDT) entails first providing a photosensitizing agent that localizes in the tumor and then activating the agent with light of a certain wavelength. Because they can improve the therapeutic result, the locations of photosensitizer localization and photodamage at the tissue or cellular level are crucial. Tumor cells' mitochondria, lysosomes, plasma membrane, nuclei, and tumor vasculature are all possible PDT targets (Veselov et al. 2022). Apoptotic, necrotic, and autophagy-associated cell death are the three main cell death processes that photodynamic treatment (PDT) can initiate. The most common type of cell death in response to PDT is usually necrosis. Heat shock proteins (HSPs), which aid in cell protection during treatment, are also expressed when PDT is administered. HSPs are essential for controlling pro-survival pathways because they attach to proteins that have been oxidatively damaged (Costley et al. 2015; Kuroki et al. 2007). A flexible class of nanomaterials, mesoporous silica nanoparticles have shown promise as carriers for photodynamic treatment (PDT) applications. Using near-infrared (NIR) irradiation, PDT can successfully treat deep-seated cancers by utilizing the optical diagnostic window of tissues (600–1000 nm), which permits deep tissue penetration. Numerous researches have used nanomaterials such as upconversion nanoparticles (UCNPs), which absorb in the near-infrared spectrum, to diagnose and cure deep-seated cancers (Ravindran Girija and Balasubramanian 2019).

2.1.4 Hormonal Therapy

Breast cancer and prostate cancer that are estrogen and progesterone receptor-positive can be effectively and non-toxically treated with hormonal therapy. The hypothalamic-pituitary-gonadal system regulates serum levels of testosterone and oestradiol. Premenopausal women make oestradiol from their ovaries, and post-menopausal women make it from the peripheral conversion of adrenal androgens via aromatase. Castration is the main form of treatment for men with prostate cancer and women with breast cancer who are premenopausal (Abraham and Staffurth 2011; Abraham and Staffurth 2020; Abraham and Staffurth 2016). Before radical surgery or radiotherapy, the size of primary cancer can be reduced by hormonal therapy (Abraham and Staffurth 2020).

2.1.5 Immunotherapy

Cancer immunotherapy is composed of approaches that use the components of the immune system or cause modification in the immune system of the host or for the cancer treatment (Dillman 2011). Tumors use several mechanisms to evade the host immune response, including induction of energy in T cells, release of immune-suppressive cytokines, and local mediators to alter the tumor environment. Immunotherapy for cancer involves the activation of T-cell response against a malignancy. Genetically modified T cells have been successful in treating B-cell lymphomas. TILs are isolated from solid tumors by surgery and expanded over several weeks ex vivo under sterile conditions to generate a sufficient number of tumor-reactive T cells (Dhar et al. 2021).

2.1.6 Targeted Therapy

By focusing on cell surface antigens, growth factors, receptors, signal transduction pathways, or gene expression modulators, targeted therapy might prevent the proliferation, metastasis, and angiogenesis of cancer cells. The two most popular forms of targeted medicine are small-molecule inhibitors and monoclonal antibodies. Despite their inability to penetrate cells, monoclonal antibodies can be coupled with harmful compounds to give targeted therapy straight to cancer cells (Shuel 2022). Direct strategies use monoclonal antibodies (MoAbs2) or small molecules that impede the target proteins' ability to communicate to change the way tumor antigens are signaled. Utilizing tumor antigens expressed on the cell surface as target devices for ligands comprising various effector molecule types, indirect techniques are based on this principle. By utilizing tumor-specific monoclonal antibodies (MoAbs) or peptide ligands that attach to receptors on tumor cells, these methods enable medications to actively target malignancies. The hyperpermeable angiogenic tumor vasculature and inefficient tumor lymphatic drainage cause "enhanced permeability and retention effects" that allow macrophages to target tumors both passively and actively (Wu et al. 2006). Cell-targeted administration can improve the efficacy and selectivity of cytotoxic medications while lowering their general toxicity. In addition to conventional liposomal and micellar formulations, developmental interest has recently shifted to a variety of nanocarrier systems. The preparation and targeting methods as well as the characteristics of various delivery systems based on liposomes, micelles, solid lipid nanoparticles, dendrimers, gold, and magnetic nanoparticles are used (Veselov et al. 2022).

2.1.7 Sonodynamic Therapy

Solid malignant tumors can be treated minimally invasively with sonodynamic therapy (SDT). When combined with a sensitizing medication, ultrasonography, and molecular oxygen, it produces cytotoxic reactive oxygen species (ROS). Reactive oxygen species (ROS) are produced during SDT, which has a number of advantages over photodynamic treatment (PDT) due to the ability of ultrasound to penetrate soft tissue and focus precisely (Costley et al. 2015). Sonodynamic therapy preferentially depends on the retention or uptake of sonosensitizer (sono-sensitizing drug) in the tumor tissues and ultrasound radiation helps in subsequent activation of drug. Sonosensitizer can be activated by deep penetration of ultrasound radiation into a small region of tumor tissues (Kuroki et al. 2007).

2.1.8 Gene Therapy

Enhancing the human genome through site-specific alterations or the correction of damaged genes in order to target therapeutic therapies is known as gene therapy. Thanks to developments in genetics and bioengineering, it is now possible to manipulate vectors to transfer extrachromosomal material to particular target cells. It can cure infectious diseases like AIDS, acquired genetic diseases like cancer, and diseases brought on by recessive gene defects. Recombinant DNA technology is the most widely used gene therapy approach. It involves inserting a healthy gene or the desired gene into a vector. This procedure entails substituting a normal gene in the patient for an aberrant gene that causes a disease (Adriouach et al. 2019).

2.1.9 Combinational Therapy

Combination chemotherapy is a therapeutic technique that brings together two or more therapeutic drugs to improve efficacy, decrease drug resistance, and provide therapeutic anticancer effects without sacrificing efficacy (Mittal et al. 2021). It is a key component of cancer treatment since it targets distinct pathways to lessen the toxicity of each medication while enabling the treatment of many malignancies simultaneously. Furthermore, combination therapy kills cancer cells by cytotoxic means while leaving healthy cells unharmed. When a caspase inhibitor and an apoptosis-inducing drug are used together, cancer treatment can have a higher therapeutic index, a more powerful cytotoxic effect, and a lower rate of resistance. In addition, combination therapy can eliminate cancer stem cells, which reduces the likelihood of relapse (Bayat Mokhtari et al. 2017).

2.1.10 Biotherapy

Biotherapy is a new idea to treat cancer by using biologicals and biological response modifiers. Many of these substances are derived from "natural" sources, originating from mammalian cells as physiological mediators of immune response and regulators of growth and maturation. With advancements in molecular biology, hybridoma technology, and computational tools, these biological substances can now be produced in large quantities and highly purified for medicinal use. The application of these biotherapeutic approaches in cancer treatment often requires expertise in immunology and molecular biology (Oldham 1986).

2.1.11 Radiotherapy

Radiotherapy is one of the most impactful management for solid tumors, but it can also damage surrounding healthy tissue. To mitigate the immediate side effects of radiotherapy, researchers have developed versatile nanoparticles capable of encapsulating drugs for slow and sustained release, minimizing immune clearance, and targeting tumor tissue. The increased permeability and retention (EPR) effect linked to the leaky blood arteries in tumors enhances tumor accumulation for these nanoparticles. Non-metal radiosensitizer silica has been employed as a coating material or as an effective nanocarrier in formulations including heavy metals for radio sensitization. Gadolinium complexes integrated into mesoporous silica nanoparticles have shown promising release profiles, demonstrating the in vitro stability of these nanoparticles (Ravindran Girija and Balasubramanian 2019).

2.1.12 Nanotherapeutics

Despite advances in chemotherapy and knowledge of tumor biology, cancer is still a common and fatal illness. The distribution of therapeutic substances in a targeted and multifunctional manner is made possible by nanotherapies. Enhancing drug solubility and stability, prolonging circulatory half-lives, minimizing off-target effects, and concentrating medications at particular target areas are all possible with nanotechnology. Drug distribution can be accomplished with a typical kind of nanodevice: polymer nanoparticles. The special characteristics of nanoparticle systems enable both active and passive tumor targeting. Nanoparticles should ideally be between 10 and 200 nm in size to take advantage of the improved permeability and EPR effect (Bronstein and Shifrina 2011). Newly developed nanotherapeutics provide a flexible framework for treating cancer that has spread. These treatments efficiently stop cancer cells from spreading from their source sites to other organs by carefully tracking and killing them. Although tumor metastasis remains a serious obstacle in the treatment of cancer, nanomedicines have demonstrated great potential in enhancing metastatic disease diagnostics and treatment (Yang et al. 2018). Nanoformulations can increase the total therapeutic index of drugs administered by conjugating chemotherapeutic chemicals to the surface of nanoparticles or encapsulating them in nano-sized carriers. The enhanced permeability and EPR effect are essential for the passive targeting of tumors, which is the main method by which nanotherapeutic platforms are given selectively, even if other mechanisms have also been proposed (Bronstein and Shifrina 2011).

3 Application of Different Types of Core-Shell Nanoconstruct in Cancer

3.1 Inorganic-Based Nanoconstruct

In order to improve these nanostructures' stability, heterogeneous doping is used. The effects of black phosphorus (BP) in biomedicine can vary depending on the metal, polymer, folic acid, albumin, and other constituents that it is conjugated with. Stimuli-responsive nanoconstructs can be made to activate in response to pH variations with near-infrared (NIR) irradiation for pH-responsive BP-based anticancer treatment. One such application has demonstrated the potential of the BP system in gene delivery. Mcl-1 is a member of the Bcl-2 group and could be used to treat cancer. Additionally, there was an increase in Mcl-1 in breast cancer cells. The development of BP nanomaterials in conjunction with PLL aimed to target Cas13a/crRNA delivery. Given that their dazzling quantum dots have been extensively employed in nanoparticulate systems to investigate particle mobility both in vitro and in vivo. The pH-responsive system is one of the stimuli-responsive systems that has the advantage of allowing the delivery system to be altered based on the internal environment. One such is the innovative pH-responsive drug delivery technology known as mesoporous silica nanoparticles (MSNs), in which the drug is bonded to the surface using pH-responsive covalent bonds. These optical nanomaterials, referred to as upconversion nanoparticles (NPs), are doped with lanthanide ions and exhibit a wide range of electronic transitions within the 4f electron shell. These NPs can upconvert a single higher-energy photon from the combination of two or more lower-energy photons. The nanoconstruct is created using metal-phenolic acid and copper ions. Tannic acid networks on the mesoporous silica-coated upconversion nanoparticles' surface. These devices allow for simultaneous medication release at the target site and real-time monitoring. These systems have anticancer medications integrated into them; hence, this nanoparticulate form aids in the effective administration of cancer therapy (Hu et al. 2019; Fernandes et al. 2023).

Furthermore, mesoporous silica nanoshells containing porphyrin molecules are coated with copper sulfide (CuS) nanoparticles (NPs) and tagged with [^{89}Zr] to produce nanostructures for better cancer imaging and treatment. These hybrid nanoconstructs improved tumor deposition, lengthened blood retention times, and were biocompatible. It is a productive method to blend imagery and therapeutic advantages inside a single nanostructure (Goel et al. 2018).

Hyaluronic acid is an excellent tumor growth and alteration agent when used as a targeted molecule for cancer cells that overexpress the CD44 receptor. Three primary responses are demonstrated by the nanoconstruct: appropriate drug release, enhanced photothermal characteristics, and effective ROS formation. The photothermal and biodistribution profiles of tumors show high nanoconstruct deposition and retention. In light of this, these nanoconstructs have enormous promise for the field of translational cancer nanomedicine (Poudel et al. 2020).

3.2 Polymer-Based Nanoconstruct

Because of their versatility in surface modification, stimulus responsiveness, and ability to mix hydrophilic and lipophilic bioactives or diagnostic chemicals, polymeric nanostructures are currently being used for cancer theranosis. One such example is the use of deformable discoidal nanostructures as a new delivery system for imaging and medicinal applications. These are made by polymerizing PEG and PLGA into a discoidal shape. Hydrophilic and hydrophobic microdomains found in these polymer matrices act as pockets for various imaging and therapeutic substances. These particles slow down the rate at which the Mononuclear Phagocyte System (MPS) sequesters them because they remain in the bloodstream for a longer amount of time (Palange et al. 2017).

Furthermore, the potential of PLGA-based nanoconstructs for radiodynamic therapy was investigated. The generation of ROS at the tumor site is the foundation of this anticancer treatment. It mostly addresses the ROS produced as a result of hypoxia, which is the tumor-induced drop in oxygen levels. In one unique method, verteporfin and perfluorooctylbromide are loaded into PLGA NPs to create nanoconstructs. Under normoxic and hypoxic circumstances, these nanoconstructs exhibit a sharp rise in ROS production. In people, this treatment has eliminated over 60% of pancreatic cancer cells, and in just 2 weeks, tumor growth was halted. These success rates highlight the potential of radiodynamic therapy-based nanoconstructs as an advanced, non-invasive treatment option for deeply located hypoxic tumors (Clement et al. 2021).

Polyurethanes (PU) nanostructures are some other polymeric nanostructures that are frequently employed in biomedical settings due to their stimuli-responsiveness and biodegradability. PU nanoconstructs are straightforward medication and cancer treatment delivery methods. The characteristics of these kinds of nanoconstructs include stimuli sensitivity, targeting, quick drug release, solubility of hydrophobic-type chemotherapy medicines, and improved efficiency. In order to achieve active targeting, they can be coupled with ligands. They are ideal nanocarriers because they are responsive to temperature, pH, stimulation, and a variety of external conditions (Gajbhiye et al. 2020).

PEGylated squalene (SQPEG)-based nanoconstructs, which are lipid-based, have also been employed in cancer treatment. Their typical size is 200 nm, and they have an 18% drug loading capacity. They assemble with lipophilic pyropheophorbide-a (Ppa) to produce nanoconstructs (Sun et al. 2022).

In addition to being lipid- and polymer-based, dendrimer-based nanoconstructs have other advantages as thermogenic agents. Made of poly(amidoamine), dendrimers (PAMAM) are a family of synthetic macromolecules with well-organized internal structures and a variety of functional groups. They have been employed extensively in cancer nanomedicine applications. One of the main drawbacks of first-generation dendrimers is their tiny size, which limits their ability to load drugs and makes them more resistant to passive tumor targeting due to longer retention times and greater permeability. Additionally, they lack stimuli-responsiveness; hence, super-structured dendrimeric nanoconstructs (SDNs) were created to get around all of these restrictions (Adriouach et al. 2019).

3.3 Dendrimer-Based Nanoconstruct

Dendrimers are nano-sized structures characterized by tree-like branches or arms, playing an increasingly significant role in anticancer therapies and diagnostic imaging applications. The biggest influence on the chemical and physical characteristics of dendrimers comes from their structure. They proliferate from the core shell, which then interacts with monomers that contain either two dormant or one reactive molecule. Due to their distinct properties, including their high compatibility with biological systems, well-defined spherical structure, and hyperbranching, dendrimers find extensive use in a variety of fields, including biomedicine and medicine. In particular, a broad range of pharmaceuticals can be incorporated into the three-dimensional structure of dendrimers to generate physiologically active drug conjugates (Mittal et al. 2021).

Pharmaceutical technologists face a number of challenges when designing liver-targeted systems. Phosphorylated PAMAM dendrimers used in dentistry clinical treatments have great potential (Siafaka et al. 2016). A PAMAM dendrimer is a spherical, highly branched macromolecule that can encapsulate active chemicals and stabilize metal nanoparticles, such as gold nanoparticles. An investigation looked into the potential for theranosis of dendrimer-gold hybrid structures loaded with curcumin. Generation five poly(amidoamine) (PAMAM) dendrimers were combined with PEGylated amine-terminated $AuCl_4^-$ ions to create a dendrimer-gold hybrid structure [118]. Curcumin was combined with the MUC-1 aptamer to produce the finished hybrid system. The results demonstrated greater cellular cytotoxicity in HT29 and C26 cells compared to the non-targeted technique, underscoring its potential for tumor imaging using CT scans and cancer therapy (Alibolandi et al. 2018; Bronstein and Shifrina 2011).

Another work showed that hyperbranched PAMAM dendrimers coupled with F3 peptide and based on unimolecular micelles may be synthesized to specifically target nucleolins that are overexpressed in MDA-MB-231 cells (Mishra et al. 2023). MDA-MB-231 cells demonstrated enhanced PAMAM micelle absorption with F3 attachment (PAMAM-DOX-F3). The ^{64}Cu was chelated to micelles for positron emission tomography (PET) imaging in order to track their pharmacokinetic behavior. Comparing ^{64}Cu-PAMAM-DOX to ^{64}Cu-PAMAM-F3, serial PET imaging showed that the latter accumulated more slowly, ineffectively, and permanently in MDA-MB-231 tumors. The features of distribution in various organs and tissues were very similar (Yang et al. 2018).

4 Challenges in the Use of Nanoconstruct

Biomolecules encapsulated in nanocarriers, such as lipoproteins, hormones, and albumins, exhibit hyper-permeation and longer retention, which results in the enhanced permeability and retention (EPR) effect. Leaky vasculature, increased transcytosis of endothelial macromolecules, and insufficient lymphatic drainage inside the tumor's interstitial space are all characteristics of solid tumors that undergo these processes [122]. Effective renal clearance and little extravasation into healthy tissues are necessary for selective tumor deposition based on the enhanced permeability and retention (EPR) effect. The EPR effect was produced by the nanoconstructs because of their smaller particle size, which was less than the vascular permeability of the tumor (600–800 nm). The targeting of NPs is influenced by the nanoconstructs' larger particle size compared to renal filtration (5–6 nm) and blood capillary pores (6–12 nm) (Choi et al. 2011).

Differences in tumor type and stage can lead to the development of tumor interstitium with unique extracellular matrix (ECM) components such as fibrin, fibronectin, hyaluronan, collagen, and proteoglycans along with varying tumor parenchyma and stromal cell compositions. The distribution and transportation of nanoconstructs within the tumor may be impacted by changes in pressure ranges and physical rigidity, which could have further effects. Their medicinal effectiveness (Swartz and Lund 2012; Perry et al. 2017). Prostate cancer and pancreatic ductal carcinoma are classified as low-level EPR tumors, while Kaposi sarcoma, hepatocellular carcinoma, renal cell carcinoma, and cancer of the head and neck are recognized as high-level EPR cancers (Regev et al. 2005).

Beyond limitations of the effect of EPR, other variables, such as physical and biological barriers to delivering a required amount of nanoconstructs, impede their distribution and deposition in tumors (Bae and Park 2011). A therapeutic drug's dose to the tumor's intended spot. Nanoconstructs experience a number of processes following intravenous distribution, including protein adsorption, particle transport, aggregation, shear force destruction, and hydrolysis. Events like these have an impact on how many nanoconstructs eventually make it to the tumor. The impact of these events on the tumor accumulation of nanoconstructs may differ based on the physicochemical properties of the nanoconstructs. An additional false impression that arises from the preclinical stage of in vivo model selection provides a hurdle. Studies carried out in vivo using cancer models developed from animals do not exhibit similarities with human tumors (Bae and Park 2011). In humans, the ratio of body mass to tumor mass is typically lower than that observed in animal models. It follows that the amount of nanoconstructs that likely reach human tumor locations is less than the necessary therapeutic range (Gould et al. 2015; Lazarovits et al. 2015).

5 Summary

Nanostructures are made up of ligands and nanoparticles that can move loaded cargo to the desired location of action. Nanostructures that may be used for both therapeutic and diagnostic reasons have been created using a variety of nanoparticulate platforms. Dealing with the side effects of cancer treatments, such as nonspecific medication distribution, toxicity, and uncontrolled release rate, is the main objective of nanoconstructions. Nanostructures based on polymers, including poly-l-lysine, poly(ε-caprolactone), poly (lactic-co-glycolic acid), and poly(propylene), are used to cure cancer. Among the many advantages of using polymers in cancer treatment include improved drug release, solubility, bioavailability, degradation, and reduced toxicity. To achieve effective cancer therapy, a medication or gene release system with exceptional loading capacity that can precisely target tumors while minimizing harm to healthy cells must be developed or engineered. The few typical categories that the targeting systems fall into are active, passive, magnetic targeting, and stimuli-responsive. Nanoconstruct design relies heavily on the intricate processes connected to thermomechanical engineering, which can be comprehended through computational models. For effective cancer treatment, various therapeutic approaches are used but the major strategies are basically based on surgery through the removal of the tumor mass. Some other therapeutic approaches like chemotherapies, radiotherapies, biotherapies, etc. also made an effective contribution to the uncontrolled growth of malignant cells.

6 Conclusion

Over the past years, the development of nanocarriers has provided consistent and ongoing insights that have shown a revolutionary change in cancer imaging, diagnosis, and treatment. Research has shown that the key to nanocarriers' success as a delivery platform is their capacity to be refined and optimized in aspects such as size, shape, composition, biocompatibility, drug payloads, biodegradability, and cellular interaction as well as entry into the tissue affected by tumor. It is claimed that by offering "an all-encompassing" cancer therapy strategy, nanoconstructs such as multi-shell\core-shell nanoconstruct, nanostars, nanoshells, magnetic, nanoflares, polymeric, and lipidic nanoconstructs have surpassed the previous nanomedicine systems. Multi-shell\core-shell nanoconstruct with a focus on how well they can handle different aspects of cancer treatment are discussed in this chapter along with their role in different diseases. Nanoparticles are ideal platforms for drug delivery due to their permeability, size, and enhanced ability to load drugs. The surface functionalization of nanoplatforms can enhance their functionality and make them more stimuli-responsive. This chapter also explored various therapeutic approaches for cancer diagnosis and treatment. Future research should focus on advancing nanostructures for targeted cancer therapy.

References

Abid N, Khan AM, Shujait S, Chaudhary K, Ikram M, Imran M, et al. Synthesis of nanomaterials using various top-down and bottom-up approaches, influencing factors, advantages, and disadvantages: a review. Adv Colloid Interf Sci. 2022;300:102597.

Abraham J, Staffurth J. Hormonal therapy for cancer. Medicine. 2011;39(12):723–7.

Abraham J, Staffurth J. Hormonal therapy for cancer. Medicine. 2016;44(1):30–3.

Abraham J, Staffurth J. Hormonal therapy for cancer. Medicine. 2020;48(2):103–7.

Adriouach S, Vorobiev V, Trefalt G, Allémann E, Lange N, Babič A, et al. Squalene-PEG: pyropheophorbide-a nanoconstructs for tumor theranostics. Biology. 2019;15(1):243–51.

Akanda MMH. Development of solid lipid nanoparticles for sustained release of anticancer agents. London: University of Greenwich; 2015.

Akanda M, Getti G, Nandi U, Mithu MS, Douroumis DJ. Bioconjugated solid lipid nanoparticles (SLNs) for targeted prostate cancer therapy. Int J Pharm. 2021;599:120416.

Alibolandi M, Hoseini F, Mohammadi M, Ramezani P, Einafshar E, Taghdisi SM, et al. Curcumin-entrapped MUC-1 aptamer targeted dendrimer-gold hybrid nanostructure as a theranostic system for colon adenocarcinoma. Int J Pharm. 2018;549(1-2):67–75.

Allen TM, Cullis PR. Drug delivery systems: entering the mainstream. Science. 2004;303(5665):1818–22.

Amjad MT, Chidharla A, Kasi A. Cancer chemotherapy. 2020.

Bae YH, Park KJ. Targeted drug delivery to tumors: myths, reality and possibility. Amsterdam: Elsevier; 2011. p. 198–205.

Bae KH, Chung HJ, Park TG. Nanomaterials for cancer therapy and imaging. Cell. 2011;31:295–302.

Banstola A, Duwa R, Emami F, Jeong J-H, Yook S. Enhanced caspase-mediated abrogation of autophagy by temozolomide-loaded and panitumumab-conjugated poly (lactic-co-glycolic acid) nanoparticles in epidermal growth factor receptor overexpressing glioblastoma cells. Mol Pharm. 2020;17(11):4386–400.

Barchet TM, Amiji MM. Challenges and opportunities in CNS delivery of therapeutics for neurodegenerative diseases. Expert Opin Drug Deliv. 2009;6(3):211–25.

Bayat Mokhtari R, Homayouni TS, Baluch N, Morgatskaya E, Kumar S, Das B, et al. Combination therapy in combating cancer. Oncotarget. 2017;8(23):38022–43.

Behranvand N, Nasri F, Zolfaghari Emameh R, Khani P, Hosseini A, Garssen J, et al. Chemotherapy: a double-edged sword in cancer treatment. Cancer Immunol Immunother. 2022;71(3):507–26.

Bilal M, Barani M, Sabir F, Rahdar A, Kyzas GZ. Nanomaterials for the treatment and diagnosis of Alzheimer's disease: an overview. NanoImpact. 2020;20:100251.

Black KC, Wang Y, Luehmann HP, Cai X, Xing W, Pang B, et al. Radioactive ^{198}Au-doped nanostructures with different shapes for in vivo analyses of their biodistribution, tumor uptake, and intratumoral distribution. ACS Nano. 2014;8(5):4385–94.

Bronstein LM, Shifrina ZB. Dendrimers as encapsulating, stabilizing, or directing agents for inorganic nanoparticles. Chem Rev. 2011;111(9):5301–44.

Cardoso BR, Roberts BR, Bush AI, Hare DJ. Selenium, selenoproteins and neurodegenerative diseases. Metallomics. 2015;7(8):1213–28.

Cervadoro A, Coclite A, Di Mascolo D, Ferreira M, Palange A, Palomba R, et al. Smart nanoconstructs for theranostics in cancer and cardiovascular diseases. In: Core-shell nanostructures for drug delivery and theranostics. Amsterdam: Elsevier; 2018. p. 297–321.

Chatterjee K, Sarkar S, Jagajjanani Rao K, Paria S. Core/shell nanoparticles in biomedical applications. Adv Colloid Interf Sci. 2014;209:8–39.

Chaturvedi VK, Singh A, Singh VK, Singh MP. Cancer nanotechnology: a new revolution for cancer diagnosis and therapy. Curr Drug Metab. 2019;20(6):416–29.

Chen F, Yan T-H, Bashir S, Liu JL. Energy TAiR. In: Synthesis of nanomaterials using top-down methods. Amsterdam: Elsevier; 2022. p. 37–60.

Choi CHJ, Zuckerman JE, Webster P, Davis ME. Targeting kidney mesangium by nanoparticles of defined size. Proc Natl Acad Sci USA. 2011;108(16):6656–61.

Clement S, Guller A, Mahbub SB, Goldys EM. Oxygen-carrying polymer nanoconstructs for radiodynamic therapy of deep hypoxic malignant tumors. Biomedicine. 2021;9(3):322.

Costley D, McEwan C, Fowley C, McHale AP, Atchison J, Nomikou N, et al. Treating cancer with sonodynamic therapy: a review. Int J Hyperth. 2015;31(2):107–17.

Cui Y, Hui W, Wang H, Wang L, Chen C. Preparation and characterization of Fe_3O_4 composite particles. Sci China Ser B Chem. 2004;47:152–8.

Cui Q, Zhu S, Yan Y, Ye Q, Ziener U, Cao ZJ, et al. Superparamagnetic Fe_3O_4/poly (N-isopropyl acrylamide) nanocomposites synthesized in inverse miniemulsions: magnetic and particle properties. J Nanosci Nanotechnol. 2015;15(6):4608–18.

Cursaru L. Surface-engineered magnetic nanoparticle platforms for cancer imaging and therapy. Acc Chem Res. 2011;44:883–92.

Debela DT, Muzazu SG, Heraro KD, Ndalama MT, Mesele BW, Haile DC, et al. New approaches and procedures for cancer treatment: current perspectives. SAGE Open Med. 2021;9:20503121211034366.

Dhar R, Seethy A, Singh S, Pethusamy K, Srivastava T, Talukdar J, et al. Cancer immunotherapy: recent advances and challenges. J Cancer Res Ther. 2021;17(4):834–44.

Dillman RO. Cancer immunotherapy. Cancer Biother Radiopharm. 2011;26(1):1–64.

Duwa R, Banstola A, Emami F, Jeong J-H, Lee S, Yook SJ, et al. Cetuximab conjugated temozolomide-loaded poly (lactic-co-glycolic acid) nanoparticles for targeted nanomedicine in EGFR overexpressing cancer cells. J Drug Deliv Sci Technol. 2020;60:101928.

Edelman R, Assaraf YG, Slavkin A, Dolev T, Shahar T, Livney YD. Developing body-components-based theranostic nanoparticles for targeting ovarian cancer. Pharmaceutics. 2019;11(5):216.

El-Toni AM, Habila MA, Labis JP, Othman ZA, Alhoshan M, Elzatahry AA, et al. Design, synthesis and applications of core–shell, hollow core, and nanorattle multifunctional nanostructures. Nanoscale. 2016;8(5):2510–31.

Erathodiyil N, Ying JY. Functionalization of inorganic nanoparticles for bioimaging applications. Acc Chem Res. 2011;44(10):925–35.

Fernandes NB, Nayak Y, Garg S, Nayak UY. Multifunctional engineered mesoporous silica/inorganic material hybrid nanoparticles: theranostic perspectives. Coord Chem Rev. 2023;478:214977.

Gajbhiye KR, Chaudhari BP, Pokharkar VB, Pawar A, Gajbhiye VJ. Stimuli-responsive biodegradable polyurethane nano-constructs as a potential triggered drug delivery vehicle for cancer therapy. Int J Pharm. 2020;588:119781.

Gawande MB, Goswami A, Asefa T, Guo H, Biradar AV, Peng D-L, et al. Core–shell nanoparticles: synthesis and applications in catalysis and electrocatalysis. Chem Soc Rev. 2015;44(21):7540–90.

Gazeau F, Lévy M, Wilhelm C. Optimizing magnetic nanoparticle design for nanothermotherapy. 2008.

Ghosh Chaudhuri R, Paria S. Core/shell nanoparticles: classes, properties, synthesis mechanisms, characterization, and applications. Chem Rev. 2012;112(4):2373–433.

Goel S, Ferreira CA, Chen F, Ellison PA, Siamof CM, Barnhart TE, et al. Activatable hybrid nanotheranostics for tetramodal imaging and synergistic photothermal/photodynamic therapy. Adv Mater. 2018;30(6):1704367.

Gonçalves MC, Martins MB. Multifunctional core-shell nanostructures. Nanomedicine. 2014;4:84–110.

Gong Y-k, Winnik FM. Strategies in biomimetic surface engineering of nanoparticles for biomedical applications. Nanoscale. 2012;4(2):360–8.

Gould SE, Junttila MR, de Sauvage FJ. Translational value of mouse models in oncology drug development. Nat Med. 2015;21(5):431–9.

Gupta AS. Nanomedicine approaches in vascular disease: a review. Nanomedicine. 2011;7(6):763–79.

Gupta J, Fatima MT, Islam Z, Khan RH, Uversky VN, Salahuddin P. Nanoparticle formulations in the diagnosis and therapy of Alzheimer's disease. Int J Biol Macromol. 2019;130:515–26.

He C, Ahmed T, Abbasi AZ, Li LY, Foltz WD, Cai P, et al. Multifunctional bioreactive-nanoconstructs for sensitive and accurate MRI of cerebrospinal fluid pathology and intervention of Alzheimer's disease. Nano Today. 2020;35:100965.

Hoang Thi TT, Nguyen Tran D-H, Bach LG, Vu-Quang H, Nguyen DC, Park KD, et al. Functional magnetic core-shell system-based iron oxide nanoparticle coated with biocompatible copolymer for anticancer drug delivery. Pharmaceutics. 2019;11(3):120.

Hu K, Shi Y, Jiang W, Han J, Huang S, Jiang X. Lactoferrin conjugated PEG-PLGA nanoparticles for brain delivery: preparation, characterization and efficacy in Parkinson's disease. Int J Pharm. 2011;415(1-2):273–83.

Hu F, Liu B, Chu H, Liu C, Li Z, Chen D, et al. Real-time monitoring of pH-responsive drug release using a metal-phenolic network-functionalized upconversion nanoconstruct. Nanoscale. 2019;11(18):9201–6.

Izadiyan Z, Shameli K, Miyake M, Teow S-Y, Peh S-C, Mohamad SE, et al. Green fabrication of biologically active magnetic core-shell Fe_3O_4/Au nanoparticles and their potential anticancer effect. Mater Sci Eng C. 2019;96:51–7.

Jose S, Cinu TA, Sebastian R, Shoja M, Aleykutty N, Durazzo A, et al. Transferrin-conjugated docetaxel–PLGA nanoparticles for tumor targeting: influence on MCF-7 cell cycle. Polymers. 2019;11(11):1905.

Kashkooli FM, Soltani M, Souri MJ. Controlled anti-cancer drug release through advanced nano-drug delivery systems: static and dynamic targeting strategies. J Control Release. 2020;327:316–49.

Kundu BK, Pragti, Carlton Ranjith WA, Shankar U, Kannan RR, Mobin SM, et al. Cancer-targeted chitosan–biotin-conjugated mesoporous silica nanoparticles as carriers of zinc complexes to achieve enhanced chemotherapy in vitro and in vivo. ACS Appl Bio Mater. 2021;5(1):190–204.

Kuo Y-C, Chen Y-C. Targeting delivery of etoposide to inhibit the growth of human glioblastoma multiforme using lactoferrin-and folic acid-grafted poly (lactide-co-glycolide) nanoparticles. Int J Pharm. 2015;479(1):138–49.

Kuroki M, Hachimine K, Abe H, Shibaguchi H, Kuroki M, Maekawa S-I, et al. Sonodynamic therapy of cancer using novel sonosensitizers. Anticancer Res. 2007;27(6A):3673–7.

Lazarovits J, Chen YY, Sykes EA, Chan WC. Nanoparticle–blood interactions: the implications on solid tumour targeting. Chem Commun. 2015;51(14):2756–67.

Li DC, Zhang ZL, Zhang SL. Preparation and characterization of Fe_3O_4@Au nanoparticles used as precursor of ferrofluids. Adv Mater Res. 2011;177:415–7.

Li N, Wang Z, Zhang Y, Zhang K, Xie J, Liu Y, et al. Curcumin-loaded redox-responsive mesoporous silica nanoparticles for targeted breast cancer therapy. Artif Cells Nanomed Biotechnol. 2018;46(sup2):921–35.

Ma X, Li Y, Hussain I, Shen R, Yang G, Zhang KJAM. Core–shell structured nanoenergetic materials: preparation and fundamental properties. Adv Mater. 2020;32(30):2001291.

Madhubala V, Nagarajan C, Baskaran P, Raguraman V, Kalaivani T. Influences of superparamagnetic Fe_3O_4@Ag core-shell nanoparticles on the growth inhibition of Huh-7 cells. Mater Today Commun. 2023;35:106139.

Mbatha LS, Chepkoech Maiyo F, Singh M. Dendrimer functionalized folate-targeted gold nanoparticles for luciferase gene silencing in vitro: a proof of principle study. Acta Pharma. 2019;69(1):49–61.

Menath J, Eatson J, Brilmayer R, Andrieu-Brunsen A, Buzza DMA, Vogel N. Defined core–shell particles as the key to complex interfacial self-assembly. PNAS. 2021;118(52):e2113394118.

Mishra S, Bhatt T, Kumar H, Jain R, Shilpi S, Jain V. Nanoconstructs for theranostic application in cancer: challenges and strategies to enhance the delivery. Front Pharmacol. 2023;14:1101320.

Mittal P, Saharan A, Verma R, Altalbawy FMA, Alfaidi MA, Batiha GE, et al. Dendrimers: a new race of pharmaceutical nanocarriers. Biomed Res Int. 2021;2021:8844030.

Moni MM, Begum M, Uddin MS, Ashraf GM. Deciphering the role of nanoparticle-based treatment for Parkinson's disease. Curr Drug Metab. 2021;22(7):550–60.

Moradi S, Najjar R, Hamishehkar H, Lotfi AJ. Triple-responsive drug nanocarrier: magnetic core-shell nanoparticles of Fe_3O_4@ poly (N-isopropylacrylamide)-grafted-chitosan, synthesis and in vitro cytotoxicity evaluation against human lung and breast cancer cells. J Drug Deliv Sci Technol. 2022;72:103426.

Murthy SK. Nanoparticles in modern medicine: state of the art and future challenges. Int J Nanomedicine. 2007;2(2):129–41.

Nam J, Won N, Bang J, Jin H, Park J, Jung S, et al. Surface engineering of inorganic nanoparticles for imaging and therapy. Adv Drug Deliv Rev. 2013;65(5):622–48.

Nguyen TT, Dung Nguyen TT, Vo TK, Tran N-M-A, Nguyen MK, Van Vo T, et al. Nanotechnology-based drug delivery for central nervous system disorders. Biomed Pharmacother. 2021a;143:112117.

Nguyen TT, Nguyen TTD, Vo TK, Nguyen MK, Van Vo T, Van Vo G, et al. Nanotechnology-based drug delivery for central nervous system disorders. Biomed Pharmacother. 2021b;143:112117.

Oldham RK. Biotherapy: the fourth modality of cancer treatment. J Cell Physiol Suppl. 1986;129(S4):91–9.

Palange AL, Palomba R, Rizzuti IF, Ferreira M, Decuzzi P. Deformable discoidal polymeric nanoconstructs for the precise delivery of therapeutic and imaging agents. Mol Ther. 2017;25(7):1514–21.

Palma AS, Casadei BR, Lotierzo MC, de Castro RD, Barbosa LRS. A short review on the applicability and use of cubosomes as nanocarriers. Biophys Rev. 2023;15(4):553–67.

Pan Y, Ma S, Cao K, Zhou S, Zhao A, Li M, et al. Therapeutic approaches targeting cancer stem cells. J Cancer Res Ther. 2018;14(7):1469–75.

Patel J, Amrutiya J, Bhatt P, Javia A, Jain M, Misra AJ. Targeted delivery of monoclonal antibody conjugated docetaxel loaded PLGA nanoparticles into EGFR overexpressed lung tumour cells. J Microencapsul. 2018;35(2):204–17.

Perona R, Sanchez-Perez IJC, Oncology T. Signalling pathways involved in clinical responses to chemotherapy. Clin Transl Oncol. 2007;9:625–33.

Perry JL, Reuter KG, Luft JC, Pecot CV, Zamboni W, DeSimone JM. Mediating passive tumor accumulation through particle size, tumor type, and location. Nano Lett. 2017;17(5):2879–86.

Poudel K, Banstola A, Tran TH, Thapa RK, Gautam M, Ou W, et al. Hyaluronic acid wreathed, trio-stimuli receptive and on-demand triggerable nanoconstruct for anchored combinatorial cancer therapy. Carbohydr Polym. 2020;249:116815.

Pramanik A, Xu Z, Ingram N, Coletta PL, Millner PA, Tyler AI, et al. Hyaluronic-acid-tagged cubosomes deliver cytotoxics specifically to cd44-positive cancer cells. Mol Pharm. 2022;19(12):4601–11.

Rajkumar S, Prabaharan MJC, Biointerfaces SB. Multi-functional core-shell Fe_3O_4@Au nanoparticles for cancer diagnosis and therapy. Colloids Surf B Biointerfaces. 2019;174:252–9.
Ravindran Girija A, Balasubramanian S. Theragnostic potentials of core/shell mesoporous silica nanostructures. Nanotheranostics. 2019;3(1):1–40.
Regev R, Yeheskely-Hayon D, Katzir H, Eytan GD. Transport of anthracyclines and mitoxantrone across membranes by a flip-flop mechanism. Biochem Pharmacol. 2005;70(1):161–9.
Regmi RK. Properties and biomedical applications of magnetic nanoparticles. 2011.
Sabio RM, Meneguin AB, Ribeiro TC, Silva RR, Chorilli MJ. New insights towards mesoporous silica nanoparticles as a technological platform for chemotherapeutic drugs delivery. Int J Pharm. 2019;564:379–409.
Saw WS, Anasamy T, Foo YY, Kwa YC, Kue CS, Yeong CH, et al. Delivery of nanoconstructs in cancer therapy: challenges and therapeutic opportunities. Adv Ther. 2021;4(3):2000206.
Sharma A. Magnetic and stimuli-responsive nanoparticles for environmental and biomedical applications. 2021.
Shuel SL. Targeted cancer therapies: clinical pearls for primary care. Can Fam Physician. 2022;68(7):515–8.
Siafaka PI, Üstündağ Okur N, Karavas E, Bikiaris DN. Surface modified multifunctional and stimuli responsive nanoparticles for drug targeting: current status and uses. Biomater Sci. 2016;17(9):1440.
Skourtis D, Stavroulaki D, Athanasiou V, Fragouli PG, Iatrou H. Nanostructured polymeric, liposomal and other materials to control the drug delivery for cardiovascular diseases. Pharmaceutics. 2020;12(12):1160.
Sperling RA, Parak WJ. Surface modification, functionalization and bioconjugation of colloidal inorganic nanoparticles. Philos Trans A Math Phys Eng Sci. 1915;2010(368):1333–83.
Srinivasarao M, Low PS. Ligand-targeted drug delivery. Chem Rev. 2017;117(19):12133–64.
Sriramoju B, Kanwar RK, Kanwar JR. Nanomedicine based nanoparticles for neurological disorders. Curr Med Chem. 2014;21(36):4154–68.
Sun T, Xu J, Chen T, Tu C, Zhu L, Yan DJBS. A self-amplified ROS-responsive chemodrug–inhibitor conjugate for multi-drug resistance tumor therapy. Biomater Sci. 2022;10(4):997–1007.
Swartz MA, Lund AW. Lymphatic and interstitial flow in the tumour microenvironment: linking mechanobiology with immunity. Nat Rev Cancer. 2012;12(3):210–9.
Tripathy S, Rodrigues J, Shimpi NG. Top-down and bottom-up approaches for synthesis of nanoparticles. Mater Res Proc. 2023;145:92–130.
Veselov VV, Nosyrev AE, Jicsinszky L, Alyautdin RN, Cravotto G. Targeted delivery methods for anticancer drugs. Cancer. 2022;14(3):622.
Viswanadh MK, Vikas JA, Reddy Adena SK, Mehata AK, Priya V, et al. Formulation and in vivo efficacy study of cetuximab decorated targeted bioadhesive nanomedicine for non-small-cell lung cancer therapy. Nanomedicine. 2020;15(24):2345–67.
Waris A, Ali A, Khan AU, Asim M, Zamel D, Fatima K, et al. Applications of various types of nanomaterials for the treatment of neurological disorders. Nanomaterials. 2022;12(13):2140.
Wei Q-Y, Xu Y-M, Lau AT. Recent progress of nanocarrier-based therapy for solid malignancies. Cancer. 2020;12(10):2783.
Wu H-C, Chang D-K, Huang C-TJ. Targeted therapy for cancer. J Cancer Mol. 2006;2(2):57–66.
Yadav A, Singh S, Sohi H, Dang S. Advances in delivery of chemotherapeutic agents for cancer treatment. AAPS PharmSciTech. 2022;23:1–14.
Yan X, Scherphof GL, Kamps JA. Liposome opsonization. J Liposome Res. 2005;15(1-2):109–39.
Yang J, Lu W, Xiao J, Zong Q, Xu H, Yin Y, et al. A positron emission tomography image-guidable unimolecular micelle nanoplatform for cancer theranostic applications. Acta Biomater. 2018;79:306–16.
Yang B, Huang L, Zhou Z, Yin S, Xi MJ. Paclitaxel loaded hyaluronic acid polymerized nanoparticles designed for ovarian cancer therapy. J Biomed Nanotechnol. 2023;19(12):2045–59.

Yao Y, Zhou Y, Liu L, Xu Y, Chen Q, Wang Y, et al. Nanoparticle-based drug delivery in cancer therapy and its role in overcoming drug resistance. Front Mol Biosci. 2020;7:193.

Yeamsuksawat T, Zhao H, Liang JJMTC. Characterization and antimicrobial performance of magnetic Fe_3O_4@Chitosan@Ag nanoparticles synthesized via suspension technique. Mater Today Commun. 2021;28:102481.

Yildiz A, Bayramol DV, Atav R, Ağirgan AÖ, Kurç MA, Ergünay U, et al. Synthesis and characterization of Fe_3O_4@Cs@Ag nanocomposite and its use in the production of magnetic and antibacterial nanofibrous membranes. Appl Surf Sci. 2020;521:146332.

Zhao Z, Ukidve A, Krishnan V, Mitragotri SJ. Effect of physicochemical and surface properties on in vivo fate of drug nanocarriers. Adv Drug Deliv Rev. 2019;143:3–21.

7 Core-Shell Nanoconstructs in Cancer Biosensing: Techniques, Applications, and Fabrication Strategies

Rahul Pokale, S. P. Rachana, Anoushka Mukharya, Viola Colaco, Gaurisha Alias Resha Ramnath Naik, Amrita Arup Roy, Srinivas Mutalik, Namdev Dhas, Ritu Kudarha, and Jayvadan K. Patel

Abstract

Biosensors have emerged as a promising tool for cancer diagnosis and monitoring due to their ability to rapidly and sensitively detect cancer biomarkers. Biosensors can be classified based on the type of biorecognition element and transducer used, with common techniques including electrochemical, optical, and piezoelectric biosensors that can detect a variety of cancer biomarkers. Core-shell nanoparticles, consisting of a core material coated with a shell, have shown great potential in biosensing applications for cancer due to their unique physicochemical properties, such as high surface area, tunable optical/electrical properties, and enhanced catalytic activity, which can significantly improve the sensitivity, selectivity, and stability of biosensors. Various synthetic approaches, such as sol-gel, hydrothermal, co-precipitation, seed-mediated growth, solvothermal, reverse microemulsion, layer-by-layer assembly, and electrochemical deposition, can be employed to fabricate core-shell nanostructures with tailored properties for biosensing applications, and the choice of core and shell materials, as well as the fabrication method, can be optimized to enhance the biorecognition, transduction, and signal amplification capabilities of the biosensor. This chapter will provide a comprehensive overview of the diverse biosensing techniques for cancer detection, the role of core-shell nanostructures in improving biosensing performance, and the strategies for the efficient fabrication of core-shell nanoconstructs for enhanced cancer biosensing applications.

R. Pokale · S. P. Rachana · A. Mukharya · V. Colaco · G. A. R. R. Naik · A. A. Roy
S. Mutalik · N. Dhas · R. Kudarha
Department of Pharmaceutics, Manipal College of Pharmaceutical Sciences, Manipal Academy of Higher Education, Manipal, Karnataka, India

J. K. Patel (✉)
Viesain Pharma LLC, GA, USA

Faculty of Pharmacy, Sankalchand Patel University, Visnagar, Gujarat, India

J. K. Patel et al. (eds.), *Core-Shell Nano Constructs for Cancer Theragnostic*,
https://doi.org/10.1007/978-981-96-3025-7_7

Keywords

Core-shell nanoparticles · Biosensors · Diagnosis · Nanotechnology · Biosensing · Cancer

1 Introduction

Core-shell nanoparticles are a versatile class of nanomaterials that consist of a core material encased in a distinct shell material. These nanoparticles can be categorized based on their core and shell compositions, which can be organic or inorganic and exist in solid, liquid, or gaseous states. The core provides essential properties, while the shell offers stability and functionality, enhancing the overall performance of the nanoparticle. This unique structure allows core-shell nanoparticles to exhibit combined properties of core and shell materials, making them particularly advantageous for biomedical applications, including cancer biosensing (Dhas et al. 2018). Core-shell nanoparticles offer significant advantages in biological applications, such as reduced cytotoxicity, enhanced stability, improved dispersibility, and efficient conjugation of biomolecules. Encapsulating a potentially toxic core with a benign material enhances the biocompatibility of nanoparticles, reducing their harmful effects and making them safer for use in medical applications. The shell also provides increased thermal and chemical stability, preventing the degradation of core materials susceptible to environmental changes. Hydrophilic coatings on hydrophobic cores improve dispersibility in biological systems, making them suitable for drug delivery and other biomedical applications. Furthermore, the shell can be modified to allow better attachment of biomolecules, facilitating targeted delivery and interaction with specific cellular receptors (Jenjob et al. 2020).

A biosensor is an analytical instrument that translates a biological reaction into an electrical signal. Detecting cancer biomarkers early can significantly reduce mortality rates and improve treatment outcomes. Designing bio-responsive shells over core-shell nanoparticles provides a larger surface for interaction, enhancing sensitivity. By leveraging their inherent magnetic, electro-sensitive, and optical properties, core-shell nanoparticles can improve the performance of various biosensor types, including piezoelectric, amperometric, and optical sensors. For example, magnetic core-shell nanoparticles can serve as Magnetic Resonance Imaging contrast agents, while luminescent silica particles can be used for optical cancer diagnostics. The ability to modify core-shell particles with biomolecules, such as folic acid, enhances their delivery into cancer cells via receptor-mediated endocytosis, improving the effectiveness of targeted therapies (Chatterjee et al. 2014). Hence, core-shell nanoparticles offer a promising approach to cancer biosensing due to their unique structure and multifunctional capabilities. Leveraging the combined properties of core and shell materials, core-shell nanoparticles can significantly improve the sensitivity, specificity, and overall performance of biosensors in cancer diagnostics and therapy.

2 Various Types of Biosensing in Cancer

2.1 Electrochemical Biosensors in Cancer Detection

Electrochemical biosensors are highly valued in the medical field for their affordability, quick response times, ease of use, potential for miniaturization, and ability to deliver high sensitivity and selectivity with minimal detection limits. These biosensors are composed of three primary components: a signal transducer, a three-electrode electrochemical system, and a biorecognition element (Chang et al. 2019; Zhang et al. 2020). They detect electrical signal changes resulting from electrochemical reactions at the electrode surface, which are then measured and recorded to quantify the presence of target molecules. Various biorecognition elements are employed in these biosensors to identify cancer biomarkers. These include traditional biological molecules such as antibodies and enzymes and synthetic molecules like aptamers, DNA sequences, and peptides (Gavas et al. 2021; Khanmohammadi et al. 2020). Each element is designed to precisely recognize cancer-related targets, essential for effective detection and monitoring.

Potentiometric and amperometric sensors are the most used electrochemical biosensors. Potentiometric sensors measure changes in electrical potential resulting from the interaction between the target biomolecule and the biorecognition element. A notable example is the light-addressable potentiometric sensor, which integrates a phage-based recognition system to detect cancer biomarkers such as human phosphatase of regenerating liver-3 and the MDA-MB-231 breast cancer cell line, demonstrating exceptional sensitivity in cancer detection. (Jia et al. 2007). On the other hand, amperometric biosensors measure the current generated from oxidation or reduction reactions that occur when a voltage is applied between two electrodes. (Wang 2006). These sensors are particularly effective for detecting specific genetic sequences associated with cancer, such as the BRCA1 and BRCA2 mutations linked to hereditary breast cancer. (Wang and Kawde 2001).

Advanced fabrication methods, such as square wave stripping voltammetry, field-effect transistors, cyclic voltammetry, electrochemical impedance spectroscopy, and square wave voltammetry, enable rapid biosensing for different types of cancer. These methods enhance sensitivity and allow for the detection of low-concentration analytes. Electrochemical impedance spectroscopy, a label-free and nondestructive monitoring technique, analyzes dielectric properties in response to varying frequencies, aiding in detecting biomarkers like epidermal growth factor receptors (Elshafey et al. 2013). Electrochemical biosensors have been successfully used in immunoassays and protein arrays for cancer biomarker detection. For example, a sandwiched immunosensor using square wave stripping voltammetry can monitor human epidermal growth factor receptor 2 in breast cancer cells (Zhu et al. 2013). An ssDNA probe functionalized with gold nanoparticles provides high sensitivity for detecting Chronic Lymphocytic Leukemia (Ensafi et al. 2011). The sandwich-type electrochemical immunosensor employs a dual recognition strategy. First, the prostate-specific antigen (PSA) biomarker binds to the quantum dots functionalized on the graphene surface. Then, a secondary antibody is added, forming a sandwich complex that was detected electrochemically (Yang et al. 2011). A label-free electrochemical impedimetric immunosensor enhances sensitivity through the synergistic effects of gold nanoparticles and protein G for epidermal growth factor receptor detection (Elshafey et al. 2013).

Despite their advantages, electrochemical biosensors face challenges repeatedly maintaining stability and using biorecognition elements. Fabrication requires considerable time and skilled personnel. Efforts to overcome these challenges include simplifying fabrication procedures and utilizing nanomaterial-assisted bio-fabrication to improve performance and reliability (Chang et al. 2014). Electrochemical biosensors play a crucial role in cancer detection and monitoring, with ongoing advancements promising to enhance their application and effectiveness in clinical settings.

2.2 Optical Biosensors in Cancer Detection

Optical biosensors are advanced light-based sensors designed to detect biological analytes by measuring changes in specific wavelengths of light. These sensors consist of a light source, optical components to shape and direct the light beam, a modulating agent, a modified sensing head, and a photodetector. (Mehrotra 2016). The transducers used in optical biosensors can be based on luminescence, fluorescence, colorimetry, or interferometry, converting variations in light wavelengths or surface plasmon resonance into electrical or digital readouts. (Tothill 2009). Optical biosensors have demonstrated remarkable adaptability across diverse biosensing applications, thanks to their user-friendly operation, capacity for simultaneous detection of multiple analytes, and seamless integration with automated microfluidic systems (Calabretta et al. 2021). These sensors leverage the intricate optical properties of nanomaterials, including surface-enhanced Raman scattering, localized surface plasmon resonance, and surface plasmon resonance. When functionalized with bioluminescent, fluorescent, or chemiluminescent probes, optical biosensors unlock a vast array of biosensing strategies. This versatility allows for the sensitive and specific detection of a wide range of analytes, from small molecules to large biomolecules and even whole cells. Hence, optical biosensors have notable applications in cancer diagnostics, as they offer sensitive, versatile, and non-invasive diagnostic tools, making them invaluable in medical diagnostics and treatment monitoring.

2.2.1 Surface-Enhanced Raman Scattering-Based Biosensors in Cancer Detection

Raman spectroscopy is practical in differentiating cancerous cells and diagnosing diseases like Alzheimer's, diabetes, and atherosclerosis. However, it suffers from weak signals, requiring long acquisition times or high-power lasers, which can damage biological samples, and autofluorescence interference complicates signal detection. Surface-enhanced Raman Scattering addresses these issues by amplifying weak Raman signals and is used to image and quantify highly sensitive cancer cell markers. For example, antibody-conjugated gold nanospheres have enabled multiplexed imaging of breast cancer cell lines (Lee et al. 2014). It has also been employed for non-invasive cancer detection, such as using silver nanoparticles to monitor nasopharyngeal cancer in blood plasma (Feng et al. 2010). Recent surface-enhanced Raman scattering advancements include detecting methylated DNA and its derivatives using gold nanoparticles modified with methylation-sensitive antibodies, achieving detection limits of 0.2 pg/μL (Ouyang et al. 2017). Magnetic nanoparticles-based surface-enhanced Raman scattering biosensors have detected miRNA at 0.3 fM in cancer cell RNA extracts (Pang et al. 2016).

2.2.2 Localized Surface Plasmon Resonance/Surface Plasmon Resonance-Based Biosensors in Cancer Detection

Optical biosensors leveraging surface plasmon resonance and localized surface plasmon resonance are crucial for detecting nucleic acid cancer biomarkers. (Zeng et al. 2014). In surface plasmon resonance-based sensors, biomolecules immobilized on a gold surface interact with cancer biomarkers, causing mass and refractive index changes and resulting in a surface plasmon resonance signal shift. (Guo 2012). Localized Surface Plasmon Resonance sensors use metallic nanoparticles like gold and silver to enhance sensitivity through sharp spectral absorption and scattering peaks. (Mayer and Hafner 2011). These peaks vary with the nanoparticles' composition, shape, size, and environmental changes. (Haes et al. 2004; Liu et al. 2007; Murray et al. 2009). For instance, Nguyen and Sim (2015) developed a localized surface plasmon resonance biosensor with peptide nucleic acid-conjugated gold nanoparticles to detect PIK3CA gene mutations and methylation in circulating tumor DNA (ctDNA).

The trend toward label-free, reusable nanomaterials has enhanced cancer biosensors' sensitivity and multiplexing capabilities. Aćimović et al. (Aćimović et al. 2014) Designed a microfluidic platform with multiple sensing sites for real-time prostate-specific antigen and human alpha-fetoprotein monitoring, achieving detection limits up to 500 pg/mL. These advancements highlight the potential of surface plaque resonance and localized surface plaque resonance-based biosensors in cancer diagnostics, offering susceptible, non-invasive, and real-time detection methods.

2.2.3 Fluorescent Biosensors in Cancer Detection

Fluorescence involves exciting fluorophores with shorter-wavelength photons, causing them to emit longer-wavelength photons after vibrational relaxation, known as Stoke's shift. This process includes the excitation of electrons, followed by radiative or nonradiative emission, intersystem crossing, vibrational relaxation, and internal conversion (Chinen et al. 2015). Conventional fluorophores face limitations in biosensing due to shorter lifetimes, low molar absorption coefficients, narrow absorption spectra, and susceptibility to photodegradation, resulting in lower signal-to-noise ratios in assays.

Fluorescent biosensors serve as imaging agents in cancer and drug discovery, offering insights into enzyme regulation at the cellular level. Green Fluorescent Protein -based and genetically encoded Förster Resonance Energy Transfer biosensors are crucial. These biosensors integrate fluorescent probes enzymatically, chemically, or genetically via receptors, transducing signals to detect ions, metabolites, and protein biomarkers with high sensitivity in complex solutions. They play crucial roles in studying gene expression, protein localization, signal transduction, and cell cycle dynamics and diagnosing diseases such as arthritis, inflammation, cardiovascular disorders, neurodegenerative diseases, infections, cancer, and metastasis. They are essential for detecting the progression of diseases, assessing treatment responses, directing surgical procedures, and identifying biomarkers early in molecular diagnostics. A genetically encoded Förster Resonance Energy Transfer biosensor has been employed to monitor Bcr-Abl kinase activity in cancer cells, correlating with disease status in chronic myeloid leukemia and predicting responses to therapy, including the emergence of drug-resistant cells (Mehrotra 2016). Hence, fluorescent biosensors are indispensable in biomedical research and clinical applications, enhancing understanding of disease mechanisms and facilitating targeted therapies.

2.2.4 Chemiluminescence-Based Biosensors in Cancer Detection

Chemiluminescence is a light emission process where excited electrons return to the ground state in molecules, occurring during chemical reactions. Chemiluminescence-based biosensors utilize this phenomenon by measuring light emitted when a biomarker binds to recognition elements like antibodies. These biosensors offer high detectability and straightforward measurement without nonspecific signals, making them advantageous (Roda et al. 2016). An immuno-sensor was developed for the carcinoembryonic antigen with a detection range of 0–50 ng/mL and a sensitive limit of 5.0 pg/mL, employing a portable luminometer (Qu et al. 2013). Another biosensor utilized flow injection chemiluminescent immunoassay, achieving a linear range of 100 pg/mL to 1000 ng/mL for carcinoembryonic antigen detection, enhanced by gold nanoparticle labeling (Hao and Ma 2012). An optical biosensor using chemiluminescence resonance energy transfer with graphene quantum dots for CA125 detection, achieving a wide linear range of 0.1–600 U/mL and a low detection limit of 0.05 U/mL (Al-Ogaidi et al. 2014). Additionally, a chip-based optical sensor capable of simultaneously detecting 12 tumor markers based on chemiluminescence demonstrates promise for multiplex biomarker detection in cancer (Sun et al. 2004). These examples highlight the versatility and potential of chemiluminescence-based biosensors in biomedical diagnostics and research.

2.2.5 Optofluidic Ring Resonator-Based Biosensors in Cancer Detection

An optofluidic ring resonator is a sophisticated sensing platform that combines microfluidics and optical ring resonators. Total internal reflection within the ring resonator causes light to travel as whispering gallery modes in an optofluidic ring resonator. When molecules bind to the optofluidic ring resonator, the effective refractive index changes, causing a spectral shift in the whispering gallery modes, which serves as the response signal (Fan et al. 2008). This technology offers several advantages, including label-free, rapid, portable, and multiplex detection capabilities, making it a versatile platform for a wide range of sensing applications (Suter et al. 2010). One application of optofluidic ring resonator technology is in detecting cancer biomarkers. An Optofluidic Ring Resonator (OFRR)-based biosensor for detecting CA15-3, a breast cancer biomarker, in serum samples. This demonstrated the potential of optofluidic ring resonators in clinical diagnostics. (Zhu et al. 2009). Hence, optofluidic ring resonator-based techniques represent significant advancements in cancer biomarker detection. These technologies provide high sensitivity and specificity, rapid and label-free detection, and the capability for multiplexing, making them valuable tools in clinical diagnostics and cancer research.

2.3 Calorimetric-Based Biosensors in Cancer Detection

All chemical and biological reactions involve heat exchange, leading to the development of calorimetric-based biosensing devices. Initially used in enzyme-based sensors, calorimetric transduction has expanded to DNA, cell, and immunosensors. This technique measures temperature changes from reactions between a biorecognition element and an analyte, correlating these changes to reactant consumption or product formation. Calorimetric devices typically use thermistors (metal oxide) or thermophiles (ceramic semiconductors). Key advantages include stability, increased sensitivity, and miniaturization, making them suitable for label-free biomolecule interaction screening (Maskow et al. 2012). Calorimetric biosensors are less prevalent in cancer diagnostics than other biosensors, but nanotechnology has broadened their applications. These biosensors measure exothermic processes, using temperature changes from enzymatic reactions to determine analyte concentration through enthalpy changes. Although not widely used for cancer diagnosis and prognosis, calorimetric biosensors have demonstrated some cancer-detecting capabilities.

2.4 Mass-Based Biosensors in Cancer Detection

2.4.1 Piezoelectric Biosensors

Piezoelectric devices, sensitive to nanoscale mass changes, are valuable in bioanalytical applications. Piezoelectric quartz crystals are optimized for analytical use with gold or silver electrodes. Applying a potential difference generates an electric field, causing mechanical oscillations at a characteristic frequency. Mass changes on the piezoelectric quartz crystals' surface cause detectable frequency shifts, allowing analyte detection at nanogram levels. Nanoparticle enhancements enable ultrasensitive detection at attomolar levels. Piezoelectric sensors have several important characteristics, such as ease of use, on-site detection, and ultrasensitivity. They are commonly employed in DNA hybridization techniques or sandwich immunoassays for cancer diagnostics, including the detection of BRCA-1 gene mutations in breast cancer (Abdul Rasheed and Sandhyarani 2016). Piezoelectric sensors also target specific cancer cell receptors and monitor cancer biomarkers in real time. Despite their advantages, challenges remain in functionalizing biochemicals and improving biosensor robustness, cost-effectiveness, and user-friendliness.

Continuous advancements include developing acoustic wave biosensors, piezoelectric microcantilever sensors, and novel piezoelectric transducers for multiplexed and label-free detection.

2.4.2 Surface Acoustic Wave Biosensors and Microcantilever Biosensors

Surface acoustic wave biosensors are mass sensors that generate acoustic waves via a transducer when an electric signal is applied. These waves are then converted back into an electrical signal. Any mass change on the device surface shifts the frequency of the output signal, allowing the detection of binding events on the sensing surface (Länge et al. 2013). This mechanism enables the sensitive detection of various analytes with high precision. Similarly, microcantilever biosensors are gaining importance for detecting biological and chemical agents. When biomolecules such as antibodies or antigens bind to the cantilever surface, the resulting change in resonance frequency and amplitude of the cantilever serves as the sensor signal (Buchapudi et al. 2011). These sensors are beneficial for detecting cancer biomarkers and other critical biological indicators with high sensitivity and precision. Figure 7.1 depicts all available biosensors used in cancer diagnosis for a better visual understanding.

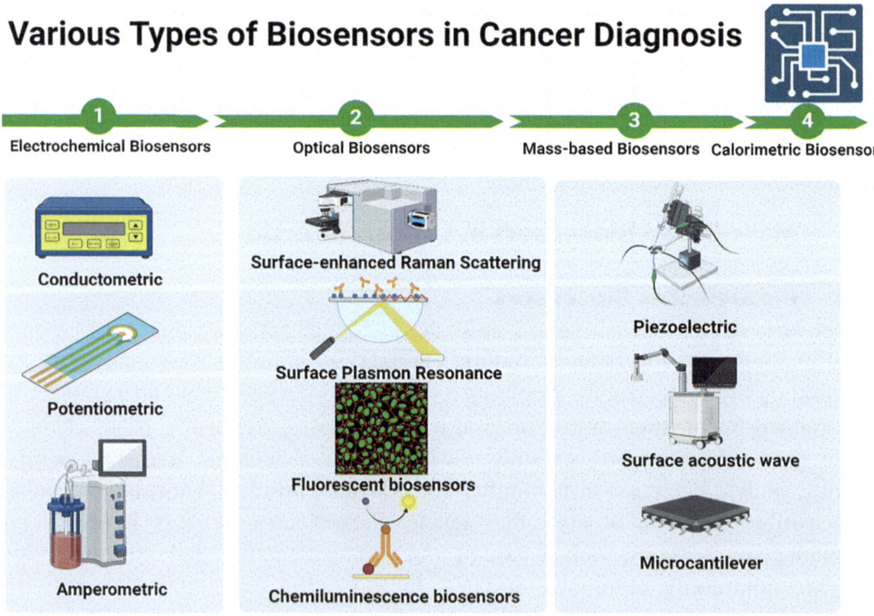

Fig. 7.1 Various types of biosensors in cancer detection

3 Role of Core-Shell Nanoconstructs in Biosensing for Cancer Detection

3.1 Breast Cancer Detection

As the second major cause of cancer-related mortality in developed nations, breast cancer is one of the most prevalent cancers among women globally. Improved diagnostic precision and tailored therapies have been made possible by developments in tumor biology and prediction biomarkers. Accurate identification of biomarkers such as calreticulin, human epidermal growth factor receptor 2 (HER2), and CA15-3 is essential for early diagnosis and treatment, improving the prognosis and quality of life of patients with breast cancer. Despite these advances, few studies have used core-shell nanoparticles for biosensing breast cancer markers.

One study used gold/silver core-shell nanoparticles to detect the breast cancer marker CA15-3. These nanoparticles, with a gold core and silver shell, were conjugated with antibodies to bind CA15-3 antigens selectively. The mercury ions etched the silver shell, causing a visible color change from earthy yellow to pinkish purple, enabling visual detection of CA15-3 in the 30–300 U/mL range. Additionally, the nanoparticles were modified onto electrodes to provide visual and precise electrochemical quantification of CA15-3. This dual-method approach offers a comprehensive means of detecting CA15-3, combining visual and electrochemical techniques for accurate and efficient assessment (Li et al. 2023). A highly sensitive electrochemical immunosensor for detecting the breast cancer biomarker HER2 was developed using core-shell gold@palldium-silver dog-bone-like nanorods and gold@platinum-rhodium nanorods. The gold@palldium-silver dog-bone-like nanorods were synthesized via a seed-assisted method, enhanced signal strength, and antibody loading. The gold@platinum-rhodium nanorods provided abundant active sites for capturing secondary antibodies and amplifying electrochemical responses. This sandwich-type immunosensor demonstrated a broad detection range of 0.001–100 ng/mL and a low detection limit of 0.25 pg/mL. It showed excellent reproducibility, specificity, and stability, performing well in human serum samples, highlighting its potential for clinical diagnostics of HER2 (Zhang et al. 2023).

A novel label-free immunosensor was developed for detecting calreticulin, a breast cancer biomarker, using core-shell iron oxide nanoparticles@gold-nanoparticles@poly(3-thiophenemalonic acid) polymer nanoparticles on an indium tin oxide electrode. These nanoparticles, with a novel conjugated polymer as the immobilization matrix, enabled high protein loading and oriented antibody immobilization. The sensor exhibited excellent repeatability, reproducibility, stability, and reusability, with 94.05–106.62% recovery rates in human serum samples, indicating high accuracy and no significant matrix effect. This design offers a promising tool for clinically detecting calreticulin and other biomarkers (Aydın et al. 2022). Another novel electrochemical nanobiosensor was presented for detecting the breast cancer biomarker miR-155 using Iron@silver core-shell nanoparticles. This sensor demonstrated a detection range of 0.5 fM to 1.0 nM with a detection limit of 0.15 fM, high selectivity, and reproducibility. The sensor effectively detected miR-155 in human serum samples, highlighting its potential for clinical applications in microRNA marker detection without sample extraction or amplification (Yazdanparast et al. 2020).

The use of core-shell nanoparticles in biosensing breast cancer biomarkers represents a significant advancement in cancer diagnostics, offering the potential for early and accurate detection, thereby improving patient outcomes.

3.2 Cervical Cancer Detection

High-risk human papillomavirus (HR-HPV) types 16 and 18 are the main cause of cervical cancer, which is the leading cause of cancer-related mortality among women worldwide, particularly in underdeveloped countries. Given the high incidence and fatality rates of cervical cancer, the development of core-shell nanoparticle-based screening techniques for early diagnosis is essential. Recent advances in biosensors using core-shell nanoparticles for detecting cervical cancer biomarkers have shown promising results.

Using screen-printed carbon electrodes modified with iron oxide-gold core-shell nanoparticles, a label-free biosensor was created to detect HPV-16. Iron oxide-gold core-shell nanoparticles were used to coat the screen-printed carbon electrodes, and single-strand DNA (ssDNA) probes thiolated with specific HPV DNA sequences were used to functionalize nanoparticles. Ferrocyanide ions were used as the redox indicator, and hybridization events were observed using differential pulse voltammetry and cyclic voltammetry. The modified electrodes showed a significant reduction in redox current after hybridization, attributed to the attraction of the redox probe to the free guanine bases in ssDNA. The biosensor achieved optimal performance with a detection limit of 0.1 nM and a sensitivity of 2.4 μA/nM, showcasing its efficiency for HPV-16 detection (Rasouli et al. 2023).

A novel surface-enhanced Raman scattering (SERS) strategy was developed for detecting miRNA-21, a biomarker for cervical cancer, using gold@5,5′-dithiobis (2-nitrobenzoic acid)@Silver core-shell nanoparticles combined with duplex-specific nuclease signal amplification (DSNSA). The gold@5,5′-dithiobis (2-nitrobenzoic acid)@Silver core-shell nanoparticles were attached to iron oxide magnetic nanoparticles via a capture probe targeting miRNA-21. Upon the presence of miRNA-21, heteroduplexes formed and were cleaved by duplex-specific nuclease, releasing Gold@5,5′-dithiobis (2-nitrobenzoic acid)@silver core-shell nanoparticles from the iron oxide nanoparticles, allowing miRNA-21 to rehybridize and trigger further signal amplification. This method demonstrated high sensitivity with a detection limit of 0.084 fM and effective single-base mismatch distinction, making it suitable for early cervical cancer diagnosis (Xu et al. 2020).

A magnetic focus lateral flow biosensor was designed to detect valosin-containing protein, a cervical cancer biomarker. The magnetic focus lateral flow biosensor utilized iron oxide core–gold shell nanoprobes that concentrated target proteins via magnetic attraction, significantly enhancing capture efficiency. This system achieved a detection limit as low as 25 fg/mL of purified valosin-containing protein in phosphate-buffered saline within 45 min, employing enzyme-based colorimetric signal amplification for visual recognition. This approach represents a 106-fold improvement over conventional lateral flow systems and demonstrated the detection of valosin-containing protein at 16 pg/mL in protein mixtures extracted from cervical cancer tissue (Ren et al. 2019).

The use of core-shell nanoparticles in biosensors for cervical cancer biomarkers offers significant advancements in sensitivity, specificity, and detection limits. These innovative approaches utilizing iron oxide-gold and gold@silver core-shell nanoparticles enhance the detection capabilities for HPV DNA, miRNA-21, and protein biomarkers like valosin-containing protein, providing promising tools for early diagnosis and improving clinical outcomes for cervical cancer patients.

3.3 Lung Cancer Detection

Lung cancer is the most common malignant tumor worldwide and is challenging to diagnose early due to nonspecific symptoms, leading to high mortality rates. Improved patient outcomes depend on a timely and accurate diagnosis, and core-shell nanoparticle-based biosensors have shown promise in the identification of lung cancer biomarkers (Lütfi Yola et al. 2021). Nickel@silver core-shell nanoparticles anchored on carbon nanofibers have been used to develop a surface-enhanced, label-free, sensitive salivary sensor platform for lung cancer pre-diagnosis. Nickel@silver/carbon nanofibers were fabricated via electrospinning, followed by chemical reduction and transmetallation, ensuring a homogeneous distribution of bimetallic nickel@silver nanoparticles on the carbon nanofibers surface. The surface-enhanced Raman spectroscopy substrate demonstrated exceptional enhancement (up to 10^7) with a detection limit of 10^{-12}, effectively analyzing salivary imidazole compounds relevant to lung cancer detection. This non-invasive, cost-effective platform holds promise for early-stage lung cancer screening (Sunil et al. 2023). Copper oxide @ Silicon core-shell nanoparticles synthesized via a sol-gel process were used to detect l-cysteine sensitively. The copper oxide @silicon-modified glassy carbon electrode exhibited excellent electrocatalytic activity, supported by high conductivity and a large surface area. Additionally, these nanoparticles showed enhanced antibacterial activity and promising anticancer effects on lung cancer cells (A549) without harming normal cells. This highlights copper oxide@silicon core-shell nanoparticles as versatile materials for biosensing and therapeutic applications in lung cancer detection and treatment (Gowtham et al. 2023).

A high-performance chemo-resistive sensor was developed to detect volatile organic compounds using core-shell nanostructures of iron oxide magnetic nanoparticles and poly(3,4-ethylene dioxythiophene). The iron oxide magnetic nanoparticles were synthesized with polymerized ionic liquids acting as linkers to a couple of magnetic nanoparticles and poly(3,4-ethylenedioxythiophene). This poly(3,4-ethylene dioxythiophene)–polymerized ionic liquids-modified iron oxide hybrid material served as a sensing channel for volatile organic compounds detection, showing high sensitivity (down to 1 ppm) and low noise. The sensor effectively detected volatile organic compounds like acetone, a biomarker in the exhaled breath of lung cancer patients, indicating its potential for early lung cancer diagnosis through volatile organic compounds biomarker detection (Tung et al. 2015). A new electrochemical immunosensor has been developed to detect CYFRA 21-1, a crucial biomarker for non-small cell lung cancer. This sensor features a platform made of a silicon nitride and molybdenum disulfide composite on multi-walled carbon nanotubes, while core-shell magnetic mesoporous silica nanoparticles coated with gold nanoparticles act as signal amplifiers. The immunosensor showed a linear detection range of 0.01–1.0 pg/mL and an impressively low detection limit of 2.00 fg/mL, with high selectivity and stability. This innovative approach offers a sensitive, rapid, and sustainable method for early lung cancer diagnosis, outperforming conventional techniques (Lütfi Yola et al. 2021).

Core-shell nanoparticle-based biosensors offer significant advancements in the sensitive and specific detection of lung cancer biomarkers. These innovative approaches utilizing core-shell nanoparticles enhance the detection capabilities for various biomarkers, providing promising tools for early diagnosis and improving clinical outcomes for lung cancer patients.

3.4 Prostate Cancer Detection

Prostate cancer is the second most common cancer in men worldwide, and an early diagnosis significantly improves patient outcomes, but current methods often lack the necessary sensitivity, accuracy, and specificity. Improved diagnostic tools with high selectivity, sensitivity, and stability are needed. Biosensors offer a promising solution with low cost, high sensitivity, specificity, and portability. Prostate-specific antigen (PSA) is widely used among various prostate cancer biomarkers. (Singh et al. 2023). Recent studies have shown that core-shell nanoparticle-based biosensors are highly effective for prostate cancer diagnosis.

A practical chemiluminescence biosensing platform was developed for detecting PSA using bimetallic gold-silver core-shell nanoparticles. These nanoparticles were synthesized via a simple green aqueous phase reduction method, exhibiting remarkable catalytic activity in the luminol-potassium ferricyanide chemiluminescence system. Anti-detecting prostate-specific antigen antibodies modified the gold@silver nanoparticles, allowing for specific detecting prostate-specific antigen capture. The chemiluminescence signal quenching degree was proportional to prostate-specific antigen concentration, with a detection range of 0.1 pg/mL to 100 ng/mL and a detection limit of 0.047 pg/mL. This method demonstrated excellent performance in human serum samples, highlighting its potential for prostate-specific antigen detection and prostate cancer biosensing (Zhang et al. 2018).

An advanced electrochemical biosensor was developed using graphene-poly(3-aminobenzoic acid) nanocomposite for electrode modification and porous-hollowed-silver-gold core-shell nanoparticles for signal amplification. The graphene-poly(3-aminobenzoic acid) modified electrodes significantly increased sensing response due to reduced impedance and increased probe binding sites. The porous-hollowed-silver-gold core-shell nanoparticles enhanced electrochemical current threefold compared to gold nanoparticles, leading to a detection limit of 0.13 pg/mL. This method demonstrated high sensitivity and specificity, suggesting its potential for early prostate cancer diagnosis (Pothipor et al. 2019). A label-free electrochemical immunosensor was developed using dendritic core-shell gold-palladium@gold nanocrystals. These nanocrystals were synthesized using a green growth-directing agent and exhibited enhanced catalytic activity. The immunosensor showed excellent performance with a detection range of 0.1–50 ng/mL and a detection limit of 0.078 ng/mL. The core-shell gold-palladium@gold nanocrystals improved stability, reproducibility, and selectivity, making this approach promising for prostate cancer diagnostics (Wang et al. 2018).

Utilizing nitrogen-doped graphene quantum dots to functionalize core-shell gold nanoparticles, a novel sandwich-type electrochemical immunosensor was created for the detection of PSA. Nanocomposites in the shape of echinoidea were made to identify secondary antibodies. These nanocomposites exhibited synergistic electrocatalytic activity, significantly amplifying the electrochemical signal. The immunosensor demonstrated a wide detection range from 0.01 pg/mL to 100 ng/mL with a detection limit of 0.003 pg/mL. It showed high sensitivity, selectivity, and stability, underscoring the potential of core-shell nanoparticles in advancing prostate cancer diagnostic (Yang et al. 2018). Hence, Core-shell nanoparticle-based biosensors have shown significant efficacy in the sensitive and specific detection of prostate cancer biomarkers, providing promising tools for early diagnosis and improving clinical outcomes for prostate cancer patients.

3.5 Liver Cancer Detection

Liver cancer poses a significant challenge due to its high mortality rate and late-stage diagnosis, which often results in poor prognosis and survival. This is primarily due to liver cancer's gradual development and a lack of identifiable early signs, making early detection challenging. Research has focused on developing more sensitive and accurate biosensing techniques to improve early detection and treatment. Core-shell nanoparticles have emerged as promising candidates for liver cancer biosensing due to their enhanced sensitivity, stability, and specificity. (Park et al. 2023).

Recent advancements include a novel biosensor utilizing upconversion@polydopamine core-shell nanoparticles developed for biosensing and imaging intracellular cytochrome c. This biosensor employs upconversion@polydopamine as an internal reference and fluorescence quencher and a Cy3-modified aptamer for ratiometric measurement of cytochrome c. The sensor exhibits high selectivity, detecting cytochrome c at a limit of 20 nM in buffer. It demonstrates low cytotoxicity, making it suitable for monitoring cytochrome c-mediated cell apoptosis pathways in complex biological samples. Future improvements in the synthesis of upconversion nanoparticles and aptamer loading efficiency could further enhance its sensitivity, advancing its potential application in liver cancer biosensing (Ma et al. 2017).

Another innovative approach involves silver/silica dioxide core-shell nanoparticles used in surface-enhanced Raman scattering to immunize human α-fetoprotein, a critical biomarker in hepatocellular carcinoma. This method employs polyclonal antibody-modified silver/silica dioxide as Raman tags and monoclonal antibody-modified silica-coated magnetic nanoparticles for immobilization and separation. The sandwich immunoassay detects human α-fetoprotein concentrations up to 0.12 μg/ml with a detection limit of 11.5 pg/ml. The robust stability of silica-coated silver Raman tags and the straightforward magnetic separation process enhance the method's sensitivity, simplicity, and cost-effectiveness compared to traditional assays (Gong et al. 2007).

Gold-silica alloy core-shell nanoparticles have been utilized for the sensitive detection of α-fetoprotein and telomerase key biomarkers linked to hepatocellular carcinoma. These nanoparticles serve as stable carriers for antibodies, facilitating efficient electron transfer and enable surface-enhanced Raman scattering detection using a three-dimensional hierarchical plasmonic nanoarchitecture label-free. Functionalized with 4-mercaptobenzoic acid and Nile blue, these nanoparticles enhance the electromagnetic field and generate amplified Raman signals upon binding to protein biomarkers. The method demonstrates a linear dynamic range of 0.2–22 ng/mL for α-fetoprotein and telomerase, with detection limits of 0.1 ng/mL and 0.06 ng/mL, respectively. Serum analysis shows good agreement with standard enzyme-linked immunosorbent assays, indicating its potential for medical diagnostics and disease monitoring (Wang et al. 2016).

An innovative core-shell hybrid nanomaterial combining surface-active maghemite nanoparticles and chromate ions has been developed for the electrochemical biosensing of polyamines in liver cancer tissues. The surface-active maghemite nanoparticles@chromate ions complexes exhibit improved charge transfer properties and enhanced electrocatalytic activity. By immobilizing bovine serum amino oxidase on the surface-active maghemite nanoparticles@chromate ions surfaces, a biologically active bio-nano-conjugate surface-active maghemite nanoparticles@chromate ions- bovine serum amino oxidase was developed. This conjugate enabled the development of a reagentless electrochemical biosensor capable of differentiating between tumor tissues and healthy liver based on polyamine levels in human liver extracts from biopsies. This approach highlights the potential of surface-active maghemite nanoparticles@chromate ions -based biosensors for sensitive liver cancer diagnostics (Magro et al. 2015).

Core-shell nanoparticles represent a cutting-edge advancement in liver cancer biosensing, significantly improving sensitivity, stability, and specificity. These nanoparticles enable the detection of crucial biomarkers and cellular processes, paving the way for earlier diagnosis and better treatment outcomes. Continued research and optimization of these biosensing techniques promise to enhance their application potential, contributing to more effective liver cancer diagnostics and monitoring.

3.6 Pancreatic Cancer Detection

A highly lethal cancer, pancreatic ductal adenocarcinoma is distinguished by an abundance of stroma, marked resistance to radiation and chemotherapy, and an advanced stage at which 80% of patients appear, making resection impossible. Patients frequently miss the optimal window to receive therapy since they lack any clear early signs, which contributes to a just 9% 5-year survival rate. For pancreatic cancer patients to have better treatment outcomes and prognoses, early identification and diagnosis are essential. Core-shell nanoparticles have been successfully used to identify and biosensitize biomarkers for pancreatic cancer (Pang et al. 2019).

A study developed a dual-surface-enhanced Raman scattering biosensor for the detection of microRNA-10b, a pancreatic cancer biomarker, in exosome and residual plasma samples. The biosensor was assisted by a duplex-specific nuclease. Iron oxide@silver-DNA-gold@silver@ 5,5′-Dithiobis (2-nitrobenzoic acid) conjugates are used in the biosensor. The DSN enzyme specifically cleaves the DNA probe upon hybridization with target microRNA-10b, releasing the surface-enhanced Raman scattering tags and initiating surface-enhanced Raman scattering intensity quenching. Through the use of dual-surface improved Raman scattering enhancement and cyclic signal amplification, single-base mismatch recognition can reach a detection limit of 1 aM. The iron oxide@silver nanoparticles serve as both capture and surface-enhanced Raman scattering substrates, while gold core@silver shell@5,5′-Dithiobis (2-nitrobenzoic acid) nanostructures act as surface-enhanced Raman scattering tags. The assay benefits from the magnetic concentration of iron oxide@silver, creating "hot spots" that enhance Raman signals. The one-step, one-pot process simplifies the detection without extensive washing or hybridization steps, making it suitable for clinical diagnostics. This biosensor accurately distinguishes pancreatic ductal adenocarcinoma from chronic pancreatitis and normal controls by quantifying microRNA-10b levels in plasma and exosome samples. This method enhances sensitivity and specificity and simplifies the diagnostic process, offering a promising tool for clinical applications and point-of-care diagnostics (Pang et al. 2019).

3.7 Brain Cancer Detection

Glioma is characterized by its challenging diagnosis and treatment, necessitating sensitive detection and accurate diagnosis for effective clinical treatments. Liquid biopsy is non-invasive, quick, and easy to use, it has a lot of potential for diagnosing cancer. Tumor markers found in biological fluids, such as miRNAs, can be used to guide this process. MiRNA-182 has been recognized as a noteworthy marker for the diagnosis and prognosis monitoring of gliomas because of its high specificity and regulatory function in the invasion and proliferation of glioma cells.

To address this, a study developed a magnetic covalent organic framework nanosphere-based fluorescent miRNA biosensor. These magnetic covalent organic framework nanospheres, featuring iron oxide cores and high-crystalline organic framework shells, were prepared using a monomer-mediated in-situ interface growth strategy. The biosensor employs hairpin DNA probes for sensitive detection of miRNA-182, achieving a detection limit of 20 fM and a linearity range of 0.1–10 pM, in blood samples. It demonstrated stability and precision, effectively quantifying miRNA-182 levels in glioma patient serum levels significantly higher than in healthy individuals and decreased post-surgery.

Additionally, a proof-of-concept capillary chip system was developed to visualize miRNA-182 in microsamples, providing a robust glioma diagnosis and prognosis method. The in-situ growing approach mediated by monomers produced uniform magnetic covalent organic framework nanospheres with well-ordered micropores and rapid magnetic separation. This biosensor leverages fluorescence quenching or amplification from DNA interactions on the magnetic covalent organic framework, enabling sensitive quantification of miRNA-182 even in complex blood samples. The magnetic covalent organic framework-based biosensor shows significant potential for accurate glioma diagnosis and prognosis, particularly its ability to monitor miRNA-182 concentrations before and after surgery. The capillary chip system further enhances its clinical utility for glioma detection. The development and application of a magnetic covalent organic framework nanosphere-based fluorescent miRNA biosensor for the sensitive detection of miRNA-182 offers a promising tool for early glioma diagnosis and effective prognosis monitoring (Liang et al. 2022).

3.8 Circulating Tumor Cells Detection

Tumor metastasis accounts for approximately 90% of cancer-related mortality. Since the concept of rare circulating tumor cells (CTCs) was first introduced by Ashworth in 1869, their significance has been increasingly recognized, particularly over the past decade. CTCs originate from primary tumor cells, enter the peripheral blood, and act as precursors for metastasis, having been identified across various cancer types, including breast, prostate, lung, and colorectal cancer. As a form of liquid biopsy, CTCs are promising biomarkers for cancer, as they can potentially represent the global features of cancer characteristics. The sensitive analysis and specific identification of CTCs in blood samples are crucial for early diagnosis, prognosis, and evaluation of treatment efficacy. The ability to detect and analyze CTCs has become an important tool in cancer management, providing insights into their biology and behavior to guide early diagnosis, treatment selection, and monitoring of disease progression. The study of CTCs has significantly advanced our understanding of the metastatic process. The integration of CTC analysis into clinical practice holds the potential to improve early detection, enable personalized treatment strategies, and ultimately enhance outcomes for cancer patients. However, the challenge lies in the extremely low abundance of circulating tumor cells in the blood (a few to hundreds per mL) amidst many hematologic cells (10^9 cells per mL) (Sun et al. 2021). Recent advancements have focused on developing efficient methods for capturing and detecting circulating tumor cells using core-shell nanoparticles. One such approach involves using core-shell electrospun nanofibrous membranes combined with hyaluronic acid-functionalized graphene quantum dots.

The nanofibrous membranes, developed from polycaprolactone and hyaluronic acid through co-axial electrospinning, capture about 75% of cancer cells via hyaluronic acid -CD44 receptor interactions, which are overexpressed on cancer cells. Graphene quantum dots-hyaluronic acid serves as a fluorescent tag for detection, where the photoluminescence intensity decreases as more cancer cells are captured, indicating higher adsorption of graphene quantum dots-hyaluronic acid to CD44 receptors. This method demonstrates high selectivity and specificity for cancer diagnosis, with significant changes in fluorescent intensity for cancer cells but not for non-cancerous cells (Asghari and Mahmoudifard 2023).

Another innovative method is the dual-mode MRI/FL diagnosis based on a magnetic core-shell aptasensor. This aptasensor, fluorescent aptasensor-doped polymer, employs magnetic nanoparticles coated with silicon dioxide@carbon for detecting circulating tumor cells and imaging tumors. It uses a fluorescent-labeled polymer aptamer targeting the MUC1 protein on cancer cells, linked to nanoparticles via EcoR1-responsive ssDNA pairing for stable cell capture and enzyme-cleavable release. The core-shell iron oxide nanoparticles, coated with silicon dioxide -doped polymer, allow for susceptible circulating tumor cell detection and accurate Magnetic Resonance/Fluorescence imaging, showing promise for clinical tumor diagnosis and liquid biopsy. (Wang et al. 2022). Additionally, a method using core-shell plasmonic nanorods with tunable nanogaps has been developed for selective and quantitative detection of circulating tumor cells by surface-enhanced Raman scattering. Gold nanorods are decorated with 4-mercaptopyridine to form a sacrificial Ag shell, replaced with HAuCl4 to prepare a gold–silver@gold core-shell structure. Functionalized with aptamers specific to circulating tumor cells, these nanorods can detect as few as 20 MCF-7 cells (breast cancer cells) in a blood-mimicking fluid using surface-enhanced Raman scattering. The nanorods also intercalate hydrophobic drugs like Doxorubicin, adding drug delivery capabilities. This versatile platform is promising for biosensing, imaging, and therapy applications. (Zhang et al. 2017). Hence, core-shell nanoparticles offer a sophisticated and efficient approach for the sensitive detection and capture of circulating tumor cells. These advancements hold significant potential for improving cancer diagnosis, prognosis, and treatment monitoring.

3.9 Colon Cancer Detection

Recent studies have explored the use of core-shell nanoparticles for diagnosing colon cancer, focusing on detecting colorectal cancer biomarkers. Two notable approaches have been introduced, each showcasing the potential of core-shell nanoparticles in biosensing for colorectal cancer. A novel molecularly imprinted polymer-surface-enhanced Raman scattering sensing platform has been developed to detect the biomarker NDKB associated with colorectal cancer. This platform utilizes gold core-silver shell nanoparticles as surface-enhanced Raman scattering sensing substrates, anchored on a silicon wafer and covered uniformly with surface-imprinted polymers. The imprinted polymers ensure specificity by recognizing NDKB, which blocks the access of Raman reporter molecules to the surface-enhanced Raman scattering sensing-active substrate, thereby reducing Raman signals. This approach achieves selective and sensitive detection of NDKB in serum, with an impressive detection limit of 0.82 pg/mL. The biosensor offers a nondestructive, rapid, and quantitative assay suitable for clinical diagnosis and prognosis in colorectal cancer. Integrating molecularly imprinted polymers with high-sensitivity surface-enhanced Raman scattering technology represents a significant advancement in biosensing, promising enhanced capabilities for early colorectal cancer detection and tumor progression monitoring (Lu et al. 2023).

Another research effort introduces a surface-enhanced Raman scattering biosensor using core-shell nanoparticles to detect colorectal cancer-related miRNAs (miR-21 and miR-31). The biosensor features gold nanocages@gold nanoparticles, hierarchical clusters, and silver magnetic nanoparticles for sample enrichment. This approach promises sensitive colorectal cancer detection via miRNA monitoring, with the potential for monitoring disease progression and treatment response. Future work includes expanding multi-miRNA detection capabilities and assessing clinical applicability in colorectal cancer management. (Wu et al. 2024).These studies demonstrate the potential of core-shell nanoparticles, particularly gold core-silver shell nanoparticles and gold nanocages@gold nanoparticles, in enabling sensitive and selective detection of colorectal cancer biomarkers. The integration of core-shell nanoparticles with advanced biosensing techniques holds promise for enhancing early colorectal cancer detection and monitoring of disease progression and treatment response. Figure 7.2 illustrates the application of core-shell nanoparticles in detecting and monitoring various cancers. Table 7.1 provides a comprehensive overview of core-shell nanoparticles used to detect different types of cancers. It lists the types of cancer, the core materials, the shell materials, the biomarkers detected, and the inferences drawn from the biosensing studies.

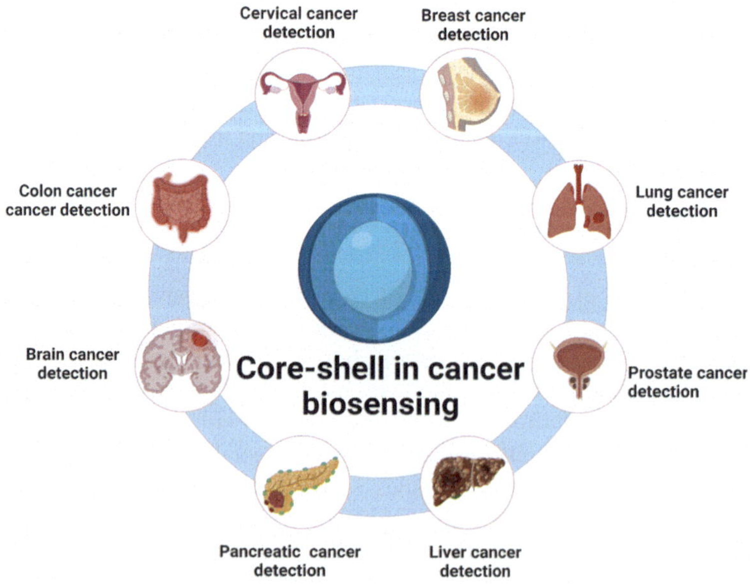

Fig. 7.2 Role of core-shell nanoparticles in biosensing for various cancers

Table 7.1 Overview of core-shell nanoparticles used to detect different types of cancers

Type of cancer	Core-shell nanoparticles	Core	Shell	Biomarker detected	Inference	Reference
Breast cancer	Gold/silver core-shell	Gold	Silver	CA15-3	Visual and electrochemical detection of CA15-3 in the range of 30–300 U/mL	(Li et al. 2023)
	Gold@palladium-silver dog-bone-like nanorods, gold@platinum-rhodium nanorods	Gold	Palladium-silver, platinum-rhodium	HER2	Highly sensitive electrochemical immunosensor for HER2 detection (0.001–100 ng/mL, 0.25 pg/mL detection limit)	(Zhang et al. 2023)
	Iron oxide nanoparticles@gold nanoparticles@poly(3-thiophenemalonic acid) polymer	Iron oxide	Gold, polymer	Calreticulin	Label-free immunosensor for calreticulin detection with high accuracy and no significant matrix effect in human serum	(Aydın et al. 2022)
	Iron@silver core-shell	Iron	Silver	miR-155	Electrochemical nanobiosensor for miR-155 detection (0.5 fM to 1.0 nM, 0.15 fM detection limit) in human serum without sample extraction or amplification	(Yazdanparast et al. 2020)
Cervical cancer	Iron oxide-gold core-shell	Iron oxide	Gold	HPV-16 DNA	Achieved a detection limit of 0.1 nM and sensitivity of 2.4 μA/nM for HPV-16 detection	(Rasouli et al. 2023)
	Gold@5,5′-Dithiobis(2-nitrobenzoic acid)@silver, iron oxide magnetic nanoparticles	Gold, iron oxide	5,5′-Dithiobis(2-nitrobenzoic acid), silver	miRNA-21	Demonstrated a detection limit of 0.084 fM and single-base mismatch distinction	(Xu et al. 2020)
	Iron oxide core–gold shell	Iron oxide	Gold	Valosin-containing protein	Achieved a detection limit of 25 fg/mL of purified valosin-containing protein and 16 pg/mL in protein mixtures extracted from cervical cancer tissue	(Ren et al. 2019)

(continued)

Table 7.1 (continued)

Type of cancer	Core-shell nanoparticles	Core	Shell	Biomarker detected	Inference	Reference
Lung cancer	Nickel@silver, carbon nanofibers	Nickel	Silver	Salivary imidazole compounds	Surface-enhanced Raman spectroscopy salivary sensor platform for non-invasive, cost-effective early-stage lung cancer screening (detection limit of 10–12)	(Sunil et al. 2023)
	Copper oxide, silicon	Copper oxide	Silicon	L-cysteine	Sensitive detection using a modified glassy carbon electrode with enhanced antibacterial activity and anticancer effects on lung cancer cells	(Gowtham et al. 2023)
	Iron oxide, poly(3,4-ethylenedioxythiophene)	Iron oxide	Poly(3,4-ethylenedioxythiophene)	Volatile organic compounds (acetone)	High-performance chemo-resistive sensor for early lung cancer diagnosis through volatile organic compounds biomarker detection (sensitivity down to 1 ppm)	(Tung et al. 2015)
	Silicon nitride–molybdenum disulfide, multi-walled carbon nanotubes, magnetic mesoporous silica nanoparticles@gold nanoparticles	Silicon nitride, multi-walled carbon nanotubes, magnetic mesoporous silica	Molybdenum disulfide, gold	CYFRA 21-1	Novel electrochemical immunosensor for non-small cell lung cancer detection with a linear range of 0.01–1.0 pg/mL and a detection limit of 2.00 fg/mL	(Lütfi Yola et al. 2021)

Prostate cancer	Gold-silver	Gold	Silver	Prostate-specific antigen	Chemiluminescence biosensing platform with a detection range of 0.1 pg/mL to 100 ng/mL and a detection limit of 0.047 pg/mL. Demonstrated excellent performance in human serum samples	(Zhang et al. 2018)
	Porous-hollowed-silver-gold	Silver	Gold	Prostate-specific antigen	Advanced electrochemical biosensor using graphene-poly(3-aminobenzoic acid) nanocomposite and porous-hollowed-silver-gold core-shell nanoparticles for signal amplification. Achieved a detection limit of 0.13 pg/mL	(Pothipor et al. 2019)
	Dendritic core-shell gold-palladium@gold	Gold-palladium	Gold	Prostate-specific antigen	Label-free electrochemical immunosensor with a detection range of 0.1–50 ng/mL and a detection limit of 0.078 ng/mL. Exhibited improved stability, reproducibility, and selectivity	(Wang et al. 2018)
	Gold nanoparticles functionalized with nitrogen-doped graphene quantum dots	Gold	Nitrogen-doped graphene quantum dots	Prostate-specific antigen	Sandwich-type electrochemical immunosensor with a wide detection range from 0.01 pg/mL to 100 ng/mL and a detection limit of 0.003 pg/mL. Showed high sensitivity, selectivity, and stability	(Yang et al. 2018)

(continued)

Table 7.1 (continued)

Type of cancer	Core-shell nanoparticles	Core	Shell	Biomarker detected	Inference	Reference
Liver cancer	Upconversion@ polydopamine	Upconversion	Polydopamine	Cytochrome C	Biosensor for monitoring cytochrome c-mediated cell apoptosis pathways in complex biological samples. Exhibits high selectivity, detecting cytochrome c at a limit of 20 nM in buffer.	(Ma et al. 2017)
	Silver/silica dioxide	Silver	Silica dioxide	Human α-fetoprotein	Surface-enhanced Raman scattering immunoassay with a detection range of up to 0.12 μg/ml and a detection limit of 11.5 pg/ml. Enhances sensitivity, simplicity, and cost-effectiveness compared to traditional assays.	(Gong et al. 2007)
	Gold-silica alloy	Gold	Silica	α-Fetoprotein, telomerase	Label-free, three-dimensional hierarchical plasmonic nanoarchitecture for surface-enhanced Raman scattering detection. Demonstrates a linear dynamic range of 0.2–22 ng/mL and detection limits of 0.1 ng/mL and 0.06 ng/mL for α-fetoprotein and telomerase, respectively.	(Wang et al. 2016)
	Surface-active maghemite nanoparticles@chromate ions	Maghemite	Chromate ions	Polyamines	Electrochemical biosensor for differentiating between tumor tissues and healthy liver based on polyamine levels in human liver extracts from biopsies.	(Magro et al. 2015)

Pancreatic cancer	Iron oxide@silver, gold@silver@5,5′-Dithiobis (2-nitrobenzoic acid)	Iron oxide, gold	Silver, 5,5′-Dithiobis (2-nitrobenzoic acid)	MicroRNA-10b	Duplex-specific nuclease-assisted dual-surface-enhanced Raman scattering biosensor for detecting microRNA-10b in exosome and residual plasma samples. Achieves a detection limit of 1 aM with single-base mismatch recognition.	(Pang et al. 2019)
Brain cancer	Magnetic covalent organic framework nanospheres	Iron oxide	Covalent organic framework	miRNA-182	Fluorescent miRNA biosensor for sensitive detection of miRNA-182 in blood samples. Achieves a detection limit of 20 fM and a linearity range of 0.1–10 pM. Demonstrates stability, precision, and the ability to quantify miRNA-182 levels in glioma patient serum.	(Liang et al. 2022)
Circulating tumor cells	Magnetic iron oxide nanoparticles, silicon dioxide@carbon	Iron oxide	Silicon dioxide, carbon	MUC1 protein	Dual-mode MRI/fluorescence aptasensor for detecting circulating tumor cells and imaging tumors. Uses a fluorescent-labeled polymer aptamer targeting the MUC1 protein on cancer cells.	(Wang et al. 2022)
	Gold nanorods, gold-silver@gold	Gold	Silver, gold	Aptamer-specific targets	Surface-enhanced Raman scattering-based detection of circulating tumor cells. Gold nanorods are decorated with 4-mercaptopyridine to form a sacrificial ag shell, replaced with HAuCl4 to prepare a gold–silver@gold core-shell structure. Functionalized with aptamers specific to circulating tumor cells.	(Zhang et al. 2017)

(continued)

Table 7.1 (continued)

Type of cancer	Core-shell nanoparticles	Core	Shell	Biomarker detected	Inference	Reference
Colorectal cancer	Gold, silver	Gold	Silver	NDKB	Molecularly imprinted polymer-surface enhanced Raman scattering sensing platform for selective and sensitive detection of NDKB in serum (detection limit of 0.82 pg/mL). Suitable for clinical diagnosis and prognosis.	(Lu et al. 2023)
	Gold nanocages, gold nanoparticles, magnetic silver nanoparticles	Gold, silver	Gold	miR-21, miR-31	Surface-enhanced Raman scattering biosensor for sensitive detection of colorectal cancer-related miRNAs. Potential for monitoring disease progression and treatment response.	(Wu et al. 2024)

4 Strategies to Fabricate Core-Shell Nanoconstructs for Efficient Biosensing Application

4.1 Chemical Synthesis Strategies

4.1.1 Sol-Gel Process

The sol-gel process is a versatile technique for synthesizing core-shell nanoparticles, particularly for cancer biosensing applications. This method involves the hydrolysis and condensation of metal alkoxides, such as tetraethyl orthosilicate, or inorganic salts like sodium silicate. In this process, alkoxysilane precursors are emulsified in water with a hydrophobic liquid, leading to hydrolysis and condensation reactions that form silica shells around the hydrophobic cores (Jenjob et al. 2020). For instance, silica-coated silver nanoparticles are synthesized by chemically reducing silver nanoparticles using trisodium citrate as a capping and reducing agent. This is followed by coating them with silica through a modified Stober method (sol-gel). This involves mixing the silver nanoparticles in ethanol, ammonia, and deionized water, with tetraethyl orthosilicate added gradually during sonication and agitation to form the silica-coated nanoparticles. These core-shell nanoparticles fabricate electrochemical immunosensors for detecting cancer biomarkers like the carcinoembryonic antigen (CEA). The combination of silver and silica enhances the sensitivity and selectivity of the immunosensor, making it an efficient and economically viable tool for cancer diagnosis, requiring only simple equipment and low reaction temperatures (Singh et al. 2021).

4.1.2 Hydrothermal Synthesis

Hydrothermal synthesis is a versatile bottom-up approach for producing core-shell nanoparticles with tailored properties and compositions. It involves heterogeneous reactions in aqueous solutions under high temperature and pressure inside a sealed vessel or autoclave, allowing precise control over nanoparticle size, shape, and structure. This method results in nanoparticles with high surface area, tailorable surface characteristics, and protective capabilities for sensitive components. Its simple equipment and nontoxic solvents like ethanol and deionized water make them ideal for applications like cancer biosensing. For instance, surrounded by a mesoporous silica shell, silver core, and potentially another layer of silver on the outside, three core-shell nanoparticles with enhanced Surface-enhanced Raman spectroscopy properties are synthesized using hydrothermal. This facile process yields nanoparticles with superior and consistent Surface-enhanced Raman spectroscopy activity. (Jiang et al. 2016). Similarly, ternary nanocomposites like manganese dioxide@ nickel ferrite with a flower-like core-shell structure are prepared through a two-step hydrothermal approach for electrochemical detection of the anti-neoplastic drug Chlorambucil (73).

4.1.3 Co-Precipitation Method

The co-precipitation method effectively fabricates core-shell nanoparticles with tailored properties for biomedical applications. This method involves precipitation of precursors from a solution, forming core-shell structures. For instance, ultra-small superparamagnetic magnetite nanoparticles and magnetite and zinc sulfide core-shell nanocomposites are synthesized using a reflux-assisted co-precipitation method, with ethylenediaminetetraacetic acid (EDTA) as a capping agent. Magnetite nanoparticles are first synthesized at 75 °C for 30 min, then dispersed in 2-methoxyethanol, and combined with zinc acetate and EDTA. Sodium sulfide is added to form the ZnS shell, and the mixture is refluxed at various temperatures. These nanoparticles, exhibit superparamagnetic behavior, making them suitable for MRI and hyperthermia applications (Rashidi Dafeh et al. 2019). Another example includes gold-coated iron oxide nanoparticles, synthesized by co-precipitating ferrous and ferric ions in a robust alkaline solution, followed by gold coating using sodium citrate as a reducing agent. These nanoparticles are used as carriers in immunoassays, demonstrating their capability to immobilize antibodies for biomedical applications (Ahmadi et al. 2014). The co-precipitation method provides precise control over nanoparticle size and composition, making it valuable for producing core-shell nanoparticles with specific properties for various biomedical uses.

5 Physical Methods

5.1 Sputtering

Sputtering is a vacuum-based physical vapor deposition process that deposits films and nanoparticles by ejecting atoms from a target material through ion bombardment. This process involves generating plasma, typically with argon gas, accelerating ions toward the target, and then transporting the ejected atoms to deposit onto a substrate. There are two types of sputtering: simple sputtering and magnetron sputtering, the latter using magnets to enhance the deposition rate and prevent target overheating. Direct Current sputtering suits conductive materials, while Radio Frequency sputtering can deposit insulating materials. Introducing reactive gases allows for reactive sputtering, which deposits compound materials. (Dhand et al. 2015).

For example, copper nanoparticles were synthesized by depositing copper on polystyrene spheres using ion sputtering, followed by calcination at 900 °C to form copper nanoparticles. These copper nanoparticles demonstrate excellent surface-enhanced Raman spectroscopy properties, making them ideal for biosensing (Kaushal et al. 2020). Another example involves silver-titanium dioxide nanowires, where titanium dioxide nanowires were formed on a titanium sheet through hydrothermal treatment, followed by Ag deposition using magnetron sputtering. These titanium dioxide nanowires were functionalized with 3’-NH2-ssDNA and used as substrates for Surface-enhanced Raman spectroscopy-based miRNA-21 detection. The silver nanoparticles uniformly coat the titanium dioxide nanowires, enhancing their Surface-enhanced Raman spectroscopy sensitivity (Peng et al. 2019). Sputtering allows precise control over the size and composition of core-shell nanoparticles, making it a valuable method for developing advanced materials for biomedical applications.

5.2 Laser Ablation

Laser Ablation of Microparticles produces nanoparticles by continuously ablating microparticles in a flowing aerosol with high-power laser pulses. This process produces nanoparticles with relatively small ranges of size and high yields from a variety of materials, including dielectrics, semiconductors, and metals. During Laser ablation, a high-energy laser pulse ablates a microparticle, generating plasma and a subsequent shockwave that vaporizes the particle. This is followed by the condensation of nanoparticles, resulting in particles with mean radii from 1.5 to 10 nm, depending on the aerosol gas type and pressure (Gallardo et al. 2009).

For example, a laser-assisted methodology was developed for synthesizing silica@gold core-satellite nanocomposites without chemical reduction steps. Silica and gold nanoparticles are modified by organosilane (3-Aminopropyl)trimethoxysilane molecules and assembled into core-satellite structures. This geometry generates a plasmonic absorption feature within the biological transparency window, making these nanocomposites suitable for biomedical applications. The methodology includes the laser ablation of silica micropowder and a gold target in aqueous solutions, followed by chemical modification and linking of the nanoparticles (Al-Kattan et al. 2021). Another study utilized pulsed laser ablation in liquid to form silver shell/gold core bimetallic nanoparticles. Researchers created bimetallic structures with controlled optical properties by irradiating a colloidal mixture of silver and gold nanoparticles with a pulsed neodymium laser. The process involves transferring laser energy to gold nanoparticles, which are cores for reducing silver ions (Mohebi et al. 2024). Hence, laser ablation methods for core-shell nanoparticles offer precise control over nanoparticle size and composition, making them valuable for applications in biomedical and other advanced fields.

6 Seed-Mediated Growth

Seed-mediated growth is a versatile and efficient technique for synthesizing core-shell nanoparticles, extensively used in cancer biosensing due to its ability to control nanoparticle size and morphology. Here, we discuss three applications utilizing this method for cancer biomarker detection. Core-shell gold-silver nanoparticles were synthesized using gold nanoparticles as seeds. Gold nanoparticles less than 12 nm were produced via the Turkevich method, which involves the chemical reduction of chloroauric acid using trisodium citrate at elevated temperatures. The gold nanoparticles were then used as seeds for the growth of silver shells by adding silver nitrate and trisodium citrate. These nanoparticles were employed in an impedimetric immunosensor for label-free detection of the cancer antigen CA125. The sensor exhibited a linear response to CA125 concentrations from 1 to 150 IU/mL and demonstrated reasonable specificity and stability, making it suitable for clinical applications without requiring preprocessing steps (Raghav and Srivastava 2015).

Gold-palladium core-shell nanoparticles, synthesized via seed-mediated growth, exhibit significantly enhanced catalytic activity, particularly for hydrogen peroxide reduction. Gold nanoparticle seeds were used to grow palladium shells by adding dihydrogen tetrachloropalladate and ascorbic acid under controlled conditions. The resulting gold-palladium nanoparticles, ranging from 30–40 nm, were used in a sandwich electrochemical immunosensor to detect CA125. The large active surface area and excellent catalytic activity of gold-palladium nanoparticles improved the sensor's sensitivity and specificity. The immunosensor demonstrated an excellent linear range and satisfactory selectivity for CA125 detection, showcasing its potential for broader biomarker detection applications. (Guo et al. 2013).

Gold@bimetallic-palladium-gold dog-bone-like nanorods core-shell for HER2 detection and gold@platinum and rhodium nanorods were prepared using a seed-assisted strategy. The gold@bimetallic-palladium-gold dog-bone-like served as a sensing platform to capture primary antibodies, while gold@platinum and rhodium nanorods were used to load secondary antibodies for signal amplification. The resulting sandwich-type electrochemical immunosensor was designed to analyze the HER2 biomarker quantitatively. The sensor exhibited a wide linear range of 0.001–100 ng/mL and a low detection limit (LOD) of 0.25 pg/mL. It demonstrated good reproducibility, specificity, and stability, making it applicable for clinical diagnosis of various biomarkers (Zhang et al. 2023). The seed-mediated growth method enables the synthesis of core-shell nanoparticles with tailored properties for cancer biosensing. These nanoparticles enhance the sensitivity, specificity, and overall performance of immunosensors, making them valuable tools for clinical diagnostics.

7 Microemulsion Method

The microemulsion method is versatile and efficient for synthesizing core-shell nanoparticles with precise control over shell thickness and surface functionality. This makes them highly suitable for cancer biosensing and other biomedical applications. Functionalized fluorescent core-shell nanoparticles serve as fluorescent labels in fluoroimmunoassays for biomarkers such as IL-6. The synthesis follows a water-in-oil (W/O) microemulsion method where surfactant TX-100, cyclohexane (oil phase), and n-hexanol (cosurfactant) are mixed, followed by the addition of a Rubpy dye solution. Tetraethyl orthosilicate initiates the hydrolysis reaction, and $NH_3{\cdot}H_2O$ is added, followed by 3-aminopropyltriethoxysilane to introduce amino groups on the nanoparticle surface. The microemulsion is broken with acetone, and the particles are centrifuged and washed. This method enhances fluorescence and sensitivity for IL-6 detection due to delayed 3-aminopropyltriethoxysilane addition, introducing more amino groups on the surface. (Hun and Zhang 2007).

Upconversion@polydopamine core@shell nanoparticles are used for biosensing and imaging cytochrome c in living cells. The preparation involves creating a W/O microemulsion and coating upconversion nanoparticles with bioinspired polydopamine. These nanoparticles provide multimodal imaging capabilities and allow for intracellular analysis of biomarkers by serving as fluorescence quenching agents and enabling ratiometric quantitative measurement of cytochrome c (Ma et al. 2017).

Gold-silica core-shell nanoparticles are synthesized for immunotargeting and theranostic applications using either citrate reduction followed by a modified Stöber method or oleylamine reduction followed by a reverse microemulsion method. Igepal surfactant forms micelles in cyclohexane, stabilizing hydrophobic gold-oleylamine nanoparticles and allowing silane precursors to condense on nanoparticle surfaces. The silica shell thickness can be tailored from ~2.5 to 14 nm by adjusting the molecular weight of Igepal and tetraethyl orthosilicate concentration, providing precise control over the functional properties of core-shell nanoparticles for targeted delivery and multimodal imaging. (Nallathamby et al. 2016).

Superparamagnetic iron oxide@silica core-shell nanoparticles are designed for cell imaging and other biomedical applications. Hydrophobic iron oxide nanoparticles are initially synthesized and then coated with a thin silica shell using tetraethyl orthosilicate and hydrolysis in a reverse microemulsion. The silica surface can be further functionalized with polyethylene glycol or amine groups, enhancing stability and biocompatibility (Zhao et al. 2014). This method ensures uniform shell formation and excellent stability in various aqueous solutions, making these nanoparticles suitable for multifunctional biomedical applications.

8 Layer-by-Layer Assembly

Layer-by-layer self-assembly is a highly effective and adaptable method for progressively depositing oppositely charged elements to create ultrathin films. The layer-by-layer technique was first presented by Iler in 1966 and refined by Decher in the 1990s. It entails subjecting a charged substrate to alternating solutions of polycations and polyanions to cause complementary compounds to adsorb on the substrate surface. The concentration of the coating substance, the ionic strength, the temperature, the pH, the washing and drying conditions, and other parameters can all be changed to get the required number of layers by repeating this process several times. Under mild conditions, the approach may produce homogenous nanocoatings with perfect structural control using a variety of materials, including natural and manufactured enzymes, polymer gels, metallic oxides, polyelectrolytes, clays, and peptides (Liu et al. 2019).

In cancer biosensing, layer-by-layer self-assembly can fabricate core-shell nanoparticles for detecting biomarkers like CA125 and CYFRA 21-1. For instance, a label-free electrochemical immunoassay for detecting CYFRA 21-1 employs poly(ε-caprolactone)-b-poly(ethylene oxide) block copolymer. This involves forming an antibody-antigen-antibody sandwich structure on a gold electrode, enabling precise detection of CYFRA 21-1 with a linear concentration range from 1.0 pg mL − 1 to 10 ng mL − 1 and a detection limit of 0.125 pg mL − 1 (Zhang et al. 2021). This method enhances the sensor's sensitivity and specificity, making it a powerful tool for cancer diagnostics.

9 Solvothermal Method

The solvothermal method, analogous to the hydrothermal treatment but using organic solvents instead of water, is beneficial for synthesizing materials that are not stable or soluble in water. This method is particularly useful in developing core-shell nanoparticles for cancer biosensing. Multiferroic magnetoelectric nanoparticles, for example, are promising tools for cancer diagnostics due to their unique electric and magnetic field interactions. Nagecetti et al. used 30-nm core-shell multiferroic magnetoelectric nanoparticles, synthesized by the solvothermal method as probes, demonstrating that cells incubated with these nanoparticles exhibited significant changes in NMR absorption spectra compared to those without multiferroic magnetoelectric nanoparticles. This effect, linked to the magnetoelectric properties of multiferroic magnetoelectric nanoparticles, was not observed with ordinary magnetic nanoparticles. The use of multiferroic magnetoelectric nanoparticles in biosensor systems highlights the broad potential for their integration into clinical practice. Key challenges in developing such systems include enhancing sensitivity, stability, reproducibility, reliability, and economic feasibility. Despite these challenges, creating multiparametric biosensor systems for cancer diagnostics remains crucial. These systems can facilitate large-scale point-of-care testing, enabling comprehensive health assessments and formulating optimal patient treatment or preventive strategies (Nagesetti et al. 2017).

10 Electrochemical Deposition

The electrochemical deposition synthesis of core-shell nanoparticles involves selectively depositing a shell material onto pre-formed core nanoparticles, allowing precise control over shell thickness and producing surfactant-free nanoparticles directly on electrodes. This method enhances the sensitivity and signal amplification of biosensors used in cancer detection. For example, a label-free electrochemical biosensor for detecting serum thrombomodulin, an endothelial glycoprotein related to tumor progression, was developed. Graphene nanosheets in nafion solution were applied to a gold electrode, and silver-silver oxide nanoparticles were immobilized using a single electrochemical deposition process. Through amino silver affinity, the thrombomodulin antibody was connected to the modified electrode. While silver-silver oxide nanoparticles functioned as redox probes and signal indicators, graphene/silver oxide nanoparticles offered stability and special electrical properties. The biosensor showed a limit of detection (LOD) of 31.5 pg/mL and a detection range of 0.1–20 ng/mL (Yang et al. 2012).

Similarly, an immunosensor for CEA detection was developed by electrodepositing gold nanoparticles on a Prussian, Blue-modified glassy carbon electrode. Functionalized with primary anti-CEA antibody and bovine serum albumin, a sandwich assay with HRP-labeled secondary anti-CEA antibody on a multi-gold nanoparticle shell/chitosan-protected graphene nanocore amplified the analytical

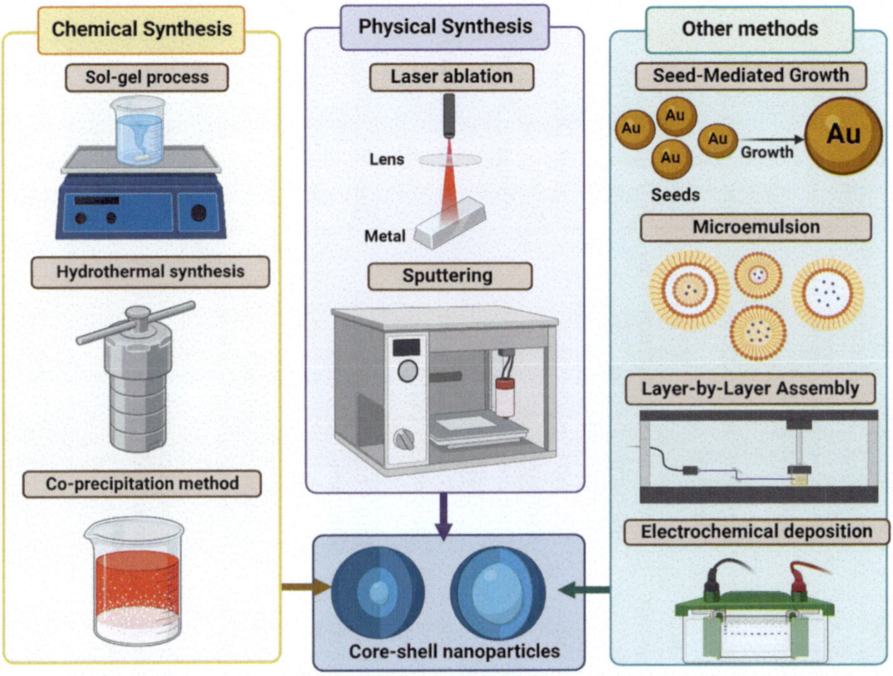

Fig. 7.3 Various strategies to fabricate core-shell nanoparticles for efficient biosensing application

signal. This sensor successfully detected CEA biomarkers in cancer patient serum with a high correlation to reference methods (Zhong et al. 2010).

These examples highlight the effectiveness of electrochemical deposition for fabricating core-shell nanoparticles in biosensors, offering high sensitivity and low detection limits for cancer biomarkers like prostate-specific antigen, epithelial cell adhesion molecule, CEA, tumor-specific growth factor, mucin 1, and thrombomodulin. Figure 7.3 illustrates various synthesis methods used for the fabrication of core-shell nanoparticles, and Table 7.2 details various core-shell nanoparticles, describing their synthesis methods, characteristics, and applications.

11 Conclusion

Core-shell nanoparticles have emerged as a versatile platform for enhancing the performance of various cancer biosensing techniques. These nanostructures can be tailored to improve the sensitivity and selectivity of electrochemical, optical, calorimetric, and mass-based biosensors for detecting various cancer biomarkers and cells. In electrochemical biosensors, core-shell nanoparticles can increase the surface area for enzyme immobilization, improve electron transfer, and amplify the electrochemical signal, enabling the sensitive detection of various cancer biomarkers. These biomarkers include cancer antigen 15-3, a protein found on the surface of

Table 7.2 Various core-shell nanoparticles, describing their synthesis methods, characteristics, and applications

Core-shell nanoparticles	Method of synthesis	Description	Applications	References
Silver@silica dioxide	Sol-gel process	Chemically reduced silver nanoparticles coated with silica using the modified Stober method	Electrochemical immunosensor for CEA detection	(Singh et al. 2021)
Silver@mesoporous silica@silver	Hydrothermal synthesis	High temperature and pressure aqueous reactions for precise control over nanoparticle properties	Enhanced SERS properties for biosensing	(Jiang et al. 2016)
Manganese dioxide@ nickel ferrite	Two-step hydrothermal synthesis	High temperature and pressure aqueous reactions for precise control over nanoparticle properties	Electrochemical detection of the anti-neoplastic drug Chlorambucil	(73)
Magnetite@zinc sulphide	Co-precipitation method	Reflux-assisted co-precipitation with EDTA as a capping agent	MRI and hyperthermia applications	(Rashidi Dafeh et al. 2019)
Gold@iron oxide	Co-precipitation method	Co-precipitation of Fe^{2+} and Fe^{3+} in alkaline solution, followed by gold coating	Immunoassays for antibody immobilization	(Ahmadi et al. 2014)
Copper nanoparticles	Sputtering	Deposition of cu on polystyrene spheres using ion sputtering, followed by calcination	Surface-enhanced Raman spectroscopy (SERS)	(Kaushal et al. 2020)
Silver/titanium dioxide nanowires	Sputtering	Hydrothermal formation of titanium dioxide nanowires on titanium sheet, followed by ag deposition using magnetron sputtering	SERS-based miRNA-21 detection	(Peng et al. 2019)
Silica@gold core-satellite	Laser ablation	Laser ablation of silica micropowder and gold target in aqueous solution, followed by chemical modification	Biomedical applications with plasmonic absorption feature	(Al-Kattan et al. 2021)
Silver shell/gold core bimetallic	Laser ablation	Pulsed laser ablation in liquid to form bimetallic structures by irradiating a colloidal mixture of silver and gold nanoparticles	Biomedical applications with controlled optical properties	(Mohebi et al. 2024)

Gold-silver nanoparticles	Seed-mediated growth	Gold nanoparticles are used as seeds for the growth of silver shells. Gold nanoparticles are produced via the Turkevich method; ag shells are grown by adding silver nitrate and trisodium citrate.	Impedimetric immunosensor for label-free detection of the cancer antigen CA125. Shows a linear response from 1 to 150 IU/mL.	(Raghav and Srivastava 2015)
Gold@pallidium nanoparticles	Seed-mediated growth	Gold nanoparticles are used as seeds for the growth of palladium shells by adding dihydrogen tetrachloropalladate and ascorbic acid under controlled conditions.	Sandwich electrochemical immunosensor for detecting CA125. Exhibits high sensitivity and specificity for CA125 detection.	(Guo et al. 2013)
Gold@palladium-silver dog-bone-like nanorods	Seed-mediated growth	Gold nanoparticles serve as seeds for the growth of palladium-silver dog-bone-like nanorods. The nanorods capture primary antibodies and load secondary antibodies for signal amplification.	An electrochemical immunosensor detects the HER2 biomarker with a wide linear range (0.001–100 ng/mL) and a low LOD of 0.25 pg/mL.	(Zhang et al. 2023)
Fluorescent Rubpy@ silica dioxide nanoparticles	Water-in-oil (W/O) microemulsion + functionalization	Surfactant TX-100, cyclohexane (oil phase), and n-hexanol (cosurfactant) form a microemulsion. Rubpy dye solution was added, followed by tetraethyl orthosilicate for hydrolysis and $NH_3{\cdot}H_2O$ and 3-aminopropyltriethoxysilane for surface functionalization. Microemulsion broke with acetone.	Enhanced fluorescence for sensitive IL-6 biomarker detection	(Hun and Zhang 2007)
Upconversion@ polydopamine core-shell NPs	Water-in-oil (W/O) microemulsion + Polydopamine coating	Upconversion is synthesized and then coated with bioinspired polydopamine using a W/O microemulsion method.	Multimodal imaging and quantitative cytochrome c detection in living cells.	(Ma et al. 2017)
Gold@silica dioxide core-shell NPs	Citrate reduction + modified Stöber method or reverse microemulsion	Gold nanoparticles are synthesized via citrate reduction. Silica oxide shell formed using a modified Stöber method or reverse microemulsion with Igepal surfactant and tetraethyl orthosilicate.	Targeted delivery and multimodal imaging with tunable silica shell thickness	(Nallathamby et al. 2016)

(continued)

Table 7.2 (continued)

Core-shell nanoparticles	Method of synthesis	Description	Applications	References
Iron oxide @silica dioxide core-shell NPs	Water-in-oil (W/O) microemulsion + tetraethyl orthosilicate hydrolysis	Hydrophobic iron oxide nanoparticles are synthesized and coated with a silica dioxide shell using tetraethyl orthosilicate hydrolysis in a reverse microemulsion setup.	Superparamagnetic properties for imaging and functionalized surfaces for drug delivery	(Zhao et al. 2014)
Poly(ε-caprolactone)-b-poly(ethylene oxide) Core-shell	Layer-by-layer self-assembly	Alternating adsorption of poly(ethylene oxide) and other molecules on a core.	Enhances sensitivity and specificity for detecting cancer biomarkers.	(Zhang et al. 2021)
Multiferroic magnetoelectric nanoparticles	Solvothermal synthesis	Organic solvents are used to create multiferroic magnetoelectric nanoparticles with both magnetic and electric properties.	Enhanced NMR absorption spectra for cancer cell analysis.	(Nagesetti et al. 2017)
Gold@graphene/silver-silver oxide	Electrochemical deposition	Gold core provides a stable base; graphene nanosheets offer high surface area; silver-silver oxide acts as redox probes and signal indicators.	Label-free electrochemical biosensor for detecting serum thrombomodulin with a detection range of 0.1–20 ng/mL and a limit of detection (LOD) of 31.5 pg/mL.	(Yang et al. 2012)
Gold@Prussian blue	Electrochemical deposition	Gold core provides a stable base; Prussian blue acts as a redox mediator for enhanced signal detection	Immunosensor for CEA detection, showing high correlation to reference methods	(Zhong et al. 2010)

breast cancer cells; human epidermal growth factor receptor 2, a protein involved in the growth and division of breast cancer cells; and Calreticulin, a protein with altered levels in various cancers. Core-shell nanoparticles have also been used in biosensors for the detection of microRNA-155, a small non-coding RNA molecule linked to breast cancer, human papillomavirus type 16 DNA, the DNA of a virus that can cause cervical cancer, and prostate-specific antigen, a protein marker for prostate cancer. Additionally, these nanostructures have been employed in the detection of other cancer biomarkers, such as MicroRNA-21, a regulator of gene expression associated with various cancers, cytokeratin-19 Fragment, a protein fragment released by lung cancer cells, and human α-Fetoprotein, a protein marker for liver cancer. Core-shell nanoparticles have also been utilized in biosensors for the detection of microRNA-10b and microRNA-182, small non-coding RNA molecules linked to breast cancer metastasis and progression, respectively, as well as MUC1 protein, a protein involved in cell signaling and adhesion with altered levels in breast, ovarian, and other cancers. Furthermore, these nanostructures have been employed in detecting valosin-containing protein, a protein with altered levels in lung cancer, and Nucleoside Diphosphate Kinase B, an enzyme with altered levels in various cancers, including lung cancer. The versatility and enhanced performance of core-shell nanoparticles in cancer biosensing make them a promising platform for developing advanced, sensitive, and selective cancer diagnostic tools. These nanostructures can be fabricated using various chemical synthesis methods, such as sol-gel, hydrothermal, and co-precipitation, as well as physical methods, including sputtering and laser ablation, and other techniques such as seed-mediated growth, microemulsion method, layer-by-layer assembly, solvothermal method, and electrochemical deposition. Hence, core-shell nanoparticles' versatility and enhanced performance in cancer biosensing make them a promising platform for developing advanced, sensitive, and selective cancer diagnostic tools.

References

Abdul Rasheed P, Sandhyarani N. Quartz crystal microbalance genosensor for sequence specific detection of attomolar DNA targets. Anal Chim Acta. 2016;905:134–9. https://doi.org/10.1016/j.aca.2015.11.033.

Aćimović SS, Ortega MA, Sanz V, Berthelot J, Garcia-Cordero JL, Renger J, Maerkl SJ, Kreuzer MP, Quidant R. LSPR chip for parallel, rapid, and sensitive detection of cancer markers in serum. Nano Lett. 2014;14:2636–41. https://doi.org/10.1021/nl500574n.

Ahmadi A, Shirazi H, Pourbagher N, Omidfar K. Synthesis and characterization of core-shell Au Fe oxide nanocomposites and their application for detecting immunological interaction. Monoclon Antib Immunodiagn Immunother. 2014;33:74–9. https://doi.org/10.1089/mab.2014.0006.

Al-Kattan A, Tselikov G, Metwally K, Popov AA, Mensah S, Kabashin AV. Laser ablation-assisted synthesis of plasmonic Si@Au core-satellite nanocomposites for biomedical applications. Nano. 2021;11:592. https://doi.org/10.3390/nano11030592.

Al-Ogaidi I, Gou H, Aguilar ZP, Guo S, Melconian AK, Al-kazaz AKA, Meng F, Wu N. Detection of the ovarian cancer biomarker CA-125 using chemiluminescence resonance energy transfer to graphene quantum dots. Chem Commun. 2014;50:1344–6. https://doi.org/10.1039/C3CC47701K.

Asghari S, Mahmoudifard M. The detection of the captured circulating tumor cells on the core-shell nanofibrous membrane using hyaluronic acid-functionalized graphene quantum dots. J Biomed Mater Res B Appl Biomater. 2023;111:1121–32. https://doi.org/10.1002/jbm.b.35219.

Aydın EB, Aydın M, Sezgintürk MK. Determination of calreticulin using Fe3O4@AuNPs core-shell functionalized with PT(COOH)2 polymer modified electrode: a new platform for the impedimetric biosensing of cancer biomarkers. Sensors Actuators B Chem. 2022;367:132099. https://doi.org/10.1016/j.snb.2022.132099.

Buchapudi KR, Huang X, Yang X, Ji H-F, Thundat T. Microcantilever biosensors for chemicals and bioorganisms. Analyst. 2011;136:1539–56. https://doi.org/10.1039/C0AN01007C.

Calabretta MM, Montali L, Lopreside A, Fragapane F, Iacoangeli F, Roda A, Bocci V, D'Elia M, Michelini E. Ultrasensitive on-field luminescence detection using a low-cost silicon photomultiplier device. Anal Chem. 2021;93:7388–93. https://doi.org/10.1021/acs.analchem.1c00899.

Chang K, Pi Y, Lu W, Wang F, Pan F, Li F, Jia S, Shi J, Deng S, Chen M. Label-free and high-sensitive detection of human breast cancer cells by aptamer-based leaky surface acoustic wave biosensor array. Biosens Bioelectron. 2014;60:318–24. https://doi.org/10.1016/j.bios.2014.04.027.

Chang J, Wang X, Wang J, Li H, Li F. Nucleic acid-functionalized metal–organic framework-based homogeneous electrochemical biosensor for simultaneous detection of multiple tumor biomarkers. Anal Chem. 2019;91:3604–10. https://doi.org/10.1021/acs.analchem.8b05599.

Chatterjee K, Sarkar S, Jagajjanani Rao K, Paria S. Core/shell nanoparticles in biomedical applications. Adv Colloid Interf Sci. 2014;209:8–39. https://doi.org/10.1016/j.cis.2013.12.008.

Chinen AB, Guan CM, Ferrer JR, Barnaby SN, Merkel TJ, Mirkin CA. Nanoparticle probes for the detection of cancer biomarkers, cells, and tissues by fluorescence. Chem Rev. 2015;115:10530–74. https://doi.org/10.1021/acs.chemrev.5b00321.

Dhand C, Dwivedi N, Jun Loh X, Ying ANJ, Kumar Verma N, Beuerman RW, Lakshminarayanan R, Ramakrishna S. Methods and strategies for the synthesis of diverse nanoparticles and their applications: a comprehensive overview. RSC Adv. 2015;5:105003–37. https://doi.org/10.1039/C5RA19388E.

Dhas NL, Raval NJ, Kudarha RR, Acharya NS, Acharya SR. Chapter 9: Core–shell nanoparticles as a drug delivery platform for tumor targeting. In: Grumezescu AM, editor. Inorganic frameworks as smart nanomedicines. William Andrew Publishing; 2018. p. 387–448. https://doi.org/10.1016/B978-0-12-813661-4.00009-2.

Elshafey R, Tavares AC, Siaj M, Zourob M. Electrochemical impedance immunosensor based on gold nanoparticles–protein G for the detection of cancer marker epidermal growth factor receptor in human plasma and brain tissue. Biosens Bioelectron. 2013;50:143–9. https://doi.org/10.1016/j.bios.2013.05.063.

Ensafi AA, Taei M, Rahmani HR, Khayamian T. Sensitive DNA impedance biosensor for detection of cancer, chronic lymphocytic leukemia, based on gold nanoparticles/gold modified electrode. Electrochim Acta. 2011;56:8176–83. https://doi.org/10.1016/j.electacta.2011.05.124.

Fan X, White IM, Shopova SI, Zhu H, Suter JD, Sun Y. Sensitive optical biosensors for unlabeled targets: a review. Anal Chim Acta. 2008;620:8–26. https://doi.org/10.1016/j.aca.2008.05.022.

Feng S, Chen R, Lin J, Pan J, Chen G, Li Y, Cheng M, Huang Z, Chen J, Zeng H. Nasopharyngeal cancer detection based on blood plasma surface-enhanced Raman spectroscopy and multivariate analysis. Biosens Bioelectron. 2010;25:2414–9. https://doi.org/10.1016/j.bios.2010.03.033.

Gallardo I, Hoffmann K, Keto JW. CdSe & ZnS core/shell nanoparticles generated by laser ablation of microparticles. Appl Phys A Mater Sci Process. 2009;94:65–72. https://doi.org/10.1007/s00339-008-4921-4.

Gavas S, Quazi S, Karpiński TM. Nanoparticles for cancer therapy: current progress and challenges. Nanoscale Res Lett. 2021;16:173. https://doi.org/10.1186/s11671-021-03628-6.

Gong J-L, Liang Y, Huang Y, Chen J-W, Jiang J-H, Shen G-L, Yu R-Q. Ag/SiO2 core-shell nanoparticle-based surface-enhanced Raman probes for immunoassay of cancer marker using silica-coated magnetic nanoparticles as separation tools. Biosens Bioelectron. 2007;22:1501–7. https://doi.org/10.1016/j.bios.2006.07.004.

Gowtham SM, Dhivya R, Muthulakshmi L, Sureshkumar S, Ashraf M, Pandi M, Mayandi J, Annaraj J, Sagadevan S. Environmentally benign and biocompatible CuO@Si core-shell nanoparticles: as electrochemical l-cysteine sensor, antibacterial and anti-lung cancer agents. Ceram Int. 2023;49:10023–31. https://doi.org/10.1016/j.ceramint.2022.11.182.

Guo X. Surface plasmon resonance based biosensor technique: a review. J Biophotonics. 2012;5:483–501. https://doi.org/10.1002/jbio.201200015.

Guo A, Wu D, Ma H, Zhang Y, Li H, Du B, Wei Q. An ultrasensitive enzyme-free electrochemical immunosensor for CA125 using Au@Pd core–shell nanoparticles as labels and platforms for signal amplification. J Mater Chem B. 2013;1:4052–8. https://doi.org/10.1039/C3TB20574F.

Haes AJ, Zou S, Schatz GC, Van Duyne RP. A nanoscale optical biosensor: the long range distance dependence of the localized surface Plasmon resonance of Noble metal nanoparticles. J Phys Chem B. 2004;108:109–16. https://doi.org/10.1021/jp0361327.

Hao M, Ma Z. An ultrasensitive chemiluminescence biosensor for carcinoembryonic antigen based on autocatalytic enlargement of Immunogold Nanoprobes. Sensors. 2012;12:17320–9. https://doi.org/10.3390/s121217320.

Hun X, Zhang Z. Functionalized fluorescent core-shell nanoparticles used as a fluorescent labels in fluoroimmunoassay for IL-6. Biosens Bioelectron. 2007;22:2743–8. https://doi.org/10.1016/j.bios.2007.01.022.

Jenjob R, Phakkeeree T, Crespy D. Core–shell particles for drug-delivery, bioimaging, sensing, and tissue engineering. Biomater Sci. 2020;8:2756–70. https://doi.org/10.1039/C9BM01872G.

Jia Y, Qin M, Zhang H, Niu W, Li X, Wang L, Li X, Bai Y, Cao Y, Feng X. Label-free biosensor: a novel phage-modified light addressable potentiometric sensor system for cancer cell monitoring. Biosens Bioelectron. 2007;22:3261–6. https://doi.org/10.1016/j.bios.2007.01.018.

Jiang T, Wang X, Zhou J, Chen D, Zhao Z. Hydrothermal synthesis of Ag@MSiO2@Ag three core–shell nanoparticles and their sensitive and stable SERS properties. Nanoscale. 2016;8:4908–14. https://doi.org/10.1039/C6NR00006A.

Kaushal S, Nanda SS, Samal S, Yi DK. Strategies for the development of metallic-nanoparticle-based label-free biosensors and their biomedical applications. Chembiochem. 2020;21:576–600. https://doi.org/10.1002/cbic.201900566.

Khanmohammadi A, Aghaie A, Vahedi E, Qazvini A, Ghanei M, Afkhami A, Hajian A, Bagheri H. Electrochemical biosensors for the detection of lung cancer biomarkers: a review. Talanta. 2020;206:120251. https://doi.org/10.1016/j.talanta.2019.120251.

Länge K, Gruhl FJ, Rapp M. Surface acoustic wave (SAW) biosensors: coupling of sensing layers and measurement. In: Jenkins G, Mansfield CD, editors. Microfluidic diagnostics: methods and protocols. Totowa, NJ: Humana Press; 2013. p. 491–505. https://doi.org/10.1007/978-1-62703-134-9_31.

Lee S, Chon H, Lee J, Ko J, Chung BH, Lim DW, Choo J. Rapid and sensitive phenotypic marker detection on breast cancer cells using surface-enhanced Raman scattering (SERS) imaging. Biosens Bioelectron. 2014;51:238–43. https://doi.org/10.1016/j.bios.2013.07.063.

Li L, Zhang W, Chen H, Zhao Z, Wang M, Chen J. Visual and electrochemical determination of breast cancer marker CA15-3 based on etching of Au@Ag core/shell nanoparticles. Int J Electrochem Sci. 2023;18:100123. https://doi.org/10.1016/j.ijoes.2023.100123.

Liang D, Zhang X, Wang Y, Huo T, Qian M, Xie Y, Li W, Yu Y, Shi W, Liu Q, Zhu J, Luo C, Cao Z, Huang R. Magnetic covalent organic framework nanospheres-based miRNA biosensor for sensitive glioma detection. Bioact Mater. 2022;14:145–51. https://doi.org/10.1016/j.bioactmat.2021.11.033.

Liu X, Atwater M, Wang J, Huo Q. Extinction coefficient of gold nanoparticles with different sizes and different capping ligands. Colloids Surf B Biointerfaces. 2007;58:3–7. https://doi.org/10.1016/j.colsurfb.2006.08.005.

Liu T, Wang Y, Zhong W, Li B, Mequanint K, Luo G, Xing M. Biomedical applications of layer-by-layer self-assembly for cell encapsulation: current status and future perspectives. Adv Healthc Mater. 2019;8:1800939. https://doi.org/10.1002/adhm.201800939.

Lu Y, Liu Q, Fu B, Li P, Xu W. Label-free MIP-SERS biosensor for sensitive detection of colorectal cancer biomarker. Talanta. 2023;258:124461. https://doi.org/10.1016/j.talanta.2023.124461.

Lütfi Yola M, Atar N, Özcan N. A novel electrochemical lung cancer biomarker cytokeratin 19 fragment antigen 21-1 immunosensor based on Si 3 N 4 /MoS 2 incorporated MWCNTs and core–shell type magnetic nanoparticles. Nanoscale. 2021;13:4660–9. https://doi.org/10.1039/D1NR00244A.

Ma L, Liu F, Lei Z, Wang Z. A novel upconversion@polydopamine core@shell nanoparticle based aptameric biosensor for biosensing and imaging of cytochrome *c* inside living cells. Biosens Bioelectron. 2017;87:638–45. https://doi.org/10.1016/j.bios.2016.09.017.

Magro, M, Bonaiuto, E, Baratella, D, Jakubec, P, Sconcerle, E, Terzo, M, Miotto, G, Macone, A, Agostinelli, E, Fasolato, S, Tucek, J, Zboril, R, Vianello, F. Iron oxide-chromate core-shell nanostructure for the electrochemical biosensing of polyamines in liver cancer tissues; 2015.

Maskow T, Wolf K, Kunze W, Enders S, Harms H. Rapid analysis of bacterial contamination of tap water using isothermal calorimetry. Thermochim Acta. 2012;543:273–80. https://doi.org/10.1016/j.tca.2012.06.002.

Mayer KM, Hafner JH. Localized surface plasmon resonance sensors. Chem Rev. 2011;111:3828–57. https://doi.org/10.1021/cr100313v.

Mehrotra P. Biosensors and their applications—a review. J Oral Biol Craniofac Res. 2016;6:153–9. https://doi.org/10.1016/j.jobcr.2015.12.002.

Mohebi E, AdibAmini S, Sari AH, Dorranian D. Formation of Agshell/Aucore bimetallic nanoparticles by pulsed laser ablation method: effect of colloidal/solution concentration. Plasmonics. 2024;19:75–95. https://doi.org/10.1007/s11468-023-01976-w.

Murray WA, Auguié B, Barnes WL. Sensitivity of localized surface plasmon resonances to bulk and local changes in the optical environment. J Phys Chem C. 2009;113:5120–5. https://doi.org/10.1021/jp810322q.

Nagesetti A, Rodzinski A, Stimphil E, Stewart T, Khanal C, Wang P, Guduru R, Liang P, Agoulnik I, Horstmyer J, Khizroev S. Multiferroic coreshell magnetoelectric nanoparticles as NMR sensitive nanoprobes for cancer cell detection. Sci Rep. 2017;7:1610. https://doi.org/10.1038/s41598-017-01647-x.

Nallathamby PD, Hopf J, Irimata LE, McGinnity TL, Roeder RK. Preparation of fluorescent Au–SiO2 core–shell nanoparticles and nanorods with tunable silica shell thickness and surface modification for immunotargeting. J Mater Chem B. 2016;4:5418–28. https://doi.org/10.1039/C6TB01659F.

Nguyen AH, Sim SJ. Nanoplasmonic biosensor: detection and amplification of dual bio-signatures of circulating tumor DNA. Biosens Bioelectron. 2015;2014(67):443–9. https://doi.org/10.1016/j.bios.2014.09.003.

Ouyang L, Hu Y, Zhu L, Cheng GJ, Irudayaraj J. A reusable laser wrapped graphene-Ag array based SERS sensor for trace detection of genomic DNA methylation. Biosens Bioelectron. 2017;92:755–62. https://doi.org/10.1016/j.bios.2016.09.072.

Pang Y, Wang C, Wang J, Sun Z, Xiao R, Wang S. Fe3O4@Ag magnetic nanoparticles for microRNA capture and duplex-specific nuclease signal amplification based SERS detection in cancer cells. Biosens Bioelectron. 2016;79:574–80. https://doi.org/10.1016/j.bios.2015.12.052.

Pang Y, Wang C, Lu L, Wang C, Sun Z, Xiao R. Dual-SERS biosensor for one-step detection of microRNAs in exosome and residual plasma of blood samples for diagnosing pancreatic cancer. Biosens Bioelectron. 2019;130:204–13. https://doi.org/10.1016/j.bios.2019.01.039.

Park S, Cho E, Chueng S-TD, Yoon J-S, Lee T, Lee J-H. Aptameric fluorescent biosensors for liver cancer diagnosis. Biosensors. 2023;13:617. https://doi.org/10.3390/bios13060617.

Peng L, Zhou J, Liang Z, Zhang Y, Petti L, Jiang T, Gu C, Yang D, Mormile P. SERS-based sandwich bioassay protocol of miRNA-21 using Au@Ag core–shell nanoparticles and a Ag/TiO 2 nanowires substrate. Anal Methods. 2019;11:2960–8. https://doi.org/10.1039/C9AY00573K.

Pothipor C, Wiriyakun N, Putnin T, Ngamaroonchote A, Jakmunee J, Ounnunkad K, Laocharoensuk R, Aroonyadet N. Highly sensitive biosensor based on graphene–poly (3-aminobenzoic acid) modified electrodes and porous-hollowed-silver-gold nanoparticle labelling for prostate cancer detection. Sensors Actuators B Chem. 2019;296:126657. https://doi.org/10.1016/j.snb.2019.126657.

Qu S, Liu J, Luo J, Huang Y, Shi W, Wang B, Cai X. A rapid and highly sensitive portable chemiluminescent immunosensor of carcinoembryonic antigen based on immunomagnetic separation in human serum. Anal Chim Acta. 2013;766:94–9. https://doi.org/10.1016/j.aca.2012.12.043.

Raghav R, Srivastava S. Core–shell gold–silver nanoparticles based impedimetric immunosensor for cancer antigen CA125. Sensors Actuators B Chem. 2015;220:557–64. https://doi.org/10.1016/j.snb.2015.05.108.

Rashidi Dafeh S, Iranmanesh P, Salarizadeh P. Fabrication, optimization, and characterization of ultra-small superparamagnetic Fe3O4 and biocompatible Fe3O4@ZnS core/shell magnetic nanoparticles: ready for biomedicine applications. Mater Sci Eng C. 2019;98:205–12. https://doi.org/10.1016/j.msec.2018.12.147.

Rasouli E, Basirun WJ, Johan MR, Rezayi M, Mahmoudian MR, Poenar DP. Electrochemical DNA-nano biosensor for the detection of cervical cancer-causing HPV-16 using ultrasmall Fe3O4-Au core-shell nanoparticles. Sens Bio Sens Res. 2023;40:100562. https://doi.org/10.1016/j.sbsr.2023.100562.

Ren W, Mohammed SI, Wereley S, Irudayaraj J. Magnetic focus lateral flow sensor for detection of cervical cancer biomarkers. Anal Chem. 2019;91:2876–84. https://doi.org/10.1021/acs.analchem.8b04848.

Roda A, Mirasoli M, Michelini E, Di Fusco M, Zangheri M, Cevenini L, Roda B, Simoni P. Progress in chemical luminescence-based biosensors: a critical review. Biosens Bioelectron. 2016;30th Anniversary(76):164–79. https://doi.org/10.1016/j.bios.2015.06.017.

Singh P, Katkar PK, Patil UM, Bohara RA. A robust electrochemical immunosensor based on core–shell nanostructured silica-coated silver for cancer (carcinoembryonic-antigen-CEA) diagnosis. RSC Adv. 2021;11:10130–43. https://doi.org/10.1039/D0RA09015H.

Singh B, Ma S, Hara TO, Singh S. Nanomaterials-based biosensors for the detection of prostate cancer biomarkers: recent trends and future perspective. Adv Mater Technol. 2023;8:2201860. https://doi.org/10.1002/admt.202201860.

Sun Z, Fu X, Zhang L, Yang X, Liu F, Hu G. A protein chip system for parallel analysis of multi-tumor markers and its application in cancer detection. Anticancer Res. 2004;24:1159–66.

Sun Z-F, Chang Y, Xia N. Recent development of nanomaterials-based cytosensors for the detection of circulating tumor cells. Biosensors. 2021;11:281. https://doi.org/10.3390/bios11080281.

Sunil N, Unnathpadi R, Pullithadathil B. Label-free SERS salivary biosensor based on Ni@Ag core–shell nanoparticles anchored on carbon nanofibers for prediagnosis of lung cancer. ACS Appl Nano Mater. 2023;6:11334–50. https://doi.org/10.1021/acsanm.3c01379.

Suter JD, Howard DJ, Shi H, Caldwell CW, Fan X. Label-free DNA methylation analysis using opto-fluidic ring resonators. Biosens Bioelectron. 2010;26:1016–20. https://doi.org/10.1016/j.bios.2010.08.050.

Tothill IE. Biosensors for cancer markers diagnosis. Semin Cell Dev Biol. 2009;20:55–62. https://doi.org/10.1016/j.semcdb.2009.01.015.

Tung TT, Losic D, Park SJ, Feller J-F, Kim T. Core-shell nanostructured hybrid composites for volatile organic compound detection. Int J Nanomedicine. 2015;10:203–14. https://doi.org/10.2147/IJN.S88305.

Wang J. Electrochemical biosensors: towards point-of-care cancer diagnostics. Biosens Bioelectron. 2006;21:1887–92. https://doi.org/10.1016/j.bios.2005.10.027.

Wang J, Kawde A-N. Pencil-based renewable biosensor for label-free electrochemical detection of DNA hybridization. Anal Chim Acta. 2001;431:219–24. https://doi.org/10.1016/S0003-2670(00)01318-0.

Wang S, Qin Y, Zou Z. Determination of liver cancer biomarkers by surface-enhanced Raman scattering using gold-silica nanoparticles. Anal Lett. 2016;49:1209–20. https://doi.org/10.1080/00032719.2015.1098656.

Wang R, Liu W-D, Wang A-J, Xue Y, Wu L, Feng J-J. A new label-free electrochemical immunosensor based on dendritic core-shell AuPd@Au nanocrystals for highly sensitive detection of prostate specific antigen. Biosens Bioelectron. 2018;99:458–63. https://doi.org/10.1016/j.bios.2017.08.010.

Wang Y, Huo T, Du Y, Qian M, Lin C, Nie H, Li W, Hao T, Zhang X, Lin N, Huang R. Sensitive CTC analysis and dual-mode MRI/FL diagnosis based on a magnetic core-shell aptasensor. Biosens Bioelectron. 2022;215:114530. https://doi.org/10.1016/j.bios.2022.114530.

Wu J, Li S, Ma Y, Zhi W, Chen T, Huang X, Huang C, Zhou X, Zhang P, Zhang Y, Zheng G, Wang Z, Zhong X, Cai H, Wang W, Sun P, Zhou H. 3D hierarchic interfacial assembly of Au nanocage@Au along with IS-AgMNPs for simultaneous, ultrasensitive, reliable, and quantitative SERS detection of colorectal cancer related miRNAs. Biosens Bioelectron. 2024;248:115993. https://doi.org/10.1016/j.bios.2023.115993.

Xu W, Zhao A, Zuo F, Khan R, Hussain HMJ, Chang J. Au@Ag core-shell nanoparticles for microRNA-21 determination based on duplex-specific nuclease signal amplification and surface-enhanced Raman scattering. Microchim Acta. 2020;187:384. https://doi.org/10.1007/s00604-020-04330-7.

Yang Q, Gong X, Song T, Yang J, Zhu S, Li Y, Cui Y, Li Y, Zhang B, Chang J. Quantum dot-based immunochromatography test strip for rapid, quantitative and sensitive detection of alpha fetoprotein. Biosens Bioelectron. 2011;30:145–50. https://doi.org/10.1016/j.bios.2011.09.002.

Yang Y-C, Dong S-W, Shen T, Jian C-X, Chang H-J, Li Y, He F-T, Zhou J-X. A label-free amperometric immunoassay for thrombomodulin using graphene/silver-silver oxide nanoparticles as a immobilization matrix. Anal Lett. 2012;45:724–34. https://doi.org/10.1080/00032719.2011.653896.

Yang Y, Yan Q, Liu Q, Li Y, Liu H, Wang P, Chen L, Zhang D, Li Y, Dong Y. An ultrasensitive sandwich-type electrochemical immunosensor based on the signal amplification strategy of echinoidea-shaped Au@Ag-Cu2O nanoparticles for prostate specific antigen detection. Biosens Bioelectron. 2018;99:450–7. https://doi.org/10.1016/j.bios.2017.08.018.

Yazdanparast S, Benvidi A, Azimzadeh M, Tezerjani MD, Ghaani MR. Experimental and theoretical study for miR-155 detection through resveratrol interaction with nucleic acids using magnetic core-shell nanoparticles. Microchim Acta. 2020;187:479. https://doi.org/10.1007/s00604-020-04447-9.

Zeng S, Baillargeat D, Ho H-P, Yong K-T. Nanomaterials enhanced surface plasmon resonance for biological and chemical sensing applications. Chem Soc Rev. 2014;43:3426–52. https://doi.org/10.1039/C3CS60479A.

Zhang Y, Yang P, Habeeb Muhammed MA, Alsaiari SK, Moosa B, Almalik A, Kumar A, Ringe E, Khashab NM. Tunable and linker free nanogaps in core–shell plasmonic nanorods for selective and quantitative detection of circulating tumor cells by SERS. ACS Appl Mater Interfaces. 2017;9:37597–605. https://doi.org/10.1021/acsami.7b10959.

Zhang W-S, Cao J-T, Dong Y-X, Wang H, Ma S-H, Liu Y-M. Enhanced chemiluminescence by Au-Ag core-shell nanoparticles: a general and practical biosensing platform for tumor marker detection. J Lumin. 2018;201:163–9. https://doi.org/10.1016/j.jlumin.2018.03.075.

Zhang X, Yu Y, Shen J, Qi W, Wang H. Design of organic/inorganic nanocomposites for ultrasensitive electrochemical detection of a cancer biomarker protein. Talanta. 2020;212:120794. https://doi.org/10.1016/j.talanta.2020.120794.

Zhang Y, Wang X, Fang X, Yuan X, Yang H, Kong J. Label-free electrochemical immunoassay for detecting CYFRA 21-1 using poly(ε-caprolactone)-*b*-poly(ethylene oxide) block copolymer. Microchem J. 2021;165:106119. https://doi.org/10.1016/j.microc.2021.106119.

Zhang J-X, Lv C-L, Tang C, Wang A-J, Mei L-P, Song P, Feng J-J. Sandwich-type ultrasensitive immunosensing of breast cancer biomarker based on core-shell au@PdAg dog-bone-like nanostructures and au@PtRh nanorods. Sensors Actuators B Chem. 2023;382:133497. https://doi.org/10.1016/j.snb.2023.133497.

Zhao X, Zhao H, Yuan H, Lan M. Multifunctional superparamagnetic Fe3O4@SiO2 Core/Shell nanoparticles: design and application for cell imaging. J Biomed Nanotechnol. 2014;10:262–70. https://doi.org/10.1166/jbn.2014.1641.

Zhong Z, Wu W, Wang D, Wang D, Shan J, Qing Y, Zhang Z. Nanogold-enwrapped graphene nanocomposites as trace labels for sensitivity enhancement of electrochemical immunosensors in clinical immunoassays: carcinoembryonic antigen as a model. Biosens Bioelectron. 2010;25:2379–83. https://doi.org/10.1016/j.bios.2010.03.009.

Zhu H, Dale PS, Caldwell CW, Fan X. Rapid and label-free detection of breast cancer biomarker CA15-3 in clinical human serum samples with Optofluidic ring resonator sensors. Anal Chem. 2009;81:9858–65. https://doi.org/10.1021/ac902437g.

Zhu Y, Chandra P, Shim Y-B. Ultrasensitive and selective electrochemical diagnosis of breast cancer based on a hydrazine–Au nanoparticle–aptamer bioconjugate. Anal Chem. 2013;85:1058–64. https://doi.org/10.1021/ac302923k.

Characterization and Evaluation Techniques for Core-Shell Nanoconstructs for Cancer Theragnostics

8

Ashish Kumar Parashar, Gaurav Kant Saraogi, and Vandana Arora Sethi

Abstract

Nanotechnology has emerged as a promising field in cancer theragnostics, offering novel approaches for early detection, targeted drug delivery, and personalized treatment. Core-shell nanoconstructs, in particular, have garnered significant attention due to their unique structural and functional properties that enable enhanced specificity, efficiency, and versatility in cancer management. The integration of advanced imaging techniques, such as quantum dots and gold nanoparticles, further enhances the diagnostic capabilities of core-shell nanoconstructs, enabling real-time monitoring and precise evaluation of their therapeutic potential. This chapter provides a comprehensive overview of the characterization and evaluation techniques employed to assess the performance of core-shell nanoconstructs for cancer theragnostics. Techniques such as electron microscopy, spectroscopy, dynamic light scattering, and zeta potential analysis are discussed for structural and morphological characterization. Additionally, advanced imaging methods like fluorescence imaging, MRI, and PET/CT are explored for their role in evaluating the in vivo diagnostic potential of these constructs. This chapter also discusses the various analytical methods used to assess the physicochemical properties, targeting capabilities, drug loading and release kinetics, and in vitro and in vivo efficacy of these nanoplatforms. The integration of these characterization techniques enables the development of highly efficient and safe nanoconstructs for cancer diagnosis and treatment, paving the way for personalized and precision oncology. Finally, this chapter explores the challenges and considerations in characterizing and evaluating core-shell nanoparticles.

A. K. Parashar (✉) · V. A. Sethi
Lloyd Institute of Management and Technology, Greater Noida, Uttar Pradesh, India

G. K. Saraogi
Sri Aurobindo Institute of Pharmacy, Indore, MP, India

J. K. Patel et al. (eds.), *Core-Shell Nano Constructs for Cancer Theragnostic*,
https://doi.org/10.1007/978-981-96-3025-7_8

Keywords

Core-shell nanoconstructs · Cancer theragnostics · Characterization · Evaluation · Targeting · Drug delive.ry · Imaging · Zeta potential

1 Introduction to Core-Shell Nanoconstructs

Core-shell nanoconstructs are a class of nanoparticles composed of a core material surrounded by a shell layer (Jang et al. 2004). These nanostructures are designed to enhance specific properties such as stability, biocompatibility, and targeted delivery for various biomedical applications, particularly in cancer theragnostics. The core of a core-shell nanoconstruct often serves as a reservoir for therapeutic agents, while the shell can be functionalized to improve targeting and imaging capabilities (Ho et al. 2015). This unique architecture allows for the integration of multiple functionalities into a single nanoparticle, making them highly versatile for cancer theragnostics (Chen et al. 2014). In cancer theragnostics, core-shell nanoconstructs play a crucial role in combining therapeutic and diagnostic capabilities (Tang et al. 2022).

The core can be loaded with therapeutic agents, such as drugs or genes, while the shell can be modified with targeting ligands to enhance specificity toward cancer cells (Li et al. 2014). Additionally, the shell can be designed to respond to specific stimuli, such as pH or temperature changes, enabling controlled release of the therapeutic payload at the tumor site. Moreover, the core-shell architecture can incorporate imaging agents, such as fluorescent dyes or magnetic nanoparticles, within the shell or core (Zhang 2021). This allows for real-time monitoring of the nanoparticle's biodistribution, cellular uptake, and therapeutic efficacy, which is essential for personalized cancer treatment (Shen et al. 2018).

The versatility of core-shell nanoconstructs lies in their ability to be synthesized using various materials, including polymers, lipids, silica, and metals (Dimov et al. 2021). The choice of materials depends on the specific application and the desired properties of the nanoparticle. For instance, polymeric core-shell nanoconstructs offer excellent biocompatibility and biodegradability, while metallic nanoparticles provide unique optical and magnetic properties for imaging and therapy (Mishra et al. 2021).

One of the key advantages of core-shell nanoconstructs is their ability to respond to various environmental stimuli, such as temperature, pH, and chemical cues. This responsiveness is often attributed to the shell layer, which can be engineered to undergo structural changes in response to these stimuli, effectively altering the properties and behavior of the entire nanostructure (Kularatne et al. 2022). The core-shell architecture also offers protection and stability to the core component, shielding it from potential degradation or unwanted interactions. Additionally, the high surface area-to-volume ratio inherent to nanoscale materials, combined with the customizable shell layer, enables core-shell nanoconstructs to be tailored for specific applications, such as drug delivery, catalysis, and environmental remediation (Fig. 8.1) (Chhour et al. 2014; Parashar 2021a).

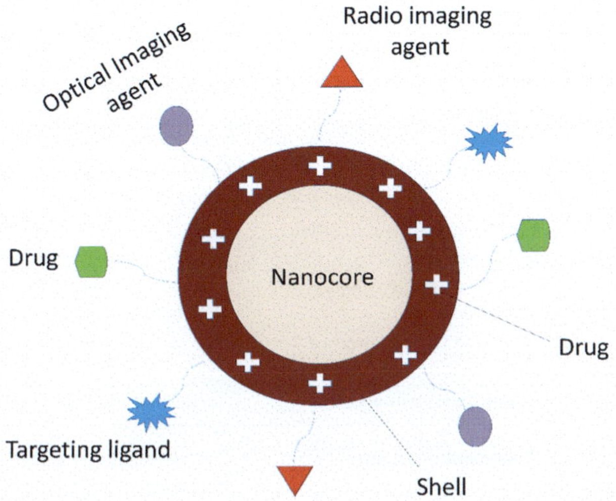

Fig. 8.1 Illustration of core-shell nanoconstructs with different functional moieties

This chapter will discuss the various characterization techniques used to understand the structural, chemical, and functional properties of core-shell nanoconstructs.

2 Characterization Techniques

Characterization of core-shell nanoparticles involves several advanced techniques that assess their physical, chemical, and biological properties. The following techniques are used for the characterization of theranostic core-shell nanoparticles (Table 8.1).

2.1 Structural Characterization

Characterizing the structure of core-shell nanoconstructs designed for cancer theragnostics is crucial to understanding their behavior and effectiveness (Pulumati et al. 2023).

Importance of Structural Characterization

- **Confirming Core-Shell Formation:** Techniques like transmission electron microscopy (TEM) help visualize the core and shell components, confirming the successful synthesis of the desired structure.
- **Size and Morphology:** The size and shape of the nanoconstruct influence their biodistribution, cellular uptake, and targeting ability. Dynamic light scattering (DLS) and TEM provide insights into these parameters.

Table 8.1 Techniques used for the characterization of nanotheranostic core-shell nanoparticles

Category	Techniques/parameters	Purpose
Structural characterization	Transmission electron microscopy	Size and shape
	Scanning electron microscopy	Size, shape, and morphology
	X-ray diffraction	Crystal structure, size, and composition
Chemical characterization	Fourier-transform infrared spectroscopy	Formation, functionalization, and potential interactions
	Energy-dispersive X-ray spectroscopy	Elemental composition and distribution
Physical characterization	Dynamic light scattering	Size, size distribution, and stability
	Zeta potential analysis	Surface charge, charge distribution, and stability
In vitro evaluation	Therapeutic efficacy	Cytotoxicity, cell viability, and drug targeting
	Imaging capabilities	Cellular uptake and imaging
In vivo evaluation	Biodistribution and pharmacokinetics	Track the fate of the nanoconstructs
	Therapeutic efficacy	Achieving optimal therapeutic outcomes

- **Surface Properties:** The surface charge and functional groups on the shell impact how the nanoconstruct interacts with the biological environment, including blood proteins and target cells. Techniques like Zeta Potential measurements and X-ray Photoelectron Spectroscopy are useful here.
- **Composition and Purity:** Determining the elemental composition and ensuring the absence of impurities is vital for safety and efficacy. Energy-dispersive X-ray spectroscopy (EDX) coupled with TEM or X-ray photoelectron spectroscopy (XPS) can provide this information.
- **Stability and Degradation:** Understanding how the nanoconstruct behaves in biological conditions, including its stability and degradation profile, is essential for predicting its performance (Cressoni et al. 2023).

2.1.1 Transmission Electron Microscopy (TEM)

Transmission electron microscopy stands as a cornerstone technique for characterizing core-shell nanoconstructs in cancer theragnostics. Its ability to visualize the internal structure of materials at the nanoscale makes it indispensable for understanding the morphology and composition of these intricate systems (Nandwana et al. 2015).

TEM works by passing a beam of electrons through a thin sample. The interaction of electrons with the sample generates a transmitted electron signal, which is then magnified and focused to create an image. The varying electron densities within the core-shell nanoconstruct create contrast in the image, allowing researchers to distinguish between the core and shell materials (Schwartz et al. 2023; Miao et al. 2024). This enables the direct visualization of the core-shell architecture, confirming the successful synthesis of the desired structure (Fig. 8.2).

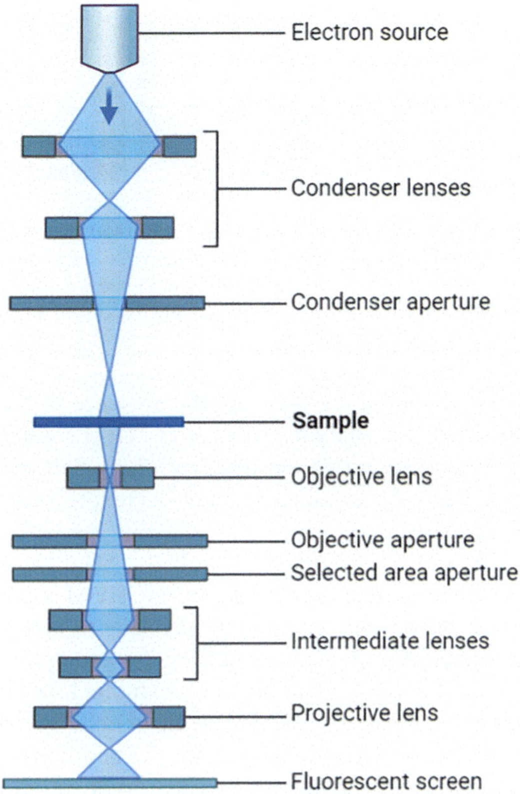

Fig. 8.2 Working principal of transmission electron microscopy

Beyond simply confirming the core-shell structure, TEM provides valuable quantitative data. By analyzing TEM images, researchers can accurately determine the size and shape of both the core and the shell (Brodusch et al. 2021). This information is crucial for understanding the nanoconstruct's behavior in biological systems. For instance, size influences how easily the nanoconstruct can navigate biological barriers and penetrate tumor tissues. Additionally, the shape of the nanoconstruct can impact its cellular uptake and interaction with target cells (Patamia et al. 2023; Tang et al. 2014).

Furthermore, advances in TEM, such as energy-dispersive X-ray spectroscopy, allow for elemental analysis to be performed simultaneously with imaging. This enables researchers to map the distribution of different elements within the core-shell nanoconstruct, providing insights into the composition and purity of each component (Wang 2012). This is particularly valuable for theragnostic applications, where the core and shell may be composed of different materials with distinct functionalities, such as a therapeutic agent in the core and an imaging agent in the shell (Wang et al. 2013; Parashar 2021b).

TEM provides critical insights into the structural features that govern the targeting efficiency of core-shell nanoconstructs. TEM imaging directly visualizes the core-shell morphology, confirming the successful encapsulation of the therapeutic

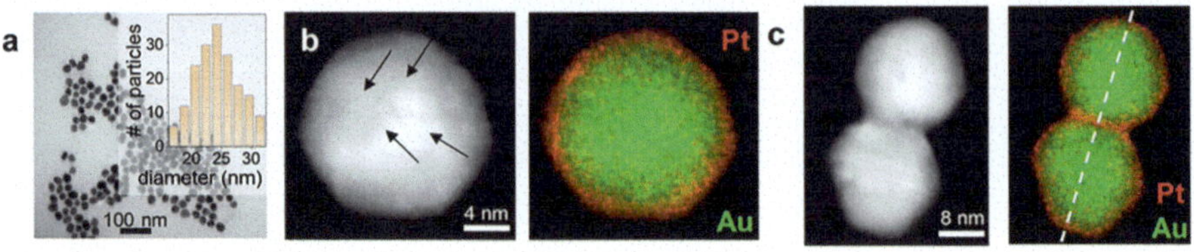

Fig. 8.3 Characterization of Pt/Au core-shell nanoparticles: TEM images of Pt/Au core-shell nanoparticles after synthesis (**a**) and the particle size distribution (inset): HAADF micrographs (**b**, **c**) with the corresponding EDX maps of Pt and Au. The arrows indicate the characteristic twinning observed also for pristine gold nanoparticles. (From: https://doi.org/10.1038/s42004-022-00680-w (Ledendecker et al. 2022))

payload within the core and the presence of the targeting ligand on the shell. For instance, in a study by Wang et al., TEM confirmed the formation of gold nanostar cores coated with a mesoporous silica shell functionalized with folic acid for targeted delivery to folate receptor-overexpressing tumor cells (Huang et al. 2017; Hu et al. 2016).

The size and shape of core-shell nanoconstructs significantly influence their biodistribution and tumor penetration. TEM analysis provides quantitative data on these parameters, enabling researchers to optimize the design for enhanced tumor accumulation (Jiang et al. 2022). For example, Ni et al. used TEM to demonstrate that smaller-sized nanoparticles exhibited superior tumor penetration compared to larger counterparts. TEM imaging of cells treated with targeted nanoconstructs can reveal their cellular uptake pathways and intracellular fate. This information is crucial for understanding the mechanism of action and optimizing therapeutic efficacy (Ni et al. 2022). A study by Zhang et al. employed TEM to demonstrate the receptor-mediated endocytosis of Her2-targeted core-shell nanoparticles in breast cancer cells (Zhang et al. 2016). TEM imaging of tumor tissue sections can provide direct evidence of the nanoconstructs' ability to penetrate deep into the tumor mass and accumulate at the target site. This is particularly valuable for evaluating the efficacy of nanoconstructs designed to overcome biological barriers within the tumor microenvironment (Fig. 8.3).

Advancements in Transmission Electron Microscopy for Analyzing Core-Shell Nanoconstructs in Cancer Theragnostics

Transmission electron microscopy has long been a cornerstone technique for characterizing core-shell nanoconstructs. However, recent advancements are pushing the boundaries of TEM analysis, providing even deeper insights into the complexities of these nanomaterials for cancer theragnostics (Sun et al. 2022).

A. **Cryogenic Transmission Electron Microscopy:**

Traditional TEM requires sample preparation techniques that can sometimes alter the native structure of delicate core-shell nanoconstructs (Sun et al. 2014). Cryo-TEM overcomes this limitation by imaging samples at cryogenic temperatures, preserving their structure and allowing for the visualization of:

- **Soft Matter Shells:** Visualizing lipid-based or polymer-based shells in their native hydrated state, which is crucial for understanding their behavior in biological environments.
- **Dynamic Processes:** Observing dynamic processes like drug encapsulation, release, and degradation in real-time.

B. **In Situ Liquid-Phase TEM:**

This technique allows researchers to study the behavior of core-shell nanoconstructs in a liquid environment, mimicking physiological conditions (Du et al. 2017). This enables the visualization of:

- **Dynamic Interactions:** Observing how nanoconstructs interact with biological molecules, cells, and tissues in real time.
- **Stimuli-Responsive Behavior:** Studying how nanoconstructs respond to environmental stimuli, such as pH changes or enzyme activity, which is crucial for designing targeted drug delivery systems (Yameen et al. 2016).

C. **Correlative Light and Electron Microscopy:**

CLEM combines the high-resolution imaging capabilities of TEM with the molecular specificity of fluorescence microscopy (Keevend et al. 2020). This powerful combination allows researchers to:

- **Track Nanoconstructs:** Following the journey of fluorescently labeled nanoconstructs within cells and tissues, providing insights into their biodistribution and targeting efficiency.
- **Correlate Structure and Function:** Correlating the structural information obtained from TEM with the functional information obtained from fluorescence imaging leads to a more comprehensive understanding of the nanoconstruct's behavior (Yan et al. 2022).

D. **Advanced Image Analysis and Tomography:**

Developments in image analysis software and tomographic reconstruction techniques are enabling researchers to extract more quantitative data from TEM images, including:

- **3D Visualization:** Reconstructing three-dimensional (3D) models of core-shell nanoconstructs, providing a more complete understanding of their morphology and internal structure (Arteaga Cardona et al. 2023).
- **Quantitative Analysis:** Measuring parameters such as core size, shell thickness, and drug loading with greater accuracy and precision.

These advancements in TEM analysis are providing unprecedented insights into the structure, function, and behavior of core-shell nanoconstructs for cancer theragnostics. As these techniques continue to evolve, they will undoubtedly play an increasingly important role in the development of more effective and personalized cancer therapies (Wang et al. 2018a).

2.1.2 Scanning Electron Microscopy (SEM)

While transmission electron microscopy excels at visualizing the internal structure of core-shell nanoconstructs, scanning electron microscopy offers complementary information about their surface morphology, composition, and interactions with biological systems (Lamvik 2000). Scanning electron microscopy is a powerful imaging technique that utilizes a focused beam of electrons to scan the surface of a sample, generating high-resolution images with detailed topographical information. At the heart of the SEM is an electron gun, typically a tungsten filament or a field emission source, which generates a beam of high-energy electrons. The electrons are accelerated toward the sample by a high voltage (typically ranging from 1 to 30 kV), gaining significant kinetic energy (Santagata et al. 2009; Conti et al. 2022).

A series of electromagnetic lenses (condenser and objective lenses) focus the electron beam into a fine spot (a few nanometers in diameter) and direct it onto the sample's surface. Scanning coils deflect the electron beam in a raster pattern, systematically scanning it across the sample's surface. As the focused electron beam interacts with the sample's surface, it generates various signals, including low-energy electrons emitted from the sample's surface, providing topographical information; high-energy electrons reflected back from the sample, providing information about the sample's composition and density (Reeves 1994); and characteristic X-rays emitted from the sample, providing information about its elemental composition (Fig. 8.4).

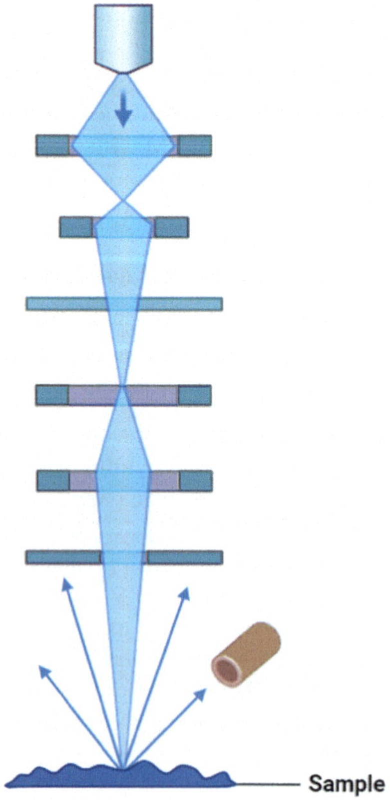

Fig. 8.4 Working principal of scanning electron microscopy

Specialized detectors capture the emitted signals. The detected signals are amplified and processed to generate an image. The processed signals are displayed on a monitor, creating a grayscale image where brighter areas correspond to stronger signal intensities. SEM can achieve magnifications up to several hundred thousand times with a resolution down to a few nanometers, allowing for the visualization of fine surface details (Sato et al. 2019). SEM provides a large depth of field, resulting in images with a three-dimensional appearance, which is particularly useful for examining samples with complex topographies. When equipped with an X-ray photoelectron spectroscopy (EDS) detector, SEM enables elemental analysis, providing valuable information about the sample's composition (Jones 2009).

SEM contributes to the characterization of the following nanomaterials for cancer theragnostics:

A. **Surface Morphology and Size Analysis:**
   - **High-Resolution Imaging:** SEM provides detailed images of the nanoconstruct's surface, revealing features like surface roughness, porosity, and the presence of any surface modifications. This is crucial for understanding how the nanoconstruct will interact with its biological environment.
   - **Size and Shape Determination:** SEM allows for the accurate measurement of the nanoconstruct's size and shape distribution, which are critical parameters influencing their biodistribution, cellular uptake, and therapeutic efficacy (Pragatisheel 2020).

B. **Elemental Composition Analysis:**
   - **Energy-Dispersive X-ray Spectroscopy:** When coupled with SEM, EDS can determine the elemental composition of the core-shell nanoconstruct. This is essential for confirming the presence of the desired elements in the core and shell materials and for detecting any impurities (Newbury 2012).

C. **Assessing Surface Modifications and Functionalization:**
   - **Visualizing Ligand Conjugation:** SEM can be used to confirm the successful conjugation of targeting ligands, antibodies, or other biomolecules to the surface of the nanoconstruct. This is crucial for verifying the functionality of targeted drug delivery systems (Lin et al. 2014).

D. **Investigating Nanoconstruct–Cell Interactions:**
   - **Visualizing Cellular Uptake:** SEM can image the interaction of core-shell nanoconstructs with cells, providing insights into the mechanisms of cellular uptake, such as endocytosis or membrane fusion.
   - **Observing Cellular Responses:** SEM can also be used to observe the effects of nanoconstructs on cell morphology, providing information about their biocompatibility and potential therapeutic effects (Francia et al. 2022).

E. **Correlative Microscopy:**
   - **Combining SEM with Other Techniques:** SEM data can be combined with information obtained from other techniques, such as TEM or fluorescence

microscopy, to provide a more comprehensive understanding of the nanoconstruct's properties and behavior (Lange et al. 2021).

Lee et al. utilized SEM to confirm the successful coating of liposomes with hyaluronic acid, a targeting ligand for CD44 receptors overexpressed on breast cancer cells (Lee et al. 2018). SEM images revealed a uniform Hyaluronic acid (HA) layer on the liposome surface, indicating successful core-shell formation. The study demonstrated enhanced drug delivery to breast cancer cells in vitro, highlighting the potential of HA-coated liposomes for targeted therapy.

Kim et al. employed SEM to characterize the morphology and structure of iron oxide core-gold shell nanoparticles (Jang et al. 2014). SEM images confirmed the successful formation of a core-shell structure, with a distinct iron oxide core surrounded by a continuous gold shell. This unique structure enabled both magnetic hyperthermia (induced by the iron oxide core) and photothermal therapy (mediated by the gold shell), offering a synergistic therapeutic approach.

Zhang et al. used SEM to assess the surface morphology and polyethylene glycol coating efficiency of mesoporous silica nanoparticles (Zhang et al. 2021). SEM images revealed a smooth and uniform Polyethylene glycol (PEG) layer on the nanoparticle surface, indicating successful surface modification. The study demonstrated that PEGylation enhanced the nanoparticles' biocompatibility and circulation time, improving their suitability for drug delivery applications.

Chen et al. utilized SEM to visualize the interaction of gold nanostars with glioblastoma cells (Geng et al. 2014). SEM images revealed the attachment and internalization of nanostars by the cells. The study demonstrated that upon laser irradiation, the nanostars generated localized heat, leading to effective ablation of glioblastoma cells (Fig. 8.5).

Advancements in scanning electron microscopy are transforming the analysis of core-shell nanoconstructs for cancer theragnostics, providing unprecedented insights into their structural and functional properties (Fan et al. 2015). Environmental SEM allows for imaging of delicate biological samples in their natural state, enabling visualization of nanoconstruct–cell interactions and dynamic processes. Correlative microscopy, integrating SEM with techniques like TEM and confocal microscopy, provides a multidimensional understanding by correlating surface features with internal structure and functional data (Hull et al. 2014). Cryo-SEM preserves the native state of samples, minimizing artifacts and enabling the study of dynamic processes like drug release with high temporal resolution. Furthermore, advanced image analysis software and machine learning algorithms are automating data extraction and analysis, enabling quantification of nanoconstruct characteristics and accelerating research progress. These advancements in SEM technology are driving the development of more effective and targeted cancer therapies using core-shell nanoconstructs (Parashar and Sethi 2024; Bardhan 2022).

2.1.3 X-Ray Diffraction (XRD)

X-ray diffraction plays a crucial role in characterizing core-shell nanoconstructs designed for cancer theragnostics, providing valuable information about their crystal structure, size, and composition. X-ray diffraction relies on the interaction of

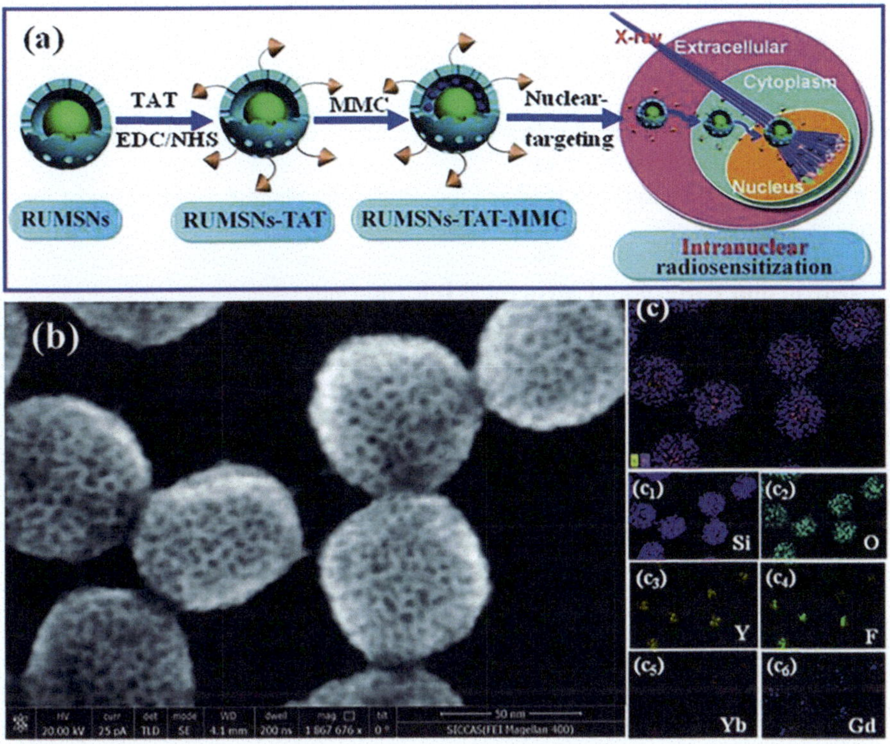

Fig. 8.5 (**a**) Schematic illustration of intranuclear radiosensitization. (**b**) Scanning electron microscopy (SEM) image of rattle-structured upconversion core/mesoporous silica nanotheranostics (RUMSNs). (**c**) Scanning transmission electron microscopy (STEM) image and the corresponding element mappings of the RUMSNs: (c1) Si; (c2) O; (c3) Y; (c4) F; (c5) Yb; (c6) Gd. (From: https://doi.org/10.1039/c4sc03080j (Ettlinger et al. 2020))

X-rays with the ordered arrangement of atoms within a crystalline material. This interaction produces a diffraction pattern, which acts as a fingerprint, revealing crucial information about the material's structure (Chung 2000; Clearfield and Bhuvanesh 2005). The working principle of X-ray diffraction involves the following:

A. **X-ray Interaction with Crystalline Lattice:**

When a beam of X-rays strikes a crystalline material, it interacts with the electrons of the atoms within the crystal lattice. This interaction causes the X-rays to scatter in various directions.

B. **Constructive Interference and Bragg's Law:**

Within the crystal, atoms are arranged in a periodic and repeating three-dimensional pattern. When the scattered X-rays from different parallel planes of atoms interfere constructively, they produce a diffracted beam (Birkholz 2005). This constructive interference occurs only when the path difference between the scattered waves is an integer multiple of the X-ray wavelength, a condition described by Bragg's Law:

$$n\lambda = 2\mathrm{d}\sin\theta$$

where n is an integer (order of diffraction)

λ is the wavelength of the X-rays

d is the interplanar spacing between the diffracting planesθ is the angle of incidence and diffraction

C. **Diffraction Pattern and Structural Information:**

The diffracted beams are detected and recorded as a diffraction pattern, typically plotted as intensity versus the diffraction angle (2θ). Each crystalline material produces a unique diffraction pattern based on its specific atomic arrangement and interplanar spacings (Ranu et al. 2024). By analyzing the position, intensity, and width of the diffraction peaks, we can determine the following:

- **Crystal Structure:** The arrangement of atoms within the unit cell.
- **Lattice Parameters:** The dimensions and angles of the unit cell.
- **Crystallite Size:** The average size of the crystalline domains.
- **Strain:** The presence of defects or stress within the crystal lattice.
- **Phase Identification:** Comparing the pattern with standard databases allows for identifying the crystalline phases present in the sample.

In essence, X-ray diffraction acts as a powerful tool to probe the internal structure of materials by analyzing the scattering pattern of X-rays caused by their interaction with the ordered arrangement of atoms within a crystal (Fig. 8.6) (Kvick 2017).

XRD patterns reveal the characteristic diffraction peaks associated with specific crystalline phases present in the core and shell materials. By analyzing the position and intensity of these peaks, we can:

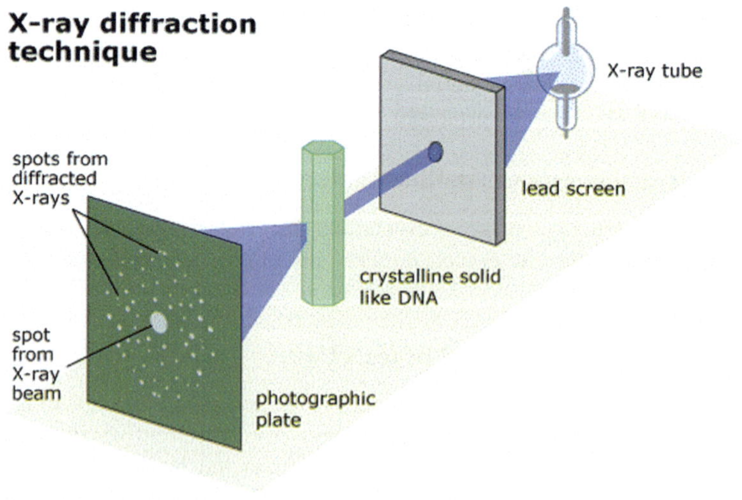

Fig. 8.6 Working principal of X-ray diffraction

- Identify the crystalline structure of the core and shell materials.
- Confirm the formation of a core-shell structure, as opposed to a simple mixture of the individual components.
- Detect the presence of any unintended crystalline phases that may arise during synthesis (Li et al. 2004; Wang et al. 2020).

The width of the diffraction peaks in an XRD pattern is inversely proportional to the crystallite size. By applying Scherrer's equation to the peak broadening, researchers can estimate the average crystallite size of the core and shell materials. Peak broadening can also arise from strain within the crystal lattice, which can be caused by defects or interfacial stress between the core and shell. The position of the diffraction peaks is sensitive to the composition of the material (Jo et al. 2022). By comparing the obtained XRD pattern with standard reference patterns, we can:

- Determine the elemental composition of the core and shell materials.
- Assess the phase purity of the nanoconstructs, identifying any impurities or secondary phases present.

XRD can be used to monitor structural changes in core-shell nanoconstructs upon surface functionalization or exposure to different environments. For example, we can track changes in crystal structure, size, or strain after conjugating targeting ligands or therapeutic agents to the surface and exposing the nanoconstructs to physiological conditions, such as pH or temperature changes (Macchione and Strumia 2023; Diez-Pascual and Rahdar 2022).

XRD is primarily sensitive to crystalline materials and may not provide detailed information about amorphous phases. It provides an average measurement over a large number of nanoconstructs and may not capture heterogeneity within the sample. Despite these limitations, XRD remains a powerful technique for characterizing core-shell nanoconstructs in cancer theragnostics, providing essential information about their structure, size, and composition, which are crucial for understanding their behavior and optimizing their performance (Venderley et al. 2022).

Rostek et al. investigated the effectiveness of X-ray powder diffraction in characterizing PVP-stabilized palladium, gold, and bimetallic Pd-Au core-shell nanoparticles (He et al. 2020). The researchers successfully synthesized nanoparticles with average core diameters of 5–6 nm and overall diameters of 7–8 nm for the core-shell structures, indicating a shell thickness of 1–2 nm. Ansari et al. carried out powder X-ray diffraction (XRD) pattern using PAN alytical X'Pert X-ray diffractometer equipped with Ni filter and CuKa (k = 1.5405 A̓) radiations (Li et al. 2018). The XRD patterns in Fig. 8.7 depict all the possible characteristic peaks of core, core shell, and core-shell/SiO_2. These sharp XRD patterns indicated that the core, core shell, and core-shell/SiO2 NCs have hexagonal phase structure and high crystallinity. Moreover, no additional impurity phases were observed in the XRD pattern, confirming the phase purity and homogeneously distribution of Ce3+ and Tb3+ ions into the NaGdF4 host. Furthermore, the intensity of XRD patterns of the core-shell NCs is slightly higher than that of core NCs. This may be due to increased grain size and crystallinity after NaGdF4 shell formation around the core.

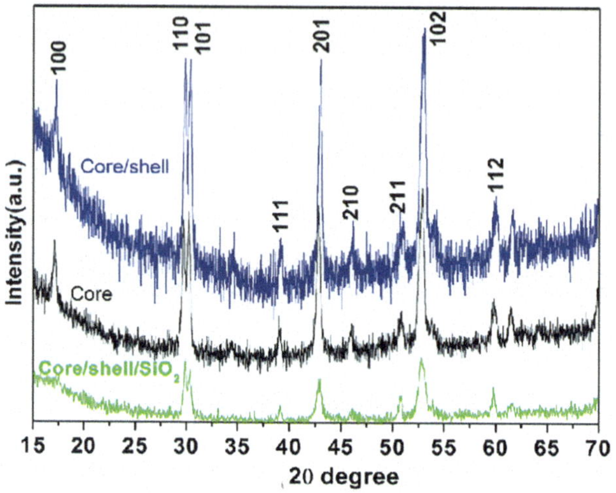

Fig. 8.7 X-ray diffraction pattern of core, core shell, and core-shell/SiO 2 NCs. (From: https://doi.org/10.1016/j.arabjc.2017.10.008 (Li et al. 2018))

2.2 Chemical Characterization

Chemical characterization of core-shell nanoconstructs in cancer theragnostics is essential for understanding their structural and functional properties, which directly influence their efficacy in targeted drug delivery and imaging (Ansari et al. 2020). Core-shell nanoconstructs typically consist of a hard nanoparticle core surrounded by a soft shell of biomolecular ligands or polymers, which can be tailored to enhance interaction with cancer cells. Functionalization with targeting ligands or therapeutic agents can be analyzed through techniques like Fourier-transform infrared spectroscopy (FTIR) and energy-dispersive X-ray spectroscopy (EDX) confirming the successful attachment of biomolecules (Yong et al. 2019).

2.2.1 Fourier-Transform Infrared Spectroscopy (FTIR)

Fourier-transform infrared spectroscopy provides valuable insights into the composition and chemical interactions within core-shell nanoconstructs designed for cancer theragnostics. By analyzing the characteristic vibrational frequencies of molecular bonds, FTIR spectroscopy helps researchers understand the formation, functionalization, and potential interactions of these nanostructures with biological systems (Fernández-González et al. 2023).

A. **Identifying Core and Shell Materials:**

FTIR spectra exhibit distinct absorption bands corresponding to specific functional groups present in the core and shell materials. For instance, the presence of a gold core can be confirmed by the characteristic Au-S stretching vibration when thiol-containing ligands are used for stabilization. Similarly, the shell composition, such as polymers like PEG or silica, can be identified through their unique vibrational fingerprints (Cimatu and Baldelli 2008).

B. **Verifying Surface Functionalization:**

FTIR plays a crucial role in verifying the successful conjugation of targeting ligands or therapeutic agents to the surface of core-shell nanoconstructs. The appearance of new absorption bands corresponding to the functional groups of the conjugated molecules confirms their attachment and provides information about the chemical bonds formed during functionalization (Shen et al. 2024).

C. **Studying Drug Loading and Release:**

For drug delivery applications, FTIR can be employed to study drug encapsulation within the core or shell of the nanoconstruct. The appearance of characteristic drug peaks after loading and their subsequent disappearance or reduction in intensity upon release provide valuable information about loading efficiency and release kinetics.

D. **Assessing Biomolecular Interactions:**

FTIR spectroscopy can also provide insights into the interactions between core-shell nanoconstructs and biomolecules, such as proteins or cell membranes. Changes in peak position, intensity, or shape upon interaction can reveal information about binding affinities, conformational changes, and potential mechanisms of cellular uptake (Wang et al. 2018b).

FTIR spectroscopy serves as a versatile tool for characterizing core-shell nanoconstructs in cancer theragnostics. By providing information about composition, functionalization, drug loading, and biomolecular interactions, FTIR contributes significantly to the development and optimization of these nanomaterials for targeted imaging and therapy.

Working Principle of Fourier-Transform Infrared Spectroscopy

FTIR spectroscopy exploits the interaction of infrared radiation with matter to identify and analyze the chemical composition of a sample. It relies on the principle that molecules absorb specific frequencies of IR light that correspond to the vibrations of their chemical bonds (Movasaghi et al. 2008).

- **Infrared Radiation and Molecular Vibrations:**

  Every molecule vibrates at specific frequencies, much like a musical instrument has specific resonant frequencies. When a beam of IR radiation passes through a sample, the molecules absorb energy from the radiation at frequencies that match their vibrational modes. These vibrational modes include stretching, bending, and twisting of the chemical bonds within the molecule.

- **Absorption Spectrum and Molecular Fingerprint:**

  The absorbed frequencies are recorded as an absorption spectrum, which plots the amount of IR radiation absorbed as a function of frequency or wavenumber. Each type of bond absorbs IR radiation at a unique set of frequencies, creating a distinct pattern in the spectrum. This pattern acts as a "molecular fingerprint" that can be used to identify the functional groups and, ultimately, the molecules present in the sample (Fig. 8.8).

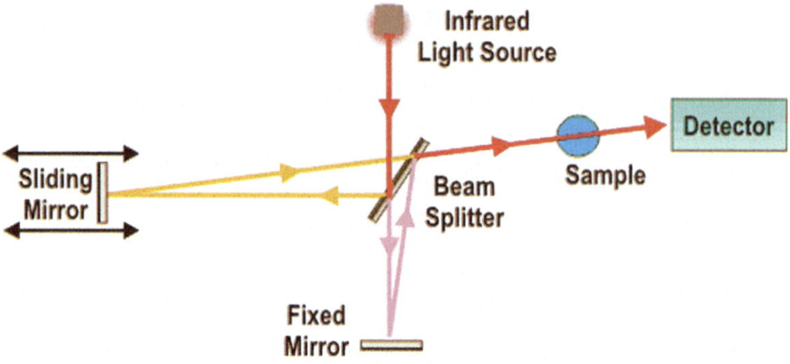

Fig. 8.8 Working principle of Fourier-transform infrared spectroscopy

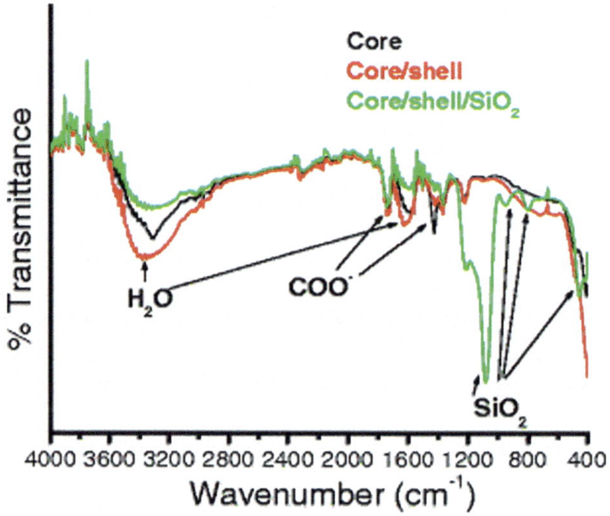

Fig. 8.9 FTIR spectra of core, core shell, and core-shell/SiO2 NPs. (From https://doi.org/10.1007/s10895-016-1824-1 (Ansari et al. 2016))

- **Fourier Transform and Interferometer.**

  Instead of using a prism or grating to separate individual frequencies of IR light like traditional dispersive IR instruments, FTIR spectroscopy employs an interferometer. The interferometer produces an interferogram, which encodes all the absorbed frequencies simultaneously. A mathematical technique called Fourier transform is then used to decode the interferogram and generate the familiar IR absorption spectrum (Fig. 8.9).

  Fourier-transform infrared spectroscopy (FTIR) offers significant advantages for the characterization of core-shell nanoconstructs in cancer theragnostics, primarily by providing detailed insights into their chemical composition and functionalization. This technique is particularly useful for identifying the presence of specific functional groups on the surface of nanoparticles, which are critical for their interaction with biological systems. For instance, FTIR can detect the characteristic absorption bands associated with various chemical bonds, such as Si-O, C=O, and N-H, allowing researchers to confirm the successful attachment of therapeutic agents or targeting ligands to the shell of the nanoconstructs. This information is vital for optimizing the design of nanoparticles for targeted drug delivery and imaging applications (Diebolder et al. 2018).

Moreover, FTIR can reveal changes in the chemical environment of the core-shell structures during synthesis and functionalization. By comparing the FTIR spectra of the raw materials and the synthesized nanoparticles, researchers can assess the effectiveness of the coating process and the stability of the resulting constructs (Biswal et al. 2012). For example, studies have shown that the intensity of specific peaks diminishes or shifts after the formation of the core-shell structure, indicating successful encapsulation or surface modification. This capability to monitor chemical interactions and modifications in real-time enhances the understanding of how these nanoconstructs behave in biological settings, which is crucial for their application in cancer theragnostics (Nica et al. 2023).

Additionally, FTIR can be employed to study the interactions between core-shell nanoconstructs and biomolecules, such as proteins or nucleic acids, which are essential for evaluating their biocompatibility and efficacy in therapeutic applications. By analyzing the spectral changes upon exposure to biological environments, researchers can gain insights into the binding mechanisms and stability of the nanoparticles in physiological conditions. FTIR spectroscopy serves as a powerful tool in the characterization of core-shell nanoconstructs, facilitating the development of effective and safe theragnostic agents for cancer treatment (Dung et al. 2021).

2.2.2 Energy-Dispersive X-Ray Spectroscopy (EDX)

Energy-dispersive X-ray spectroscopy (EDX) is a vital analytical technique for the characterization of core-shell nanoconstructs in cancer theragnostics, providing detailed insights into their elemental composition and distribution. This technique operates on the principle that each element emits characteristic X-rays when excited by a beam of electrons or X-rays. By analyzing the emitted X-rays, EDX can identify the specific elements present in the core and shell materials of the nanoconstructs, which is essential for understanding their functional properties and interactions in biological systems (Pattammattel et al. 2020).

In the context of core-shell nanoconstructs, EDX is particularly beneficial for determining the thickness and uniformity of the shell layer surrounding the core. For instance, quantitative EDX line profiles can be employed to analyze the elemental distribution across the nanostructure, allowing researchers to assess how effectively the shell encapsulates the core material. This information is crucial for optimizing the design of nanoparticles for targeted drug delivery, as variations in shell thickness can significantly influence drug release rates and biodistribution. Studies have demonstrated that EDX can provide precise atomic-scale quantitative analysis, enabling the evaluation of shell thickness down to atomic layers, which is particularly important for applications requiring high precision, such as in cancer theragnostics (Edmonds et al. 2016; Feng et al. 2017).

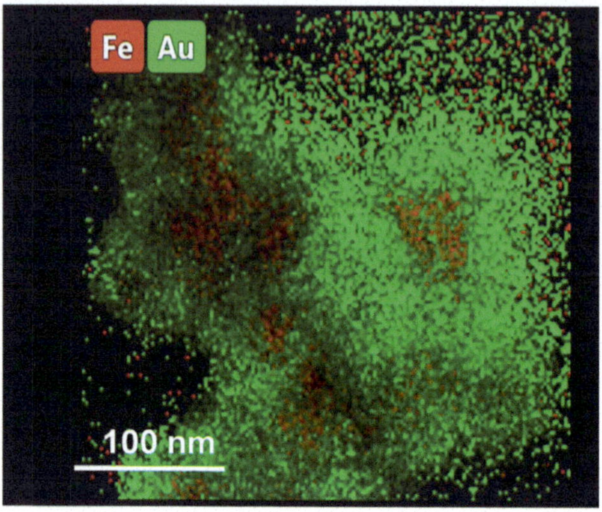

Fig. 8.10 Energy-dispersive X-ray spectroscopy in characterization of core-shell nanoconstructs in cancer theragnostics. (https://doi.org/10.3390/pharmaceutics13030416 [93])

EDX can be combined with other imaging techniques, such as scanning transmission electron microscopy (STEM), to enhance the characterization process. By correlating the elemental composition obtained from EDX with high-resolution structural images from STEM, researchers can gain a comprehensive understanding of the nanoconstructs' morphology and composition. This multimodal approach allows for a more thorough investigation of how the core and shell materials interact and function together, ultimately contributing to the development of more effective theragnostic agents for cancer treatment (Li et al. 2022). Overall, EDX is an indispensable tool in the characterization of core-shell nanoconstructs, facilitating advancements in their application in cancer theragnostics (Fig. 8.10).

A. **Working Principle of Energy-Dispersive X-ray Spectroscopy**

Energy-dispersive X-ray spectroscopy is an analytical technique used to determine the elemental composition of a material. It works by analyzing the X-ray emissions produced when a high-energy beam of electrons interacts with the sample. A focused beam of high-energy electrons, typically generated by an electron microscope, is directed at the sample. When these electrons strike the atoms within the sample, they can knock out an electron from the inner shells of the atom, leaving a vacancy. An electron from a higher energy level then fills this vacancy, releasing energy in the form of an X-ray photon. The energy of this X-ray photon is characteristic of the difference in energy levels between the two electron shells involved and is unique to the element from which it originated (Carlson and Krause 1965).

An EDS detector measures the energy of these emitted X-rays. By analyzing the energy and intensity of the X-ray peaks in the spectrum, the elemental composition of the sample can be determined. Each peak in the spectrum corresponds to a specific element present in the sample, and the intensity of the peak is proportional to the concentration of that element (Cesareo 2010).

B. **Key Features of EDS:**

- EDS can detect a wide range of elements, typically from beryllium to uranium (U).
- EDS provides excellent spatial resolution, allowing for elemental analysis of very small areas within a sample, often on the nanometer scale.
- EDS can provide quantitative information about the elemental composition, allowing researchers to determine the relative amounts of different elements present in a sample (Burke et al. 2020).

Several studies have utilized energy-dispersive X-ray spectroscopy (EDX) to characterize core-shell nanoconstructs in cancer theragnostics, highlighting its effectiveness in determining the elemental composition and structural properties.

Shin Inamoto et al. analyzed palladium-platinum (Pd-Pt) core-shell nanoparticles using EDX in conjunction with scanning transmission electron microscopy (STEM) (Hayakawa et al. 2022). The researchers developed a quantitative method to ascertain the thickness of the platinum shell, revealing that it corresponded to two atomic layers. This precise atomic-scale analysis demonstrated the capability of EDX to provide critical information about the structural integrity and composition of core-shell nanoparticles, which are essential for their application in catalytic processes and potentially in therapeutic contexts (Schwartz et al. 2024).

Ansari et al. focused on the characterization of core-shell nanorods and silica nanoparticles using EDX (Awada et al. 2019). This study emphasized the ability of EDX to differentiate between the core and shell materials, providing insights into the elemental distribution and confirming the successful synthesis of the core-shell structures. By mapping the elemental composition, researchers could assess the uniformity and effectiveness of the shell in encapsulating the core, which is vital for optimizing drug delivery systems in cancer therapy.

These studies illustrate the significance of EDX in the characterization of core-shell nanoconstructs, enabling researchers to evaluate their composition, structural properties, and potential interactions in biological environments. The information gained from EDX analyses is crucial for advancing the development of effective theragnostic agents in cancer treatment, ensuring that the designed nanoconstructs meet the required specifications for targeted therapy and imaging applications (Sobhana et al. 2023).

2.3 Physical Characterization

2.3.1 Dynamic Light Scattering (DLS)

Dynamic light scattering plays a crucial role in characterizing core-shell nanoconstructs designed for cancer theragnostics by providing essential information about their size, size distribution, and stability (Dibaba et al. 2022). These parameters are critical for understanding the behavior of nanoparticles in biological systems, influencing their circulation time, cellular uptake, and overall therapeutic efficacy. DLS measures the hydrodynamic diameter of the nanoparticles by analyzing the fluctuations in scattered light intensity caused by their Brownian motion in solution. By comparing the size of the core material along with that of the core-shell

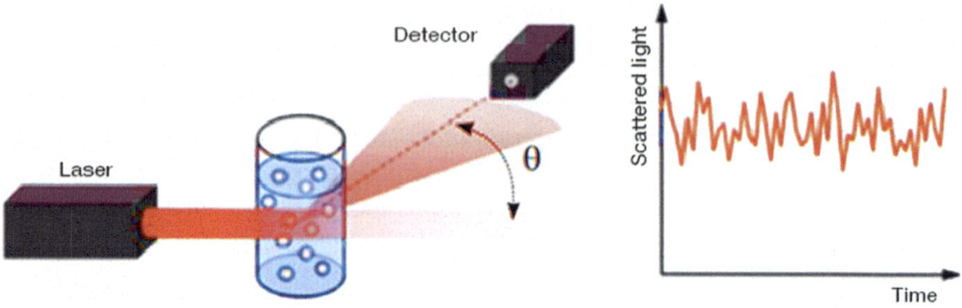

Fig. 8.11 Working principle of dynamic light scattering

nanoconstruct, researchers can estimate the thickness of the shell layer, which is particularly important for drug delivery applications. Moreover, DLS helps monitor the stability of these nanoconstructs over time, as changes in size distribution can indicate aggregation or degradation, ultimately impacting their efficacy and safety in therapeutic applications (Sarode et al. 2020).

Dynamic light scattering is based on the principle of Brownian motion and its effect on the scattering of light. When light interacts with particles in a solution, it gets scattered in all directions. The intensity of this scattered light constantly fluctuates due to the random movement of the particles, known as Brownian motion. Smaller particles move more rapidly, leading to faster fluctuations in scattered light intensity, while larger particles exhibit slower fluctuations (Volpe et al. 2014).

DLS instruments analyze these fluctuations in scattered light intensity over time. By calculating the autocorrelation function of the scattered light intensity signal, DLS can determine the diffusion coefficient of the particles in the solution. The Stokes-Einstein equation then relates this diffusion coefficient to the hydrodynamic diameter of the particles, providing information about their size and size distribution (Anton Paar 2024).

In essence, DLS utilizes the relationship between particle movement, light scattering, and time to characterize the size and size distribution of particles in a solution, making it a valuable tool in various fields, including nanotechnology, colloid science, and biophysics (Fig. 8.11).

Several studies have utilized dynamic light scattering (DLS) to characterize core-shell nanoconstructs for cancer theragnostics, providing insights into their size, stability, and functional capabilities. Barui et al. focused on Gd-Mn doped ZnO core-shell nanoparticles, which were characterized using DLS to assess their size and stability (Du et al. 2020). The nanoconstructs demonstrated high bio- and hemocompatibility, making them suitable for therapeutic applications against pancreatic cancer. DLS measurements indicated significant size distribution data that correlated with the particles' efficacy in biological systems. Aswathy et al. highlighted the potential of core-shell mesoporous silica nanoparticles as theragnostic agents. DLS was employed to evaluate the size and stability of these nanoparticles, which possess multifunctional properties for drug delivery and imaging. The study emphasized the importance of size control in enhancing drug loading capacity and targeting efficiency in cancer therapy.

Caro et al. investigated $Fe3O4$-Au core-shell nanoparticles designed for multimodal imaging and photothermal therapy (Caro et al. 2021). DLS was used to analyze the mean hydrodynamic diameter and ζ-potential, which provided insights into the stability and potential interactions of the nanoparticles in biological environments. This characterization was crucial for assessing their effectiveness in imaging and therapeutic applications. Huang et al. reported on mesoporous silica-based triple-modal imaging probes that incorporated DLS for size characterization (Huang et al. 2017). The study demonstrated how DLS measurements could provide insights into the stability and size distribution of nanoparticles, which are essential for their application in non-invasive imaging techniques for tumor metastasis detection.

Recent advancements in dynamic light scattering have significantly enhanced its capabilities for analyzing core-shell nanoconstructs in cancer theragnostics, enabling more detailed and insightful characterization (Doan et al. 2021). One notable advancement is the development of high-resolution DLS instruments, which offer improved size resolution and sensitivity, allowing for the detection of subtle changes in size distribution, particularly important for characterizing complex core-shell structures and monitoring their stability over time. Furthermore, the integration of DLS with other analytical techniques, such as depolarized DLS and multi-angle DLS, provides comprehensive information about particle shape, surface properties, and interactions with the surrounding environment, crucial for understanding their behavior in biological systems. These advancements, coupled with sophisticated data analysis algorithms, have made DLS an indispensable tool for optimizing the design and performance of core-shell nanoconstructs for targeted drug delivery, imaging, and other theragnostic applications (Herlan and Bräse 2020).

2.3.2 Zeta Potential Analysis

Zeta potential analysis is a crucial technique for characterizing core-shell nanoconstructs in cancer theragnostics, as it provides valuable insights into their stability and their interactions with biological systems. By measuring the electrical potential at the slipping plane of a particle, zeta potential analysis reveals the surface charge and charge distribution of core-shell nanoparticles (Gaikwad et al. 2020). This information is essential for predicting their stability in solution, as particles with high zeta potential values (either positive or negative) tend to repel each other, preventing aggregation and maintaining their desired size distribution. Moreover, understanding the surface charge of these nanoconstructs is crucial for predicting their interactions with biological components, such as proteins, cells, and tissues, which can influence their biodistribution, targeting efficiency, and therapeutic efficacy (Zein et al. 2020). Zeta potential analysis, therefore, plays a vital role in optimizing the design and performance of core-shell nanoconstructs for targeted drug delivery, imaging, and other theragnostic applications.

When a solid surface comes into contact with a liquid, an electrical double layer forms at the interface. This double layer consists of two oppositely charged layers:

- **Surface Charge:** The surface of the particle acquires a charge due to ionization, ion adsorption, or other chemical processes.
- **Counterion Layer:** Ions of opposite charge from the surrounding liquid accumulate near the surface, attracted by the electrostatic forces. This layer is further divided into two parts:

- **Stern Layer:** A tightly bound layer of ions directly adsorbed onto the surface.
- **Diffuse Layer:** A more loosely associated layer of ions extending further into the liquid.

The boundary between the immobile Stern layer and the mobile diffuse layer is called the slipping plane. The electrical potential at this plane is the zeta potential. It represents the potential difference between the bulk liquid and the layer of fluid that moves with the particle when an electric field is applied (Souza et al. 2023). Zeta potential is typically measured indirectly by determining the electrophoretic mobility of the particles. This involves applying an electric field across the dispersion and observing the movement of the charged particles toward the oppositely charged electrode. The velocity of particle movement under the influence of the electric field is called electrophoretic mobility. It is directly proportional to the zeta potential and inversely proportional to the viscosity of the medium (Libretexts 2013).

A. **Instrumentation and Measurement Techniques:**

Zeta potential analyzers measure zeta potential and employ one of the following techniques:

- **Laser Doppler Velocimetry:** This technique uses a laser beam to illuminate the moving particles. The scattered light from the particles undergoes a Doppler shift, which is proportional to their velocity. By analyzing the frequency shift, the electrophoretic mobility and, consequently, the zeta potential can be determined (Fig. 8.12).
- **Phase Analysis Light Scattering:** This technique analyzes the phase shift of scattered light caused by the movement of particles in the electric field. It is particularly suitable for measuring the zeta potential of small particles and samples with low electrophoretic mobility (Han et al. 2023).

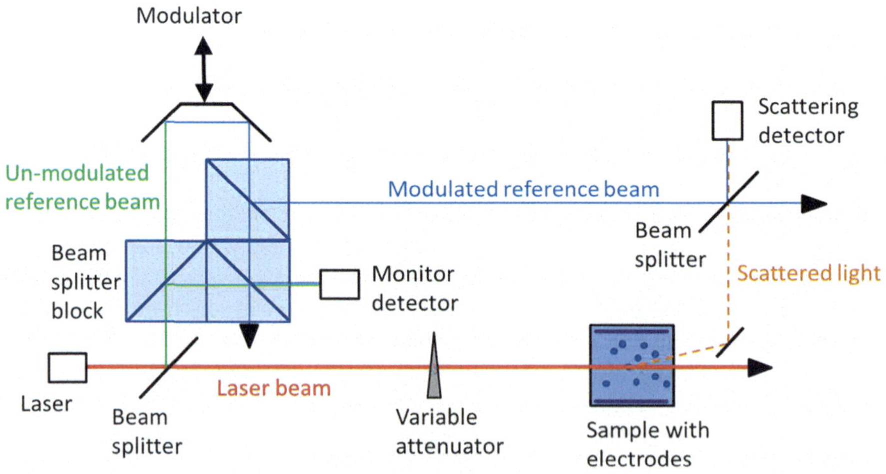

Fig. 8.12 Working principle of zeta potential analyzers

The magnitude and sign of the zeta potential provide valuable information about the stability and behavior of colloidal systems:

- **High Zeta Potential:** Indicates a strong electrostatic repulsion between particles, leading to a stable dispersion resistant to aggregation.
- **Low Zeta Potential:** Suggests weak electrostatic repulsion, making the dispersion prone to aggregation, flocculation, or sedimentation.

By measuring the surface charge of these nanoparticles, zeta potential provides valuable insights into their stability, colloidal behavior, and potential interactions in biological systems (Feng et al. 2020). Huang et al. reported on mesoporous silica-based triple-modal imaging probes designed for non-invasive diagnosis of tumor metastasis (Huang et al. 2017). Zeta potential analysis was employed to characterize the surface charge of these nanoparticles, which was an important parameter for evaluating their stability and potential interactions with biological molecules. Tian et al. investigated the zeta potential of HC-SiO2(PPVA/Cs)n core-shell nanoparticles, which are designed for biomedical applications. The zeta potential measurements provided insights into the surface charge of these nanoparticles, which is a crucial factor in determining their stability and potential interactions with cells and tissues. Chatterjee et al. focused on Prussian blue (PB) core coated by a compact ZIF-8 shell (core-shell dual-MOFs, CSD-MOFs) for cancer theragnostics (Sahoo et al. 2022). Zeta potential analysis was used to characterize the surface charge of these nanoparticles, which is an important parameter for understanding their stability and potential interactions in biological systems (Fig. 8.13).

Advancements in zeta potential analysis have significantly enhanced the characterization of core-shell nanoconstructs for cancer theragnostics, enabling

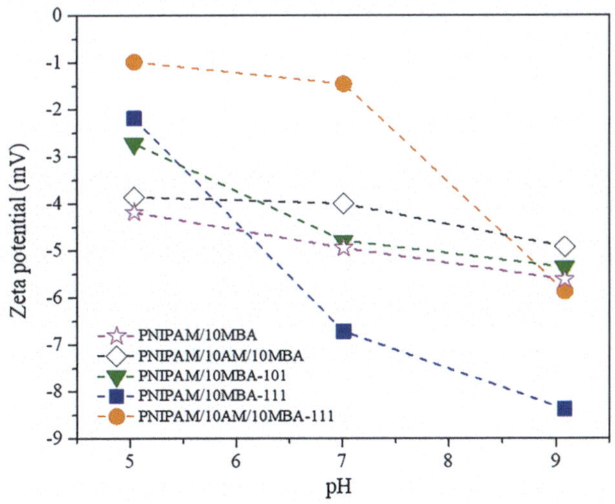

Fig. 8.13 Zeta potential of the polymeric core particles and hybrid core-shell particles versus pH. (From: https://doi.org/10.1007/s10853-019-03317-x (Cao-Luu et al. 2019))

researchers to gain deeper insights into their stability and surface properties (Yang et al. 2019). Modern zeta potential analyzers offer improved sensitivity and resolution, allowing for the detection of subtle changes in surface charge, crucial for understanding the complex interactions between core-shell nanoparticles and biological systems. Furthermore, the development of techniques like electrophoretic light scattering and tunable resistive pulse sensing has enabled the measurement of zeta potential in complex biological media, providing more accurate and relevant information about nanoparticle behavior in vivo (Buskermolen et al. 2022). These advancements, coupled with sophisticated data analysis algorithms, have made zeta potential analysis an indispensable tool for optimizing the design and performance of core-shell nanoconstructs for targeted drug delivery, imaging, and other theragnostic applications.

3 Evaluation of Core-Shell Nanoconstructs

The evaluation of core-shell nanoconstructs is a critical step in determining their suitability and efficacy for various applications, particularly in the biomedical field. This evaluation process goes beyond simply characterizing their physicochemical properties and delves into understanding their behavior, performance, and potential risks in complex biological systems (Vawhal et al. 2023).

3.1 In Vitro Evaluation of Core-Shell Nanoconstructs: Therapeutic Efficacy and Imaging Capabilities

In vitro evaluation using cell culture models is a crucial step in assessing the potential of core-shell nanoconstructs for cancer theragnostics. This stage focuses on evaluating both the therapeutic efficacy and imaging capabilities of these nanostructures:

3.1.1 Therapeutic Efficacy

This aspect focuses on determining how effectively the core-shell nanoconstructs deliver their therapeutic payload and impact cancer cell behavior:

- **Cytotoxicity and Cell Viability Assays:** These assays are fundamental for evaluating the inherent toxicity of the nanoconstructs on various cancer cell lines. Common assays include MTT, MTS, and flow cytometry, which measure cell viability and metabolic activity. This helps determine the safe dosage range for further experiments (Tasso et al. 2020).
- **Targeted Drug Delivery and Release:** The ability of the nanoconstructs to selectively target cancer cells and release their therapeutic payload in a controlled manner is crucial. This can be assessed by conjugating targeting ligands to the nanoconstruct surface and evaluating their binding affinity and specificity to cancer cells. Drug release kinetics can be studied under different conditions mimicking the tumor microenvironment (e.g., acidic pH, presence of specific enzymes) (Adhikari et al. 2015).

- **Mechanism of Action:** Understanding how the delivered therapy exerts its effect on cancer cells is essential. This might involve investigating pathways like apoptosis (programmed cell death), autophagy (cellular self-degradation), or cell cycle arrest. Techniques like Western blotting, flow cytometry, and gene expression analysis can be employed (Fucikova et al. 2020).
- **Combination Therapy Studies:** Core-shell nanoconstructs can be designed to deliver multiple therapeutic agents or combine with other treatment modalities like chemotherapy or photodynamic therapy. In vitro studies can assess the synergistic effects of these combinations (Mohammadniaei et al. 2018).

3.1.2 Imaging Capabilities

This aspect focuses on evaluating the potential of core-shell nanoconstructs as contrast agents for cancer imaging:

- **Cellular Uptake and Imaging:** The ability of the nanoconstructs to be internalized by cancer cells and generate a detectable signal is crucial. This can be visualized using techniques like confocal microscopy and flow cytometry, employing fluorescently labeled nanoconstructs (Reichel et al. 2019).
- **Imaging Sensitivity and Specificity:** The sensitivity of the imaging signal generated by the nanoconstructs and their ability to differentiate between cancerous and healthy cells are crucial parameters. This can be assessed by comparing the signal intensity in different cell lines and under different conditions.
- **Multimodal Imaging Potential:** Core-shell nanoconstructs can be designed to be detectable by multiple imaging modalities, offering complementary information. In vitro studies can evaluate their performance in different imaging setups, such as fluorescence imaging, magnetic resonance imaging, or computed tomography (Liu et al. 2019).

By combining these in vitro evaluations of therapeutic efficacy and imaging capabilities, researchers can gain a comprehensive understanding of the potential of core-shell nanoconstructs for cancer theragnostics. This information is crucial for optimizing their design and selecting the most promising candidates for further in vivo studies and eventual clinical translation.

3.2 In Vivo Evaluation of Core-Shell Nanoconstructs

In vivo evaluation represents a critical step in the development of core-shell nanoconstructs for cancer theragnostics, bridging the gap between promising in vitro results and potential clinical applications. This stage involves testing the nanoconstructs in living organisms, typically animal models, to assess their behavior, efficacy, and safety in a complex biological environment (Stern et al. 2010).

3.2.1 Biodistribution and Pharmacokinetics

- **Biodistribution Studies:** These studies aim to track the fate of the nanoconstructs within the body after administration. By labeling the nanoconstructs with fluorescent tags or radioisotopes, researchers can visualize and quantify their accumulation in different organs and tissues over time using techniques like fluorescence imaging or positron emission tomography. This information is crucial for understanding whether the nanoconstructs reach the target tumor site effectively and if there is any unwanted accumulation in healthy organs (Zhang et al. 2022).

- **Pharmacokinetic Analysis:** This aspect focuses on studying the absorption, distribution, metabolism, and excretion profile of the nanoconstructs. By analyzing their concentration in blood and other bodily fluids over time, researchers can determine crucial parameters like half-life, clearance rate, and volume of distribution. This information helps optimize dosing regimens and predict potential drug interactions.

3.2.2 Therapeutic Efficacy

- **Tumor Models:** Various animal models, such as xenograft or syngeneic models, are used to mimic human cancers. These models allow researchers to evaluate the ability of the nanoconstructs to inhibit tumor growth, reduce tumor volume, and improve survival rates compared to control groups or standard treatments.
- **Treatment Regimens:** Different administration routes (e.g., intravenous, intratumoral) and dosing schedules are tested to determine the most effective approach for delivering the nanoconstructs and achieving optimal therapeutic outcomes.
- **Combination Therapies:** In vivo studies are crucial for evaluating the efficacy of core-shell nanoconstructs in combination with other therapies, such as chemotherapy, radiotherapy, or immunotherapy. This allows researchers to assess potential synergistic effects and develop more effective treatment strategies (Shen et al. 2022).

3.2.3 Toxicity and Safety

- **Toxicity Studies:** Assessing the potential toxicity of nanoconstructs in vivo is paramount. Animals are closely monitored for any adverse effects, such as weight loss, behavioral changes, or organ dysfunction. Blood tests and histological analysis of major organs are performed to detect any signs of toxicity.
- **Immunogenicity:** The potential of nanoconstructs to trigger an immune response is carefully evaluated. This involves monitoring for inflammatory markers and analyzing immune cell populations.
- **Long-Term Effects:** Studies may be conducted to assess the long-term fate and potential chronic toxicity of nanoconstructs, especially if they are designed for repeated administration or have components with known long-term effects (Nehoff et al. 2014; Lappas 2014).

3.2.4 Imaging Applications

- **In Vivo Imaging:** The ability of the nanoconstructs to enhance tumor visualization in living animals is assessed using various imaging modalities, such as fluorescence imaging, MRI, PET, or CT. This helps evaluate their sensitivity, specificity, and potential for early disease detection or image-guided surgery.

- **Theragnostic Applications:** In vivo studies are crucial for evaluating the combined therapeutic and diagnostic capabilities of core-shell nanoconstructs. This involves monitoring tumor response to therapy while simultaneously tracking the nanoconstructs' distribution and accumulation within the tumor (Xu et al. 2024; Parashar et al. 2023).

In vivo evaluation remains an indispensable step in the development of core-shell nanoconstructs for cancer theragnostics. By providing valuable insights into their behavior, efficacy, and safety in a complex biological environment, these studies pave the way for translating promising nanomedicine approaches into safe and effective clinical applications.

4 Challenges and Considerations in Characterizing and Evaluating Core-Shell Nanoparticles

Core-shell nanoparticles, with their intricate structure and diverse functionalities, present unique challenges when it comes to characterization and evaluation. Accurately assessing their properties is crucial for understanding their behavior, optimizing their design, and ensuring their safe and effective use.

4.1 Characterization Challenges

4.1.1 Structural Complexity

Determining the exact core size, shell thickness, and overall morphology can be challenging. Techniques like transmission electron microscopy provide valuable insights, but achieving high-resolution images of these nanoscale structures, especially with complex core-shell architectures or very thin shells, can be difficult.

- **Compositional Heterogeneity:** Variations in the composition and distribution of elements within the core and shell can significantly impact the nanoparticle's properties. Techniques like X-ray photoelectron spectroscopy and energy-dispersive X-ray spectroscopy are essential for elemental analysis, but accurately quantifying elements at such small scales and resolving compositional gradients within the nanoparticle remains challenging.
- **Surface Chemistry:** The surface chemistry of core-shell nanoparticles, particularly the shell layer, plays a crucial role in their stability, biocompatibility, and interactions with their surroundings. Characterizing surface ligands, functional groups, and charge density requires sensitive techniques like Fourier-transform infrared spectroscopy, zeta potential measurements, and surface-enhanced Raman spectroscopy.
- **Dynamic Nature:** Core-shell nanoparticles can exhibit dynamic behavior in different environments, such as changes in size, shape, or surface properties. Characterizing these dynamic changes requires time-resolved measurements and techniques that can probe the nanoparticle's behavior under relevant conditions (Baer 2018).

4.2 Evaluation Challenges

- **In Vitro-In Vivo Correlation:** Bridging the gap between in vitro findings and in vivo behavior is crucial. Factors like protein corona formation, interactions with biological barriers, and clearance mechanisms can significantly influence the nanoparticle's fate and efficacy in living organisms.
- **Reproducibility and Batch-to-Batch Variability:** Synthesizing core-shell nanoparticles with consistent properties across different batches can be challenging. Variations in synthesis parameters can lead to inconsistencies in size, shape, composition, and surface chemistry, affecting reproducibility and comparability of results.
- **Long-Term Stability and Degradation:** Understanding the long-term stability and degradation pathways of core-shell nanoparticles is essential, especially for biomedical applications. Factors like storage conditions, exposure to biological fluids, and potential degradation products need to be carefully considered.
- **Safety and Toxicity:** Thoroughly evaluating the potential toxicity of core-shell nanoparticles is paramount. This involves assessing both short-term and long-term effects, considering factors like dose, route of administration, and potential accumulation in organs (Sharma et al. 2021).

4.3 Addressing the Challenges

- **Combining Multiple Characterization Techniques:** Employing a combination of complementary techniques provides a more comprehensive understanding of the nanoparticle's properties.
- **Developing Standardized Protocols:** Establishing standardized protocols for synthesis, characterization, and evaluation can improve reproducibility and comparability of results.
- **Utilizing Advanced Imaging and Spectroscopy:** Leveraging advanced imaging techniques like cryo-TEM and high-resolution spectroscopy methods can provide more detailed insights into the nanoparticle's structure and composition.
- **Developing Robust In Vitro and In Vivo Models:** Employing relevant and well-characterized biological models that mimic the intended application is crucial for evaluating the nanoparticle's behavior and efficacy.
- **Considering Ethical Aspects:** Adhering to ethical guidelines and conducting thorough risk assessments are essential when working with nanomaterials.

By acknowledging these challenges and adopting rigorous characterization and evaluation strategies, researchers can advance the development of safe and effective core-shell nanoparticles for various applications, from medicine to energy and beyond.

5 Conclusion

In conclusion, the successful development and implementation of core-shell nanoconstructs for cancer theragnostics critically depend on a multifaceted approach to characterization and evaluation. Employing a synergistic combination of advanced imaging, spectroscopy, in vitro, and in vivo techniques is essential to fully understand their physicochemical properties, biological behavior, and potential for safe and effective clinical translation. Bridging the gap between in vitro findings and in vivo performance, ensuring reproducibility and batch-to-batch consistency, and prioritizing safety by design are paramount considerations. As we continue to refine our characterization and evaluation strategies, integrating high-throughput methods, computational modeling, and personalized approaches, we pave the way for a future where core-shell nanoconstructs revolutionize cancer diagnosis and treatment, offering new hope for patients worldwide.

References

Adhikari C, Das A, Chakraborty A. Controlled release of a sparingly water-soluble anticancer drug through pH-responsive functionalized gold-nanoparticle-decorated liposomes. Chem Int 2015;16(4):866–871. Available from: https://doi.org/10.1002/cphc.201402748.

Ansari AA, Parchur AK, Kumar B, Rai SB. Influence of shell formation on morphological structure, optical and emission intensity on aqueous dispersible NaYF4:Ce/Tb nanoparticles. J Fluoresc [Internet] 2016;26(4):1151–1159. Available from: https://doi.org/10.1007/s10895-016-1824-1.

Ansari AA, Khan S, Aldalbahi A, Parchur AK, Kumar B, Kumar A, et al. In-vitro cytotoxicity evaluation of surface design luminescent lanthanide core/shell nanocrystals. Arab J Chem [Internet] 2020;13(1):1259–1270. Available from: https://doi.org/10.1016/j.arabjc.2017.10.008.

Anton Paar. The principles of dynamic light scattering [Internet]. Anton Paar. [cited 2024 Aug 16]. Available from: https://wiki.anton-paar.com/pl-pl/zasady-dynamicznego-rozpraszania-swiatla.

Arteaga Cardona F, Jain N, Popescu R, Busko D, Madirov E, Arús BA, et al. Preventing cation intermixing enables 50% quantum yield in sub-15 nm short-wave infrared-emitting rare-earth based core-shell nanocrystals. Nat Commun [Internet]. 2023;14(1). Available from: https://doi.org/10.1038/s41467-023-40031-4.

Awada H, Al Samad A, Laurencin D, Gilbert R, Dumail X, El Jundi A, et al. Controlled anchoring of iron oxide nanoparticles on polymeric nanofibers: easy access to core@shell organic–inorganic nanocomposites for magneto-scaffolds. ACS Appl Mater Interfaces [Internet]. 2019;11(9):9519–9529. Available from: https://doi.org/10.1021/acsami.8b19099.

Baer DR. The chameleon effect: characterization challenges due to the variability of nanoparticles and their surfaces. Front Chem [Internet] 2018;6:145. Available from: https://doi.org/10.3389/fchem.2018.00145.

Bardhan N. Nanomaterials in diagnostics, imaging and delivery: Applications from COVID-19 to cancer. MRS Commun [Internet]. 2022 [cited 2024 Aug 16];12(6):1119–39. Available from: https://doi.org/10.1557/s43579-022-00257-7

Birkholz, M. Principles of X-ray diffraction [Internet]. In: Thin film analysis by X-ray scattering. Wiley; 2005. p. 1–40. Available from: https://doi.org/10.1002/3527607595.ch1.

Biswal D, Peeples BN, Spence DD, Peeples C, Bell CN, Pradhan AK. Synthesis and characterization of functionalized magnetic nanoparticles. In: Varadan VK, editor. SPIE proceedings. SPIE; 2012.

Brodusch N, Brahimi SV, Barbosa De Melo E, Song J, Yue S, Piché N, et al. Scanning electron microscopy versus transmission electron microscopy for material characterization: a comparative study on high-strength steels. Scanning [Internet] 2021;2021:1–19. Available from: https://doi.org/10.1155/2021/5511618.

Burke MG, Volpe L, Scenini F, Prestat E. Correlative microscopy: elucidating the mechanisms of SCC in structural alloys in PWR environments. Microsc Microanal [Internet] 2020;26(S2):168–170. Available from: https://doi.org/10.1017/s1431927620013665.

Buskermolen AD, Lin Y-T, van Smeden L, van Haaften RB, Yan J, Sergelen K, et al. Continuous biomarker monitoring with single molecule resolution by measuring free particle motion. Nat Commun [Internet] 2022;13(1). Available from: https://doi.org/10.1038/s41467-022-33487-3.

Cao-Luu N-H, Pham Q-T, Yao Z-H, Wang F-M, Chern C-S. Synthesis and characterization of PNIPAM microgel core–silica shell particles. J Mater Sci [Internet] 2019;54(10):7503–7516. Available from: https://doi.org/10.1007/s10853-019-03317-x.

Carlson TA, Krause MO. Atomic readjustment to vacancies in theKandLShells of argon. Phys Rev [Internet] 1965;137(6A):A1655–A1662. Available from: https://doi.org/10.1103/physrev.137.a1655.

Caro C, Gámez F, Quaresma P, Páez-Muñoz JM, Domínguez A, Pearson JR, et al. Fe3O4-Au core-shell nanoparticles as a multimodal platform for *In Vivo* imaging and focused photothermal therapy. Pharmaceutics [Internet] 2021;13(3):416. Available from: https://doi.org/10.3390/pharmaceutics13030416.

Cesareo R. X-ray fluorescence spectrometry. In: Ullmann's encyclopedia of industrial chemistry. Weinheim: Wiley-VCH Verlag GmbH & Co. KGaA; 2010.

Chen F, Ehlerding EB, Cai W. Theranostic nanoparticles. J Nucl Med [Internet] 2014;55(12):1919–1922. Available from: https://doi.org/10.2967/jnumed.114.146019.

Chhour P, Gallo N, Cheheltani R, Williams D, Al-Zaki A, Paik T, et al. Nanodisco balls: control over surface versus core loading of diagnostically active nanocrystals into polymer nanoparticles. ACS Nano [Internet] 2014;8(9):9143–9153. Available from: https://doi.org/10.1021/nn502730q

Chung DDL. Structure determination, X-ray diffraction for [Internet]. In: Encyclopedia of analytical chemistry. Wiley; 2000. Available from: https://doi.org/10.1002/9780470027318.a6804.

Cimatu K, Baldelli S. Chemical imaging of corrosion: sum frequency generation imaging microscopy of cyanide on gold at the Solid−Liquid interface. J Am Chem Soc [Internet] 2008;130(25):8030–8037. Available from: https://doi.org/10.1021/ja8011676.

Clearfield A, Bhuvanesh N. X-ray powder diffraction [Internet]. In: Encyclopedia of inorganic chemistry. Wiley; 2005. Available from: https://doi.org/10.1002/0470862106.ia332.

Conti S, Remuzzi G, Benigni A, Tomasoni S. Imaging the kidney with an unconventional scanning electron microscopy technique: analysis of the subpodocyte space in diabetic mice. Int J Mol Sci [Internet] 2022;23(3):1699. Available from: https://doi.org/10.3390/ijms23031699.

Cressoni C, Vurro F, Milan E, Muccilli M, Mazzer F, Gerosa M, et al. From nanothermometry to bioimaging: lanthanide-activated KY3F10 nanostructures as biocompatible multifunctional tools for nanomedicine. ACS Appl Mater Interfaces [Internet]. 2023;15(9):12171–12188. Available from: https://doi.org/10.1021/acsami.2c22000.

Dibaba ST, Xie Y, Xi W, Bednarkiewicz A, Ren W, Sun L. Nd3+-sensitized upconversion nanoparticle coated with antimony shell for bioimaging and photothermal therapy *In Vitro* using single laser irradiation. J Rare Earths [Internet] 2022;40(6):862–869. Available from: https://doi.org/10.1016/j.jre.2021.05.015.

Diebolder P, Vazquez-Pufleau M, Bandara N, Mpoy C, Raliya R, Thimsen E, et al. Aerosol-synthesized siliceous nanoparticles: impact of morphology and functionalization on biodistribution. Int J Nanomedicine [Internet] 2018;13:7375–7393. Available from: https://doi.org/10.2147/ijn.s177350.

Diez-Pascual AM, Rahdar A. Functional nanomaterials in biomedicine: current uses and potential applications. ChemMedChem [Internet] 2022;17(16). Available from: https://doi.org/10.1002/cmdc.202200142.

Dimov IP, Tous C, Li N, Häfeli UO, Martel S, Soulez G. Future advances in diagnosis and drug delivery in interventional radiology using MR imaging-steered theranostic iron oxide nanoparticles. J Vasc Interv Radiol [Internet] 2021;32(9):1292–1295.e1. Available from: https://doi.org/10.1016/j.jvir.2021.05.027.

Doan VHM, Nguyen VT, Mondal S, Vo TMT, Ly CD, Vu DD, et al. Fluorescence/photoacoustic imaging-guided nanomaterials for highly efficient cancer theragnostic agent. Sci Rep [Internet]. 2021;11(1). Available from: https://doi.org/10.1038/s41598-021-95660-w.

Du JS, Chen P-C, Dravid VP, Mirkin CA. Using STEM to probe the in-situ dynamics of multimetallic nanoparticles grown in polymer nanoreactors. Microsc Microanal [Internet]. 2017 [cited 2024 Aug 16];23(S1):872–3. Available from: https://academic.oup.com/mam/article/23/S1/872/6945116?login=false.

Du Y, Yang C, Li F, Liao H, Chen Z, Lin P, et al. Core–shell–satellite nanomaces as remotely controlled self-fueling Fenton reagents for imaging-guided triple-negative breast cancer-specific therapy. Small [Internet] 2020;16(31). Available from: https://doi.org/10.1002/smll.202002537.

Dung NT, Linh NTN, Chi DL, Hoa NTH, Hung NP, Ha NT, et al. Optical properties and stability of small hollow gold nanoparticles. RSC Adv [Internet]. 2021;11(22):13458–13465. Available from: https://doi.org/10.1039/d0ra09417j.

Edmonds S, Volpe A, Shmeeda H, Parente-Pereira AC, Radia R, Baguña-Torres J, et al. Exploiting the metal-chelating properties of the drug cargo for *In Vivo* positron emission tomography imaging of liposomal nanomedicines. ACS Nano [Internet]. 2016;10(11):10294–10307. Available from: https://doi.org/10.1021/acsnano.6b05935.

Ettlinger R, Moreno N, Ziółkowska N, Ullrich A, Krug von Nidda H-A, Jirák D, et al. *In vitro* studies of Fe3O4-ZIF-8 core–shell nanoparticles designed as potential theragnostics. Part Part Syst Charact [Internet] 2020;37(12). Available from: https://doi.org/10.1002/ppsc.202000185.

Fan W, Shen B, Bu W, Zheng X, He Q, Cui Z, et al. Design of an intelligent sub-50 nm nuclear-targeting nanotheranostic system for imaging guided intranuclear radiosensitization. Chem Sci [Internet] 2015;6(3):1747–1753. Available from: https://doi.org/10.1039/c4sc03080j.

Feng L, Yang D, He F, Gai S, Li C, Dai Y, et al. A core–shell-satellite structured Fe3O4@g-C3N4–UCNPs–PEG for T1/T2-weighted dual-modal MRI-guided photodynamic therapy. Adv Healthc Mater [Internet] 2017;6(18). Available from: https://doi.org/10.1002/adhm.201700502.

Feng Y, Kilker SR, Lee Y. Surface charge (zeta-potential) of nanoencapsulated food ingredients. In: Characterization of nanoencapsulated food ingredients. Elsevier; 2020. p. 213–41.

Fernández-González A, Obaya Á, Chimeno-Trinchet C, Fontanil T, Badía-Laíño R. Viability of ABO blood typing with ATR-FTIR spectroscopy [Internet]. Preprints. 2023. Available from: https://doi.org/10.20944/preprints202307.2081.v1

Francia V, Reker-Smit C, Salvati A. Mechanisms of uptake and membrane curvature generation for the internalization of silica nanoparticles by cells. Nano Lett [Internet] 2022;22(7):3118–3124. Available from: https://doi.org/10.1021/acs.nanolett.2c00537.

Fucikova J, Kepp O, Kasikova L, Petroni G, Yamazaki T, Liu P, et al. Detection of immunogenic cell death and its relevance for cancer therapy. Cell Death Dis [Internet] 2020;11(11): 1013. Available from: https://doi.org/10.1038/s41419-020-03221-2.

Gaikwad SS, Avari J, Patil ML. Zeta potential as a diagnostic tool to determine the angina risk. In: Apolipoproteins, triglycerides and cholesterol. IntechOpen; 2020.

Geng F, Xing JZ, Chen J, Yang R, Hao Y, Song K, et al. Pegylated glucose gold nanoparticles for improved in-vivo bio-distribution and enhanced radiotherapy on cervical cancer. J Biomed Nanotechnol [Internet] 2014;10(7):1205–1216. Available from: https://doi.org/10.1166/jbn.2014.1855.

Han Y, Ding L, Wang Y, Zheng H, Fang L. Shape discrimination of individual aerosol particles using light scattering. Sensors (Basel) [Internet]. 2023;23(12):5464. Available from: https://doi.org/10.3390/s23125464

Hayakawa Y, Furuya M, Tahara H, Kosuge Y, Kimura T, Sugawa K, et al. Modulation technique of localized surface plasmon resonance of palladium nanospheres by coating with titanium dioxide shell for application to photothermal therapy agent. Nanoscale Res Lett [Internet] 2022;17(1). Available from: https://doi.org/10.1186/s11671-022-03697-1.

He DS, Huang Y, Myers BD, Isheim D, Fan X, Deng Y, et al. Nanosized monoatomic palladium metallic glass [Internet]. arXiv [cond-mat.mtrl-sci]. 2020 [cited 2024 Aug 16]. Available from: http://arxiv.org/abs/2004.01851

Herlan C, Bräse S. Lanthanide conjugates as versatile instruments for therapy and diagnostics. Dalton Trans [Internet] 2020;49(8):2397–2402. Available from: https://doi.org/10.1039/c9dt04851k.

Ho L-C, Hsu C-H, Ou C-M, Wang C-W, Liu T-P, Hwang L-P, et al. Unibody core-shell smart polymer as a theranostic nanoparticle for drug delivery and MR imaging. Biomaterials [Internet]. 2015;37:436–446. Available from: https://doi.org/10.1016/j.biomaterials.2014.10.006.

Hu Y, Wang R, Wang S, Ding L, Li J, Luo Y, et al. Multifunctional Fe3O4 @ Au core/shell nanostars: a unique platform for multimode imaging and photothermal therapy of tumors. Sci Rep [Internet]. 2016;6(1) Available from: https://www.nature.com/articles/srep28325.pdf.

Huang H, Yang D-P, Liu M, Wang X, Zhang Z, Zhou G, et al. pH-sensitive Au–BSA–DOX–FA nanocomposites for combined CT imaging and targeted drug delivery. Int J Nanomedicine [Internet]. 2017;12:2829–43. Available from: https://www.dovepress.com/terms.php.

Hull LC, Farrell D, Grodzinski P. Highlights of recent developments and trends in cancer nanotechnology research—view from NCI Alliance for Nanotechnology in Cancer. Biotechnol Adv [Internet] 2014;32(4):666–678. Available from: https://doi.org/10.1016/j.biotechadv.2013.08.003

Ionic mobility and electrophoresis [Internet]. Chemistry LibreTexts. Libretexts; 2013 [cited 2024 Aug 16]. Available from: https://chem.libretexts.org/Bookshelves/Analytical_Chemistry/Supplemental_Modules_(Analytical_Chemistry)/Instrumentation_and_Analysis/Capillary_Electrophoresis/Ionic_Mobility_and_Electrophoresis.

Jang J, Li XL, Oh JH. Facile fabrication of polymer and carbon nanocapsules using polypyrrole core/shell nanomaterialsElectronic Supplementary Information (ESI) available: Experimental details for the synthesis and carbonization of PPy nanocapsules, instrumentation, XRD pattern and Raman spectrum. See http://www.rsc.org/suppdata/cc/b3/b316881f/. Chem Commun (Camb) [Internet]. 2004 [cited 2024 Aug 16];(7):794. Available from: https://pubs.rsc.org/en/content/articlelanding/2004/CC/b316881f

Jang H, Kim Y-K, Huh H, Min D-H. Facile synthesis and intraparticle self-catalytic oxidation of dextran-coated hollow Au–Ag nanoshell and its application for chemo-thermotherapy. ACS Nano [Internet]. 2014;8(1):467–475. Available from: https://doi.org/10.1021/nn404833b.

Jiang X, Lee M, Xia J, Luo T, Liu J, Rodriguez M, et al. Two-stage SN38 release from a core–shell nanoparticle enhances tumor deposition and antitumor efficacy for synergistic combination with immune checkpoint blockade. ACS Nano [Internet]. 2022;16(12):21417–21430. Available from: https://doi.org/10.1021/acsnano.2c09788.

Jo H, Wi DH, Lee T, Kwon Y, Jeong C, Lee J, et al. Revealing 3-dimensional core-shell interface structures at the single-atom level [Internet]. arXiv [cond-mat.mtrl-sci]. 2022. Available from: http://arxiv.org/abs/2207.06677

Jones IP. Electron microscopy for engineers. IOP Conf Ser Mater Sci Eng [Internet]. 2009;4:012001. Available from: https://doi.org/10.1088/1757-899x/4/1/012001

Keevend K, Krummenacher R, Kungas E, Gerken LRH, Gogos A, Stiefel M, et al. Correlative cathodoluminescence electron microscopy: Immunolabeling using rare-earth element doped nanoparticles. Small [Internet]. 2020;16(44). Available from: https://doi.org/10.1002/smll.202004615.

Kularatne RN, Crist RM, Stern ST. The future of tissue-targeted lipid nanoparticle-mediated nucleic acid delivery. Pharmaceuticals (Basel) [Internet] 2022;15(7):897. Available from: https://doi.org/10.3390/ph15070897.

Kvick Å. X-ray diffraction, materials science applications. In: Encyclopedia of Spectroscopy and Spectrometry. Elsevier; 2017. p. 648–655.

Lamvik MK. Scanning electron microscopy in analysis of surfaces [Internet]. In: Encyclopedia of analytical chemistry. Wiley; 2000. Available from: https://doi.org/10.1002/9780470027318.a2512.

Lange F, Agüi-Gonzalez P, Riedel D, Phan NTN, Jakobs S, Rizzoli SO. Correlative fluorescence microscopy, transmission electron microscopy and secondary ion mass spectrometry (CLEM-SIMS) for cellular imaging. PLoS One [Internet] 2021;16(5):e0240768. Available from: https://doi.org/10.1371/journal.pone.0240768.

Lappas CM. Immunotoxicology of nanomaterials [Internet]. In: Handbook of nanotoxicology, nanomedicine and stem cell use in toxicology. Wiley; 2014. p. 99–110. Available from: https://doi.org/10.1002/9781118856017.ch5.

Ledendecker M, Paciok P, Osowiecki WT, Pander M, Heggen M, Göhl D, et al. Engineering gold-platinum core-shell nanoparticles by self-limitation in solution. Commun Chem [Internet] 2022;5(1). Available from: https://doi.org/10.1038/s42004-022-00680-w.

Lee SY, Ko S-H, Shim J-S, Kim D-D, Cho H-J. Correction to "tumor targeting and lipid rafts disrupting hyaluronic acid-cyclodextrin-based nanoassembled structure for cancer therapy". ACS Appl Mater Interfaces [Internet]. 2018;10(50):44197–44198. Available from: https://doi.org/10.1021/acsami.8b19910.

Li J, Zeng H, Sun S, Liu JP, Wang ZL. Analyzing the structure of CoFe−Fe3O4 Core−Shell nanoparticles by electron imaging and diffraction. J Phys Chem B [Internet] 2004;108(37):14005–14008. Available from: https://doi.org/10.1021/jp047602f.

Li W-P, Liao P-Y, Su C-H, Yeh C-S. Formation of oligonucleotide-gated silica shell-coated Fe3O4-Au core–shell nanotrisoctahedra for magnetically targeted and near-infrared light-responsive theranostic platform. J Am Chem Soc [Internet]. 2014;136(28):10062–10075. Available from: https://doi.org/10.1021/ja504118q.

Li Z, Li Z, Sun L, Du B, Wang Y, Zhao G, et al. Core–shell bi2Se3@mSiO2-PEG as a multifunctional drug-delivery nanoplatform for synergistic Thermo-chemotherapy with infrared thermal imaging of cancer cells. Part Part Syst Charact [Internet]. 2018;35(3). Available from: https://doi.org/10.1002/ppsc.201700337.

Li J, Huang D, Cheng R, Figueiredo P, Fontana F, Correia A, et al. Multifunctional biomimetic nanovaccines based on photothermal and weak-immunostimulatory nanoparticulate cores for the immunotherapy of solid tumors. Adv Mater [Internet]. 2022;34(9). Available from: https://doi.org/10.1002/adma.202108012.

Lin E-W, Boehnke N, Maynard HD. Protein–polymer conjugation via ligand affinity and photoactivation of glutathione S-transferase. Bioconjug Chem [Internet] 2014;25(10):1902–1909. Available from: https://doi.org/10.1021/bc500380r.

Liu Y, Gong CS, Lin L, Zhou Z, Liu Y, Yang Z, et al. Core-shell metal-organic frameworks with fluorescence switch to trigger an enhanced photodynamic therapy. Theranostics [Internet]. 2019;9(10):2791–2799. Available from: https://doi.org/10.7150/thno.34740.

Macchione MA, Strumia MC. Stimuli-responsive nanosystems as smart nanotheranostics. In: Advanced nanoformulations. Elsevier; 2023. p. 363–96.

Miao C, Shen Y, Lang Y, Li H, Gong Y, Liu Y, et al. Biomimetic nanoparticles with enhanced rapamycin delivery for autism spectrum disorder treatment via autophagy activation and oxidative stress modulation. Theranostics [Internet]. 2024;14(11):4375–4392. Available from: https://doi.org/10.7150/thno.95614.

Mishra S, Tripathi RM, Sinha OP. Recent developments in detoxification of organic pollutants using CdS-based nanocomposites. Nano Biomed Eng [Internet] 2021;13(2). Available from: https://doi.org/10.5101/nbe.v13i2.p95-108.

Mohammadniaei M, Lee T, Bharate BG, Yoon J, Choi HK, Park S-J, et al. Bifunctional Au@bi2Se3 core–shell nanoparticle for synergetic therapy by SERS-traceable AntagomiR delivery and photothermal treatment. Small [Internet]. 2018;14(38). Available from: https://doi.org/10.1002/smll.201802934.

Movasaghi Z, Rehman S, ur Rehman DI. Fourier transform infrared (FTIR) spectroscopy of biological tissues. Appl Spectrosc Rev [Internet] 2008;43(2):134–179. Available from: https://doi.org/10.1080/05704920701829043.

Nandwana V, De M, Chu S, Jaiswal M, Rotz M, Meade TJ, et al. Theranostic magnetic nanostructures (MNS) for cancer. In: Cancer treatment and research. Cham: Springer International Publishing; 2015. p. 51–83.

Nehoff H, Parayath NN, Taurin S, Greish K. *In Vivo*evaluation of acute and chronic nanotoxicity [Internet]. In: Handbook of nanotoxicology, nanomedicine and stem cell use in toxicology. Wiley; 2014. p. 65–86. Available from: https://doi.org/10.1002/9781118856017.ch3.

Newbury DE. Energy dispersive spectrometry [Internet]. In: Characterization of materials. Wiley; 2012. p. 1–26. Available from: https://doi.org/10.1002/0471266965.com087.pub2.

Ni R, Huang L, Li Z, Zhang W, Wang Y, Shen Y, et al. Multifunctional ROS-responsive and TME-modulated lipid-polymer hybrid nanoparticles for enhanced tumor penetration. Int J

Nanomedicine [Internet]. 2022;17:5883–5897. Available from: https://doi.org/10.2147/ijn.s383517.

Nica V, Marino A, Pucci C, Şen Ö, Emanet M, De Pasquale D, et al. Cell-membrane-coated and cell-penetrating peptide-conjugated trimagnetic nanoparticles for targeted magnetic hyperthermia of prostate cancer cells. ACS Appl Mater Interfaces [Internet] 2023;15(25):30008–30028. Available from: https://doi.org/10.1021/acsami.3c07248.

Parashar AK. Synthesis and characterization of temozolomide loaded theranostic quantum dots for the treatment of brain glioma. J Med Pharm Allied Sci [Internet] 2021a;10(3):2778–2782. Available from: https://doi.org/10.22270/jmpas.v10i3.1073.

Parashar AK. Synthesis and characterization of ligand anchored poly propyleneiminedendrimers for the treatment of brain glioma. J Med Pharm Allied Sci [Internet]. 2021b;10(3):2784–2789. Available from: https://doi.org/10.22270/jmpas.v10i3.1084.

Parashar AK, Sethi VA. Polymer nanoparticles for cancer immunotherapy: preparation methods and applications. Nat Cell Sci [Internet] 2024;000(000):000–000. Available from: https://doi.org/10.61474/ncs.2024.00012.

Parashar AK, Verma KK, Kumar R, Arora V. A concise review of carbon dots and their pharmaceutical and biomedical applications. Recent Adv Drug Deliv Formul [Internet] 2023;17(3):183–192. Available from: https://doi.org/10.2174/0126673878237423230919070049

Patamia V, Zagni C, Fiorenza R, Fuochi V, Dattilo S, Riccobene PM, et al. Total bio-based material for drug delivery and iron chelation to fight cancer through antimicrobial activity. Nanomaterials (Basel) [Internet] 2023;13(14):2036. Available from: https://doi.org/10.3390/nano13142036.

Pattammattel A, Tappero R, Ge M, Chu YS, Huang X, Gao Y, et al. High-sensitivity nanoscale chemical imaging with hard x-ray nano-XANES. Sci Adv [Internet] 2020;6(37). Available from: https://doi.org/10.1126/sciadv.abb3615.

Pragatisheel PJ. Silver nanostructures, chemical synthesis methods, and biomedical applications. In: Nanotechnology in the life sciences. Cham: Springer International Publishing; 2020. p. 281–303.

Pulumati A, Pulumati A, Dwarakanath BS, Verma A, Papineni RVL. Technological advancements in cancer diagnostics: improvements and limitations. Cancer Rep [Internet] 2023;6(2):e1764. Available from: https://doi.org/10.1002/cnr2.1764.

Ranu R, Kadam SL, Gade VK, Desarada SV, Yewale MA, Chavan KB. Comparative microstructural analysis of V2O5nanoparticles via x-ray diffraction (XRD) technique. Nanotechnology [Internet] 2024;35(43):435701. Available from: https://doi.org/10.1088/1361-6528/ad67e7.

Reeves C. The uses of scanning electron microscopy for studying semiconductor devices. Int J Electron [Internet] 1994;77(6):919–928. Available from: https://doi.org/10.1080/00207219408926111.

Reichel D, Tripathi M, Butte P, Saouaf R, Perez JM. Tumor-activatable clinical nanoprobe for cancer imaging. Nanotheranostics [Internet] 2019;3(2):196–211. Available from: https://doi.org/10.7150/ntno.34921.

Sahoo P, Kundu S, Roy S, Sharma SK, Ghosh J, Mishra S, et al. Fundamental understanding of the size and surface modification effects on r1, the relaxivity of Prussian blue nanocube@m-SiO2: a novel targeted chemo-photodynamic theranostic agent to treat colon cancer. RSC Adv [Internet]. 2022;12(38):24555–70. Available from: https://pubs.rsc.org/en/content/articlepdf/2022/ra/d2ra03995h.

Santagata F, Yang CK, Creemer JF, French PJ, Sarro PM. Single wafer surface micromachined field emission electron source. In: 2009 IEEE 22nd international conference on micro electro mechanical systems. IEEE; 2009.

Sarode A, Annapragada A, Guo J, Mitragotri S. Layered self-assemblies for controlled drug delivery: a translational overview. Biomaterials [Internet] 2020;242(119929):119929. Available from: https://doi.org/10.1016/j.biomaterials.2020.119929.

Sato M, Konomi M, Tamochi R, Ishikawa T. Development ofFE-SEMTechnologies for life science fields [Internet]. In: Biological field emission scanning electron microscopy. Wiley; 2019. p. 25–51. Available from: https://doi.org/10.1002/9781118663233.ch3.

Schwartz J, Di ZW, Jiang Y, Manassa J, Pietryga J, Qian Y, et al. Imaging 3D chemistry at 1 nm resolution with fused multi-modal electron tomography. 2023; 15(1):3555. Available from: https://doi.org/10.48550/ARXIV.2304.12259

Schwartz J, Di ZW, Jiang Y, Manassa J, Pietryga J, Qian Y, et al. Imaging 3D chemistry at 1 nm resolution with fused multi-modal electron tomography. Nat Commun [Internet]. 2024;15(1):3555. Available from: https://doi.org/10.1038/s41467-024-47558-0.

Sharma S, Parveen R, Chatterji BP. Toxicology of nanoparticles in drug delivery. Curr Pathobiol Rep [Internet] 2021;9(4):133–144. Available from: https://doi.org/10.1007/s40139-021-00227-z.

Shen Z, Song J, Zhou Z, Yung BC, Aronova MA, Li Y, et al. Dotted core–shell nanoparticles for T1-weighted MRI of tumors. Adv Mater [Internet] 2018;30(33). Available from: https://doi.org/10.1002/adma.201803163.

Shen C-L, Liu H-R, Lou Q, Wang F, Liu K-K, Dong L, et al. Recent progress of carbon dots in targeted bioimaging and cancer therapy. Theranostics [Internet] 2022;12(6):2860–2893. Available from: https://doi.org/10.7150/thno.70721.

Shen S, Guo X, Wu L, Wang M, Wang X, Kong F, et al. Dual-core@shell-structured Fe3O4–NaYF4@TiO2nanocomposites as a magnetic targeting drug carrier for bioimaging and combined chemo-sonodynamic therapy. J Mater Chem B Mater Biol Med [Internet]. 2014 [cited 2024 Aug 16];2(35):5775–84. Available from: https://pubs.rsc.org/en/content/articlelanding/2014/TB/C4TB00841C.

Sobhana S, Sarathy NP, Karthikeyan L, Shanthi K, Vivek R. Ultra-small NIR-responsive nanotheranostic agent for targeted photothermal ablation induced damage-associated molecular patterns (DAMPs) from post-PTT of tumor cells activate immunogenic cell death. Nanotheranostics [Internet]. 2023;7(1):41–60. Available from: https://doi.org/10.7150/ntno.76720

Souza TC, Costa AFS, Vinhas GM, Sarubbo LA. Synthesis of iron oxides and influence on final sizes and distribution in bacterial cellulose applications [Internet]. Preprints 2023. Available from: https://doi.org/10.20944/preprints202307.0915.v1

Stern ST, Hall JB, Yu LL, Wood LJ, Paciotti GF, Tamarkin L, et al. Translational considerations for cancer nanomedicine. J Control Release [Internet] 2010;146(2):164–174. Available from: https://doi.org/10.1016/j.jconrel.2010.04.008.

Sun X, Huang X, Yan X, Wang Y, Guo J, Jacobson O, et al. Chelator-free 64cu-integrated gold nanomaterials for positron emission tomography imaging guided photothermal cancer therapy. ACS Nano [Internet]. 2014;8(8):8438–8446. Available from: https://doi.org/10.1021/nn502950t.

Sun C-K, Wang Y-H, Chen Y-L, Lu T-Y, Chen H-Y, Pan S-C, et al. Fabrication of an Au-doped Cu/Fe oxide-polymer core–shell nanoreactor with chemodynamic and photodynamic dual effects as potential cancer therapeutic agents. Sci Rep [Internet] 2022;12(1). Available from: https://doi.org/10.1038/s41598-022-23002-5.

Tang L, Yang X, Yin Q, Cai K, Wang H, Chaudhury I, et al. Investigating the optimal size of anticancer nanomedicine. Proc Natl Acad Sci U S A [Internet]. 2014;111(43):15344–15349. Available from: https://doi.org/10.1073/pnas.1411499111.

Tang L, Li J, Pan T, Yin Y, Mei Y, Xiao Q, et al. Versatile carbon nanoplatforms for cancer treatment and diagnosis: strategies, applications and future perspectives. Theranostics [Internet]. 2022;12(5):2290–2321. Available from: https://doi.org/10.7150/thno.69628.

Tasso M, Lago Huvelle MA, Diaz Bessone I, Picco AS. Toxicity assessment of nanomaterials. In: Nanomedicine and nanotoxicology. Cham: Springer International Publishing; 2020. p. 383–446.

Vawhal PK, Jadhav SB, Kaushik S, Panigrahi KC, Nayak C, Urmee H, et al. Coumarin-based sulfonamide derivatives as potential DPP-IV inhibitors: pre-ADME analysis, toxicity profile, computational analysis, and *In Vitro* enzyme assay. Molecules [Internet]. 2023;28(3):1004. Available from: https://doi.org/10.3390/molecules28031004.

Venderley J, Mallayya K, Matty M, Krogstad M, Ruff J, Pleiss G, et al. Harnessing interpretable and unsupervised machine learning to address big data from modern X-ray diffraction. Proc Natl Acad Sci U S A [Internet] 2022;119(24). Available from: https://doi.org/10.1073/pnas.2109665119.

Volpe G, Volpe G, Gigan S. Brownian motion in a speckle light field: tunable anomalous diffusion and selective optical manipulation. Sci Rep [Internet]. 2014 [cited 2024 Aug 16];4(1):1–6. Available from: https://www.nature.com/articles/srep03936

Wang Y. High-resolution 3DImaging and material analysis with TransmissionX-ray microscopy and nano-CT [Internet]. In: Characterization of materials. Wiley; 2012. p. 1–10. Available from: https://doi.org/10.1002/0471266965.com133.

Wang F, Pauletti GM, Wang J, Zhang J, Ewing RC, Wang Y, et al. Dual surface-functionalized Janus nanocomposites of polystyrene/Fe3O4@SiO2 for simultaneous tumor cell targeting and stimulus-induced drug release. Adv Mater [Internet]. 2013;25(25):3485–3489.Available from: https://doi.org/10.1002/adma.201301376

Wang H, Mu Q, Revia R, Wang K, Tian B, Lin G, et al. Iron oxide-carbon core-shell nanoparticles for dual-modal imaging-guided photothermal therapy. J Control Release [Internet]. 2018a;289:70–78. Available from: https://doi.org/10.1016/j.jconrel.2018.09.022.

Wang Y-C, Rhéaume É, Lesage F, Kakkar A. Synthetic methodologies to gold nanoshells: an overview. Molecules [Internet] 2018b;23(11):2851. Available from: https://doi.org/10.3390/molecules23112851.

Wang B-X, Zheng H, Krivyakina E, Chmaissem O, Lopes PP, Lynn JW, et al. Synthesis and characterization of bulk Nd1-x Sr x NiO_2 and Nd1-x Sr x NiO_3. Phys Rev Mater [Internet] 2020;4(8). Available from: https://doi.org/10.1103/PhysRevMaterials.4.084409.

Xu L, Zhu J, Rong L, Yang H, Wang B, Lu S, et al. Osteoblast-specific down-regulation of NLRP3 inflammasome by aptamer-functionalized liposome nanoparticles improves bone quality in postmenopausal osteoporosis rats. Theranostics [Internet]. 2024;14(10):3945–3962. Available from: https://doi.org/10.7150/thno.95423.

Yameen B, Wu J, Vilos C, Whyte A, Werstler D, Pollit L, et al. Stimuli-responsive nanotheranostics. In: Nanotheranostics for personalized medicine. World Scientific; 2016. p. 267–96.

Yan Y, Chen B, Yin Q, Wang Z, Yang Y, Wan F, et al. Dissecting extracellular and intracellular distribution of nanoparticles and their contribution to therapeutic response by monochromatic ratiometric imaging. Nat Commun [Internet]. 2022;13(1). Available from: https://doi.org/10.1038/s41467-022-29679-6.

Yang G, Liu Y, Wang H, Wilson R, Hui Y, Yu L, et al. Bioinspired core–shell nanoparticles for hydrophobic drug delivery. Angew Chem Int Ed Engl [Internet] 2019;58(40):14357–14364. Available from: https://doi.org/10.1002/anie.201908357

Yong T, Zhang X, Bie N, Zhang H, Zhang X, Li F, et al. Tumor exosome-based nanoparticles are efficient drug carriers for chemotherapy. Nat Commun [Internet]. 2019;10(1). Available from: https://doi.org/10.1038/s41467-019-11718-4.

Zein R, Sharrouf W, Selting K. Physical properties of nanoparticles that result in improved cancer targeting. J Oncol [Internet] 2020;2020:1–16. Available from: https://doi.org/10.1155/2020/5194780.

Zhang Q. Recent progress of polymer micelles used as anti-cancer multifunctional nanocarriers. J Phys Conf Ser [Internet] 2021;1885(3):032049. Available from: https://doi.org/10.1088/1742-6596/1885/3/032049.

Zhang J, Zhang X, Liu G, Chang D, Liang X, Zhu X, et al. Intracellular trafficking network of protein nanocapsules: endocytosis, exocytosis and autophagy. Theranostics [Internet]. 2016;6(12):2099–2113. Available from: https://doi.org/10.7150/thno.16587

Zhang D, Zhang W, Wu X, Li Q, Mu Z, Sun F, et al. Dual modal imaging-guided drug delivery system for combined chemo-photothermal melanoma therapy. Int J Nanomedicine [Internet]. 2021;16:3457–3472. Available from: https://doi.org/10.2147/ijn.s306269.

Zhang C, Tu W, Chen X, Xu B, Li X, Hu C, et al. Intelligent design of polymer nanogels for full-process sensitized radiotherapy and dual-mode computed tomography/magnetic resonance imaging of tumors. Theranostics [Internet]. 2022;12(7):3420–3437. Available from: https://doi.org/10.7150/thno.70346.

Part II

Organic/Organic Core Shell Nanoconstructs

Polymer/polymer Core-Shell Nanoconstructs for Cancer Theragnostics

9

Gaurav Tiwari, K. Kranthi Kumar, Madhusmruti Khandai, Shashi Ravi Suman Rudrangi, and Namdev Dhas

Abstract

The development of innovative nanomaterials for cancer detection has received a lot of attention in recent years. Because of their special qualities and adaptability, polymer core-shell nanoconstructs have become one of the most promising options among them. The review starts off with a summary of the various polymer kinds that are utilized as core and shell materials, emphasizing their benefits and drawbacks. The synthesis of polymer core-shell nanoconstructs using different techniques is then covered, including layer-by-layer assembly, template synthesis, and emulsion polymerization. Polymer core-shell nanoconstructs, which combine therapeutic and diagnostic functions into a single nanostructure, have emerged as promising platforms for cancer theragnostic applications. Development, synthesis, and potential uses of polymer core-shell nanoconstructs for cancer detection and therapies have advanced recently, and this review summarizes these developments. A number of benefits come with the core-shell

G. Tiwari (✉)
PSIT-Pranveer Singh Institute of Technology (Pharmacy), Kanpur, Uttar Pradesh, India
e-mail: dr.gauravtiwari@psit.in

K. K. Kumar
College of Pharmaceutical Sciences, Dayananda Sagar University Bengaluru, Bengaluru, India

M. Khandai
Royal College of Pharmacy and Health Sciences, Ganjam, India

S. R. S. Rudrangi
Department of Product Development and Commercialisation, Kent Pharma UK Limited, Ashford, Kent, UK

N. Dhas
Department of Pharmaceutics, Manipal College of Pharmaceutical Sciences, Manipal Academy of Higher Education (MAHE), Manipal, Karnataka, India

J. K. Patel et al. (eds.), *Core-Shell Nano Constructs for Cancer Theragnostic*,
https://doi.org/10.1007/978-981-96-3025-7_9

architecture, such as improved stability, regulated drug release kinetics, and the capacity to combine numerous imaging and therapeutic agents at once. Several polymerization methods, including layer-by-layer assembly, polymer blending, and emulsion polymerization, have been used to precisely control the size, shape, and surface characteristics of these nanoconstructs. Moreover, surface modification techniques allow for the targeted destruction of cancer cells, increasing therapeutic efficacy and reducing off-target effects. Additionally, to produce synergistic effects and overcome multidrug resistance, therapeutic agents such as nucleic acids, photothermal agents, and chemotherapeutic drugs can be conjugated or encapsulated onto the nanoconstructs. Polymer core-shell nanoconstructs have the potential to revolutionize personalized cancer therapy through precise diagnosis, targeted drug delivery, and real-time treatment efficacy monitoring in the clinic. Before being widely used in clinical settings, issues like biocompatibility, systemic toxicity, and regulatory approval must be resolved. All things considered, polymer core-shell nanostructures offer a flexible platform for developing cancer therapeutics and enhancing patient outcomes.

Keywords

Types of core material · Types of shell material · Specific Surface Functionalization Strategies · Future perspectives

1 Introduction

As one of the primary causes of morbidity and death globally, cancer demands the creation of novel therapeutic and diagnostic strategies. Polymer core-shell nanoconstructs have become a flexible platform for cancer theragnostics, with the ability to provide tailored treatment plans, improved imaging, and targeted drug delivery. Because of the unique design of these nanoconstructs—a core substance surrounded by a shell layer—cancer may be precisely and successfully managed by integrating therapeutic and diagnostic components into a single unit. A variety of therapeutic chemicals, including photothermal agents, nucleic acids, and chemotherapeutic medications, may be put into the cores of polymer core-shell nanoconstructs to reduce systemic toxicity and give tailored therapy to cancer cells (Rosenholm et al. 2010). The shell layer can also be engineered to facilitate surface functionalization for targeted delivery, control drug release kinetics, and offer stability. Polymer core-shell nanoconstructs are excellent candidates for theragnostic applications in cancer treatment due to their multifunctional nature. In this review, we emphasize the latest developments in polymer core-shell nanoconstruct design, synthesis, and applications for cancer theragnostic applications. We go over the different polymerization methods that are utilized to create these nanostructures, along with the approaches taken for surface modification and targeting. Additionally, we investigate the incorporation of diagnostic modalities into polymer core-shell nanoconstructs for precise cancer detection and treatment response monitoring, including computed tomography (CT), positron emission tomography (PET), magnetic resonance imaging

(MRI), and fluorescence imaging (Coll et al. 2011). All things considered, polymer core-shell nanoconstructs are a promising method for developing cancer theragnostics, with the potential to lead to more individualized treatment plans and better patient outcomes. Regulatory approval, systemic toxicity, and biocompatibility are a few of the issues that need to be resolved before they can reach their full clinical potential. The purpose of this review is to shed light on the state of polymer core-shell nanoconstructs for cancer theragnostics as of right now and to point out potential future developments in this fascinating field (Lai et al. 2013).

The use of nanomedicines in the identification and management of diseases is very broad. According to Bae et al. (2011), nanotechnology has made a significant contribution to cancer treatment and offers a novel paradigm for addressing problems with current chemotherapeutic agents. It is a non-invasive therapy that enhances human wellness and can be applied as a molecular tool for specific molecular medical interventions. The field of nanoscale and nanotechnology research is rapidly evolving. The development of nanosystems for use in the medical field has led to new difficulties in the design of smart materials. Creating new systems that can administer medicinal agents to patients in a less risky and more efficient manner is the primary objective in this area. Applications for nanoconstructs are found in a number of disciplines, involving material science, energy, electronics, and healthcare. Nanoconstructs, for instance, can be created for the administration of medication, imaging, and diagnosis in the medical field. The delivery of medications is one common use for nanoconstructs. Drug delivery to precise target sites within the body can be facilitated by designing nanoparticles, which can enhance drug efficiency and minimize adverse reactions (Tang et al. 2012; Lou et al. 2008).

2 Types of Core Material

1. **Magnetic Core**

The choice of magnetic core material for core-shell nanoconstructs depends on the specific application and desired properties. Here are some commonly used magnetic core materials in the development of core-shell nanoconstructs:

- **Superparamagnetic Iron Oxide (SPION):** Iron oxide nanoparticles, mainly magnetite (Fe_3O_4) or maghemite (γ-Fe_2O_3), make up its composition. The magnetic moments of SPIONs align with an external magnetic field, but when the field vanishes, they revert to a random state. This phenomenon is known as super-paramagnetism. This feature is useful for a number of applications, such as imaging and magnetic targeting. Pharmaceutical substances can be loaded into SPIONs, and when an external magnetic field is applied, their magnetic characteristics enable targeted drug delivery (Tang et al. 2012; Lou et al. 2008; Wan and Zhao 2007; Lee and Char 2009). To increase the core's stability and biological compatibility, compatible coatings like lipids or polymers can be applied. When exposed to an alternating magnetic field, it can produce heat—a condition called magnetic hyperthermia. This characteristic is investigated for medicinal uses, especially in the therapy of cancer, where targeted heating can be produced to destroy cancer cells (Mandal and Kruk 2012).

- **Cobalt Nanoparticles:** Cobalt nanoparticles are made up of cobalt metal in the nanoparticulate arrangement. Cobalt is a ferromagnetic material, and cobalt nanoparticles show magnetic features. They are therefore appropriate for uses that take advantage of magnetic conduct. Because of their magnetic qualities, they can be utilized as contrast materials in imaging techniques like magnetic resonance imaging (MRI). Their ability to image is improved when they are incorporated into nanoconstructs. Theranostics, or combining treatment and diagnostics, and drug delivery are two biomedical applications of cobalt nanoparticles that are currently being investigated (Tang et al. 2012; Tan et al. 2005). They are especially intriguing for targeted drug delivery in cancer treatments when subjected to an external magnetic field due to their magnetic characteristics.
- **Cobalt Nanoparticles:** Nickel metal in nanoparticulate form creates nickel nanoparticles. It is possible to create nickel nanoparticles that are nanoscale in size, usually with a diameter of a couple to multiple tens of nanometers. Nickel nanoparticles have the ability in biomedical imaging because of their ability to function as MRI contrast agents because of their magnetic characteristics. Nickel nanoparticles can be utilized for magnetic hyperthermia, just like other magnetic nanoparticles. They produce heat in the presence of an alternating magnetic field, that can be used for targeted hyperthermic therapy, especially for the management of cancer. Applications of nickel nanoparticles in medicine, such as drug delivery and imaging, are currently being researched (Lou et al. 2008; Li et al. 2010). They are especially intriguing for targeted drug delivery in the presence of an external magnetic field due to their magnetic characteristics.
- **Iron Nanoparticles:** Iron metal in nanoparticulate form makes up the majority of iron nanoparticles. Due to their distinct magnetic characteristics, iron oxide nanoparticles like magnetite (Fe_3O_4) and maghemite (γ-Fe_2O_3) are frequently employed. The magnetic properties of iron nanoparticles, especially iron oxide nanoparticles, make them useful as contrast agents in magnetic resonance imaging. They improve the tissue's visibility during imaging. When placed in an alternating magnetic field, it can produce heat, which makes it appropriate for magnetic hyperthermia in cancer treatment. These nanostructures are investigated for a number of imaging modalities, such as photoacoustic and fluorescence imaging. For use in biomedicine, iron nanoparticles—particularly iron oxide nanoparticles—are being actively researched. They are taken into consideration for therapy, imaging, and targeted drug delivery (Lou et al. 2008; Wan and Zhao 2007; Lee and Char 2009; Mandal and Kruk 2012; Tan et al. 2005; Li et al. 2010; Li et al. 2012).
- **Manganese Ferrite Nanoparticles:** Manganese (Mn), iron (Fe), and oxygen (O) make up manganese ferrite nanoparticles. $MnFe_2O_4$ is the chemical formula. A ferrimagnetic substance with magnetic properties is manganese ferrite. It is intriguing for use in magnetic and electronic devices since it also possesses semi-

conducting properties. Because of their magnetic characteristics, manganese ferrite nanoparticles may be investigated as MRI contrast agents. Their application could improve medical diagnostic imaging features. It might be used as a fundamental component of medication delivery systems. Targeted drug delivery with an external magnetic field is made possible by the magnetic properties. Its semiconducting qualities make it useful for sensors and electronic devices (Wan and Zhao 2007; Blas et al. 2008). Applications in domains like magnetoelectronics and spintronics might be possible.

2. Metallic Core Materials

Metallic core materials play a crucial role in the development of core-shell nanoconstructs, offering unique physical and chemical properties that make them valuable for various biomedical applications. The following are some commonly used metallic core materials for core-shell nanoconstructs:

- **Gold Nanoparticles (AuNPs):** Gold atoms are arranged in nanoscale dimensions to form gold nanoparticles. They can take on a variety of forms, such as cubes, spheres, rods, and more. The sizes of AuNPs can be produced in a range of 1 to 100 nanometers, and both size and shape have a significant impact on their characteristics. In imaging methods like computed tomography (CT) and photoacoustic imaging, AuNPs can act as contrast agents. Because of its ability to convert light into heat, photothermal therapy can be used to treat cancer (Lee and Char 2009; Kim et al. 2008). When used in biosensors for biomolecule detection, it exhibits high sensitivity, quick detection, and flexible surface functionalization. When used in Photodynamic therapy (PDT) for cancer treatment, AuNPs can improve photosensitizer effectiveness.
- **Silver Nanoparticles (AgNPs):** When using silver as a core material for nanoconstructs, silver nanoparticles (AgNPs) or silver-based nanostructures are integrated into a larger structure for particular uses. Surface plasmon resonance (SPR), one of its special optical qualities, can be used for a number of purposes, including imaging and sensing. Due to their well-known antimicrobial qualities, silver nanoparticles are useful for use in medical devices, wound dressings, and antibacterial coatings. Because silver nanoparticles can increase contrast, they can be used as contrast agents in X-ray, computed tomography (CT), and photoacoustic imaging procedures. It could be the main component of medication delivery systems. They have the ability to encapsulate medications and enable targeted delivery to particular tissues or cells (Lee and Char 2009; Mandal and Kruk 2012; Tan et al. 2005; Li et al. 2010; Li et al. 2012; Blas et al. 2008; Kim et al. 2008; Vasani et al. 2011).
- **Platinum Nanoparticles (PtNPs):** Platinum nanoparticles (PtNPs) have special qualities that can be applied to a variety of applications when used as nanoconstruct core materials. PtNPs efficiently convert chemical energy into electrical energy in fuel cells by acting as catalysts. They can be included in nanostructures

to improve the efficiency of catalysis. PtNPs can be produced by sonochemical techniques, colloidal techniques, and chemical reduction. The size, form, and surface characteristics of the nanoparticles can be altered using these techniques. It can facilitate effective catalysis in chemical reactions by serving as the fundamental component of catalytic nanoconstructs. In the development of fuel cells and other energy conversion devices, this is especially crucial. Because of their paramagnetic characteristics, PtNPs can be used as contrast agents in imaging methods like magnetic resonance imaging (MRI). It can be applied to prodrugs based on platinum (Mandal and Kruk 2012; Hu et al. 2000). This method increases the therapy's selectivity by using PtNPs' catalytic properties to cause inactive prodrugs to transform into their active form only in cancer cells.

- **Titanium Dioxide Nanoparticles (TiO_2NPs):** The properties of titanium dioxide (TiO_2) nanoparticles make them suitable as a nanoconstruct core material; they have been thoroughly researched for a variety of applications. In photodynamic therapy, titanium dioxide nanoparticles can function as photosensitizers. Reactive oxygen species (ROS) produced by TiO_2 in response to light exposure cause localized cytotoxic effects on cancer cells. It can be added to nanoconstructs to increase the specificity of photodynamic therapy by delivering photosensitizers to cancer cells in a targeted manner. In photoacoustic imaging, TiO_2 nanoparticles can be employed as contrast agents by absorbing light and generating acoustic signals (Tan et al. 2005; Fu et al. 2003). This imaging technique can yield important data for the detection and tracking of cancer. As radiosensitizers, it has been investigated. TiO_2 can increase the production of reactive oxygen species when exposed to ionizing radiation, which can increase. When exposed to ionizing radiation, TiO_2 can enhance the generation of reactive oxygen species, increasing the effectiveness of radiotherapy against cancer cells (Rao and Lopez 2000).

3. **Polymeric Core material**

Polymeric core materials in core-shell nanoconstructs offer a versatile platform for various biomedical applications, including drug delivery, imaging, and therapy. These materials provide advantages such as biocompatibility, tunable properties, and the ability to encapsulate or conjugate therapeutic agents. Here are some common polymeric core materials for core-shell nanoconstructs:

- **Poly(lactic-co-glycolic acid) (PLGA):** Lactic-co-glycolic acid, or poly (lactic-co-glycolic acid) acid, is a biocompatible and biodegradable polymer that has been extensively employed as a building block for the creation of nanoconstructs for a range of uses, including cancer treatment (Tan et al. 2005; Li et al. 2010). Anticancer medications can be delivered and encapsulated using it. This maximizes the drug's effectiveness for therapy while reducing its negative impact on healthy tissues by enabling the controlled release of the medication. Additionally, it has the ability to encapsulate several medications, allowing combination therapy to have synergistic effects on cancer cells. To increase PLGA nanoparticles' specificity for cancer cells, targeting ligands like peptides or antibodies can be added. By reducing off-target effects, this targeted delivery raises the therapeutic index overall. For a variety of imaging modalities (such as computed tomography, magnetic resonance imaging, or fluorescence imaging), it can be loaded with imaging agents like fluorophores or contrast agents. This permits imaging and therapy (theranostics) to be done concurrently. Cancer vaccines can be delivered via PLGA nanoconstructs. To elicit an immune response directed against cancer cells, antigens or tumor-associated antigens can be encapsulated in PLGA particles (Liu et al. 2014; Saint-Cricq et al. 2015; Li et al. 2016).

- **Polyethylene Glycol (PEG):** In the field of nanotechnology, polyethylene glycol (PEG) is a versatile and extensively used polymer, especially in the creation of nanoconstructs for cancer therapy. PEG has various qualities that make it appropriate for use in nanoconstructs, which are nanoscale structures intended to deliver therapeutic agents to particular targets within the body. Biocompatibility is essential for reducing side effects and guaranteeing the security of nanoconstructs in cancer treatment. PEG chains are attached to the surface of nanoconstructs through a process called PEGylation, which gives them stealth characteristics. By extending the duration of nanoconstructs' circulation in the bloodstream and enhancing their accumulation at the tumor site through the enhanced permeability and retention (EPR) effect, stealthiness aids in immune system evasion (Li et al. 2012; Hu et al. 2018). Effective encapsulation and delivery of hydrophobic anticancer drugs are facilitated by hydrophilic characteristics. PEG-based nanoconstructs can be designed to provide controlled and sustained drug release. Controlled release enhances the therapeutic efficacy of anticancer drugs, reduces systemic toxicity, and maintains therapeutic drug levels over an extended period (Wang et al. 2017).
- **Polyvinyl Alcohol (PVA):** Polyvinyl Alcohol (PVA) is a synthetic polymer that has found applications in various fields, including the development of nanoconstructs for cancer therapy. PVA is generally considered biocompatible, making it suitable for biomedical applications. Biocompatibility is crucial for minimizing adverse reactions and ensuring the safety of nanoconstructs in the biological environment. It can be utilized to encapsulate various drugs, including chemotherapeutic agents. Encapsulation protects the drug payload, improves its solubility, and facilitates controlled release, enhancing the therapeutic efficacy against cancer cells. PVA-based nanoconstructs can be engineered to provide controlled and sustained drug release. Controlled release contributes to maintaining therapeutic drug levels, reducing systemic toxicity, and improving the overall effectiveness of cancer treatment (Blas et al. 2008; Liu et al. 2017; Chen et al. 2015; Sun et al. 2018; Yu et al. 2013). Targeted drug delivery improves the selectivity of the nanoconstructs, ensuring that therapeutic agents are delivered primarily to cancer cells, reducing off-target effects.
- **Polyethyleneimine (PEI):** The cationic polymer polyethyleneimine, or PEI, has attracted a lot of attention in the field of nanomedicine, especially as a possible building block for nanoconstructs used in cancer treatment. Because of the amino

groups, it has a positive charge. The cationic charge makes it possible for negatively charged materials, like cell membranes or nucleic acids, to interact electrostatically, which promotes internalization and uptake by cells. Due to its effectiveness in delivering nucleic acids like DNA or RNA, it has been thoroughly investigated. Gene delivery has multiple applications in cancer therapy, such as immunomodulation, gene replacement, and silencing (Kim et al. 2008; Ding et al. 2014). Therapeutic agents can be released under controlled conditions thanks to drug encapsulation, which increases drug efficacy while reducing adverse reactions. pH-responsive drug release can be particularly beneficial in the acidic tumor microenvironment, enhancing the specificity of drug delivery to cancer cells. Its surface modification enhances the specificity of nanoconstructs, enabling targeted drug delivery or imaging to cancer cells (Kievit et al. 2009).

- **Polydopamine:** Due to its special qualities—such as adhesiveness, biocompatibility, and the capacity to encapsulate and deliver therapeutic agents—polydopamine (PDA), a versatile polymer derived from dopamine, has drawn interest in the field of nanomedicine. Because of its powerful adhesive qualities, PDA can stick to a variety of surfaces, including the surfaces of nanoparticles. When it comes to drug delivery applications, adhesive properties help to stabilize and coat nanoparticles, improving their performance. It is capable of effectively encasing and delivering a wide range of medications, including chemotherapy agents that provide controlled and sustained drug release, enhancing therapeutic efficacy and mitigating side effects. It is also biodegradable under certain conditions. In order to reduce long-term accumulation and guarantee the safe removal of PDA-based nanoconstructs from the body, biodegradability is crucial (Kim et al. 2008; Kievit et al. 2010; Stephen et al. 2016).
- **Poly(l-lysine):** A positively charged synthetic polymer called poly(l-lysine) (PLL) has been studied for possible uses in the creation of nanoconstructs for cancer treatment. Because of the amino groups in its structure, PLL has a positive charge. The positive charge makes it easier for negatively charged cellular constituents to interact with one another, which may improve cancer cells' ability to absorb and internalize cells. PLL has been applied to the delivery of DNA and RNA, among other nucleic acids. This is pertinent to gene therapy techniques, in which delivering genetic material to cancer cells is an essential component of the therapeutic regimen (Vasani et al. 2011; Wang et al. 2012). Targeting ligands, imaging agents, and therapeutic agents can all be integrated into PLL-based nanoconstructs to create multifunctional structures. A comprehensive approach to cancer treatment is made possible by multifunctionality, which combines imaging, diagnosis, and therapy into a single unit. Its imaging powers are helpful in visualizing tumors, enabling early detection and monitoring treatment responses (Lee et al. 2012).
- **Polymeric Micelles:** Amphiphilic block copolymers are the building blocks of polymeric micelles, which are nanoscale structures that have shown promise as cancer medication delivery vehicles. These micelles are useful for effectively delivering therapeutic agents to cancer cells because they have a number of benefits. Amphiphilic block copolymers with hydrophobic and hydrophilic segments

combine to form polymeric micelles. With the hydrophilic shell acting as a barrier, this structure enables the encapsulation of hydrophobic drugs in the core, improving their solubility and stability. Drug release that is regulated and prolonged can be achieved with polymeric micelles. Drug efficacy is increased, side effects are decreased, and therapeutic concentrations are sustained for a longer amount of time with controlled release (Vasani et al. 2011; Hu et al. 2000; Wan et al. 2013). Improved solubility is essential for creating a wider variety of therapeutic agents and increasing the number of medications that can be used to treat cancer. Its Active targeting enhances the specificity of drug delivery, improving accumulation at the tumor site while minimizing exposure to healthy tissues. Passive targeting can also occur through the enhanced permeability and retention (EPR) effect.

- **Poly(Lactic Acid) (PLA):** Amphiphilic block copolymers are the building blocks of polymeric micelles, which are nanoscale structures that have shown promise as cancer medication delivery vehicles. These micelles are useful for effectively delivering therapeutic agents to cancer cells because they have a number of benefits. Amphiphilic block copolymers with hydrophobic and hydrophilic segments combine to form polymeric micelles. With the hydrophilic shell acting as a barrier, this structure enables the encapsulation of hydrophobic drugs in the core, improving their solubility and stability. Drug release that is regulated and prolonged can be achieved with polymeric micelles (Fu et al. 2003; Lee et al. 2016). Drug efficacy is increased, side effects are decreased, and therapeutic concentrations are sustained for a longer amount of time with controlled release. Improved solubility is essential for creating a wider variety of therapeutic agents and increasing the number of medications that can be used to treat cancer. It can also take advantage of the EPR effect for passive targeting of tumor tissues. Passive targeting enhances the accumulation of nanoconstructs at the tumor site, improving drug delivery efficiency. Its immunomodulation may complement cancer therapy by stimulating or modulating the immune system to recognize and eliminate cancer cells (Quang et al. 2018).
- **Chitosan:** The natural polysaccharide chitosan, which is derived from chitin, has drawn a lot of interest in the field of nanomedicine due to its possible uses in cancer treatment drug delivery systems (Rao and Lopez 2000). Due to its biocompatibility and biodegradability, chitosan can be safely removed from the body without causing any negative side effects. For the nanoconstruct to eventually break down more easily, be compatible with biological systems, and have the least amount of toxicity, it must be biocompatible. It can interact with mucosal surfaces because of its mucoadhesive characteristics. Both hydrophobic and hydrophilic medications can be effectively encapsulated by chitosan. The payload is safeguarded, its solubility is improved, and controlled and prolonged drug release is made possible by drug encapsulation. It is possible to add targeting ligands to chitosan nanoconstructs to enable targeted recognition of cancer cells. Targeted drug delivery enhances the selectivity of the nanoconstructs, improving the efficiency of drug delivery to cancer cells while minimizing effects on healthy tissues. Chitosan improves the stability of drugs and protects them from degrada-

tion (Hu et al. 2000; Rao and Lopez 2000; Quang et al. 2018). Chitosan is amenable to various modifications, allowing for the incorporation of functional groups, ligands, or imaging agents.

4. **Carbon-Based Core Material**

Carbon-based nanoconstructs have shown promise in various applications in cancer therapy and imaging. The unique properties of carbon nanomaterials make them versatile for different functionalities (Wijagkanalan et al. 2011; Boyd et al. 2006; Leiro et al. 2015; Lee et al. 2004; Kwon et al. 2001).

- **Carbon Nanotubes (CNTs):** CNTs can be used to create drug delivery nanoconstruct cores. Because of their large surface area, anticancer medications can be loaded and released under controlled conditions. CNTs can be functionalized to improve their targeting and biocompatibility.
- **Graphene Oxide (GO):** Therapeutic agents can be encapsulated and delivered to cancer cells via GO-based nanoconstructs. Because of its distinct structure and functional groups, GO offers sites for drug conjugation as well as controlled drug release.
- **Graphene and Carbon Nanotubes:** Both materials have superior photothermal qualities. They produce heat in response to near-infrared (NIR) light, which can result in localized hyperthermia (Liu et al. 2014). This can be used in photothermal therapy to destroy cancer cells selectively.
- **Carbon Nanodots:** For fluorescence imaging, carbon nanodots, which are tiny carbon nanoparticles, can be employed as imaging agents. They are appropriate for applications involving cancer imaging due to their size, biocompatibility, and fluorescence characteristics.
- **Graphene-Based Imaging Agents:** Magnetic resonance imaging (MRI) and computed tomography (CT) are two imaging applications that can benefit from the functionalization of graphene and graphene oxide with contrast agents.
- **Carbon Nanotubes for Gene Delivery:** In gene therapy applications, CNTs can be used to deliver genetic material. They can deliver nucleic acids into cells for therapeutic purposes in the treatment of cancer because of their special structure.
- **Carbon Nanomaterials as Radiosensitizers:** The potential of carbon nanomaterials, particularly carbon nanotubes (CNTs), as radiosensitizers has been investigated. They have the ability to increase ionizing radiation absorption, enhancing radiotherapy's ability to kill cancer cells.
- **Carbon Dots for PDT:** Another kind of carbon-based nanomaterial is carbon dots, which can be utilized in PDT. When activated by light, they release reactive oxygen species, which aid in the targeted destruction of cancer cells (Saint-Cricq et al. 2015).
- **Multifunctional Carbon Nanoconstructs:** By combining therapeutic, imaging, and drug delivery functions onto a single platform, carbon nanoconstructs can be developed for combination therapies.

5. **Silica Core Material**

Silica-based hybrid materials have gained attention as promising nanoconstruct core materials in the field of cancer therapy and imaging. These materials often combine the unique properties of silica with other components, such as organic polymers or inorganic nanoparticles, to achieve enhanced functionalities (Lemoine et al. 1996; Danhier 2016; Porsch et al. 2013).

- **Mesoporous Silica Nanoparticles (MSNs):** Anticancer medications can be loaded into mesoporous silica nanoparticles due to their clearly defined pores. MSNs' large surface area and adjustable pore size allow for controlled drug release, enhancing the effectiveness of treatment.
- **Polymer-Silica Hybrids:** By combining silica and polymers, hybrid nanoconstructs can provide the structural stability of silica with the biocompatibility of polymers. This combination offers controlled release and improves drug-loading capabilities.
- **Surface Functionalization:** To specifically recognize cancer cells, silica-based nanoconstructs can be functionalized with targeting ligands (such as peptides or antibodies) (Hu et al. 2018). This focused strategy lessens side effects while increasing the effectiveness of medication delivery.
- **Fluorescent Silica Nanoparticles:** Fluorescence imaging uses silica nanoparticles doped with fluorescent dyes as imaging agents (Wang et al. 2017; Liu et al. 2017). By altering their surface, these nanoconstructs can be used to specifically image cancer cells.
- **Multimodal Imaging:** Magnetic and gold nanoparticles, for example, are two examples of imaging agents that can be incorporated into silica hybrids to enable multimodal imaging. This allows for the provision of complementary information during a single imaging session.
- **Gold-Silica Nanoshells:** Photothermal therapy can be designed using silica-gold hybrid nanoconstructs, like nanoshells. In order to kill cancer cells, the gold component absorbs light and transforms it into heat. This causes localized hyperthermia.
- **Silica Nanoparticles for Gene Delivery:** Nucleic acids, such as DNA or siRNA, can be delivered using silica-based nanoconstructs for gene therapy applications in cancer treatment.
- **Biodegradable Silica Hybrids:** A few silica-based hybrids are made to be biodegradable. This means that the material will break down gradually and reduce long-term exposure to it (Chen et al. 2015) (Fig. 9.1).

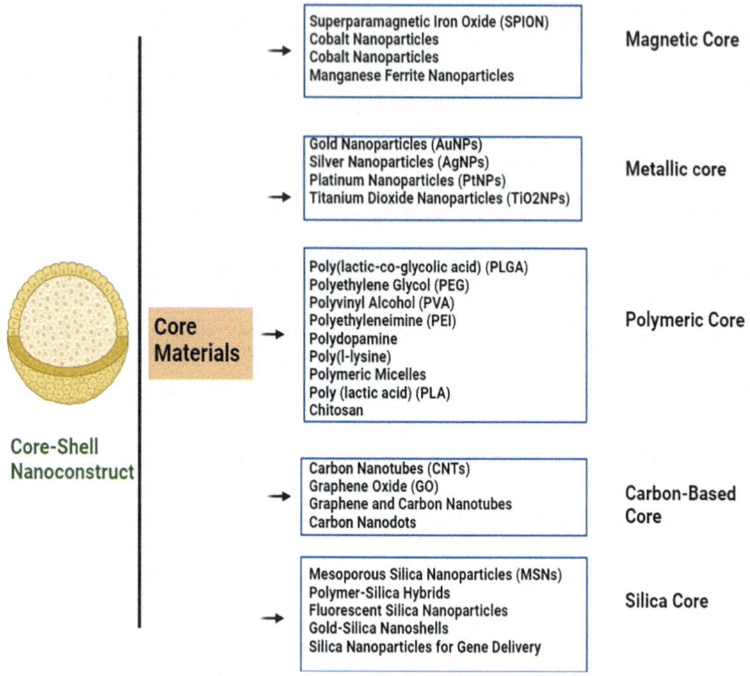

Fig. 9.1 Various core materials in a core-shell nanoconstruct

3 Types of Shell Material

1. **Lipid Shell**

Lipid-based nanoconstructs, particularly liposomes, are being extensively studied for their potential applications in cancer diagnostics. Liposomes offer a flexible and efficient platform for theragnostics, which is the integration of therapeutic and diagnostic functions into a single platform. Because lipid materials are typically biocompatible, they can be used in biological systems. This is important for drug delivery systems because the nanostructure needs to interact with tissues and cells without causing harm. They can contain a variety of hydrophilic and hydrophobic anticancer medications in their aqueous core or lipid bilayers. This reduces the likelihood of systemic side effects while enabling targeted drug delivery to cancer cells. To specifically recognize and bind to cancer cells, targeting ligands, such as antibodies or peptides, can be added to the liposome surface. This improves the medication delivery's selectivity to tumor sites (Sun et al. 2018; Yu et al. 2013; Ding et al. 2014; Kievit et al. 2009; Kievit et al. 2010). Imaging agents, such as radionuclides for nuclear imaging, fluorophores for fluorescence imaging, or contrast agents for magnetic resonance imaging, can be loaded into liposomes. This makes it possible to image the tumor non-invasively, which helps with diagnosis and tracking the effectiveness of treatment. Certain liposomes are engineered to discharge their cargo in reaction to the acidic milieu found within tumors. The therapeutic efficacy can be increased by using this pH sensitivity to trigger drug release within the tumor. They can be altered to have a longer bloodstream circulation period by adding polyethylene glycol (PEG) or other covert coatings (Yu et al. 2013; Stephen et al. 2016). Because of this stealth effect, the immune system's quick clearance is slowed down, improving the likelihood of reaching the tumor site.

2. **Polymer Shell**

Polymers find frequent usage as shell materials in nanoconstructs for a range of purposes, such as imaging, diagnostics, and drug delivery. Polymer shells stabilize the cargo that is enclosed and shield it from deterioration and premature release. For the therapeutic payload to be effective, this stability is especially crucial. It is possible to engineer polymers so that they can perform several functions inside of one nanostructure. For instance, a polymer shell may concurrently carry targeting ligands, imaging agents, and therapeutic drugs. Numerous polymers, including polyethylene glycol (PEG), poly(lactic-co-glycolic acid) (PLGA), and polyethyleneimine (PEI), are biocompatible and are utilized in cancer therapeutics. Applications in the intricate biological environments of cancer tissues depend on this. Numerous therapeutic agents, such as nucleic acids, chemotherapeutic medications, and other targeted therapies, can be delivered and encapsulated using them. The therapeutic efficacy of the payload can be improved by engineering controlled-release mechanisms. Fluorophores, radionuclides, and contrast agents for magnetic resonance imaging (MRI) can all be encapsulated in polymers (Ding et al. 2014; Wan et al. 2013). This facilitates non-invasive imaging of the tumor, which helps with diagnosis, tracks the effectiveness of treatment, and directs additional therapeutic interventions. They enable the fusion of several functions into a single nanoconstruct. An instance of a polymer-based nanoconstruct is capable of carrying three different drugs at once: a therapeutic drug, imaging agents for diagnosis, and targeting ligands for targeted delivery. Certain polymers react to particular stimuli, like changes in pH, temperature, or enzymatic activity. The therapeutic efficacy can be increased by creating nanoconstructs that release their payload within the tumor microenvironment in a controlled manner by taking advantage of this responsiveness (Kievit et al. 2009). The stealth properties that are imparted by PEGylation—the addition of PEG to the polymer shell—reduce immune system clearance and prolong the duration of circulation in the bloodstream. To guarantee adequate accumulation at the tumor site, this is crucial. They offer an adaptable framework for customizing nanoconstructs to fit the needs of particular cancer subtypes or patient types. This enhances the idea of personalized medicine by enabling theragnostics to be tailored for best results (Lee et al. 2016).

3. **Protein Shell**

Proteins are naturally occurring biomolecules, so they are biocompatible and well-tolerated in biological systems. This characteristic is essential for uses in therapeutics and medicine. They can help design nanoscale systems for a variety of uses, such as medication delivery, imaging, and diagnostics, by acting as superior shell materials for nanoconstructs. They can be chosen or engineered to specifically target particular tissues or cells (Kievit et al. 2010). Nanoconstructs can become more specific for cancer cells or other disease targets by adding targeting ligands into the protein shell. Certain proteins possess innate self-assembly characteristics that enable them to generate distinct nanostructures on their own. This characteristic makes it easier to fabricate nanoconstructs and gives you control over their size and composition. They can contain a variety of medicinal substances, such as nucleic acids, chemotherapeutic medications, or other bioactive compounds. This reduces the likelihood of systemic side effects while enabling targeted drug delivery to cancer cells (Quang et al. 2018). They can also behave similarly to natural biological processes when interacting with biological molecules. This may be helpful for the delivery of drugs, as the protein shell may help cancer cells recognize, internalize, and release therapeutics under controlled conditions (Stephen et al. 2016). Certain proteins react to changes in pH, temperature, or other environmental cues by exhibiting stimuli-responsive behavior. This characteristic can be used to create nanoconstructs that release their payload within the tumor microenvironment in a regulated manner. Protein engineering's adaptability makes it possible to tailor nanoconstructs for particular cancer types or patient profiles, which adds to the idea of personalized medicine (Wijagkanalan et al. 2011).

4. **Peptides Shell**

Peptides are made up of short chains of amino acids, and they are becoming more and more recognized for their potential as shell materials in nanoconstructs. This is especially true in the field of nanomedicine, where they can be used for imaging and drug delivery. Peptides are useful for medical applications, such as cancer therapeutics, because they are frequently biocompatible and well-tolerated by biological systems. Numerous peptides have the ability to self-assemble into nanostructures on their own. This characteristic gives control over the size, shape, and structure of nanoconstructs and makes their fabrication easier. Through the recognition of distinct biomarkers or receptors on the cell surface, they can be engineered to specifically target cancer cells. By doing this, off-target effects are decreased, and drug delivery or imaging's specificity is improved. Certain stimuli, such as variations in pH, temperature, or enzyme activity, cause certain peptides to react. This reactivity can be used to create controlled-release therapeutic agent-releasing nanoconstructs in the tumor microenvironment. Imaging agents, such as fluorophores, radionuclides, or contrast agents, can be added to peptides through modification (Wang et al. 2012). This makes it possible to image the tumor noninvasively, which helps with diagnosis and tracking the effectiveness of treatment.

Since many of them are biodegradable by nature, it is possible to create nanoconstructs that decompose organically over time. This is especially useful for systems that do not need to stay in the body for an extended period of time. Peptide design's adaptability makes it possible to customize treatment for particular cancer types or patient profiles, which advances the idea of personalized medicine. Different therapeutic agents, such as nucleic acids, chemotherapeutic medications, or other bioactive molecules, can be encapsulated in peptide-based nanoconstructs. This makes possible targeted medication delivery to cancer cells while minimizing systemic side effects (Lee et al. 2012; Wan et al. 2013; Lee et al. 2016; Quang et al. 2018; Wijagkanalan et al. 2011; Boyd et al. 2006; Leiro et al. 2015).

5. **Dendrimers Shell**

With their highly branched structure and well-defined structure, dendrimers provide exact control over surface functionalities, size, and shape. It is useful to have this structural precision when creating nanoconstructs with particular characteristics. Multivalent interactions are possible because of the numerous functional groups that are usually present on their surface. This characteristic can be used to attach imaging or diagnostic agents, improve targeting, and increase drug-loading capacity. They can contain a variety of medicinal substances, such as nucleic acids, chemotherapeutic medications, or other bioactive compounds. Therapeutic results may be enhanced by the regulated and prolonged release of medications from dendrimer-based nanoconstructs. Numerous dendrimers have good biocompatibility, including polyamidoamine (PAMAM) and polypropyleneimine (PPI) dendrimers (Wan et al. 2013; Leiro et al. 2015). Notwithstanding, the possibility of cytotoxicity persists, and adjustments can be implemented to enhance safety characteristics. They are readily functionalized using polymers, imaging agents, or targeting ligands. This surface modification makes it possible to create multifunctional nanoconstructs with imaging capabilities and the ability to target cancer cells. Imaging agents, such as radionuclides, fluorophores, or contrast agents for magnetic resonance imaging (MRI), can be added to dendrimers. This makes it possible to image the tumor non-invasively, which helps with diagnosis and tracking the effectiveness of treatment. Their multivalent surface functional groups frequently show effective cellular uptake. This feature is essential for targeting cancer cells with therapeutic payloads (Lee et al. 2016).

6. **Inorganic Material Shell**

Nanoconstructs are stabilized and made durable by inorganic materials like silica, gold, or iron oxide nanoparticles. During circulation and delivery, this stability is beneficial for preserving the construct's structural integrity. Their unique properties and functionalities have led to their exploration as shell materials in nanoconstructs for cancer therapeutics. Some inorganic materials are suitable for diagnostic use because they naturally have imaging properties (Quang et al. 2018). For magnetic resonance imaging (MRI), iron oxide nanoparticles can be used as contrast agents, and gold nanoparticles have unique optical properties. They can be made to transport therapeutic payloads to particular cancerous tissues or cells. Targeting ligands functionalize the surface to improve drug delivery specificity. A controlled release of medication within the tumor microenvironment is possible thanks to the engineering of certain inorganic materials to react to external stimuli, such as temperature or pH changes. It is simple to alter the surface of inorganic nanoparticles to increase targeting efficiency, decrease immunogenicity, or improve biocompatibility. These modifications can be accomplished by applying surface ligands or polymer coatings. Inorganic materials can be made more biocompatible by carefully choosing them and modifying their surfaces, which will ensure that they have fewer negative effects on biological systems. By combining different therapeutic agents with imaging capabilities, inorganic nanoconstructs can enable combination therapies that address multiple cancer aspects at once (Wijagkanalan et al. 2011). Inorganic materials can have their surfaces modified, such as with polyethylene glycol (PEG) coatings, to lengthen their bloodstream circulation times and promote tumor accumulation.

7. **Polysaccharide Shell**

The potential of natural polysaccharides as shell materials in nanoconstructs has been studied. These polysaccharides come from a variety of sources, including bacteria, algae, and plants. Natural polysaccharides that are well-tolerated and biocompatible by the human body include hyaluronic acid, alginate, and chitosan. This characteristic is crucial for uses in cancer therapeutics. Their biodistribution profiles are often advantageous, facilitating the secure and effective transportation of nanoconstructs to intended locations (Boyd et al. 2006). It is simple to alter the surface of polysaccharide-based nanoconstructs to improve their characteristics. For example, conjugation with other molecules or chemical modifications can increase the stability or efficiency of targeting. They react to temperature changes and other environmental stimuli. In the tumor microenvironment, this receptivity can be used to release drugs under controlled conditions. Therapeutic agents such as proteins, nucleic acids, and chemotherapy drugs can be encapsulated in polysaccharide shells. Therapeutic payloads can now be released with precision and control thanks to this. They can be loaded with contrast or fluorophore agents for non-invasive tumor imaging, which can help with diagnosis and track the effectiveness of treatment (Leiro et al. 2015). Polysaccharide-based nanoconstructs can have their surfaces modified with polyethylene glycol (PEGylation) or other covert coatings to decrease immune system recognition and prolong their bloodstream circulation. Certain polysaccharides show immunomodulatory characteristics that could affect the immune system. In order to improve the immune system's function in cancer treatment, this aspect can be investigated (Fig. 9.2).

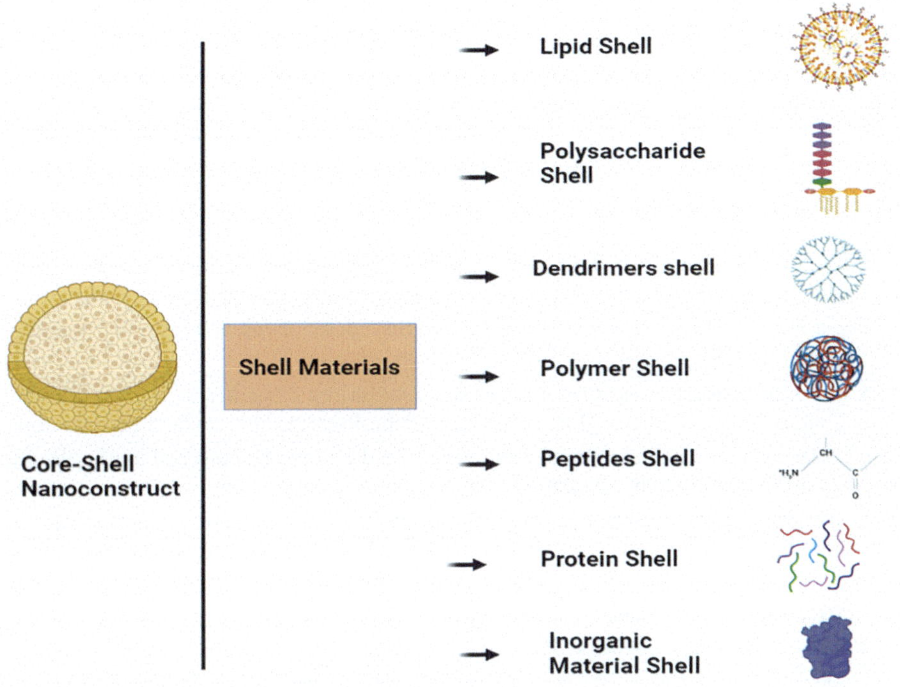

Fig. 9.2 Shell materials for core-shell nanoconstruct

4 Specific Surface Functionalization Strategies

Polymer core-shell nanostructures have become highly adaptable platforms for diverse biomedical uses, such as tissue engineering, drug delivery, and imaging. These nanoconstructs' surface functionalization is essential for improving their stability, biocompatibility, and targeting effectiveness. We address particular surface functionalization techniques used in polymer core-shell nanoconstructs for biomedical uses in this review. Covalent conjugation, in which functional groups are added to the surface of nanoconstructs to enable the attachment of targeting ligands and imaging agents, is among the primary tactics. Copper-catalyzed azide-alkyne cycloaddition (CuAAC) and other click chemistry reactions provide a focused and effective way to functionalize surfaces. The process of layer-by-layer (LBL) assembly facilitates the incorporation of multiple layers of functional molecules and polyelectrolytes by providing precise control over the surface coating. Polymer brush coatings create a dense and stable surface modification, whereas hydrophilic coatings increase biocompatibility and reduce non-specific interactions (Boyd et al. 2006; Leiro et al. 2015; Lee et al. 2004; Kwon et al. 2001; Lemoine et al. 1996). It is possible to create pH-responsive coatings that will release medication in response to pH variations, like those that occur in the tumor microenvironment. Another tactic to improve the stability and cellular uptake of nanoconstructs is surface charge modification. All things considered; these surface functionalization techniques allow polymer core-shell nanoconstructs to be tailored for particular biomedical uses. It is anticipated that additional investigations in this field will produce novel nanoconstructs with enhanced characteristics and functions, thereby increasing their potential applications in clinical practice and biomedical research (Danhier 2016). Several strategies can be used to functionalize the surface of these nanostructures, which are as follows:

- **Covalent Conjugation**: The nanoconstructs' surface can be modified by adding functional groups like amines, carboxylates, or thiols. For targeted cell targeting, these functional groups can then be employed to covalently attach targeting ligands, such as aptamers, peptides, or antibodies.
- **Click Chemistry**: A variety of molecules, including targeting ligands and imaging agents, can be utilized to selectively functionalize the surface of nanoconstructs through click chemistry reactions like the copper-catalyzed azide-alkyne cycloaddition (CuAAC) reaction.
- **Layer-by-Layer Assembly**: In LBL assembly, polyelectrolytes are progressively adsorbed onto the surface of nanoconstructs (Leiro et al. 2015). Targeting ligands and other functional molecules can be added to the surface coating using this technique, which also gives exact control over its composition and thickness.
- **Polymer Brush Coatings**: Polymer brushes, which are composed of tightly packed polymer chains that are grafted onto nanoconstructs, can be employed to alter the surface characteristics and improve stability. To further functionalize the polymer chains, functional groups can be added.
- **Hydrophilic Coatings**: These coatings have the potential to enhance the biocompatibility and stability of nanoconstructs (Lee et al. 2004). To lessen non-specific interactions with proteins and cells, hydrophilic polymers, like polyethylene glycol (PEG), can be used to functionalize the surface and create these coatings.
- **pH-Responsive Coatings**: These coatings have the ability to release medication in reaction to pH variations, like those that occur in the tumor microenvironment. The surface of nanoconstructs can be coated with these coatings by applying pH-responsive polymers, like poly(acrylic acid) (PAA).
- **Surface Charge Modification**: Stability and cellular uptake of nanoconstructs can be increased by modifying their surface charge. Controlling the surface charge can be accomplished by functionalizing the surface with charged molecules, like polyelectrolytes (Li et al. 2012; Liu et al. 2014).

5 Biomedical Applications

Core-shell polymer nanoconstructs have demonstrated significant potential for use in biomedical applications, especially in the area of cancer therapeutics. These nanoconstructs, which are made up of a shell layer encircling a core material, provide a flexible platform for the simultaneous delivery of imaging contrast agents and therapeutic agents, allowing for targeted therapy and accurate diagnosis (Kwon et al. 2001; Lemoine et al. 1996; Danhier 2016; Porsch et al. 2013). The biomedical uses of polymer core-shell nanoconstructs in cancer therapeutics are covered here.

- **Targeted Drug Delivery**: Targeted drug delivery is among the main uses for polymer core-shell nanoconstructs. Chemotherapeutic drugs can be loaded into the nanoconstruct's core, and targeting ligands that bind and recognize particular receptors on cancer cells can functionalize the shell. By enhancing drug accumulation at the tumor site, this targeted approach lowers off-target effects and increases effectiveness in therapy.
- **Imaging**: A variety of imaging modalities, including fluorescence, magnetic resonance imaging (MRI), and positron emission tomography (PET), can be incorporated into polymer core-shell nanoconstructs. These imaging agents enable non-invasive tumor imaging and treatment response monitoring by being integrated into the nanoconstruct's core or applied to its surface.
- **Theranostics**: Theranostics, the concept of combining therapeutic and diagnostic functions in a single nanoconstruct, is a potent tool for treating cancer. Therapeutic agents can be delivered and imaging features can be combined to create polymer core-shell nanoconstructs, which enable real-time drug delivery and treatment response monitoring.
- **Multimodal Imaging**: This type of imaging, in which several imaging modalities are applied concurrently to provide a more complete picture of the tumor, can be achieved by adding several imaging agents into the core-shell nanoconstruct (Lemoine et al. 1996). Better treatment planning and diagnosis accuracy may result from this.
- **Personalized Medicine**: Polymer core-shell nanoconstructs are modular, allowing for customization according to the unique features of the patient and the tumor. Personalized therapy methods that optimize the effectiveness of therapy and reduce adverse reactions can be achieved by adjusting the nanoconstruct's composition, size, and surface characteristics.
- **Tissue Engineering**: By simulating the extracellular matrix and fostering tissue regeneration, scaffolds made of nanoconstructs can be employed in tissue engineering. In order to improve tissue repair, they can also deliver growth factors or other bioactive molecules. Bioactive molecules, including cytokines and growth factors, can be delivered by nanoconstructs to facilitate tissue regeneration (Danhier 2016). These molecules can have a beneficial effect on the healing process and the functionality of the engineered tissue when released under controlled conditions.
- **Vaccines**: By boosting antigen delivery, encouraging immune cell uptake of antigen, and offering sustained antigen release, nanoconstructs can increase the effectiveness of vaccines. Additionally, adjuvants can be delivered through them to strengthen immune responses. Adjuvants and multiple antigens can be delivered in a single formulation using nanoconstructs, which makes vaccination administration easier and increases vaccine compliance (Porsch et al. 2013) (Fig. 9.3).

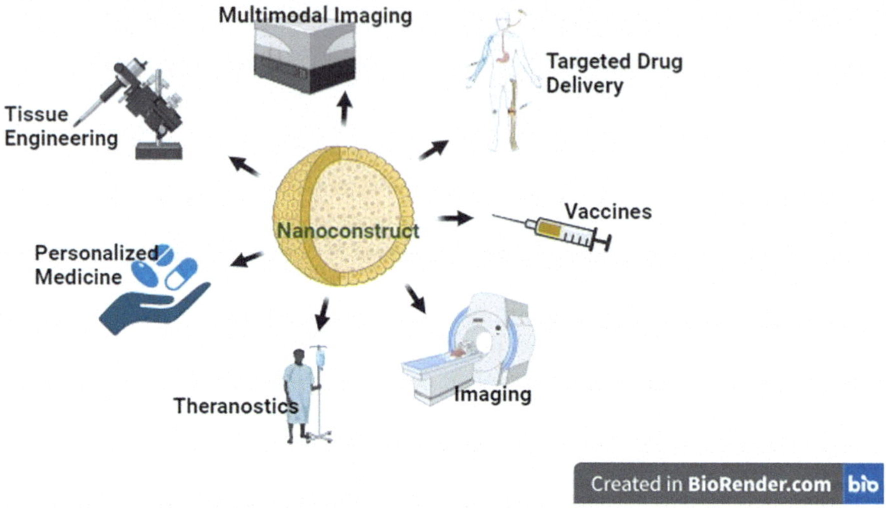

Fig. 9.3 Various biomedical applications of core-shell nanoconstruct

6 Conclusion and Future Perspectives

Drug delivery, image processing, and tissue engineering are just a few of the medical uses for which polymer core-shell nanoconstructs have shown great promise. Improved stability, regulated drug release kinetics, and the capacity to combine several functions into a single nanostructure are just a few benefits of the novel core-shell architecture. The strategies employed for surface functionalization are essential in customizing the characteristics of polymer core-shell nanoconstructs for distinct uses. Targeting ligands, imaging agents, and therapeutic molecules can be incorporated thanks to the exact control over surface properties made possible by covalent conjugation, layer-by-layer assembly, and polymer brush coatings. Future developments in polymer core-shell nanoconstruct research include exploring biodegradable materials for elevated security and biocompatibility, improving targeting strategies for increased specificity and efficiency, and creating multifunctional nanoconstructs with multiple therapeutic and diagnostic applications. A major obstacle still facing the clinical translation of polymer core-shell nanoconstructs is scale-up, manufacturing, and regulatory approval issues. To sum up, polymer core-shell nanoconstructs have the potential to transform tissue engineering, imaging, drug delivery, and clinical practice. They also hold great promise for advancing biomedical research. Novel nanoconstructs with enhanced qualities and functionalities are anticipated to be developed as a result of ongoing research and innovation in this field, ultimately helping patients and enhancing healthcare outcomes.

References

Blas H, Save M, Pasetto P, et al. Elaboration of monodisperse spherical hollow particles with ordered mesoporous silica shells via dual latex/surfactant templating: radial orientation of mesopore channels. Langmuir. 2008;24:13132–7. https://doi.org/10.1021/la801897k.

Boyd BJ, Kaminskas LM, Karellas P, et al. Cationic poly-l-lysine dendrimers: pharmacokinetics, biodistribution, and evidence for metabolism and bioresorption after intravenous administration to rats. Mol Pharm. 2006;3:614–27. https://doi.org/10.1021/mp060032e.

Chen Q, Wang X, Wang C, et al. Drug-induced self-assembly of modified albumins as nanotheranostics for tumor-targeted combination therapy. ACS Nano. 2015;9:5223–33. https://doi.org/10.1021/acsnano.5b00640.

Coll C, Mondragón L, Martínez-Máñez R, et al. Enzyme-mediated controlled release systems by anchoring peptide sequences on mesoporous silica supports. Angew Chem Int Ed. 2011;50:2138–40. https://doi.org/10.1002/anie.201004133.

Danhier F. To exploit the tumor microenvironment: since the EPR effect fails in the clinic, what is the future of nanomedicine? J Control Release. 2016;244:108–21. https://doi.org/10.1016/j.jconrel.2016.11.015.

Ding M, Zeng X, He X, et al. Cell internalizable and intracellularly degradable cationic polyurethane micelles as a potential platform for efficient imaging and drug delivery. Biomacromolecules. 2014;15:2896–906. https://doi.org/10.1021/bm500506v.

Fu Q, Rao GR, Ista LK, et al. Control of molecular transport through stimuli-responsive ordered mesoporous materials. Adv Mater. 2003;15:1262–6. https://doi.org/10.1002/adma.200305165.

Hu TJ, Gao J, Wu C. Swelling and shrinking of poly(N Isopropylacrylamide) chains adsorbed on the surface of polystyrene nanoparticles. J Macromol Sci Phys B. 2000;39:407–14. https://doi.org/10.1081/mb-100100395.

Hu DR, Chen LJ, Qu Y, et al. Oxygen-generating hybrid polymeric nanoparticles with encapsulated doxorubicin and chlorin e6 for trimodal imaging-guided combined chemo-photodynamic therapy. Theranostics. 2018;8:1558–74. https://doi.org/10.7150/thno.22989.

Kievit FM, Gunn JW, Lee D. PEI-PEG-chitosan copolymer coated iron oxide nanoparticles for safe gene delivery: synthesis, complexation, and transfection. Adv Funct Mater. 2009;19:2244–51. https://doi.org/10.1002/adfm.200801844.

Kievit FM, Veiseh O, Fang C, et al. Chlorotoxin labeled magnetic nanovectors for targeted gene delivery to glioma. ACS Nano. 2010;4:4587–94. https://doi.org/10.1021/nn1008512.

Kim J, Kim H, Lee N, et al. Multifunctional uniform nanoparticles composed of a magnetite nanocrystal core and a mesoporous silica shell for magnetic resonance and fluorescence imaging and for drug delivery. Angew Chem Int Ed Engl. 2008;47:8438–41. https://doi.org/10.1002/anie.200802469.

Kwon HY, Lee JY, Choi SW, et al. Preparation of PLGA nanoparticles containing estrogen by emulsification-diffusion method. Colloids Surf A Physicochem Eng Asp. 2001;182:123–30. https://doi.org/10.1016/s0927-7757(00)00825-6.

Lai CH, Lai NC, Chuang YJ, et al. Trivalent galactosyl-functionalized mesoporous silica nanoparticles as a target-specific delivery system for boron neutron capture therapy. Nanoscale. 2013;5:9412–8. https://doi.org/10.1039/c3nr02594b.

Lee H, Char K. Morphological changes from silica tubules to hollow spheres controlled by the intermolecular interactions within block copolymer micelle templates. ACS Appl Mater Interfaces. 2009;1:913–20. https://doi.org/10.1021/am900026s.

Lee SJ, Jeong JR, Shin SC, et al. Nanoparticles of magnetic ferric oxides encapsulated with poly(D,L latide-co-glycolide) and their applications to magnetic resonance imaging contrast agent. J Magn Magn Mater. 2004;272:2432–3. https://doi.org/10.1016/j.jmmm.2003.12.416.

Lee SJ, Muthiah M, Lee HJ, et al. Synthesis and characterization of magnetic nanoparticle-embedded multi-functional polymeric micelles for MRI-guided gene delivery. Macromol Res. 2012;20:188–96. https://doi.org/10.1007/s13233-012-0023-4.

Lee SY, Yang CY, Peng CL, et al. A theranostic micelleplex co-delivering SN-38 and VEGF siRNA for colorectal cancer therapy. Biomaterials. 2016;86:92–105. https://doi.org/10.1016/j.biomaterials.2016.01.068.

Leiro V, Garcia JP, Tomás H, et al. The present and the future of degradable dendrimers and derivatives in theranostics. Bioconjug Chem. 2015;26:1185–97. https://doi.org/10.1021/bc5006224.

Lemoine D, Francois C, Kedzierewicz F, et al. Stability study of nanoparticles of poly(e-caprolactone), poly(D,L-lactide) and poly(D,L-lactide-co-glycolide). Biomaterials. 1996;17:2191–7. https://doi.org/10.1016/0142-9612(96)00049-x.

Li J, Liu J, Wang D, et al. Interfacially controlled synthesis of hollow mesoporous silica spheres with radially oriented pore structures. Langmuir. 2010;26:12267–72. https://doi.org/10.1021/la101225j.

Li CY, Ma C, Wang F, et al. Preparation and biomedical applications of core -shell silica/magnetic nanoparticle composites. J Nanosci Nanotechnol. 2012;12:2964–72. https://doi.org/10.1166/jnn.2012.6428.

Li M, Li L, Zhan C, et al. Core-shell nanostars for multimodal therapy and imaging. Theranostics. 2016;6:2306–13. https://doi.org/10.7150/thno.15843.

Liu J, Detrembleur C, Pauw-Gillet MD, Mornet S, et al. Heat-triggered drug release systems based on mesoporous silica nanoparticles filled with a maghemite core and phase-change molecules as gatekeepers. Mater Chem B. 2014;2:59–70. https://doi.org/10.1039/c3tb21229g.

Liu J, Chen Q, Zhu W, et al. Nanoscale-coordination-polymer-shelled manganese dioxide composite nanoparticles: a multistage redox/pH/H_2O_2-responsive cancer theranostic nanoplatform. Adv Funct Mater. 2017;27:1605926. https://doi.org/10.1002/adfm.201605926.

Lou XW, Archer LA, Yang Z. Hollow micro/nanostructures: synthesis and applications. Adv Mater. 2008;20:3987–4019. https://doi.org/10.1002/adma.200800854.

Mandal M, Kruk M. Family of single-micelle templated organosilica hol low nanospheres and nanotubes synthesized through adjustment of organo silica surfactant ratio. Chem Mater. 2012;24:123–32. https://doi.org/10.1021/cm202136r.

Porsch C, Zhang Y, Östlund Å, et al. In vitro evaluation of non-protein adsorbing breast cancer theranostics based on 19F-polymer containing nanoparticles. Part Part Syst Charact. 2013;30:381–90. https://doi.org/10.1002/ppsc.201300018.

Quang HV, Chang C, Song P, et al. Caveolae-mediated mesenchymal stem cell labelling by PSS-coated PLGA PFOB nano-contrast agent for MRI. Theranostics. 2018;8:2657–71. https://doi.org/10.7150/thno.23206.

Rao GV, Lopez GP. Encapsulation of poly(N-isopropyl acrylamide) in silica: a stimuli-responsive porous hybrid material that incorporates molecular nano-valves. Adv Mater. 2000;12:1692–5. https://doi.org/10.1002/1521-4095(200011)12:22<1692::aid-adma1692>3.0.co;2-c.

Rosenholm JM, Sahlgren C, Lindén M. Towards multifunctional, targeted drug delivery systems using mesoporous silica nanoparticles opportunities & challenges. Nanoscale. 2010;2:1870–3. https://doi.org/10.1039/c0nr00156b.

Saint-Cricq P, Deshayes S, Zink JI, et al. Magnetic field activated drug delivery using thermodegradable azo-functionalised PEG-coated core-shell mesoporous silica nanoparticles. Nanoscale. 2015;7:13168–72. https://doi.org/10.1039/c5nr03777h.

Stephen ZR, Dayringer CJ, Lim JJ, et al. Approach to rapid synthesis and functionalization of iron oxide nanoparticles for high gene transfection. ACS Appl Mater Interfaces. 2016;8:6320–8. https://doi.org/10.1021/acsami.5b10883.

Sun X, Zhang M, Du R, et al. A polyethyleneimine-driven self-assembled nanoplatform for fluorescence and MR dual-mode imaging guided cancer chemotherapy. Chem Eng J. 2018;350:69–78. https://doi.org/10.1016/j.cej.2018.05.157.

Tan B, Lehmler H, Vyas SM, et al. Fluorinated-surfactant-templated synthesis of hollow silica particles with a single layer of mesopores in their shells. Adv Mater. 2005;17:2368–71. https://doi.org/10.1002/adma.200500344.

Tang F, Li L, Chen D. Mesoporous silica nanoparticles: synthesis, biocompatibility and drug delivery. Adv Mater. 2012;24:1504–34. https://doi.org/10.1002/adma.201104763.

Vasani RB, McInnes SJ, Cole MA, et al. Stimulus-responsiveness and drug release from porous silicon films ATRP-grafted with poly(N-isopropylacrylamide). Langmuir. 2011;27:7843–53. https://doi.org/10.1021/la200551g.
Wan Y, Zhao D. On the controllable soft-templating approach to mesoporous silicates. Chem Rev. 2007;107:2821–60. https://doi.org/10.1002/chin.200743247.
Wan Q, Xie L, Gao L, et al. Self-assembled magnetic theranostic nanoparticles for highly sensitive MRI of minicircle DNA delivery. Nanoscale. 2013;5:744–52. https://doi.org/10.1039/c2nr32438e.
Wang C, Ravi S, Martinez GV, et al. Dual-purpose magnetic micelles for MRI and gene delivery. J Control Release. 2012;163:82–92. https://doi.org/10.1016/j.jconrel.2012.04.030.
Wang L, Li D, Hao Y, et al. Gold nanorod – based poly (lactic-co-glycolic acid) with manganese dioxide core – shell structured multifunctional nanoplatform for cancer theranostic applications. Int J Nanomedicine. 2017;12:3059–75. https://doi.org/10.2147/ijn.s128844.
Wijagkanalan W, Kawakami S, Hashida M. Designing dendrimers for drug delivery and imaging: pharmacokinetic considerations. Pharm Res. 2011;28:1500–19. https://doi.org/10.1007/s11095-010-0339-8.
Yu J, Li X, Luo Y, et al. Poly(ethylene glycol) shell-sheddable magnetic nanomicelle as the carrier of doxorubicin with enhanced cellular uptake. Colloids Surf B Biointerfaces. 2013;107:213–9. https://doi.org/10.1016/j.colsurfb.2013.02.009.

Polymer/Lipid Core-Shell Nanoconstructs for Cancer Theragnostic

10

Viola Colaco, Anoushka Mukharya, Amrita Arup Roy, Gaurisha Alias Resha Ramnath Naik, Rahul Pokale, Ritu Kudarha, Srinivas Mutalik, and Namdev Dhas

Abstract

Cancer, a prevalent global health issue, is commonly treated with surgery, chemotherapy, radiation therapy, and hormonal therapy, often resulting in severe side effects and recurrence. The advent of nanoparticles has revolutionized drug delivery by enhancing the antitumor therapy efficacy and overcoming multidrug resistance. These nanocarriers encapsulate drugs, enabling targeted delivery and mitigating the limitations of pure drug forms. Polymeric nanoparticles (PNPs) have gained significant interest due to their small size, controlled drug release, and high stability. Similarly, lipid-based nanocarriers offer comparable benefits but face challenges such as structural instability and short circulation time. To address these limitations, polymer-lipid hybrid nanoparticles (PLNs) have been developed, combining the advantages of both polymeric and lipid-based systems. Various polymers such as polylactic-co-glycolic acid (PLGA), polylactic acid (PLA), polycaprolactone (PCL), poly(vinyl alcohol) (PVA) chitosan and hyaluronic acid along with natural and synthetic phospholipids, are utilized in PLN fabrication. Surface functionalization using aptamer, antibodies, ligands, and PEGylation enhances the cancer-targeting properties of these nanoparticles. The drug release mechanism of PLNs involves diffusion across the lipid bilayer and erosion of the polymer core, influenced by factors such as drug solubility, polymer-drug interactions, polymer degradation rate, and particle size. This chapter discusses the types and fabrication processes of PLNs, the polymers and lipids used, surface functionalization, drug release mechanisms, and their biomedical applications in cancer treatment.

V. Colaco · A. Mukharya · A. A. Roy · G. A. R. R. Naik · R. Pokale
R. Kudarha · S. Mutalik · N. Dhas (✉)
Department of Pharmaceutics, Manipal College of Pharmaceutical Sciences, Manipal Academy of Higher Education, Manipal, India

J. K. Patel et al. (eds.), *Core-Shell Nano Constructs for Cancer Theragnostic*,
https://doi.org/10.1007/978-981-96-3025-7_10

Keywords

Polymeric nanoparticles · Lipid nanoparticles · Polymer-lipid hybrid nanoparticles · Drug delivery · Targeted therapy · Surface functionalization

1 Introduction

Cancer is a major health issue worldwide that is often treated with surgery, chemotherapy, radiation therapy, targeted therapy, and hormone therapy. However, chemotherapy and radiation therapy frequently result in severe adverse effects and recurrence (Gavas et al. 2021; Jovčevska and Muyldermans 2020; Park et al. 2018). The development of nanoparticles resulted in nanotherapeutic drug delivery, which enhances drug delivery while combating antitumor MDR through combination therapy (Dhas et al. 2024a; Gavas et al. 2021; Palazzolo et al. n.d.). The particles with a diameter of less than 100 nm are known as nanoparticles and are composed of lipids, polymers, metals, and organic and inorganic materials (Carrasco-Esteban et al. 2021; Datta et al. 2025a). The dimensions and composition of the nanoparticles closely resemble biological structures and molecules, providing functional characteristics beneficial for both in vitro and in vivo research. These nanoparticles act as carriers that are constructed from biodegradable materials, allowing the encapsulation of drugs for targeted delivery, thus mitigating the limitations associated with drugs in their pure form (Angolkar et al. 2023; Colaco et al. 2024a; Dhas et al. 2024b; Yusuf et al. 2023).

Polymeric nanoparticles (PNPs) consist of solid nanoparticles, polymeric micelles, polymer-drug conjugates, and polyplexes. The PNPs have drawn interest due to their distinct characteristics like small particle size, which can be modified with specific targeting to improve therapeutic efficacy and wide diagnostic and drug delivery applications (Abasian et al. 2020; Jain et al. 2023). They exhibit controlled release of hydrophobic drugs while shielding the cargo from the external environment. They also have low polydispersity and are highly stable (Crucho and Barros 2017; Zielińska et al. 2020). Lipid-based nanocarriers like liposomes, nanostructured lipid carriers, and solid lipid nanoparticles have gained attention in drug delivery and cancer treatment (Colaco et al. 2024b; Xu et al. 2022). Lipid nanoparticles (LNPs) can encapsulate hydrophobic and hydrophilic drugs, leading to reduced toxicity, regulated release from the carrier, drug pharmacological action, and improved permeability, potentially due to increased membrane fluidity (García-Pinel et al. 2019; Sivadasan et al. 2021; Xu et al. 2022). These formulations can influence the intestinal environment, increasing drug transfer to the lymphatic system and interfering with enterocyte-based transport pathways. Recently, polymer–lipid hybrid nanoparticles (PLNs) have been formulated to amalgamate the benefits of PNPs and liposomes while addressing their limitation, such as structural instability, short circulation period, and leakage of materials. PLNs have a composite profile that includes therapeutic efficacy, biomimetic features, and biodegradability, offering a viable approach to drug delivery strategies (Datta et al. 2025b; Hadinoto et al. 2013;

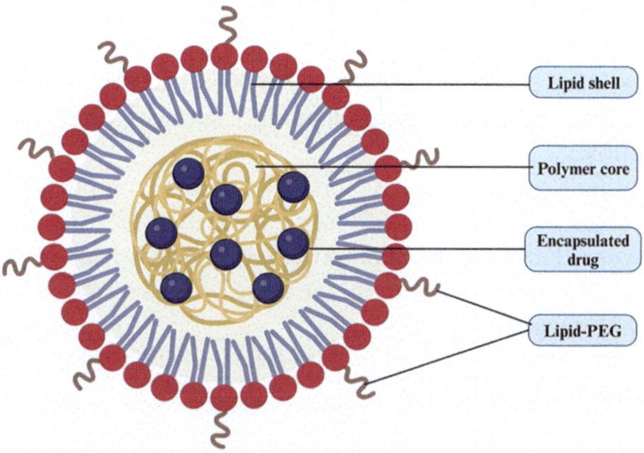

Fig. 10.1 Structure of PLNs, consisting of a drug encapsulated in the polymer core, a lipid shell, and a lipid-PEG layer at the outer layer

Sivadasan et al. 2021). According to research findings, these PLNs exhibit prolonged systemic circulation and drug release for at least 24 h. Moreover, biocompatible PLNs are efficiently degraded and eliminated by the body within 72 h (Parveen et al. 2023). Figure 10.1 depicts the structure of the PLN.

1.1 Polymer-Lipid Hybrid Nanoparticles

The PLNs are novel types of particles comprising a polymeric core encapsulated within a lipid shell. These nanoparticles possess a synergistic blend of benefits from both polymeric and lipid-based systems, enabling encapsulation of both hydrophilic and hydrophobic drugs due to their core-shell structure (Bandi et al. 2023; Dhas et al. 2022; Sivadasan et al. 2021; Wilhelm Romero et al. 2021). The structure of drug-loaded polymeric-lipid core-shell nanoparticles, as shown in Fig. 10.1, contains three distinct components: the core, the lipid shell layer, and the lipid-PEG layer. The core comprises a polymer matrix and the hydrophobic drug complex, having high drug-loading efficacy. The polymeric core is enveloped by single or multiple lipid layers, augmenting bioavailability and facilitating in drug release from the core matrix (Dave et al. 2019; Mukherjee et al. 2019; Sivadasan et al. 2021). The inner lipid layer functions as a structural barricade, reducing the loss of entrapped drugs during formulation and preventing diffusion of water inward avoiding polymer degradation and sustained release kinetics. Further, the outer lipid-polymeric layer functions as a diffusion barrier and extends the nanoparticles' circulation time in the systemic circulation. The interstitial space between the polymeric core and lipid shell is filled with an aqueous layer. The outer layer of PLNs is made up of lipid-PEG coating the lipid layer and thus promotes stability, increases circulation time, and aids in steric stabilization (Chaudhary et al. 2018; Hadinoto et al. 2013; Jose et al. 2018).

Advantages of PLNs (Abasian et al. 2020; Jain et al. 2023; Sivadasan et al. 2021)

1. The PLNs have the combined benefit of PNPs and LNPs, such as suitable drug release profiles, enhanced drug loading, surface functionalization, and biocompatibility.
2. The use of stimuli-responsive polymers helps in the controlled release of drugs along with enhanced therapeutic efficacy.
3. The lipid layer offers structural integrity and prevents leakage of the contents from the core and water penetration, ensuring continuous resistance and maximum structural integrity.
4. The PLNs demonstrate enhanced in vivo efficacy when compared to individual PNPs and LNPs.

Limitations of PLNs (Dave et al. 2019; Shah et al. 2022; Sivadasan et al. 2021)

1. Difficulty in optimizing the lipid/polymer ratio within nanoparticles hinders the translational rate from laboratory research to clinical practical applications.
2. The changes in the ratio of lipid/polymer composition for long-term storage can lead to leakage of the drug, phase separation, agglomeration, and destabilization.
3. The loading of hydrophilic into the PLN matrix is limited.
4. The fabrication of PLNs requires additional steps which contribute to higher costs, potentially leading to patient unacceptance.

2 Types of PLNs

The PLNs are categorized into different groups based on the arrangement of polymer and lipid components. It includes polymer core-lipid shell hybrid nanoparticles, hollow core-shell type PLNs, biomimetic PLNs, monolithic PLNs, and polymer-caged liposomes, which are shown in Fig. 10.2.

2.1 Polymer Core-Lipid Shell Hybrid Nanoparticles

The polymeric core-lipid shell hybrid nanoparticles are the basic form of PLNs, in which a polymer core is surrounded by a single lipid layer consisting of a lipid shell and lipid-PEG. The structural advantages and biomimetic properties of the lipid contribute to the development of advanced delivery systems for the treatment of various disorders (Dave et al. 2019; Sivadasan et al. 2021). These nanoparticles help in site-specific delivery, extended encapsulation potential, and sustained release (Gajbhiye et al. 2023). Additionally, the polymer influences the drug release profile, thereby enhancing the lipid layer stability. However, the drawback of this system is encapsulation of hydrophilic drugs, to overcome this limitation a blend of polymer and lipid can be employed. In this context, amphiphilic lipids have a significant

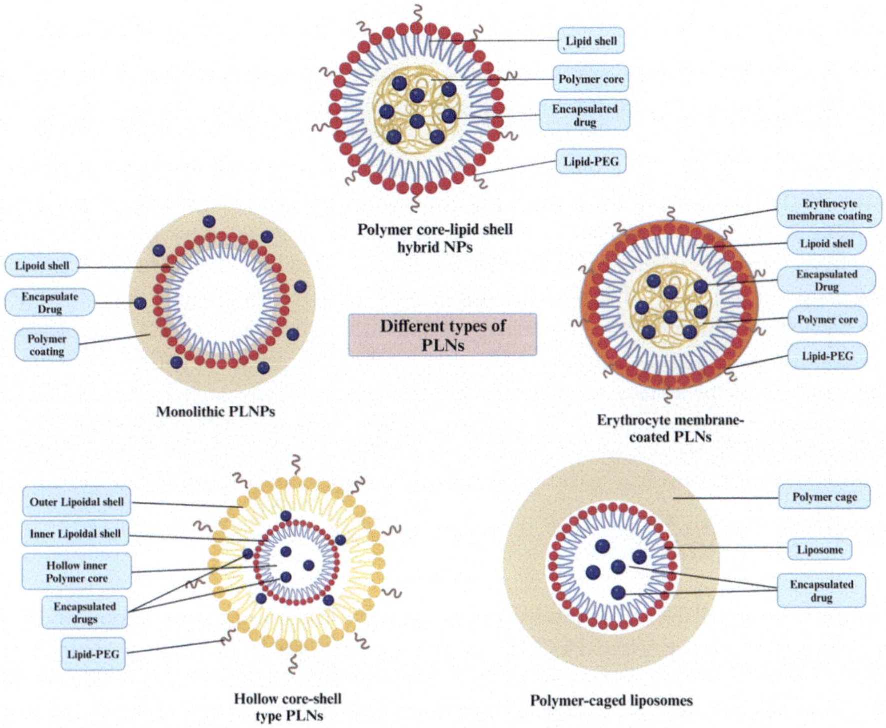

Fig. 10.2 Different types of PLNs with their structural composition

influence on the entrapment of hydrophobic drugs within the lipophilic junction and hydrophilic drugs within the bilayer. This property enables the entrapment and concurrent delivery of various hydrophobic and hydrophilic drugs (Gajbhiye et al. 2023; Mandal et al. 2013).

2.2 Hollow Core-Shell Type PLNs

The hollow core-shell type PLNs contain a hollow polymer core enveloped by a concentrated lipid layer, which is further coated with a polymer layer on the inner core and encased by a lipid coating containing lipid-PEG. The core is surrounded by one or more layers of lipid. This configuration typically amalgamates the properties of both liposomes and PNPs. The drug molecules can be dispersed in both the hollow polymer core and the space between the inner and outer lipid layers. The coating of LNPs with cationic and zwitterionic phospholipids improves the electrostatic interactions with oppositely charged polymers (Dave et al. 2019; Mandal et al. 2013). These nanoparticles reduce the susceptibility of serum nuclease, improve the delivery of genetic agents, and inhibit phagocytic uptake.

2.3 Biomimetic PLNs

The biomimetic PLNs are also known as cell membrane camouflaged PNPs. In this type, the PLNs are coated with the cellular membrane, derived from erythrocytes, platelets, or leucocytes exhibiting inherent propensity for prolonged systemic circulation and shielding of the nanoparticles from macrophage uptake (Gajbhiye et al. 2023; Hadinoto et al. 2013). These nanoparticles possess the capability to traverse membrane barriers effortlessly, enabling sustained drug delivery over extended durations through circulation in the bloodstream (Dave et al. 2019; Hu et al. 2011). The drug molecules are distributed within the polymeric core. Biodegradable PNPs can be surface functionalized by coating them with natural erythrocyte membranes, incorporating membrane lipids. This erythrocyte membrane coating mimicked endogenous erythrocytes, allowing the nanoparticles to prevent recognition by the reticuloendothelial system (RES) and prolonging circulation time (Hu et al. 2011; Wen et al. 2021).

2.4 Monolithic PLNs

The monolithic PLNs, also referred to as mixed PLNs, are characterized by the distribution of lipid or lipid-PEG layer within a polymeric core matrix containing drug molecules. This distribution results in homogeneous dispersion of lipids within the dispersed polymeric structure (Dave et al. 2019). These PLNs are effective at entrapping highly lipophilic drug molecules that would otherwise be difficult to encapsulate within the polymer. To reduce the systemic toxicity the lipid to polymer ratio can be adjusted during the manufacturing process (Cai et al. 2015; Gajbhiye et al. 2023). These nanoparticles help in targeted delivery of the chemotherapeutic agents and reduce systemic toxicity.

2.5 Polymer-Caged Liposomes

Polymer-caged liposomes involve the polymer encapsulation at the surface of liposomes to confer system stability (Dave et al. 2019). It comprises a hydrophilic core with a drug, enveloped by an outer polymeric structure. Such arrangement imparts stability to the nanoparticles, preventing drug leakage and controlled release into the systemic circulation, which may exhibit sensitivity to pH or proteases (Gajbhiye et al. 2023; Lee et al. 2007). However, the limitation of this type is the potential for polymer cages to readily detach from the liposome surface, reverting to an unstable form. Moreover, the RES may efficiently capture polymer-caged liposomes, leading to decreased uptake as the lipid layer is absent in their structure (Cao et al. 2022; Lee et al. 2010). The liposomal core can encapsulate anticancer drugs, DNA, siRNA, proteins, and peptides (Bangera et al. 2023).

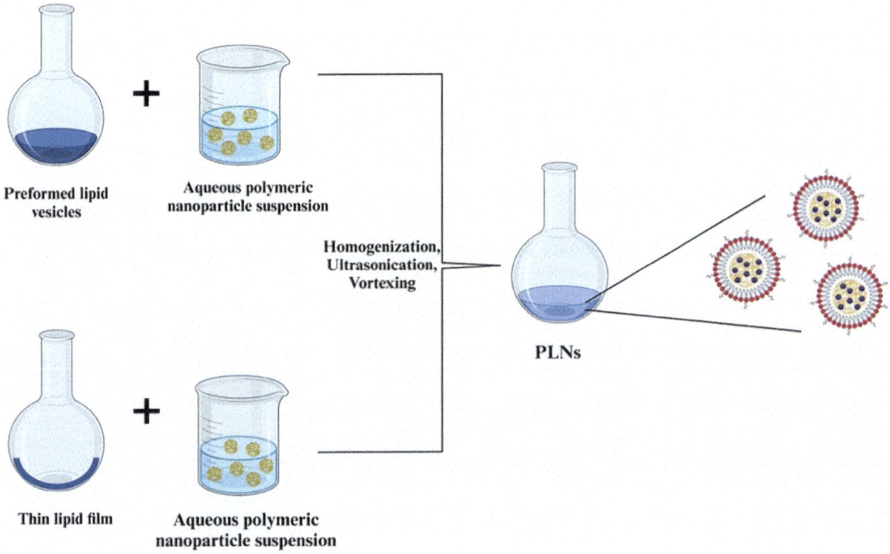

Fig. 10.3 Two-step method

3 Method of Preparation

The molecular mechanism for the formation of PLNs is undergoing investigation. The method of preparation of PLNs can be categorized into single-step method and two-step method. In the single-step method, an organic phase containing polymers is introduced into the aqueous phase which contains lipids, leading to the precipitate of the polymer and self-assemble lipid around the core as shown in Fig. 10.3. In the two-step method, the PNPs and the LNPs are formulated separately which are mixed, and due to hydrophobic interactions between polymer and lipid chain PLNs are formed (Gajbhiye et al. 2023; Mukherjee et al. 2019), as shown in Fig. 10.4.

3.1 Two-Step Method

The two-step method entails the preparation of PNPs and LNPs separately, which are then combined using external energy sources such as heat, vortexing, ultrasonication, or homogenization. The process is facilitated by hydrophobic, electrostatic, and van der Waals interactions leading to the formation of a lipid shell over the polymer core (Hadinoto et al. 2013; Jain et al. 2023; Mandal et al. 2013). This method is categorized further into conventional and non-conventional approaches. The uniformity of the LNPs, the ionic strength of the aqueous medium, and the ratio of LNPs to PNPs are the parameters determining size homogeneity and colloidal stability in this approach (Mukherjee et al. 2019).

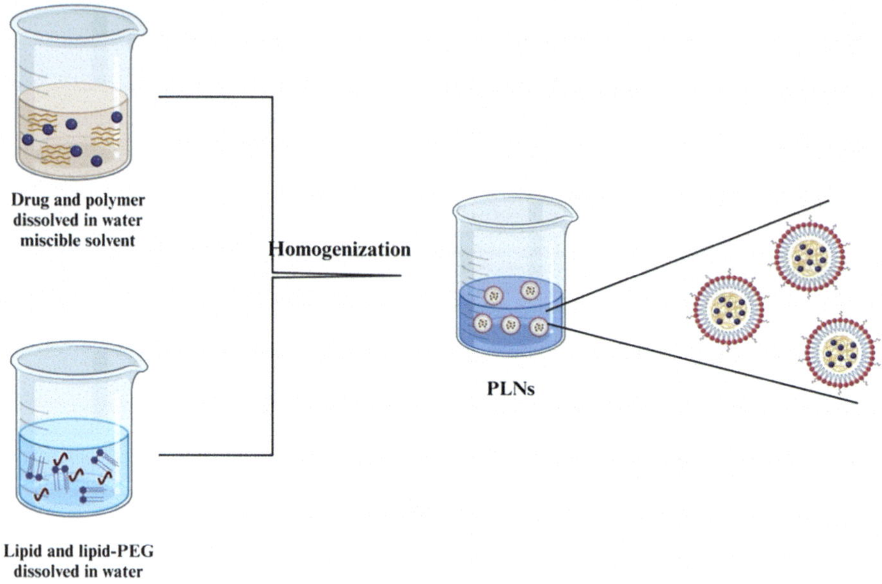

Fig. 10.4 Single-step method

The two-step preparation parameters have a significant influence on the physical characteristics of PLNs. Troutier and colleagues provided evidence of the influence of various parameters on PLN size and colloidal stability. Better monodispersity was demonstrated by the PLNs formulated from smaller and homogeneous particles, whereas aggregation was influenced by the charge on the LNPs (Troutier et al. 2005). In another study, Thevenot et al. investigated the impact of PEGylated lipids on PLN stability and found that long-chain PEG improved colloidal stability through steric stabilization. Despite lower zeta potentials, PLNs with longer PEG chains showed greater stability in salt solutions (Thevenot et al. 2007).

3.1.1 Non-conventional Approach

The non-conventional approach is mostly employed in PLN preparation on a large scale. This approach entails either spray drying or soft lithography particle molding to prepare PLNs. In spray drying the PNPs are prepared and subsequently dispersed in organic solvent containing different lipids to form a lipoidal polymeric suspension. This suspension is then spray-dried to produce PLNs (Hitzman et al. 2006; Mukherjee et al. 2019). Particle replication in nonwetting templates, a soft lithography molding technology is utilized in the formulation of PLNs for the delivery of genetic materials. First, an organic solvent is used to dissolve the polymer before cast on a polyethylene terephthalate (PET) sheet. The polymer solution fills the mold cavities when heated and solidifies on cooling, producing PNPs, which are then taken out of the mold and coated with polyvinyl alcohol using aqueous lipid solution, resulting in the formation of PLNs (Hadinoto et al. 2013; Hasan et al. 2012).

3.1.2 Conventional Approach

The conventional method is usually used to prepare PLNs on a small scale. The PLNs can be prepared through high-pressure homogenization (Fenart et al. 1999), solvent evaporation (Szczęch and Szczepanowicz 2020), and nanoprecipitation (Craparo et al. 2022). The prepared PNPs are subsequently complexed with LNPs on their surface using a rotary evaporator and electrostatically bound lipids or PEG. The resulting mixture undergoes ultrasonication or vortexing at temperatures above the lipid transition temperature needed for PLN formation, and nanoparticles are separated from the solution using centrifugation (Dave et al. 2019). After preparation, the PLN suspension is homogenized or extruded to produce the appropriate particle size.

3.2 One-Step Method

The one-step method was designed in order to overcome the drawbacks of two-step methods. The drug and polymer solution is mixed directly with the lipid solution using emulsification solvent evaporation or nanoprecipitation procedures. It has benefits like scalability, affordability, and adherence to traditional preparation techniques. Lipid acts as a stabilizer for PLNs in this method (Mukherjee et al. 2019; Shah et al. 2022).

3.2.1 Self-Assemble Nanoprecipitation Method

In the nanoprecipitation method, the polymers and drugs to be encapsulated are dissolved in an organic solvent. Simultaneously, the lipid and lipid PEG compounds are dissolved in aqueous solvent through heating at 65–70 °C yielding a homogeneously dispersed solution. Subsequently, the polymeric and drug-containing solution is introduced into the lipid solution dropwise, inducing precipitation of the polymer. During this process, the lipid or lipid-PEG molecules undergo self-assembly around the polymer core facilitated by hydrophobic interactions. Specifically, the hydrophobic tails of the lipid attach to the polymeric core with their hydrophilic heads extending outwards in the aqueous environment, thus resulting in the formation of stabilized PLNs. Further, the PLN suspension was centrifuged to remove extra lipids and polymer and the solvent used was evaporated (Dave et al. 2019; Hadinoto et al. 2013). When lipid–PEG is utilized, the lipid tail extends into the lipid layer, while the PEG head is exposed to the aqueous environment, serving as a steric stabilizer (Su et al. 2011) (Fig. 10.5).

To overcome the limitations of nanoprecipitation such as reproducibility, scalability, and variations in physicochemical properties Tahir et al., developed a microfluidic coflow nanoprecipitation technique for the preparation of PLNs. PLGA was dissolved in acetonitrile along with sorafenib which served as inner fluid and lecithin and 1,2-Distearoyl-sn-glycero-3-phosphorylethanolamine -PEG 2000 in a 2:3 ratio dissolved in ethanol solution served as outer fluid. Both fluids were injected into the microfluidic device to form PLNs. After collection, the PLNs were

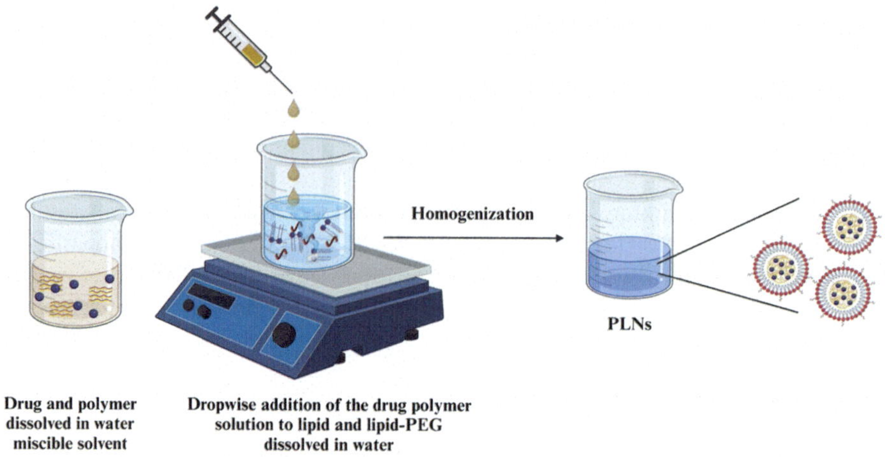

Fig. 10.5 Schematic representation of nanoprecipitation method

self-assembled by stirring. The resultant particles had spherical shape with particle size less than equal to 250 nm and low polydispersity. (Tahir et al. 2020).

Formulation parameters in the nanoprecipitation method include lipid-to-polymer (L/P) mass ratio, determining the characteristics of nanoparticles. Elevated ratios may lead to concentration surpassing the micelle concentration, thus resulting in liposome formation in addition to PLNs, whereas reduced ratios may induce PLNs to aggregate due to inadequate lipid coating (Yang et al. 2012). Zhang et al. demonstrated the significance of lipid coating in enhancing encapsulation efficiency and drug loading in PLNs. At an optimal L/P ratio, PLNs exhibited enhanced encapsulation efficiency of Docetaxel (DTX) compared to non-hybrid PLGA and PEGylated-PLGA nanoparticles. PLNs also showed sustained drug release over 20 h, longer than PLGA and PEGylated-PLGA nanoparticles. Similar trends regarding the impact of the L/P ratio were observed in studies by Chan et al. (Chan et al. 2009; Zhang et al. 2008).

3.2.2 Modified Emulsification Solvent Evaporation Method

The modified emulsification solvent evaporation method is utilized for drugs soluble in hydrophobic solvents. This technique involves dispersing lipids in water employing agitation methods such as stirring and sonication, subsequently subjecting the mixture to heat to dissolve the organic solvent, polymer, and drug. The organic solution is then slowly added dropwise to the aqueous dispersion, followed by a probe sonicated to generate small polymeric particles encapsulated with a lipid layer, yielding spherical nanoparticles. The organic solvent is removed via an evaporator using a rotary evaporator, and the resulting solution is stirred overnight before purification through cold centrifugation and subsequent washing with distilled water (Dave et al. 2019; Hadinoto et al. 2013). This method is further categorized into single and double emulsification methods as shown in Fig. 10.6.

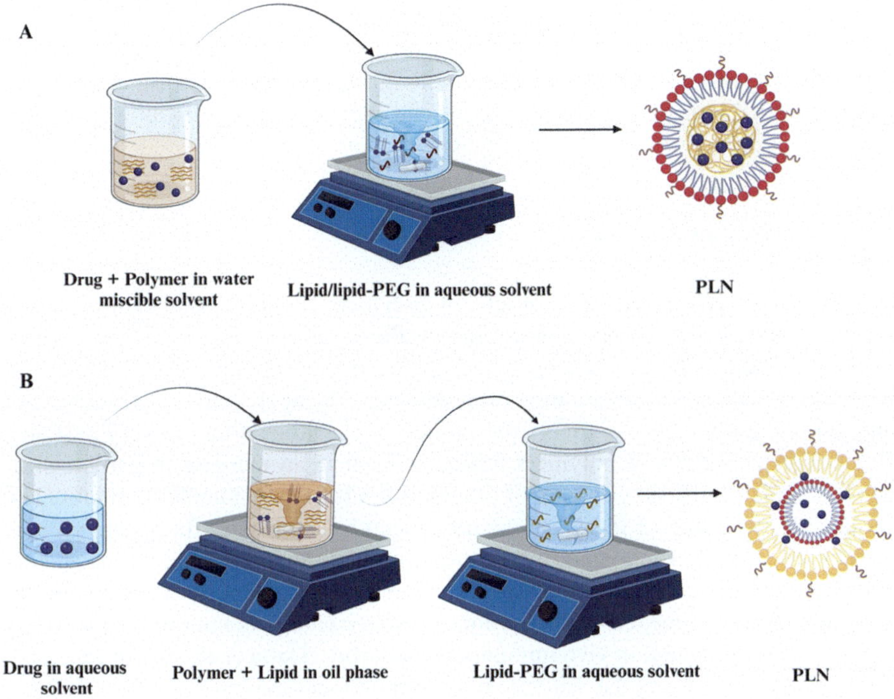

Fig. 10.6 Schematic representation of (**a**) Single emulsification method and (**b**) Double emulsification method

A single emulsification method is employed for drugs that are soluble in hydrophobic solvents. This technique fabricates an oil-in-water (o/w) emulsion by combining a water-immiscible solvent containing polymer and drug with a lipid-containing aqueous phase and is dissolved under ultrasonication or continuous stirring. Subsequently, the organic solvent is removed to prepare PLNs (Jain et al. 2023; Mukherjee et al. 2019).

The double emulsification method is used for hydrophilic drugs. In this method, emulsification of an aqueous phase containing drug in an organic phase having polymer and lipid to form a water-in-oil (w/o) mixture, which is further subjected to another emulsification process in an aqueous phase containing lipid-PEG resulting in a water-in-oil-in-water (w/o/w) emulsion, to yield PLNs. The nanoparticles formed here have a lipid and polymer layer surrounding the inner aqueous core and an outer lipid-PEG shell. This method produces larger particles compared to the single emulsification method (Mukherjee et al. 2019; Shah et al. 2022).

Formulation parameters such as the L/P ratio and lipid-PEG fraction have a considerable impact on PLN properties. Increasing the lipid-PEG to polymer ratio enhances PLN colloidal stability and size reduction, resulting in improved drug encapsulation efficiency (Mukherjee et al. 2019). Cheow et al. explored the impact of L/P ratio variations on the size and encapsulation efficiency of PLNs using

PLGA, SPC, and TPGS. PLNs exhibited aggregation at L/P ratios below 15%, but at ratios above 15%, their size reduced, with an optimal ratio of 30% yielding maximum production yield (Cheow and Hadinoto 2011). Additionally, Liu et al. found that higher L/P ratios led to smaller PLNs and influenced encapsulation efficiency by affecting lipid coverage on the polymer core (Liu et al. 2010).

4 Types of Polymers

The PLNs are composed of a variety of polymers ranging from natural, semi-synthetic, and synthetic polymers, serving as the core for PLNs coated with lipids. These polymers help in entrapping hydrophilic or hydrophobic drugs within the polymeric core, thereby facilitating drug delivery. The aggregation of polymer and intrinsic viscosity directly affects the particle size within drug delivery systems. Higher concentrations of polymers typically result in an increase in particle size, whereas higher intrinsic viscosity of polymers tends to yield smaller particles. The surface zeta potential of PLNs can be modulated by modifying functional groups, thereby affecting particle stability. Immunocompatibility assessments reveal that hybrid nanoparticles bearing methoxyl surface groups exhibit the lowest complement activation. Hence, careful selection of surface charge is imperative for ensuring in vitro stability and in vivo immunocompatibility (Datta et al. 2025b; Jain et al. 2023). Table 10.1 shows the different types of polymers used.

5 Types of Shell Materials and Their Properties

The shell of the PLNs is made up of lipids which are of natural and synthetic origin, characterized by their hydrophobic or amphiphilic nature, and are found in various compounds which include oils, fatty acids, waxes, and steroids. The phospholipids are often utilized for the surface engineering of PNPs (Jain et al. 2023). Cholesterol is the major lipid used for the preparation of liposomes to improve the rigidity and stability (Nsairat et al. 2022).

The bilayer of the cell membrane is made up of glycerophospholipids. The natural phospholipids are sourced from diverse origins such as soya bean and egg yolk (Monteiro et al. 2014). These phospholipids are categorized into phosphatidylcholine (PC), phosphatidylethanolamine (PE), phosphatidylglycerol (PG), phosphatidylserine (PS), phosphatidic acid (PA) and phosphatidylinositol (PI) depending on their polar head groups (Nsairat et al. 2022). The inherent instability of natural phospholipids in comparison to synthetic lipids is attributed to the unsaturated nature of their hydrocarbon chains. Fatty acid composition analysis reveals that the egg-derived phospholipids and PCs consist primarily of stearic acid, palmitic acid, oleic acid, linoleic acid, and arachidonic acid. The predominant saturated acids are stearic in PE and PS, and palmitic in other lipids (Nsairat et al. 2022). Lipids sourced from natural or cell membranes present several advantageous characteristics, including mimicking the natural cell surface, inherent targeting ability, natural immune

Table 10.1 Types of polymers used in the preparation of PLNs

Polymer	Source	Properties	Reference
Polylactic-co-glycolic acid (PLGA)	Chemical interaction of lactic acid and glycolic acid	A hydrophobic biodegradable and biocompatible polymer with high bioavailability and flexibility in payload formulation	Alsaab et al. (2022), Pandita et al. (2015)
Polylactic acid (PLA)	Lactic acid	Its adaptability permits alterations via plasticizers, polymer blending, or additive composites. It enhances biocompatibility and thermal processing and requires less energy than petroleum-based polymers.	Casalini et al. (2019), Mukherjee et al. (2019)
Polycaprolactone (PCL)	Polymerization of the cyclic monomer of Ɛ-caprolactone.	A biocompatible, biodegradable, and bioresorbable aliphatic polyester polymer, exhibiting hydrophobicity and semi-crystalline structure. It has enhanced viscoelastic properties.	Christen and Vercesi (2020), Labet and Thielemans (2009)
Poly(vinyl alcohol) (PVA)	Hydrolysis of polyvinyl acetate	A hydrophilic thermoplastic polymer, biodegradable, biocompatible, stable to temperature variations, and non-toxic to living tissues and cells. It promotes the formation of small particles with a narrow size distribution.	Ekinci et al. (2022), Gaaz et al. (2015), Higazy et al. (2021)
Chitosan (D-glucosamine and N-acetyl-D-glucosamine)	Deacetylation of chitin yields a natural polysaccharide, a key component of fungal cell walls, crustacean and insect exoskeletons, and fish scales, derived from (1–4)-2-amino-2-deoxy-β-D-glucan.	Has pH sensitivity, biocompatibility, and bioactive properties, It is water-soluble, bioadhesive, and forms complexes with mucus, enhancing its solubility in acidic pH environments.	Mikušová and Mikuš (2021), Mohammed et al. (2017), Yanat and Schroën (2021)

(continued)

Table 10.1 (continued)

Polymer	Source	Properties	Reference
Hyaluronic acid (HA)	Synthesized by hyaluronan synthase from animal or bacterial source	Has good biocompatibility, biodegradability, high viscoelasticity, and receptor-binding properties It reduces toxicity, enhance nanocarrier dispersity in aqueous solutions, and enables the constructing tumor-targeted drug delivery systems. It acts as a natural ligand for the endocytic receptor, CD44, which is often overexpressed in cancer cells, facilitating tumor targeting.	Curcio et al. (2022), Fu et al. (2023), Huang and Huang (2018)
Polyethylenimine (PEI)	Acid-catalyzed ring-opening polymerization of the aziridine monomer,	PEI-based nanocarriers demonstrate low toxicity, the ability to deliver nucleic acids and chemotherapy drugs simultaneously, easy modification with targeting molecules, and responsiveness to external stimuli.	Chrószcz and Barszczewska-Rybarek (2020), Wang et al. (2015)
Poly(2-hydroxyethyl methacrylate) (PHEMA)	Copolymerization of HEMA with other co-monomers	It can be adjusted by cross-linking density, copolymerization, and mesoscopic pores, offering controlled pharmaceutical and protein release through enzymatically susceptible monomers or cross-linking agents.	*2 Hydroxyethyl Methacrylate—an Overview | ScienceDirect Topics* (n.d.), Pernari and Wells (2023), Zare et al. (2021)
Eudragit	Derived from methacrylic acid,	It has potential to encapsulate, increase the solubility and bioavailability of poorly soluble drugs	Xu et al. (2018)
Polyamidoamine	Michael-type addition reaction of amine compounds to bis-acrylamides	Biodegradable cationic polymers, with good water solubility, low cytotoxicity, and long-term biodegradability,	Lin and Engbersen (2008), Pontes et al. (2024)

evasion, prolonged circulating half-life, and regulated tissue distribution (Troutier et al. 2005).

Due to the inherent instability of phospholipids in lipid environments, modifications are often performed to enhance their stability by altering both polar and non-polar regions, thereby generating synthetic phospholipids (Jain et al. 2023). Synthetic lipids offer advantages such as a controlled drug release mechanism, stealth characteristics, prolonged circulation time, and surface adaptability for functional group targeting (Troutier et al. 2005). The synthetic lipids include 1,2-distear oyl-sn-glycero-3-phosphorylethanolamine (DSPE), 1,2-dimyristoyl-sn-glycero-3-phosphoethanolaminediethylene (DMPE), 1,2-dimyristoyl-sn-glycero-3-phospho choline (DMPC), 1,2-dipalmitoyl-sn-glycero-3-phosphorylethanolamine (DPPE), 1,2-dipalmitoyl-sn-glycero-3-phosphatidic acid (DPPA). 1,2-Dioleoyl-3-trimethylammonium-propane (DOTAP), 1,2-dioleoyl-sn-glycero-3-phosphoethano lamine (DOPE), 1,2-dioleoyl-sn-glycero-3-phosphocholine (DOPC), glyceryl monooleate (GMO), and glycerol monostearate (GMS).

6 Surface Functionalization and Strategies for PLNs

Surface functionalization of nanoparticles involves modifying their surface properties to confer unique characteristics in enhancing their pharmacokinetic (PK) and pharmacodynamic (PD) properties. The PK-PD profile of the drug can be affected by the lipophilicity and charge on the surface of nanoparticles. PEGylation is the common method of increasing surface hydrophilicity, which decreases cell internalization but increases plasma circulation time (Jain et al. 2023; Mandal et al. 2013; Shah et al. 2022). PEGylation has emerged as a conventional approach for fabricating long-circulating nanoparticles, minimizing plasma protein adsorption, reducing macrophage uptake, and preventing particle aggregation while providing stability through steric hindrance (Halder et al. 2024; Shah et al. 2022). Figure 10.7 depicts the various surface functionalization moieties and Table 10.2 shows the PLNs surface functionalized for different cancer treatments.

A pH-sensitive PEG coating undergoes shedding in acidic conditions, allowing fusion with cell membrane and entry into tumor cells (Clawson et al. 2011). Clawson et al. formulated PLNs containing PLGA core with lipid and lipid-PEG shell. In order to impart pH sensitivity, a lipid-(succinate)-mPEG conjugate was synthesized with the intention of providing a hydrolyzable PEG stealth layer, facilitating easy shedding at low pH. The destabilization and aggregation of hybrid nanoparticles are pH-dependent, modulated by the quality of lipid-(succinate)-mPEG incorporated into the lipid shell. Increased incorporation enhances particle stability, particularly at lower pH, broadening the pH range for engineered removal of the PEG coating, thereby enhancing the versatility of nanoparticle drug delivery systems (Clawson et al. 2011). In another study, Wang et al. fabricated pH-sensitive folate-decorated PLNs containing carboplatin and paclitaxel in PCL core and soya lecithin lipid, and PEG modification on the surface was done for cervical cancer. The PLNs showed 80% release at pH 5.5 at 24 h which was delayed at pH 7.4 to 48 h with decorated

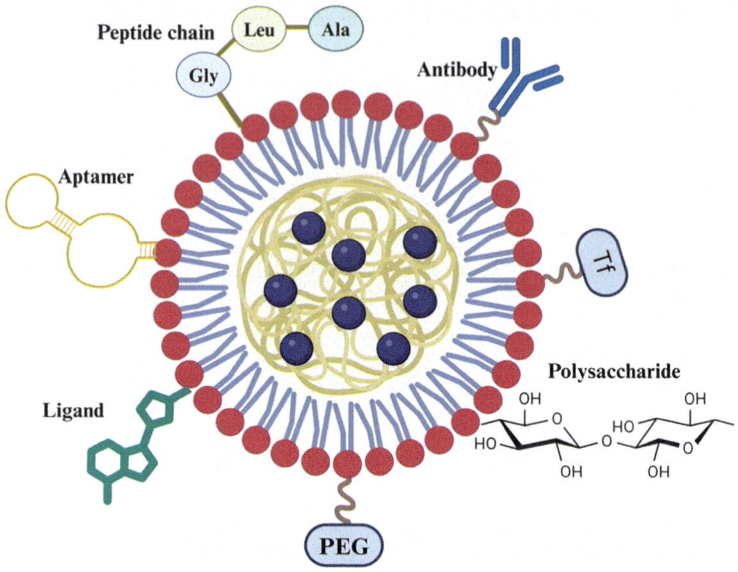

Fig. 10.7 Schematic representation of functionalization of PLNs with different types of functional moieties

FA, additionally, it showed efficient cellular uptake and significant cytotoxicity. Moreover, their substantial tumor accumulation and notable antitumor efficacy support their potential as a promising targeted therapy (Wang 2020).

Garg et al. developed co-encapsulation of methotrexate (MTX) and aceclofenac (ACL) containing PLNs consisting of PCL core and DSPE-PEG (2000), which were further functionalized with fucose for breast cancer treatment. The PLNs were fabricated using a modified single-step nanoprecipitation approach. The in vitro release studies were carried out in pH 7.4, which revealed burst release within 12 h following sustained release, additionally, the release of ACL was higher in comparison to MTX, which can be due to the higher entrapment efficiency of ACL. The anticancer activity of the coencapsulated and free drugs was investigated on MCF-7 and MDA-MBA-231 cells to determine the dependence of concentration and time. The results revealed that ACL did not show much cytotoxicity activity, whereas MTX showed cytotoxicity activity at all times. The fucose functionalized PLNs containing MTX and ACL revealed a significant cytotoxicity effect as compared to other formulations. This formulation exhibited significantly enhanced bioavailability (8–10 folds) and synergistic effects of the drug combination, resulting in superior control over tumor growth in the DMBA (7,12-dimethylbenz[a] anthracene) induced breast cancer mouse model. The PK and biodistribution studies revealed that the residence time and half-life of the drug were increased in comparison to free formulations up to 72 h post-administration. Additionally, a significant amount of MTX was detected in tumors when delivered through fucose-functionalized PLNs in comparison to non-functionalized PLNs. The PD studies revealed that the fucose

Table 10.2 Various surface functionalized PLNs for cancer treatments

Drug/s	Cancer	Polymer	Lipid	Fabrication method	Targeting ligand	Reference
DOX and gallic acid	Leukemia	PCL	DSPE-PEG	Single step nanoprecipitation	Hyaluronic acid	Shao et al. (2019)
DTX	KB cancer cells	PLGA	DSPE-PEG	Modified nanoprecipitation method	Folic acid	Dehaini et al. (2016)
Isoliquiritigenin	Breast cancer	PLGA	Lecithin and DSPE–PEG	Modified single-step nanoprecipitation	RGD	Gao et al. (2017)
Methotrexate and beta-carotene	Breast cancer	PLA	DSPE-PEG and Stearyl amine	Self-assembled nanoprecipitation technique	Fructose	Jain et al. (2017)
Erlotinib and bevacizumab	Non-small cell lung cancer	PCL	Soy phosphatidylcholine (SPC)	Sonication method	HA	Pang et al. (2020)
CIS and fluoropyrimidine (5-FU)	Esophageal adenocarcinoma	PCL	DSPE-PEG, SPC	Solvent displacement method	Trastuzumab	Fu et al. (2020)
Sorafenib	Liver cancer	Chitosan	DSPE-PEG	Emulsification solvent evaporation metho	Folic acid	Tang and Li (2019)
CIS	Ovarian carcinoma	Chitosan	Lipoid S75	Single-step ionic gelation method	Fluorescent dyes, rhodamine 123, and rhodamine-PE	Khan et al. (2019)

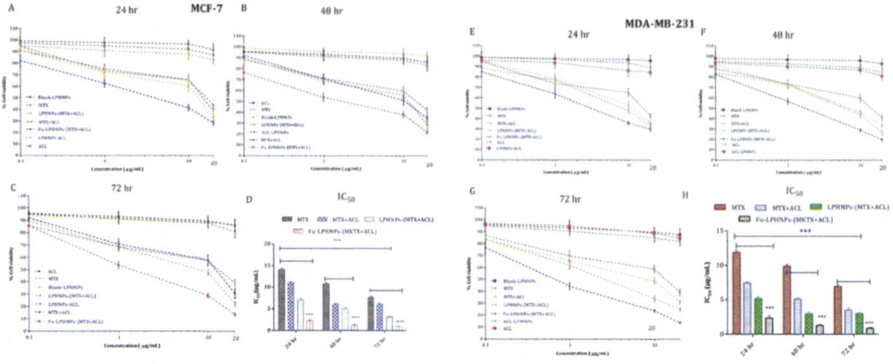

Fig. 10.8 In vitro anticancer activity performed on MCF-7 (A-D) and MDA-MBA-231 cells, with different ACL and MTX-based free and coencapsulated PLNs after 24, 48, and 72 h. IC_{50} was determined using a drug sensitivity assay. (Adapted with permission Garg et al. 2017)

functionalized PLNs containing MTX and ACL showed a significant decrease in tumor volume along with an increase in survival rate, indicating enhanced therapeutic efficacy in breast cancer treatment (Garg et al. 2017) (Fig. 10.8).

Omar et al., prepared rivastigmine containing PLNs with PLGA core, L-α-phosphatidylcholine shell, and surface modified with dextran cholic acid for brain targeting. The nanoparticles were prepared by the modified nanoprecipitation method and organic solvent was removed using dialysis. The surface modification was carried out by carbodiimide chemistry. The nanoparticles were 100 nm in size with a negative charge, which showed efficient penetration into the brain in animal models. The MTT assay showed enhanced bioavailability as compared to blank PLNs. The surface-modified nanoparticles showed prolonged circulation and enhanced AUC in plasma, improving drug bioavailability in the brain in comparison to the free drug. The in vivo pharmacokinetic studies showed sustained release of the drug with more targeting efficiency (5.4 folds) with reduced biodistribution in non-targeted organs hence limiting side effects (Omar et al. 2020).

Aptamer-functionalized PLNs are employed for the co-delivery of drugs for treating cancer (Tiwari et al. 2023a). Chen et al., designed curcumin and cabazitaxel-loaded and aptamer-functionalized PLNs for prostate cancer treatment. The core consisted of PLGA and the shell was made up of soyphosphatidylcholine (SPC) which was fabricated using a self-assembled nanoprecipitation technique. The aptamer conjugation was carried out by carbodiimide chemistry in the presence of EDC. The nanoparticle hybrid had a positive charge with a particle size of 121 nm and showed sustained release. The release of drugs from APT-CUR/CTX-PLNs occurred at a slower rate compared to CUR/CTX-PLN, revealing that the aptamer covering on the surface may act as a molecular barrier, aiding in the retention of drugs within the nanoparticles. The in vitro cell line studies were investigated on LNCaP and PC3 cells. The aptamer functionalized PLNs showed cell inhibition ability and tumor inhibition activity and also exhibited remarkable antitumor efficacy, which could be due to the high affinity of the outer lipid layer to the cell membrane. The Chou-Talalay method was used for evaluating the synergistic effect of the prepared nanoparticles containing curcumin and cabazitaxel with different ratios, from which a 2:5 ratio showed better synergistic effect. The in vivo studies of aptamer-functionalized PLNs in comparison to non-functionalized PLNs showed significantly better half-life and plasma AUC and additionally demonstrated antitumor efficiency with reduced systemic toxicity (Chen et al. 2020b).

Monoclonal antibodies (mAbs) are potent targeting ligands for drug delivery, yet their size often results in heightened nonspecific uptake by the RES and shortened circulation time of nanoparticles. Furthermore, mAbs are prone to nonspecific conjugation to nanoparticles and particle dimerization. Jain et al. reported PEGylated PLA nanoparticles containing mRNA which were coated with cationic surfactant DOTAP and functionalized with mAbs to facilitate receptor-mediated endocytic uptake in T-cell leukemia. The PLNs were fabricated using the double emulsion solvent evaporation method and the functionalization of mAbs on the surface was carried out using maleimide chemistry and incubated with thiolated antibodies. Coating of DOTAP enhanced the release of coated nanoparticles from the endosomes into the cytosol in comparison to PLA nanoparticles. The CD7 antigen receptor was selected as the targeting marker demonstrating swift internalization post ligand binding. These PEGylated PLNs exhibited stability in the bloodstream following systemic administration, facilitating mRNA delivery for protein translation within cells expressing CD7 antigen in vivo (Jain et al. 2020).

Wang and colleagues fabricated PLNs containing paclitaxel (PTX) and cisplatin (CIS), along with Arg-Gly-Asp peptide sequence (RGD peptide) for lung cancer treatment. The PLNs were formulated using a modified emulsification sonication approach where the core contained PLGA (50,50) and CIS and PTX were conjugated to DPA through EDC chemistry. Further RGD was conjugated with PTX-ss-PEG to get RGD-ss-PTX which made up the shell. The in vitro cell line studies were carried out on A549 and NCI-H1299 cells, the results showed that RGD-modified PLNs showed enhanced cellular uptake and stronger fluorescence in comparison to non-modified PLNs, whereas in NCI-H1299 cells not much difference was seen. The dual drug-loaded PLNs showed enhanced cytotoxicity and synergistic effect in addition to profound suppression in tumor growth in comparison to free drugs (Wang et al. 2018).

6.1 Strategies for Surface Functionalization

Strategies for surface functionalization of nanoparticles include methods like covalent grafting and noncovalent binding (Rouco et al. 2022). Noncovalent binding is achieved through an affinity-based receptor-ligand system and is conjugated through noncovalent interactions like electrostatic interactions, π-π stacking or attaching particular secondary ligands (Rouco et al. 2022; Thiruppathi et al. 2017). This approach involves the reversible, straightforward process of adsorption, which does not necessitate any chemical modification or activation of the functional groups on the nanoparticles (Shen et al. 2017). Weak interactions like hydrogen bonding, van der Waals forces, and electrostatic interactions are necessary for physical adsorption. Ionic bonds between the surfaces of oppositely charged nanoparticles and antibodies are necessary for ionic binding (Marques et al. 2020; Zhao et al. 2012).

In covalent grafting, the nanoparticle must contain reactive groups to be preactivated, which can be carried out using carbodiimide chemistry, maleimide chemistry, and click chemistry, to provide increased stability and repeatability (Marques et al. 2020). Carbodiimide chemistry entails the conjugation via the primary amine moiety present on antibodies, which are abundant on antibody surfaces and highly reactive with various functional groups on nanoparticles, enabling covalent binding without chemical modification (Polo et al. 2013). The carboxyl groups present on nanoparticle surfaces undergo activation facilitated by agents like 1-ethyl-3-(-3-dim ethylaminopropyl) carbodiimide (EDC), a water-soluble reagent, to react with amine groups, thus resulting in the formation of an amide bond (Marques et al. 2020). Primary amines and carboxylic or phosphate groups are coupled by EDC. The presence of N-hydroxysuccinimide (NHS) or N-hydroxysulfoxuccinimide (sulfo-NHS) in the carbodiimide cross-linking reaction enhances coupling efficiency by allowing a two-step reaction that prevents intra and intermolecular cross-linking of antibodies containing both amine and carboxyl groups (V. J. Yao et al. 2016).

Maleimide chemistry is the process by which the maleimide double bond is alkylated with sulfhydryl groups at pH values of 6.5–7.5 to form stable thioether bonds. Maleimide chemistry binds to sulfhydryl groups (-SH) of antibodies, which are present in cysteine amino acid side chains (Marques et al. 2020). Compared to primary amines, this reaction proceeds more quickly with free sulfhydryl groups. This reaction is faster with free sulfhydryl than with primary amines. Maleimide cross-linking reagents exist in two forms: homobifunctional or heterobifunctional. Homobifunctional maleimides facilitate the conversion of protein disulfide bonds into permanent thioether bonds, linking cysteine sulfhydryls permanently. Heterobifunctional reagents contain amine-reactive (such as succinimidyl ester or NHS ester) and sulfhydryl-reactive ends (Baldwin and Kiick 2011; Behrens and Liu 2014; Kolodych et al. 2015).

Click chemistry denotes the category of chemical reactions characterized by orthogonality, site-specificity, and rapid reaction rates. They are straightforward and typically require little to no purification. Click chemistry results in the formation of irreversible chemical bonds without producing cytotoxic byproducts (Marques et al. 2020).

7 Drug Release Mechanism

The PLNs exhibit unique release properties combining features of liposomes and PNPs. Release from liposomes occurs through diffusion across the bilayer, while PNPs undergo erosion followed by drug diffusion. The lipid coating delays drug release and polymer degradation by limiting water penetration and small molecule diffusion. Critical factors affecting drug release include drug solubility, interactions between drug and polymer, degradation rate of the polymer, and particle size.

Release mechanisms vary depending on whether drugs are physically encapsulated or chemically conjugated, involving compound diffusion or hydrolysis of linkages between drugs and polymer chains (Sivadasan et al. 2021). Researchers have developed various PLNs capable of tailored delivery in reaction to various stimuli such as temperature, pH, redox responsiveness, magnetic fields, enzyme responsiveness, and photo-irradiation. Fig. 10.9 depicts the various drug release mechanisms from PLNs (Garkal et al. 2023; Kulkarni et al. 2023; Mishra et al. 2023; Rajak et al. 2023; Tiwari et al. 2023b).

Zhang et al., developed pH-responsive PLNs comprising polymer core made up of poly β-amino ester (PBAE) and lipid shell containing DSPE-PEG2000 and lecithin for targeted DTX delivery, further surface functionalized with folic acid (FA/PBAE/DTX-NPs) for breast cancer treatment. The nanoparticles exhibited uniform. The release of DTX from FA/PBAE/DTX-NPs was evaluated under different pH. Notably, drug release from FA/PBAE/DTX-NPs was slower at higher pH values (pH 7.4 and pH 6.8) compared to lower pH (pH 5.5), indicating the potential activation in acidic tumor microenvironment (TME). PBAE has a tertiary amine

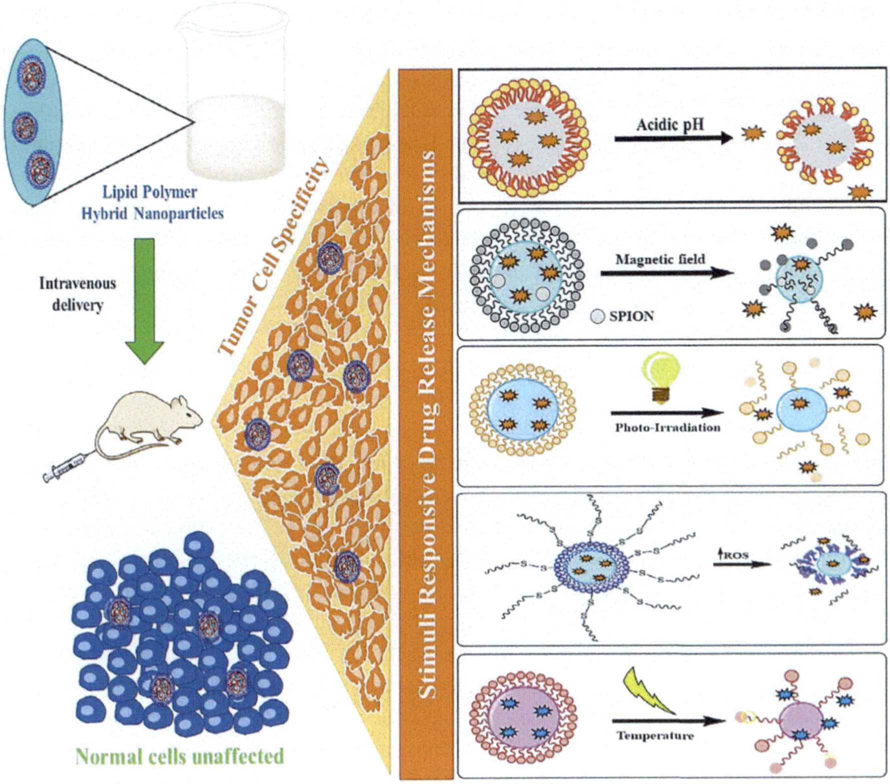

Fig. 10.9 Drug release mechanism of PLNs in conjugation of different stimuli in TME. (Adapted with permission Shah et al. 2022)

group that enhances endosomal/lysosomal outflow via the proton sponge effect, facilitating rapid drug release. Additionally, enhanced intracellular uptake, cytotoxicity, and tumor targeting, along with reduced systemic toxicity, suggest promising potential for FA/PBAE/DTX-NPs in tumor therapy (Zhang et al. 2021).

Wang et al. formulated redox-responsive PLNs for the delivery of afatinib (Afa) and Tf-modified (Tf-SS-Afa-PLNs) for treating non-small cell lung cancer. The nanoparticles were formulated by nanoprecipitation method using PLGA and lecithin. In vitro, glutathione-triggered drug release was evaluated, which revealed that with an increase in the concentration of GSH, the drug release was high (80%) as compared to a decrease in the concentration of GSH. Cell viability assay study revealed that Tf-SS-Afa-PLNs exhibited cell inhibition efficiency compared to untargeted Afa-PLNs and free drugs under hypoxic conditions, suggesting the redox sensitivity of Tf-SS-Afa-PLNs (Wang et al. 2019).

Joshy et al. developed PLNs containing nickel ferrite (NFO) using PVA and stearic acid along with PEG to facilitate the delivery of zidovudine (AZT) delivery with responsiveness to the magnetic field for controlled drug release. In vitro investigation of drug release showed that the release of AZT was enhanced in the presence of a magnetic field (80%), whereas in its absence, the release was decreased (13%) after 48 h, attributed to the influence of the magnetic field. Additionally, the investigation confirmed the loading and sustained release of AZT (Joshy et al. 2020). Deok Kong et al. formulated PLNs containing camptothecin (CPT) and iron oxide nanoparticles (IONPs) for breast cancer. IONPs were embedded within lecithin, DSPE-PEG, and PLGA. The drug release was determined using a radio frequency (RF) magnetic field. The results revealed that under an RF magnetic field, heating Fe_3O_4 and loosening the polymeric cores. Magnetic field-assisted drug release showed nearly 100% release in 48 h at 100 kHz. In vitro tests on MT2 mouse breast cancer cells revealed lower cell growth with CPT-loaded LPHNPs under RF stimulation. The platform offers simple synthesis, stability, and controlled drug release, potentially enhancing cancer treatment (Deok Kong et al. 2013) (Fig. 10.10).

The photoresponsive technique for drug release harnesses the photothermal effect, enabling controlled release either through triggered drug release or direct disintegration of the drug carrier upon light irradiation. To formulate these light-sensitive nanocarriers, their composition must incorporate photochromic moieties capable of inducing photoreaction, consequently altering polarity and shifting the hydrophilic–hydrophobic equilibrium sufficiently to cause the rupture of nanocarriers upon light exposure (Jain et al. 2023). Yao et al. developed a photoresponsive PLNs comprising a PLGA core coated with a monolayer of lecithin and a photosensitive layer (PEG-hexadecyl block polymer with a 2-nitrobenzyl linker), encapsulating DOX. The in vitro photosensitive drug release from the polymeric shell was triggered upon light sensitization and showed enhanced drug release that is, with light irradiation, it showed 76% drug release whereas in the absence of light irradiation only 10% release. The results from confocal microscopy and flow cytometry confirmed the light-regulated drug-release dynamics within cancer cells. Also in living animals, the drug release was significantly accelerated with light irradiation (Yao et al. 2017).

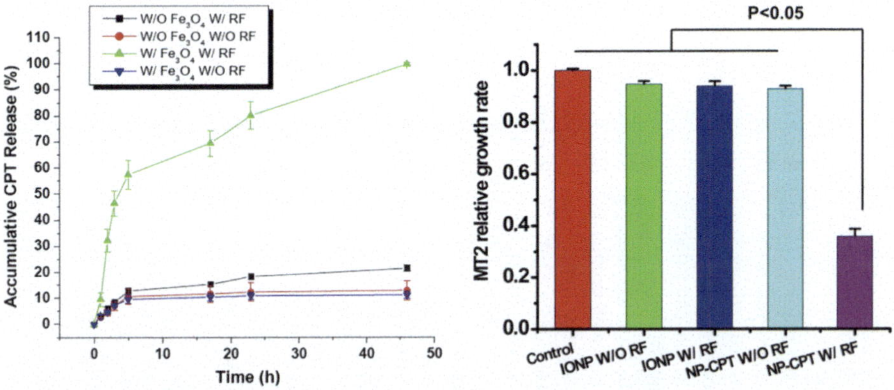

Fig. 10.10 (**a**) Drug release from PLNs containing CPT and IONPs under the influence of Rf magnetic field. The cumulative drug release was carried out for 50 h; the particles were exposed to Rf for 4 min, two times with breaks; (**b**) MTT assay carried out on MT2 cancer cell line, treated with free-CPT and loaded-CPT-PLNs, with and without Rf field activation. (Adapted with permission Deok Kong et al. 2013)

8 Biomedical Applications

The application of PLNs offers a versatile platform for drug delivery in cancer treatments, these hybrid nanoparticles offer the combined advantage of both polymer and lipids providing benefits such as stability, enhanced loading capacity, prolonged release, increased biocompatibility, and enhanced therapeutic efficacy of the drug (Gajbhiye et al. 2023; Persano et al. 2021). Furthermore, the co-administered of hydrophobic drugs with analogous therapeutic agents like nucleic acids, proteins, and peptides is feasible. Enhanced delivery can be achieved by ligating the PLNs with the cancer cell-specific ligand, facilitating their uptake. The synergistic effect of encapsulation and controlled release of cargo upon reaching cancer cells render PLNs an intriguing system for in vitro and in vivo investigations across various cancer conditions (Gajbhiye et al. 2023). The amalgamation of liposomes and PNPs within PLNs offers superior attributes that could propel the advancement of the current nanotherapeutic systems. Table 10.3 depicts the applications of PLNs in cancer treatments.

8.1 Drug Delivery

The PLNs can incorporate both hydrophobic and hydrophilic drugs, however, PLNs prepared by double emulsification approach can incorporate hydrophilic moiety whereas, there is limited encapsulation efficiency of hydrophilic drugs (Shah et al. 2022). Shafique et al., formulated DOX containing PLNs, consisting of eudragit RS-100, oleic acid, stearic acid, and ethyl cellulose (EC). The findings indicated

Table 10.3 Different PLN preparations with their targeting cancer treatment

Drug	Cancer	Polymer	Lipid	Fabrication method	Outcome	Reference
Palbociclib	Brest cancer	PLGA	L-α PC	Nanoprecipitation with modifications	The nanoparticles showed significantly improved anti-cancer efficacy, reducing cell migration, and inducing apoptosis	Rajana et al. (2024)
5-fluorouracil	Colorectal cancer	PLGA	Lipoid S 100 and DSPE-PEG2000	Nanoprecipitation method with modifications	Reduced dose-related toxicity with sustained release behaviour	Khan et al. (2024)
Exemestane	Estrogen receptor-positive breast cancer	PCL	Phospholipon 90 G	Nanoprecipitation method	Enhanced intestinal permeation, and better tumor inhibition efficacy	Rizwanullah et al. (2023)
Baicalin	Colorectal cancer	Chitosan	SPC D70SPE-PEG2000	Nanoprecipitation method	Increased biochemical characteristics, with increased efficacy	Riadi et al. (2023)
5-fluorouracil and sulforaphane	Colon cancer	PLGA	Phospholipid and DSPE-PEG 2000-NH_2	Nanoprecipitation method	Enhanced cytotoxicity and improved therapeutic efficacy	Li et al. (2022)
Ursolic acid	Pancreatic ductal adenocarcinoma cells	PLGA	SPC and DSPE-PEG 2000	Nanoprecipitation method	Enhanced cytotoxicity and excellent stability	Markowski et al. (2022)
Emodin	Breast cancer	PLGA	Soybean phospholipid	Nanoprecipitation method	Reduced epithelial-mesenchymal transition, enhancing sensitivity in turn improving drug efficacy	Zou et al. (2021)

Methotrexate	Glioblastoma	PLGA	Glyceryl tripalmitate	Solvent injection and homogenisation	Exhibits higher cytotoxicity to glioma cells with potential targeting	Bhattacharya (2021)
Anastrozole	Breast cancer	PCL	Stearic acid.	Direct emulsification solvent evaporation method	Enhanced therapeutic efficacy	Massadeh et al. (2020)
Curcumin and tamoxifen	Cancer cell lines (A2780, MDA-mb231, Ntra2, HepG2, and PC3)	Chitosan	Lecithin	Ionic gelation method	Inhibited cancer cell proliferation, growth and migration	Chen et al. (2020a)
DTX and CIS	Lung cancer	Poly(L-lactide) (5000)-poly(ethylene glycol) (2000)-maleimide	Glyceryl monostearate and lecithin	Thin-film hydration and ultrasonic dispersion method	Aptamer-conjugate nanoparticles showed enhanced cytotoxicity, synergy antitumor effect and tumor inhibition ability as compared to single drug and non-aptamer nanoparticles	Wu et al. (2020)
Plumbagin	Melanoma	PLGA	Phosphatidylcholine and DSPE-PEG2000-maleimide	One-step nanoprecipitation	Enhanced therapeutic effect in Tf-bearing nanoparticles	Sakpakdeejaroen et al. (2021)
Psoralen	Lung cancer	PLGA	Soybean phospholipids and DSPE-PEG2000	Nanoprecipitation	Enhanced cytotoxicity effect	Yuan et al. (2019)

(continued)

Table 10.3 (continued)

Drug	Cancer	Polymer	Lipid	Fabrication method	Outcome	Reference
CRISPR/Cas9 plasmids	Glioblastoma	PLGA	Lecithin	Nanoprecipitation	Efficient gene targeting with therapeutic application	Yang et al. (2021)
DOX	Breast cancer	PLGA	Lecithin, and DSPE-PEG 2000	Modified single step nano-precipitation process	Controlled delivery with enhanced particle internalization	Tahir et al. (2019)
Psoralen	Breast cancer	PLGA (50:50)	Soybean lecithin	Emulsification evaporation-low temperature solidification method	Showed sustained release with enhanced therapeutic efficiency	Huang et al. (2018)

that increasing the concentration of oleic acid and EC increases the encapsulation efficiency significantly. The in vitro studies revealed a favorable drug release profile with initial burst release of the drug followed by sustained release, wherein higher drug payload results in prolonged release time. These observations were confirmed by in vitro and in vivo studies demonstrating the sustained drug release and enhanced bioavailability. Density functional theory calculations underscore the strong affinity between the polymers and drug molecules, implying stable drug delivery (Shafique et al. 2023). In another study done by Yang et al., they formulated PLNs containing DOX and edelfosine for osteosarcoma treatment. The in vitro drug release was evaluated under different pH conditions, which revealed that the drug release was higher in acidic conditions compared to physiological pH (pH 7.4). Cellular uptake assay conducted in MG63 cells exhibited enhanced internalization and subcellular distribution of targeted nanoparticles relative to non-targeted counterpoints. MTT and caspase-3/7 activity assay demonstrated the superior anticancer efficacy of co-administration of both drugs within folate-targeted nanoparticles in inducing cancer cell apoptosis. In vivo results displayed potent tumor cell killing and significant suppression of tumor growth without observable adverse effects (Yang et al. 2020). The study conducted by Yalcin et al. formulated gemcitabine hydrochloride (GEM) containing PLNs for breast cancer. The cytotoxicity studies were conducted by MTT assay using MCF-7 and MDA-MB-231 which showed significantly reduced cell viability compared to native GEM formulations. The lower IC50 value of PLNs indicates higher effectiveness in cancer cell apoptosis. This increased cytotoxicity is attributed to the lipid shell of the PLNs, which enhances adherence to cell membranes, leading to improved intracellular accumulation of GEM and enhanced anticancer efficiency. In vivo findings showed significant antitumor capability along with low toxicity (Yalcin et al. 2020). Another study conducted by Korucu Aktas and colleagues, developed crizotinib (CL) loaded PLNs for determination of its anticancer activity in non-small cell lung cancer (NSCLS). The PLNs were prepared using the nanoprecipitation method, where the PLGA or PCL was used for the polymeric core containing CL and SPC and DSSPE-PEG 2000 as the lipid shell. The box-Behnken design was used for optimizing the prepared nanoparticles. The release of CL from PLGA and PCL showed a controlled and sustained release from the PLNs. The cytotoxicity study was conducted on NCI-H2228 cells which revealed dosed-dependent and enhanced cytotoxicity of CL. The in vitro inhibitory activity was studied using rt-PCR and ELISA, which showed a significant reduction in gene expression and protein levels especially in PLGA PLNs, indicating efficiency against lung cancer (Korucu Aktas et al. 2023) (Fig. 10.11).

8.2 Gene Delivery

Gene delivery has gotten increased attention recently. Saeed and colleagues developed siRNA containing PLNs formulated of chitosan-phthalate and lecithin nanoparticle, cross-linked with tripolyphosphate (CSP-LC-TPP NPs), and investigated its effectiveness against breast cancer cells. The nanoparticles exhibited

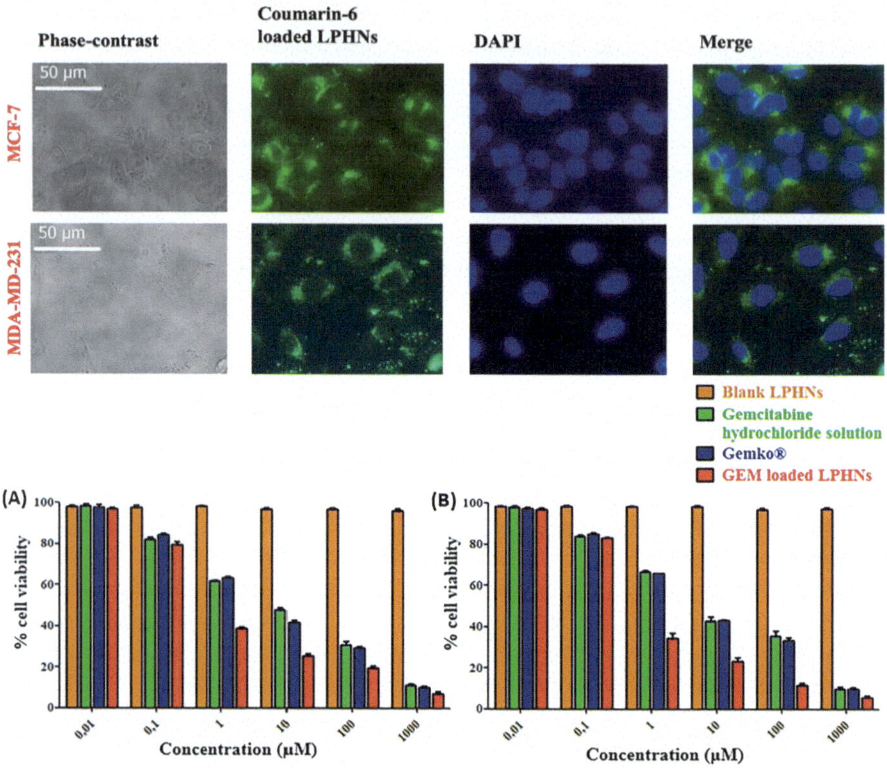

Fig. 10.11 Fluorescence imaging of PLNs containing coumarin-6; MTT assay of various nanoformulations (**a**) MCF-7 and (**b**)MDA-MB-231 cell lines. (Adapted with permission Yalcin et al. 2020)

exceptional dual-layered shells, characterized by excellent stability under physiological pH conditions and in serum solutions. Assessments through MTT assay showed that nanoparticles possess non-cytotoxic properties and can efficiently penetrate into cancer cells. These nanoparticles achieved a 44% silencing against SLUG protein in MDA-MB-453 cancer cells. Furthermore, the nanoparticles demonstrated the ability to protect siRNA in serum and resist displacement by varying concentrations of heparin (Saeed et al. 2021). In another study conducted by Wei et al., they fabricated PLNs using microfluidic technology by incorporating siRNA into PCL-PEI polymer and coated with neutral lipids (LPS NPs). The cellular uptake and internalization mechanisms of these PLNs were investigated on human prostate cancer PC-3 cells. Despite their negatively charged zeta potential, the nanoparticles exhibited high cellular uptake. The inhibition assay and fluorescence labeling revealed that the primary internalization pathway was through macropinocytosis. In vivo distribution and antitumor studies showed that both LPS and lipid/micelle/siRNA (LMS NPs) rapidly distributed into tumor tissue and major organs, with LPS NPs achieving extended and prolonged fluorescence intensity in tumors. LPS NPs showed enhanced tumor growth inhibition compared to LMS NPs, with a higher accumulation in tumor tissue and less retention in the liver. Ex vivo analysis further confirmed the stronger tumor inhibition by LPS NPs, associated with the downregulation of EGFR protein expression. These findings suggest that LPS NPs, with their stable nanostructure and efficient siRNA encapsulation, exhibit promising antitumor efficacy and biosafety in vivo (Wei et al. 2020) (Fig. 10.12).

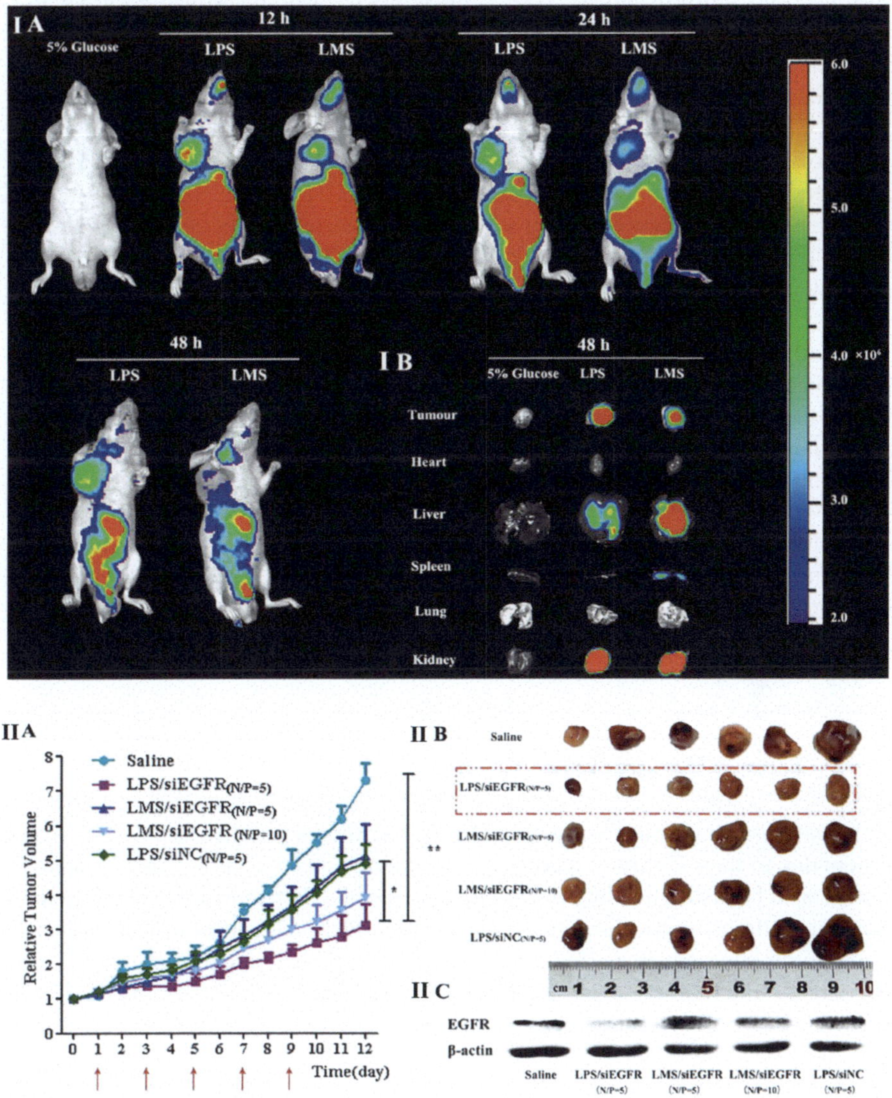

Fig. 10.12 I: Biodistribution of LPS and LMS NPs in Balb/c nude mice bearing PC-3 tumors. (**a**) In vivo imaging and (**b**) Ex vivo imaging in tumor and organs after administration; II: Antitumor activity of NPs in PC-3 tumor-bearing micein right flank, (**a**) After dose administration the change in tumor size; (**b**) Excised tumor tissues from PC-3 tumor-bearing mice after treatment; (**c**) Western-blot assay. (Adapted with permission Wei et al. 2020)

8.3 Theragnostics

Photodynamic therapy uses photosensitizer molecules along with light for treating cancer cells and is potentially safer than systemic chemotherapy. Thakur et al. developed a zinc phthalocyanine (ZnPC) and co-encapsulated with quercetin (QC) in PLNs. The in vitro drug release showed that QC and ZnPC showed a biphasic release profile, characterized by an initial burst release of the drug followed by a sustained and uniform release. The cellular uptake was investigated on MCF-7, which demonstrated significant cellular uptake efficiency, reaching up to 80% after 5 h of incubation, and showed a higher cytotoxic effect. Clathrin-mediated uptake of the nanoparticles is identified as a primary mechanism for efficient cell internalization. The in vivo studies showed that at the tumor site, there was minimal systemic toxicity with an enhanced antitumor effect (Thakur et al. 2021). Gu et al., formulated folate-modified, CIS, ICG-loaded PLNs for breast cancer. The in vitro release showed accelerated release when laser irradiated. Without laser irradiation, the nanoparticles showed an initial burst release (16.03%) within the first 12 h, with a cumulative release of 60.03% over 72 h. However, laser irradiation significantly increased the release rate, with 31.97% released by 12 h and a total release of 95.02% by 72 h. The cytotoxicity study using the MTT assay indicated that higher drug concentrations induce lower cell viability. The folate-mediated nanoparticles exhibit higher cell toxicity compared to CINPs (conventional inorganic nanoparticles), indicating good folate-targeting ability. Additionally, the combination of folate-mediated nanoparticles or CINPs with laser irradiation enhanced the apoptosis on MCF-7 cells, suggesting a synergistic effect of nanoparticles and laser irradiation. The intracellular uptake showed that the efficient internalization of folate-mediated nanoparticles was through receptor-mediated endocytosis of folate receptors. In contrast, there is no specific uptake of CINPs by either MCF-7 or A549 cells under the same conditions, indicating the targeting specificity of folate-conjugated nanoparticles to FRs (Gu et al. 2017).

9 Future Perspective

In the past decade, the development of PLNs has shown promise as a drug delivery system due to their ability to modify configuration and serve as a versatile delivery system with high drug loading potential, superior bloodstream stability, and precise targeting abilities, particularly in cancer therapy. PLNs can deliver drugs precisely while minimizing adverse effects on normal cells, perform dual drug therapy, and respond to internal or external stimuli, enhancing their therapeutic efficacy. Responsive hybrid systems, capable of adapting to environmental such as temperature and pH while maintaining biological activity, offer potential applications in targeted drug delivery, particularly in combination therapy, which is often the most effective approach for treating cancer, requiring controlled delivery. Recent advancements focus on PLNs for gene therapy and immunotherapy, expanding their potential applications and demonstrating efficacy in overcoming MDR in cancer cells while minimizing systemic toxicity. Future research should concentrate on exploring unexplored areas and optimizing PLN design to advance current therapeutics while maintaining properties and stability. Collaboration between researchers, industry professionals, and regulatory agencies is crucial to addressing challenges and facilitating clinical translation of PLNs and will be pivotal in bringing LPHNP-based products from academic labs to the market. Prioritizing the translation of PLNs into therapies aimed at increasing life expectancy and improving the quality of life for patients in the clinical market is imperative.

10 Conclusion

PLNs have demonstrated remarkable success in clinical and drug delivery applications, particularly in the field of oncology. These nanoparticles represent a promising nanocarrier class offering pharmaceutical benefits superior to both PNPs and liposomes. The PLNs prevent drug leakage and exhibit enhanced stability and surface functionalization possibilities, making them extravagant carriers advancing current nanotherapeutics. The polymeric core of PLNs contributes to encapsulation efficiency, drug uptake, and regulated drug release kinetics while the lipid shell, enhances structural integrity, stability, loading capacity, and drug permeation across cellular membranes. This topic comprehensively discusses the PLNs for cancer theragnostic covers their method of fabrication, core and shell material, surface functionalization, stimuli-responsive drug release, and applications in cancer treatment. Different arrangements of lipids and polymers result in various types of PLNs, each with specific advantages. Surface functionalization of PLNs particularly with PEGylated lipids, can alter plasma circulation time, while using the antibodies, aptamers, and ligands aids in site-specific binding and reduces adverse effects. Various materials used for PLN preparation help overcome drawbacks associated with conventional polymeric and liposomal nanocarriers. Based on the studies presented, PLNs offer a versatile platform with promising potential for improving cancer therapy. They exhibit various responsive behaviors such as pH responsiveness, redox sensitivity, magnetic field responsiveness, and photo-responsiveness, allowing controlled drug release in TME. PLNs enhance intracellular uptake, cytotoxicity, and tumor targeting, while reducing the systemic toxicity making them suitable for targeted cancer therapy. Further, the development of surface modifications, such as folic acid and transferrin, enables enhanced targeting of cancer cells, for improving the efficacy of drug delivery and treatment outcomes. Further research and development in PLNs hold promise of advancing personalized medicine approaches and improving the efficacy and safety of cancer treatment.

References

2 Hydroxyethyl methacrylate—an overview | ScienceDirect topics. (n.d.). Retrieved February 23, 2024., from https://www.sciencedirect.com/topics/pharmacology-toxicology-and-pharmaceutical-science/2-hydroxyethyl-methacrylate

Abasian P, Ghanavati S, Rahebi S, Nouri Khorasani S, Khalili S. Polymeric nanocarriers in targeted drug delivery systems: A review. Polym Adv Technol. 2020;31(12):2939–54. https://doi.org/10.1002/pat.5031.

Alsaab HO, Alharbi FD, Alhibs AS, Alanazi NB, Alshehri BY, Saleh MA, Alshehri FS, Algarni MA, Almugaiteeb T, Uddin MN, Alzhrani RM. PLGA-based nanomedicine: history of advancement and development in clinical applications of multiple diseases. Pharmaceutics. 2022;14(12):2728. https://doi.org/10.3390/pharmaceutics14122728.

Angolkar M, Paramshetti S, Halagali P, Jain V, Patil AB, Somanna P. Nanotechnological advancements in the brain tumor therapy: a novel approach. Ther Deliv. 2023; https://doi.org/10.4155/tde-2022-0035.

Baldwin AD, Kiick KL. Tunable degradation of Maleimide–Thiol adducts in reducing environments. Bioconjug Chem. 2011;22(10):1946–53. https://doi.org/10.1021/bc200148v.

Bandi SP, Datta D, Venuganti VVK. Hydrocaffeic acid-chitosan coating of gastric patch provides long-acting mucoadhesive delivery of model chemotherapeutic agent. Int J Pharm. 2023;631:122504. https://doi.org/10.1016/j.ijpharm.2022.122504.

Bangera PD, Kara DD, Tanvi K, Tippavajhala VK, Rathnanand M. Highlights on cell-penetrating peptides and polymer-lipid hybrid nanoparticle: overview and therapeutic applications for targeted anticancer therapy. AAPS PharmSciTech. 2023;24(5):124. https://doi.org/10.1208/s12249-023-02576-x.

Behrens CR, Liu B. Methods for site-specific drug conjugation to antibodies. MAbs. 2014;6(1):46–53. https://doi.org/10.4161/mabs.26632.

Bhattacharya S. Methotrexate-loaded polymeric lipid hybrid nanoparticles (PLHNPs): a reliable drug delivery system for the treatment of glioblastoma. J Exp Nanosci. 2021;16(1):344–67. https://doi.org/10.1080/17458080.2021.1983172.

Cai J, Huang H, Song W, Hu H, Chen J, Zhang L, Li P, Wu R, Wu C. Preparation and evaluation of lipid polymer nanoparticles for eradicating H. pylori biofilm and impairing antibacterial resistance in vitro. Int J Pharm. 2015;495(2):728–37. https://doi.org/10.1016/j.ijpharm.2015.09.055.

Cao Y, Dong X, Chen X. Polymer-modified liposomes for drug delivery: from fundamentals to applications. Pharmaceutics. 2022;14(4):778. https://doi.org/10.3390/pharmaceutics14040778.

Carrasco-Esteban E, Domínguez-Rullán JA, Barrionuevo-Castillo P, Pelari-Mici L, Leaman O, Sastre-Gallego S, López-Campos F. Current role of nanoparticles in the treatment of lung cancer. J Clin Transl Res. 2021;7(2):140–55.

Casalini T, Rossi F, Castrovinci A, Perale G. A perspective on polylactic acid-based polymers use for nanoparticles synthesis and applications. Front Bioeng Biotechnol. 2019;7:483145. https://doi.org/10.3389/fbioe.2019.00259.

Chan JM, Zhang L, Yuet KP, Liao G, Rhee J-W, Langer R, Farokhzad OC. PLGA–lecithin–PEG core–shell nanoparticles for controlled drug delivery. Biomaterials. 2009;30(8):1627–34. https://doi.org/10.1016/j.biomaterials.2008.12.013.

Chaudhary, Z., Ahmed, N., .ur.Rehman, A., & Khan, G. M. (2018). Lipid polymer hybrid carrier systems for cancer targeting: a review. Int J Polym Mater Polym Biomater, 67(2), 86–100. doi:https://doi.org/10.1080/00914037.2017.1300900.

Chen T-C, Chuang J-Y, Ko C-Y, Kao T-J, Yang P-Y, Yu C-H, Liu M-S, Hu S-L, Tsai Y-T, Chan H, Chang W-C, Hsu T-I. AR ubiquitination induced by the curcumin analog suppresses growth of temozolomide-resistant glioblastoma through disrupting GPX4-mediated redox homeostasis. Redox Biol. 2020a;30:101413. https://doi.org/10.1016/j.redox.2019.101413.

Chen Y, Deng Y, Zhu C, Xiang C. Anti prostate cancer therapy: aptamer-functionalized, curcumin and cabazitaxel co-delivered, tumor targeted lipid-polymer hybrid nanoparticles. Biomed Pharmacother. 2020b;127:110181. https://doi.org/10.1016/j.biopha.2020.110181.

Cheow WS, Hadinoto K. Factors affecting drug encapsulation and stability of lipid–polymer hybrid nanoparticles. Colloids Surf B: Biointerfaces. 2011;85(2):214–20. https://doi.org/10.1016/j.colsurfb.2011.02.033.

Christen M-O, Vercesi F. Polycaprolactone: how a well-known and futuristic polymer has become an innovative collagen-stimulator in esthetics. Clin Cosmet Investig Dermatol. 2020;13:31–48. https://doi.org/10.2147/CCID.S229054.

Chrószcz M, Barszczewska-Rybarek I. Nanoparticles of quaternary ammonium polyethylenimine derivatives for application in dental materials. Polymers. 2020;12(11):Article 11. https://doi.org/10.3390/polym12112551.

Clawson C, Ton L, Aryal S, Fu V, Esener S, Zhang L. Synthesis and characterization of lipid–polymer hybrid nanoparticles with pH-triggered poly(ethylene glycol) shedding. Langmuir. 2011;27(17):10556–61. https://doi.org/10.1021/la202123e.

Colaco V, Pokale R, Mukharya A, Patel JK, Pathak YV, Mutalik S, Goswami H, Dhas N. Lymphatic system: history, anatomy, physiology, challenges, and opportunities. In: Dhas N, Patel JK, Pathak YV, editors. Advanced targeting of the lymphatic system. Cham: Springer; 2024a. p. 1–21. https://doi.org/10.1007/978-3-031-64828-1_1.

Colaco V, Roy AA, Naik GARR, Mondal A, Mutalik S, Dhas N. Advancement in lipid-based nanocomposites for theranostic applications in lung carcinoma treatment. OpenNano. 2024b;15:100199. https://doi.org/10.1016/j.onano.2023.100199.

Craparo EF, Cabibbo M, Scialabba C, Giammona G, Cavallaro G. Inhalable formulation based on lipid–polymer hybrid nanoparticles for the macrophage targeted delivery of roflumilast. Biomacromolecules. 2022;23(8):3439–51. https://doi.org/10.1021/acs.biomac.2c00576.

Crucho CIC, Barros MT. Polymeric nanoparticles: a study on the preparation variables and characterization methods. Mater Sci Eng C. 2017;80:771–84. https://doi.org/10.1016/j.msec.2017.06.004.

Curcio M, Vittorio O, Bell JL, Iemma F, Nicoletta FP, Cirillo G. Hyaluronic acid within self-assembling nanoparticles: endless possibilities for targeted cancer therapy. Nanomaterials. 2022;12(16):2851. https://doi.org/10.3390/nano12162851.

Datta D, Colaco V, Alizadeh B, Bandi SP, Dhas N. 14—Micro/nanoparticles. In: Parambath A, editor. Polymers for oral drug delivery technologies. 629: Elsevier; 2025a. p. 573. https://doi.org/10.1016/B978-0-443-13774-7.00014-1.

Datta D, Colaco V, Bandi SP, Sharma H, Dhas N, Giram PS. 7—Classes/types of polymers used in oral delivery (natural, semisynthetic, synthetic), their chemical structure and general functionalities. In: Parambath A, editor. Polymers for Oral drug delivery technologies. Elsevier; 2025b. p. 263–333. https://doi.org/10.1016/B978-0-443-13774-7.00007-4.

Dave V, Tak K, Sohgaura A, Gupta A, Sadhu V, Reddy KR. Lipid-polymer hybrid nanoparticles: synthesis strategies and biomedical applications. J Microbiol Methods. 2019;160:130–42. https://doi.org/10.1016/j.mimet.2019.03.017.

Dehaini D, Fang RH, Luk BT, Pang Z, Hu C-MJ, Kroll AV, Yu CL, Gao W, Zhang L. Ultra-small lipid–polymer hybrid nanoparticles for tumor-penetrating drug delivery. Nanoscale. 2016;8(30):14411–9. https://doi.org/10.1039/C6NR04091H.

Deok Kong S, Sartor M, Jack H, C.-M., Zhang, W., Zhang, L., & Jin, S. Magnetic field activated lipid–polymer hybrid nanoparticles for stimuli-responsive drug release. Acta Biomater. 2013;9(3):5447–52. https://doi.org/10.1016/j.actbio.2012.11.006.

Dhas N, Preetha HS, Dubey A, Ravi G, Govindan I, Rama A, Naha A, Hebbar S. Factorial design-based fabrication of biopolymer-functionalized Asiatic acid-embedded liposomes: in-vitro characterization and evaluation. J Appl Pharm Sci. 2022;12(11):071–81. https://doi.org/10.7324/JAPS.2022.121108.

Dhas N, Kudarha R, Kulkarni S, Soman S, Navti PD, Kulkarni J, Roy AA, Colaco V, Raychaudhuri R, Gupta A, Pardeshi C, Bari D, Tiwari R, Patel J, Moorkoth S, Mutalik S. Nanoengineered platform-based microenvironment-triggered immunotherapy in cancer treatment. Front Biosci-Landmark. 2024a;29(10):Article 10. https://doi.org/10.31083/j.fbl2910349.

Dhas N, Kudarha R, Tiwari R, Tiwari G, Garg N, Kumar P, Kulkarni S, Kulkarni J, Soman S, Hegde AR, Patel J, Garkal A, Sami A, Datta D, Colaco V, Mehta T, Vora L, Mutalik S. Recent advancements in nanomaterial-mediated ferroptosis-induced cancer therapy: importance of molecular dynamics and novel strategies. Life Sci. 2024b;346:122629. https://doi.org/10.1016/j.lfs.2024.122629.

Ekinci M, Yeğen G, Aksu B, İlem-Özdemir D. Preparation and evaluation of poly(lactic acid)/poly(vinyl alcohol) nanoparticles using the quality by design approach. ACS Omega. 2022;7(38):33793–807. https://doi.org/10.1021/acsomega.2c02141.

Fenart L, Casanova A, Dehouck B, Duhem C, Slupek S, Cecchelli R, Betbeder D. Evaluation of effect of charge and lipid coating on ability of 60-nm nanoparticles to cross an in vitro model of the blood-brain barrier. J Pharmacol Exp Ther. 1999;291(3):1017–22. Scopus

Fu Q, Wang J, Liu H. Chemo-immune synergetic therapy of esophageal carcinoma: Trastuzumab modified, cisplatin and fluorouracil co-delivered lipid-polymer hybrid nanoparticles. Drug Deliv. 2020;27(1):1535–43. https://doi.org/10.1080/10717544.2020.1837294.

Fu C-P, Cai X-Y, Chen S-L, Yu H-W, Fang Y, Feng X-C, Zhang L-M, Li C-Y. Hyaluronic acid-based nanocarriers for anticancer drug delivery. Polymers. 2023;15(10):Article 10. https://doi.org/10.3390/polym15102317.

Gaaz TS, Sulong AB, Akhtar MN, Kadhum AAH, Mohamad AB, Al-Amiery AA. Properties and applications of polyvinyl alcohol, halloysite nanotubes and their nanocomposites. Molecules. 2015;20(12):22833–47. https://doi.org/10.3390/molecules201219884.

Gajbhiye KR, Salve R, Narwade M, Sheikh A, Kesharwani P, Gajbhiye V. Lipid polymer hybrid nanoparticles: A custom-tailored next-generation approach for cancer therapeutics. Mol Cancer. 2023;22(1):160. https://doi.org/10.1186/s12943-023-01849-0.

Gao F, Zhang J, Fu C, Xie X, Peng F, You J, Tang H, Wang Z, Li P, Chen J. iRGD-modified lipid–polymer hybrid nanoparticles loaded with isoliquiritigenin to enhance anti-breast cancer effect and tumor-targeting ability. Int J Nanomedicine. 2017;12:4147–62. https://doi.org/10.2147/IJN.S134148.

García-Pinel B, Porras-Alcalá C, Ortega-Rodríguez A, Sarabia F, Prados J, Melguizo C, López-Romero JM. Lipid-based nanoparticles: application and recent advances in cancer treatment. Nano. 2019;9(4):638. https://doi.org/10.3390/nano9040638.

Garg NK, Tyagi RK, Sharma G, Jain A, Singh B, Jain S, Katare OP. Functionalized lipid–polymer hybrid nanoparticles mediated codelivery of methotrexate and aceclofenac: a synergistic effect in breast cancer with improved pharmacokinetics attributes. Mol Pharm. 2017;14(6):1883–97. https://doi.org/10.1021/acs.molpharmaceut.6b01148.

Garkal A, Sami A, Patel J, Patel L, Dhas NL, Mehta T. Light-responsive delivery nanoplatforms in cancer theranostics. In: Site-specific cancer nanotheranostics. CRC Press; 2023.

Gavas S, Quazi S, Karpiński TM. Nanoparticles for cancer therapy: current progress and challenges. Nanoscale Res Lett. 2021;16:173. https://doi.org/10.1186/s11671-021-03628-6.

Gu L, Shi T, Sun Y, You C, Wang S, Wen G, Chen L, Zhang X, Zhu J, Sun B. Folate-modified, indocyanine green-loaded lipid-polymer hybrid nanoparticles for targeted delivery of cisplatin. J Biomater Sci Polym Ed. 2017;28(7):690–702. https://doi.org/10.1080/09205063.2017.1296347.

Hadinoto K, Sundaresan A, Cheow WS. Lipid–polymer hybrid nanoparticles as a new generation therapeutic delivery platform: a review. Eur J Pharm Biopharm. 2013;85(3, Part A):427–43. https://doi.org/10.1016/j.ejpb.2013.07.002.

Halder T, Saha B, Dhas N, Acharya S, Acharya N. Development and evaluation of multi-functionalized sialic acid conjugated asiatic acid nanoconstruct to mitigate cognitive deficits in Alzheimer's disease. Drug Dev Res. 2024;85(1):e22146. https://doi.org/10.1002/ddr.22146.

Hasan W, Chu K, Gullapalli A, Dunn SS, Enlow EM, Luft JC, Tian S, Napier ME, Pohlhaus PD, Rolland JP, DeSimone JM. Delivery of multiple siRNAs using lipid-coated PLGA nanoparticles for treatment of prostate cancer. Nano Lett. 2012;12(1):287–92. https://doi.org/10.1021/nl2035354.

Higazy IM, Mahmoud AA, Ghorab MM, Ammar HO. Development and evaluation of polyvinyl alcohol stabilized polylactide-co-caprolactone-based nanoparticles for brain delivery. J Drug Deliv Sci Technol. 2021;61:102274. https://doi.org/10.1016/j.jddst.2020.102274.

Hitzman CJ, Elmquist WF, Wattenberg LW, Wiedmann TS. Development of a respirable, sustained release microcarrier for 5-fluorouracil I: in vitro assessment of liposomes, microspheres, and lipid coated nanoparticles. J Pharm Sci. 2006;95(5):1114–26. https://doi.org/10.1002/jps.20591.

Hu C-MJ, Zhang L, Aryal S, Cheung C, Fang RH, Zhang L. Erythrocyte membrane-camouflaged polymeric nanoparticles as a biomimetic delivery platform. Proc Natl Acad Sci. 2011;108(27):10980–5. https://doi.org/10.1073/pnas.1106634108.

Huang G, Huang H. Application of hyaluronic acid as carriers in drug delivery. Drug Deliv. 2018;25(1):766–72. https://doi.org/10.1080/10717544.2018.1450910.

Huang Q, Cai T, Li Q, Huang Y, Liu Q, Wang B, Xia X, Wang Q, Whitney JCC, Cole SPC, Cai Y. Preparation of psoralen polymer–lipid hybrid nanoparticles and their reversal of multidrug resistance in MCF-7/ADR cells. Drug Deliv. 2018;25(1):1044–54. https://doi.org/10.1080/10717544.2018.1464084.

Jain A, Sharma G, Kushwah V, Garg NK, Kesharwani P, Ghoshal G, Singh B, Shivhare US, Jain S, Katare OP. Methotrexate and beta-carotene loaded-lipid polymer hybrid nanoparticles: a preclinical study for breast cancer. Nanomedicine. 2017;12(15):1851–72. https://doi.org/10.2217/nnm-2017-0011.

Jain AK, Bataille CJR, Milhas S, Miller A, Zhang J, Rabbitts TH. Immunopolymer lipid nanoparticles for delivery of macromolecules to antigen-expressing cells. ACS Appl Bio Mater. 2020;3(12):8481–95. https://doi.org/10.1021/acsabm.0c00857.

Jain S, Kumar M, Kumar P, Verma J, Rosenholm JM, Bansal KK, Vaidya A. Lipid–polymer hybrid nanosystems: a rational fusion for advanced therapeutic delivery. J Funct Biomater. 2023;14(9):Article 9. https://doi.org/10.3390/jfb14090437.

Jose C, Amra K, Bhavsar C, Momin MM, Omri A. Polymeric lipid hybrid nanoparticles: properties and therapeutic applications. Crit Rev Ther Drug Carrier Syst. 2018;35(6):555. https://doi.org/10.1615/CritRevTherDrugCarrierSyst.2018024751.

Joshy KS, Augustine R, Mayeen A, Alex SM, Hasan A, Thomas S, Chi H. NiFe2O4/poly(ethylene glycol)/lipid–polymer hybrid nanoparticles for anti-cancer drug delivery. New J Chem. 2020;44(42):18162–72. https://doi.org/10.1039/D0NJ01163K.

Jovčevska I, Muyldermans S. The therapeutic potential of nanobodies. BioDrugs. 2020;34(1):11–26. https://doi.org/10.1007/s40259-019-00392-z.

Khan MM, Madni A, Torchilin V, Filipczak N, Pan J, Tahir N, Shah H. Lipid-chitosan hybrid nanoparticles for controlled delivery of cisplatin. Drug Deliv. 2019;26(1):765–72. https://doi.org/10.1080/10717544.2019.1642420.

Khan S, Madni A, Aamir MN, Khan S, Ahmad F-D, Basit A, Jan N, Shah H, Shafiq A, Anwar M. Design and evaluation of sustained-release lipid-PLGA hybrid nanoparticles for enhanced anticancer efficacy of 5-fluorouracil. Part Sci Technol. 2024;42(2):269–87. https://doi.org/10.1080/02726351.2023.2230924.

Kolodych S, Koniev O, Baatarkhuu Z, Bonnefoy J-Y, Debaene F, Cianférani S, Van Dorsselaer A, Wagner A. CBTF: new amine-to-thiol coupling reagent for preparation of antibody conjugates with increased plasma stability. Bioconjug Chem. 2015;26(2):197–200. https://doi.org/10.1021/bc500610g.

Korucu Aktas P, Baysal I, Yabanoglu-Ciftci S, Arica B. Development and in vitro evaluation of crizotinib-loaded lipid–polymer hybrid nanoparticles using box–Behnken design in non-small cell lung cancer. AAPS PharmSciTech. 2023;24(7):178. https://doi.org/10.1208/s12249-023-02634-4.

Kulkarni D, Gadade D, Kapare H, Dhas NL, Ban M. Characterization techniques for stimuli-responsive delivery nanoplatforms in cancer treatment. In: Site-specific cancer nanotheranostics. CRC Press; 2023.

Labet M, Thielemans W. Synthesis of polycaprolactone: a review. Chem Soc Rev. 2009;38(12):3484–504. https://doi.org/10.1039/B820162P.

Lee S-M, Chen H, Dettmer CM, O'Halloran TV, Nguyen ST. Polymer-caged Lipsomes: a pH-responsive delivery system with high stability. J Am Chem Soc. 2007;129(49):15096–7. https://doi.org/10.1021/ja070748i.

Lee S-M, O'Halloran TV, Nguyen ST. Polymer-caged nanobins for synergistic cisplatin-doxorubicin combination chemotherapy. J Am Chem Soc. 2010;132(48):17130–8. https://doi.org/10.1021/ja107333g.

Li S, Xu Z, Alrobaian M, Afzal O, Kazmi I, Almalki WH, Altamimi ASA, Al-Abbasi FA, Alharbi KS, Altowayan WM, Singh T, Akhter MH, Gupta M, Rahman M, Beg S. EGF-functionalized lipid–polymer hybrid nanoparticles of 5-fluorouracil and sulforaphane with enhanced bioavailability and anticancer activity against colon carcinoma. Biotechnol Appl Biochem. 2022;69(5):2205–21. https://doi.org/10.1002/bab.2279.

Lin C, Engbersen JFJ. Effect of chemical functionalities in poly(amido amine)s for non-viral gene transfection. J Control Release. 2008;132(3):267–72. https://doi.org/10.1016/j.jconrel.2008.06.022.

Liu Y, Pan J, Feng S-S. Nanoparticles of lipid monolayer shell and biodegradable polymer core for controlled release of paclitaxel: effects of surfactants on particles size, characteristics and *in vitro* performance. Int J Pharm. 2010;395(1):243–50. https://doi.org/10.1016/j.ijpharm.2010.05.008.

Mandal B, Bhattacharjee H, Mittal N, Sah H, Balabathula P, Thoma LA, Wood GC. Core–shell-type lipid–polymer hybrid nanoparticles as a drug delivery platform. Nanomedicine. 2013;9(4):474–91. https://doi.org/10.1016/j.nano.2012.11.010.

Markowski A, Jaromin A, Migdał P, Olczak E, Zygmunt A, Zaremba-Czogalla M, Pawlik K, Gubernator J. Design and development of a new type of hybrid PLGA/lipid nanoparticle as an ursolic acid delivery system against pancreatic ductal adenocarcinoma cells, Int J Molecul Sci. 2022;23(10) Scopus https://doi.org/10.3390/ijms23105536.

Marques AC, Costa PJ, Velho S, Amaral MH. Functionalizing nanoparticles with cancer-targeting antibodies: a comparison of strategies. J Control Release. 2020;320:180–200. https://doi.org/10.1016/j.jconrel.2020.01.035.

Massadeh S, Omer ME, Alterawi A, Ali R, Alanazi FH, Almutairi F, Almotairi W, Alobaidi FF, Alhelal K, Almutairi MS, Almalik A, Obaidat AA, Alaamery M, Yassin AE. Optimized polyethylene glycolylated polymer–lipid hybrid nanoparticles as a potential breast cancer treatment. Pharmaceutics. 2020;12(7):Article 7. https://doi.org/10.3390/pharmaceutics12070666.

Mikušová V, Mikuš P. Advances in chitosan-based nanoparticles for drug delivery. Int J Mol Sci. 2021;22(17):Article 17. https://doi.org/10.3390/ijms22179652.

Mishra P, Tripathi YB, Dhas NL, Garg N. Hypoxia-responsive delivery nanoplatforms in cancer theranostics. In: Site-specific cancer nanotheranostics. CRC Press; 2023.

Mohammed MA, Syeda JTM, Wasan KM, Wasan EK. An overview of chitosan nanoparticles and its application in non-parenteral drug delivery. Pharmaceutics. 2017;9(4):53. https://doi.org/10.3390/pharmaceutics9040053.

Monteiro N, Martins A, Reis RL, Neves NM. Liposomes in tissue engineering and regenerative medicine. J R Soc Interface. 2014;11(101):20140459. https://doi.org/10.1098/rsif.2014.0459.

Mukherjee A, Waters AK, Kalyan P, Achrol AS, Kesari S, Yenugonda VM. Lipid–polymer hybrid nanoparticles as a next-generation drug delivery platform: state of the art, emerging technologies, and perspectives. Int J Nanomedicine. 2019;14:1937–52. https://doi.org/10.2147/IJN.S198353.

Nsairat H, Khater D, Sayed U, Odeh F, Al Bawab A, Alshaer W. Liposomes: structure, composition, types, and clinical applications. Heliyon. 2022;8(5):e09394. https://doi.org/10.1016/j.heliyon.2022.e09394.

Omar SH, Osman R, Mamdouh W, Abdel-Bar HM, Awad GAS. Bioinspired lipid-polysaccharide modified hybrid nanoparticles as a brain-targeted highly loaded carrier for a hydrophilic drug. Int J Biol Macromol. 2020;165(Pt A):483–94. https://doi.org/10.1016/j.ijbiomac.2020.09.170.

Palazzolo S, Bayda S, Hadla M, Caligiuri I, Corona G, Toffoli G, Rizzolio F. The clinical translation of organic nanomaterials for cancer therapy: a focus on polymeric nanoparticles, micelles, liposomes and exosomes. Curr Med Chem. n.d.;25(34):4224–68.

Pandita D, Kumar S, Lather V. Hybrid poly(lactic-co-glycolic acid) nanoparticles: design and delivery prospectives. Drug Discov Today. 2015;20(1):95–104. https://doi.org/10.1016/j.drudis.2014.09.018.

Pang J, Xing H, Sun Y, Feng S, Wang S. Non-small cell lung cancer combination therapy: hyaluronic acid modified, epidermal growth factor receptor targeted, pH sensitive lipid-polymer hybrid nanoparticles for the delivery of erlotinib plus bevacizumab. Biomed Pharmacother = Biomedecine & Pharmacotherapie. 2020;125:109861. https://doi.org/10.1016/j.biopha.2020.109861.

Park W, Heo Y-J, Han DK. New opportunities for nanoparticles in cancer immunotherapy. Biomater Res. 2018;22(1):24. https://doi.org/10.1186/s40824-018-0133-y.

Parveen S, Gupta P, Kumar S, Banerjee M. Lipid polymer hybrid nanoparticles as potent vehicles for drug delivery in cancer therapeutics. Med Drug Discov. 2023;20:100165. https://doi.org/10.1016/j.medidd.2023.100165.

Pernari J, Wells LA. Poly(2-hydroxyethyl methacrylate) surface chemistry and modulus differentially modulate neutrophils and lens epithelial cells—possible implications in cellular responses to intraocular lenses. J Biomed Mater Res A. 2023;111(6):863–78. https://doi.org/10.1002/jbm.a.37537.

Persano F, Gigli G, Leporatti S. Lipid-polymer hybrid nanoparticles in cancer therapy: current overview and future directions. Nano Express. 2021;2(1):012006. https://doi.org/10.1088/2632-959X/abeb4b.

Polo E, Puertas S, Moros M, Batalla P, Guisán JM, de la Fuente JM, Grazú V. Tips for the functionalization of nanoparticles with antibodies. In: Guisan JM, editor. Immobilization of enzymes and cells. 3rd ed. Humana Press; 2013. p. 149–63. https://doi.org/10.1007/978-1-62703-550-7_11.

Pontes AP, van der Wal S, Roelofs K, Grobbink A, Creemers LB, Engbersen JFJ, Rip J. A poly(amidoamine)-based polymeric nanoparticle platform for efficient *in vivo* delivery of mRNA. Biomater Adv. 2024;156:213713. https://doi.org/10.1016/j.bioadv.2023.213713.

Rajak N, Kumar P, Dhas NL, Singh Y, Garg N. ROS-responsive delivery nanoplatforms in cancer theranostics. In: Site-specific cancer nanotheranostics. CRC Press; 2023.

Rajana N, Sandeep Chary P, Bhavana V, Deshmukh R, Dukka K, Sharma A, Kumar Mehra N. Targeted delivery and apoptosis induction of CDK-4/6 inhibitor loaded folic acid decorated lipid-polymer hybrid nanoparticles in breast cancer cells. Int J Pharm. 2024;651:123787. https://doi.org/10.1016/j.ijpharm.2024.123787.

Riadi Y, Afzal O, Geesi MH, Almalki WH, Singh T. Baicalin-loaded lipid–polymer hybrid nanoparticles inhibiting the proliferation of human colon cancer: pharmacokinetics and in vivo evaluation. Polymers. 2023;15(3):Article 3. https://doi.org/10.3390/polym15030598.

Rizwanullah M, Perwez A, Alam M, Ahmad S, Mir SR, Rizvi M, M. A., & Amin, S. Polymer-lipid hybrid nanoparticles of exemestane for improved oral bioavailability and anti-tumor efficacy: an extensive preclinical investigation. Int J Pharm. 2023;642:123136. https://doi.org/10.1016/j.ijpharm.2023.123136.

Rouco H, García-García P, Évora C, Díaz-Rodríguez P, Delgado A. Screening strategies for surface modification of lipid-polymer hybrid nanoparticles. Int J Pharm. 2022;624:121973. https://doi.org/10.1016/j.ijpharm.2022.121973.

Saeed RM, Abdullah M, Ahram M, Taha MO. Novel ellipsoid chitosan-phthalate lecithin nanoparticles for siRNA delivery. Front Bioeng Biotechnol. 2021;9:695371. https://doi.org/10.3389/fbioe.2021.695371.

Sakpakdeejaroen I, Somani S, Laskar P, Mullin M, Dufès C. Regression of melanoma following intravenous injection of plumbagin entrapped in transferrin-conjugated, lipid–polymer hybrid nanoparticles. Int J Nanomedicine. 2021;16:2615–31. https://doi.org/10.2147/IJN.S293480.

Shafique M, Ur Rehman M, Kamal Z, Alzhrani RM, Alshehri S, Alamri AH, Bakkari MA, Sabei FY, Safhi AY, Mohammed AM, Hamd MAE, Almawash S. Formulation development of lipid polymer hybrid nanoparticles of doxorubicin and its in-vitro, in-vivo and computational evaluation. Front Pharmacol. 2023;14:1025013. https://doi.org/10.3389/fphar.2023.1025013.

Shah S, Famta P, Raghuvanshi RS, Singh SB, Srivastava S. Lipid polymer hybrid nanocarriers: insights into synthesis aspects, characterization, release mechanisms, surface functionalization and potential implications. Colloid Interfac Sci Commun. 2022;46:100570. https://doi.org/10.1016/j.colcom.2021.100570.

Shao Y, Luo W, Guo Q, Li X, Zhang Q, Li J. In vitro and in vivo effect of hyaluronic acid modified, doxorubicin and gallic acid co-delivered lipid-polymeric hybrid nano-system for leukemia therapy. Drug Des Devel Ther. 2019;13:2043–55. https://doi.org/10.2147/DDDT.S202818.

Shen M, Rusling JF, Dixit CK. Site-selective orientated immobilization of antibodies and conjugates for immunodiagnostics development. Methods. 2017;116:95–111. https://doi.org/10.1016/j.ymeth.2016.11.010.

Sivadasan D, Sultan MH, Madkhali O, Almoshari Y, Thangavel N. Polymeric lipid hybrid nanoparticles (PLNs) as emerging drug delivery platform—a comprehensive review of their properties, preparation methods, and therapeutic applications. Pharmaceutics. 2021;13(8):1291. https://doi.org/10.3390/pharmaceutics13081291.

Su X, Fricke J, Kavanagh DG, Irvine DJ. In vitro and in vivo mRNA delivery using lipid-enveloped pH-responsive polymer nanoparticles. Mol Pharm. 2011;8(3):774–87. https://doi.org/10.1021/mp100390w.

Szczęch M, Szczepanowicz K. Polymeric core-shell nanoparticles prepared by spontaneous emulsification solvent evaporation and functionalized by the layer-by-layer method. Nanomaterials. 2020;10:496. https://doi.org/10.3390/nano10030496.

Tahir N, Madni A, Correia A, Rehman M, Balasubramanian V, Khan MM, Santos HA. Lipid-polymer hybrid nanoparticles for controlled delivery of hydrophilic and lipophilic doxorubicin for breast cancer therapy. Int J Nanomedicine. 2019;14:4961–74. https://doi.org/10.2147/IJN.S209325.

Tahir N, Madni A, Li W, Correia A, Khan MM, Rahim MA, Santos HA. Microfluidic fabrication and characterization of Sorafenib-loaded lipid-polymer hybrid nanoparticles for controlled drug delivery. Int J Pharm. 2020;581:119275. https://doi.org/10.1016/j.ijpharm.2020.119275.

Tang S, Li Y. Sorafenib-loaded ligand-functionalized polymer-lipid hybrid nanoparticles for enhanced therapeutic effect against liver cancer. J Nanosci Nanotechnol. 2019;19(11):6866–71. https://doi.org/10.1166/jnn.2019.16936.

Thakur NS, Mandal N, Patel G, Kirar S, Reddy YN, Kushwah V, Jain S, Kalia YN, Bhaumik J, Banerjee UC. Co-administration of zinc phthalocyanine and quercetin via hybrid nanoparticles for augmented photodynamic therapy. Nanomedicine. 2021;33:102368. https://doi.org/10.1016/j.nano.2021.102368.

Thevenot J, Troutier A-L, David L, Delair T, Ladavière C. Steric stabilization of lipid/polymer particle assemblies by poly(ethylene glycol)-lipids. Biomacromolecules. 2007;8(11):3651–60. https://doi.org/10.1021/bm700753q.

Thiruppathi R, Mishra S, Ganapathy M, Padmanabhan P, Gulyás B. Nanoparticle functionalization and its potentials for molecular imaging. Advanced Science. 2017;4(3):1600279. https://doi.org/10.1002/advs.201600279.

Tiwari G, Babu AK, Khatri C, Dhas NL. Aptamer-based targeting of nanoplatforms for cancer theranostics. In: Site-specific cancer nanotheranostics. CRC Press; 2023a.

Tiwari R, Ansari VA, Mishra J, Dhas NL. Ultrasound-responsive delivery nanoplatforms in cancer theranostics. In: Site-specific cancer nanotheranostics. CRC Press; 2023b.

Troutier A-L, Delair T, Pichot C, Ladavière C. Physicochemical and interfacial investigation of lipid/polymer particle assemblies. Langmuir. 2005;21(4):1305–13. https://doi.org/10.1021/la047659t.

Wang J. Combination treatment of cervical cancer using folate-decorated, pH-sensitive, carboplatin and paclitaxel co-loaded lipid-polymer hybrid nanoparticles. Drug Des Devel Ther. 2020;14:823–32. https://doi.org/10.2147/DDDT.S235098.

Wang X, Niu D, Hu C, Li P. Polyethyleneimine-based nanocarriers for gene delivery. Curr Pharm Des. 2015;21(42):6140–56. https://doi.org/10.2174/1381612821666151027152907.

Wang G, Wang Z, Li C, Duan G, Wang K, Li Q, Tao T. RGD peptide-modified, paclitaxel prodrug-based, dual-drugs loaded, and redox-sensitive lipid-polymer nanoparticles for the enhanced lung cancer therapy. Biomed Pharmacother. 2018;106:275–84. https://doi.org/10.1016/j.biopha.2018.06.137.

Wang J, Su G, Yin X, Luo J, Gu R, Wang S, Feng J, Chen B. Non-small cell lung cancer-targeted, redox-sensitive lipid-polymer hybrid nanoparticles for the delivery of a second-generation irreversible epidermal growth factor inhibitor—Afatinib: *in vitro* and *in vivo* evaluation. Biomed Pharmacother. 2019;120:109493. https://doi.org/10.1016/j.biopha.2019.109493.

Wei W, Sun J, Guo X-Y, Chen X, Wang R, Qiu C, Zhang H-T, Pang W-H, Wang J-C, Zhang Q. Microfluidic-based holonomic constraints of siRNA in the kernel of lipid/polymer hybrid nanoassemblies for improving stable and safe in vivo delivery. ACS Appl Mater Interfaces. 2020;12(13):14839–54. https://doi.org/10.1021/acsami.9b22781.

Wen Q, Zhang Y, Muluh TA, Xiong K, Wang B, Lu Y, Wu Z, Liu Y, Shi H, Xiao S, Fu S. Erythrocyte membrane-camouflaged gefitinib/albumin nanoparticles for tumor imaging and targeted therapy against lung cancer. Int J Biol Macromol. 2021;193:228–37. https://doi.org/10.1016/j.ijbiomac.2021.10.113.

Wilhelm Romero K, Quirós MI, Vargas Huertas F, Vega-Baudrit JR, Navarro-Hoyos M, Araya-Sibaja AM. Design of hybrid polymeric-lipid nanoparticles using curcumin as a model: preparation, characterization, and in vitro evaluation of Demethoxycurcumin and bisdemethoxycurcumin-loaded nanoparticles. Polymers. 2021;13(23):Article 23. https://doi.org/10.3390/polym13234207.

Wu R, Zhang Z, Wang B, Chen G, Zhang Y, Deng H, Tang Z, Mao J, Wang L. Combination chemotherapy of lung cancer—co-delivery of docetaxel prodrug and cisplatin using aptamer-decorated lipid–polymer hybrid nanoparticles. Drug Des Devel Ther. 2020;14:2249–61. https://doi.org/10.2147/DDDT.S246574.

Xu B, Zhang W, Chen Y, Xu Y, Wang B, Zong L. Eudragit® L100-coated mannosylated chitosan nanoparticles for oral protein vaccine delivery. Int J Biol Macromol. 2018;113:534–42. https://doi.org/10.1016/j.ijbiomac.2018.02.016.

Xu L, Wang X, Liu Y, Yang G, Falconer RJ, Zhao C-X. Lipid nanoparticles for drug delivery. Adv NanoBiomed Res. 2022;2(2):2100109. https://doi.org/10.1002/anbr.202100109.

Yalcin TE, Ilbasmis-Tamer S, Takka S. Antitumor activity of gemcitabine hydrochloride loaded lipid polymer hybrid nanoparticles (LPHNs): in vitro and in vivo. Int J Pharm. 2020;580:119246. https://doi.org/10.1016/j.ijpharm.2020.119246.

Yanat M, Schroën K. Preparation methods and applications of chitosan nanoparticles; with an outlook toward reinforcement of biodegradable packaging. React Funct Polym. 2021;161:104849. https://doi.org/10.1016/j.reactfunctpolym.2021.104849.

Yang X-Z, Dou S, Wang Y-C, Long H-Y, Xiong M-H, Mao C-Q, Yao Y-D, Wang J. Single-step assembly of cationic lipid–polymer hybrid nanoparticles for systemic delivery of siRNA. ACS Nano. 2012;6(6):4955–65. https://doi.org/10.1021/nn300500u.

Yang P, Zhang L, Wang T, Liu Q, Wang J, Wang Y, Tu Z, Lin F. Doxorubicin and edelfosine combo-loaded lipid–polymer hybrid nanoparticles for synergistic anticancer effect against drug-resistant osteosarcoma. Onco Targets Ther. 2020;13:8055–67. https://doi.org/10.2147/OTT.S259428.

Yang Q, Zhou Y, Chen J, Huang N, Wang Z, Cheng Y. Gene therapy for drug-resistant glioblastoma via lipid-polymer hybrid nanoparticles combined with focused ultrasound. Int J Nanomedicine. 2021;16:185–99. https://doi.org/10.2147/IJN.S286221.

Yao VJ, D'Angelo S, Butler KS, Theron C, Smith TL, Marchiò S, Gelovani JG, Sidman RL, Dobroff AS, Brinker CJ, Bradbury ARM, Arap W, Pasqualini R. Ligand-targeted theranostic nanomedicines against cancer. J Control Release. 2016;240:267–86. https://doi.org/10.1016/j.jconrel.2016.01.002.

Yao C, Wu M, Zhang C, Lin X, Wei Z, Zheng Y, Zhang D, Zhang Z, Liu X. Photoresponsive lipid-polymer hybrid nanoparticles for controlled doxorubicin release. Nanotechnology. 2017;28(25):255101. https://doi.org/10.1088/1361-6528/aa702a.

Yuan Y, Cai T, Callaghan R, Li Q, Huang Y, Wang B, Huang Q, Du M, Ma Q, Chiba P, Cai Y. Psoralen-loaded lipid-polymer hybrid nanoparticles enhance doxorubicin efficacy in multidrug-resistant HepG2 cells. Int J Nanomedicine. 2019;14:2207–18. https://doi.org/10.2147/IJN.S189924.

Yusuf A, Almotairy ARZ, Henidi H, Alshehri OY, Aldughaim MS. Nanoparticles as drug delivery systems: a review of the implication of nanoparticles. Physicochemical Properties on

Responses in Biological Systems Polymers. 2023;15(7):Article 7. https://doi.org/10.3390/polym15071596.
Zare M, Bigham A, Zare M, Luo H, Rezvani Ghomi E, Ramakrishna S. pHEMA: an overview for biomedical applications. Int J Mol Sci. 2021;22(12):6376. https://doi.org/10.3390/ijms22126376.
Zhang L, Chan JM, Gu FX, Rhee J-W, Wang AZ, Radovic-Moreno AF, Alexis F, Langer R, Farokhzad OC. Self-assembled lipid–polymer hybrid nanoparticles: a robust drug delivery platform. ACS Nano. 2008;2(8):1696–702. https://doi.org/10.1021/nn800275r.
Zhang H, Dong S, Zhang S, Li Y, Li J, Dai Y, Wang D. pH-responsive lipid polymer hybrid nanoparticles (LPHNs) based on poly (β-amino ester) as a promising candidate to resist breast cancers. J Drug Deliv Sci Technol. 2021;61:102102. https://doi.org/10.1016/j.jddst.2020.102102.
Zhao X, Pan F, Garcia-Gancedo L, Flewitt AJ, Ashley GM, Luo J, Lu JR. Interfacial recognition of human prostate-specific antigen by immobilized monoclonal antibody: effects of solution conditions and surface chemistry. J R Soc Interface. 2012;9(75):2457–67. https://doi.org/10.1098/rsif.2012.0148.
Zielińska A, Carreiró F, Oliveira AM, Neves A, Pires B, Venkatesh DN, Durazzo A, Lucarini M, Eder P, Silva AM, Santini A, Souto EB. Polymeric nanoparticles: production, characterization, toxicology and ecotoxicology. Molecules. 2020;25(16):Article 16. https://doi.org/10.3390/molecules25163731.
Zou T, Lan M, Liu F, Li L, Cai T, Tian H, Cai Y. Emodin-loaded polymer-lipid hybrid nanoparticles enhance the sensitivity of breast cancer to doxorubicin by inhibiting epithelial–mesenchymal transition. Cancer Nanotechnol. 2021;12(1):22. https://doi.org/10.1186/s12645-021-00093-9.

Lipid/Polymer Core-Shell Nanoconstructs for Cancer Theragnostic

11

Amrita Arup Roy, Gaurisha Alias Resha Ramnath Naik, Rahul Pokale, Viola Colaco, Anoushka Mukharya, Ritu Kudarha, Namdev Dhas, and Srinivas Mutalik

Abstract

Lipid-polymer hybrid nanoparticles (LPHNPs) represent an advanced platform for drug delivery and cancer theragnostics, combining lipid cores and polymer shells. Solid lipid cores, such as stearic acid and glyceryl monostearate, provide controlled drug release, high drug loading capacity, and enhanced bioavailability. Charged lipid cores, incorporating cationic or anionic lipids like DOPE and PS, improve cellular uptake and targeted delivery. The hybrid cores integrate the structural integrity and prolonged circulation of polymers like PLGA and PEG with the drug encapsulation benefits of lipids. LPHNPs are prepared using methods like nanoprecipitation, emulsification-solvent evaporation, double emulsion, microfluidics, and supercritical fluid techniques, each offering unique advantages. Polymer coatings such as PEG, PLGA, chitosan, and PCL enhance biocompatibility, stability, and controlled release. Surface functionalization strategies, including ligand conjugation, antibody attachment, and stimuli-responsive coatings, enable targeted drug delivery, improved efficacy, and reduced off-target effects. In cancer theranostics, LPHNPs enable targeted drug delivery, multimodal imaging, theranostic capabilities, controlled release, and biomarker detection. These nanoconstructs encapsulate chemotherapeutic drugs and imaging agents, allowing precise tumor targeting, real-time imaging, and treatment monitoring.

A. A. Roy · G. A. R. R. Naik · R. Pokale · V. Colaco · A. Mukharya
R. Kudarha · N. Dhas · S. Mutalik (✉)
Department of Pharmaceutics, Manipal College of Pharmaceutical Sciences, Manipal Academy of Higher Education, Manipal, India
e-mail: ss.mutalik@manipal.edu

J. K. Patel et al. (eds.), *Core-Shell Nano Constructs for Cancer Theragnostic*,
https://doi.org/10.1007/978-981-96-3025-7_11

In this chapter about LPHNPs, we provide a multifaceted approach to cancer treatment, combining therapeutic and diagnostic functionalities to improve efficacy and patient outcomes, representing a significant advancement in precision oncology.

Keywords

Lipid-polymer hybrid nanoparticles (LPHNPs) · Drug delivery · Cancer theranostics · Solid lipid cores · Charged lipid cores · Hybrid cores

1 Introduction

Cancer remains a leading cause of mortality worldwide, prompting significant research into nanomedicines for therapeutic intervention. According to the World Health Organization (WHO), cancer was responsible for approximately ten million deaths in 2020 (de Martel et al. 2020). Despite advancements, only a few nanomedicines have gained FDA approval, with some, like Doxil, showing suboptimal efficacy due to various factors, including tumor heterogeneity and the challenges of in vivo behavior (O'Brien et al. 2004). Tumor accumulation of nanomedicines can vary widely among patients and tumor types, complicating therapeutic outcomes (Gupta et al. 2024). Furthermore, barriers such as abnormal tumor vasculature and dense extracellular matrix hinder optimal drug delivery (Minchinton and Tannock 2006).

Current animal models often fail to accurately mimic human conditions, making predictions about nanomedicine behavior challenging (Eliasof et al. 2013). Combining diagnosis with therapy, known as theranostics, offers a promising solution (Kim et al. 2017). Lipidic nanomedicines, with their biocompatibility and safety, are particularly well-suited for theranostic applications. By co-encapsulating imaging agents and drugs, lipidic nanocarriers enable real-time monitoring of drug delivery and therapeutic response (Bukhari et al. 2021).

Various lipidic structures, including nanoemulsions, liposomes (Fernandes et al. 2021), solid lipid nanoparticles (SLNs) (da Rocha et al. 2020), and nanostructured lipid carriers (NLCs), offer versatile platforms for theranostic applications. Imaging agents compatible with conventional diagnostic techniques, such as MRI, CT, PET, and SPECT, enhance the diagnostic capabilities of lipidic nanomedicines (Haider et al. 2020). The integration of imaging with nanomedicine-based therapy holds great potential for overcoming the hurdles in cancer treatment (Colaco et al. 2024). By simultaneously assessing drug delivery, release, and distribution, theranostic lipidic nanomedicines offer a promising strategy for personalized and effective cancer therapy, marking a paradigm shift in cancer treatment approaches (Hegde et al. 2023).

Lipid-polymer hybrid nanoparticles (LPHNs) represent an innovative approach in drug delivery systems, harnessing the advantages of both liposomes and polymeric nanoparticles (Naik et al. 2024). This hybrid design enables the combination of desirable properties from lipids and polymers, addressing the limitations encountered with traditional nanocarriers (Persano et al. 2021). One significant advantage of LPHNs lies in their versatility and adaptability, allowing for the incorporation of various therapeutic agents, including small molecule drugs, nucleic acids, and proteins, within their structure. This flexibility in cargo accommodation enhances their potential for targeted drug delivery, therapy, and diagnostic applications. The advantages of LPHNPs also include solubilizing hydrophobic drugs, combining various drug types, and reducing exposure of host organs to toxic substances. The incorporation of biomimetic lipids or PEG enhances stability and circulation time, making LPHNPs promising for dual drug delivery and nucleic acid therapeutics. Surface modifications allow for active targeting and stimuli-responsive drug release (Krishnamurthy et al. 2015; Zangabad et al. 2018).

The synthesis of LPHNs involves the integration of lipids and polymers through various techniques such as nanoprecipitation, emulsification, or self-assembly. These methods enable precise control over particle size, surface properties, and drug encapsulation efficiency, contributing to the tunable characteristics of LPHNs. LPHNPs combine the biomimetic properties of lipids with the mechanistic advantages of polymers. The synthesis of LPHNPs involves a blend of natural, synthetic, or semi-synthetic polymers and lipids, offering diverse properties such as size, shape, surface charge, and responsiveness to stimuli (Zhang and Zhang 2010). The shell of LPHNPs can be composed of biodegradable polymers like polylactic acid (PLA) or polycaprolactone (PCL), while lipids such as phosphatidylcholine (PC) and cholesterol form the lipidic core encapsulating the drug (Anderson and Shive 1997; Fang et al. 2014). This combination ensures biocompatibility and facilitates drug release at the target site. Polymer-caged liposome nanoparticles (NPs) offer solutions to challenges faced by conventional liposomes, such as susceptibility to physiological conditions and premature cargo leakage. These NPs feature a hydrophilic core containing the payload, enveloped by a polymeric structure. This design enhances stability, prevents drug leakage, and enables controlled release through pH or protease sensitivity. However, drawbacks include the potential dissociation of polymer cages and increased susceptibility to reticuloendothelial system (RES) uptake. Synthesis methodologies involve one-step or two-step procedures, including the synthesis of monolithic hybrid systems, polymer-core lipid-shell NPs, hollow-core lipid-polymer-lipid NPs, and biomimetic LPHNPs (Bochicchio et al. 2021). Techniques include nanoprecipitation, emulsification-solvent evaporation, double emulsion, and membrane coating. Polymer-caged liposome NPs are synthesized by decorating liposomes with hydrophobic group-anchored polymers and water-loving polymers. This structure can be achieved by dropping a polymer solution into a liposome dispersion or by functionalizing liposomes with cross-linked polymers, enhancing stability but potentially affecting payload release control (Mohanty et al. 2020).

One of the key features of LPHNs is their stimuli-responsive behavior, which enables controlled drug release in response to specific triggers. Endogenous stimuli, such as variations in pH, redox potential, hypoxia, or the presence of reactive oxygen species (ROS) in the tumor microenvironment, can be exploited to trigger drug release selectively at the target site while minimizing off-target effects (Lee et al. 2007; Mohanty et al. 2022).

LPHNPs exhibit superior characteristics such as controlled drug release, enhanced biocompatibility, and improved structural integrity compared to conventional nanocarriers. The integration of lipids and polymers allows for precise modulation of drug release kinetics and enhanced targeting capabilities, addressing limitations associated with burst release and reticuloendothelial system uptake (Lee et al. 2010). Additionally, exogenous stimuli, including light, temperature, magnetic fields, and ultrasound, offer external control over drug release kinetics, allowing for spatiotemporal modulation of therapeutic effects. This precise control enhances the therapeutic efficacy and minimizes systemic toxicity associated with conventional drug delivery systems (Mandal et al. 2013).

Various types of stimuli-responsive LPHNs have been developed, each tailored to exploit specific triggers for drug release. For example, pH-responsive LPHNs undergo structural changes in acidic tumor microenvironments, leading to cargo release. Similarly, redox-responsive LPHNs utilize the differential redox potential between intracellular and extracellular environments for triggered drug release within cells (Duan et al. 2020).

Despite the promising advantages of LPHNs, several challenges remain, including scalability, reproducibility, and long-term stability. Furthermore, the translation of these innovative nanocarriers from preclinical studies to clinical applications necessitates rigorous evaluation of safety, efficacy, and biocompatibility. In conclusion, LPHNs represent a promising platform for advanced drug delivery, therapy, and diagnosis, offering tailored solutions to overcome the limitations of traditional nanocarriers. With ongoing research and development efforts addressing current challenges, LPHNs hold tremendous potential for revolutionizing personalized medicine and improving patient outcomes in various disease contexts.

2 Types of Core Materials and Their Properties

Lipid polymer-based nanoparticles represent a versatile and promising platform for drug delivery applications in the field of pharmaceuticals. These nanoparticles offer numerous advantages, including biocompatibility, controlled release, and the ability to encapsulate a wide range of drugs, both hydrophobic and hydrophilic. A key aspect of lipid/polymer-based nanoparticles is their core material, which plays a critical role in determining their properties and performance as drug carriers. In this chapter, we will delve into the various types of core materials utilized in lipid-based nanoparticles and their corresponding properties. Specifically, we will explore solid lipid cores, charged lipid cores, and lipid/polymer hybrid cores. Each type of core material offers distinct advantages and characteristics, making them suitable for different drug delivery applications (Dhas 2021).

Understanding the composition and properties of these core materials is essential for designing lipid/polymer-based nanoparticles tailored to specific pharmaceutical needs. By harnessing the unique attributes of each core material, researchers can develop nanoparticles with enhanced stability, drug loading capacity, controlled release kinetics, and targeting capabilities. This knowledge is pivotal for advancing the field of lipid/polymer-based drug delivery and developing innovative solutions for delivering therapeutics to target sites within the body (Nangare et al. 2022).

Lipid-core and polymer-shell nanoparticles represent a sophisticated drug delivery system that combines the unique properties of lipids and polymers to achieve optimal therapeutic outcomes. Let's delve deeper into the properties and advantages of utilizing such nanoparticles:

2.1 Solid Lipid Core

2.1.1 Composition

Solid lipid nanoparticles (SLNs) are typically composed of biocompatible lipids such as triglycerides, fatty acids, or waxes. These lipids are solid at room temperature and form the core matrix of the nanoparticles. Triglycerides, commonly used in SLNs, consist of glycerol linked to three fatty acid chains. The fatty acids can vary in chain length and saturation, influencing the properties of the SLNs (Satapathy et al. 2021).

2.1.2 Properties

Stability The solid lipid core provides remarkable stability to the encapsulated drug, shielding it from environmental factors such as light, heat, and oxidation. This stability ensures the integrity of the drug payload throughout storage and transportation, minimizing degradation and preserving therapeutic efficacy (Badilli et al. 2018).

Controlled Release Kinetics The solid matrix of SLNs facilitates controlled release kinetics, allowing for sustained and prolonged drug release over time. This controlled release profile is advantageous for maintaining therapeutic concentrations of the drug at the target site, enhancing efficacy while reducing side effects associated with rapid drug release (Newton and Kaur 2019).

High Encapsulation Capacity SLNs exhibit a high encapsulation capacity, meaning they can efficiently load and retain both hydrophobic and hydrophilic drugs within their lipid core. This feature is particularly beneficial for enhancing the bioavailability of poorly soluble drugs, improving their therapeutic outcomes (Deshpande et al. 2017).

Biocompatibility and Low Toxicity SLNs are biocompatible and exhibit low toxicity, making them suitable for various pharmaceutical formulations. The use of biocompatible lipids ensures compatibility with biological systems, reducing the risk of adverse reactions or immune responses when administered to patients (Dahiya and Dahiya 2022).

2.2 Charged Lipid Core

2.2.1 Composition

Charged lipids play a crucial role in the formulation of lipid-based nanoparticles with a core-shell structure. Various types of charged lipids are utilized to create nanoparticles tailored for specific applications.

Zwitterionic Lipids These lipids have both positive and negative charges within the same molecule, resulting in a neutral overall charge. Phosphatidylcholine (PC) and phosphatidylethanolamine (PE) are examples of zwitterionic lipids commonly used in nanoparticle formulations (Meng and Grimm 2021).

Cationic Lipids Cationic lipids have a positive charge, making them particularly useful for promoting cellular uptake of nanoparticles through electrostatic interactions with negatively charged cell membranes. Examples of cationic lipids include dioleoyltrimethylammonium propane (DOTAP) and 1,2-dioleoyl-3-trimethylammonium-propane (DOTMA) (Sun and Lu 2023).

Anionic Lipids Anionic lipids carry a negative charge and can be incorporated into nanoparticles to modulate surface charge and interactions with biological components. Examples of anionic lipids include phosphatidylserine (PS) and phosphatidylglycerol (PG) (Schmidpeter et al. 2022).

Neutral Lipids While not charged, neutral lipids such as lecithin (phosphatidylcholine) are essential components of lipid-based nanoparticles. They contribute to the stability and structure of nanoparticles while also influencing drug encapsulation and release properties (Rod-in et al. 2020).

2.2.2 Properties

Modulation of Surface Charge Incorporating charged lipids allows for the modulation of nanoparticle surface charge, influencing interactions with biological membranes, cells, and proteins. For instance, cationic lipids can enhance cellular uptake of nanoparticles, facilitating drug delivery to target cells.

Enhanced Drug Encapsulation Charged lipids contribute to the formation of stable nanoparticles with improved drug encapsulation efficiency. The electrostatic interactions between charged lipids and drug molecules help to retain drugs within the nanoparticle core, enhancing their stability and bioavailability (Sánchez et al. 2020).

Targeted Drug Delivery Tailoring the surface charge of nanoparticles using charged lipids enables targeted drug delivery strategies. By promoting interactions with specific cell types or tissues, charged nanoparticles can achieve targeted drug accumulation and therapeutic effects while minimizing off-target effects.

Functionalization Possibilities Charged lipids also facilitate the incorporation of functional moieties onto the nanoparticle surface. This allows for the conjugation of targeting ligands, imaging agents, or therapeutic molecules, expanding the capabilities of lipid-based nanoparticles for imaging, diagnostics, and targeted therapy applications (Zhang et al. 2023).

2.3 Lipid-Polymer Hybrid Core

Lipid-core and polymer-shell nanoparticles offer a versatile platform for drug delivery, providing enhanced stability, controlled release kinetics, and improved drug encapsulation efficiency. The incorporation of a lipid core with a polymer shell harnesses the synergistic properties of both materials, making them valuable tools for formulating tailored drug delivery systems with diverse pharmaceutical applications (Table 11.1).

2.3.1 Composition

Lipid-polymer hybrid nanoparticles are sophisticated nanostructures that integrate a core composed of a blend of polymers and lipids, resulting in a unique core-shell architecture. The polymer component of these nanoparticles encompasses a variety of biodegradable and biocompatible materials, including but not limited to poly(lactic-co-glycolic acid) (PLGA), polyethylene glycol (PEG), chitosan, and polylactic acid (PLA). On the other hand, the lipid component typically includes phospholipids such as phosphatidylcholine or fatty acids like stearic acid (Mukherjee et al. 2019).

2.3.2 Properties

The hybrid nature of lipid-polymer nanoparticles endows them with a myriad of advantageous properties for drug delivery applications. Firstly, the lipid component enhances biocompatibility, allowing for safe interaction with biological systems. Additionally, lipids play a crucial role in facilitating efficient drug encapsulation within the nanoparticles, protecting the drug payload from degradation and enhancing its stability. Meanwhile, the polymer component contributes to the structural integrity of the nanoparticles, ensuring their stability during storage and transportation. Moreover, polymers enable controlled drug release, enabling tunable release kinetics tailored to specific therapeutic requirements (Rezigue 2020).

Furthermore, lipid-polymer hybrid nanoparticles offer versatility and customization options. Their core-shell architecture permits the incorporation of targeting ligands, imaging agents, or other functional moieties on the nanoparticle surface, enabling precise and targeted drug delivery to desired sites within the body. This targeted drug delivery capability enhances therapeutic efficacy while minimizing off-target effects (Table 11.1).

Table 11.1 Lipid/polymer core-shell nanoconstructs and their application

Drug encapsuled	Polymer shell	Lipid core	Particle size	EE/DL	Applications	References
Hydrophilic/ hydrophobic drug	Chitosan, polyvinyl alcohol, poly(acrylic acid)	Dipalmitoyl phosphatidylcholine (DPPC) and Dicetyl phosphate (DCP)	5–6 μm	–	Mucoadhesion	Takeuchi et al. (1994)
Doxorubicin	Polyvinylalcohol and Hydroxypropylmethylcellulose	Egg phosphatidylcholine	104 ± 9 nm	92%	Anticancer	Takeuchi et al. (1999)
Hydrophilic/ hydrophobic drug	Polysaccharides (chitosan, alginate and pectin)	Egg-phosphatidylcholine/ DOTAP	140 nm	–	Dental drug delivery	Pistone et al. (2017)
5(6)-carboxyfluorescein	Poly(L-lysine) hydrobromide (PLL)	Egg yolk L-α-phosphatidylcholine	57 ± 9 nm	–	Increase resistance in biological media	Hermal et al. (2020)
Hydrophilic/ hydrophobic drug	Eudragit S100	Egg phosphatidylcholine	5–10 μm	–	Gastrointestinal (GI) targeting	Barea et al. (2010)
Leuprolide acetate	Chitosan	Lecithin	54 nm	62.48%	Improve GI stablilty	Guo et al. (2003)
Doxorubicin and pEGFP	PLGA	Folic acid coated polymeric liposome	435.9 ± 8.7	–	Gene co-delivery system for cancer treatment	Wang et al. (2010)
Doxorubicin	Poly(ethylene glycol)	DPPC/DPPE	121 ± 1	–	Tumor targeting	Qhattal et al. (2014)
Doxorubicin-HCl	mPEG-P(HPMA-g-his)	DPPC	96.2 ± 2.1	81.3 ± 15.14%	On targeted cancer therapy	Chiang and Lo (2014)

3 Types of Shell Materials and Their Properties

Polymeric shell materials are indispensable components in enhancing the performance of nanoparticles across various applications. They serve as protective coatings around the core of nanoparticles, offering several advantages crucial for their functionality. One of the primary benefits of polymeric shells is their ability to improve biocompatibility, ensuring compatibility with biological systems (Zielińska et al. 2020). Additionally, polymeric shells provide anchor sites for molecular linkages, facilitating the attachment of targeting ligands or therapeutic agents, which is essential for targeted drug delivery and other biomedical applications. Moreover, these shells offer protection against oxidation, safeguarding the integrity and functionality of the core material from environmental factors (Singh and Bhateria 2021).

Polymeric shell materials possess inherent properties that further contribute to their significance in nanoparticle design. They exhibit high thermal stability, which enhances the overall stability of the nanoparticle structure, making them suitable for various applications. Furthermore, the solubility of the shell material influences the dispersion and behavior of the nanoparticle, impacting its performance in drug delivery, bioimaging, and other applications. Importantly, polymeric shells can mitigate the toxicity associated with the core material, ensuring the safety of nanoparticles for biomedical use (Dhas et al. 2018).

In drug delivery, tailored polymeric nanocapsules effectively load, protect, and release bioactive substances, enabling targeted drug delivery with enhanced efficacy. For bioimaging applications, engineered polymeric shells enable precise visualization of specific biological targets, contributing to advances in diagnostic imaging techniques (Deng et al. 2020). Additionally, in the field of solar cells, polymeric shells play a crucial role in enhancing the efficiency of solar cell materials, thereby improving their performance in energy conversion applications (Table 11.2).

Synthesis and characterization of nanoparticles with polymeric shells involve various techniques, including nanoprecipitation, emulsion-diffusion, double emulsification, and layer-by-layer assembly. These methods allow for the controlled deposition of polymer layers around the nanoparticle core, resulting in the formation of well-defined polymeric shell structures with tailored properties. By carefully selecting and engineering polymer coatings, researchers can tailor the properties of nanoparticles to meet the specific requirements of drug delivery, imaging, and other therapeutic modalities, paving the way for innovative and effective biomedical interventions (Wang et al. 2022).

3.1 Polyethylene Glycol (PEG)

3.1.1 Properties of PEG-Coated Nanoparticles

Biocompatibility PEG (polyethylene glycol) is highly biocompatible, making it suitable for use in medical applications without causing adverse reactions or immune responses in the body.

Table 11.2 Lipid/polymer core-shell nanoconstructs and their characterization and application

Drug encapsuled	Polymer shell	Particle size	Zeta Potential	EE/DL	Applications	References
Memantine hydrochloride (MEM)	PEG	152.6 ± 0.5 nm	− 22.4 mV	–	MEM–PEG–PLGA nanoparticles effectively target Alzheimer's disease, crossing the blood-brain barrier and reducing β-amyloid plaques.	Sánchez-López et al. (2018)
Rapamycin	PLGA	200–500 nm	−18.27 ± 0.52 mV to −11.22 ± 0.29 mV	1.17 ± 0.52% to 54.83 ± 0.29%	PLGA-based nanoformulations exhibit enhanced anticancer efficacy against glioma cells.	Escalona-Rayo et al. (2019)
Turmeric oil	Chitosan/Alginate	184.8 ± 14.8 nm	−21.8 ± 1.1 mV	70.3 ± 1.3% and 3.4 ± 1.3%	TO-CS/Alg-NCs with poloxamer 407 were prepared, exhibiting 200 nm size, sustained drug release, and enhanced cytotoxicity, promising for cancer therapy.	San et al. (2022)
Quercetin (QUE)	Poly(ε-caprolactone) (PCL)	212 nm	−11 mV	–	QUE-loaded lipid-core nanocapsules (LNC) with PCL as the polymer wall demonstrated sustained antioxidant activity, highlighting PCL's role in formulation stability and potential for nanostructured product development.	Weiss-Angeli et al. (2012)
Combretastatin A4 (CA4) and Doxorubicin (DOX)	PLGA	394.7 nm	–	–	PLGA-based core-shell nanoparticles allow sequential release of CA4 and DOX for cancer combination therapy, offering tunable drug release profiles tailored to clinical needs.	Cao et al. (2014)

Non-immunogenicity PEG does not trigger immune reactions, ensuring that PEG-coated nanoparticles can be safely administered to patients without eliciting immune responses that could compromise treatment efficacy (Pham Le Khanh et al. 2022).

Resistance to protein adsorption PEG has a strong resistance to protein adsorption, preventing proteins from binding to the surface of nanoparticles and thereby reducing the risk of opsonization and clearance by the immune system.

Enhanced circulation time PEGylation of nanoparticles extends their circulation time in the bloodstream, allowing for prolonged exposure to target tissues and enhanced therapeutic effects.

3.1.2 Applications of PEG-Coated Nanoparticles

Drug delivery PEG-coated nanoparticles are widely used in drug delivery systems to encapsulate and deliver various therapeutics, including chemotherapeutic agents, proteins, and nucleic acids. The PEG coating improves the stability and pharmacokinetics of drugs, leading to enhanced therapeutic outcomes and reduced side effects.

Imaging modalities PEG-coated nanoparticles are employed in imaging techniques such as magnetic resonance imaging (MRI) and positron emission tomography (PET) to improve contrast and targeting efficiency. The presence of PEG enhances the nanoparticles' visibility in imaging studies, providing accurate diagnostic information.

Anti-fouling properties The anti-fouling properties of PEG-coated nanoparticles make them suitable for long-term circulation in biological fluids, maintaining their integrity and ensuring sustained drug release at the target site. This property enhances therapeutic efficacy and patient compliance (Raina et al. 2021).

3.2 Poly(Lactic-co-Glycolic Acid) (PLGA)

3.2.1 Properties

PLGA is a biodegradable and biocompatible copolymer composed of lactic acid and glycolic acid monomers, making it suitable for biomedical applications. The ratio of lactic acid to glycolic acid in PLGA can be adjusted to tailor the degradation kinetics and mechanical properties of the polymer, allowing for the controlled release of encapsulated drugs over a specified period. PLGA nanoparticles possess a high loading capacity for drugs due to their porous structure and can protect sensitive compounds from degradation in biological environments (Dhas and Mehta 2021a). PLGA is nontoxic and does not induce inflammatory responses, further enhancing its biocompatibility and safety profile. Surface modification of PLGA nanoparticles is feasible, enabling the conjugation of targeting ligands, imaging agents, or stimuli-responsive moieties for controlled drug release (Makadia and Siegel 2011).

3.2.2 Applications

PLGA nanoparticles have diverse applications in drug delivery systems, including cancer therapy, vaccine delivery, and tissue engineering. In cancer therapy, PLGA nanoparticles are used to encapsulate chemotherapeutic agents, allowing for sustained drug release and improved therapeutic efficacy while minimizing systemic toxicity. They can also deliver nucleic acid-based therapeutics such as siRNA or mRNA for targeted gene therapy. In vaccine delivery, PLGA nanoparticles are employed to encapsulate antigens and adjuvants, enhancing antigen stability, immunogenicity, and controlled release kinetics for prolonged immune response. Furthermore, PLGA nanoparticles play a crucial role in tissue engineering by delivering growth factors, cells, or bioactive molecules to promote tissue regeneration and repair. The ability to modify PLGA nanoparticle surfaces with ligands specific to target cells or tissues enables precise targeting and enhanced therapeutic outcomes (Elmowafy et al. 2019).

3.3 Chitosan

3.3.1 Properties

Chitosan is a naturally occurring polysaccharide derived from chitin, which is abundant in the exoskeletons of crustaceans such as shrimp and crabs. It possesses several unique properties that make it an attractive material for biomedical applications. Firstly, chitosan is mucoadhesive, meaning it has the ability to adhere to mucosal surfaces such as those found in the gastrointestinal tract, nasal passages, and oral cavity. This mucoadhesive property allows for prolonged residence time of chitosan-based formulations at the site of administration, enhancing drug absorption and bioavailability. Secondly, chitosan exhibits a positive charge under acidic conditions, such as those found in the stomach or inflamed tissues. This positive charge enables chitosan to interact with negatively charged cell membranes, promoting cellular uptake and intracellular delivery of encapsulated therapeutic agents. Additionally, chitosan is biocompatible, biodegradable, nontoxic, and immunologically inert, making it safe for use in humans and minimizing the risk of adverse reactions (Aranaz et al. 2021).

3.3.2 Applications

Chitosan-coated nanoparticles have a wide range of applications in drug delivery, wound healing, and gene delivery. In oral drug delivery, chitosan nanoparticles are used to improve the bioavailability of poorly water-soluble drugs by enhancing their solubility and intestinal absorption. The mucoadhesive properties of chitosan enable targeted drug delivery to specific regions of the gastrointestinal tract, reducing systemic exposure and potential side effects. Furthermore, chitosan nanoparticles can protect encapsulated drugs from enzymatic degradation in the gastrointestinal environment, leading to sustained release and prolonged therapeutic effects. In nasal drug delivery, chitosan-based formulations are employed to deliver drugs directly to the nasal mucosa, bypassing first-pass metabolism and achieving rapid onset of action (Dhas and Mehta 2020). Chitosan is also utilized in wound healing applications due to its hemostatic properties, ability to promote tissue regeneration, and antimicrobial activity. Lastly, chitosan nanoparticles are explored for gene delivery purposes, where they can efficiently deliver nucleic acids such as siRNA or plasmid DNA to target cells, offering potential for gene therapy and genetic engineering applications (Muzamil et al. 2022).

3.4 Gelatin

3.4.1 Properties

Gelatin nanoparticles are highly versatile carriers with exceptional properties that make them ideal for a range of biomedical applications. Derived from collagen, the main structural protein in the extracellular matrix of connective tissues, gelatin is biocompatible and biodegradable, making it well-suited for medical use. Its biodegradability ensures that gelatin nanoparticles can undergo enzymatic degradation under physiological conditions, eliminating concerns about long-term accumulation and toxicity. Moreover, gelatin's natural origin reduces the risk of immunogenic responses or adverse reactions in the body. One of the key advantages of gelatin nanoparticles is their tunable properties. By modifying parameters such as molecular weight, crosslinking density, and surface functionalization, researchers can tailor the characteristics of gelatin nanoparticles. For instance, altering the degree of crosslinking can influence the degradation rate of nanoparticles, allowing for controlled drug release kinetics. Surface modification can also be employed to modify the surface charge, hydrophobicity, or targeting ligands of gelatin nanoparticles, enhancing their specificity and efficacy in various applications (Song et al. 2019).

Furthermore, gelatin nanoparticles exhibit good stability, enabling the encapsulation and protection of sensitive therapeutic agents such as drugs, proteins, or nucleic acids. Their small size and high surface area-to-volume ratio facilitate efficient drug loading and release, enhancing therapeutic outcomes while minimizing systemic side effects. Overall, gelatin nanoparticles offer a versatile and biocompatible platform for drug delivery, tissue engineering, and medical imaging applications (Madkhali 2023).

3.4.2 Applications

In drug delivery, gelatin nanoparticles serve as effective carriers for encapsulating and delivering therapeutic agents to target tissues or cells. They can be loaded with a wide range of drugs, including small molecules, peptides, and nucleic acids, and release them in a controlled manner, improving treatment efficacy and reducing dosing frequency. Gelatin nanoparticles are particularly valuable for targeted drug delivery, as their surface can be functionalized with ligands that recognize specific receptors or biomarkers on target cells, enabling precise localization and enhanced therapeutic outcomes. In tissue engineering, gelatin nanoparticles play a crucial role as scaffolds for cell growth and tissue regeneration. Their biocompatibility and ability to support cell adhesion, proliferation, and differentiation make them suitable for engineering artificial tissues or promoting wound healing. By mimicking the extracellular matrix environment, gelatin nanoparticles create a favorable microenvironment for cell growth and tissue repair, making them valuable tools in regenerative medicine (Jiang et al. 2023).

Moreover, gelatin nanoparticles can be functionalized with contrast agents for use in medical imaging modalities such as magnetic resonance imaging (MRI) or computed tomography (CT). These contrast-enhanced nanoparticles enable noninvasive visualization of biological structures, disease markers, or drug distribution in vivo, providing valuable diagnostic information and aiding in treatment monitoring (Yasmin et al. 2017).

3.5 Poly(Lactic Acid) (PLA)

3.5.1 Properties

PLA is a biodegradable polyester derived from renewable resources such as corn starch or sugarcane, making it an environmentally friendly material. Its biodegradability is a key property, as PLA undergoes hydrolytic degradation in physiological conditions, breaking down into lactic acid. This lactic acid is a natural metabolite that can be easily metabolized and eliminated from the body, reducing concerns about long-term accumulation or toxicity. PLA nanoparticles exhibit excellent controlled drug release kinetics, allowing for sustained and predictable release of encapsulated drugs. The release profile can be tailored by adjusting parameters such as the molecular weight of PLA and the composition of the nanoparticles. PLA is biocompatible and nontoxic, making it suitable for various biomedical applications. Its compatibility with biological systems minimizes adverse reactions or immune responses, ensuring safe use in medical settings. PLA nanoparticles also demonstrate good stability, preserving the integrity of encapsulated drugs during storage and transportation. These properties collectively make PLA a versatile and attractive material for drug delivery systems and other biomedical applications (Elmowafy et al. 2019).

3.5.2 Applications

PLA nanoparticles find extensive applications in drug delivery systems for various therapeutic areas, including cancer therapy, infectious diseases, and inflammatory disorders. Their biodegradability and biocompatibility make them ideal carriers for delivering drugs to target sites while minimizing systemic toxicity. PLA nanoparticles can be engineered to achieve sustained release formulations, improving patient compliance and therapeutic outcomes. Additionally, they can encapsulate a wide range of drugs, including hydrophobic and hydrophilic compounds, enhancing their versatility in pharmaceutical formulations. PLA nanoparticles can be surface-modified with targeting ligands or stimuli-responsive polymers to enhance site-specific drug delivery. Targeted delivery systems improve drug accumulation at diseased tissues or cells, increasing therapeutic efficacy and reducing off-target effects. PLA nanoparticles can also be functionalized to respond to specific stimuli such as pH, temperature, or enzymes, enabling controlled drug release in response to physiological conditions (Xiao et al. 2010).

3.6 Polyethyleneimine (PEI)

3.6.1 Properties

PEI is a cationic polymer known for its high density of amino groups, which gives it a positive charge. This property enables efficient complexation with negatively charged molecules such as nucleic acids (DNA and RNA), forming stable polyplexes. The proton sponge effect of PEI plays a crucial role in endosomal escape following cellular uptake. This effect involves the buffering capacity of PEI, which helps to disrupt endosomal membranes and release the polyplexes into the cytoplasm. As a result, PEI nanoparticles exhibit robust nucleic acid delivery capabilities, protecting genetic material from enzymatic degradation and facilitating their intracellular delivery.

PEI nanoparticles can condense genetic material into nanosized complexes, protecting it during transit and enhancing cellular uptake. The size and charge of the PEI-based complexes can be tailored to optimize transfection efficiency and minimize cytotoxicity. Additionally, PEI exhibits good biocompatibility and low immunogenicity, making it suitable for biomedical applications (Zhao and Zhou 2022).

3.6.2 Applications

PEI-coated nanoparticles are widely utilized in gene therapy, siRNA delivery, and DNA vaccination due to their efficient transfection capabilities. They are particularly valuable for delivering genetic material into a wide range of cell types, including primary cells and stem cells, enabling genetic manipulation and gene expression modulation. PEI nanoparticles can be engineered to incorporate targeting ligands or stimuli-responsive moieties, allowing for site-specific delivery and controlled release of nucleic acids.

In gene therapy, PEI-based nanoparticles play a crucial role in delivering therapeutic genes to target cells, offering potential treatments for genetic disorders and other diseases. In siRNA delivery, PEI facilitates the delivery of small interfering RNAs to silence specific genes, providing a therapeutic approach for various diseases. Moreover, in DNA vaccination, PEI nanoparticles can deliver DNA vaccines to immune cells, eliciting immune responses against pathogens or cancer cells (Ihm et al. 2015).

3.7 Polycaprolactone (PCL)

Polycaprolactone (PCL) is a biodegradable polyester that has garnered significant attention in the field of nanotechnology, particularly in the context of lipid/polymer core-shell nanoconstructs. PCL offers several advantageous properties that make it an attractive choice for such nanoconstructs.

3.7.1 Properties

Biodegradability Polycaprolactone (PCL) is a biodegradable polyester, which means it can be broken down by natural processes in the body over time. This property is highly beneficial in biomedical applications as it allows for controlled degradation of the polymer shell, ensuring the safe elimination of the nanoparticle from the body once its therapeutic function is complete (Łukasiewicz et al. 2021).

3.7.2 Applications

Controlled Drug Release PCL's biodegradability makes it an excellent choice for the polymer shell in nanoparticles designed for controlled drug release. The gradual degradation of PCL allows for sustained and controlled release of encapsulated drugs over an extended period. This property is particularly useful in drug delivery systems where a steady concentration of the drug at the target site is desired.

Tissue Engineering PCL-based polymer shells in nanoparticles are widely used in tissue engineering applications. The biodegradable nature of PCL enables the gradual degradation of the polymer shell, providing temporary structural support for cell growth and tissue regeneration. PCL-based nanoparticles can be loaded with growth factors or therapeutic agents to enhance tissue regeneration and repair (Abdullah et al. 2020).

Scaffold for Cell Culture PCL polymer shells can serve as scaffolds for cell culture in vitro. The biocompatible and biodegradable properties of PCL allow cells to adhere, proliferate, and differentiate within the nanoparticle scaffold. This application is valuable in tissue engineering research and regenerative medicine.

Targeted Delivery Systems PCL-based polymer shells can be functionalized with targeting ligands or surface modifications to achieve targeted drug delivery. By decorating the surface of PCL nanoparticles with specific molecules, they can selectively target and bind to cells or tissues of interest, enhancing the efficacy and specificity of drug delivery (Aminu and Audu 2023).

3.8 Polyvinyl Alcohol (PVA)

Polyvinyl alcohol (PVA) in the polymer shell of lipid/polymer core-shell nanoconstructs offers advantages in biocompatibility, film-forming ability, drug stabilization, controlled release formulations, and biomedical imaging, making it a versatile material in nanomedicine and drug delivery systems.

3.8.1 Properties

Biocompatibility Polyvinyl alcohol (PVA) is a biocompatible polymer, meaning it is well-tolerated by biological systems without causing significant adverse reactions. This property is crucial in biomedical applications, where the nanoparticle's interaction with biological tissues must be minimized to ensure safety and efficacy.

Film-forming Ability PVA has excellent film-forming properties, allowing it to create a stable and protective coating around the lipid core of nanoparticles. This coating enhances the stability of the nanoparticles, preventing premature drug release and degradation during storage and transportation (Morimune et al. 2012).

3.8.2 Applications

Drug Delivery Systems PVA-based polymer shells are commonly used in drug delivery systems to encapsulate and protect therapeutic agents. The biocompatibility and film-forming ability of PVA ensure the safe delivery of drugs to target sites, providing controlled release and improved bioavailability.

Stabilization of Lipid Cores PVA's film-forming properties help stabilize the lipid core of nanoparticles, preventing aggregation and maintaining the structural integrity of the nanoparticles. This stabilization is essential for ensuring consistent drug encapsulation and controlled release kinetics (Rivera-Hernández et al. 2021).

Controlled Release Formulations PVA-based polymer shells can be engineered to achieve specific drug release profiles, such as sustained or triggered release. By modulating the properties of the PVA coating, researchers can design nanoparticles with tailored release kinetics to meet therapeutic requirements.

Biomedical Imaging PVA-coated nanoparticles can also be utilized in biomedical imaging applications. The stable and biocompatible nature of PVA makes it suitable for incorporating imaging contrast agents or fluorescent dyes into the polymer shell, enabling noninvasive imaging and diagnostics (Xu et al. 2019).

3.9 Pluronic F-68

Pluronic F-68 offers valuable properties such as amphiphilicity, emulsifying ability, stability enhancement, biocompatibility, and controlled release capabilities in the polymer shell of lipid/polymer core-shell nanoconstructs. These properties make it a versatile material for various biomedical and drug-delivery applications.

3.9.1 Properties

Amphiphilic Nature Pluronic F-68 is an amphiphilic block copolymer composed of hydrophilic poly(ethylene oxide) (PEO) and hydrophobic poly(propylene oxide) (PPO) blocks. This amphiphilic nature allows Pluronic F-68 to interact with both hydrophobic and hydrophilic components, making it suitable for forming stable polymer shells around lipid-core nanoparticles.

Emulsifying and Stabilizing Agent Pluronic F-68 possesses emulsifying properties, enabling the formation of stable emulsions and nanostructures. When used in the polymer shell of nanoparticles, it acts as a stabilizing agent, preventing the aggregation and precipitation of nanoparticles during formulation and storage.

Biocompatibility Pluronic F-68 is biocompatible and nontoxic, making it suitable for biomedical applications. Its safety profile allows for the use of Pluronic F-68-coated nanoparticles in drug delivery systems without causing significant adverse effects on biological systems (Khaliq et al. 2023).

3.9.2 Applications

Drug Delivery Systems Pluronic F-68 is commonly used in the polymer shell of nanoparticles for drug delivery applications. It stabilizes the nanoparticles, enhances their biocompatibility, and provides controlled release properties. The amphiphilic nature of pluronic F-68 also helps in solubilizing hydrophobic drugs within the polymer shell.

Enhanced Stability Pluronic F-68 improves the stability of lipid/polymer core-shell nanoconstructs by preventing aggregation and maintaining the structural integrity of the nanoparticles. This enhanced stability is crucial for ensuring consistent drug encapsulation, controlled release kinetics, and prolonged circulation time in the body (Paris et al. 2023).

Biomedical Imaging Pluronic F-68-coated nanoparticles can be utilized for biomedical imaging applications. The polymer shell can be engineered to incorporate imaging contrast agents or fluorescent dyes, allowing for non-invasive imaging and diagnostics. Pluronic F-68's emulsifying and stabilizing properties contribute to the development of effective imaging agents.

Controlled Release Formulations Pluronic F-68 can be used to design nanoparticles with specific drug release profiles, such as sustained release or stimuli-responsive release. By adjusting the composition and properties of the polymer shell, researchers can tailor the release kinetics of encapsulated drugs to achieve desired therapeutic outcomes.

3.10 Alginate

Alginate polymer in lipid-core/polymer-shell hybrid nanoparticles offers valuable properties such as biocompatibility, hydrophilicity, gel-forming ability, and controlled drug release for various biomedical and drug delivery applications.

3.10.1 Properties

Biocompatibility Alginate is a naturally derived biopolymer extracted from brown algae. It is biocompatible and nontoxic, making it suitable for biomedical applications without causing significant adverse effects on biological systems.

Hydrophilicity Alginate is highly hydrophilic, allowing it to interact with water molecules and form stable aqueous solutions. This property is beneficial for incorporating hydrophilic drugs or biomolecules into the polymer shell of lipid-core/polymer-shell hybrid nanoparticles.

Gel-Forming Ability Alginate has the unique property of forming hydrogels in the presence of divalent cations such as calcium ions. This gel-forming ability can be utilized to create crosslinked structures within the polymer shell, enhancing the stability and mechanical properties of the nanoparticles.

Ionotropic Gelation Alginate can undergo ionotropic gelation when exposed to calcium ions, leading to the formation of a crosslinked network. This process can be used to control the release of encapsulated drugs from the polymer shell by modulating the crosslinking density.

3.10.2 Applications

Drug Delivery Systems Alginate-based polymer shells in lipid-core/polymer-shell nanoparticles are utilized for drug delivery applications. The hydrophilic nature of alginate allows for the encapsulation of water-soluble drugs, while its gel-forming ability enables controlled drug release over time.

Targeted Delivery Alginate polymer shells can be functionalized with targeting ligands or antibodies to achieve targeted drug delivery. The surface modification of nanoparticles enhances their affinity for specific cell types or tissues, improving therapeutic efficacy and reducing off-target effects.

Biomedical Imaging Alginate-coated lipid-core/polymer-shell nanoparticles can incorporate imaging contrast agents or fluorescent dyes for biomedical imaging applications. The polymer shell provides stability and protection to the imaging agents, allowing for noninvasive imaging and diagnostics.

Cell Encapsulation and Tissue Engineering Alginate is commonly used for cell encapsulation and tissue engineering applications. In lipid-core/polymer-shell hybrid nanoparticles, alginate-based polymer shells can encapsulate cells or growth factors, supporting cell viability, proliferation, and tissue regeneration.

4 Method for Preparation

4.1 Emulsification/Solvent Evaporation Method

The emulsification method for producing lipid-core/polymer-shell nanoparticles involves several steps to ensure the successful formation of these versatile nanocarriers.

4.1.1 Preparation of Lipid Core and Polymer Solution

This method involves the emulsification of a lipid core with a polymer solution to form lipid-core/polymer-shell nanoparticles.

Lipids and drugs, while the polymer solution consists of polymers like poly(lactic-co-glycolic acid) (PLGA) dissolved in an organic solvent. Initially, the lipid core and polymer solution are mixed to form an oil-in-water emulsion, where the lipid core is dispersed as droplets in the aqueous phase. The emulsion is then subjected to evaporation or extraction of the organic solvent, leading to the formation of solid lipid-core/polymer-shell nanoparticles with the drug encapsulated within as portrayed in Fig. 11.1 (Szczęch and Szczepanowicz 2020).

4.2 Nanoprecipitation Method

Nanoprecipitation is a simple and rapid method for preparing lipid-core/polymer-shell nanoparticles. It involves the direct injection of a lipid core solution (e.g., lipids dissolved in an organic solvent) into an aqueous phase containing a polymer (e.g., PLGA) and surfactants. The rapid mixing of the two phases results in the formation of nano-sized droplets, followed by the diffusion of the organic solvent into the aqueous phase. This leads to the precipitation of lipid-core/polymer-shell nanoparticles with the drug encapsulated inside, as depicted in Fig. 11.2. Emulsification/Solvent Evaporation Method one of the methods used for the preparation of Lipid/Polymer core-shell nanoconstructs.

(Xu et al. 2023).

4.3 Double Emulsion Method (Water-in-Oil-in-Water, W/O/W)

The double emulsion method is particularly useful for encapsulating hydrophilic drugs or water-soluble compounds within lipid-core/polymer-shell nanoparticles. Initially, a water-in-oil emulsion (W/O) is formed by emulsifying an aqueous phase containing the drug with a lipid solution in an organic solvent. This results in the formation of drug-containing droplets dispersed in the oil phase. The W/O emulsion

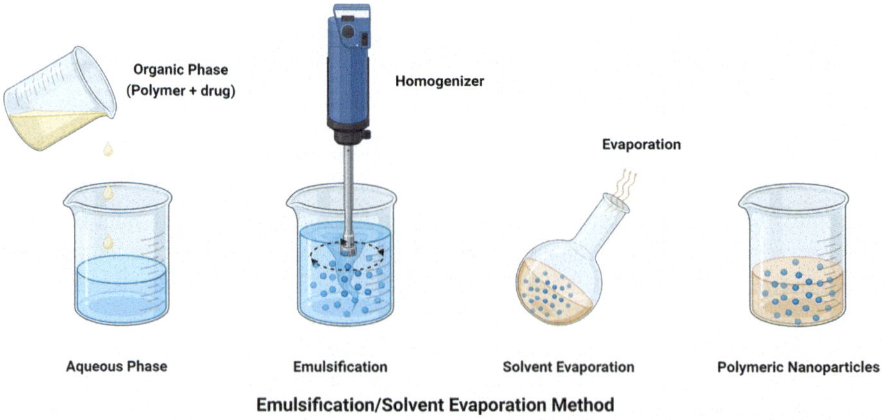

Fig. 11.1 Emulsification/solvent evaporation method for the preparation of lipid/polymer core-shell nanoconstructs

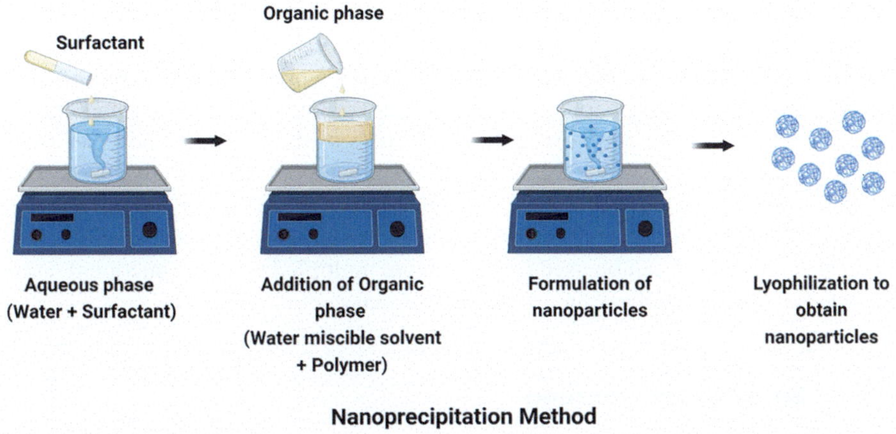

Fig. 11.2 Nanoprecipitation method for the preparation of lipid/polymer core-shell nanoconstructs

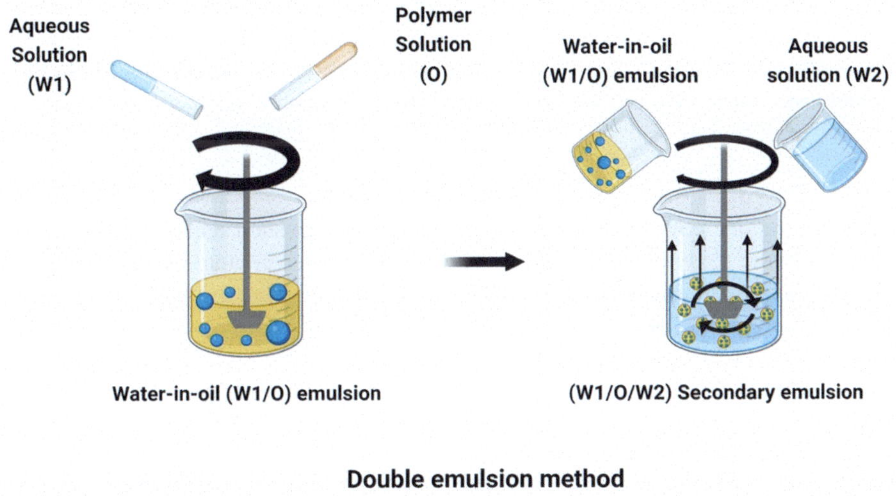

Fig. 11.3 Double emulsion method for the preparation of lipid/polymer core-shell nanoconstructs

is then emulsified into a second aqueous phase containing a polymer and surfactants to form the double emulsion (W/O/W). Subsequent solvent evaporation or extraction leads to the formation of nanoparticles lipid-core/polymer-shell nanoparticles with a double emulsion structure (Iqbal et al. 2015) (Fig. 11.3).

4.4 Layer-by-Layer Assembly Method

The layer-by-layer assembly method involves the sequential deposition of lipid core and polymer shell materials onto a solid substrate or template. This method allows precise control over the composition and thickness of the lipid-core/polymer-shell nanoparticles by nanoparticles alternating layers of lipid and polymer materials through techniques such as electrostatic interactions, adsorption, or chemical bonding. Layer-by-layer assembly offers flexibility in tailoring precise the properties of the nanoparticles and can be used to incorporate multiple layers or functional components for specific application interactions (Amasya et al. 2020).

4.5 Spray Drying Method

Spray drying is employed for the production of dry powder formulations of lipid-core/polymer-shell nanoparticles, particularly for inhalation or oral delivery applications. In this method, a solution or suspension containing lipid and polymer materials is atomized into fine droplets using a spray nozzle. The droplets are then rapidly dried in a hot air stream, leading to the formation of solid lipid-core/polymer-shell nanoparticles. The precise control of the drying conditions, including temperature, humidity, and airflow rate, is crucial in achieving the desired particle size and morphology.

4.5.1 Powder Collection and Characterization

The resulting dry powder containing lipid-core/polymer-shell nanoparticles is collected from the spray drying chamber. The powder is then characterized using various analytical techniques such as scanning electron microscopy (SEM), dynamic light scattering (DLS), X-ray diffraction (XRD), and Fourier-transform infrared spectroscopy (FTIR) to assess particle size, morphology, crystallinity, chemical composition, and stability.

4.5.2 Application Considerations

The dry powder formulations of lipid-core/polymer-shell nanoparticles are particularly suited for pulmonary or oral administration as elaborated in Fig. 11.4. Each method route offers unique advantages and can be optimized based on factors such as the physicochemical properties of the materials, desired nanoparticle characteristics (e.g., size, drug loading, release profile), and intended route of administration (Sengel-Turk et al. 2024).

5 Specific Surface Functionalization Strategies

Specific surface functionalization strategies for lipid/polymer core-shell nanoconstructs play a crucial role in enhancing their performance and efficacy in biomedical applications (Fig. 11.5). These strategies involve modifying the outer surface of the

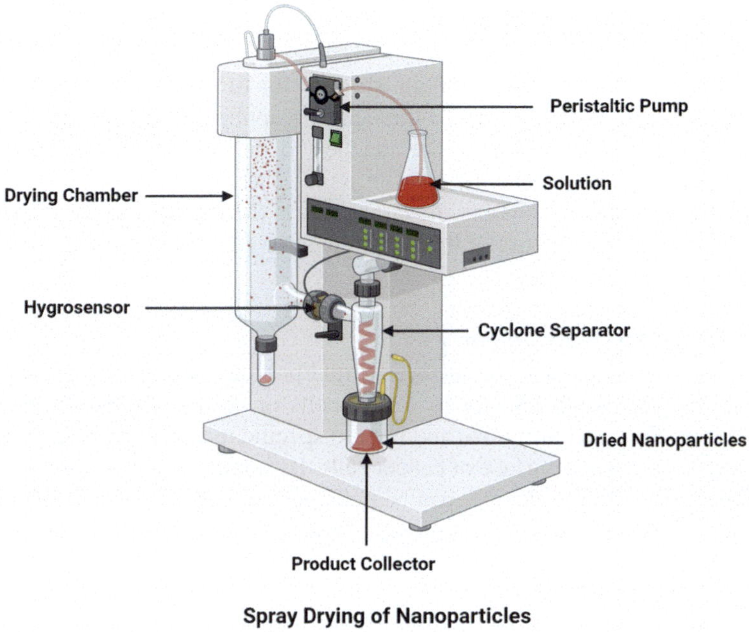

Fig. 11.4 Spray drier used for the preparation of lipid/polymer core-shell nanoconstructs

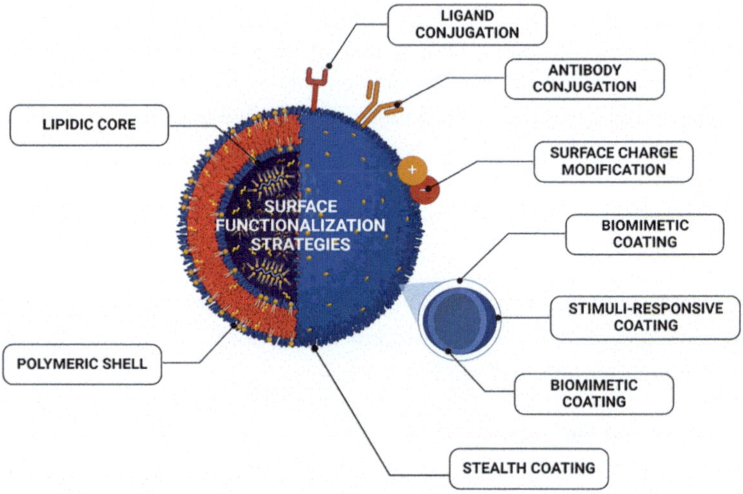

Fig. 11.5 Surface functionalization strategies for lipid/polymer core-shell nanoconstructs

nanoparticles to impart specific properties or functionalities tailored to the intended application (Gajbhiye et al. 2023). Below are some detailed surface functionalization strategies commonly used for lipid/polymer core-shell nanoconstructs:

5.1 Ligand Conjugation

Ligand conjugation represents a crucial strategy for functionalizing lipid/polymer core-shell nanoconstructs for cancer theragnostic applications. Ligands, including peptides, antibodies, or aptamers, are attached to the surface of nanoparticles to enable targeted delivery to specific cells or tissues, thereby enhancing the efficiency of drug delivery while minimizing off-target effects (Mohanty et al. 2020). Here's a detailed elaboration on the use of ligand conjugation for functionalizing lipid/polymer core-shell nanoconstructs:

5.1.1 Targeted Drug Delivery

The attachment of ligands to the surface of lipid/polymer core-shell nanoconstructs allows for targeted drug delivery to cancer cells or tissues. Ligands are selected based on their ability to recognize and bind to specific receptors or biomarkers that are overexpressed on the surface of cancer cells. By conjugating targeting ligands to the nanoparticles, drug molecules can be delivered directly to tumor cells, enhancing therapeutic efficacy while minimizing systemic side effects (Zhang et al. 2014).

5.1.2 Selection of Ligands

Various types of ligands can be employed for targeted drug delivery, including peptides, antibodies, or aptamers. Peptides are short amino acid sequences that can specifically target cell surface receptors involved in cancer progression. Antibodies are highly specific proteins that can recognize and bind to antigens present on the surface of cancer cells with high affinity. Aptamers are single-stranded nucleic acids that can fold into specific three-dimensional structures to bind to target molecules with high selectivity. The choice of ligand depends on the specific characteristics of the target cancer cells and the desired mode of action (Agrawal et al. 2015).

5.1.3 Surface Functionalization

Ligands are typically conjugated to the surface of lipid/polymer core-shell nanoconstructs through covalent or noncovalent interactions. Surface functionalization methods allow for precise control over the density and orientation of ligands on the nanoparticle surface, ensuring optimal targeting efficiency. Strategies such as click chemistry, bioconjugation, or avidin-biotin interactions are commonly employed for ligand conjugation (Shah et al. 2022).

5.1.4 Enhanced Targeting Efficiency

The conjugation of targeting ligands to lipid/polymer core-shell nanoconstructs enhances the targeting efficiency of the nanoparticles to cancer cells. Targeted nanoparticles can selectively bind to receptors or biomarkers on the surface of cancer cells, leading to increased cellular uptake and accumulation of therapeutic payloads within the tumor microenvironment. This targeted approach improves drug delivery to tumor cells while minimizing exposure to healthy tissues, thereby reducing systemic toxicity and side effects (Mehandole et al. 2023).

5.1.5 Personalized Cancer Therapy

Ligand-conjugated lipid/polymer core-shell nanoconstructs offer a promising platform for personalized cancer therapy. By selecting specific ligands that target unique molecular signatures of individual tumors, these nanoparticles can be tailored to deliver customized therapeutic payloads to different types of cancer cells. This personalized approach holds great potential for improving treatment outcomes and overcoming drug resistance in cancer patients (Mukherjee et al. 2019).

5.2 Antibody Conjugation

Lipid/polymer core-shell nanoconstructs represent an innovative approach for cancer theragnostic applications, particularly when combined with antibody conjugation. Antibodies or antibody fragments can be precisely attached to the surface of nanoparticles, enabling targeted delivery to cells expressing specific antigens or receptors. This conjugation enhances the nanoparticles' ability to recognize and bind to diseased cells, such as cancer cells, while minimizing interaction with healthy tissues. By leveraging the specificity of antibodies, these nanoconstructs can effectively deliver therapeutic payloads directly to the tumor site, enhancing treatment efficacy and minimizing off-target effects. Additionally, the integration of imaging agents into these nanoconstructs allows for real-time monitoring of treatment response, enabling personalized cancer therapy. This targeted delivery strategy holds significant promise for improving patient outcomes in oncology, as it enables tailored treatment regimens based on individual tumor characteristics and patient responses (Marques et al. 2023).

5.3 Surface Charge Modification

By altering the surface charge of nanoparticles through the incorporation of positively or negatively charged molecules, their interactions with biological molecules or cell membranes can be finely tuned. This modulation of surface charge influences various aspects of nanoparticle behavior, including cellular uptake, biodistribution, and overall therapeutic efficacy. When designed for cancer theragnostic purposes, lipid/polymer core-shell nanoconstructs can be engineered to carry diagnostic imaging agents and therapeutic payloads. By modifying the surface charge, these nanoconstructs can enhance their targeting specificity to cancer cells while minimizing interactions with healthy tissues. Positively charged nanoparticles may exhibit increased cellular uptake due to electrostatic interactions with negatively charged cell membranes, facilitating efficient delivery of therapeutic agents into cancer cells. Conversely, negatively charged nanoparticles may show improved circulation stability and reduced nonspecific uptake by reticuloendothelial system (RES) organs, potentially leading to enhanced tumor accumulation via the enhanced permeability and retention (EPR) effect (Campani et al. 2018).

Furthermore, surface charge modification can influence the immune response elicited by nanoconstructs, impacting their immunogenicity and biocompatibility. By carefully selecting surface charge-modifying agents, such as cationic or anionic polymers, the immunological profile of lipid/polymer core-shell nanoconstructs can be tailored to minimize adverse reactions and promote safe administration in vivo.

5.4 Stimuli-Responsive Coatings

Incorporating stimuli-responsive polymers or coatings onto the nanoparticle surface enables controlled drug release in response to specific environmental cues, such as changes in pH, temperature, or enzymatic activity (Dhas et al. 2021). Stimuli-responsive nanoconstructs offer spatiotemporal control over drug release, enhancing therapeutic outcomes and minimizing side effects (Xiao et al. 2023).

5.5 Biomimetic Coatings

Biomimetic coatings represent a cutting-edge approach in the functionalization of lipid/polymer core-shell nanoconstructs for cancer theragnostic applications. These coatings are designed to mimic the surface properties of biological membranes or structures, thereby enhancing biocompatibility and improving interactions with biological systems (Carmona-Ribeiro 2019). Here's a detailed elaboration on the use of biomimetic coatings for functionalizing lipid/polymer core-shell nanoconstructs:

5.5.1 Mimicking Biological Membranes

Biomimetic coatings aim to replicate the composition and structure of biological membranes, which play crucial roles in cellular interactions and signaling processes. By mimicking the surface properties of cell membranes, biomimetic coatings can enhance the biocompatibility of lipid/polymer core-shell nanoconstructs and promote favorable interactions with biological systems (Dhas et al. 2022).

5.5.2 Lipid Bilayers

One approach to biomimetic coating involves the formation of lipid bilayers on the surface of lipid/polymer core-shell nanoconstructs. Lipid bilayers consist of two layers of amphiphilic lipid molecules arranged in a double membrane structure, similar to the lipid bilayer of cell membranes. These lipid bilayers can provide a biocompatible interface that mimics the cell membrane, allowing for enhanced cellular uptake and interaction with biological targets (Shah et al. 2022).

5.5.3 Cell Membrane-Derived Vesicles

Another strategy for biomimetic coating involves the use of cell membrane-derived vesicles, such as exosomes or liposomes, to coat the surface of lipid/polymer core-shell nanoconstructs. These vesicles are obtained from the membrane of donor cells and retain the surface properties and biomolecules present on the cell membrane.

Coating lipid/polymer nanoconstructs with cell membrane-derived vesicles can improve stability, reduce immune recognition, and promote specific targeting of biological tissues or cells.

5.5.4 Improved Stability and Reduced Immune Recognition

Biomimetic coatings, such as lipid bilayers or cell membrane-derived vesicles, can improve the stability of lipid/polymer core-shell nanoconstructs by providing a protective layer that shields the nanoparticles from environmental factors and immune recognition. The biomimetic nature of these coatings allows for enhanced biocompatibility and reduced immunogenicity, making the nanoconstructs suitable for in vivo applications.

5.5.5 Enhanced Cellular Interactions

Biomimetic coatings can also facilitate enhanced interactions between lipid/polymer core-shell nanoconstructs and biological systems, including cells and tissues. By mimicking the surface properties of biological membranes, these coatings can promote cellular uptake, internalization, and intracellular delivery of therapeutic payloads or imaging agents. This enhanced cellular interaction is particularly advantageous for cancer theragnostic applications, where precise targeting and delivery to tumor cells are critical.

5.6 Stealth Coatings

Stealth coatings play a crucial role in the functionalization of lipid/polymer core-shell nanoconstructs for cancer theragnostic applications. These coatings are designed to minimize recognition and clearance by the immune system, thereby prolonging circulation time and improving nanoparticle accumulation at the target site. One of the most commonly used stealth coatings is polyethylene glycol (PEG), which is known for its ability to impart stealth properties to nanoparticles. Below is a detailed elaboration on the functionalization of lipid/polymer core-shell nanoconstructs with stealth coatings for cancer theragnostic purposes:

5.6.1 Surface Modification with PEG

Polyethylene glycol (PEG) is a hydrophilic polymer that can be conjugated to the surface of lipid/polymer core-shell nanoconstructs to create a stealth coating. PEGylation involves the covalent attachment of PEG chains to the nanoparticle surface, forming a protective layer that shields the nanoparticles from recognition and clearance by the immune system. The hydrophilic nature of PEG creates a hydration layer around the nanoparticles, reducing protein adsorption and opsonization, which are key mechanisms of immune recognition and clearance (Fam et al. 2020).

5.6.2 Prolonged Circulation Time

The stealth coating provided by PEGylation extends the circulation time of lipid/polymer core-shell nanoconstructs in the bloodstream. By evading recognition and clearance by the immune system, PEGylated nanoconstructs can circulate for longer periods, allowing for increased accumulation at the target site, such as tumor tissues. This prolonged circulation time enhances the therapeutic efficacy of the nanoconstructs by ensuring sustained exposure to the target cells.

5.6.3 Enhanced Tumor Accumulation

The stealth properties conferred by PEGylation also facilitate enhanced accumulation of lipid/polymer core-shell nanoconstructs at the tumor site. This phenomenon, known as the enhanced permeability and retention (EPR) effect, is mediated by the leaky vasculature and impaired lymphatic drainage characteristic of solid tumors. PEGylated nanoconstructs can passively accumulate in the tumor tissue through fenestrations in the tumor vasculature, leading to higher drug concentrations at the target site (Campani et al. 2018).

5.6.4 Improved Therapeutic Efficacy

The stealth coating provided by PEGylation not only enhances tumor accumulation but also improves the therapeutic efficacy of lipid/polymer core-shell nanoconstructs. By minimizing premature clearance and systemic distribution, PEGylated nanoconstructs can deliver therapeutic payloads, such as chemotherapeutic drugs or imaging agents, more effectively to the tumor tissue. This targeted delivery approach reduces off-target effects and systemic toxicity while maximizing the therapeutic benefits for cancer patients.

5.7 Molecules

Biorecognition molecules play a crucial role in the design of lipid/polymer core-shell nanoconstructs for cancer theragnostic applications. These molecules, such as aptamers and folate, offer specific targeting abilities by recognizing and binding to molecular markers overexpressed on cancer cells, facilitating the selective delivery of therapeutic agents or imaging probes. Below is a detailed elaboration on the utilization of biorecognition molecules in lipid/polymer core-shell nanoconstructs for cancer theragnostic purposes:

5.7.1 Aptamers

Aptamers are short, single-stranded DNA or RNA molecules that can bind to specific target molecules with high affinity and specificity. In the context of cancer theragnostic nanoconstructs, aptamers can be selected to recognize cancer-specific biomarkers, such as cell surface receptors or tumor-associated antigens (Tiwari et al. 2023). By conjugating aptamers to the surface of lipid/polymer core-shell nanoconstructs, these nanoparticles can be engineered to selectively target cancer cells while sparing healthy tissues. Aptamer-targeted nanoconstructs can deliver therapeutic agents, such as chemotherapeutic drugs or nucleic acid therapeutics, directly to cancer cells, enhancing treatment efficacy and reducing off-target effects (Liu et al. 2014).

5.7.2 Folate

Folate receptors are often overexpressed on the surface of various cancer cells, making them attractive targets for cancer theragnostic applications. Folate molecules can be conjugated to the surface of lipid/polymer core-shell nanoconstructs to exploit this receptor-mediated targeting mechanism. Folate-targeted nanoconstructs can efficiently bind to cancer cells via folate receptor-mediated endocytosis, allowing for the selective delivery of imaging agents or therapeutic payloads. This targeted approach enables precise tumor localization and improved imaging contrast or therapeutic efficacy (Zhang et al. 2015).

5.7.3 Dual-Targeting Strategies

In some cases, a combination of biorecognition molecules targeting different molecular markers may be employed to enhance the specificity and efficacy of lipid/polymer core-shell nanoconstructs for cancer theragnostic applications. For example, dual-targeting nanoconstructs may utilize both aptamers and folate ligands to target different surface receptors overexpressed on cancer cells, further improving tumor cell recognition and uptake. This dual-targeting strategy can enhance the selectivity and precision of cancer theragnostic nanoconstructs, leading to improved diagnostic accuracy and therapeutic outcomes (Essa et al. 2020).

5.7.4 Imaging and Therapeutic Integration

Biorecognition molecules can also be coupled with imaging probes or therapeutic agents within lipid/polymer core-shell nanoconstructs to enable simultaneous diagnostic imaging and therapeutic intervention. For instance, aptamer-conjugated nanoconstructs can be loaded with imaging agents, such as fluorescent dyes or contrast agents, for real-time visualization of tumor localization and biodistribution. Additionally, therapeutic payloads, such as chemotherapeutic drugs or nucleic acid-based therapeutics, can be incorporated into the nanoconstructs to exert targeted anticancer effects while minimizing systemic toxicity (Bukhari et al. 2021).

By incorporating these surface functionalization strategies, lipid/polymer core-shell nanoconstructs can be precisely engineered for various biomedical applications, including drug delivery, imaging, and diagnostics. The choice of surface modification strategy depends on the specific requirements of the application and the desired properties of the nanoparticles.

6 Biomedical Application

Lipid/polymer core-shell nanoconstructs have emerged as a highly promising platform for cancer theragnostic applications due to their versatile nature and multifunctional capabilities. These nanoconstructs consist of a lipid core surrounded by a polymer shell, allowing for precise control over drug encapsulation, surface functionalization, and therapeutic payload delivery (Fig. 11.6).

Here, we delve into the detailed biomedical applications of lipid/polymer core-shell nanoconstructs for cancer theragnostic purposes:

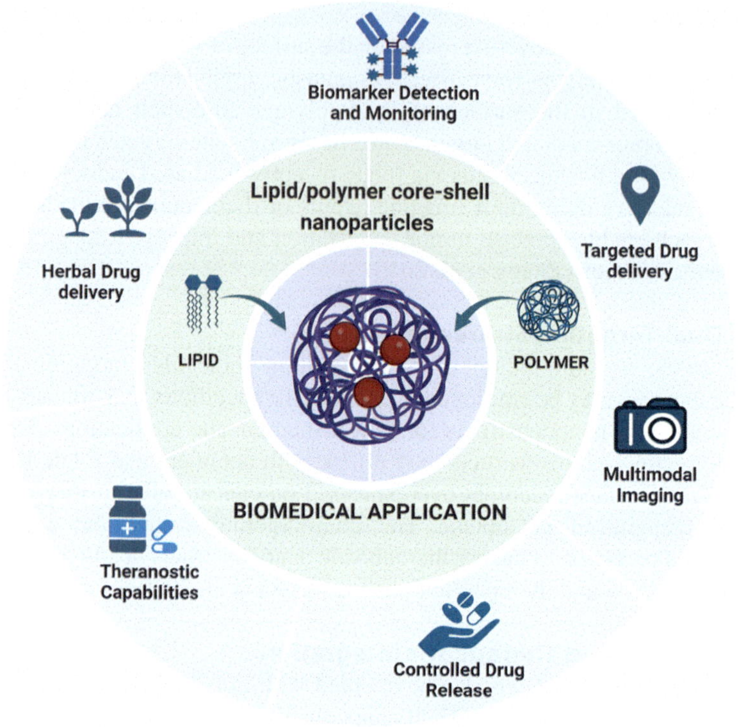

Fig. 11.6 Biomedical application of lipid/polymer core-shell nanoconstructs

6.1 Targeted Drug Delivery

The lipid core of these nanoconstructs serves as an ideal carrier for the encapsulation of chemotherapeutic drugs, shielding them from degradation and facilitating their delivery to tumor sites. By modifying the surface of the polymer shell with targeting ligands, such as antibodies, peptides, or aptamers, lipid/polymer core-shell nanoconstructs can selectively bind to cancer-specific receptors or antigens, enabling targeted drug delivery to malignant cells while minimizing off-target effects on healthy tissues (Dhas and Mehta 2021b). This targeted approach enhances therapeutic efficacy and reduces systemic toxicity, improving patient outcomes in cancer treatment (Sengel-Turk et al. 2024).

Doxorubicin (DOX) and indocyanine green (ICG) were encapsulated in lipid-core polymer-shell nanoparticles (DINPs) using a single-step sonication method, ensuring good monodispersity and stability. These DINPs displayed enhanced temperature response and faster DOX release under laser irradiation, along with prolonged retention time in tumors. The fluorescence of DOX and ICG within DINPs allowed for visualization of subcellular location in vitro and metabolic distribution in vivo. Combining DINPs with laser irradiation synergistically induced apoptosis in DOX-sensitive and DOX-resistant cancer cells, effectively suppressing tumor growth. Importantly, a single dose of DINPs with laser irradiation showed no tumor recurrence, highlighting the potential of these nanoparticles in targeted cancer imaging and chemo-photothermal therapy (Zheng et al. 2013).

Jason and his colleagues engineered nanoscale liposomal polymeric gels (nLGs) to address the immunoinhibitory properties of the tumor microenvironment. These nLGs are designed to deliver a combination of a TGF-β inhibitor and IL-2 in a sustained manner to enhance the effectiveness of immunotherapy. By encapsulating these agents within lipid-core polymer shells, we have developed a novel approach to combatting the challenges posed by the tumor microenvironment in conventional immunotherapy.

The utilization of nLGs loaded with a TGF-β inhibitor and IL-2 has shown promising results in preclinical studies. These studies have demonstrated that the nLGs can significantly delay tumor growth and increase the survival rate of tumor-bearing mice. Moreover, they have been shown to enhance the activity of natural killer cells and promote the infiltration of intratumoral-activated CD8+ T cells, which are crucial components of the antitumor immune response.

This innovative approach not only addresses the immunosuppressive nature of the tumor microenvironment but also activates both innate and adaptive immune responses against the tumor. The sustained release of the TGF-β inhibitor and IL-2 from the nLGs ensures a prolonged and targeted effect, leading to enhanced therapeutic outcomes compared to conventional immunotherapy approaches (Park et al. 2012).

6.2 Multimodal Imaging

Lipid/polymer core-shell nanoconstructs can be engineered to incorporate various imaging contrast agents, including fluorescent dyes, magnetic nanoparticles, or radioactive isotopes, within the lipid core or polymer shell. This allows for multimodal imaging of tumors using techniques such as optical imaging, magnetic resonance imaging (MRI), computed tomography (CT), and positron emission tomography (PET). By providing real-time visualization of tumor morphology, vascularization, and metabolic activity, these nanoconstructs enable accurate tumor localization, characterization, and monitoring of treatment response, facilitating personalized cancer care (Gajbhiye et al. 2023).

Hybrid nanocarriers, termed Lipid-Polymer NPs (LiPoNs), are designed to bridge the gap between synthetic and biological identity, enhancing the clinical translation of nanomedicine. The fabrication process involves utilizing a coupled hydrodynamic flow focusing (cHFF) technique to precisely control the time scales of solvent exchange and the coupling of polymer nanoprecipitation with lipid self-assembly. This simultaneous control guides the formation of LiPoNs, which feature a core-shell structure comprising a polymeric chitosan core enveloped in a lipid bilayer. One of the key advantages of LiPoNs is their ability to co-encapsulate Gd-DTPA for magnetic resonance imaging (MRI) and Irinotecan/Atto 633 compounds for optical imaging. This enables multimodal imaging, allowing researchers to obtain complementary information from different imaging modalities.

LiPoNs are characterized by their monodisperse nature, with an average size of 77 nm, ensuring structural integrity in various environmental conditions. Moreover, they exhibit high biocompatibility, making them suitable for biomedical applications. Preliminary studies have demonstrated the potential of Irinotecan-loaded LiPoNs in enhancing drug delivery and therapeutic efficacy against U87 MG cancer cells. These promising results pave the way for further exploration and optimization of LiPoNs as versatile nanocarriers for imaging and therapeutic purposes in cancer research and beyond (Roffo et al. 2022).

Researchers developed a core-interlayer-shell structure Fe3O4@mSiO2@lipid-PEG-methotrexate nanoparticle (FMLM) for multimodal imaging-guided combination therapy. The Fe3O4 core enables magnet-stimulated drug release and magnetic resonance imaging, while the mSiO2 layer encapsulates doxorubicin (Dox) for chemotherapy. The lipid-PEG shell provides cellular targeting and water dispersibility, along with the loading of photosensitizer zinc phthalocyanine (ZnPc) for near-infrared fluorescence imaging and photodynamic therapy. Both in vitro and in vivo studies demonstrated enhanced tumor accumulation, cellular uptake, anticancer activity, and reduced side effects compared to free drug formulations. This versatile nanoplatform offers dual-modal imaging and combined chemo-photodynamic cancer therapy (Liu et al. 2017).

6.3 Theranostic Capabilities

By combining diagnostic and therapeutic functionalities within a single platform, lipid/polymer core-shell nanoconstructs enable theranostic applications in cancer management. These nanoconstructs can simultaneously deliver therapeutic agents to cancer cells while providing real-time imaging of tumor progression and treatment response. This integrated approach allows clinicians to monitor treatment efficacy, predict therapeutic outcomes, and tailor treatment regimens based on individual patient responses, ultimately improving patient outcomes and prognosis (Bukhari et al. 2021).

The researchers present an innovative approach for in vivo combined-modality imaging using multifunctional drug delivery nanoparticles. These nanoparticles, referred to as nanosomes, are designed with a dextran core encapsulating doxorubicin, iron oxide for magnetic resonance imaging (MRI) contrast enhancement, and BODIPY for fluorescence imaging. The integration of these components into a single nanoparticle system allows for simultaneous therapeutic drug delivery and imaging capabilities. To enhance their in vivo performance, the nanosomes are surface-modified with polyethylene glycol (PEG) and incorporate acetylated lipids that are ultraviolet cross-linked for increased physical stability. This surface modification and lipid composition contribute to prolonged circulation time, reduced clearance by the immune system, and enhanced tumor accumulation through the enhanced permeability and retention (EPR) effect.

A key aspect of the research is the development and utilization of a rodent dorsal skinfold window chamber model. This model enables both MRI and noninvasive optical imaging of nanoparticle accumulation within tumors, providing valuable insights into their distribution and targeting efficiency. The tumors within the window chamber are genetically labeled with DsRed-2, allowing for the co-localization of MR images, tumor fluorescence (red), and nanoparticle fluorescence (blue).

Furthermore, the researchers extended their imaging capabilities to orthotopic pancreatic tumors expressing DsRed-2. By employing iron oxide-dextran liposomes for MRI and fluorescence imaging, they demonstrated the deep tissue imaging potential of these multifunctional nanoparticles. This comprehensive imaging approach enables them to visualize the biodistribution, tumor targeting, and therapeutic efficacy of the nanosomes in real time, providing valuable information for future clinical translation and personalized medicine applications (Erten et al. 2010).

6.4 Controlled Drug Release

The polymer shell of lipid/polymer core-shell nanoconstructs can be engineered to respond to specific stimuli present in the tumor microenvironment, such as pH, temperature, or enzymatic activity. This enables controlled and triggered drug release at the tumor site, enhancing therapeutic efficacy while minimizing systemic side effects. Additionally, spatiotemporal control over drug release reduces the risk of drug resistance and improves drug bioavailability, leading to enhanced tumor regression and prolonged survival in cancer patients (Barba et al. 2019).

Core-shell NPs are intricately designed, comprising a poly(D, L-lactide-co-glycolide) hydrophobic core, a soybean lecithin monolayer, and a poly(ethylene glycol) shell. The synthesis methodology involved a tailored nanoprecipitation technique combined with self-assembly processes to ensure the desired NP characteristics. Throughout the study, researchers delved into understanding the nuanced parameters influencing the biological and physicochemical attributes of these core-shell NPs. These include stability, size distribution, drug release kinetics, and cytotoxicity profiles. Of particular interest was the encapsulation of docetaxel, a model chemotherapy drug, within these NPs, as it allowed them to investigate how variations in lipid coverage impact drug release behavior.

Findings revealed that factors such as the lipid/polymer mass ratio and lipid/lipid-PEG molar ratio significantly influenced NP stability, with implications on size uniformity and drug release kinetics. Devised scalable processes for NP formulation, purification, and long-term storage, ensuring reproducibility and reliability in large-scale production scenarios. To gauge the biocompatibility and therapeutic potential of these NPs, they conducted extensive in vitro evaluations using two human cell lines, HeLa and HepG2. The results showcased not only the biocompatibility of these particles but also their ability to effectively deliver and release docetaxel in a controlled manner, showcasing their promise as a viable controlled-release drug delivery system (Chan et al. 2009).

6.5 Biomarker Detection and Monitoring

Functionalization of lipid/polymer core-shell nanoconstructs with targeting ligands allows for selective recognition and binding to cancer-specific biomarkers present in the tumor microenvironment or circulating in biological fluids. By detecting and quantifying these biomarkers, lipid/polymer core-shell nanoconstructs enable early cancer diagnosis, prognostic assessment, and monitoring of disease progression. This facilitates timely intervention, patient stratification, and personalized treatment planning, optimizing clinical outcomes and patient survival in cancer therapy (Bukhari et al. 2021).

The recent study conducted by the research team focused on the development of lipid-core/polymer-shell hybrid nanoparticles (HNPs) as potential drug carriers. Utilizing vitamin E acetate (VEA) as the oily core component and a tailor-made amphiphilic polymer as the wrapping shell, the researchers aimed to create HNPs with enhanced drug delivery capabilities. To assess the formation, composition, and stability of the HNPs, a dual-color labeling strategy was employed. The oil phase was labeled using a newly developed green fluorogenic BODIPY tracker, while the polymer was covalently attached with a red-emitting rhodamine. Advanced analytical techniques such as dual-color electrophoresis gel analysis, dynamic light scattering (DLS), and electron microscopy were utilized for comprehensive characterization.

The study revealed that an optimal ratio of 20 wt% polymer content resulted in stable HNPs with desirable properties, including high loading capacity and surface functionality. Moreover, by varying the composition of the polymeric shell, different types of HNPs were generated, each with distinct characteristics and functionalities.

In summary, lipid/polymer core-shell nanoconstructs offer a comprehensive and integrated approach to cancer diagnosis and therapy, leveraging their multifunctional capabilities, tunable properties, and biocompatibility to address the complex challenges associated with cancer management. Their potential for targeted drug delivery, multimodal imaging, theranostic applications, controlled drug release, and biomarker detection holds great promise in advancing precision oncology and revolutionizing cancer treatment paradigms (Bou et al. 2020).

7 Future Perspectives

Despite the significant progress in the development of lipid/polymer core-shell nanoconstructs for cancer theragnostic applications, several challenges and opportunities remain on the horizon. Future research efforts should focus on advancing the design, fabrication, and characterization of nanoconstructs to optimize their performance and clinical translation. This includes exploring novel materials, engineering strategies, and synthesis methods to enhance the stability, biocompatibility, and targeting efficiency of core-shell nanoconstructs (Gajbhiye et al. 2023).

Moreover, there is a need for comprehensive preclinical and clinical studies to evaluate the safety, efficacy, and pharmacokinetics of lipid/polymer core-shell nanoconstructs in vivo. Long-term toxicity assessments, pharmacodynamic studies, and biodistribution analyses are essential to ensure the safety and efficacy of these nanoconstructs in clinical settings. Additionally, large-scale clinical trials are warranted to validate the clinical utility and effectiveness of lipid/polymer core-shell nanoconstructs in diverse patient populations and cancer types.

Furthermore, the integration of emerging technologies, such as artificial intelligence, machine learning, and microfluidics, holds great promise for advancing the field of cancer theragnostics. These technologies can facilitate the rapid screening of drug candidates, prediction of treatment responses, and optimization of treatment regimens, leading to more personalized and effective cancer care (Burkett et al. 2023).

As a result, lipid/polymer core-shell nanoconstructs represent a paradigm-shifting approach to cancer theragnostics, offering a multifunctional platform for targeted drug delivery, imaging, and therapeutic monitoring. With continued research and innovation, these nanoconstructs have the potential to revolutionize cancer diagnosis, treatment, and patient outcomes, ushering in a new era of precision oncology.

8 Conclusion

Lipid/polymer core-shell nanoconstructs represent a cutting-edge technology with multifaceted applications in cancer theragnostic interventions. These nanoconstructs combine the advantageous properties of lipid cores and polymer shells, offering a versatile platform for targeted drug delivery, imaging, and therapeutic monitoring in cancer treatment. By harnessing the unique features of both lipid and polymer components, these nanoconstructs can overcome the limitations of conventional cancer therapies and pave the way for more effective and personalized treatment strategies.

The integration of lipid cores provides several benefits, including high drug loading capacity, biocompatibility, and the ability to encapsulate both hydrophobic and hydrophilic drugs. Lipid cores can protect encapsulated drugs from premature degradation and clearance, thereby enhancing their stability and bioavailability. Moreover, lipid-based nanoconstructs offer tunable release kinetics, allowing for sustained and controlled drug release at the tumor site, which is critical for achieving optimal therapeutic outcomes.

The polymer shell component plays a pivotal role in enhancing the stability, biocompatibility, and targeting specificity of the nanoconstructs. Polymer shells can be engineered to modulate the pharmacokinetics and biodistribution of encapsulated drugs, thereby improving their accumulation at the tumor site while minimizing off-target effects. Additionally, surface functionalization of polymer shells with targeting ligands, such as antibodies or peptides, enables precise recognition and binding to cancer cells, further enhancing the selectivity and efficacy of drug delivery.

The synergistic combination of lipid cores and polymer shells in core-shell nanoconstructs enables synergistic therapeutic and diagnostic functionalities, making them ideal candidates for cancer theragnostic applications. These nanoconstructs can simultaneously deliver therapeutic agents, such as chemotherapeutic drugs, nucleic acids, or photothermal agents, while also providing real-time imaging and monitoring of tumor response to treatment. This integrated approach allows for personalized treatment regimens tailored to individual patient characteristics and disease profiles, leading to improved therapeutic outcomes and reduced adverse effects.

References

Abdullah T, Gauthaman K, Mostafavi A, Alshahrie A, Salah N, Morganti P, Chianese A, Tamayol A, Memic A. Sustainable drug release from polycaprolactone coated chitin-lignin gel fibrous scaffolds. Sci Rep. 2020;10:1–12. https://doi.org/10.1038/s41598-020-76971-w.

Agrawal U, Chashoo G, Sharma PR, Kumar A, Saxena AK, Vyas SP. Tailored polymer–lipid hybrid nanoparticles for the delivery of drug conjugate: dual strategy for brain targeting. Colloids Surf B: Biointerfaces. 2015;126:414–25. https://doi.org/10.1016/j.colsurfb.2014.12.045.

Amasya G, Bakar Ates F, Wintgens V, Amiel C. Layer by layer assembly of core-corona structured solid lipid nanoparticles with β-cyclodextrin polymers. Int J Pharm. 2020;592:119994. https://doi.org/10.1016/j.ijpharm.2020.119994.

Aminu N, Audu MM. Chapter 3- Polycaprolactone-based nanoparticles for advanced therapeutic applications. In: Hasnain MS, Nayak AK, Aminabhavi TM, editors. Polymeric nanosystems. Academic Press; 2023. p. 37–84. https://doi.org/10.1016/B978-0-323-85656-0.00003-6.

Anderson JM, Shive MS. Biodegradation and biocompatibility of PLA and PLGA microspheres. Adv Drug Deliv Rev, Biodegradable Microspheres/Therapeutic Peptide Delivery. 1997;28:5–24. https://doi.org/10.1016/S0169-409X(97)00048-3.

Aranaz I, Alcántara AR, Civera MC, Arias C, Elorza B, Heras Caballero A, Acosta N. Chitosan: an overview of its properties and applications. Polymers (Basel). 2021;13:3256. https://doi.org/10.3390/polym13193256.

Badilli U, Gumustas M, Uslu B, Ozkan SA. Chapter 9—Lipid-based nanoparticles for dermal drug delivery. In: Grumezescu AM, editor. Organic materials as smart nanocarriers for drug delivery. William Andrew Publishing; 2018. p. 369–413. https://doi.org/10.1016/B978-0-12-813663-8.00009-9.

Barba AA, Bochicchio S, Dalmoro A, Caccavo D, Cascone S, Lamberti G. Chapter 10—Polymeric and lipid-based systems for controlled drug release: an engineering point of view. In: Grumezescu AM, editor. Nanomaterials for drug delivery and therapy. William Andrew Publishing; 2019. p. 267–304. https://doi.org/10.1016/B978-0-12-816505-8.00013-8.

Barea MJ, Jenkins MJ, Gaber MH, Bridson RH. Evaluation of liposomes coated with a pH responsive polymer. Int J Pharm. 2010;402:89–94. https://doi.org/10.1016/j.ijpharm.2010.09.028.

Bochicchio S, Lamberti G, Barba AA. Polymer–lipid pharmaceutical nanocarriers: innovations by new formulations and production technologies. Pharmaceutics. 2021;13:198. https://doi.org/10.3390/pharmaceutics13020198.

Bou S, Wang X, Anton N, Bouchaala R, Klymchenko AS, Collot M. Lipid-core/polymer-shell hybrid nanoparticles: synthesis and characterization by fluorescence labeling and electrophoresis. Soft Matter. 2020;16:4173–81. https://doi.org/10.1039/d0sm00077a.

Bukhari SI, Imam SS, Ahmad MZ, Vuddanda PR, Alshehri S, Mahdi WA, Ahmad J. Recent progress in lipid nanoparticles for cancer theranostics: opportunity and challenges. Pharmaceutics. 2021;13:840. https://doi.org/10.3390/pharmaceutics13060840.

Burkett BJ, Bartlett DJ, McGarrah PW, Lewis AR, Johnson DR, Berberoğlu K, Pandey MK, Packard AT, Halfdanarson TR, Hruska CB, Johnson GB, Kendi AT. A review of theranostics: perspectives on emerging approaches and clinical advancements. Radiol Imaging Cancer. 2023;5:e220157. https://doi.org/10.1148/rycan.220157.

Campani V, Giarra S, De Rosa G. Lipid-based core-shell nanoparticles: evolution and potentialities in drug delivery. Open Nano. 2018;3:5–17. https://doi.org/10.1016/j.onano.2017.12.001.

Cao Y, Wang B, Wang Y, Lou D. Polymer-controlled core–shell nanoparticles: a novel strategy for sequential drug release. RSC Adv. 2014;4:30430–9. https://doi.org/10.1039/C4RA03610G.

Carmona-Ribeiro AM. Biomimetic nanomaterials from the assembly of polymers, lipids, and surfactants. In: Surfactants and detergents. Intech Open; 2019. https://doi.org/10.5772/intechopen.84618.

Chan JM, Zhang L, Yuet KP, Liao G, Rhee J-W, Langer R, Farokhzad OC. PLGA-lecithin-PEG core-shell nanoparticles for controlled drug delivery. Biomaterials. 2009;30:1627–34. https://doi.org/10.1016/j.biomaterials.2008.12.013.

Chiang Y-T, Lo C-L. pH-responsive polymer-liposomes for intracellular drug delivery and tumor extracellular matrix switched-on targeted cancer therapy. Biomaterials. 2014;35:5414–24. https://doi.org/10.1016/j.biomaterials.2014.03.046.

Colaco V, Roy AA, Naik GARR, Mondal A, Mutalik S, Dhas N. Advancement in lipid-based nanocomposites for theranostic applications in lung carcinoma treatment. Open Nano. 2024;15:100199. https://doi.org/10.1016/j.onano.2023.100199.

da Rocha MCO, da Silva PB, Radicchi MA, Andrade BYG, de Oliveira JV, Venus T, Merker C, Estrela-Lopis I, Longo JPF, Báo SN. Docetaxel-loaded solid lipid nanoparticles prevent tumor growth and lung metastasis of 4T1 murine mammary carcinoma cells. J Nanobiotechnol. 2020;18:43. https://doi.org/10.1186/s12951-020-00604-7.

Dahiya S, Dahiya R. Chapter 6—Potential of colloidal carriers for nanocosmeceutical applications. In: Das MK, editor. Nanocosmeceuticals. Academic Press; 2022. p. 169–208. https://doi.org/10.1016/B978-0-323-91077-4.00002-8.

Deng S, Gigliobianco MR, Censi R, Di Martino P. Polymeric nanocapsules as nanotechnological alternative for drug delivery system: current status, challenges and opportunities. Nanomaterials (Basel). 2020;10:847. https://doi.org/10.3390/nano10050847.

Deshpande A, Mohamed M, Daftardar SB, Patel M, Boddu SHS, Nesamony J. Chapter 12—Solid lipid nanoparticles in drug delivery: opportunities and challenges. In: Mitra AK, Cholkar K, Mandal A, editors. Emerging nanotechnologies for diagnostics, drug delivery and medical devices, micro and nano technologies. Boston: Elsevier; 2017. p. 291–330. https://doi.org/10.1016/B978-0-323-42978-8.00012-7.

Dhas N. Development and characterization of core shell nanoparticles for the treatment of Alzheimer's disease using non invasive approach. University; 2021.

Dhas N, García MC, Kudarha R, Pandey A, Nikam AN, Gopalan D, Fernandes G, Soman S, Kulkarni S, Seetharam RN, Tiwari R, Wairkar S, Pardeshi C, Mutalik S. Advancements in cell membrane camouflaged nanoparticles: a bioinspired platform for cancer therapy. J Control Release. 2022;346:71–97. https://doi.org/10.1016/j.jconrel.2022.04.019.

Dhas N, Kudarha R, Pandey A, Nikam AN, Sharma S, Singh A, Garkal A, Hariharan K, Singh A, Bangar P, Yadhav D, Parikh D, Sawant K, Mutalik S, Garg N, Mehta T. Stimuli responsive and receptor targeted iron oxide based nanoplatforms for multimodal therapy and imaging of cancer: conjugation chemistry and alternative therapeutic strategies. J Control Release. 2021;333:188–245. https://doi.org/10.1016/j.jconrel.2021.03.021.

Dhas N, Mehta T. Intranasal delivery of chitosan decorated PLGA core/shell nanoparticles containing flavonoid to reduce oxidative stress in the treatment of Alzheimer's disease. J Drug Deliv Sci Technol. 2021a;61:102242. https://doi.org/10.1016/j.jddst.2020.102242.

Dhas N, Mehta T. Intranasal delivery of chitosan decorated PLGA core/shell nanoparticles containing flavonoid to reduce oxidative stress in the treatment of Alzheimer's disease. J Drug Deliv Sci Technol. 2021b;61:102242. https://doi.org/10.1016/j.jddst.2020.102242.

Dhas N, Mehta T. Cationic biopolymer functionalized nanoparticles encapsulating lutein to attenuate oxidative stress in effective treatment of Alzheimer's disease: a non-invasive approach. Int J Pharm. 2020;586:119553. https://doi.org/10.1016/j.ijpharm.2020.119553.

Dhas NL, Raval NJ, Kudarha RR, Acharya NS, Acharya SR. Chapter 9—Core–shell nanoparticles as a drug delivery platform for tumor targeting. In: Grumezescu AM, editor. Inorganic frameworks as smart nanomedicines. William Andrew Publishing; 2018. p. 387–448. https://doi.org/10.1016/B978-0-12-813661-4.00009-2.

Duan Y, Dhar A, Patel C, Khimani M, Neogi S, Sharma P, Kumar NS, Vekariya RL. A brief review on solid lipid nanoparticles: part and parcel of contemporary drug delivery systems. RSC Adv. 2020;10:26777–91. https://doi.org/10.1039/D0RA03491F.

Eliasof S, Lazarus D, Peters CG, Case RI, Cole RO, Hwang J, Schluep T, Chao J, Lin J, Yen Y, Han H, Wiley DT, Zuckerman JE, Davis ME. Correlating preclinical animal studies and human clinical trials of a multifunctional, polymeric nanoparticle. Proc Natl Acad Sci. 2013;110:15127–32. https://doi.org/10.1073/pnas.1309566110.

Elmowafy EM, Tiboni M, Soliman ME. Biocompatibility, biodegradation and biomedical applications of poly(lactic acid)/poly(lactic-co-glycolic acid) micro and nanoparticles. J Pharm Investig. 2019;49:347–80. https://doi.org/10.1007/s40005-019-00439-x.

Erten A, Wrasidlo W, Scadeng M, Esener S, Hoffman RM, Bouvet M, Makale M. Magnetic resonance and fluorescence imaging of doxorubicin-loaded nanoparticles using a novel in vivo model. Nanomed Nanotechnol Biol Med. 2010;6:797–807. https://doi.org/10.1016/j.nano.2010.06.005.

Escalona-Rayo O, Fuentes-Vázquez P, Jardon-Xicotencatl S, García-Tovar CG, Mendoza-Elvira S, Quintanar-Guerrero D. Rapamycin-loaded polysorbate 80-coated PLGA nanoparticles: optimization of formulation variables and *in vitro* anti-glioma assessment. J Drug Deliv Sci Tech. 2019;52:488–99. https://doi.org/10.1016/j.jddst.2019.05.026.

Essa ML, El-Kemary MA, Ebrahem Saied EM, Leporatti S, Nemany Hanafy NA. Nano targeted therapies made of lipids and polymers have promising strategy for the treatment of lung cancer. Materials (Basel). 2020;13:5397. https://doi.org/10.3390/ma13235397.

Fam SY, Chee CF, Yong CY, Ho KL, Mariatulqabtiah AR, Tan WS. Stealth coating of nanoparticles in drug-delivery systems. Nano. 2020;10 https://doi.org/10.3390/nano10040787.

Fang D-L, Chen Y, Xu B, Ren K, He Z-Y, He L-L, Lei Y, Fan C-M, Song X-R. Development of lipid-Shell and polymer core nanoparticles with water-soluble salidroside for anti-cancer therapy. Int J Mol Sci. 2014;15:3373–88. https://doi.org/10.3390/ijms15033373.

Fernandes M, Lopes I, Magalhães L, Sárria MP, Machado R, Sousa JC, Botelho C, Teixeira J, Gomes AC. Novel concept of exosome-like liposomes for the treatment of Alzheimer's disease. J Control Release. 2021;336:130–43. https://doi.org/10.1016/j.jconrel.2021.06.018.

Gajbhiye KR, Salve R, Narwade M, Sheikh A, Kesharwani P, Gajbhiye V. Lipid polymer hybrid nanoparticles: a custom-tailored next-generation approach for cancer therapeutics. Mol Cancer. 2023;22:160. https://doi.org/10.1186/s12943-023-01849-0.

Guo J, Ping Q, Jiang G, Huang L, Tong Y. Chitosan-coated liposomes: characterization and interaction with leuprolide. Int J Pharm. 2003;260:167–73. https://doi.org/10.1016/S0378-5173(03)00254-0.

Gupta A, Kulkarni S, Soman S, Saha M, Kulkarni J, Rana K, Dhas N, Ayesha Farhana S, Kumar Tiyyagura P, Pandey A, Moorkoth S, Mutalik S. Breaking barriers in cancer management: the promising role of microsphere conjugates in cancer diagnosis and therapy. Int J Pharm. 2024;665:124687. https://doi.org/10.1016/j.ijpharm.2024.124687.

Haider N, Fatima S, Taha M, Rizwanullah M, Firdous J, Ahmad R, Mazhar F, Khan MA. Nanomedicines in diagnosis and treatment of cancer: an update. Curr Pharm Des. 2020;26:1216–31. https://doi.org/10.2174/1381612826666200318170716.

Hegde AR, Paul M, Kumbham S, Roy AA, Ahmad SF, Parekh H, Biswas S, Mutalik S. Ameliorative anticancer effect of dendrimeric peptide modified liposomes of letrozole: *in vitro* and *in vivo* performance evaluations. Int J Pharm. 2023;648:123582. https://doi.org/10.1016/j.ijpharm.2023.123582.

Hermal F, Frisch B, Specht A, Bourel-Bonnet L, Heurtault B. Development and characterization of layer-by-layer coated liposomes with poly(L-lysine) and poly(L-glutamic acid) to increase their resistance in biological media. Int J Pharm. 2020;586:119568. https://doi.org/10.1016/j.ijpharm.2020.119568.

Ihm JE, Krier I, Lim JM, Shim S, Han DK, Hubbell JA. Improved biocompatibility of polyethylenimine (PEI) as a gene carrier by conjugating urocanic acid: in vitro and in vivo. Macromol Res. 2015;23:387–95. https://doi.org/10.1007/s13233-015-3047-8.

Iqbal M, Valour J-P, Fessi H, Elaissari A. Preparation of biodegradable PCL particles via double emulsion evaporation method using ultrasound technique. Colloid Polym Sci. 2015;293:861–73. https://doi.org/10.1007/s00396-014-3464-9.

Jiang X, Du Z, Zhang X, Zaman F, Song Z, Guan Y, Yu T, Huang Y. Gelatin-based anticancer drug delivery nanosystems: a mini review. Front Bioeng Biotechnol. 2023;11:1158749. https://doi.org/10.3389/fbioe.2023.1158749.

Khaliq NU, Lee J, Kim S, Sung D, Kim H. Pluronic F-68 and F-127 based nanomedicines for advancing combination cancer therapy. Pharmaceutics. 2023;15:2102. https://doi.org/10.3390/pharmaceutics15082102.

Kim M, Jang J, Cha C. Carbon nanomaterials as versatile platforms for theranostic applications. Drug Discov Today, SI: Therap Nanomater. 2017;22:1430–7. https://doi.org/10.1016/j.drudis.2017.05.004.

Krishnamurthy S, Vaiyapuri R, Zhang L, Chan JM. Lipid-coated polymeric nanoparticles for cancer drug delivery. Biomater Sci. 2015;3:923–36. https://doi.org/10.1039/C4BM00427B.

Lee S-M, Chen H, Dettmer CM, O'Halloran TV, Nguyen ST. Polymer-caged lipsomes: a pH-responsive delivery system with high stability. J Am Chem Soc. 2007;129:15096–7. https://doi.org/10.1021/ja070748i.

Lee S-M, O'Halloran TV, Nguyen ST. Polymer-caged nanobins for synergistic cisplatin–doxorubicin combination chemotherapy. J Am Chem Soc. 2010;132:17130–8. https://doi.org/10.1021/ja107333g.

Liu G, Ma J, Li Y, Li Q, Tan C, Song H, Cai S, Chen D, Hou Z, Chen Q, Zhu X. Core-interlayer-shell Fe3O4@mSiO2@lipid-PEG-methotrexate nanoparticle for multimodal imaging and multistage targeted chemo-photodynamic therapy. Int J Pharm. 2017;521:19. https://doi.org/10.1016/j.ijpharm.2017.01.068.

Liu Q, Jin C, Wang Y, Fang X, Zhang X, Chen Z, Tan W. Aptamer-conjugated nanomaterials for specific cancer cell recognition and targeted cancer therapy. NPG Asia Mater. 2014;6:e95. https://doi.org/10.1038/am.2014.12.

Łukasiewicz S, Mikołajczyk A, Błasiak E, Fic E, Dziedzicka-Wasylewska M. Polycaprolactone nanoparticles as promising candidates for nanocarriers in novel nanomedicines. Pharmaceutics. 2021;13:191. https://doi.org/10.3390/pharmaceutics13020191.

Madkhali OA. Drug delivery of gelatin nanoparticles as a biodegradable polymer for the treatment of infectious diseases: perspectives and challenges. Polymers. 2023;15:4327. https://doi.org/10.3390/polym15214327.

Makadia HK, Siegel SJ. Poly Lactic-co-Glycolic Acid (PLGA) as biodegradable controlled drug delivery carrier. Polymers (Basel). 2011;3:1377–97. https://doi.org/10.3390/polym3031377.

Mandal B, Bhattacharjee H, Mittal N, Sah H, Balabathula P, Thoma LA, Wood GC. Core–shell-type lipid–polymer hybrid nanoparticles as a drug delivery platform. Nanomed Nanotechnol Biol Med. 2013;9:474–91. https://doi.org/10.1016/j.nano.2012.11.010.

Marques AC, Costa PC, Velho S, Amaral MH. Lipid nanoparticles functionalized with antibodies for anticancer drug therapy. Pharmaceutics. 2023;15:216. https://doi.org/10.3390/pharmaceutics15010216.

de Martel C, Georges D, Bray F, Ferlay J, Clifford GM. Global burden of cancer attributable to infections in 2018: a worldwide incidence analysis. Lancet Glob Health. 2020;8:e180–90. https://doi.org/10.1016/S2214-109X(19)30488-7.

Mehandole A, Walke N, Mahajan S, Aalhate M, Maji I, Gupta U, Mehra NK, Singh PK. Core–shell type lipidic and polymeric nanocapsules: the transformative multifaceted delivery systems. AAPS Pharm Sci Tech. 2023;24:50. https://doi.org/10.1208/s12249-023-02504-z.

Meng N, Grimm D. Membrane-destabilizing ionizable phospholipids: novel components for organ-selective mRNA delivery and CRISPR–Cas gene editing. Sig Transduct Target Ther. 2021;6:1–3. https://doi.org/10.1038/s41392-021-00642-z.

Minchinton AI, Tannock IF. Drug penetration in solid tumours. Nat Rev Cancer. 2006;6:583–92. https://doi.org/10.1038/nrc1893.

Mohanty A, Uthaman S, Park I-K. Chapter 12—Lipid–polymer hybrid nanoparticles as a smart drug delivery platform. In: Gajbhiye V, Gajbhiye KR, Hong S, editors. Stimuli-responsive nanocarriers. Academic Press; 2022. p. 319–49. https://doi.org/10.1016/B978-0-12-824456-2.00012-6.

Mohanty A, Uthaman S, Park I-K. Utilization of polymer-lipid hybrid nanoparticles for targeted anti-cancer therapy. Molecules. 2020;25:4377. https://doi.org/10.3390/molecules25194377.

Morimune S, Nishino T, Goto T. Poly(vinyl alcohol)/graphene oxide nanocomposites prepared by a simple eco-process. Polym J. 2012;44:1056–63. https://doi.org/10.1038/pj.2012.58.

Mukherjee A, Waters AK, Kalyan P, Achrol AS, Kesari S, Yenugonda VM. Lipid–polymer hybrid nanoparticles as a next-generation drug delivery platform: state of the art, emerging technologies, and perspectives. Int J Nanomedicine. 2019;14:1937–52. https://doi.org/10.2147/IJN.S198353.

Muzamil S, Rather G, Patrício A, Haq Z, Sheikh AA, Zahoor M, Shah MZUH, Singh H, Khan A, Imtiyaz S, Ahmad B, Nabi S, Rakhshan R, Hassan S, Fonte P. Chitosan nanoparticles: a versatile platform for biomedical applications. Materials. 2022;15 https://doi.org/10.3390/ma15196521.

Naik Ga RR, Roy AA, Mutalik S, Dhas N. Unleashing the power of polymeric nanoparticles—creative triumph against antibiotic resistance: a review. Int J Biol Macromol. 2024;278:134977. https://doi.org/10.1016/j.ijbiomac.2024.134977.

Nangare SN, Patil PO, Tade RS, Patil AG, Deshmukh PK, Pandey A, Khan ZG, Mutalik S, Jadhav NR, Patel JK, More MP, Bari SB. Polymeric nanoplatforms for the targeted treatment of prostate cancer. In: Padhi S, Behera A, Lichtfouse E, editors. Polymeric nanoparticles for the treatment of solid tumors. Cham: Springer; 2022. p. 427–71. https://doi.org/10.1007/978-3-031-14848-4_16.

Newton AMJ, Kaur S. Chapter 9—Solid lipid nanoparticles for skin and drug delivery: methods of preparation and characterization techniques and applications. In: Grumezescu AM, editor. Nanoarchitectonics in biomedicine. William Andrew Publishing; 2019. p. 295–334. https://doi.org/10.1016/B978-0-12-816200-2.00015-3.

O'Brien MER, Wigler N, Inbar M, Rosso R, Grischke E, Santoro A, Catane R, Kieback DG, Tomczak P, Ackland SP, Orlandi F, Mellars L, Alland L, Tendler C, CAELYX Breast Cancer Study Group. Reduced cardiotoxicity and comparable efficacy in a phase III trial of pegylated liposomal doxorubicin HCl (CAELYX/Doxil) versus conventional doxorubicin for first-line treatment of metastatic breast cancer. Ann Oncol. 2004;15:440–9. https://doi.org/10.1093/annonc/mdh097.

Paris JL, Gaspar R, Coelho F, De Beule PAA, Silva BFB. Stability criterion for the assembly of core–shell lipid–polymer–nucleic acid nanoparticles. ACS Nano. 2023;17:17587–94. https://doi.org/10.1021/acsnano.3c07204.

Park J, Wrzesinski SH, Stern E, Look M, Criscione J, Ragheb R, Jay SM, Demento SL, Agawu A, Licona Limon P, Ferrandino AF, Gonzalez D, Habermann A, Flavell RA, Fahmy TM. Combination delivery of TGF-β inhibitor and IL-2 by nanoscale liposomal polymeric gels enhances tumour immunotherapy. Nat Mater. 2012;11:895–905. https://doi.org/10.1038/nmat3355.

Persano F, Gigli G, Leporatti S. Lipid-polymer hybrid nanoparticles in cancer therapy: current overview and future directions. Nano Ex. 2021;2:012006. https://doi.org/10.1088/2632-959X/abeb4b.

Pham Le Khanh H, Nemes D, Rusznyák Á, Ujhelyi Z, Fehér P, Fenyvesi F, Váradi J, Vecsernyés M, Bácskay I. Comparative investigation of cellular effects of polyethylene glycol (PEG) derivatives. Polymers (Basel). 2022;14:279. https://doi.org/10.3390/polym14020279.

Pistone S, Rykke M, Smistad G, Hiorth M. Polysaccharide-coated liposomal formulations for dental targeting. Int J Pharm. 2017;516:106–15. https://doi.org/10.1016/j.ijpharm.2016.11.028.

Qhattal HSS, Hye T, Alali A, Liu X. Hyaluronan polymer length, grafting density, and surface poly (ethylene glycol) coating influence in vivo circulation and tumor targeting of hyaluronan-grafted liposomes. ACS Nano. 2014;8:5423–40. https://doi.org/10.1021/nn405839n.

Raina N, Singh AK, Islam A. Biological implications of polyethylene glycol and PEGylation: therapeutic approaches based on biophysical studies and protein structure-based drug design tools. In: Singh SK, editor. Innovations and implementations of computer aided drug discovery strategies in rational drug design. Singapore: Springer; 2021. p. 273–94. https://doi.org/10.1007/978-981-15-8936-2_11.

Rezigue M. Lipid and polymeric nanoparticles: drug delivery applications. In: Krishnan A, Chuturgoon A, editors. Integrative nanomedicine for new therapies. Cham: Springer; 2020. p. 167–230. https://doi.org/10.1007/978-3-030-36260-7_7.

Rivera-Hernández G, Antunes-Ricardo M, Martínez-Morales P, Sánchez ML. Polyvinyl alcohol based-drug delivery systems for cancer treatment. Int J Pharm. 2021;600:120478. https://doi.org/10.1016/j.ijpharm.2021.120478.

Rod-in W, Monmai C, Shin I, You S, Park WJ. Neutral lipids, glycolipids, and phospholipids, isolated from sandfish (Arctoscopus japonicus) eggs, exhibit anti-inflammatory activity in LPS-stimulated RAW264.7 cells through NF-κB and MAPKs pathways. Mar Drugs. 2020;18:480. https://doi.org/10.3390/md18090480.

Roffo F, Ponsiglione AM, Netti PA, Torino E. coupled Hydrodynamic Flow Focusing (cHFF) to Engineer Lipid–Polymer Nanoparticles (LiPoNs) for multimodal imaging and theranostic applications. Biomedicines. 2022;10:438. https://doi.org/10.3390/biomedicines10020438.

San HHM, Alcantara KP, Bulatao BPI, Chaichompoo W, Nalinratana N, Suksamrarn A, Vajragupta O, Rojsitthisak P, Rojsitthisak P. Development of turmeric oil—loaded chitosan/alginate nanocapsules for cytotoxicity enhancement against breast cancer. Polymers. 2022;14:1835. https://doi.org/10.3390/polym14091835.

Sánchez A, Mejía SP, Orozco J. Recent advances in polymeric nanoparticle-encapsulated drugs against intracellular infections. Molecules. 2020;25:3760. https://doi.org/10.3390/molecules25163760.

Sánchez-López E, Ettcheto M, Egea MA, Espina M, Cano A, Calpena AC, Camins A, Carmona N, Silva AM, Souto EB, García ML. Memantine loaded PLGA PEGylated nanoparticles for Alzheimer's disease: in vitro and in vivo characterization. J Nanobiotechnol. 2018;16:32. https://doi.org/10.1186/s12951-018-0356-z.

Satapathy MK, Yen T-L, Jan J-S, Tang R-D, Wang J-Y, Taliyan R, Yang C-H. Solid Lipid Nanoparticles (SLNs): an advanced drug delivery system targeting brain through BBB. Pharmaceutics. 2021;13:1183. https://doi.org/10.3390/pharmaceutics13081183.

Schmidpeter PAM, Wu D, Rheinberger J, Riegelhaupt PM, Tang H, Robinson CV, Nimigean CM. Anionic lipids unlock the gates of select ion channels in the pacemaker family. Nat Struct Mol Biol. 2022;29:1092–100. https://doi.org/10.1038/s41594-022-00851-2.

Sengel-Turk CT, Paksoy AO, Alpturk O. The state of the art in core–shell-type lipid–polymer hybrid nanocarriers and beyond. Polym Bull. 2024;81:4771–800. https://doi.org/10.1007/s00289-023-04951-x.

Shah S, Famta P, Raghuvanshi RS, Singh SB, Srivastava S. Lipid polymer hybrid nanocarriers: insights into synthesis aspects, characterization, release mechanisms, surface functionalization and potential implications. Colloid Interf Sci Commun. 2022;46:100570. https://doi.org/10.1016/j.colcom.2021.100570.

Singh R, Bhateria R. Core–shell nanostructures: a simplest two-component system with enhanced properties and multiple applications. Environ Geochem Health. 2021;43:2459–82. https://doi.org/10.1007/s10653-020-00766-1.

Song X, Gan K, Qin S, Chen L, Liu X, Chen T, Liu H. Preparation and characterization of general-purpose gelatin-based co-loading flavonoids nano-core structure. Sci Rep. 2019;9:6365. https://doi.org/10.1038/s41598-019-42909-0.

Sun D, Lu Z-R. Structure and function of cationic and ionizable lipids for nucleic acid delivery. Pharm Res. 2023;40:27–46. https://doi.org/10.1007/s11095-022-03460-2.

Szczęch M, Szczepanowicz K. Polymeric core-shell nanoparticles prepared by spontaneous emulsification solvent evaporation and functionalized by the layer-by-layer method. Nanomaterials (Basel). 2020;10:496. https://doi.org/10.3390/nano10030496.

Takeuchi H, Kojima H, Yamamoto H, Kawashima Y. Enhanced tumour accumulation of doxorubicin with polymer-coated liposomes in rats. Pharm Pharmacol Commun. 1999;5:445–8. https://doi.org/10.1211/146080899128735144.

Takeuchi H, Yamamoto H, Niwa T, Hino T, Kawashima Y. Mucoadhesion of polymer-coated liposomes to rat intestine in vitro. Chem Pharm Bull. 1994;42:1954–6. https://doi.org/10.1248/cpb.42.1954.

Tiwari G, Babu AK, Khatri C, Dhas NL. Aptamer-based targeting of nanoplatforms for cancer theranostics. In: Site-specific cancer nanotheranostics. CRC Press; 2023.

Wang C, Park MJ, Yu H, Matsuyama H, Drioli E, Shon HK. Recent advances of nanocomposite membranes using layer-by-layer assembly. J Membr Sci. 2022;661:120926. https://doi.org/10.1016/j.memsci.2022.120926.

Wang H, Zhao P, Su W, Wang S, Liao Z, Niu R, Chang J. PLGA/polymeric liposome for targeted drug and gene co-delivery. Biomaterials. 2010;31:8741–8. https://doi.org/10.1016/j.biomaterials.2010.07.082.

Weiss-Angeli V, Poletto FS, de Marco SL, Salvador M, da Silveira NP, Guterres SS, Pohlmann AR. Sustained antioxidant activity of quercetin-loaded lipid-core nanocapsules. J Nanosci Nanotechnol. 2012;12:2874–80. https://doi.org/10.1166/jnn.2012.5770.

Xiao R, Zhou G, Wen Y, Ye J, Li X, Wang X. Recent advances on stimuli-responsive biopolymer-based nanocomposites for drug delivery. Compos Part B. 2023;266:111018. https://doi.org/10.1016/j.compositesb.2023.111018.

Xiao RZ, Zeng ZW, Zhou GL, Wang JJ, Li FZ, Wang AM. Recent advances in PEG–PLA block copolymer nanoparticles. Int J Nanomedicine. 2010;5:1057–65. https://doi.org/10.2147/IJN.S14912.

Xu L, Wang X, Yang G, Zhao Z, Weng Y, Li Y, Liu Y, Zhao C-X. Development of a concentration-controlled sequential nanoprecipitation for making lipid nanoparticles with high drug loading. Aggregate. 2023;4:e369. https://doi.org/10.1002/agt2.369.

Xu X, Zhao J, Wang M, Wang L, Yang J. 3D printed polyvinyl alcohol tablets with multiple release profiles. Sci Rep. 2019;9:12487. https://doi.org/10.1038/s41598-019-48921-8.

Yasmin R, Shah M, Khan SA, Ali R. Gelatin nanoparticles: a potential candidate for medical applications. Nanotechnol Rev. 2017;6:191–207. https://doi.org/10.1515/ntrev-2016-0009.

Zangabad PS, Mirkiani S, Shahsavari S, Masoudi B, Masroor M, Hamed H, Jafari Z, Taghipour YD, Hashemi H, Karimi M, Hamblin MR. Stimulus-responsive liposomes as smart nanoplatforms for drug delivery applications. Nanotechnol Rev. 2018;7:95–122. https://doi.org/10.1515/ntrev-2017-0154.

Zhang L, Zhu D, Dong X, Sun H, Song C, Wang C, Kong D. Folate-modified lipid–polymer hybrid nanoparticles for targeted paclitaxel delivery. Int J Nanomedicine. 2015;10:2101–14. https://doi.org/10.2147/IJN.S77667.

Zhang L, Zhang L. Lipid–polymer hybrid nanoparticles: synthesis, characterization and applications. Nano LIFE. 2010;01:163–73. https://doi.org/10.1142/S179398441000016X.

Zhang Y, Huang Y, Li S. Polymeric micelles: nanocarriers for cancer-targeted drug delivery. AAPS PharmSciTech. 2014;15:862–71. https://doi.org/10.1208/s12249-014-0113-z.

Zhang Z, Feng Z, Zhao X, Jean D, Yu Z, Chapman ER. Functionalization and higher-order organization of liposomes with DNA nanostructures. Nat Commun. 2023;14:5256. https://doi.org/10.1038/s41467-023-41013-2.

Zhao C, Zhou B. Polyethyleneimine-based drug delivery systems for cancer theranostics. J Funct Biomater. 2022;14:12. https://doi.org/10.3390/jfb14010012.

Zheng M, Yue C, Ma Y, Gong P, Zhao P, Zheng C, Sheng Z, Zhang P, Wang Z, Cai L. Single-step assembly of DOX/ICG loaded lipid–polymer nanoparticles for highly effective chemo-photothermal combination therapy. ACS Nano. 2013;7:2056–67. https://doi.org/10.1021/nn400334y.

Zielińska A, Carreiró F, Oliveira AM, Neves A, Pires B, Venkatesh DN, Durazzo A, Lucarini M, Eder P, Silva AM, Santini A, Souto EB. Polymeric nanoparticles: production, characterization. Toxicol Ecotoxicol Molecul. 2020;25:3731. https://doi.org/10.3390/molecules25163731.

Part III

Inorganic/Organic Core shell Nanoconstructs

12 Magnetic/Organic Core-Shell Nanoconstructs for Cancer Theranostic

Gaurisha Alias Resha Ramnath Naik, S. P. Rachana, Amrita Arup Roy, Rahul Pokale, Viola Colaco, Anoushka Mukharya, Ritu Kudarha, Srinivas Mutalik, Namdev Dhas, and Deepanjan Datta

Abstract

Magnetic-organic core-shell nanoconstructs, which integrate therapeutic and diagnostic functions onto a sole platform, provide a revolutionary approach to cancer theranostics. These nanoparticles, which have an organic shell for improved biocompatibility and targeted drug delivery and a magnetic core for enhanced imaging capabilities, have shown significant potential to improve treatment precision through real-time therapeutic efficacy monitoring while minimizing systemic toxicity. Their synthesis involves a variety of techniques that could be tuned at one end to attain important properties such as size and surface charge and, on the other end, functionalization for efficient tumor targeting and drug release. Different magnetic core materials like metals and their oxides offer unique properties for imaging (e.g., MRI) and treatment applications (e.g., hyperthermia). Surface functionalization for the attachment of targeting ligands that enhance delivery specificity complements the enhanced stability and functionality conferred by the organic shell materials. In such a way, nanoconstructs are a hybrid material that already suggests many different applications—from chemotherapeutic, phototherapeutic, and ultrasound-responsive strategies—the promise they hold in revolutionizing both cancer diagnosis and therapy through new design and synthesis methods.

G. A. R. R. Naik · S. P. Rachana · A. A. Roy · R. Pokale · V. Colaco · A. Mukharya
R. Kudarha · S. Mutalik · N. Dhas · D. Datta (✉)
Department of Pharmaceutics, Manipal College of Pharmaceutical Sciences, Manipal Academy of Higher Education (MAHE), Manipal, Udupi, India
e-mail: deepanjan.datta@manipal.edu

J. K. Patel et al. (eds.), *Core-Shell Nano Constructs for Cancer Theragnostic*,
https://doi.org/10.1007/978-981-96-3025-7_12

Keywords

Cancer · Core-shell nanoparticles · Metallic core · Organic shell · Theranostic nanoparticles

1 Introduction

Cancer is a major global health concern that necessitates the development of novel therapeutic and diagnostic strategies. The combination of targeted delivery and improved imaging capabilities along with strong therapeutic effects makes metal-organic core-shell nanoparticles a promising platform for cancer theranostics (Fernandes 2023). These nanoparticles are generally comprised of organic shells that are often engineered to have particular characteristics like stability, controlled release, and biocompatibility, encapsulating a metal core usually made of gold, silver, or iron (Dhas et al. 2018). The organic shell provides a flexible and adaptable surface for interaction with biological processes, while the metal core of the nanoparticle provides a strong and solid base.

Metal-core organic-shell nanoparticles are an impressive tool for cancer theranostics due to their distinct features. They are developed to target specific cancer cells, which makes it possible to administer pharmaceuticals precisely and improve imaging capabilities. Additionally, they can be functionalized with other biomolecules to improve targeted specificity and lessen adverse effects, such as peptides or antibodies (Zhao et al. 2016). Moreover, it is desirable to construct the organic shell such that medicinal chemicals are released in a controlled manner, reducing systemic toxicity and optimizing therapeutic effectiveness (Mishra et al. 2023). Metal-organic core-shell nanoparticles have demonstrated an abundance of promise in the field of cancer diagnosis and treatment. They are relevant for noninvasive tumor progression and therapy response monitoring using imaging modalities such as positron emission tomography (PET), computed tomography (CT), and magnetic resonance imaging (MRI) (Saeb et al. 2021; Wang et al. 2018). Additionally, these nanoparticles are often designed to deliver various therapeutics, including chemotherapeutic drugs, gene therapy vectors, and immunomodulatory agents, to enhance cancer treatment outcomes.

Metal-organic core-shell nanoparticles are a versatile platform that can be tailored for targeted, controlled, and combination cancer therapy while also enabling multimodal imaging and overcoming drug resistance (Dhas et al. 2018; Kashyap et al. 2023). Their unique properties make them a promising approach to improving cancer treatment outcomes. Metal-organic core-shell nanoparticles offer several key benefits for the treatment of cancer:

1. Targeted drug delivery: The nanoparticles target cancer cells by functionalizing the organic shell with biomolecules like antibodies or peptides. This enhances specificity and reduces off-target effects.

2. Controlled drug release: The organic shell is engineered to release the drug in a controlled manner in response to tumor microenvironment stimuli like acidic pH or high hydrogen peroxide levels. This minimizes systemic toxicity and maximizes therapeutic efficacy.
3. Synergistic combination therapy: The core-shell structure allows the co-delivery of multiple therapeutic agents like chemotherapeutic drugs and photosensitizers. This enables synergistic combination therapy, which is more effective than single agents.
4. Multimodal imaging: The metal core can be used for various imaging modalities like MRI, CT, and PET, enabling noninvasive monitoring of tumor progression and treatment response.
5. Overcoming drug resistance: Nanoparticle-based drug delivery can help overcome multidrug resistance, a major challenge in cancer treatment.
6. Biocompatibility and stability: The nanoparticles can be designed to be biocompatible and have enhanced stability compared to free drugs.

We aim to provide a comprehensive overview of the current state of metal/organic core-shell nanoparticles for cancer theranostics, highlighting their types, method of fabrication, surface functionalization, and applications in the diagnosis and treatment of cancer. It will confer challenges and future directions for developing these nanoparticles, emphasizing their ability to transform cancer management and enhance outcomes for patients.

2 Methods of Synthesis

The selection of the fabrication method depends on the desired properties of the metal/organic core/shell nanoparticles, such as size, shape, composition, and stability. Each method has its advantages and disadvantages, and researchers often combine multiple methods to achieve the desired nanoparticle properties (Büyüktiryaki et al. 2022). Table 12.1 summarizes the key methods for synthesizing metal-organic core-shell nanoparticles.

3 Types of Magnetic Core Materials and Their Properties

3.1 Magnetic Core

Magnetic core-shell nanoparticles (MNPs) have garnered substantial interest in recent years due to their distinctive properties and wide-ranging applications. These nanoparticles comprise a magnetic core, often made of gold, silver, or platinum, encapsulated within a shell material such as iron oxide, cobalt ferrite, or manganese oxides. The shell serves to protect the core from oxidation and enhance biocompatibility, making these nanoparticles suitable for various biomedical applications (López-Ortega 2023). Recent advancements have focused on synthesizing

Table 12.1 Key methods for synthesizing metal-organic core-shell nanoparticles with advantages and disadvantages

Method	Description	Advantages	Disadvantages
Micro-emulsion synthesis	Uses micro-emulsion to create a stable environment for nanoparticle synthesis. Metal core formed first, then organic shell deposited.	Produces nanoparticles with narrow size distribution.	Requires careful control of micro-emulsion composition and conditions.
Hydrothermal synthesis	Uses high temperatures and pressures to synthesize nanoparticles. Metal core formed first, then organic shell deposited through hydrothermal reaction.	Allows for synthesis of nanoparticles with specific shapes and sizes.	Requires specialized equipment and careful control of temperature and pressure.
Solvothermal synthesis	Similar to hydrothermal synthesis but uses a solvent instead of water.	Allows for fabrication of nanoparticles with specific shapes and sizes.	Requires careful control of solvent composition and conditions.
Thermal decomposition	Involves thermal decomposition of a metal precursor to form the metal core, which is then coated with an organic shell.	Produces nanoparticles with high crystallinity and narrow size distribution.	Requires high temperatures and careful control of reaction conditions.
Co-precipitation	Involves the simultaneous precipitation of the metal core and the organic shell from a solution.	Simple and cost-effective method.	Produces nanoparticles with a wide size distribution.
Seed-mediated growth	Uses pre-formed nanoparticles (seeds) as a template for the growth of the metal core and the organic shell.	Allows for the synthesis of nanoparticles with specific shapes and sizes.	Requires careful control of seed composition and growth conditions.
Layer-by-layer methods	Involves the sequential deposition of layers of the metal core and the organic shell through various techniques such as electrochemical deposition or chemical vapor deposition.	Allows for precise control over the composition and structure of the nanoparticles.	Requires specialized equipment and multiple deposition steps.
Laser ablation	Uses a laser to ablate a metal target, creating nanoparticles that are then coated with an organic shell.	Produces nanoparticles with high purity and minimal impurities.	Requires specialized equipment and careful control of laser parameters.
Mussel-inspired Polydopamine coating	Uses mussel-inspired polydopamine (PDA) to coat the metal core, providing a versatile strategy for constructing well-defined core-shell nanohybrids.	Simple, efficient, and environmentally friendly method.	Requires careful control of PDA coating conditions.

core-shell nanoparticles with specific features, such as combining magnetostrictive and piezoelectric phases to produce a solid magnetoelectric effect. The property of this material has great potential for utilization in biomedical applications such as tissue imaging, drug delivery, and therapeutics (Fiocchi et al. 2022). The MNPS are versatile for therapeutics such as targeted drug delivery and the magnetic hyperthymia of cancer treatment. These nanoparticles are crucial for cancer theranostics because they provide a multifunctional platform for both diagnosis and therapy. Because of their distinct physicochemical properties, synergistic effects, and interfacial heterojunctions of MNPs contribute to their excellent performance in cancer therapy (Baldea et al. 2023). MNPs have various therapeutic applications, including drug delivery, carrying out bioactive substances into cells, and utilizing magnetic hyperthermia in cancer treatment (Bian et al. 2021). Furthermore, The application of magnetic nanoparticles in cancer diagnosis and therapy has shown promising potential, enhancing diagnostic imaging sensitivities and accuracy while reducing the adverse effects of chemotherapy drugs (Adamiano et al. 2018).

3.2 Types of Magnetic Core Nanoparticles

Magnetic core nanoparticles consist of a magnetic metal or metal oxide core surrounded by a protective shell layer. The core material determines the magnetic properties of the material, while the shell enhances stability, functionality, and adaptability for various applications. These nanoparticles can broadly be classified into two main categories: Class I, which includes metallic core nanoparticles, and Class II, which includes metallic oxide core nanoparticles.

3.2.1 Class I: Metal Core

Metallic core nanoparticles, such as gold, silver, platinum, palladium, and copper, depicted in Fig. 12.1, exhibit unique properties like localized surface plasmon resonance, making them suitable for cancer theranostics. These nanoparticles can generate reactive oxygen species like hydroxyl radicals and singlet oxygen, crucial for

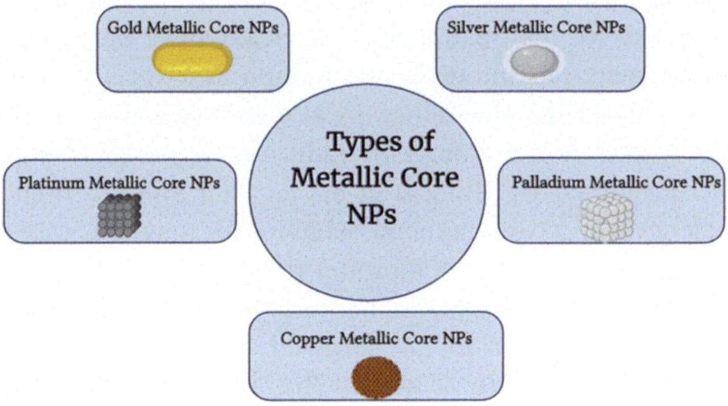

Fig. 12.1 Types of metallic core nanoparticles

inducing cell apoptosis in the tumor microenvironment. Core-shell nanoparticles, particularly those with noble metal cores or shells, are extensively researched for targeted photodynamic therapy due to their biocompatibility and tunable surface plasmonic resonance properties (Matlou and Abrahamse 2021). Metal-based nanoparticles have shown significant potential in cancer therapy by enhancing radiosensitivity and acting as tumor-selective radiosensitizers, which could have clinical applications in radiotherapy (Kumar et al. 2023b). Furthermore, an innovative approach to cancer diagnosis and treatment has been provided by synthesizing core-shell plasmonic nanomaterials with effective chemo-dynamic properties and dual-mode imaging capabilities. Due to their unique properties, metallic core nanoparticles play a crucial role in cancer theranostics. These nanoparticles, such as gold nanoparticles (AuNPs) and silver nanoparticles (AgNPs), offer advantages like high surface area to volume ratio, surface plasmon resonance, and light scattering, making them ideal for various applications in cancer diagnosis and therapeutics (Gupta et al. n.d.). Gold nanoparticles are preferred for their nontoxic nature, plasmonic properties, and ability to facilitate real-time monitoring and noninvasive imaging of diseases, thereby enhancing early detection and treatment precision (Kumar et al. 2023a). Additionally, metallic nanoparticles have been investigated for their surface chemistry and shape to integrate cancer diagnostics with treatment techniques, allowing for precise imaging, effective treatment delivery, and real-time therapeutic efficacy monitoring (Rabaan et al. 2022). The optical properties of metallic nanoparticles, such as gold nanoparticles, can be leveraged to induce cell apoptosis by generating reactive oxygen species, making them valuable for cancer treatment (Kumar et al. 2023b). Several types of metallic core materials nanoparticles used in cancer theranostics are described below.

3.2.1.1 Gold Metallic Core Nanoparticles (AuNPs)

Gold metallic core nanoparticles have garnered significant attention due to their distinctive properties, including variable surface chemistry, strong biocompatibility, and distinct optical properties. AuNPs are widely used in cancer theranostics due to their excellent biocompatibility, ease of synthesis, and ability to functionalize targeted drug delivery and imaging. They can be used in the targeted delivery of drugs, photodynamic therapy, and photothermal therapy (Hu et al. 2017a). AuNPs serve as beneficial therapeutics for the treatment of cervical cancer because their combination with magnetic materials allows accumulation at a specific region utilizing a magnetic field for efficient localized therapy at tumor sites. Hence, Fe_3O_4@Au NPs with remarkable properties for synergistic radio-photothermal therapy against cervical cancer were reported by Rui Hu et al., in their studies: high photothermal conversion efficiency, superparamagnetic behavior for magnetic targeting, and synergistic therapeutic effects. When combining radiotherapy and photothermal therapy, it is enhanced by an external magnetic field. Fe_3O_4@Au NPs show promise as a multifunctional nano platform for the targeted and effective radio-photothermal treatment of cervical cancer (Hu et al. 2017b). The core-shell structure of gold-coated magnetic nanoparticles offers enhanced stability, superior biocompatibility, and surface reactivity, making them stand out among other nanoparticles. When radiation and photothermal therapy are combined with an external magnetic field, Fe_3O_4@Au NPs demonstrate excellent photothermal conversion efficiency, superparamagnetic behavior for magnetic targeting, and synergistic effects that increase their therapeutic benefits (Robinson et al. 2010).

AuNPs have emerged as potential tools for cancer treatment due to their unique properties. They can functionalize with various molecules, such as antibodies or targeting ligands, to specifically bind to cancer cells and deliver therapeutic agents directly to them. These nanoparticles can also enhance the efficacy of chemotherapy drugs by increasing their uptake in cancer cells and reducing their toxicity to healthy cells. Additionally, AuNPs can act as photothermal agents, converting light energy into heat, which can be used to selectively destroy cancer cells while sparing healthy tissue (Zhang 2020).

Surface modifications enable gold nanoparticles (AuNPs) to selectively target cancer cells while sparing healthy tissues, enhancing treatment efficacy, and minimizing side effects. AuNPs can induce programmed cell death (apoptosis) in cancer cells, inhibiting their growth and spread. Enhanced drug delivery facilitated by AuNPs can improve the bioavailability and distribution of chemotherapeutic agents within tumor tissues, enhancing treatment outcomes. Studies have shown the effectiveness of AuNPs in reducing tumor growth and increasing cancer cell death, making them promising agents for cancer therapy (Yaqoob et al. 2020). Compared to conventional treatments, AuNPs offer reduced toxicity to healthy tissues, minimizing adverse effects. This improved safety profile enhances patient quality of life during treatment.

AuNPs pave the way for personalized cancer medicine by enabling tailored therapies specific to individual patients. Their versatility allows for targeted and precise treatment delivery. However, regulatory hurdles and approval processes challenge the clinical translation of gold nanoparticles (Abed et al. 2019). Addressing these barriers is crucial to realizing their therapeutic potential in cancer treatment. Long-term safety and efficacy data are essential to ascertain the stability and performance of gold nanoparticles over extended periods. Robust research is needed to ensure their clinical viability.

3.2.1.2 Silver Metallic Core Nanoparticles (AgNPs)

Silver metallic core nanoparticles (AgNPs) are nanoparticles composed of silver with a metallic core structure, typically ranging from 1 to 100 nanometres in size. These nanoparticles exhibit distinct physicochemical properties that are beneficial in various fields, including drug delivery, antimicrobial applications, imaging technology, and coatings for biomedical devices (Zhang et al. 2016). Due to their distinct physical, chemical, and optical properties, gold nanoparticles are attractive for various biomedical applications. They possess antimicrobial and anticancer properties, which can be attributed to the release of silver ions or the formation of radical species after they are taken up by living cells or bacteria (Aksoy et al. 2023). AgNPs have been synthesized through various methods, including biological, chemical, and physical processes, and have been utilized in various medical applications, including antimicrobial and anticancer therapy, wound repair, bone healing, vaccine adjuvants, and eco-friendly approaches. They display antimicrobial properties, making them effective against various animal pathogens like *Staphylococcus aureus*, *Escherichia coli*, and *Bacillus abortus*, potentially aiding in combating antibiotic resistance (Villalobos Gutiérrez et al. 2023). Studies have additionally examined altering the surface of AgNPs and mixing them with other metals and biomaterials to enhance their specific functions and minimize their cytotoxicity. These developments may result in a more excellent range of AgNPs employed in therapeutic and industrial applications. One research conducted by Kabir et al. synthesized AgNPs with advantageous properties for biomedical purposes, including antimicrobial and anticancer applications. Their synthesis involves reducing silver ions via diverse methods, including plant extract utilization. AgNPs demonstrate anticancer effects by triggering apoptosis, generating reactive oxygen species (ROS), and causing DNA damage in cancer cells (Kabir et al. 2023).

Moreover, they serve as adjuvants for immunizations and antidiabetic therapies and stimulate bone regeneration and wound healing. AgNPs have demonstrated efficacy in the treatment of wound infections, diabetes, and various malignancies, including cervical, breast, lung, and colon cancer. Current research suggests that gold nanoparticles are effective and well-tolerated in individuals despite concerns about their toxicity. However, further investigation is necessary to assess their biological safety profile (Abdel-Fattah and Ali 2018).

The potential of AgNPs to generate reactive oxygen species, inhibit cancer cell proliferation, cause cell death, and inhibit tumor growth has been extensively studied. AgNPs exhibit antiangiogenic properties by preventing VEGF-induced angiogenesis in both in vitro and in vivo conditions. Additionally, AgNPs have been shown to alter P-glycoprotein (Pgp) activity, enhancing chemotherapeutic efficiency against drug-resistant cancer cells. This highlights the remarkable potential of AgNPs as combinational partners. Furthermore, the production of double-stranded DNA breaks and chromosomal instability, which triggers the start of apoptotic execution, contributes to the genotoxicity of AgNPs. This mode of action implies that AgNPs and a broad range of DNA-targeting cancer treatments may be mutually beneficial (Abdel-Fattah and Ali 2018). These nanoparticles are remarkable due to their ability to target specific body areas, possess an elevated surface area-to-volume ratio, and exhibit biocompatibility with living tissues. These characteristics enable them to effectively transport cancer drugs directly to tumor sites, minimize side effects, and enhance drug stability. AgNPs produced from natural sources, such as the leaf extract of *Artocarpus heterophyllus*, have shown promising results in eliminating cancer cells, indicating their potential for cancer treatment.

3.2.1.3 Platinum Metallic Core Nanoparticles (PtNPs)

Platinum metallic core nanoparticles (PtNPs) are a type of metallic core-containing platinum nanoparticle. They possess several unique properties that make them valuable in cancer theranostics. PtNPs have been shown to have antioxidant properties, which can help reduce oxidative stress and damage caused by free radicals in cancer cells. Furthermore, PtNPs are cytotoxic to various cancer cells, including breast, lung, and ovarian cancer cells. They can also be functionalized with targeting ligands to enhance their ability to target specific cancer cells and tissues. PtNPs have emerged as promising agents in cancer theranostics, offering a multifaceted approach to cancer treatment. PtNPs can be functionalized with biomolecules like antibodies, DNA, RNA, and proteins to enhance their anticancer effects through ROS generation, apoptosis induction, and photothermal effects (Villalobos Gutiérrez et al. 2023).

Additionally, PtNPs have proven to be effective at inhibiting the proliferation of tumor cells in sonodynamic therapy by acting as sonosensitizers through cavitation and the production of reactive oxygen species (ROS) without causing hyperthermia (Zahraie et al. 2023). These unique physicochemical properties of PtNPs make them versatile for various applications, including nanomedicine, nanocatalysis, and environmental remediation, showcasing their potential in cancer therapy and photothermal treatments (Yerpude et al. 2023).

Furthermore, PtNPs and other metallic core nanoparticles have been explored for their theranostics capabilities, combining diagnostic and therapeutic functions for personalized cancer treatment, focusing on their unique properties like X-ray attenuation and near-infrared activity. Targeted delivery and real-time monitoring can be achieved by encapsulating Pt drugs within polymeric nanosystems, enhancing antitumor efficacy and enabling imaging-guided cancer therapy (Liu et al. 2023). According to Almarzoug et al., PtNPs are effective in cancer therapy due to their biocompatibility, photothermal antitumor performance, and ability to enhance radiotherapy. These nanoplatforms can be designed to target specific cancer cells and tissues, and their surface chemistry can be modified to improve their therapeutic efficacy. Additionally, PtNPs exhibit excellent electrocatalytic activity, which can be leveraged to mediate innovative electrodynamic therapy.

PtNPs are thermally stable, making them ideal for photothermal therapy. Additionally, they are biocompatible and nontoxic at therapeutically relevant concentrations; PtNPs have shown significant promise in combating cancer by inducing apoptosis in cancer cells. These nanoparticles can trigger programmed cell death by causing DNA damage, disrupting mitochondrial function, increasing the levels of proteins that promote apoptosis, and activating specific enzymes such as caspase 3 (Almarzoug et al. 2020). Furthermore, platinum nanoparticles can make cancer cells more sensitive to radiation therapy when combined with palladium, thereby enhancing the cell-killing effects (Klebowski et al. 2022). Overall, the activation of caspase-3 plays a crucial role in the process of cancer cell death caused by platinum nanoparticles, making them promising candidates for developing effective anticancer treatments.

3.2.1.4 Palladium Metallic Core Nanoparticles (PdNPs)

Palladium nanoparticles (PdNPs) have garnered significant attention due to their unique properties and wide applications, including in catalysis, sensing, and medicine. PdNPs are used in cancer theranostics due to their ability to deliver medications and ligand targeting. They can used for photothermal therapy and targeted drug delivery (Joudeh et al. 2022). These metallic nanoparticles can be functionalized with various molecules for enhanced targeted drug delivery and imaging applications. PdNPs are available in various forms, including nanoparticles, nanoshells, and nanorods, offering promising cancer diagnosis and treatment opportunities. Numerous synthesis methods have been explored to efficiently produce PdNPs, with environmentally friendly approaches, such as using phytochemicals for biogenic synthesis, gaining attention. Because of its unique properties and versatility, Pd has attracted much attention from researchers who are applying it to fabricate magnetic materials. Pd is a platinum group metal with remarkable corrosion resistance, thermal stability, and catalytic activity, which makes it a desirable option for many applications Vedyagin et al., The incorporation of palladium into magnetic materials, such as alloys and composites, has demonstrated the potential to enhance their performance and expand their applications in various industries.

PdNPs possess distinct properties that make them suitable for cancer theranostics. For photothermal therapy, they efficiently absorb near-infrared (NIR) light and convert it into heat. They hold significant potential for cancer therapy due to their biocompatibility. Additionally, PdNPs serve as contrast agents in imaging modalities such as MRI, CT, and PA imaging. They enable precise drug delivery to cancer cells, enhancing the effectiveness of therapies like radiation and chemotherapy (Liu et al. 2020).

3.2.1.5 Copper Metallic Core Nanoparticles (CuNPs)

Copper metallic core nanoparticles are implemented in cancer theranostics due to their unique physicochemical properties, including optical, thermal, and electrical conductivity. They can be applied to targeted delivery of drugs and photothermal therapy. Due to their multifunctional properties, CuNPs have emerged as promising agents in cancer theranostics. CuNPs can act as antioxidants, antimicrobials, and anticancer agents, inducing apoptosis, inhibiting cell proliferation, and preventing tumor angiogenesis (Tyagi et al. 2023). These nanoparticles release labile copper pools that trigger antioxidant responses and metal homeostasis alterations within cells, influencing cancer therapy outcomes. The development and application of Cu-based nanomaterials have significantly advanced cancer diagnosis and therapy, with Cu playing a crucial role in physiological metabolism and cancer development (Bonet-Aletá et al. 2023). Furthermore, the unique properties of copper, such as its involvement in oxidative stress and energy metabolism, make it a valuable element in cancer diagnostics and antitumor therapy, potentially modulating cancer cell survival through various mechanisms. Additionally, utilizing copper clusters as carriers in nano drug delivery systems can enhance the combined effect of chemo/chemo dynamic/photodynamic therapy on cancer, demonstrating excellent biosafety and tumor suppression capabilities (Aishajiang et al. 2023).

CuNPs have emerged as a promising therapeutic strategy in cancer treatment due to their distinct characteristics that facilitate the selective targeting and elimination of tumor cells. These nanoparticles comprise a copper-centered core enveloped by a shell material, typically a polymer or silica, enabling effective drug loading, precise delivery, and improved compatibility with biological systems. The magnetic properties of the copper core contribute to targeted delivery and inherent anticancer capabilities, exploiting copper's essential role as a trace element capable of triggering oxidative stress and apoptosis in cancerous cells. Moreover, the core-shell configuration allows for integrating therapeutic substances like chemotherapy drugs or photosensitizers, offering a comprehensive treatment approach for enhanced efficacy. The magnetic characteristics of the copper core empower the nanoparticles to be directed and accumulated at the tumor location through an external magnetic field. CuNPs have shown great potential in improving cancer treatment and diagnosis. They work in various ways, such as causing oxidative stress, which leads to increased production of reactive oxygen species (ROS) and cell death in cancer cells (Ghasemi et al. 2022). CuNPs can also be used to deliver drugs for cancer therapy, increasing the effectiveness of chemotherapy and other treatments.

Additionally, they can induce a specific type of cell death called cuproptosis in cancer cells, which helps suppress tumors (Zhang et al. 2023). By combining with other types of cell death, like ferroptosis and apoptosis, CuNPs offer a promising approach to treating cancer. Xu and colleagues developed CuNPs loaded with a substance that inhibits copper chaperones, which can enhance the effects of chemotherapy by increasing copper levels inside cells and promoting ROS production, ultimately improving cancer treatment and stimulating an immune response against cancer cells (Xu et al. 2023).

3.2.2 Class II: Metal Oxide Core

Metal oxide core materials depicted in Fig. 12.2, specifically ferrites, have garnered significant attention due to their multifunctional properties from core-shell nanoparticles. Ferrites, a class of magnetic oxides, exhibit remarkable structural, electrical, and magnetic properties, making them valuable in various technological applications such as electronics, biomedicine, and energy systems. These materials are

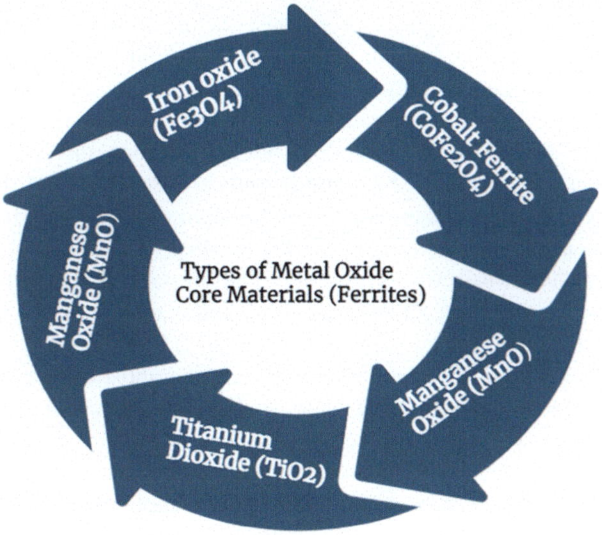

Fig. 12.2 Types of metal oxide core materials (ferrites)

prepared using diverse methods, ranging from solid-state reactions to advanced wet chemical processes, impacting their properties significantly (Mélinon et al. 2014). Metal oxide nanoparticles are widely used in various applications due to their unique properties and potential for tailoring their properties through size and shape control. Here are some examples of metal oxide nanoparticles and their applications. Metal oxide core materials, specifically ferrites, have been extensively studied for their potential in cancer theranostics (Mokhosi et al. 2022). Ferrites comprise ferric oxide and one or more oxides of other metals, such as manganese, copper, iron, cobalt, or titanium. These nanomaterials exhibit unique magnetic properties and high safety for human use, making them promising candidates for various cancer therapies. Ferrite nanoparticles can catalyze the Fenton reaction, producing reactive oxygen species (ROS) that induce the death of tumor cells. The acidic microenvironment of tumors enhances this mechanism and can be further improved by combining ferrite nanoparticles with drugs or other energy fields (Garcia-Muñoz et al. 2020). Ferrite nanoparticles can be activated by an external alternating magnetic field, generating heat that can used to kill cancer cells. The nanoparticles' specific absorption rate (SAR) is a crucial factor in determining their effectiveness for this magnetic hyperthermia application. Ferrite nanoparticles can be functionalized with agents or coatings to enhance their retention in affected tissues and serve as magnetic resonance imaging (MRI) contrast agents. This allows for targeted delivery of anticancer agents and the monitoring of treatment efficacy (McHenry and Laughlin 2022). Overall, metal oxide core nanoparticles present a potential avenue for targeted cancer treatment through selective cell interaction and efficient drug delivery mechanisms.

3.2.2.1 Iron Oxide (Fe_3O_4)

Iron oxide (Fe_3O_4), also known as magnetite, is a highly researched material due to its unique combination of magnetic properties, chemical inertness, and biocompatibility. These characteristics make it an attractive material for various applications, including biomedical and technological uses. Different applications of magnetic Fe_3O_4 material have been incorporated into various practices like magnetic resonance imaging (MRI), anti-radar material, cancer treatment, and drug delivery systems (Pourmadadi et al. 2022). Iron compounds are metallic, and the type of its crystal structure determines its properties, including magnetic properties. Iron oxide compounds include magnetite (Fe_3O_4), hematite (Fe_2O_3), and maghemite (Fe_2O_3), which are composed of iron with different oxidation states. Due to their high magnetic moment and stability, Fe_3O_4 NPs are employed in magnetic utilization, such as magnetic relaxation imaging, storage, and retrieval gadgets (Zhang et al. n.d.).

Fe_3O_4 has shown significant potential in cancer theranostics due to its magnetic properties and biocompatibility due to its multifunctional capabilities, including magnetic resonance imaging, drug delivery, hyperthermia treatment, and immunotherapy (Revathy et al. 2023). These nanoparticles can be synthesized through various methods and easily surface-modified for specific applications, such as targeting overexpressed biomarkers in breast cancer for precise drug delivery and imaging. The biological effects of Fe_2O_4 are attributed to their core iron oxide properties inducing reactive oxygen species, modulating cellular metabolism, and their coatings interacting with cell receptors, offering innovative therapeutic opportunities (Halder et al. 2022). In this case, by regulating the size distribution of Fe_3O_4 NPs, we can achieve maximum magnetic resonance imaging of different modes to improve their capabilities in cancer diagnosis (Zhao et al. 2020). Overall, using iron oxide nanoparticles in cancer theranostics offers a beneficial pathway toward improving novel and individualized therapy for cancer diseases. Cancerous lesions are essential in public health, and there is a tremendous effort to develop an excellent nano platform for proper diagnosis and treatment.

3.2.2.2 Cobalt Ferrite ($CoFe_2O_4$)

Cobalt ferrite ($CoFe_2O_4$) metal oxide core materials have been widely investigated for their use as magnetic nanoparticles in cancer theranostics. The size of cobalt ferrite nanoparticles can tailored by the synthesis method, which affects their magnetic properties and biological performance (Barani et al. 2022). Smaller nanoparticles with a narrower distribution exhibit improved cellular uptake and enhanced anticancer activity compared to their larger counterparts. Cobalt ferrite nanoparticles' unique structural, magnetic, and physicochemical properties make them highly versatile and promising candidates for cancer theranostics applications, including targeted drug delivery, magnetic hyperthermia, and MRI contrast enhancement.

In their studies, Rotunjanu et al., provided that within the course of the investigations, Cobalt ferrite ($CoFe_2O_4$) nanoparticles are promising in cancer theranostics because of magnetic as well as structure characteristics. Recent studies report that doping other elements, such as dysprosium (Dy), increases the anticancer property of the $CoFe_2O_4$ nanoparticles through the mitochondria apoptotic cancer cell death pathway. These nanoparticles possess considerable antiproliferative activity and are accompanied by high selectivity. They can thereby become efficient agents against cancer cells. Moreover, they have been used as drug delivery systems and as a suitable candidate in magnetic hyperthermia cancer therapy (Rotunjanu et al. 2023). The synthesis methods employed can result in nanoparticles with varying lattice constants, internal stresses, and cation inversion values, affecting their overall properties (Khalili Najafabad et al. 2023). Cobalt ferrite nanoparticles are highly versatile for various applications, including magnetic hyperthermia and as drug delivery vehicles for cancer therapy.

3.2.2.3 Manganese Oxide (MnO)

Manganese oxide (MnO) metal oxide core nanoparticles have been extensively studied for their potential in cancer theranostics due to their unique properties and applications. These nanoparticles can significantly amplify magnetic resonance imaging (MRI) signals at the tumor site, enhancing the sensitivity of cancer recognition and tumor margin detection. They can also be used as contrast agents for MRI, improving the visibility of cancer cells and tissues (Cai et al. 2019). Additionally, MnO nanoparticles are biocompatible, making them a promising platform for cancer diagnosis and treatment.

MnO nanoparticles can be synthesized using various methods and functionalized with targeting moieties to deliver therapeutic agents specifically to cancer cells. They have been used for multimodal imaging, including MRI, photoacoustic imaging (PAI), and near-infrared fluorescence imaging (NIR-FL), allowing for more comprehensive and accurate cancer diagnosis (Zhu et al. 2020). In cancer theranostics, MnO metal oxide core materials have been used for magnetic targeting, allowing for targeted delivery of therapeutic agents to specific cancer cells or tissues. They have also been used for magnetic hyperthermia, generating heat in cancer cells to enhance the efficacy of chemotherapy and radiation therapy. The unique properties of MnO metal oxide core materials, such as their high relaxation efficiency and rapid water proton exchange rate, make them an attractive platform for cancer diagnosis and treatment (Yang et al. 2017). MnO metal oxide core materials have been explored for their potential in cancer theranostics due to their unique properties and applications. They are biocompatible for MRI, targeted drug delivery, and magnetic hyperthermia (Mohammadi et al. 2023) and the versatility and potential of MnO metal oxide core materials make them a promising platform for cancer diagnosis and treatment.

3.2.2.4 Titanium Dioxide (TiO_2)

Titanium dioxide (TiO_2) metal oxide core nanoparticles have been extensively studied for their potential in cancer theranostics due to their unique properties and applications. These nanoparticles have been used in various biomedical applications, including cancer diagnosis and treatment, due to their biocompatibility, lightness, high corrosion resistance, high thermal stability, low ion release, and nonmagnetic properties. They can be used as contrast agents for magnetic resonance imaging (MRI), enhancing the visibility of cancer cells and tissues (Raja et al. 2020). Additionally, TiO_2 metal oxide core materials have been shown to have antimicrobial effects and antitumor properties and can be used for targeted drug delivery and photodynamic therapy. TiO_2 groups have been investigated based on density functional theory (DFT), which reveals unique properties such as coordination of the titanium atom and variations in the electronic structure. Furthermore, TiO_2 nanoparticles have been used for multimodal imaging, including MRI, photoacoustic imaging (PAI), and near-infrared fluorescence imaging (NIR-FL), allowing for more comprehensive and accurate cancer diagnosis (Kawassaki et al. 2021).

Titanium dioxide (TiO_2) metal oxide core nanoparticles possess unique properties that make them highly suitable for cancer theranostics. These properties include their biocompatibility, low cytotoxicity, and photocatalytic properties, which enable them to interact with biological systems and induce apoptosis in cancer cells. TiO2 nanoparticles can be used as a drug delivery vehicle, allowing for targeted delivery of therapeutic agents to cancer cells (Dai et al. 2020). Additionally, their ability to generate reactive oxygen species (ROS) and singlet oxygen ($^1O_2$) makes them practical for photodynamic and sonodynamic therapy. Furthermore, TiO2 nanoparticles can be used as a contrast agent for magnetic resonance imaging (MRI) and computed tomography (CT), enhancing the visibility of cancer cells and tissues. Their inert nature and high stability make them suitable for long-term implantation and biodegradable applications (Rezaei et al. 2022). Overall, the unique properties of

TiO_2 metal oxide core nanoparticles make them a promising platform for cancer theranostics, offering a multifaceted approach to cancer diagnosis and treatment.

3.2.2.5 Zinc Oxide (ZnO)

Zinc oxide (ZnO) metal oxide core nanoparticles have been extensively studied for their potential in cancer theranostics due to their unique properties and applications. These nanoparticles exhibit selective cytotoxicity against tumor cells, primarily due to their enhanced intracellular release of dissolved Zn^{2+}, increased reactive oxygen species (ROS) production, and mitochondrial dysfunction. They can be used as theranostic agents for imaging of cancer cells and delivery of anticancer molecules, with the ability to accumulate in the tumor microenvironment and induce apoptosis (Mousa et al. 2023). Additionally, ZnO metal oxide core nanoparticles have been used for pH-dependent or pH-triggered oral and dermal drug delivery for skin diseases, leveraging their luminescent properties for theranostic and imaging applications. The biocompatibility, low toxicity, and inexpensive nature of ZnO nanoparticles make them a promising cancer diagnosis and treatment platform with potential applications in multimodal imaging, targeted drug delivery, and photodynamic therapy (Naser et al. 2023).

Due to their unique properties and applications, ZnO metal oxide core nanoparticles have been extensively studied for their potential in cancer theranostics. One of the main advantages of using ZnO NPs in cancer theranostics is their biocompatibility and biodegradability, which reduces the risk of toxicity and side effects (Bisht and Rayamajhi 2016). Additionally, ZnO NPs exhibit selective cytotoxicity against cancer cells, primarily due to their ability to generate reactive oxygen species (ROS) and induce apoptosis. This property makes them a promising platform for targeted drug delivery, as they can used as carriers for sustained and targeted delivery of anticancer drugs into tumor cells (Anjum et al. 2021). Furthermore, ZnO NPs have luminescent properties, making them suitable for multimodal imaging applications such as magnetic resonance imaging (MRI), photoacoustic imaging (PAI), and near-infrared fluorescence imaging (NIR-FL). The low cost and toxicity of ZnO NPs also make them a promising platform for cancer theranostics.

4 Shell Materials

In core-shell nanoparticles, the shell refers to the outer layer of a material that surrounds the core. This outer layer can be solid, continuous, multi-layered, or discontinuous and is typically made of a different material than the core (Mishra et al. 2023). The shell is crucial in modifying the core's properties and enhancing the nanoparticle's overall functionality. It can provide protection, stability, and control over the release of substances from the core, making core-shell nanoparticles useful in various applications such as drug delivery, bioimaging, and solar cells (Gopal and Sudarshan 2022).

The shell materials in core-shell nanoparticles are important due to their ability to provide specific functionalities like targeting, imaging, and controlled release of the therapeutic agents. The shell nanomaterials are chosen based on their biocompatibility, stability, and inability to interact with the target site. Common nanomaterials used for shells are classified into organic and inorganic (Dhas et al. 2018). Organic nanoparticles include polymeric nanoparticles, lipidic nanoparticles, micelles, and dendrimers, whereas inorganic nanoparticles are categorized into metal-based, metallic oxide-based, silica-based, quantum dots, carbon nanotubes, and graphene-based (Kumar et al. 2020). In this section, we will focus on the organic nanomaterials used to synthesize the shell component of a core-shell nanoparticle.

4.1 Organic Shell

Organic Shell nanoparticles are a type of core-shell nanoparticle where the core is generally made up of inorganic or organic material and the shell is composed of organic material like polymers, and lipids. These organic shell nanoparticles as depicted in Fig. 12.3 can enhance the performance of inorganic nanoparticles by improving their biocompatibility, acting as anchor sites for molecular linkages, or protecting them from oxidation (Chiozzi and Rossi 2020). These nanoparticles have diverse applications in various fields, such as drug delivery, biomedical applications, bioimaging, chemistry, catalysis, and electrical applications. They can be designed to improve the chemical and thermal stability of the inorganic nanoparticles and control the release of molecules from the core (Dhas et al. 2023). A few of the characteristics of organic shell nanoparticles include functionality, reliability, efficiency, maintainability, usability, and portability. The functionality of core-shell nanoparticles refers to their ability to combine multiple properties from the core and shell, enabling them to perform specific tasks which include targeting for enhanced delivery and efficacy of therapeutics, enhancement of imaging capabilities like MRI

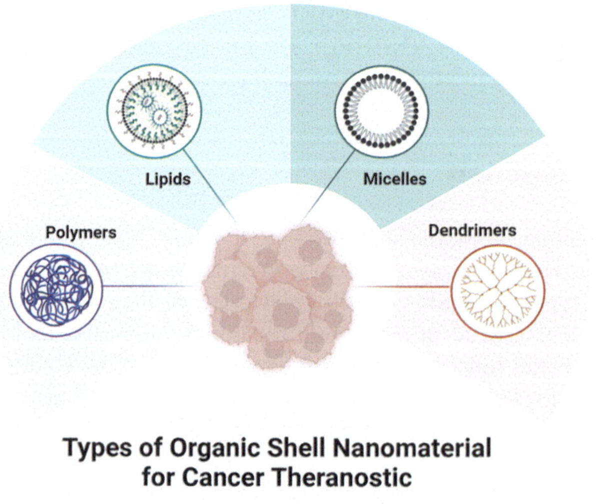

Fig. 12.3 Types of organic shell nanomaterial for cancer theranostic

or CT scans allowing for real-time monitoring of the nanoparticles in the body, controlled delivery of core material enhancing the efficacy and enhancement of radiosensitization improving the radiotherapy enhancement. Reliability refers to nanoparticles' consistency and dependability in performing their intended functions. Similarly, efficacy refers to their ability to effectively deliver therapeutics to cancer cells while minimizing the harm to the healthy tissues. Maintainability is crucial to ensure that the nanoparticles remain effective and stable over time, whereas usability and portability are crucial for ensuring effective delivery and efficacy in various clinical settings. The advantages of organic shell nanoparticles include targeted delivery, controlled release, stability and solubility, biocompatibility, bioimaging, enhanced permeability, multifunctionality, and improved efficacy (Mahdavi et al. 2020). Though there are numerous advantages, there are a few disadvantages as well. Cytotoxicity, limited stability and targeting ability, difficulty in scalability, and potential toxicity, as well as regulatory challenges.

4.1.1 Polymers

Polymers have emerged as a crucial component in developing innovative cancer theranostics, offering a versatile platform for designing and delivering targeted therapies. The choice of polymers depends on factors like drug compatibility, biodegradability, targeting ability, and release kinetic profile. It is found that synthetic polymers generally offer more tunability, while natural polymers are more biocompatible (Ali et al. 2020). These complex biomaterials can be engineered to encapsulate and release therapeutic agents, enhance their bioavailability, and facilitate their targeted delivery to cancer cells. The diverse range of synthetic and natural polymers available allows for the creation of tailored systems that can address specific challenges in cancer treatment, such as improving drug solubility, enhancing circulation times, and reducing immunogenicity. This versatility has led to the widespread adoption of polymers in cancer theranostics, where they play a vital role in enhancing treatment outcomes and improving patient quality of life.

The biocompatibility and biodegradability of polymers reduce the risk of adverse reactions, while their ability to accumulate in tumors via the enhanced permeability and retention (EPR) effect and be functionalized with specific targeting ligands enhance their targeting capabilities (Kumar et al. 2013). Additionally, polymers can be designed to provide a controlled release of therapeutic agents in response to environmental cues, such as pH changes or the presence of specific enzymes, further improving the efficacy of cancer treatments. The multimodal imaging capabilities of polymers, including enhanced computed tomography (CT) and photoacoustic (PA) contrast, enable comprehensive diagnostic information to guide and monitor therapeutic interventions. However, polymers are not without their limitations, as they can be susceptible to opsonization and rapid clearance, elicit immune responses, and face challenges in terms of stability and scalability. The biocompatibility and safety of the polymeric materials, as well as the ability to precisely control drug release and tumor targeting, must be carefully evaluated before these systems can be translated to the clinic.

4.1.2 Lipids

Lipids are also widely used as shell materials in cancer nanotheranostic systems due to their excellent biocompatibility, biodegradability, and ability to encapsulate hydrophilic and hydrophobic agents. The most commonly used lipid-based shells are phospholipids and cholesterol (Fernandez-Fernandez et al. 2023). Phospholipids like phosphatidylcholine and phosphatidylethanolamine can self-assemble into liposomal structures, which can entrap the core and therapeutic or imaging agents. Similarly, cholesterol is often incorporated in the lipid bilayer to improve the stability and rigidity of the nanoparticles (Bukhari et al. 2021). Lipid-based nanosystems include liposomes and solid lipid nanoparticles. The main benefit of using lipids as shell materials is their ability to enhance the pharmacokinetics and biodistribution of the enclosed substance through easy modification. Nonetheless, there are also challenges that include achieving the optimal size, stability, scalability, and maximizing drug loading efficiency, among others (Akanda et al. 2023; Mashaghi et al. 2013). However, despite these challenges, lipid-based nanoparticles remain a promising platform for cancer nanotheranostics due to their versatility, biocompatibility, and ability to improve the therapeutic index of the encapsulated agents.

4.1.3 Micelles

Micelles are self-assembled amphiphilic copolymers in the form of shell nanostructures that have shown great potential as theranostic agents for cancer imaging and therapy. The hydrophilic component of the copolymer forms the shell and provides stability on the aqueous medium, while the hydrophobic core is used for encapsulation of the core nanostructures like hydrophobic therapeutic and imaging agents (Upponi et al. 2018). The hydrophobic core regulates the sustained release of the entrapped entity, whereas the hydrophilic component enables efficient tumor accumulation (Gupta et al. 2024). The micelle formation occurs when the concentration of block co-polymer exceeds a critical micelle concentration (CMC). At concentrations above the CMC, the hydrophobic segments of the block copolymers begin to associate with each other to minimize contact with water molecules. The formation of micelles is driven by a decrease in the free energy of the system. The removal of the hydrophobic segments from the aqueous environment and the re-establishment of the hydrogen bond network in water reduces the overall free energy, leading to the spontaneous self-assembly of the micelles (Ali et al. 2020; Pawar et al. 2022). Polymeric micelles are promising nanocarriers that can solubilize hydrophobic drugs, provide controlled release, and passively target tumors while remaining stable in circulation due to their PEGylated shell. However, challenges remain in ensuring the safety, efficacy, and regulatory approval of these nanotheranostic systems. Further research is needed to address issues such as nontargeted biodistribution, complex fabrication, and potential toxicity (Ahmad et al. 2014; Kapare and Metkar 2020).

4.1.4 Dendrimers

Dendrimers are tree-like, hyperbranched macromolecules with well-defined, monodisperse, and highly symmetric structures. They consist of a central core, repeating branching units, and terminal functional groups (Pérez-Ferreiro et al. 2023). This unique structure allows for the encapsulation of drugs, imaging agents, and targeting ligands within the dendrimer's interior, while the surface can be modified with various functional groups (Abbasi et al. 2014). Dendrimers are typically classified by their generation, which refers to the number of branching iterations from the central core. Higher-generation dendrimers have more branching units and terminal groups, allowing for increased loading capacity and multifunctionality. Common dendrimer types used in cancer theranostics include polyamidoamine (PAMAM) and poly(propylene imine) (PPI) dendrimers (Cruz et al. 2023; Huang et al. 2023). The advantages of using dendrimers as shell materials in cancer theranostics include their well-defined structure, high drug loading capacity, and ability to incorporate multiple functionalities like targeting, imaging, and therapeutic efficacy within a single nanoparticle (Ouyang et al. 2021). However, some potential disadvantages include concerns about their biocompatibility and potential toxicity, as well as challenges in large-scale synthesis and purification (Table 12.2).

5 Specific Surface Functionalization

Surface functionalization is a crucial aspect in the development of core-shell cancer theranostic. This functionalization can enhance their properties in cancer diagnosis and therapy (Dhas et al. 2024. PEGylation is one of the most frequently used surface modifications, which involves coating the nanoparticles with polyethylene glycol (PEG). PEG is a hydrophilic polymer that improves the biocompatibility and stability of nanoparticles in biological systems (Suk et al. 2016). It also increases the circulation time of nanoparticles in the body, leading to accumulation more effectively in tumor tissues through the enhanced permeability and retention (EPR) effect. This passive targeting strategy helps to minimize off-target effects and enhance the therapeutic index of the nanoparticles (Danhier et al. 2010). Another polymer used for conjugation is polyethyleneimine (PEI), which can interact with negatively charged surfaces. These complexes can be internalized by tumor cells, where PEI helps in the endosomal escape of the drug, enhancing its therapeutic efficacy.

Another surface modification method is the conjugation of antibodies and peptides. These targeting ligands can be attached to the surface of nanoparticles to enable selective delivery to specific cell types or tissues (Hui et al. 2023). For example, antibodies against tumor-associated antigens can be used to target cancer cells, while peptides like RGD can bind to integrins overexpressed on tumor vasculature, enhancing nanoparticle accumulation in the tumor (Jeong et al. 2018; Juan et al. 2020). Stimuli-responsive modifications are another important aspect of surface modification. The nanoparticle surface can be designed to respond to specific TME stimuli-like acidic pH or overexpressed enzymes. This helps trigger drug release and activation of therapeutic effects selectively in the tumor, minimizing systemic toxicity (Karimi et al. 2016; Mura et al. 2013). Targeting ligands like folic acid, transferrin, and hyaluronic acid are often conjugated to the surface of the

Table 12.2 Summary of the studies involving the different types of organic nanomaterials as shell component

Sr. No.	Shell	Role	Core	Drug	Cell line	Animal Study	Outcome	Reference
1.	Polyethylene Glycol (PEG)	Physical stability and compatibility	Polypyrrole-coated bismuth nanohybrids	Polypyrrole	Cytotoxicity study using HUVEC and 4T1 cell line, and in-vitro photothermal therapy assessment using 4T1 cells	Biodistribution, Photothermal therapy, and toxicity assessment were done in tumor-bearing BALB/c mice	It was observed that the photoconversion efficiency of developed nanohybrids significantly increased compared to metallic bismuth nanoparticles (up to 46%). It was also found that the temperature could reach 87.3 °C after 10 min of irradiation. It was also found that the developed nanohybrids significantly enhanced CT efficiency compared to iohexol, showing potential as a theranostic nanoplatform for cancer imaging and therapy.	Yang et al. (2018)

2.	Poly lactic co glycolic acid (PLGA)	Drug delivery	Fe/FeO nanoparticles	Doxorubicin	Cytotoxicity studies and Photothermal ablation studies were performed using NIH3T3 or KB cells	In vivo MRI, fluorescence imaging, antitumor efficiency, and blood biochemistry were performed on mice with KB tumor	The in vitro studies found that the treatment groups with nanocapsules and irradiation showed better efficacy affecting cell viability than the control groups. It was also observed that there was a significant enhancement in MRI signal intensity at the tumor site after the administration of developed nanoparticles. Similarly, fluorescence intensity was also enhanced. Moreover, antitumor efficiency evaluation demonstrated a promising anti-tumor efficacy when administered with laser irradiation.	Wang et al. (2019)
3.	Poly ε-Caprolactone (PCL)	Drug delivery and stimuli triggered drug release	Fe_3O_4 NPs	Gemcitabine	Cytotoxicity studies using HFF-1 cell	MRI experiments, ex vivo, and toxicity determinations in mice bearing PC3 prostate tumors	MRI investigations showed effective in vivo distribution of nanoparticles up to 24 h post-administration, with a 5 mg/kg Fe concentration and field strength of 9.4 T. Moreover, in vitro cytotoxicity assessments in HFF-1 fibroblasts displayed favorable cell viability, as indicated by the calculated relative cell viability values and errors related to controls.	García-García et al. (2023)

(continued)

Table 12.2 (continued)

Sr. No.	Shell	Role	Core	Drug	Cell line	Animal Study	Outcome	Reference
4.	Alginate	Protective barrier for core component and for its mucoadhesive property	Shikonin and indocyanine green	Shikonin	Cytotoxicity and cell viability studies were performed on CP70 and SKOV3 cell lines	–	The study demonstrated that the developed nanoparticles exhibited a cytotoxic effect against the cell line. Similarly, IC50 values were approximately 2.08 mM for CP70 cells and 4.43 mM for SKOV3 cells, which indicated that the concentration showed higher cytotoxicity towards CP70 cells compared to SKOV3 cells.	Bai et al. (2022)
5.	Glycogen	Enhancement of accumulation in tumor tissue	Polysaccharide drug delivery systems	Polysaccharide drug delivery systems	Cytotoxicity and cell viability studies were performed on CP70 and SKOV3 cell lines	–	The study demonstrated that the developed nanoparticles exhibited a cytotoxic effect against the cell line. Similarly, IC50 values were approximately 2.08 mM for CP70 cells and 4.43 mM for SKOV3 cells which indicated that the concentration showed higher cytocxicity towards CP70 cells compared to SKOV3 cells.	Bai et al. (2022)

6.	1,2-dioleoyl-sn-glycero-3-phosphoethanolamine (DOPE)—F68 complex	To increase the water-solubility core and drug	IR780	Paclitaxel	Cellular internalization, cytotoxicity assay, cell apoptosis, cell cycle analysis and photothermal efficacies were performed in the MHCC-97H cell line	In-vivo biodistribution studies, synergistic effects of combined treatment were performed in BALB/c nude mice	The study evaluated the in vivo efficacy of PDFI nanoparticles for photothermal (PTT) and photodynamic (PDT) therapy in MHCC-97H tumor-bearing mice. PDFI nanoparticles showed lower cytotoxicity compared to DFI nanocores, indicating reduced phototoxicity of IR780. Cell apoptosis and cell cycle arrest were assessed using flow cytometry, with specified concentrations of IR780 and PTX.	Wang et al. (2017)
7.	Liposome	Enhancement of biocompatibility and cellular uptake	AuCu and AuNCs	AuCu and AuNCs	Cell internalization and fluorescence imaging experiments were performed in HeLa cells.	In vivo studies were conducted using ICR mice to evaluate the antitumor effects of nanoplatform.	The study showed that the developed nanoplatform has promising multifunctional diagnostic and therapy effects for tumors, with good biocompatibility and targeted accumulation at the tumor site leading to effective antitumor results.	Liu et al. (2021)

(continued)

Table 12.2 (continued)

Sr. No.	Shell	Role	Core	Drug	Cell line	Animal Study	Outcome	Reference
8.	Micelles composed of starch-octanoic acid (ST-OA)	Stabilizing the micelles, protecting the core components, and facilitating targeted drug delivery to the tumor site.	SPIONs and doxorubicin.	Doxorubicin	BEL-7402 cells were used in the study to evaluate the in vitro drug release behavior and cellular uptake.	Nude mice were used in the animal study to assess the in vivo antitumor activity and safety of the drug-loaded micelles.	The study showed that the developed nano-platform effectively inhibited tumor growth, with tumor volumes of 391.13 ± 48.86 mm^3 (1 mg/kg) and 391.13 ± 48.86 mm^3 (2 mg/kg) compared to the saline group. The micelles exhibited relatively lesser toxicity compared to adriamycin group, indicating efficient tumor growth inhibition with reduced harm to the heart tissue.	Jie et al. (2019)
9.	Supramolecular PEGylated dendritic systems	To provide a pH-dependent release and enhance the overall efficacy	Platinum	Platinum	Intracellular platinum delivery and antitumor utility was evaluated on A549 tumor cell line.	In vivo studies were conducted using nude mice bearing A549 tumor xenografts for assessing the therapeutic efficacy.	The study found TSPDSs more effective and less toxic than clinical cisplatin, with higher tumor targeting and 71.7% tumor inhibition (vs. 68.1%). TSPDSs also had an 8.4-fold higher maximum plasma concentration and significantly better pharmacokinetics, including a 21.67-fold increase in area under the curve (AUC), a 3.17-fold longer half-life, and a 20-fold lower blood clearance.	Li et al. (2016)

nanoparticles to enhance tumor targeting, which binds to the overexpressed receptors on cancer cells, enabling selective uptake of the nanoparticles (Danhier et al. 2010; Sriraman et al. 2014).

Similarly, imaging agents are also conjugated to the surface of nanoparticles to enhance their diagnostic capabilities for cancer theranostic. Fluorescent dyes or radionuclides can be loaded on the organic shell of the magnetic nanoparticles for diagnostic purposes, enabling multimodal imaging (Howes et al. 2010). This allows for selective delivery to tumors along with therapeutic agents resulting in improved residence time in the tumor site. This facilitates the visualization of tumors when employing multimodal imaging methods and also as nanoparticles removed from the blood circulation (Burke et al. 2017). This allows for easier logistics between contrast administration and imaging, improvement of imaging signal, and reduction of the need for multiple contrast agent dosing (Lin et al. 2021) (Table 12.3).

6 Biomedical Applications (Theranostic Application)

Magnetic core-organic shell nanosystems are revolutionizing cancer theranostics by seamlessly integrating diagnostic and therapeutic capabilities. These nanosystems consist of a magnetic core that enables advanced imaging techniques like MRI for early tumor detection and localization (Farzin et al. 2020). The organic shell can be functionalized with therapeutic agents for targeted drug delivery, minimizing side effects. Additionally, the magnetic core can generate heat for hyperthermia treatment upon exposure to an alternating magnetic field, effectively killing cancer cells (Dhas et al. 2018). This multifunctionality exemplifies the theranostic potential of these nanosystems, paving the way for personalized and effective cancer treatments. Researchers are focused on refining their design, improving targeting accuracy, and guaranteeing safety and effectiveness for clinical use. Additionally, we will delve into the studies conducted to investigate this particular aspect of multimodal therapy in this section.

6.1 Chemotheranostic Applications

The magnetic core-organic shell nanosystems in chemotherapeutics represent a notable progression in cancer theranostics (Kulkarni et al. 2023). These nanosystems facilitate improved delivery and efficacy of chemotherapy drugs by exploiting their magnetic properties for precise transport and controlled release. Functionalizing the organic shell can enhance drug solubility and stability, while the magnetic core enables accurate localization to tumor sites using external magnetic fields. This targeted approach minimizes systemic toxicity and maximizes therapeutic impact, presenting a more effective and safer treatment modality for cancer patients.

Gabal and colleagues performed a study focused on developing dual-mode MRI contrast agents utilizing superparamagnetic iron oxide nanoparticles (SPIONs) as the core and samarium ions as the coating for cancer theranostics. The developed

Table 12.3 Case studies of surface functionalization of magnetic/organic core/shell nanoconstructs

Sr. No.	Targeting Moiety	Role	Drug	Particle Size	Zeta Potential	Cell line Study	Animal Study	Outcome	Reference
1	Bovine serum albumin	Provide stability and biocompatibility to the nanoparticles in biological fluids	Doxorubicin (DOX)	41 ± 4 nm	−30 ± 3 mV	Cell viability was evaluated using human fibroblast (HF) cell line	–	PEGylated MNP-BSA nanoparticles, small in size, are promising for anticancer drug delivery and therapy. The toxicity of MNP-BSA NPs decreased when they were coated with PEG compared to the non-coated nanoparticles.	Semkina et al. (2015)
2	Polyethylenimine (PEI)	"To bind with other biomolecules and decrease the polymer's toxicity."	–	256–575 nm	More than 30.0 nV	–	–	The study found that treating PEI with succination and p-carboxybenzylation led to a reduction in the zeta potential of PEI, particularly in PEI-SA derivatives. This reduction was attributed to a decrease in the available amino groups in the polymer structure. However, the study also noted that there was no significant decrease in the binding capacity of biomolecules.	Kasprzak et al. (2015)

3	L-thyroxine	Enhancement of targeting ability.	Plasmid DNA encoding for recombinant protein IL-12	210 nm	Around 30 mV	Transfection efficiency study, cytotoxicity assessment and cellular uptake study were performed on OVCAR-3 and HepG2 cell lines.	Imaging and biodistribution studies were performed in female Balb/c mice.	The study explored the biodistribution and transfection efficiency of radiolabeled polyplexes (99mTc-bPEI/plasmid DNA) in Balb/c mice and human cancer cell lines. Significant accumulation was observed in the kidneys and bladder, indicating renal excretion, while lower levels in the liver suggested high hydrophilicity. Higher charge to polymer (C/P) ratios enhanced hIL-12 expression, with integrin receptors aiding uptake in OVCAR-3 cells. Unmodified bPEI exhibited notable cytotoxicity (40% viability at $C/P = 8$) in HepG2 cells, while lower C/P ratios improved viability (85% at $C/P = 0.25$), highlighting the potential for targeted gene delivery with varying cytotoxic effects.	Sadeghpour et al. (2018)
4	Mesothelin (MSLN) single-chain variable fragment (scFv)	To enhance specificity of nanoparticles	miR-198 mimics	153–332 nm	–	Cell binding assay, cell internalization assay, gene expression analysis and targeting efficiency were evaluated in Mia-PaCa-2 cells.	–	The study showed that developed nanoparticles effectively target and deliver therapeutic nucleic acids to MSLN-expressing pancreatic cancer cells, demonstrating enhanced binding and internalization in vitro. In vivo experiments confirmed efficient delivery to tumors, indicating potential therapeutic benefits	Lü et al. (2021)

nanoparticles were synthesized and characterized for size, morphology, drug loading, and encapsulation efficiency. The nanoparticles exhibited potential as T1-T2 dual-modal contrast agents, with longitudinal relaxivity (r1) and transverse relaxivity (r2) estimated using NMR relaxation times. In vivo investigations assessed the effectiveness of the developed nanoformulation against human breast cancer (MCF-7) cells using wound healing assays and molecular analysis. The nanoformulation exhibited enhanced cytotoxicity against MCF-7 cells compared to cisplatin alone, indicating its promising role as an inhibitory agent against breast cancer cell proliferation. The enhanced efficacy of the nanoparticle formulation is attributed to the synergistic effects of the superparamagnetic Fe_3O_4 core for magnetic targeting, the PEG-co-PMMA shell for stability and controlled drug release, and the hydrophobicity introduced by SmO2 to promote cellular adhesion and uptake. In summary, the developed nanoformulation dual-mode MRI contrast agent demonstrated improved anticancer activity against breast cancer cells compared to cisplatin alone, highlighting its potential for enhanced cancer diagnosis and treatment (Abo Gabal et al. 2024). Gracia García-García et al. developed tri-stimuli responsive nanoparticles having potential theranostic activity. The nanoparticles have a magnetite core and PCL shell with gemcitabine as a chemotherapeutic agent. The nanoparticles exhibited superparamagnetic properties and potential as T2 contrast agents for MRI. Safety was demonstrated in vitro and ex vivo. The nanoparticles showed pH- and heat-responsive drug release capabilities in vitro. In vivo, MRI and tissue analysis revealed improved tumor targeting when applying a magnetic field. This tri-stimuli responsive (magnetite/PCL)/chitosan nanoparticle system has promising theranostic applications, combining imaging and chemotherapy for tumor treatment (García-García et al. 2023).

Liu et al. conducted a study that investigates the development of magnetic core-shell S-nitrosothiols nanoparticles, specifically Fe_3O_4@S-nitrosothiols modified with hyaluronic acid (HA) and folic acid (FA), as a dual-targeting theranostic platform for cancer treatment. These nanoparticles exhibited significant cytotoxicity towards HepG2 cancer cells, with a median lethal time of approximately 14 h compared to 45 h for normal cells at a concentration of 220 μg/mL. They demonstrated superparamagnetic properties, with saturated magnetic values of 52 emu/g for Fe_3O_4 and 15 emu/g for the functionalized nanoparticles, enabling effective magnetic targeting. In vivo studies revealed improved tumor growth inhibition and survival rates in mice treated with these nanoparticles, particularly when guided by an external magnetic field, with some mice surviving beyond 40 days. The nanoparticles also released sufficient nitric oxide (NO) for effective chemotherapy, comparable to the small molecular NO donor PYRRO-NO, validating their potential for enhanced cancer therapy through targeted delivery and controlled release mechanisms. Furthermore, the study emphasizes the diagnostic capabilities of these nanoparticles, particularly in magnetic resonance imaging (MRI). The nanoparticles were characterized for their MRI potential, demonstrating significant contrast enhancement due to the magnetic properties of the Fe_3O_4 core, which allows for improved visualization of tumor sites and facilitates accurate diagnosis and monitoring of tumor progression. In vivo MRI characterization confirmed the effective accumulation of the nanoparticles in tumor tissues, providing a clear distinction between tumor and normal tissues. Biodistribution assays tracked the relative iron content in major organs and tumors, revealing that the Fe_3O_4@S-nitrosothiols-HA-FA nanoparticles exhibited superior tumor targeting performance, especially with external magnetic guidance. This dual functionality—acting as both a therapeutic agent and a diagnostic tool—highlights the nanoparticles' potential as a theranostic system, enabling simultaneous tumor imaging and targeted nitric oxide release for chemotherapy, thus enhancing the overall efficacy of cancer treatment.

6.2 Phototheranostic Applications

Magnetic core-organic shell nanosystems have emerged as a potent tool in phototherapy applications for cancer theranostics (Garkal et al. 2023). These nanosystems can convert light into heat or reactive oxygen species, enabling precise photothermal or photodynamic therapy. The magnetic core enhances targeting and retention at tumor sites, while the organic shell can be designed to respond to specific wavelengths of light. This allows for controlled and localized treatment, effectively destroying cancer cells with minimal damage to surrounding healthy tissues, and represents a significant advancement in non-invasive cancer treatment strategies.

Feng and colleagues developed.

Zhilong Xu and his colleagues performed a study focusing on the innovative development of a multifunctional nanozyme, MCPtCe6, which is engineered to enhance cancer treatment through the integration of PTT, PDT, and MRI capabilities. This nanozyme is constructed with a unique yolk-shell structure that encapsulates Fe_3O_4 nanoparticles within a carbon shell, while platinum nanoparticles are dispersed on the carbon layer. This design improves the components' stability and dispersion and synergistically enhances their therapeutic effects. MCPtCe6 exhibits an impressive PCE of 28.28%, which is significantly higher than many conventional photothermal agents. NIR irradiation can raise the temperature of tumor tissues to approximately 60 °C, effectively inducing thermal ablation of cancer cells. This thermal effect is complemented by the nanozyme's ability to generate ROS when activated by light, which is crucial for the efficacy of PDT. The combination of PTT and PDT leads to enhanced tumor cell apoptosis, providing a dual mechanism of action that significantly improves therapeutic outcomes compared to using either modality alone. In vivo studies demonstrated that treatment with MCPtCe6 resulted in substantial reductions in tumor volume in animal models, particularly when combined with laser irradiation. This highlights the potential of MCPtCe6 to inhibit tumor growth effectively and facilitate rapid tumor eradication. Furthermore, the nanozyme serves as a T2 MRI contrast agent, allowing for real-time imaging of tumor sites. This capability is vital for precise treatment planning and monitoring, enabling clinicians to visualize tumor dynamics and assess therapeutic responses during treatment. The study also emphasizes the biocompatibility of MCPtCe6, as it was administered to healthy mice at a dose of 31.25 mg/kg without significant adverse effects. Blood parameters and organ histology remained within normal ranges, indicating that the nanozyme can be safely utilized in clinical settings (Xu et al. 2022).

Shengxiang Fu et al. explored the innovative use of Ru/MnO2 nanoparticles as dual-function agents for MRI and PTT in cancer treatment. These nanoparticles were synthesized through a one-step process, resulting in a range of T1 relaxivities from 6.5 to 18.2 mM^{-1} s^{-1} at 3.0 T, with the Ru/MnO2–5 variant exhibiting the highest relaxivity of 33.3 mM^{-1} s^{-1} at 1.5 T. This significant enhancement in T1 relaxivity indicates their potential to improve MRI signal intensity, making them effective contrast agents for tumor imaging. In vivo, experiments conducted on H22 tumor-bearing mice revealed that the T1 MR signal increased significantly after injecting Ru/MnO2–5 nanoparticles, peaking at approximately 1.8 times the pre-injection level. This increase demonstrates the nanoparticles' effective targeting of tumor tissue via glucose transporters (GLUTs), which are overexpressed in various tumors, thereby enhancing the precision of tumor imaging. Furthermore, the study evaluated the photothermal activity of Ru/MnO2–5 nanoparticles, which showed a dose-dependent decrease in cell viability in H22 cells when subjected to laser irradiation. This indicates their capability to induce thermal effects that can selectively destroy cancer cells. In vivo, assessments also revealed that mice treated with Ru/MnO2–5 nanoparticles experienced significant tumor growth inhibition compared to the saline control group, highlighting the therapeutic potential of these nanoparticles in cancer treatment (Fu et al. 2024).

6.3 Ultrasound Responsive with Diagnostic Applications

Using magnetic core-organic shell nanosystems in sonodynamic therapy presents a novel and promising approach to cancer theranostics (Tiwari et al. 2023). These nanosystems utilize their magnetic properties to enhance the production of reactive species under ultrasound exposure, targeting cancer cells selectively. The organic shell can be tailored to optimize ultrasound absorption and reactive species generation. This method offers a noninvasive treatment option with the ability to penetrate deep tissues and precisely target tumor cells while sparing healthy tissues, thus broadening the scope of effective cancer treatment modalities.

Maghsoudinia and colleagues investigated the safety and effectiveness of Gd-DOTA/DOX nanodroplets. Blood tests and tissue examinations revealed no notable changes in RBCs, WBCs, platelets, or organ functions, indicating high compatibility and a hemolysis percentage of less than 1.5%. This confirms their safety for IV use. The stability of the nanodroplets was also assessed over 3 months, showing no significant changes in size (mean size of 25.5 ± 4.1 nm) or drug encapsulation efficiency (92.11 ± 3.6% of DOX trapped). The study also demonstrates that gadolinium-loaded nanodroplets exhibit a significant ultrasound-responsive effect, with a drug release rate of approximately 80% within 30 min of US exposure, compared to only 20% without ultrasound. The echogenicity of the nanodroplets increased by 3.5-fold when converted to microbubbles, enhancing their visibility during ultrasound imaging. Furthermore, the T1 relaxivity of the Gd-DOTA/DOX NDs was measured at 5.2 $mM^{-1}\ s^{-1}$, indicating a substantial improvement in MRI contrast compared to conventional gadolinium-based contrast agents. Furthermore, studies on how the nanodroplets move within the body showed that the concentration of DOX in vital organs was notably lower in Gd-DOTA/DOX nanodroplets than in free DOX, indicating targeted delivery and reduced systemic toxicity. These thorough assessments suggest that Gd-DOTA/DOX nanodroplets have the potential to be a safe and effective treatment for cancer (Maghsoudinia et al. 2022).

7 Conclusion and Future Prospective

The emergence of magnetic-organic core-shell nanoconstructs marks a transformative advancement in cancer theranostics, combining the therapeutic and diagnostic capabilities into a single, multifunctional platform. These innovative nanomaterials leverage the unique properties of magnetic nanoparticles, such as their ability to be manipulated by external magnetic fields alongside organic shells that can encapsulate therapeutic agents and facilitate targeted delivery to tumor sites. This dual functionality enhances the precision of cancer treatment and allows for real-time monitoring of therapeutic efficacy, significantly reducing the risk of adverse effects on surrounding healthy tissues. As research continues to optimize these constructs, their potential to overcome the limitations of traditional cancer therapies becomes increasingly evident, paving the way for more effective and personalized treatment strategies. The future of magnetic/organic core-shell nanoconstructs in cancer theranostics is filled with promising opportunities for innovation and advancement. A critical area for development lies in enhancing the functionalization of these nanomaterials to improve targeting specificity and reduce off-target effects. By incorporating advanced targeting ligands, such as antibodies or peptides, and designing stimuli-responsive systems that can release therapeutic agents in response to specific tumor microenvironmental cues, researchers can significantly increase the efficacy of these treatments. Additionally, the integration of multimodal imaging techniques—combining modalities like magnetic resonance imaging (MRI), fluorescence imaging, and ultrasound—can provide a comprehensive view of tumor dynamics and treatment responses, facilitating more informed clinical decisions. Moreover, the successful clinical translation of these nanoconstructs will require interdisciplinary collaboration among chemists, biologists, and clinicians to address safety, efficacy, and scalability challenges. As regulatory frameworks evolve and more clinical trials are conducted, these advanced nanomaterials have a strong potential to become integral components of cancer therapy and diagnosis. By continuing to explore and refine these technologies, we can look forward to a future where magnetic/organic core-shell nanoconstructs play a pivotal role in transforming cancer care, ultimately leading to improved patient outcomes and enhanced quality of life.

References

Abbasi E, Aval SF, Akbarzadeh A, Milani M, Nasrabadi HT, Joo SW, Hanifehpour Y, Nejati-Koshki K, Pashaei-Asl R. Dendrimers: synthesis, applications, and properties. Nanoscale Res Lett. 2014;9:247. https://doi.org/10.1186/1556-276X-9-247.

Abdel-Fattah WI, Ali GW. On the anti-cancer activities of silver nanoparticles. JABB. 2018;5 https://doi.org/10.15406/jabb.2018.05.00116.

Abed Z, Beik J, Laurent S, Eslahi N, Khani T, Davani ES, Ghaznavi H, Shakeri-Zadeh A. Iron oxide–gold core–shell nano-theranostic for magnetically targeted photothermal therapy under magnetic resonance imaging guidance. J Cancer Res Clin Oncol. 2019;145:1213–9. https://doi.org/10.1007/s00432-019-02870-x.

Abo Gabal R, shokeir D, Awadalla A, Hamam ET, Oraby AH. Design of magnetic nanoplatforms for simultaneous molecular imaging and cisplatin delivery for breast cancer cells theranostics. J Drug Deliv Sci Technol. 2024;95:105533. https://doi.org/10.1016/j.jddst.2024.105533.

Adamiano A, Iafisco M, Tampieri A. Magnetic core-shell nanoparticles. In: Core-Shell nanostructures for drug delivery and theranostics. Elsevier; 2018. p. 259–96. https://doi.org/10.1016/B978-0-08-102198-9.00009-0.

Ahmad Z, Shah A, Siddiq M, Kraatz H-B. Polymeric micelles as drug delivery vehicles. RSC Adv. 2014;4:17028–38. https://doi.org/10.1039/C3RA47370H.

Aishajiang R, Liu Z, Wang T, Zhou L, Yu D. Recent advances in cancer therapeutic copper-based nanomaterials for antitumor therapy. Molecules. 2023;28:2303. https://doi.org/10.3390/molecules28052303.

Akanda M, Mithu MSH, Douroumis D. Solid lipid nanoparticles: an effective lipid-based technology for cancer treatment. J Drug Deliv Sci Technol. 2023;86:104709. https://doi.org/10.1016/j.jddst.2023.104709.

Aksoy A, Alazragi R, Alabdali AYM, Aljazzar R, Sadi SE, Alostaz M, Hindi ME. Antibacterial activity of metallic-core gold and silver nanoparticles against some animal pathogens. Ann Anim Sci. 2023;23:473–9. https://doi.org/10.2478/aoas-2023-0008.

Ali I, Alsehli M, Scotti L, Tullius Scotti M, Tsai S-T, Yu R-S, Hsieh MF, Chen J-C. Progress in polymeric nano-medicines for theranostic cancer treatment. Polymers (Basel). 2020;12:598. https://doi.org/10.3390/polym12030598.

Almarzoug MHA, Ali D, Alarifi S, Alkahtani S, Alhadheq AM. Platinum nanoparticles induced genotoxicity and apoptotic activity in human normal and cancer hepatic cells via oxidative stress-mediated Bax/Bcl-2 and caspase-3 expression. Environ Toxicol. 2020;35:930–41. https://doi.org/10.1002/tox.22929.

Anjum S, Hashim M, Malik SA, Khan M, Lorenzo JM, Abbasi BH, Hano C. Recent advances in zinc oxide nanoparticles (ZnO NPs) for cancer diagnosis, target drug delivery, and treatment. Cancers (Basel). 2021;13:4570. https://doi.org/10.3390/cancers13184570.

Bai M-Y, Chen T-H, Wang Y-C, Lai Y-J. Transformation of theranostic alginate-based microbubbles from raspberry-like to core–shell-like microbubbles and *in vitro* studies. RSC Adv. 2022;12:31943–9. https://doi.org/10.1039/D2RA06298D.

Baldea I, Petran A, Florea A, Sevastre-Berghian A, Nenu I, Filip GA, Cenariu M, Radu MT, Iacovita C. Magnetic nanoclusters stabilized with poly [3,4-Dihydroxybenzhydrazide] as efficient therapeutic agents for cancer cells destruction. Nanomaterials. 2023;13:933. https://doi.org/10.3390/nano13050933.

Barani M, Rahdar A, Mukhtar M, Razzaq S, Qindeel M, Hosseini Olam SA, Paiva-Santos AC, Ajalli N, Sargazi S, Balakrishnan D, Gupta AK, Pandey S. Recent application of cobalt ferrite nanoparticles as a theranostic agent. Mater Today Chem. 2022;26:101131. https://doi.org/10.1016/j.mtchem.2022.101131.

Bian X, Guo T, Zhang J, Xia J, Feng X, Wang F, Lin M, Tian W. The magnetic nanomaterial biofunctions in cancer diagnosis and therapy. J Nanomater. 2021;2021:9968166. https://doi.org/10.1155/2021/9968166.

Bisht G, Rayamajhi S. ZnO nanoparticles: a promising anticancer agent. Nano. 2016;3:9. https://doi.org/10.5772/63437.

Bonet-Aletá J, Pezacki A, Oi M, Hueso JL, Santamaría J, Chang CJ. Therapeutic copper-based nanoparticles release labile copper (II) and trigger cellular responses in glutathione and NRF2 redox pathways and metal homeostasis. 2023; https://doi.org/10.26434/chemrxiv-2023-0hzxw.

Bukhari SI, Imam SS, Ahmad MZ, Vuddanda PR, Alshehri S, Mahdi WA, Ahmad J. Recent progress in lipid nanoparticles for cancer theranostics: opportunity and challenges. Pharmaceutics. 2021;13:840. https://doi.org/10.3390/pharmaceutics13060840.

Burke BP, Cawthorne C, Archibald SJ. Multimodal nanoparticle imaging agents: design and applications. Philos Trans R Soc A Math Phys Eng Sci. 2017;375:20170261. https://doi.org/10.1098/rsta.2017.0261.

Büyüktiryaki S, Keçili R, Özkütük EB, Ersöz A, Say R. Synthesis of core-shell magnetic nanoparticles. In: Thomas S, Rezazadeh Nochehdehi A, editors. Handbook of magnetic hybrid nanoalloys and their nanocomposites. Cham: Springer; 2022. p. 65–106. https://doi.org/10.1007/978-3-030-90948-2_9.

Cai X, Zhu Q, Zeng Y, Zeng Q, Chen X, Zhan Y. manganese oxide nanoparticles as MRI contrast agents in tumor multimodal imaging and therapy. IJN. 2019;14:8321–44. https://doi.org/10.2147/IJN.S218085.

Chiozzi V, Rossi F. Inorganic–organic core/shell nanoparticles: progress and applications. Nanoscale Adv. 2020;2:5090–105. https://doi.org/10.1039/D0NA00411A.

Cruz A, Barbosa J, Antunes P, Bonifácio VDB, Pinto SN. A glimpse into dendrimers integration in cancer imaging and theranostics. Int J Mol Sci. 2023;24:5430. https://doi.org/10.3390/ijms24065430.

Dai T, Ren W, Wu A. Cancer theranostics of black TiO2 nanoparticles. In: TiO2 Nanoparticles. Wiley; 2020. p. 185–215. https://doi.org/10.1002/9783527825431.ch6.

Danhier F, Feron O, Préat V. To exploit the tumor microenvironment: passive and active tumor targeting of nanocarriers for anti-cancer drug delivery. J Control Release. 2010;148:135–46. https://doi.org/10.1016/j.jconrel.2010.08.027.

Dhas NL, Raval NJ, Kudarha RR, Acharya NS, Acharya SR. Core–shell nanoparticles as a drug delivery platform for tumor targeting. In: Inorganic frameworks as smart nanomedicines. Elsevier; 2018. p. 387–448. https://doi.org/10.1016/B978-0-12-813661-4.00009-2.

Dhas N, Neyyar S, Garkal A, Kudarha R, Patel J, Mutalik S, Mehta T. Biomedical applications of nanocarriers in nasal delivery. In: Pathak YV, Yadav HKS, editors. Nasal drug delivery: formulations, developments, challenges, and solutions. Cham: Springer; 2023. p. 101–26. https://doi.org/10.1007/978-3-031-23112-4_7.

Dhas N, Kudarha R, Tiwari R, Tiwari G, Garg N, Kumar P, Kulkarni S, Kulkarni J, Soman S, Hegde AR, Patel J, Garkal A, Sami A, Datta D, Colaco V, Mehta T, Vora L, Mutalik S. Recent advancements in nanomaterial-mediated ferroptosis-induced cancer therapy: importance of molecular dynamics and novel strategies. Life Sci. 2024;346:122629. https://doi.org/10.1016/j.lfs.2024.122629.

Farzin A, Etesami SA, Quint J, Memic A, Tamayol A. Magnetic nanoparticles in cancer therapy and diagnosis. Adv Healthc Mater. 2020;9:e1901058. https://doi.org/10.1002/adhm.201901058.

Fernandes DA. Review on metal-based theranostic nanoparticles for cancer therapy and imaging. Technol Cancer Res Treat. 2023;22:15330338231191493. https://doi.org/10.1177/15330338231191493.

Fernandez-Fernandez A, Manchanda R, Kumari M. Lipid-engineered nanotherapeutics for cancer management. Front Pharmacol. 2023;14:1125093. https://doi.org/10.3389/fphar.2023.1125093.

Fiocchi S, Chiaramello E, Marrella A, Suarato G, Bonato M, Parazzini M, Ravazzani P. Modeling of core-shell magneto-electric nanoparticles for biomedical applications: effect of composition, dimension, and magnetic field features on magnetoelectric response. PLoS One. 2022;17:e0274676. https://doi.org/10.1371/journal.pone.0274676.

Fu S, Cai Z, Gu H, Lui S, Ai H, Song B, Wu M. Rutin-coated ultrasmall manganese oxide nanoparticles for targeted magnetic resonance imaging and photothermal therapy of malignant tumors. J Colloid Interface Sci. 2024;670:499–508. https://doi.org/10.1016/j.jcis.2024.05.067.

García-García G, Caro C, Fernández-Álvarez F, García-Martín ML, Arias JL. Multi-stimuli-responsive chitosan-functionalized magnetite/poly(ε-caprolactone) nanoparticles as theranostic platforms for combined tumor magnetic resonance imaging and chemotherapy. Nanomedicine. 2023;52:102695. https://doi.org/10.1016/j.nano.2023.102695.

Garcia-Muñoz P, Fresno F, de la Peña O'Shea VA, Keller N. Ferrite materials for photoassisted environmental and solar fuels applications. Top Curr Chem (Z). 2020;378:1–56. https://doi.org/10.1007/s41061-019-0270-3.

Garkal A, Sami A, Patel J, Patel L, Dhas NL, Mehta T. Light-responsive delivery nanoplatforms in cancer theranostics. In: Site-specific cancer nanotheranostics. CRC Press; 2023.

Ghasemi P, Shafiee G, Ziamajidi N, Abbasalipourkabir R. Copper nanoparticles induce apoptosis and oxidative stress in SW-480 human colon cancer cell line. Biol Trace Elem Res. 2022; https://doi.org/10.21203/rs.3.rs-1885906/v1.

Gopal L, Sudarshan T. Core shell particles. Surf Eng. 2022;38:675–9. https://doi.org/10.1080/02670844.2022.2170550.

Gupta D, Roy P, Sharma R, Kasana R, Rathore P, Gupta TK. Recent nanotheranostic approaches in cancer research. Clin Exp Med. 2024;24:8. https://doi.org/10.1007/s10238-023-01262-3.

Gupta P, Mishra K, Mittal AK, Handa N, Paul MK. Current expansion of silver and gold nanomaterials towards cancer theranostics: development of therapeutics. Curr Nanosci. n.d.;20:356–72. https://doi.org/10.2174/1573413719666230503144904.

Halder J, Pradhan D, Biswasroy P, Rai VK, Kar B, Ghosh G, Rath G. Trends in iron oxide nanoparticles: a nano-platform for theranostic application in breast cancer. J Drug Target. 2022;30:1055–75. https://doi.org/10.1080/1061186X.2022.2095389.

Howes P, Green M, Bowers A, Parker D, Varma G, Kallumadil M, Hughes M, Warley A, Brain A, Botnar R. Magnetic conjugated polymer nanoparticles as bimodal imaging agents. J Am Chem Soc. 2010;132:9833–42. https://doi.org/10.1021/ja1031634.

Hu R, Zheng M, Wu J, Li C, Shen D, Yang D, Li L, Ge M, Chang Z, Dong W. Core-shell magnetic gold nanoparticles for magnetic field-enhanced radio-photothermal therapy in cervical cancer. Nanomaterials. 2017a;7:111. https://doi.org/10.3390/nano7050111.

Hu R, Zheng M, Wu J, Li C, Shen D, Yang D, Li L, Ge M, Chang Z, Dong W. Core-shell magnetic gold nanoparticles for magnetic field-enhanced radio-photothermal therapy in cervical cancer. Nanomaterials. 2017b;7:111. https://doi.org/10.3390/nano7050111.

Huang T, Li G, Guo Y, Zhang G, Shchabin D, Shi X, Shen M. Recent advances in PAMAM dendrimer-based CT contrast agents for molecular imaging and theranostics of cancer. Sens Diagn. 2023;2:1145–57. https://doi.org/10.1039/D3SD00101F.

Hui L, Chen W, Najlah M. The challenges to develop antibody-conjugated nanomedicine products. Nano TransMed. 2023;2:100018. https://doi.org/10.1016/j.ntm.2023.100018.

Jeong W, Bu J, Kubiatowicz LJ, Chen SS, Kim Y, Hong S. Peptide–nanoparticle conjugates: a next generation of diagnostic and therapeutic platforms? Nano Converg. 2018;5:38. https://doi.org/10.1186/s40580-018-0170-1.

Jie L, Lang D, Kang X, Yang Z, Du Y, Ying X. Superparamagnetic iron oxide nanoparticles/doxorubicin-loaded starch-octanoic micelles for targeted tumor therapy. J Nanosci Nanotechnol. 2019;19:5456–62. https://doi.org/10.1166/jnn.2019.16548.

Joudeh N, Saragliadis A, Koster G, Mikheenko P, Linke D. Synthesis methods and applications of palladium nanoparticles: a review. Front Nanotechnol. 2022;4 https://doi.org/10.3389/fnano.2022.1062608.

Juan A, Cimas FJ, Bravo I, Pandiella A, Ocaña A, Alonso-Moreno C. Antibody conjugation of nanoparticles as therapeutics for breast cancer treatment. Int J Mol Sci. 2020;21:6018. https://doi.org/10.3390/ijms21176018.

Kabir MF, Rahman MZ, Ferdousy J, Ullah AKM a, Razzakul IM, Rahman MM. Green synthesis of size-controlled silver nanoparticles and their anti-cancer potentiality. 2023; https://doi.org/10.21203/rs.3.rs-2694046/v1.

Kapare HS, Metkar SR. Micellar drug delivery system: a review. Pharmaceutical Resonance. 2020;2

Karimi M, Ghasemi A, Zangabad PS, Rahighi R, Basri SMM, Mirshekari H, Amiri M, Pishabad ZS, Aslani A, Bozorgomid M, Ghosh D, Beyzavi A, Vaseghi A, Aref AR, Haghani L, Bahrami

S, Hamblin MR. Smart micro/nanoparticles in stimulus-responsive drug/gene delivery systems. Chem Soc Rev. 2016;45:1457–501. https://doi.org/10.1039/C5CS00798D.

Kashyap BK, Singh VV, Solanki MK, Kumar A, Ruokolainen J, Kesari KK. Smart nanomaterials in cancer theranostics: challenges and opportunities. ACS Omega. 2023;8:14290–320. https://doi.org/10.1021/acsomega.2c07840.

Kasprzak A, Popławska M, Bystrzejewski M, Łabędź O, Grudziński IP. Conjugation of polyethylenimine and its derivatives to carbon-encapsulated iron nanoparticles. RSC Adv. 2015;5:85556–67. https://doi.org/10.1039/C5RA17912B.

Kawassaki RK, Romano M, Dietrich N, Araki K. Titanium and iron oxide nanoparticles for cancer therapy: surface chemistry and biological implications. Front Nanotechnol. 2021;3 https://doi.org/10.3389/fnano.2021.735434.

Khalili Najafabad B, Attaran N, Barati M, Mohammadi Z, Mahmoudi M, Sazgarnia A. Cobalt ferrite nanoparticle for the elimination of CD133+CD44+ and CD44+CD24−, in breast and skin cancer stem cells, using non-ionizing treatments. Heliyon. 2023;9:e19893. https://doi.org/10.1016/j.heliyon.2023.e19893.

Klebowski B, Stec M, Depciuch J, Panek A, Krzempek D, Komenda W, Gałuszka-Bulaga A, Pajor-Swierzy A, Baran J, Parlinska-Wojtan M. Improving the effect of cancer cells irradiation with X-rays and high-energy protons using bimetallic palladium-platinum nanoparticles with various nanostructures. Cancers. 2022;14:5899. https://doi.org/10.3390/cancers14235899.

Kulkarni D, Gadade D, Kapare H, Dhas NL, Ban M. Characterization techniques for stimuli-responsive delivery nanoplatforms in cancer treatment. In: Site-specific cancer nanotheranostics. CRC Press; 2023.

Kumar KS, Kumar VB, Paik P. Recent advancement in functional core-shell nanoparticles of polymers: synthesis, physical properties, and applications in medical biotechnology. J Nanopart. 2013;2013:672059. https://doi.org/10.1155/2013/672059.

Kumar R, Mondal K, Panda PK, Kaushik A, Abolhassani R, Ahuja R, Rubahn H-G, Mishra YK. Core–shell nanostructures: perspectives towards drug delivery applications. J Mater Chem B. 2020;8:8992–9027. https://doi.org/10.1039/D0TB01559H.

Kumar A, Das N, Rayavarapu RG. Role of tunable gold nanostructures in cancer nanotheranostics: implications on synthesis, toxicity, clinical applications and their associated opportunities and challenges. J Nanotheranost. 2023a;4:1–34. https://doi.org/10.3390/jnt4010001.

Kumar S, Shukla MK, Sharma AK, Jayaprakash GK, Tonk RK, Chellappan DK, Singh SK, Dua K, Ahmed F, Bhattacharyya S, Kumar D. Metal-based nanomaterials and nanocomposites as promising frontier in cancer chemotherapy. MedComm. 2023b;4:e253. https://doi.org/10.1002/mco2.253.

Li Y, Li Y, Zhang X, Xu X, Zhang Z, Hu C, He Y, Gu Z. Supramolecular PEGylated dendritic systems as pH/redox dual-responsive theranostic nanoplatforms for platinum drug delivery and NIR imaging. Theranostics. 2016;6:1293–305. https://doi.org/10.7150/thno.15081.

Lin X, O'Reilly Beringhs A, Lu X. Applications of nanoparticle-antibody conjugates in immunoassays and tumor imaging. AAPS J. 2021;23:43. https://doi.org/10.1208/s12248-021-00561-5.

Liu Y, Li J, Chen M, Chen X, Zheng N. Palladium-based nanomaterials for cancer imaging and therapy. Theranostics. 2020;10:10057–74. https://doi.org/10.7150/thno.45990.

Liu X, Yang Y, Wang X, Liu X, Cheng H, Wang P, Shen Y, Xie A, Zhu M. Self-assembled Au_4Cu_4/Au_{25} NCs@liposome tumor nanotheranostics with PT/fluorescence imaging-guided synergetic PTT/PDT. J Mater Chem B. 2021;9:6396–405. https://doi.org/10.1039/D1TB01092A.

Liu W, Li X, Wang T, Xiong F, Sun C, Yao X, Huang W. Platinum drug-incorporating polymeric nanosystems for precise cancer therapy. Small. 2023;19:2208241. https://doi.org/10.1002/smll.202208241.

López-Ortega A. Core–shell magnetic nanoparticles. Nanomaterials. 2023;13:822. https://doi.org/10.3390/nano13050822.

Lü J-M, Liang Z, Liu D, Zhan B, Yao Q, Chen C. Two antibody-guided Lactic-co-Glycolic Acid-Polyethylenimine (LGA-PEI) nanoparticle delivery systems for therapeutic nucleic acids. Pharmaceuticals (Basel). 2021;14:841. https://doi.org/10.3390/ph14090841.

Maghsoudinia F, Akbari-Zadeh H, Aminolroayaei F, Birgani FF, Shanei A, Samani RK. Ultrasound responsive Gd-DOTA/doxorubicin-loaded nanodroplet as a theranostic agent for magnetic resonance image-guided controlled release drug delivery of melanoma cancer. Eur J Pharm Sci. 2022;174:106207. https://doi.org/10.1016/j.ejps.2022.106207.

Mahdavi Z, Rezvani H, Moraveji MK. Core–shell nanoparticles used in drug delivery-microfluidics: a review. RSC Adv. 2020;10:18280–95. https://doi.org/10.1039/D0RA01032D.

Mashaghi S, Jadidi T, Koenderink G, Mashaghi A. Lipid nanotechnology. Int J Mol Sci. 2013;14:4242–82. https://doi.org/10.3390/ijms14024242.

Matlou GG, Abrahamse H. Metallic core-shell nanoparticles as drug delivery vehicles in targeted photodynamic therapy of cancer. In: Chakraborti S, editor. Handbook of oxidative stress in cancer: therapeutic aspects. Singapore: Springer; 2021. p. 1–17. https://doi.org/10.1007/978-981-16-1247-3_208-1.

McHenry ME, Laughlin DE. Structure and chemistry of ferrites and related oxide systems. In: Modern ferrites. Wiley; 2022. p. 21–45. https://doi.org/10.1002/9781118971499.ch3.

Mélinon P, Begin-Colin S, Duvail JL, Gauffre F, Boime NH, Ledoux G, Plain J, Reiss P, Silly F, Warot-Fonrose B. Engineered inorganic core/shell nanoparticles. Phy Rep. 2014;543:163–97. https://doi.org/10.1016/j.physrep.2014.05.003.

Mishra S, Bhatt T, Kumar H, Jain R, Shilpi S, Jain V. Nanoconstructs for theranostic application in cancer: challenges and strategies to enhance the delivery. Front Pharmacol. 2023;14:1101320. https://doi.org/10.3389/fphar.2023.1101320.

Mohammadi Z, Montazerabadi A, Irajirad R, Attaran N, Abedi H, Mousavi Shaegh SA, Sazgarnia A. Optimization of cobalt ferrite magnetic nanoparticle as a theranostic agent: MRI and hyperthermia. Magn Reson Mater Phy. 2023;36:749–66. https://doi.org/10.1007/s10334-023-01072-4.

Mokhosi SR, Mdlalose W, Nhlapo A, Singh M. Advances in the synthesis and application of magnetic ferrite nanoparticles for cancer therapy. Pharmaceutics. 2022;14:937. https://doi.org/10.3390/pharmaceutics14050937.

Mousa AB, Moawad R, Abdallah Y, Abdel-Rasheed M, Zaher AMA. Zinc oxide nanoparticles promise anticancer and antibacterial activity in ovarian cancer. Pharm Res. 2023;40:2281–90. https://doi.org/10.1007/s11095-023-03505-0.

Mura S, Nicolas J, Couvreur P. Stimuli-responsive nanocarriers for drug delivery. Nature Mater. 2013;12:991–1003. https://doi.org/10.1038/nmat3776.

Naser SS, Ghosh B, Simnani FZ, Singh D, Choudhury A, Nandi A, Sinha A, Jha E, Panda PK, Suar M, Verma SK. Emerging trends in the application of green synthesized biocompatible ZnO nanoparticles for translational paradigm in cancer therapy. J Nanotheranost. 2023;4:248–79. https://doi.org/10.3390/jnt4030012.

Ouyang Z, Gao Y, Shen M, Shi X. Dendrimer-based nanohybrids in cancer photomedicine. Mater Today Bio. 2021;10:100111. https://doi.org/10.1016/j.mtbio.2021.100111.

Pawar A, Kamdi V, Alaspure A, Gangane P. Recent updates on polymeric micelles: A review. IJPSRR. 2022:37–52. https://doi.org/10.47583/ijpsrr.2022.v73i01.010.

Pérez-Ferreiro M, Abelairas AM, Criado A, Gómez IJ, Mosquera J. Dendrimers: exploring their wide structural variety and applications. Polymers (Basel). 2023;15:4369. https://doi.org/10.3390/polym15224369.

Pourmadadi M, Rahmani E, Shamsabadipour A, Mahtabian S, Ahmadi M, Rahdar A, Díez-Pascual AM. Role of iron oxide (Fe2O3) nanocomposites in advanced biomedical applications: a state-of-the-art review. Nanomaterials. 2022;12:3873. https://doi.org/10.3390/nano12213873.

Rabaan AA, Bukhamsin R, AlSaihati H, Alshamrani SA, AlSihati J, Al-Afghani HM, Alsubki RA, Abuzaid AA, Al-Abdulhadi S, Aldawood Y, Alsaleh AA, Alhashem YN, Almatouq JA, Emran TB, Al-Ahmed SH, Nainu F, Mohapatra RK. Recent trends and developments in multifunctional nanoparticles for cancer theranostics. Molecules. 2022;27:8659. https://doi.org/10.3390/molecules27248659.

Raja G, Cao S, Kim D-H, Kim T-J. Mechanoregulation of titanium dioxide nanoparticles in cancer therapy. Mater Sci Eng C Mater Biol Appl. 2020;107:110303. https://doi.org/10.1016/j.msec.2019.110303.

Revathy R, Sajini T, Augustine C, Joseph N. Iron-based magnetic nanomaterials: sustainable approaches of synthesis and applications. Result Eng. 2023;18:101114. https://doi.org/10.1016/j.rineng.2023.101114.

Rezaei M, Khoshgard K, Mohammadi S. Recent applications of titanium dioxide nanoparticles as cancer theranostic agents. Int J Basic Sci Med. 2022;7:147–56. https://doi.org/10.34172/ijbsm.2022.26.

Robinson I, Tung LD, Maenosono S, Wälti C, Thanh NTK. Synthesis of core-shell gold coated magnetic nanoparticles and their interaction with thiolated DNA. Nanoscale. 2010;2:2624–30. https://doi.org/10.1039/C0NR00621A.

Rotunjanu S, Racoviceanu R, Mioc A, Milan A, Negrea-Ghiulai R, Mioc M, Marangoci NL, Şoica C. Newly synthesized CoFe2−xDyxO4 (x = 0; 0.1; 0.2; 0.4) nanoparticles reveal promising anticancer activity against melanoma (A375) and breast cancer (MCF-7) cells. Int J Mol Sci. 2023;24:15733. https://doi.org/10.3390/ijms242115733.

Sadeghpour H, Khalvati B, Entezar-Almahdi E, Savadi N, Hossaini Alhashemi S, Raoufi M, Dehshahri A. Double domain polyethylenimine-based nanoparticles for integrin receptor mediated delivery of plasmid DNA. Sci Rep. 2018;8:6842. https://doi.org/10.1038/s41598-018-25277-z.

Saeb MR, Rabiee N, Mozafari M, Verpoort F, Voskressensky LG, Luque R. Metal–Organic Frameworks (MOFs) for cancer therapy. Materials (Basel). 2021;14:7277. https://doi.org/10.3390/ma14237277.

Semkina A, Abakumov M, Grinenko N, Abakumov A, Skorikov A, Mironova E, Davydova G, Majouga AG, Nukolova N, Kabanov A, Chekhonin V. Core–shell–corona doxorubicin-loaded superparamagnetic Fe3O4 nanoparticles for cancer theranostics. Colloids Surf B: Biointerfaces. 2015;136:1073–80. https://doi.org/10.1016/j.colsurfb.2015.11.009.

Sriraman SK, Aryasomayajula B, Torchilin VP. Barriers to drug delivery in solid tumors. Tissue Barriers. 2014;2:e29528. https://doi.org/10.4161/tisb.29528.

Suk JS, Xu Q, Kim N, Hanes J, Ensign LM. PEGylation as a strategy for improving nanoparticle-based drug and gene delivery. Adv Drug Deliv Rev, Non-antigenic regulators of targeting for imaging and therapy. 2016;99:28–51. https://doi.org/10.1016/j.addr.2015.09.012.

Tiwari R, Ansari VA, Mishra J, Dhas NL. Ultrasound-responsive delivery nanoplatforms in cancer theranostics. In: Site-Specific Cancer Nanotheranostics. CRC Press; 2023.

Tyagi PK, Arya A, Mazumder AM, Tyagi S. Development of copper nanoparticles and their prospective uses as antioxidants, antimicrobials, anticancer agents in the pharmaceutical sector. prnano. 2023;6:1048–65. https://doi.org/10.33218/001c.83932.

Upponi JR, Jerajani K, Nagesha DK, Kulkarni P, Sridhar S, Ferris C, Torchilin VP. Polymeric micelles: theranostic co-delivery system for poorly water-soluble drugs and contrast agents. Biomaterials. 2018;170:26–36. https://doi.org/10.1016/j.biomaterials.2018.03.054.

Villalobos Gutiérrez PT, Muñoz Carrillo JL, Sandoval Salazar C, Viveros Paredes JM, Gutiérrez Coronado O. Functionalized metal nanoparticles in cancer therapy. Pharmaceutics. 2023;15:1932. https://doi.org/10.3390/pharmaceutics15071932.

Wang D, Zhang S, Zhang T, Wan G, Chen B, Xiong Q, Zhang J, Zhang W, Wang Y. Pullulan-coated phospholipid and Pluronic F68 complex nanoparticles for carrying IR780 and paclitaxel to treat hepatocellular carcinoma by combining photothermal therapy/photodynamic therapy and chemotherapy. IJN. 2017;12:8649–70. https://doi.org/10.2147/IJN.S147591.

Wang X, Li H, Chen G. Core-shell nanoparticles for cancer imaging and therapy; 2018. p. 143–75. https://doi.org/10.1016/B978-0-08-102198-9.00006-5.

Wang Z, Ju Y, Ali Z, Yin H, Sheng F, Lin J, Wang B, Hou Y. Near-infrared light and tumor microenvironment dual responsive size-switchable nanocapsules for multimodal tumor theranostics. Nat Commun. 2019;10:4418. https://doi.org/10.1038/s41467-019-12142-4.

Xu Z, Chen J, Li Y, Hu T, Fan L, Xi J, Han J, Guo R. Yolk-shell Fe3O4@carbon@platinum-Chlorin e6 nanozyme for MRI-assisted synergistic catalytic-photodynamic-photothermal tumor therapy. J Colloid Interface Sci. 2022;628:1033–43. https://doi.org/10.1016/j.jcis.2022.08.006.

Xu W, Wang Y, Hou G, Wang J, Wang T, Qian J, Suo A. Tumor microenvironment responsive hollow Nanoplatform for triple amplification of oxidative stress to enhance cuproptosis-based synergistic cancer therapy. Adv Healthc Mater. 2023;12:2202949. https://doi.org/10.1002/adhm.202202949.

Yang G, Xu L, Chao Y, Xu J, Sun X, Wu Y, Peng R, Liu Z. Hollow MnO2 as a tumor-microenvironment-responsive biodegradable nano-platform for combination therapy favoring antitumor immune responses. Nat Commun. 2017;8:902. https://doi.org/10.1038/s41467-017-01050-0.

Yang S, Li Z, Wang Y, Fan X, Miao Z, Hu Y, Li Z, Sun Y, Besenbacher F, Yu M. Multifunctional Bi@PPy-PEG core–shell nanohybrids for dual-modal imaging and photothermal therapy. ACS Appl Mater Interfaces. 2018;10:1605–15. https://doi.org/10.1021/acsami.7b17838.

Yaqoob SB, Adnan R, Rameez Khan RM, Rashid M. Gold, silver, and palladium nanoparticles: a chemical tool for biomedical applications. Front Chem. 2020;8 https://doi.org/10.3389/fchem.2020.00376.

Yerpude ST, Potbhare AK, Bhilkar P, Rai AR, Singh RP, Abdala AA, Adhikari R, Sharma R, Chaudhary RG. Biomedical, clinical and environmental applications of platinum-based nanohybrids: an updated review. Environ Res. 2023;231:116148. https://doi.org/10.1016/j.envres.2023.116148.

Zahraie N, Haghighi H, Salehi F, Daneshvar F, Tamaddon P, Sattarahmady N. Pulsed sonodynamic therapy of melanoma cancer cells using nanoparticles of and mesoporous platinum. Ultrasound Med Biol. 2023;49:2160–8. https://doi.org/10.1016/j.ultrasmedbio.2023.06.011.

Zhang Z. Research Progress of gold Core-shell structured nanoparticles in tumor therapy. J Phys Conf Ser. 2020;1699:012007. https://doi.org/10.1088/1742-6596/1699/1/012007.

Zhang X-F, Liu Z-G, Shen W, Gurunathan S. Silver nanoparticles: synthesis, characterization, properties, applications, and therapeutic approaches. Int J Mol Sci. 2016;17:1534. https://doi.org/10.3390/ijms17091534.

Zhang J, Li P, Wang T, Li J, Yun K, Zhang X, Yang X. A copper nanocluster-based multifunctional nanoplatform for augmented chemo/chemodynamic/photodynamic combination therapy of breast cancer. Pharmacol Res. 2023;187:106632. https://doi.org/10.1016/j.phrs.2022.106632.

Zhang J, Komarek AC, Jin, M, Wang X, Jia Y, Li W, Hu Z, Peng W, Tjeng LH, Deng Z, Yu R, Zhang S, Liu M, Yang Y, Lin H, Chen C-T, Li X, Zhu J, Jin C. Crystal structure and properties of iron-based spin-chain compound Ba9Fe3Se15. n.d.

Zhao H-X, Zou Q, Sun S-K, Yu C, Zhang X, Li R-J, Fu Y-Y. Theranostic metal–organic framework core–shell composites for magnetic resonance imaging and drug delivery †electronic supplementary information (ESI) available: materials, experimental details and characterization. Chem Sci. 2016;7:5294–301. https://doi.org/10.1039/c6sc01359g.

Zhao S, Yu X, Qian Y, Chen W, Shen J. Multifunctional magnetic iron oxide nanoparticles: an advanced platform for cancer theranostics. Theranostics. 2020;10:6278–309. https://doi.org/10.7150/thno.42564.

Zhu D, Zhu X-H, Ren S-Z, Lu Y-D, Zhu H-L. Manganese dioxide (MnO 2) based nanomaterials for cancer therapies and theranostics. J Drug Target. 2020;29:1–37. https://doi.org/10.1080/1061186X.2020.1815209.

Nonmagnetic Inorganic/Organic Core-Shell Nanoconstructs for Cancer Theranostics

13

Sandhya Vasanth, Alima Misiriya, Fathima Thahsin, and Sneh Priya

Abstract

Nonmagnetic inorganic/organic core-shell nanoconstructs are the advanced, versatile platform in treating cancer as theranostics. Theranostics is an emerging tool in cancer treatment; it is a combination technique in which a drug is loaded in a suitable formulation, which can diagnose images, and the drug will target the cells. The core is made of nonmagnetic inorganic material such as mesoporous silica, gold, quantum, cerium oxide, silica, titanium dioxide, and calcium phosphate; and the shells are made of organic materials such as liposomes, polymer, dendrimer, albumin, chitosan, cyclodextrin, and protein-based nanoparticles. In addition to their versatility, these nanoconstructs are low in toxicity, stable, biocompatible, and environmentally friendly, ensuring safety in their use. They have excellent imaging techniques and inbuilt fluorescence properties, which help in photothermal imaging to diagnose cancer cells. Inorganic nonmagnetic cores such as gold and quantum have photothermal therapy, which can synergistically treat cancer cells. The organic shell can undergo surface modification such as stimuli-responsive, protein-peptide binding, targeting ligand, grafting and chemical electrostatic interaction, and treatment with the cancer cells. The organic shell can also target the tumor cells with ligands such as antibodies and folic acid and conjugate to the shell for active targeting by stimuli that are responsive to release drugs in a controlled manner. The applications are dual imaging, targeted drug delivery, photothermal therapy, thermos-chemotherapy, enhanced biocompatibility, real-time monitoring, and multimodal theranostic. However, the stud-

S. Vasanth (✉) · A. Misiriya · F. Thahsin
Department of Pharmaceutics, Yenepoya Pharmacy College and Research Centre, Yenepoya (Deemed to be University), Mangalore, Karnataka, India

S. Priya
Department of Pharmaceutics, Nitte Gulabi Shetty Institute of Pharmaceutical Science, Nitte (Deemed to be University), Mangalore, Karnataka, India

J. K. Patel et al. (eds.), *Core-Shell Nano Constructs for Cancer Theragnostic*,
https://doi.org/10.1007/978-981-96-3025-7_13

ies have proved precision and efficient results in preclinical studies, and challenges such as addressing the regulatory scalability from batch to batch that need to be focused on for implementation in a clinical study. This chapter briefs the types, surface modification, properties, and application of cancer theranostics of core-shell nanoconstructs.

Keywords

Nonmagnetic inorganic core · Organic shell · Core-shell nanoconstruct · Theranostic · Surface modification

Abbreviation

API	Active pharmaceutical ingredients
AuNPs	Gold nanoparticles
CaP	Calcium phosphate
CDs	Cyclodextrins
CeO_2 NPs	Cerium oxide nanoparticles
HA	Hyaluronic acid
MPS	Mononuclear phagocyte system
MSNs	mesoporous silica nanoparticles
NPs	Nanoparticles
PDT	Photodynamic therapy
PTT	Photothermal therapy
QDs	Quantum dots
ROS	Reactive oxygen species
TiO_2	Titanium dioxide

1 Introduction

Cancer is a complex disease with a proliferation of cells, which is far higher than normal incidence. Early diagnosis becomes challenging mainly because most cancer cells are asymptomatic at early stages and diagnosed at more advanced stages when treatment becomes tougher and the prognosis poorer. The heterogeneity of cancer, consisting of numerous kinds, mutations, and behaviors, making it even harder to approach tailored therapies. Presently available standard treatments include chemotherapy, radiation therapy, or surgery. These are generally efficient but carry significant toxicity side effects because they lack specificity for cancer cells. They work on the principles of those areas in the body in which cells proliferate fast, killing both cancerous and normal cells, thus causing systemic toxicity. The development of cancer-specific therapies is important for the improvement of patient outcomes. Targeted therapies selectively target cancer cells based on their unique molecular characteristics, potentially improving efficacy and reducing side effects. Advances in molecular profiling and our understanding of cancer biology pave the way for more personalized and targeted approaches (Dai et al. 2017). Nanotechnology has greatly aided cancer treatment and offers a novel approach to problems associated with current chemotherapy drugs (Bae et al. 2011). Theranosis is a crucial weapon in the fight to integrate precision medicine—which combines diagnostics and therapy—into clinical practice. Precision medicine targets disorders at the molecular level with imaging agents. It has been demonstrated that cancer theranosis is very beneficial for both physicians and patients. The primary goals of nanotechnology in treating cancer cells are to distribute payloads effectively and selectively without causing hazardous side effects and to track the effectiveness of noninvasively delivered therapeutics over an extended period with reliability. Targeted, nontargeted, or stimulus-responsive theranostic agents can be created by using this nanotechnology. Combinations of diagnostic and therapeutic functionalities within a single platform make theranostic carriers promising for cancer treatment. The carriers usually contain a targeting ligand, an antibody or peptide attached to a nanoparticle or other delivery vehicle. The ligand is engineered to bind specifically to cell surface receptors, often overexpressed in cancer cells, thereby allowing for targeted delivery of diagnostic or therapeutic agents. In the present scenario, ligands and nanoparticles (NPs) are combined to create nanoconstructs, a potentially effective cancer treatment method. The drug can be supplied actively or in response to stimuli and has a straightforward shape and stability. The adverse effects of traditional cancer treatments, such as toxicity and unwanted biodistribution, are mitigated using nanoconstructs (Mishra et al. 2023a). Core and shell nanoconstruct, two components with biomolecular ligands, target drug delivery and diagnosis overexpressing cancer cells in control (Dam et al. 2014). The active core and shell compartment for drug delivery are found in nanosystems. Nanoparticle-based chemotherapy has several advantages over traditional chemotherapy: improved drug delivery, targeted therapy, and reduced systemic toxicity. Incorporating chemotherapeutic agents in nanoparticles can increase drug solubility, permeability, and stability, prolong release, and target the tumor site (Allen and Cullis 2004). Thus, this targeted approach minimizes side effects on other tissues, reducing side effects and improving efficacy. Moreover, nanoparticles can be designed to achieve a controlled release of drugs, enhancing treatment efficacy (Schmidt 1990). The nanostructure should be engineered to hold a range of drugs within or on their surface, making them the perfect instrument for combination therapy (Başağaoğlu et al. 2013). Core/shell nanoparticles can be categorized based on the mass of the particle or the two or more different materials that make up the core. With the various combinations of materials and elements on Earth, it might be challenging to enumerate them all. Because of this, dividing the components that make core/shell nanoparticles into organic and inorganic categories would seem fairer. When examining the core and shell in this instance, four possible combinations exist: inorganic-core and inorganic-shell, inorganic-core and organic-shell, organic-core and inorganic-shell, and organic-core and organic-shell nanoconstruct are available for cancer theranostics (Oldenburg et al. 1998). While coatings often offer advantages such as decreased cytotoxicity, improved biocompatibility, cytocompatibility, dispersibility, and chemical and thermal stability (Li et al. (2020), they also regulate certain compounds' release from the core (Liu et al. 2010; Sounderya and Zhang 2010). Core-shell nanoconstructs form a versatile platform for several cancer-related applications, including drug delivery, imaging, and catalysis. Incorporating therapeutic or diagnostic agents in the core is feasible, while the shell provides protection, enhanced biocompatibility, and surface modification for targeting. The applications examined by researchers cover various fields, including medicine, catalysis, and electronics. This chapter discusses the nonmagnetic inorganic cores with organic shells (Singh et al. 2020a).

Inorganic nanoparticles contain metallic oxides or have an organic shell around their inorganic core, for the most part. However, the definition of "inorganic nanoparticle" is broader than this and covers a much larger variety of materials and structures. Inorganic nanoparticles can be formed with different materials such as metals, metal oxides, non-metals, and other inorganic compounds. They can be formed with varied structures, such as simple spheres, rods, tubes, complex core shells, or even multicomponent architectures. The key defining characteristic of inorganic nanoparticles is their size within the 1–100 nm scale. The nanoparticles of the inorganic core have been widely developed for diagnostics, treatment, phototherapy, and immunotherapy for cancer. Some commonly used nonmagnetic inorganic cores in biomedical applications are mesoporous silica nanoparticles (MSNs), quantum dots (QDs), and gold nanoparticles (AuNPs) (Zhao et al. 2017). Organic shells, typically composed of polymers, carbonaceous sugars, and similar materials, improve compatibility with biological and cellular systems, particularly in applications related to the human body. Furthermore, they safeguard and ensure the metal core's oxidation resistance and provide locations for supplementary molecular connections. The control of the coating's thickness during the synthesis process is closely related to the nanoparticle's final chemical and physical properties.

2 Types of Core Material and Their Properties

Various nonmagnetic inorganic core nanoconstructs are used in cancer theranostics. These constructs often serve as carriers for therapeutic and imaging agents.

2.1 Mesoporous Silica Nanoparticles (MSNs)

MSNs are nonmagnetic inorganic core nanomaterials developed as pore channels within silica particles. These materials have a pore size ranging from 2 nm to 50 nm. They are mainly employed for biological applications and drug carriers. These materials have a larger surface area and pore volume, which makes them suitable for loading medicines or natural substances into the pore channels. These pore sizes are adjustable and regulate the mesoporous structure to release the drug in a controlled manner, which helps dissolve API (active pharmaceutical ingredients), inhibiting the crystallization of API or drugs loaded in this core.

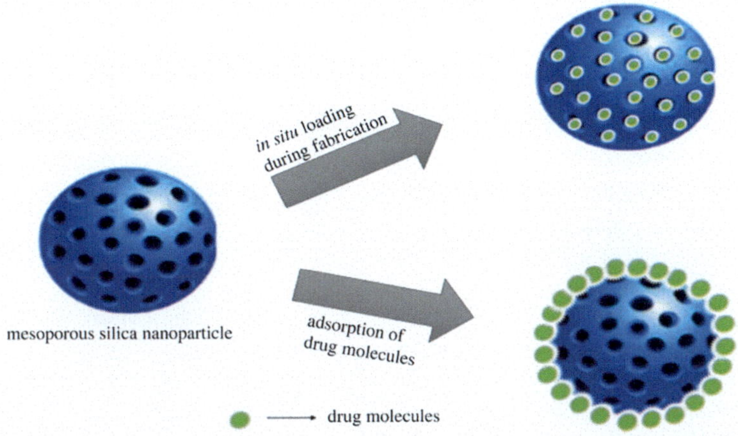

Fig. 13.1 Drug fabrication in MSNs is done using adsorption and in situ methods (Tella et al. 2022)

Additionally, the high density of silanol on the external and inner surfaces of MSNs allows for easy modification, which can lead to improved control over drug loading. Because of this property, the MSNs can be easily modified for precise, accurate, controlled, and targeted delivery of drugs or diagnostic agents to the required area, increasing the effectiveness of therapy or diagnosis and reducing the harmful effects. MSNs have gained popularity as excellent carriers because of their chemical, thermal, and mechanical stability for a wide range of pH and temperature in physiological circumstances and their controlled release kinetics. The surface modification of silica nanoparticles is functional and applicable, especially in drug delivery. Changes in the surface properties allow hybrid, multifunctional nanoparticles to be tailored to specific applications. For example, the loading and delivery of hydrophobic drugs can be enhanced by surface modification with hydrophobic materials (Narayan et al. 2018). Due to the favorable interaction of a hydrophobic surface with the hydrophobic drugs, improving their encapsulation efficiency and stability within the nanoparticle. Also, surface modification can improve nanoparticles' biocompatibility, dispersibility, and targeting ability. The untimely discharge of drugs can be restricted while the administered dosage can be increased (Fig. 13.1).

2.2 Gold Nanoparticles (AuNPs)

AuNPs are microscopic gold particles ranging from 1 to 100 nm, and their color varies from red to purple, depending on their size. It is possible to synthesize AuNPs with desired dimensions and structures, making them applicable to various purposes. The size, shape, and subsequent optical, electronic, and catalytic characteristics of the gold nanoparticles are controlled in a typical preparation. Spheres, rods, cubes, and triangles are common shapes with their distinct properties. This level of

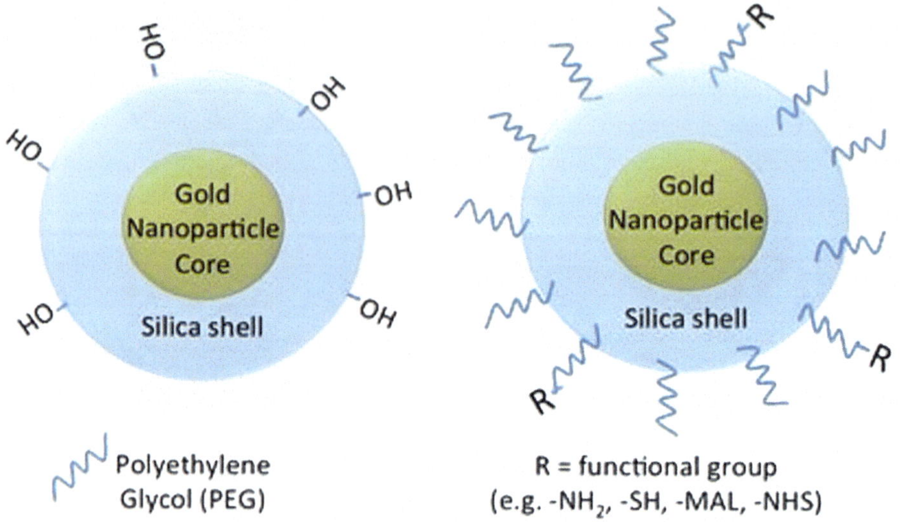

Fig. 13.2 Hydrogel-encapsulated mesoporous silica-coated gold nanoshells (Kim et al. 2019)

accuracy in controlling size and structure makes gold nanoparticles very useful in biomedicine, electronics, and catalysis. The synthesis of AuNPs is created using different methods involving physics, chemistry, and biology. The method used for synthesizing gold nanoparticles chemical uses various chemicals and solvents, gold chloride, and trisodium citrate for the citrate reduction process. Bio-nanoparticle synthesis is a recently developed and environmentally friendly topic with great potential. This method uses different species of plants and bacteria to make AuNPs. The microbes can help break down gold ions by releasing enzymes needed to make AuNPs. AuNPs have attracted attention from various research disciplines, particularly biopharmaceutical. AuNPs have unique optical properties that can be changed, easy production and modification, and precise control over the particles' physicochemical properties. They also have unique electronic and SPR properties. The distinct physical and chemical features of AuNP make it an excellent framework for various applications in therapeutics, diagnostics, biomarkers, drug delivery, and imaging (Arvizo et al. 2010). Furthermore, AuNPs of diverse geometries, including nanospheres, nanorods, nanoshells, and nanocages, have been created. AuNPs of varying shapes exhibit distinct features crucial in fulfilling certain activities across multiple domains. The subsequent text provides an elaborate account of the many configurations of AuNPs (Fig. 13.2).

2.3 Quantum Dots

Quantum dots (QDs) are inorganic nonmagnetic core nanoconstructs with semiconductor unit nanoparticles with distinctive optical and electronic properties that make them effective in cancer theranostics. They are inorganic, nonmagnetic core

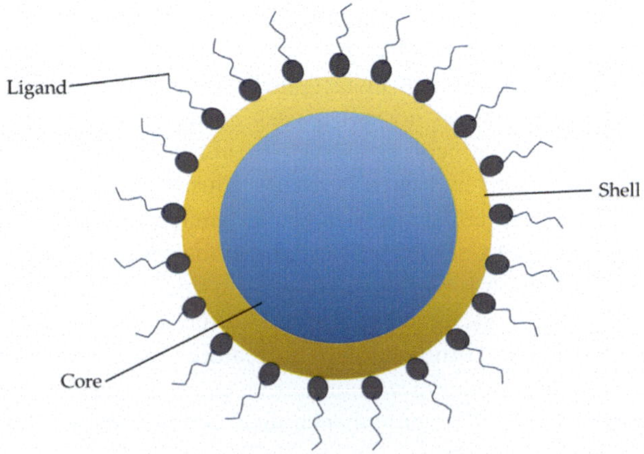

Fig. 13.3 Quantum dot (QD) nanoparticle core (Hamidu et al. 2023)

nanoconstructs in cancer theranostics. Core material is semiconductor nanocrystals (e.g., cadmium selenide, lead sulphide). QDs have a size-dependent bandgap, leading to tenable emission wavelengths. This size tunability allows for the engineering of QDs with specific emission spectra suitable for various imaging and diagnostic applications. QDs exhibit intense and stable fluorescence, making them excellent probes for fluorescence imaging. Their broad excitation and narrow emission spectra enable multiplexed imaging using multiple QDs with distinct emission wavelengths. QDs are highly photostable, resisting degradation and fading under continuous illumination. This property is crucial for long-term imaging applications. QDs have a high molar extinction coefficient, allowing for efficient light absorption and emission. This property enhances their sensitivity as imaging probes. QDs can be easily surface-modified for interaction with specific receptors of cancer cells, such as antibodies, peptides, or targeting ligands, to improve their biocompatibility, specificity, and functionality for cancer targeting (Shabbir et al. 2023). QDs can be used for different imaging modes, including fluorescence imaging, photoacoustic imaging, and, in some cases, MRI. This multimodal imaging capability allows for comprehensive cancer diagnostics (Fig. 13.3).

2.4 Cerium Oxide Nanoparticles (CeO_2 NPs)

Cerium oxide nanoparticles (CeO_2 NPs), known as ceria nanoparticles, have shown promising potential in various biomedical applications, including cancer theranostics. Here are detailed properties of cerium oxide nanoparticles used as inorganic, nonmagnetic core nanoconstructs in cancer theranostics. The core material is cerium oxide (CeO_2). CeO_2 NPs exhibit antioxidant behaviors, acting as a catalyst for oxidation and reduction reactions. This property is particularly relevant in cancer theranostics, where oxidative stress plays a crucial role. CeO_2 NPs can switch between

Ce^{3+} and Ce^{4+} oxidation states, allowing them to scavenge reactive oxygen species (ROS). This property is beneficial in mitigating oxidative stress associated with cancer cells. CeO_2 NPs have been investigated for their potential radioprotective effects. They may help protect normal tissues during radiation therapy, making them valuable in cancer treatment. CeO_2 NPs could scavenge various ROS, including superoxide radicals and hydrogen peroxide. This ROS-scavenging property is relevant for mitigating oxidative stress-induced damage in cancer cells. The anti-inflammatory properties of CeO_2 NPs contribute to their potential as therapeutics in cancer, where inflammation is often associated with tumor progression (Singh et al. 2020b). CeO_2 NPs can be designed to respond to the acidic tumor microenvironment, enabling pH-triggered drug release. This feature enhances therapeutic agents' specificity and controlled release within cancer cells. Surface modifications and functionalization of CeO_2 NPs can enhance their cellular uptake, improving their efficiency as carriers for therapeutic agents.

2.5 Silica Nanoparticles

Silica is generally considered biocompatible and has a long history of use in various biomedical applications. It is inert and does not elicit significant immune responses. The core material is (SiO_2). Silica nanoparticles can be engineered to have controlled sizes, typically ranging from 5–200 nm. This tunability is crucial for drug delivery and imaging applications. The high surface area of silica nanostructure allows for the attachment of various functional groups, enhancing their versatility in binding targeting ligands and therapeutic agents (Selvarajan et al. 2020). Silica nanoparticles can be easily surface-modified with multiple proteins, antibodies, polymers, and targeting ligands to improve their specificity and biocompatibility. Surface modification facilitates the loading of therapeutic agents and imaging contrast agents, enabling a multifunctional platform for cancer theranostics.

2.6 Titanium Dioxide (TiO_2)

Titanium dioxide (TiO_2), also known as titanium (IV) oxide or titania, is a nonmagnetic inorganic substance that has gained scientific attention due to its photoactivity. TiO_2 generates various ROS when exposed to UV radiation in water. The ability to create ROS and trigger cell death is utilized in photodynamic therapy (PDT) to treat several conditions, including psoriasis and cancer. Titanium dioxide nanoparticles were examined for their use as photosensitizing agents in treating tumors. TiO_2 nanoparticles and their composites and combinations with other chemicals or biomolecules can effectively be photosensitizers in photodynamic therapy (PDT) (Ziental et al. 2020). Furthermore, different chemical molecules can be attached to TiO_2 nanoparticles, resulting in hybrid materials. Nanostructures can enhance light absorption, making them suitable for focused therapy in medicine. Various titanium dioxide methods were experimented with to enhance effective anticancer and antibacterial treatments (Fig. 13.4).

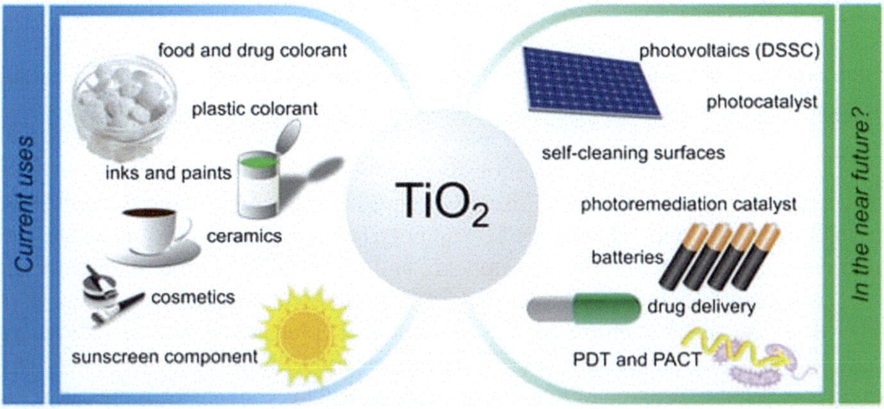

Fig. 13.4 TiO_2 applications (Ziental et al. 2020)

2.7 Calcium Phosphate (CaP)

Carbonate apatite, a type of synthetic calcium phosphate, closely resembles the inorganic mineral found in natural bone. It is known for its excellent biocompatibility and biodegradability and is mainly used in orthopedic implants and bone tissue engineering. Their pH-dependent solubility, easy production, and functionalization make them valuable for cancer treatment. Biodegradable nanoparticles are commonly favored for cancer treatment due to their predictable body clearance pathways and mechanisms, making them safer clinical options. The size, charge, shape, content, and surface chemistry of CaP nanoparticles influence the internalization of nanomedicines. The dose of bio-agents and their functions are influenced by the cellular entry mechanism and their ultimate intracellular location. Due to the abundance of organelles in the cell, the delivery mechanism must target the exact spot accurately to achieve the intended outcome. Future studies on how cells transport their components to specific cellular compartments could enhance drug development for more effective tumor treatment.

3 Organic Shell

3.1 Liposomes

Nanoparticles with a lipid shell and a solid core encapsulated within a single particle structure have recently been developed. The shell of the formulation might consist of either a single layer or many layers of lipids, depending on the formulation procedure and the specific core-shell combinations. These solid cores serve as a means of support, offering mechanical stability, regulated morphology, biodegradability, an enhanced surface area-to-volume ratio, and a limited size range (Akbarzadeh et al. 2013).

3.2 Polymer

Polymeric nanoconstructs are used in cancer theranostics because they can change their surface, respond to stimuli, and hold bioactive or diagnostic chemicals that are both water- and fat-loving. An example of this is deformable discoidal nanoconstructs, which are utilized as an innovative method of administration for imaging and therapeutic objectives. These discoidal shapes are formed through the polymerization of poly(ethylene glycol) (PEG) and poly(lactide-co-glycoside (PLGA). The polymer matrices contain both hydrophobic and hydrophilic microdomains, which serve as compartments for different imaging and medicinal substances. These particles decrease quick capture by the mononuclear phagocyte system (MPS) by remaining in the bloodstream longer. The utilization of PLGA-based nanoconstructs was investigated to provide radiodynamic treatment. This cancer treatment relies on generating ROS, specifically at the tumor's location. Its primary function is focused on tumor-induced hypoxia, which decreases the presence of low oxygen levels, leading to the generation of ROS. There is one. A unique method involves the use of nanoconstructs composed of PLGA nanoparticles. That contain verteporfin and perfluorooctylbromide. When exposed to both normal oxygen levels (normoxic) and low oxygen levels (hypoxic), under certain circumstances, these nanoconstructs rapidly surge ROS generation. This therapy has resulted in the death of approximately 60% of pancreatic cancer cells studied and inhibited the proliferation of tumors within a fortnight. The success rates demonstrate the efficacy of nanoconstructs derived by radiodynamic treatment. Improved and less invasive therapy for deeply situated hypoxic tumors (Perumal et al. 2022; Zielińska et al. 2020).

3.3 Dendrimer

They are highly branched macromolecules. They are mainly used as an anticancer therapy and imaging contrast agents. Modifying dendrimer's peripheral groups enables the identification of antibody-dendrimer, peptide-dendrimer conjugates, or dendritic boxes that encapsulate guest molecules. It helps to produce multiple interactions with biological receptor sites, which is mainly helpful in designing antiviral therapeutic agents. It shows high solubility, reduced systemic toxicity, increased half-life, excellent stability, and enhanced permeation and retention. The dendrimers exhibit a spherical shape and possess a high degree of branching. Macromolecules such as gold nanoparticles can stabilize metal nanoparticles by enclosing bioactive substances. A study investigated the possibility of a curcumin-loaded dendrimer-gold hybrid for theranosis. By synthesizing poly(amidoamine) of the fifth generation AuCl4-terminated dendrimers functionalized with PEGylated amines negative ions, a hybrid structure combining dendrimers and gold was created and ultimate amalgamation was achieved. The system was equipped with curcumin conjugated to the MUC-1 aptamer. The results demonstrated heightened cellular cytotoxicity in HT29 and C26 cells compared to the nontargeted system and established. The potential lies in applying both cancer therapy and CT scan-based tumor imaging (Zielińska et al. 2020).

3.4 Albumin Nanoparticles

The albumin shell plays a crucial role in cancer theranostic applications. It provides a stealth coating for enhanced biocompatibility and acts as a targeting ligand for efficient tumor accumulation. The albumin shell also preserves the native form of proteins, preventing subsequent adsorption cascades of other proteins and influencing cellular uptake. Additionally, albumin-based nanocarriers have shown great promise in constructing multifunctional theranostic platforms, combining diagnostic and therapeutic functions. Furthermore, albumin-core nanogels with a folic acid functionalized hyperbranched amylopectin shell have been developed for oral delivery of cancer therapeutics. These nanogels exhibit stability, specificity, and enhanced cellular uptake in folate receptor-positive cancer cells (Dam et al. 2014). Overall, the albumin shell in cancer theranostics offers improved biocompatibility, targeting capabilities, and drug delivery efficiency, making it valuable in developing novel theranostic platforms. Albumin-based theranostic platforms responsive to tumor microenvironments can be developed. The unique features of tumor microenvironments, such as reduced pH, hypoxia, high levels of ROS, and over-expression of specific enzymes, can be utilized as responsive targets for albumin-based theranostic agents. The engineering of such platforms requires careful design of surface chemistry, self-assembly techniques, and genetic modification of albumin (Abbasi et al. 2014).

3.5 Chitosan Nanoparticles

Chitosan-based nanoparticles have shown promise in cancer theranostics, specifically in developing drug delivery systems for tumor targeting and imaging. These nanoparticles have a unique core-shell structure, with chitosan forming the shell. The chitosan shell provides stability to the nanoparticles and allows for controlled release of therapeutic agents. Additionally, chitosan-based theranostics have been found to have better cellular imaging capability, tumor-targeted drug release, and multimodal therapeutic efficiencies. The chitosan shell also contributes to the biocompatibility, biodegradability, and long-circulating capabilities of the nanoparticles, making them suitable for in vivo applications. Furthermore, chitosan and its derivatives can be easily modified or functionalized with other bioactive molecules to enhance their physicochemical and biological properties. Overall, chitosan-based nanoparticles hold great potential for cancer theranostics, offering improved drug delivery and imaging capabilities for advanced cancer therapy (Hassanin and Elzoghby 2020).

3.6 Cyclodextrin Nanoparticles

Cyclodextrins (CDs) have been extensively studied for their potential in cancer theranostics. CDs are versatile polymers that can be chemically modified for drug delivery, making them suitable for improving the physicochemical properties of antineoplastic agents. CD-based nanoparticles have been developed as carriers for therapeutic and diagnostic agents, showing promise in cancer treatment. Polydopamine-coated magnetic nanoparticles functionalized with cyclodextrin have demonstrated high performance in combined chemo- and photothermal therapy of liver cancer. Nonmagnetic nanocomposites with a shell of SiO_2 encapsulated within cyclodextrin have shown potential for fluorescence imaging, cancer cell-targeting, and hydrophobic drug delivery. CD-based nanoplatforms have also been developed as stimuli-responsive systems for cancer treatment and theranostics, with the ability to respond to various stimuli such as pH and temperature. These studies highlight the clinical significance and potential of cyclodextrin-based systems in cancer theranostics (Mohammed et al. 2017).

3.7 Protein-Based Nanoparticles

The protein shell of cancer theranostic agents plays a crucial role in their functionality and effectiveness. Protein shells, such as human serum albumin and bovine serum albumin, have encapsulated various nanoparticles for multimodal imaging and therapy. These protein-shell nanoparticles offer advantages such as enhanced bioavailability, protection of the core nanoparticles, and the ability to bind targeting groups. Additionally, protein shells can be functionalized to optimize their therapeutic outcomes, such as inhibiting cancer migration and inducing cytotoxic stress. The combination of protein shells with different core nanoparticles allows for the development of intelligent nanoaggregates that can overcome limitations such as premature leakage of photosensitizers and lack of oxygen in hypoxic cancer cells. Overall, the protein shell of cancer theranostic agents is a critical component that enables targeted imaging and therapy for improved clinical outcomes (Gadade and Pekamwar 2020).

3.8 Properties of Nonmagnetic Inorganic-Organic Core-Shell Nanoconstructs

All the nonmagnetic inorganic-organic core shells used Au, SiO_2 are low toxic, stable, biocompatible and eco-friendly. The optical properties of most inorganic nonmagnetic cores are optimum. Gold nanoparticle surface plasmon resonance has excellent optical imaging in surface-enhanced Raman scattering (SERS) and photothermal imaging. Quantum dots have fluorescence properties in turn, and offer strong and stable signals for fluorescence imaging. Titanium nonmagnetic inorganic-organic core-shell nanoconstructs can provide multimodal imaging by incorporating different types of organic shells. For example, gold-silica core-shell nanoparticles can combine the optical properties of gold with the porous structure of silica, enabling both optical and CT imaging. Quantum dot-based nanoconstructs can provide fluorescence imaging alongside other modalities like CT or photoacoustic imaging.

The organic shell can easily undergo surface modification with ligands to target tumor cells. Drug encapsulation within the inorganic core will protect drugs from the degradation microenvironment of tumor cells and enable controlled release at the tumor site.

Nonmagnetic core-shell nanoconstructs are also designed with organic shells that change pH and temperature sensitivity properties.

The photothermal therapy (PTT) properties of nonmagnetic core-shell nanoconstructs are higher than those of magnetic core nanoconstructs. For example, AuNPs in the core can absorb near-infrared light and convert it into heat, enabling PTT to selectively ablate cancer cells. The organic shell can enhance the stability and biocompatibility of the nanoparticles during PTT.

These nanoconstructs can offer a synergistic approach to cancer treatment by combining PTT with drug delivery. The heat energy generated by nonmagnetic metals such as gold and quantum will enhance the drug's release into the tumor cell and improve the effectiveness of chemotherapy.

The organic shell can be easily modified to monitor the real-time drug release into the tumor microenvironment. By incorporating pH-sensitive fluorescent dyes, the shell can indicate changes in pH as the nanoconstructs reach the tumor. These core-shell nanoconstructs have low immunogenicity, reducing the risk of immune responses when introduced into the body (Figs. 13.5 and 13.6).

4 Specific Surface Functionalization Strategies

4.1 Chemical Treatment

Nonmagnetic inorganic, organic core-shell nanoconstructs are surface-modified using silane coupling agents by chemical treatments. It will improve core-shell dispersion stability in various liquid media. Plueddemann et al. reported the chemical treatment or the concept of silane used as a coupling agent (Dai et al. 2017). Research has continued to advance the compatibility between inorganic core and organic shell surfaces and improve the nanostructure's optical, kinetic, and electric properties. The modified nanostructure will behave differently within organic shell solvents or polymer matrices, showing better dispersion. Surface modifications can also be achieved through reactions with metal alkoxides, alkyl or aryl isocyanates, and epoxides. Recent studies have shown that because of surface modification, the mechanical and optical protective properties of polyurethane composite coatings and the dispersion stability of TiO_2 nanoparticles in organic solvents have been improved.

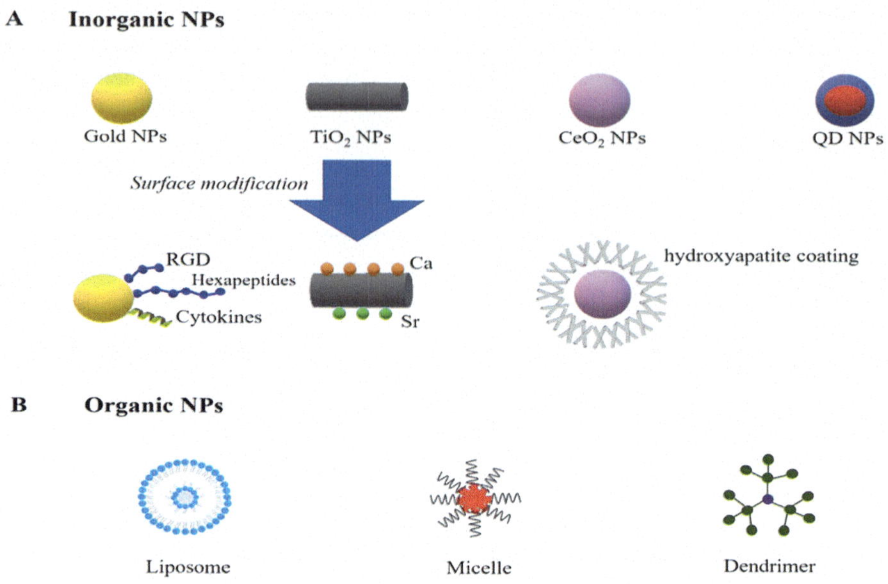

Fig. 13.5 Overview of nonmagnetic inorganic NP core and organic shell used for nanoconstructs (Cai et al. 2022)

4.2 Grafting of Polymers

Core and shell nanoconstructs are colloidal systems capable of encapsulating or embedding the polymer with API for their structure or adsorbing it onto their shells. Grafting of polymers will increase the physiochemical characteristics, including stability within biological fluids, functionalizing their coats, and modulating the polymer deprivation and leakage of entrapped active drugs concerning specific stimuli. Polymers such as polystyrene, poly(methacrylate), polyacrylates and polyacrylamide(non-biodegradable), and poly (D,L-glycolide) (PLG), poly (D,L-lactide) (PLA), and poly(lactide-co-glycoside) (PLGA), (Karlsson et al. 2018) were used to develop the core-shell nanoconstructs. However, non-biodegradable polymers show chronic toxicity and inflammatory reactions, whereas biodegradable polymers show decreased or acute systemic toxicity and are highly biocompatible (Dinarvand et al. 2011). The metabolization and elimination process of biodegradable polymers first involves degrading polymers into monomers and oligomers via a normal pathway, whereby polymers are eliminated from the body (Hong et al. 2020). The demonstration on the controlled release of macromolecules using polymer by Langer and Folkman in 1976 initiated the development of drug delivery system for tumor therapy via an antiangiogenic (Gagliardi et al. 2021). These delivery systems can also be given in the form of combination drugs showing dual release. Examples include PLGA nanoparticles coated with PEGylated lipids, which initially release combretastatin as an anti-angiogenesis agent and then control the release of doxorubicin acting as a chemotherapeutic agent from the PLGA core.

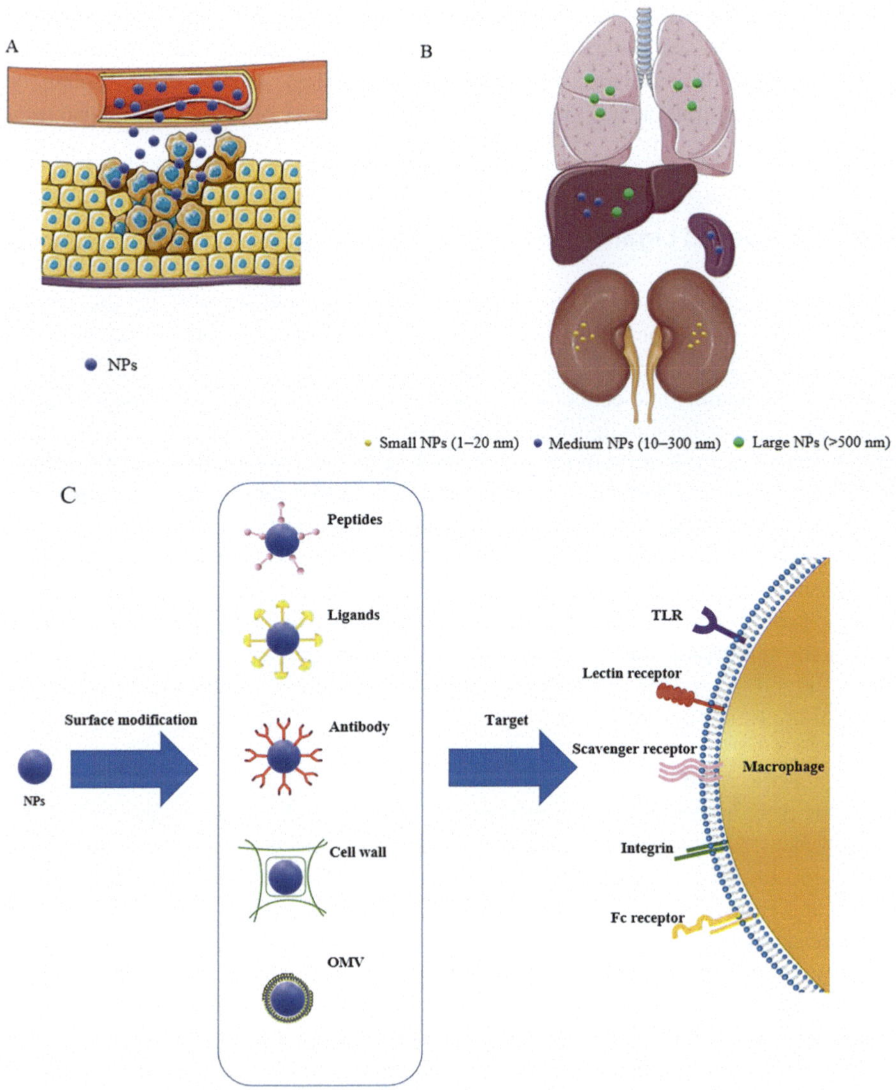

Fig. 13.6 Summary of macrophage-targeting nanotechnologies. (**a**) Passive macrophage-targeting. NPs accumulate in tissues through the vascular leakage. (**b**) Passive macrophage-targeting. NPs with different sizes preferentially accumulate in different organs. (**c**) Active macrophage-targeting. With various surface modification methods, NPs can specially target macrophages via recognition by receptors on macrophage membrane (Cai et al. 2022)

PEGylated squalene (SQPEG)–based nanoconstructs are used to treat cancer therapy. These nanoparticles formed by assembling lipophilic pyropheophorbide-a (Ppa) showed more excellent fluorescence quenching, i.e., 99.99%. The commonly used biodegradable polymer is polyurethane core-shell nanoconstructs for biomedical applications since they are stimuli-responsive and biodegradable. Grafting with polymer core-shell nanoconstructs has advantages such as sustaining and controlling drug release, the solubility of lipophilic chemo drugs, enhanced efficiency, and stimuli sensitivity. These provide targeted delivery and are also used in cancer treatment (Mishra et al. 2023b).

4.3 Targeted Ligand

Even though conventional chemotherapies, radiation, and surgery are the golden standard for cancer, they have improved the survival rate in cancer patients to a certain extent. The drawbacks include the nonspecific tissue distribution of the medication to the healthy and cancer cells, severe side effects, including damaged health cells, directly insufficient distribution to the cancer cells or organ, and rapid elimination from the systemic circulation. It makes the therapy ineffective. Therefore, anticancer drugs are delivered, which could ensure that they release, sustain, control, target, and effectively to a particular site, limiting the toxicity to maintain a patient's safety and quality of life. The research focused on the different approaches to applying efficient ligands, enzymes, or proteins that efficiently bind target-specific tumor cell receptors. The ligands commonly used are peptides, transferrin, folic acid, hyaluronic acid, and antibodies. These are employed in single or combination ligands by anticancer drugs to enhance intracellular uptake and target selectivity of the tumor cell receptor.

4.3.1 Folic Acid

Folic acid is a purine and a glutamate moiety with vitamin B9, linked by p-aminobenzoic acid. Folate receptors undergo endocytosis for the transport of folate into tumor cells. The different isoforms of folate receptors are FR-α, FR-β, FR- δ, and FR-γ. FR-α regularizes tumor cell growth, regulation, and signaling, and an increase of FR-α is mostly in 40% of human cancers, such as the ovary, breast, prostate, uterus, kidney, brain, and lungs. About 25% of cancer is due to the overexpression of FR.-β. Among these two FR-α & FR-β, the cancer cells have high affinities, which leads to uncontrolled tumor cell proliferation. Since the folate receptor is a tumor marker, it enhances the tumor-selective activity. The present model promotes cancer cell-selective core-shell nanoconstruct to deliver drugs targeting folic acid-conjugated molecules to FR-α, leading to receptor-mediated endocytosis. The drug carriers, such as liposomes and micelles, are also conjugated with folic acid to facilitate the tumor-specific delivery of anticancer drugs such as doxorubicin, paclitaxel, 5-fluorouracil, etc. (Bajracharya et al. 2022; Yu et al. 2010)

4.3.2 Hyaluronic Acid (HA)

Hyaluronic acid (HA) is a polysaccharide that consists of alternate units of N-acetyl-D-glucosamine and D-glucuronic acid linked by 1-β-3 and 1-β-4 glycosidic bonds. HA occurs in the human body's extracellular matrix (ECM) in sodium hyaluronate. It is majorly present in the skin and synovial fluid. Fibroblasts, chondrocytes, and synoviocytes are responsible for the synthesis of HA. Specific alterations in the ECM are observed due to the interaction of malignant cells with HA, which leads to tumor cell activity and cell motility, essential in metastatic progression. It is a group catalyst CD44 (cluster-determinant 44 receptor), a transmembrane glycoprotein. The isoform of CD44, i.e., CD44s, appears in healthy tissues, whereas CD44s is overexpressed in tumor cells. Hence, CD44 is also considered a biomarker in tumor cell diagnosis, mainly for the head, neck, bowel, breast, colorectal, prostate, pancreas, and bowel. CD44 receptor has an elevated affinity for the HA ligand. Hence, HA catalyst is a desirable protein for encapsulating inorganic-organic in core-shell nanoconstruct drugs. The surface-modified drug can target and select cancer cells for treatment and diagnosis via endocytosis between cancer cells and HA-CD44-mediated.

Examples include the covalent bonded encapsulated HA and paclitaxel (PTX) nanoconstructs, which increased the selective delivery of PTX to cancer cells. Cai et al. studied HA-doxrubicin (DOX) nanoconstruct in rodent breast tumor models. The surface modification of DOX with HA could deliver DOX to cancer cells and inhibit cells, which showed an excellent anticancer effect by inhibiting breast cancer progression in rodent breast cancer models, reducing cardiac toxicity, and reducing toxic effects in healthy tissues. Another technique is coating or embedding the nanoconstructs with HA to convert into HA surface layer targeting CD44 delivery, improving biocompatibility, blood circulation time, and active targeting capacity. For example, HA surface treatment with curcumin-loaded zinc nanoconstructs will effectively deliver curcumin into CD44-overexpressed CT26 tumor cells, thereby increasing the drug distribution in tumors and enhancing the antitumor effect (Bajracharya et al. 2022; Yu et al. 2010; Kim et al. 2018; Michalczyk et al. 2022).

4.3.3 Peptides

A peptide is a linear chain of amino acids comprising less than 50 amino acids stabilized by disulfide bonds. These are from various sources that show antitumor effects, such as natural sources:

1. Plant-derived: *Ganoderma lucidum* polysaccharide peptide (Gl-PP) is antiangiogenic.
2. Animal-derived: Atrial natriuretic peptide.
   (a) Peptides derived from animal proteins: Angiotensin and growth-inhibitory peptide (GIP) derived from α-fetoprotein.
   (b) Marine source peptides: Jaspamide and somocystinamide A (mediate apoptosis) and aplidin (causes cell cycle arrest).
   (c) Microbial peptides: Mycobacteria-derived muramyl dipeptide (MDP), FK565, and *Streptomyces*-derived bestatin (Luo et al. 2019).

Tumor-homing-tumor-targeting peptides are peptide ligands that are utilized to deliver drugs to tumors. They are developed to be tumor cell-specific to increase nanoparticle internalization into tumor cells (Bajracharya et al. 2022). These peptides attach to overexpressed molecules, and some may promote or inhibit signaling pathways in cancer cells or tumor tissue by binding to them. Peptides that target aberrant cellular signaling pathways are oncogenic signaling pathways, which affect cancer cell activity, apoptosis escape, and tumor cell proliferation (Cai et al. 2010). Therapeutic peptides are classified into two types: cell-targeting peptides (CTP) and cell-permeable peptides. Cell targeting peptides bond to molecular markers (cell membrane proteins) on the targeted cell, delivering conjugates to a specific cell type while protecting all other healthy cells from the often-toxic effects. They act either at the cell membrane or by binding to the molecular target, causing the peptide-therapeutic complex to be internalized. Cell-permeable peptides interact with charged components of the cell membrane. The positively charged peptide can preferentially target cancer cells since their outer cell membranes are negatively charged compared to normal cells (Karami Fath et al. 2022). A peptide-drug conjugate should have a tumor-homing peptide, a linker, and a cytotoxic agent, with the homing peptide being target-specific. The linker aids in conjugating medicines or cytotoxic agents to the tumor-homing peptide (Abbasi et al. 2014). Examples include the somatostatin receptor 2 (SSTR2) binding octreotide, conjugated to doxorubicin via a cleavable disulphide link and used to treat pituitary, pancreatic, and breast cancers (Li et al. 2021).

4.3.4 Transferrin

Transferrin (Tf) is a glycoprotein that spans the cell membrane and plays a crucial role in the transportation and control of iron within human cells. The substance attaches to the transferrin receptor (TfR) and facilitates the movement of iron inside cells. Transferrin undergoes a structural alteration upon binding to iron, which is essential for its specific identification by the transferrin receptor. Because iron is a crucial protein cofactor in vital life processes like cell proliferation and growth, numerous tumor cells exhibit persistent and highly expressed TfR on their surface. Compared to normal cells, the transferrin receptor is overexpressed in common tumors such as pancreatic, colon, and bladder cancer.

TfR expression is correlated with the malignancy of tumors, being more prominent in breast cancer, glioma, lung adenocarcinoma, chronic lymphocytic leukemia, and highly malignant liver cancer with straightforward metastasis. Researchers have shown interest in using it as an effective targeting technique for delivering medications to tumor cells. TfR can provide medicines to cancer cells or inhibit the receptor's regular activity, causing cancer cell death. There are two forms of TfR: TfR1 and TfR2. TfR1 controls iron absorption and cellular proliferation, while TfR2 is in the liver. TfR1's high affinity to Tf and its overexpression in cancer cells make it a promising target for enhancing the efficacy of chemotherapy. Curcumin was found to promote autophagy and death in several tumor cells by suppressing TfR1 expression, suggesting its potential as a TfR1 inhibitor for cancer treatment (Vadevoo et al. 2023; Lelle et al. 2015).

4.4 Stimuli-Responsive Targeting

PEGylation encapsulates the antitumor drug's surface in nanoconstructs, providing a stealth effect. It is a widely used strategy to increase blood circulation time in tumor cells, leading to the accumulation of antitumor drugs in nanoconstructs through enhanced penetrability and retention effect. The drawback of PEGylation is that it disturbs the tumor cell merging of nanoconstructs, resulting in incomplete therapeutic effectiveness. This process is known as the PEG dilemma. It can be overcome by applying the shell-separable nanoconstructs, which will separate the PEG shell from the nanoconstructs in chosen tumor tissues in retort to the local environment (stimuli-responsive targeting). It promotes cellular uptake and enhances antitumor efficacy. Stimuli-responsive targeting enhances controlled release and promotes tumor accumulation at tumor microenvironments or intracellular spaces of cancer cells by responding to external or internal stimuli. External stimuli (temperature, magnetic field, light, etc.) and endogenous (internal) stimuli, such as pH, enzymes, oxidative stress, etc., can affect nanoconstructs' effectiveness. Light in the red and near-infrared (NIR) region (650–950 nm) is considered an external stimulus and used in stimuli-responsive drug delivery systems, which are used in photothermal and photodynamic therapy, thereby improving chemotherapeutic effects. This targeting is also used to overcome multidrug resistance in cancer treatment (Hong et al. 2023; Shen et al. 2018).

4.5 Bioinspired Membrane-Coated Nanosystems

This approach involves using cell-derived membranes to cover nanoparticles. These systems offer benefits such as biocompatibility and safety due to the presence of natural cell membranes. Encapsulating nanoparticles with these membranes allows for extended presence in the body without eliciting an immunological reaction. The membrane-coated nanosystems have low protein corona formation and do not experience a loss of targeting ability seen in covalently conjugated ligands. Additionally, using membranes derived from activated immune cells or bacteria can have secondary roles in regulating immune responses in the tumor microenvironment. Several membrane sources include blood cells, such as erythrocytes, platelets, and leukocytes (Desai et al. 2023).

(a) Cancer cells
(b) Stem cells
(c) Extracellular vesicles
(d) Viral capsids
(e) Bacteria: Bacterial cell membrane, outer membrane vesicles (OMV) present in gram-negative bacteria (Zhu et al. 2019).

Li et al. researched using erythrocyte membrane-coated upconversion nanoparticles (UCNPs) for pre-targeted multimodality imaging of triple-negative breast cancer. They integrated UCNPs with erythrocyte-derived membranes. The tumor absorption was increased by adding DSPE-PEG-FA molecules to the cell membranes. They created a platform utilizing positron emission tomography (PET) in conjunction with MRI and UCL imaging to improve the detection of deep tissues. PET imaging was conducted in 4T1 breast cancer-bearing mice using a pre-targeting technique and "in vivo" click chemistry. The blood biochemistry profile, hemocompatibility testing, and histologic examination indicated favorable in vivo biocompatibility of the nanosystem (Mi 2020).

5 Application

5.1 Dual Imaging Modalities

The inorganic core, often made of materials like gold (Au) or quantum dots (QDs), provides contrast for imaging techniques such as magnetic resonance imaging (MRI) or computed tomography (CT). The organic shell can be functionalized with organic dyes or fluorophores for optical imaging, providing complementary information to MRI or CT. This dual imaging capability allows for better visualization of tumors and monitoring of treatment response.

5.2 Targeted Drug Delivery

The organic shell can be tailored to have specific ligands or antibodies for targeting cancer cells. This targeting ability enhances the accumulation of therapeutic agents at the tumor site, improving efficacy and reducing side effects. The inorganic core can be loaded with drugs, such as chemotherapeutic agents or gene therapy vectors. The core protects the drugs from degradation and provides controlled release, enhancing their therapeutic potential (Chiozzi and Rossi 2020).

5.3 Photothermal Therapy (PTT)

Inorganic cores like gold nanoparticles strongly absorb near-infrared (NIR) light. When exposed to NIR light, they generate heat, which can selectively ablate cancer cells. The organic shell can enhance the stability and biocompatibility of the nanoparticles during PTT. Additionally, the shell can be functionalized with targeting ligands for specific cancer cell destruction.

5.4 Thermo-chemotherapy

Combined with PTT, inorganic-organic core-shell nanoconstructs can deliver chemotherapy drugs to the tumor site. The heat generated by the inorganic core improves drug release and penetration into the tumor, synergizing the effects of chemotherapy and PTT.

5.5 Enhanced Biocompatibility

The organic shell provides a biocompatible surface, reducing potential cytotoxicity and improving circulation time in the body. Inorganic-organic nanoconstructs can be designed to evade the immune system, minimizing clearance and maximizing accumulation at the tumor site.

5.6 Real-Time Monitoring

The organic shell can be engineered to respond to specific stimuli, such as pH or enzyme activity in the tumor microenvironment. This "smart" behavior allows for real-time monitoring of drug release and treatment efficacy. Some inorganic cores exhibit unique properties under certain conditions, such as changes in magnetic or optical properties. Monitoring these changes can provide valuable information during treatment.

5.7 Multimodal Theranostics

Nanoconstructs can offer multimodal imaging and therapy by combining different inorganic cores and organic shells. For example, a single nanoconstruct could provide MRI, optical imaging, PTT, and targeted drug delivery in a single platform.

6 Conclusion

In conclusion, nonmagnetic inorganic-organic core-shell nanoconstructs represent a promising avenue for cancer theranostics, offering a synergistic combination of biocompatibility, targeted drug delivery, imaging capabilities, and controlled therapy. Their multifunctional nature and ability to respond to the tumor microenvironment make them valuable tools in personalized medicine, paving the way for more effective and tailored cancer treatments with reduced side effects. Continued research and development in this field hold the potential to translate these nanoconstructs into clinically viable solutions for improved cancer diagnosis and therapy.

References

Abbasi E, Aval SF, Akbarzadeh A, Milani M, Nasrabadi HT, Joo SW, Hanifehpour Y, Nejati-Koshki K, Pashaei-Asl R. Dendrimers: synthesis, applications, and properties. Nanoscale Res Lett. 2014;9:10.

Akbarzadeh A, Rezaei-Sadabady R, Davaran S, Joo SW, Zarghami N, Hanifehpour Y, Samiei M, Kouhi M, Nejati-Koshki K. Liposome: classification, preparation, and applications. Nanoscale Res Lett. 2013;8:1–9.

Allen TM, Cullis PR. Drug delivery systems: entering the mainstream. Science. 2004;3:1818–22.

Arvizo R, Bhattacharya R, Mukherjee P. Gold nanoparticles: opportunities and challenges in nanomedicine. Expert Opin Drug Deliv. 2010;7(6):753–63.

Bae KH, Chung HJ, Park TG. Nanomaterials for cancer therapy and imaging. Mol Cells. 2011;31:295–302.

Bajracharya R, Song JG, Patil BR, Lee SH, Noh HM, Kim DH, Kim GL, Seo SH, Park JW, Jeong SH, Lee CH. Functional ligands for improving anticancer drug therapy: current status and applications to drug delivery systems. Drug Deliv. 2022;29(1):1959–70.

Başağaoğlu H, Allwein S, Succi S, Dixon H, Carrola JT, Stothoff S. Two- and three-dimensional lattice Boltzmann simulations of particle migration in microchannels. Microfluid Nanofluid. 2013;15:785–96.

Cai SA, Thati S, Bagby TR, Diab HM, Davies NM, Cohen MS, Forrest ML. Localized doxorubicin chemotherapy with a biopolymeric nanocarrier improves survival and reduces toxicity in xenografts of human breast cancer. J Control Drug Deliv Release. 2010;146:212–8.

Cai D, Gao W, Li Z, Zhang Y, Xiao L, Xiao Y. Current development of nano-drug delivery to target macrophages. Biomedicine. 2022;10(5):1203. https://doi.org/10.3390/biomedicines10051203.

Chiozzi V, Rossi F. Inorganic–organic core/shell nanoparticles: progress and applications. Nanoscale Adv. 2020;2:5090–105.

Dai W, Wang X, Song G, Liu T, He B, Zhang H, et al. Combination antitumor therapy with targeted dual-nanomedicines. Adv Drug Deliv Rev. 2017;115:23–45.

Dam DHM, Lee RC, Odom TW. Improved in vitro efficacy of gold nanoconstructs by increased loading of G-quadruplex aptamer. Nano Lett. 2014;14:2843–8.

Desai N, Rana D, Pande S, Salave S, Giri J, Benival D, Kommineni N. "Bioinspired" membrane-coated nanosystems in cancer theranostics: a comprehensive review. Pharmaceutics. 2023;15(6):1677.

Dinarvand R, Sepehri N, Manoochehri S, Rouhani H, Atyabi F. Polylactide-co-glycolide nanoparticles for controlled delivery of anticancer agents. Int J Nanomedicine. 2011;27:877–95.

Gadade DD, Pekamwar SS. Cyclodextrin based nanoparticles for drug delivery and theranostics. Adv Pharm Bull. 2020;10(2):166.

Gagliardi A, Giuliano E, Venkateswararao E, Fresta M, Bulotta S, Awasthi V, Cosco D. Biodegradable polymeric nanoparticles for drug delivery to solid tumors. Front Pharmacol. 2021;12:601–26.

Hamidu A, Pitt WG, Husseini GA. Recent breakthroughs in using quantum dots for cancer imaging and drug delivery purposes. Nanomaterials. 2023;13(18):2566. https://doi.org/10.3390/nano13182566.

Hassanin I, Elzoghby A. Albumin-based nanoparticles: a promising strategy to overcome cancer drug resistance. CDR. 2020;3(4):930.

Hong S, Choi DW, Kim HN, Park CG, Lee W, Park HH. Protein-based nanoparticles as drug delivery systems. Pharmaceutics. 2020;12(7):604.

Hong L, Li W, Li Y, Yin S. Nanoparticle-based drug delivery systems targeting cancer cell surfaces. RSC Adv. 2023;13(31):21365–82.

Karami Fath M, Babakhaniyan K, Zokaei M, Yaghoubian A, Akbari S, Khorsandi M, Soofi A, Nabi-Afjadi M, Zalpoor H, Jalalifar F, Azargoonjahromi A. Anti-cancer peptide-based therapeutic strategies in solid tumors. CMBL. 2022;27(1):33.

Karlsson J, Vaughan HJ, Green JJ. Biodegradable polymeric nanoparticles for therapeutic cancer treatments. Annu Rev Chem Biomol Eng. 2018;9:105–27.

Kim JH, Moon MJ, Kim DY, Heo SH, Jeong YY. Hyaluronic acid-based nanomaterials for cancer therapy. Polymers. 2018;10(10):1133.

Kim BS, Chen Y-T, Srinoi P, Marquez MD, Lee TR. Hydrogel-encapsulated mesoporous silica-coated gold nanoshells for smart drug delivery. Int J Mol Sci. 2019;20:3422. https://doi.org/10.3390/ijms20143422.

Lelle M, et al. Octreotide-mediated tumor-targeted drug delivery via a cleavable doxorubicin–peptide conjugate. Mol Pharm. 2015;12:4290–300.

Li M, Fang H, Liu Q, Gai Y, Yuan L, Wang S, Li H, Hou Y, Gao M, Lan X. Red blood cell membrane-coated upconversion nanoparticles for pretargeted multimodality imaging of triple-negative breast cancer. Biomater Sci. 2020;8:1802–14.

Li CM, Haratipour P, Lingeman RG, Perry JJ, Gu L, Hickey RJ, Malkas LH. Novel peptide therapeutic approaches for cancer treatment. Cells. 2021;10(11):2908.

Liu HL, Hou P, Zhang WX, Wu JH. Colloid surface. Physicochem Eng Aspect. 2010;356:21–7.

Luo Z, Dai Y, Gao H. Development and application of hyaluronic acid in tumor targeting drug delivery. Acta Pharm Sin B. 2019;9(6):1099–112.

Mi P. Stimuli-responsive nanocarriers for drug delivery, tumour imaging, therapy and theranostics. Theranostics. 2020;10(10):4557.

Michalczyk M, Humeniuk E, Adamczuk G, Korga-Plewko A. Hyaluronic acid as a modern approach in anticancer therapy-review. Int J Mol Sci. 2022;24(1):103.

Mishra S, Bhatt T, Kumar H, Jain R, Shilpi S, Jain V. Nanoconstructs for theranostic application in cancer: challenges and strategies to enhance the delivery. Front Pharmacol. 2023a;14:1–18.

Mishra S, Bhatt T, Kumar H, Jain R, Shilpi S, Jain V. Nanoconstructs for theranostic application in cancer: challenges and strategies to enhance the delivery. Front Pharmacol. 2023b;14:1101320.

Mohammed MA, Syeda JT, Wasan KM, Wasan EK. An overview of chitosan nanoparticles and its application in non-parenteral drug delivery. Pharmaceutics. 2017;9(4):53.

Narayan R, Nayak UY, Raichur AM, Garg S. Mesoporous Silica Nanoparticles: A Comprehensive Review on Synthesis and Recent Advances. Pharmaceutics. 2018;10(3):118. https://doi.org/10.3390/pharmaceutics10030118

Oldenburg SJ, Averitt RD, Westcott SL, Halas NJ. Chem Phys Lett. 1998;288:243–7.

Perumal S, Atchudan R, Lee W. A review of polymeric micelles and their applications. Polymers. 2022;14(12):2510.

Schmidt PC. Encyclopedia of pharmaceutical technology. New York: Marcel Dekker; 1990. https://doi.org/10.1002/pauz.19900190309.

Selvarajan V, Obuobi S, Pui Lai Rachel E. Silica nanoparticles—a versatile tool for the treatment of bacterial infections. Front Chem. 2020;8:602.

Shabbir H, Wojnicki M. Recent Progress of Non-Cadmium and Organic Quantum Dots for Optoelectronic Applications with a Focus on Photodetector Devices. Electronics. 2023;12(6):1327. https://doi.org/10.3390/electronics12061327

Shen Y, Li X, Dong D, Zhang B, Xue Y, Shang P. Transferrin receptor 1 in cancer: a new sight for cancer therapy. Am J Cancer Res. 2018;8(6):916.

Singh S, Dhawan A, Karhana S, Bhat M, Dinda AK. Quantum dots: an emerging tool for point-of-care testing. Micromachines. 2020a;11:1058.

Singh KRB, Nayak V, Sarkar T, Singh RP. Cerium oxide nanoparticles: properties, biosynthesis and biomedical application. RSC Adv. 2020b;10:27194–214.

Sounderya N, Zhang Y. Recent Pat Biomed Eng. 2010;1:34–42.

Tella JO, Adekoya JA, Ajanaku KO. Mesoporous silica nanocarriers as drug delivery systems for anti-tubercular agents: a review. R Soc Open Sci. 2022;9(6):220013.

Vadevoo SM, Gurung S, Lee HS, Gunassekaran GR, Lee SM, Yoon JW, Lee YK, Lee B. Peptides as multifunctional players in cancer therapy. Exp Mol Med. 2023;1:1.

Yu BO, Tai HC, Xue W, Lee LJ, Lee RJ. Receptor-targeted nanocarriers for therapeutic delivery to cancer. Mol Membr Biol. 2010;27(7):286–98.

Zhao T, Nguyen N-T, Xie Y, Sun X, Li Q, Li X. Inorganic nanocrystals functionalized mesoporous silica nanoparticles: fabrication and enhanced bio-applications. Front Chem. 2017;5:118.

Zhu Y, Chen C, Cao Z, Shen S, Li L, Li D, Wang J, Yang X. On-demand PEGylation and dePEGylation of PLA-based nanocarriers via amphiphilic mPEG-TK-Ce6 for nanoenabled cancer chemotherapy. Theranostics. 2019;9(26):8312.

Zielińska A, Carreiró F, Oliveira AM, Neves A, Pires B, Venkatesh DN, Durazzo A, Lucarini M, Eder P, Silva AM, Santini A. Polymeric nanoparticles: production, characterization, toxicology and ecotoxicology. Molecules. 2020;25(16):3731.

Ziental D, Czarczynska-Goslinska B, Mlynarczyk DT, Glowacka-Sobotta A, Stanisz B, Goslinski T, Sobotta L. Titanium dioxide nanoparticles: prospects and applications in medicine. Nanomaterials. 2020;10(2):387. https://doi.org/10.3390/nano10020387.

Part IV

Organic/Inorganic Core Shell Nanoconstructs

Advances in Organic/Magnetic Core–Shell Nanoconstructs for Cancer Theragnostics

14

Mohit Angolkar, Sharanya Paramshetti, Madhuchandra Kenchegowda, Riyaz Ali M. Osmani, K. M. Asha Spandana, Amarjitsing Rajput, and K. Trideva Sastri

Abstract

Cancer diagnosis and treatment have significantly advanced in recent years, resulting in improved survival rates. Conventional methods such as surgery, radiotherapy, and chemotherapy, along with newer approaches like hormonal therapy and immunotherapy, have achieved notable success. However, challenges like post-surgery relapses, potential adverse effects, limited benefits to specific patient groups, and low response rates remain major concerns. Despite technological advancements, cancer continues to pose a global challenge, with GLOBOCAN reporting 19.9 million cases and 9.7 million deaths worldwide in 2023. This chapter explores the innovative use of organic/magnetic core–shell nanoconstructs in cancer theranostics—a field combining therapy and diagnostic functions within a single platform. These sophisticated drug delivery systems merge therapeutic actions with imaging modalities like magnetic resonance imaging (MRI), near-infrared (NIR) imaging, magnetic particle imaging (MPI), and photoacoustic imaging (PAI). By incorporating polymers and magnetic nanoparticles (MNPs), these nanoconstructs offer targeted drug delivery, enhanced imaging, and hyperthermia treatment, thus improving treatment precision and efficacy. The physical properties of MNPs allow them to act as both

M. Angolkar · S. Paramshetti · M. Kenchegowda · K. M. Asha Spandana · K. Trideva Sastri
Department of Pharmaceutics, JSS College of Pharmacy, JSS Academy of Higher Education & Research (JSS AHER), Mysuru, Karnataka, India
e-mail: asha@jssuni.edu.in; ktridevasastri@jssuni.edu.in

R. A. M. Osmani (✉)
Department of Pharmaceutics, College of Pharmacy, King Khalid University (KKU), Al Faraa, Abha, Saudi Arabia

A. Rajput
Department of Pharmaceutics, Poona College of Pharmacy, Bharti Vidyapeeth Deemed University, Erandwane, Pune, Maharashtra, India

J. K. Patel et al. (eds.), *Core-Shell Nano Constructs for Cancer Theragnostic*,
https://doi.org/10.1007/978-981-96-3025-7_14

imaging probes and drug delivery systems, offering real-time monitoring and personalized treatment options. This chapter delves into the design strategies, surface modifications, and functionalization of magnetic core–shell nanoconstructs, highlighting their biomedical applications, challenges, and future prospects in the realm of cancer theranostics.

Keywords

Cancer theranostics · Magnetic nanoconstructs · Core–shell nanoparticles · Biosensors

1 Introduction

The past few years have seen significant advancements in cancer diagnosis and treatment, resulting in higher survival rates. Traditional methods like surgery, radiotherapy, and chemotherapy, along with newer approaches such as hormonal therapy and immunotherapy, have achieved notable success in treating cancer. Despite these advancements, there are still considerable challenges, including the risk of relapse after surgery, the adverse side effects of therapies, limited effectiveness for certain patient groups, and generally low response rates (Roy Chowdhury et al. 2016; Farkona et al. 2016; Ben Djemaa et al. 2018).

Despite advancements in technology and research, cancer continues to be a significant global health challenge. According to current GLOBOCAN data, there were 19.9 million new cancer cases worldwide in 2023, leading to 9.7 million deaths across all cancer types (Cancer Today 2024). Cancer, characterized by multifactorial neoplastic diseases and high mortality rates, holds profound significance in the medical, economic, and social sectors. This is coupled with increased patient expectations and stricter treatment demands. Over time, drug delivery systems (DDSs) have become increasingly sophisticated, integrating therapeutic actions with imaging modalities like near-infrared (NIR) imaging, magnetic resonance imaging (MRI), photoacoustic imaging (PAI), and magnetic particle imaging (MPI). This integration, known as theranostics, enables more precise treatment delivery, enhancing efficacy. To create effective theranostic drug delivery systems, it is necessary to design vehicles that can simultaneously adsorb and release drugs, provide imaging capabilities, and offer additional therapeutic benefits. Combining polymers with magnetic nanoparticles (MNPs) is particularly advantageous. Polymers can serve as drug reservoirs and platforms for further functionalization, such as targeting of cells or for imaging purposes, while MNPs allow MPI, hyperthermia, and MRI, which are cancer therapy methods that can be synergistically combined with drug therapy.

Magnetic nanoparticles have recently emerged as multifunctional platforms in the nanotheranostic field, offering both imaging and therapeutic capabilities (Shi et al. 2017; Singh and Sahoo 2014). The physical characteristics of these nanosystems make them valuable both as imaging probes for the diagnosis of cancer as well as DDSs. This dual functionality is particularly advantageous for monitoring real-time therapeutic responses and supporting targeted therapy regimens and personalized medicine. Due to the magnetic susceptibility of these nanoparticles (NPs), they are oriented and controlled using an external magnetic field. These nanoparticles can be made from various materials, each exhibiting unique magnetic effects based on their spin and orbital characteristics. Their magnetic susceptibility is defined by how the nanoparticles' magnetization changes in response to an external magnetic field (Buschow et al. 2001).

Also, these MNPs can be configured in different constructs, such as single-phase particles, oriented chain arrays, core–shell nanoparticles with two phases and a polymer coat, or multicore nanoparticles (Angelakeris 2017). Among the various magnetic constructs utilized for multifunctional applications in cancer theranostics, magnetic nanoparticles are typically engineered with a magnetic core, a shell, and a polymer coating to enable targeting and therapeutic capabilities (Singh and Sahoo 2014; Kudr et al. 2017). The selection of this structure is driven by the intrinsic properties of the magnetic core, which allow it to function as an agent for hyperthermia therapy, MRI contrast agents, and controlled DDS. The components of the shell are crucial for protecting the core and addressing immunogenicity and biocompatibility challenges. Adding a polymer coating prevents nanoparticle aggregation, extends their half-life, facilitates controlled release of drugs, and provides functional groups for biomolecule conjugation. This allows for the incorporation of functionalization agents like chemotherapeutic and biotherapeutic agents (including proteins or nucleic acids), targeting agents, photosensitizing agents, and fluorescent agents. These additions enable applications in photodynamic therapy (PDT) or photothermal therapy (PTT) and enhance other cellular trafficking functions (Revia and Zhang 2016). This chapter provides an overview of magnetic core–shell nanoconstructs for cancer theranostics, emphasizing their design strategies, surface modifications, and functionalization. It explores the biomedical applications of core–shell MNPs, discusses the challenges associated with their use, and considers potential future prospects in the field.

2 Core–Shell Nanoconstructs

Core–shell nanoconstructs consist of a core material coated with another material. In biological applications, they offer significant advantages over simple nanoparticles, improving properties like reduced cytotoxicity; enhanced dispersibility, cytocompatibility, and biocompatibility; improved ability to conjugate with other bioactive substances; and increased chemical and thermal stability (Law et al. 2008; Sounderya and Zhang 2008). To elaborate more: (i) When nanoconstructs are inherently toxic, they can harm tissues and organs of the host. Coating the toxic core with a benign material significantly reduces toxicity and enhances biocompatibility. Additionally, the shell can improve the core material's properties. For example, in semiconductor materials, a shell can enhance optical properties and photostability. (ii) Hydrophilicity is crucial for dispersing nanoparticles in biological (aqueous) systems. Increasing bio-dispersibility, biocompatibility, and cytocompatibility make core–shell nanoparticles a viable alternative to conventional drug delivery vehicles. These nanoconstructs are also attractive to researchers due to the ease of synthesis. When the core material is hydrophobic, coating it with a hydrophilic material can solve issues of dispersibility and compatibility. (iii) Conjugating biomolecules to nanoconstruct surfaces is essential for many biological applications. Often, the core material may not easily bind to certain biomolecules. Coating the core with a suitable biocompatible material facilitates this conjugation, expanding the range of potential bio-applications. (iv) Core materials may be prone to thermal or chemical changes when exposed to their environment. Thus, the stability of these materials can be enhanced by coating these materials with some inert material. Therefore, core–shell nanoconstructs hold greater promise for biological applications compared to single-component nanoconstructs. Table 14.1 offers an overview of various core–shell nanoconstructs, detailing various core materials and their respective applications.

2.1 Coating of Magnetic Nanoparticles

A technique for in situ surface polymerization was devised in order to coat magnetic nanoparticles using polypyrrole and poly(3,4-ethylenedioxythiophene):poly(4-styrene sulfonate) (PEDOT:PSS) (Yan et al. 2017). The conjugated polymer enables NIR absorbance, and the nanoparticles' robust multimodal imaging abilities are employed through MRI and PAI. Additionally, effective hyperthermia treatment in mice resulted in successful tumor ablation.

Polyarabic acid, a prominent constituent of gum acacia, is biocompatible and assists in penetrating cell membranes. By coating magnetic nanoparticles with polyarabic acid and subsequently functionalizing and loading with doxorubicin (DOX), theranostic nanosystems were developed. These systems exhibited exceptional cell penetration, uptake of DOX, and pH-sensitive drug release (DOX) in breast cancer cells (Patitsa et al. 2017). Additionally, the nanoparticles exhibited minimal cytotoxicity, excellent biocompatibility, and contrast characteristics that are comparable to conventional commercial agents. Applying coatings of chitosan, dextran, and polyethylene glycol (PEG) provided improved control over the properties of the polymer coat, thereby enhancing applications in both hyperthermia and MRI (Zahraei et al. 2016).

A nanotheranostic nanosystem was engineered by conjugating magnetite nanoparticles with cyclodextrin-based nanosponges (CDNSs), which were subsequently functionalized using folic acid (FA) (Gholibegloo et al. 2019). Curcumin was encapsulated within the lipophilic cavity of the CD, and an external magnetic field was utilized to direct the nanosystem toward the site of the tumor. The acidic environment of the tumor microenvironment then triggered the controlled drug release.

A core–shell magneto-fluorescent nanogel was formulated, featuring iron oxide nanoparticle (IONP) cores surrounded by a shell with photoluminescence. Co-macromers consisting of PEG–maleic acid–glycine were cross-linked together

Table 14.1 Overview of organic core–shell nanoconstructs on the basis of various core materials and their applications

Core/shell material	Surface functionalization	Application	Reference
PEG/PCL	Biomarkers	Drug delivery	Li et al. (2008)
Polycaprolactone/dextran	Biomarkers, lectin		Rodrigues et al. (2003)
Cholesterol/chitosan	MPEG		Jang et al. (2010)
Ferrite-impregnated acrylonitrile/acrylamide	Various biomarkers and stabilizers		Sahiner and Ilgin (2010)
Chitosan/cholesterol	Folate, PEG		Wang et al. (2010)
Polystyrene/Poly(butyl-2-cyanoacrylate)	Thioflavin receptors		Siegemund et al. (2006)
PLGF-PLAF/PLEOF	Various biomarkers and stabilizers		He et al. (2008)
PMMA/PEI	Lactate, aspartate, and biomarkers		Feng and Li (2007)
PLGA/PEG	Folate and various biomarkers and stabilizers	Drug targeting	Kim et al. (2010)
Chitosan/polyacrylamide	Gene insertion and associated targeted molecules	Gene transfection	Chen et al. (2005)
PAEA128-b-PS40/DNA	Gene insertion and associated targeted molecules		Zhang et al. (2010)
PGLA/DNA-functionalized glycol/chitosan	Folate and other biomarkers		Lee et al. (2010)
Polylysine/PELGE	Gene insertion and associated targeted molecules		Nie et al. (2007)
Histidine and lysine oligopeptide/DNA	Gene insertion and associated targeted molecules		Wiradharma et al. (2008)
L-aspartate/PEI	Gene insertion and associated targeted molecules		Yu et al. (2009)
Oligopeptide/DNA-PEG	Gene insertion and associated targeted molecules		Harada-Shiba et al. (2002)
N-isopropylmethacrylamide/N,N'-methylenebisacrylamide	EGFR, peptide recognition tag	Drug targeting and gene silencing	Dickerson et al. (2010)

(continued)

Table 14.1 (continued)

Core/shell material	Surface functionalization	Application	Reference
Fe_3O_4 embedded in poly(DL-lactide) (PLA)/polyvinyl alcohol (PVA)	–	Ultrasound, MRI	Yang et al. (2009)
Cy5 dye/SiO_2	PEG	Optical imaging	Burns et al. (2009)
Fluorescein isothiocyanate dye/ SiO_2	–	Optical imaging and cell labeling	He et al. (2004)

to form the structure (Vijayan et al. 2019). The nanogel demonstrated high cytocompatibility, efficient cellular uptake, and capabilities for fluorescent imaging and hyperthermia treatment. In another nanotheranostic nanosystem, crosslinking was employed to link α-lactalbumin molecules together using PEG and glutaraldehyde (Delavari et al. 2019). The resultant polymer complex was subsequently conjugated to MNPs with PEI. This redox-sensitive protein complex facilitated the precise release of the chemotherapy drug DOX directly within the acidic tumor environment.

In yet another instance, a thermo-responsive fluorescent polymer (TFP) was attached to the surface of an IONP to create a biodegradable and nontoxic theranostic agent (Pandey et al. 2020). To achieve this, allylamine, poly(N-isopropylacrylamide) (PNIPAM), and a fluorescent polymer were subjected to copolymerization and then conjugated to nanoparticles to form TFP-NPs through a free radical polymerization mechanism. In vitro studies confirmed that DOX-loaded polymer-NP conjugates exhibited low cytotoxicity and high biocompatibility.

2.2 Nanoprecipitation of a Polymer and Nanoparticles

A pH-switch nanoprecipitation technique was employed to encapsulate iron oxide nanoparticles using a copolymer–drug conjugate that includes a tumor-homing peptide (iRGD) known for its specificity toward tumors (Herranz-Blanco et al. 2016). The nanosystem demonstrated enhanced lysosomal escape facilitated by poly(histidine), leveraging the proton sponge effect for polymer release into the cytoplasm. Strong magnetic responsiveness was achieved through clustering, enabling magnetic-guided treatment. Moreover, intracellular rupture of the DOX–polymer linkage facilitated the efficient delivery of DOX, leading to its accumulation of tumor cells in the nuclei.

IONPs and DOX were incorporated into a copolymer of poly(N-ε-carbobenzyloxy-L-lysine) coupled with hyaluronan to develop micelles through nanoprecipitation. These micelles were designed for targeted detection and therapy of tumor (Yang et al. 2020). The copolymer employed for coating comprised polypeptides and polysaccharides linked via disulfide bonds, which, when exposed to the higher concentrations of glutathione (GSH) present in the tissues of the tumor, facilitated the release of doxorubicin. In vitro tests conducted with HepG2 cells demonstrated accelerated release of the drug for a period of 24 h from these redox-sensitive theranostic nanoparticles. Additionally, there was enhanced intracellular uptake under conditions mimicking high glutathione concentrations found in tumor cells.

2.3 Self-Assembly in Smart Material Formation

A nanoparticle beacon was developed by coating ferrihydrite and gold nanoparticles with a flexible polymer. This polymer was specifically designed with amino-group-terminated oligonucleotides featuring terminal biotin. Conjugation of DNA and streptavidin was utilized to assemble a highly sensitive, input-switchable structure (Cherkasov et al. 2020). This controlled the accessibility of the terminal receptor group to facilitate target binding. DNA serves as a molecular trigger, initiating transformations in the smart material, enabling binding to cancer cells, and facilitating the release of the drug. Biotin's affinity toward the surface of the gold shields it from interacting until the activation of the complex by the specific input DNA, allowing interaction of biotin with streptavidin (target). These constructs exhibit a unique signal transmission pathway between input (biochemical cue) and a targeting receptor (nanoprobe) through interactions of surface polymer, promising significant future applications.

3 Surface Modifications

For MNPs to be utilized in both in vitro and in vivo settings, surface functionalization is essential to achieve the following: (i) Prevent agglomeration to maintain nanoparticle stability, (ii) ensure biocompatibility and introduce chemical functionalities for drug conjugation and attachment of targeting ligands, (iii) reduce nonspecific cell interactions, and (iv) improve the pharmacokinetics of MNPs (Gupta et al. 2007). A wide range of inorganic and organic coatings have been studied, such as PEG, dimercaptosuccinic acid (DMSA), chitosan, dextran, liposomes, silica, and gold (Jun et al. 2005; Kohler et al. 2004; Xie et al. 2007; Mornet et al. 2005; Kim et al. 2009; Pradhan et al. 2010; Wang et al. 2008; Ma et al. 2007). Coating of MNPs can be done via various techniques, such as in situ coating, post-synthesis end grafting, and post-synthesis adsorption (De et al. 2011).

Polyethylene glycol is an amphiphilic and neutral polymer that has been extensively utilized in FDA-approved pharmaceutical formulations as excipients (Fuertges and Abuchowski 1990). Coating MNPs with PEG enhances their ability to disperse in biological media and prolongs their circulating time in the bloodstream. This is because PEG-coated nanoparticles are less likely to be recognized and cleared by the reticuloendothelial system (RES) (Harris and Chess 2003). Lutz and coworkers (Lutz et al. 2006) showed the in situ coating of Fe_3O_4 MNPs with polyethylene glycol under aqueous conditions. They also achieved PEG grafting using a single-point chemical anchoring technique through various functional groups such as dopamine, phosphate derivatives, and silanes (Kohler et al. 2004; Kim et al. 2005; Xu et al. 2004). Peng and Sun described ligand exchange involving bifunctional polyethylene glycol and dopamine (Peng et al. 2006). Recently, it has been suggested that nitrodopamine is a highly stable chemical to anchor MNPs (Amstad et al. 2009). Highly stable MNPs were developed by coating Fe_3O_4 with bifunctional polyethylene glycol conjugated to nitrodopamine and terminal carboxylate groups (Jaiswal et al. 2014). Nitrodopamine was attached covalently to the surface of Fe_3O_4 at one end, leaving the terminal carboxylate open for functionalization by targeting therapeutic agents or ligands.

Dextran is frequently employed for coating MNP due to its biocompatibility and ability to engage in polar interactions such as hydrogen bonding and chelation. Adding dextran during the synthesis of Fe_3O_4 resulted in Fe_3O_4 MNPs coated with dextran (Molday and MacKenzie 1982). Further refinements of this method led to the development of clinically approved products such as ferumoxides (AMI-25) and ferumoxtran-10 (AMI-277) (Weissleder and Pittet 2008; Medarova et al. 2007; Kumar et al. 2010; DeVita and Chu 2008; Kohler et al. 2006; Sun et al. 2008). These have cores of approximately 5 nm but exhibit significant differences in the thickness of dextran coating—ferumoxtran ranges from 20 to 40 nm, whereas ferumoxides range from 80 to 150 nm. These variations contribute to differing blood circulation times, with ferumoxtran having a circulation time of around 24 h and ferumoxides around 2 h (Wang et al. 2001). Due to the nonspecific adherence of dextran molecules because of hydroxyl interactions with the Fe_3O_4 core, there is always a potential for desorption (Carmen Bautista et al. 2005). In order to prevent desorption, dextran was cross-linked chemically on the surface of the MNPs (Wunderbaldinger et al. 2002). By employing this technique, clinically approved products such as ferucarbotran and ferumoxytol have been developed (Li et al. 2005; Reimer and Balzer 2003).

Silica coating on MNPs is favored due to its aqueous stability and ease of synthesis. Hydrolyzing silica in a basic solution allows for achieving a controllable and uniform thickness of silica coating on MNPs (Ma et al. 2007; Lu et al. 2002; Pinho et al. 2010a). Anticancer drugs and fluorescent agents like fluorescein isothiocyanate (FITC) have been administered using these silica shells (Liong et al. 2008). Incorporating additional functions into the silica coating has enhanced its targeting ability and labeling capabilities. Adding (3-aminopropyl)triethoxysilane (APTES) to the silica precursors enabled the coating of silica shells with controlled thickness and primary amine groups (Aslam et al. 2007). Similarly, Lu et al. developed fluorescent MNPs as a multimodal diagnostic agent by reacting (3-aminopropyl) triethoxysilane with FITC-functionalized dyes (Lu et al. 2002). At present, ferumoxsil (AMI-121), an oral T2 contrast agent used to differentiate intestinal loops from surrounding tissues and organs, is available as silica-coated MNPs (Hahn et al. 1990).

3.1 Mitigating Phagocytosis Using Self-Peptide

The mononuclear phagocyte system (MPS) presents a significant challenge to nanoparticle delivery systems in vivo. To evade MPS recognition, most DDSs adopt a "passive stealth" strategy by incorporating polyethylene glycol or zwitterionic polymers. These polymers delay the adsorption of serum proteins onto nanovehicles, thereby reducing MPS-mediated phagocytosis. However, within the body, the cell surface protein CD47 on host cells interacts with phagocyte receptors, facilitating anti-phagocytosis. CD47 acts as a "self-peptide," aiding the mononuclear phagocyte system in distinguishing between host and foreign particles. Zhang and coworkers investigated an approach involving the assembly of biodegradable poly(lactide-co-glycolide)-PEG (PLGA-PEG) with synthetic CD47 self-peptide to interact with the SIRPα receptor expressed on phagocytes. This represents an "active stealth" strategy aimed at enhancing the anti-phagocytic properties of nanovehicles within the body (Zhang et al. 2017). IONPs can absorb anticancer drugs through the process of self-assembly into micelles, which enhances their circulation time in the bloodstream, facilitates imaging, and improves the delivery of drugs in vivo.

3.2 Modification Strategies for Enhanced Tumor Targeting

Functionalizing core–shell IONP-polymer constructs with folic acid enhances their targeting ability and uptake into the tumor cells through folic acid receptor-mediated endocytosis (Gholibegloo et al. 2019; Huang et al. 2017; Roy et al. 2017). Functionalization of NPs with cysteine and lysine derivatives has been shown to enhance ligand targeting and uptake by the tumor cells (Wei et al. 2017). A tissue-type plasminogen activator peptide was utilized to modify a magnetic construct engineered to target pancreatic cancer cells (Dobiasch et al. 2016).

3.3 Enhancing Blood–Brain Barrier Penetration with Surfactants

IONPs modified using Tween 80, PEI, and PEG, along with the application of magnetic field, facilitated in vivo penetration of the blood–brain barrier (BBB). This suggests promise for future theranostic delivery through the BBB using Tween–nanoparticle conjugates (Huang et al. 2016). Research has demonstrated that other polymers such as vitamin E TPGS, Pluronic F68, and Brij 35 have been shown to enhance passage through the BBB, facilitating access to brain tumors such as glioblastoma (Luque-Michel et al. 2019). Lyophilization has proven effective for preserving theranostics, maintaining their physical and chemical characteristics as well as their magnetic properties even after prolonged storage (Luque-Michel et al. 2019; Yang et al. 2017).

4 Biomedical Applications

Core–shell nanoconstructs are predominantly tailored for biomedical applications, leveraging their surface chemistry to enhance binding affinity with drugs, receptors, ligands, and other biomolecules. This design strategy facilitates targeted drug delivery, molecular imaging, and theranostic applications in biomedical research and clinical settings (Sahoo and Labhasetwar 2003; Gilmore et al. 2008). This advancement has spurred the synthesis of novel nanoparticles that are more compatible with biological systems compared to bulk materials. Enhanced cytocompatibility and biocompatibility increase their therapeutic potential, paving the way for developing novel DDSs with enhanced characteristics such as prolonged residence time, enhanced bioavailability, reduction in dosage frequency, and increased specificity. For instance, a bioinspired polymeric coating on a hydrophobic drug can enable controlled drug release at the target site, triggered by ion-, temperature-, or pH-specific polymer degradation. This approach ensures precise drug delivery and enhances therapeutic efficacy (Mahmud et al. 2007; Panda et al. 2008; Van Tomme et al. 2008). Core–shell nanoconstructs are widely employed in bio-imaging due to their excellent biocompatibility in comparison to simpler nanoparticles (Chen et al. 2010). This is due to the core material of nanoconstructs. In core–shell nanoconstructs, the material of the shell plays a crucial role in determining surface characteristics such as biocompatibility and the ability to conjugate bioactive materials, facilitated by reactive entities on its surface. Tuning the thickness of the shell allows for optimizing the contrast agent's properties and for binding biomolecules, enabling applications in targeted delivery of drug, biosensing, specific binding, and other biomedical functions (Pinho et al. 2010b). Figure 14.1 illustrates a schematic representation of core–shell nanoconstructs designed for diverse biomedical uses.

4.1 Core–Shell Nanoconstructs in Bio-imaging

Bio-imaging techniques have significantly enhanced healthcare by advancing disease diagnosis, treatment, and prevention through imaging technologies. MRI, ultrasound, PET, and optical imaging have become pivotal for early disease detection, the exploration of molecular biology, and the assessment of medical interventions. Recent advancements in core–shell nanostructures have further propelled the capabilities of bio-imaging techniques.

4.1.1 Magnetic Resonance Imaging and Magnetic Particle Imaging

Theranostic nanosystems utilizing IONPs are typically employed as T2 (negative) MRI contrast agents. Enhanced MRI contrast is achieved when these nanoparticles exhibit higher magnetization and efficiently load water protons within a polymeric shell, optimizing their performance as contrast agents (Yang et al. 2020; Liao et al. 2017). Polymeric NP constructs incorporating Gd^{3+} complexes or Fe^{3+}-terpyridine complexes offer highly effective T1 (positive) MRI contrast enhancement (Roy et al. 2016; Fétiveau et al. 2019; Patra et al. 2018).

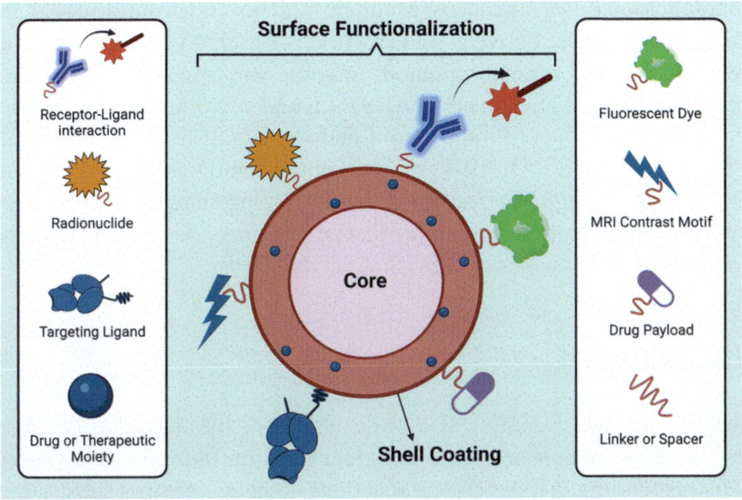

Fig. 14.1 Schematic of multifunctional nanoconstructs designed for drug delivery, molecular imaging, and therapy, which are potentially customizable for personalized diagnosis and treatments

Magnetic particle imaging, a novel imaging technique, directly measures the spatial distribution of IONPs in tissues by assessing their reaction to a magnetic field. It provides superior contrast compared to magnetic resonance imaging. A polymeric nanoparticle system has shown promising capabilities as an MPI contrast agent (Rost et al. 2020).

4.1.2 Fluorescence Imaging

Typically, fluorescence imaging is combined with magnetic resonance imaging using a polymer labeled with fluorescent markers (Pandey et al. 2020). For instance, functionalizing IONPs with a two-photon fluorescent dye-labeled polymer resulted in a nanoprobe that has the ability to quantify pancreatic beta cell mass (BCM), which serves as an indication of the onset of type 2 diabetes (Xin et al. 2020). Beta pancreatic cells' acidic environment caused dye release, which allowed for the fluorescent detection of BCM using confocal one- and two-photon microscopy. Also, the nanoprobe effectively prevented the toxic aggregation of human islet amyloid polypeptide, which is linked to beta cell degeneration in type 2 diabetes. Regarding magneto-fluorescent nanogels discussed in the earlier section, they exhibited dual emissions (green and red) when observed in HeLa cells under various excitation wavelengths, demonstrating their potential for fluorescence imaging applications in cancer cells (Vijayan et al. 2019).

4.1.3 Photoacoustic Imaging

Photoacoustic imaging (PAI) is indeed a relatively novel technique that combines optical excitation (typically laser pulses) with ultrasound detection to visualize tissues and structures (Attia et al. 2019). Combining photoacoustic imaging (PAI) with other imaging modalities enhances accuracy and provides multimodal imaging capabilities. For instance, Fetiveau and coworkers observed a robust photoacoustic signal within tumors when irradiated around 808 nm using Gd-containing Fe^{3+} [Fe^{2+} $(CN)_6$] NPs (Fétiveau et al. 2019). Lu and coworkers utilized dendrimer-stabilized nanoflowers incorporating gold and ultrasmall iron oxide nanoparticles (UIONPs) for integrated PAI, MRI, and computed tomography imaging (Lu et al. 2018). Multimodal imaging capabilities, incorporating magnetic resonance imaging, photoacoustic imaging, and fluorescence imaging, have been demonstrated with fluorescent Janus nanostructures (Song et al. 2019).

4.2 Biosensors

A biosensor is a type of analytical tool designed for the detection or analysis of biological samples by converting a biological reaction into an electrical signal. It operates by responding to biochemical reactions (such as enzyme–substrate interactions) or biomolecular interactions (like antigen–antibody binding, receptor–ligand interactions, nucleic acid–protein binding, or metal–macromolecule interactions). The device then translates these signals into electronic signals that can be quantified, amplified, and interpreted to produce understandable and actionable outputs.

A well-functioning biosensor typically includes a highly stable and specific biocatalyst that is capable of analyzing substrates through reactions that are independent of physical properties like pH, agitation, and temperature. It provides a linear, precise, accurate, and reproducible response within the desired range without needing to dilute or concentrate. The key components of a biosensor include the following: (i) Receptor: This is where the biological reaction occurs, converting the substrate into a product; (ii) Transducer: It converts the biological reactions or signals generated by the response into electrical signals that can be measured and processed; (iii) Amplifier: It enhances the electrical signal from the transducer to improve the signal-to-noise ratio and increase sensitivity; (iv) Processor: It processes the electrical signal to extract useful information, often involving algorithms or computational methods; and (v) Display unit: It presents the processed data in a human-readable format, making it comprehensible and actionable. These components work together to enable biosensors to detect and quantify specific analytes in biological samples, providing valuable insights for varied applications such as medical diagnostics, food safety, and environmental monitoring (Arora et al. 2011). The biosensor receptor typically mimics a metabolic function using biological substances. Enzymes are widely used as receptors, detecting substrates either in vitro or in vivo. Other receptors include immobile antigen/antibody pairs, nucleic acids, cell organelles containing nonspecific enzymes, or whole cells like microbes. Traditional biosensors immobilize these receptors using polymeric matrices or semipermeable membranes. Incorporating a bio-responsive shell around core–shell nanoconstructs enhances the surface area for interactions and reduces the particle size to significantly increase sensor sensitivity.

Biosensors are categorized into several types based on their transducing system, including calorimetric, optical, potentiometric, amperometric, optoelectronic, and piezoelectric. Among these, the optical, amperometric, and piezoelectric types are considered the most promising due to their heightened sensitivity and ease of design. Electronic, fluorescence, and crystallography techniques are currently the focus of extensive research in biosensor development. Nanoparticles, owing to their magnetic, optical, and electro-sensitive properties, significantly enhance the performance of these biosensor types. Field-effect transistor (FET)-based biosensors had been recognized as superior in terms of accuracy, resolution, and response time until recently.

4.3 Cancer Therapeutics

Magnetic nanoparticles are utilized in therapeutic applications as versatile tools capable of functioning as DDSs, gene carriers, and agents for thermoablation therapies like magnetic hyperthermia, PTT, and PDT (Vallabani and Singh 2018). Among therapeutic methods, hyperthermia is extensively studied. Different techniques such as microwaves, lasers, and ionizing radiation have been used for thermoablation. However, these methods can interfere with genetic materials and lack therapy selectivity, potentially damaging healthy cells. In contrast, magnetic nanoparticle-induced magnetic hyperthermia (MHT) offers externally regulated local heating focused on specific body regions, minimizing the risk of harm to healthy tissues compared to prior techniques (Bañobre-López et al. 2013).

PTT utilizes an infrared laser in order to activate light-absorbing magnetic nanoparticles, leading to a more concentrated heating effect per nanoparticle compared to MHT. However, a significant limitation is the penetration depth of incident infrared light. Therefore, combining both these thermal strategies into a single magnetic nanoparticle creates a synergistic therapeutic approach that responds both to external magnetic fields and incident light, overcoming individual technique limitations (Curcio et al. 2019).

A trimodal treatment can be achieved through a single nanoplatform by tuning magnetic nanoparticles for PTT, MHT, and PDT. Photodynamic therapy involves selective wavelength radiation that excites a photosensitizer, transferring energy to surrounding oxygen molecules to generate reactive oxygen species (ROS) and induce cell death. A nanohybrid system comprises a copper sulfide shell and multi-core iron oxide nanoparticles synthesized via the polyol technique serves as an optimized nanotheranostics nanosystem with an MHT- and MRI-responsive core and a PDT- and PTT-responsive shell. This integrated trimodal thermal nanosystem offers cumulative heating capacity, potentially enabling a low-dose benefit in cancer nanotheranostic (Curcio et al. 2019).

MNPs can function effectively as DDSs due to their enlarged surface area, which facilitates high drug loading capacities and optimized bioavailability. This capability allows for lesser administration of drug doses while enhancing the selectivity of tissues. This approach holds promise for improving therapeutic outcomes by delivering drugs more efficiently to targeted sites within the body (Gobbo et al. 2015). Moreover, chemotherapeutic agents can be incorporated into these nano-sized platforms by first coating the MNPs with molecules that provide attachment points for conjugating, complexing, or encapsulating the drug molecule. The loading capacity of MNPs for anticancer drugs has been extensively studied, with DOX being the most commonly used chemotherapeutic drug. DOX has demonstrated efficacy in treating various cancers, including breast, thyroid, lung, and ovarian cancers, among others. Its mechanism of action involves intercalation into DNA, inhibition of DNA repair mediated by topoisomerase II, and cell death through the generation of ROS (Thorn et al. 2011).

Another promising therapeutic strategy having potential applications in cancer treatment is gene delivery. Magnetic nanoparticles in conjunction with siRNA molecules are emerging as versatile nanoplatforms capable of disrupting protein translation in the cytoplasm and suppressing gene expression in tumor cells (Fig. 14.2).

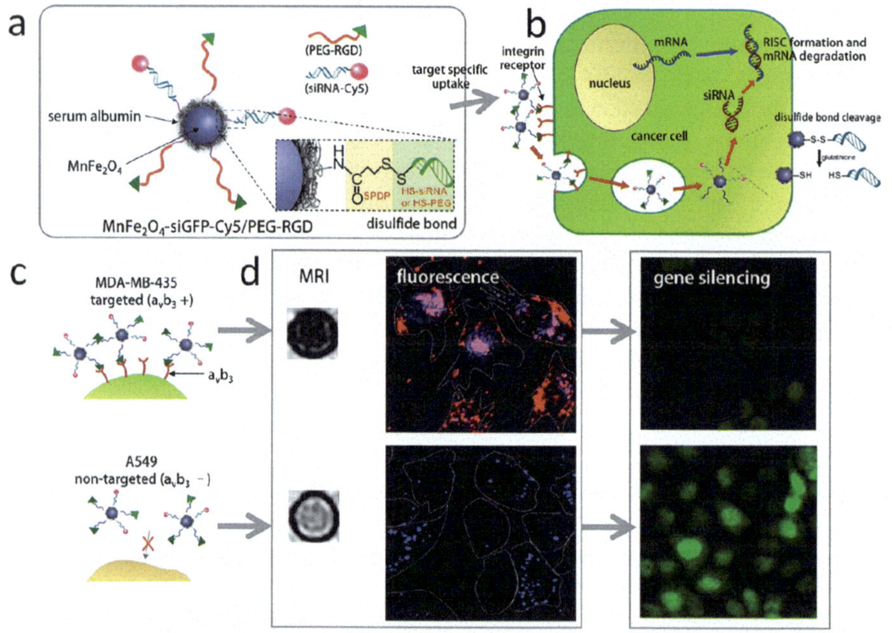

Fig. 14.2 (**a**) Diagrammatic representation of all-in-one nanoparticles of $MnFe_2O_4$-siGFP-Cy5/PEG-RGD for theranostic purposes. (**b**) Schematic representation detailing the intracellular mechanism of $MnFe_2O_4$-siGFP-Cy5/PEG-RGD NPs, starting from target-specific uptake of NPs to degradation by mRNA. (**c**) Schematic illustration of the target-specific binding of the NPs to Rυβ3 integrin positive cells. (**d**) T2-weighted magnetic resonance images of MDA-MB-435 and A549 cells posttreatment with $MnFe_2O_4$-siGFP-Cy5/PEG-RGD. Confocal microscopy images depicting the cellular distribution of $MnFe_2O_4$-siGFP-Cy5/PEG-RGD NPs. Evaluation of the gene-silencing effect of the multimodal NP systems with ($MnFe_2O_4$-siGFP-Cy5/PEG-RGD) and without a targeting moiety ($MnFe_2O_4$-siGFP-Cy5/PEG) on MDA-MB-435 and A549 cell lines. (Reprinted with permission from Yoo et al. 2011)

This approach integrates imaging and therapeutic functionalities due to the unique properties of MNPs, highlighting their potential as innovative theranostic agents (Yoo et al. 2011).

To date, there has been a growing diversity of MNPs tailored for cancer theranostics, with notable examples outlined in Table 14.2. The field of cancer nanotheranostics continues to advance, necessitating ongoing development of more cost-effective, specific, safer, and sensitive strategies to enhance the performance of MNPs in terms of efficiency and efficacy (Yang et al. 2018).

5 Challenges and Future Prospectives

Technological advancements in the synthesis, characterization, cytotoxicity, and biocompatibility of MNPs demonstrate significant progress. These advancements have enhanced the stability, quality, and application of these nano-sized systems for biomedical and theranostic purposes in diagnosis, imaging, and therapy. However, several fundamental aspects still warrant researchers' attention (Zhu et al. 2017).

Additional research is needed to delve into the biophysical mechanisms, explore diverse types of magnetic nanoparticles beyond IONPs, investigate novel constructs, and establish comprehensive nanotoxicity guidelines. Despite some MNPs entering clinical trials, the current focus remains primarily on either imaging or therapeutic applications rather than magnetic nanotheranostic systems. For instance, Feraheme® (ferumoxytol-MNPs) is used for magnetic resonance imaging but lacks authorization for cancer treatment or combined imaging and therapeutic applications. Additional studies are necessary to advance magnetic nanotheranostic systems toward clinical translation. Table 14.3 lists a few clinical trials involving MNPs for diagnosis and cancer treatment.

6 Conclusions

The field of MNPs has made significant strides, particularly in their application to nanotheranostics. Their nanoscale size, controllable physicochemical properties, responsiveness to external magnetic fields, and potential for multifunctional surface modifications make MNPs exceptional candidates for targeted imaging and drug delivery in theranostics. These attributes collectively enhance the precision and efficacy of cancer treatment. Key objectives in developing these nanosystems include achieving stability in the biological milieu, ensuring controlled release of drugs, achieving high diagnostic sensitivity, and minimizing toxicity. Robust guidance and classification systems are needed to effectively characterize and parameterize MNPs to meet these goals. Continued innovation is essential to advance the field, as underscored by ongoing research initiatives and investments. The future of cancer nanotheranostics holds promise for transformative biomedical applications, offering new strategies to address the complex challenges of cancer diagnosis and treatment. As research progresses, integrating MNPs into clinical practice is expected to increase,

Table 14.2 Multifunctional magnetic nanoparticles for imaging and therapy[a]

Magnetic nanoparticle system	Characteristic features	Diagnostic methods	Therapeutic applicability	Tumor	Ref.
Manganese-doped IONPs, coated with BSA and functionalized with a cyclic Arg-Gly-asp (cRGD) peptide and cy5 dye-labeled siRNA	Core diameter:15 nm	MRI	Inhibition of green fluorescence protein using siRNA component and disruption of receptor-mediated endocytosis targeting of tumor cells overexpression of $\alpha v\beta 3$ integrin with the RGD peptide	Breast	Lee et al. (2009), Danhier et al. (2012)
Rituximab-loaded liposome, iron oxide MNP, encapsulated with PEG	Average core size: 7–10 nm; zeta potential of −9.0 mV; PDI: 0.1–0.3; SPION-PVA encapsulation efficiency: 44.6%	MRI	Rituximab	Brain lymphoma	Saesoo et al. (2018)
cRGD-functionalized, doxorubicin-conjugated and ^{64}Cu-labeled SPION	Mean core diameter:10 nm; T2 relaxivity coefficient: 101.9 $mM^{-1}{\cdot}s^{-1}$; mean hydrodynamic size: 68 nm; loading capacity: 5.8% *w/w;* ^{64}Cu $T_{1/2}$:12.7 h	PET; MRI	Doxorubicin chemotherapy	Glioblastoma	Yang et al. (2011)
Paclitaxel-loaded, polyethylene glycol-modified liposome iron oxide (IO) MNP	Full nanoplatform size:168.3 nm; core size: 7 nm; zeta potential: −10.5 mV; PDI: 0.197; paclitaxel entrapment efficiency above 90%	MRI	Paclitaxel	Breast	Silva et al. (2019)
Indium-111-labeled, trastuzumab-doxorubicin-conjugated, and APTES-PEG-coated SPION	Mean diameter: 16 nm; radiolabeling efficiency: 97.6%; magnetization: 52 emu/g; trastuzumab conjugation capacity: 63.79%	MRI; SPECT	Antibody and chemotherapeutic agents Suppression of tumor	Breast	Zolata et al. (2015)

Gold nanorod-capped magnetite core/ mesoporous silica shell NPs	Homogenous size distribution; mean diameter: 386.6 nm; T2 relaxivity coefficient: 393.8 $mM^{-1}\cdot s^{-1}$; no reported cytotoxicity <100 μg/mL; DOX loading capacity: 30% w/w and positive therapy effect: < 39–42 °C; absorption peak: 90 nm	MRI	Doxorubicin chemotherapy; PTT	–	Ma et al. (2012)
Liposome, ADT-loaded iron oxide MNP, encapsulated with PEG	Core size of 7 nm; PDI of 0.19; final size of 211 nm; T_2* of 12.85 ms; ADT loading capacity: 49.6%	MRI	H_2S	Liver	Liu et al. (2017)
Multicore IONP with CuS shell	Mean core diameter: 25.5 nm; zeta potential: −14.1 mV at pH 7; hydrodynamic size: 156 nm; magnetization: 84 emu/g	MRI	MHT; PTT; PDT	–	Curcio et al. (2019)
Gold shell–core IONP	Mean diameter: 100 nm; T2 relaxivity coefficient: 76.2 $mM^{-1}\cdot s^{-1}$;hydrodynamic size: 179 nm	PAI; MRI	PTT	Breast	Huang et al. (2015)

[a]Abbreviations: *BSA* bovine serum albumin, *MRI* magnetic resonance imaging, *SPION* superparamagnetic iron oxide nanoparticle, *PEG* polyethylene glycol, *PAI* photoacoustic imaging, *APTES* (3-aminopropyl) triethoxysiliane, *SERS* surface-enhanced Raman spectroscopy, *cRGD* cyclic arginine–glycine–aspartate peptide, *cy5 dye* cyanine dye, *ADT* hydrophobic anethole dithiolethione, *RGD* arginyl–glycyl–aspartic acid, *NIR* near-infrared, *US* ultrasound, *T2** decay of transverse relaxation, results of spin–spin relaxation and inherent inhomogeneity of the main magnetic field

Table 14.3 A non-exhaustive list of clinical trials investigating the use of magnetic nanoparticles in the diagnosis and/or treatment of cancer[a]

Magnetic nanoparticle	Applications	Tumor site	Clinical trial	Status
Paclitaxel albumin-stabilized NPs	Chemotherapy	Breast	NCT00675259	Completed
IONP for magnetic hyperthermia	MRI and magnetic hyperthermia	Prostate	NCT02033447	Completed
SPIO nanoparticle	MRI	Pancreas	NCT00920023	Completed
USPIO NP-ferumoxytol	MRI	Head and neck	NCT01895829	Active
		Lymph node	NCT01815333	Active
		Thyroid	NCT01927887	Completed

[a]Data sourced from ClinicalTrials.gov (2024)
Abbreviations: *MRI* magnetic resonance imaging, *SPIO* superparamagnetic iron oxide, *USPIO* ultrasmall superparamagnetic iron oxide

driving advancements in personalized medicine. Advancements in organic/magnetic core–shell nanoconstructs not only promise improved therapeutic outcomes but also hold the potential to revolutionize cancer theranostics. This area represents a pivotal focus for future scientific exploration and technological innovation.

Acknowledgments The authors express their heartfelt gratitude toward the leadership and management of College of Pharmacy, King Khalid University (KKU), Al Faraa, Abha, Saudi Arabia and JSS College of Pharmacy, JSS Academy of Higher Education and Research (JSS AHER), Mysuru, Karnataka, India for providing all the necessary facilities for the completion of this piece of writing.

Conflicts of Interest The authors declare no competing interests.

References

Amstad E, Gillich T, Bilecka I, Textor M, Reimhult E. Ultrastable iron oxide nanoparticle colloidal suspensions using dispersants with catechol-derived anchor groups. Nano Lett. 2009;9(12):4042–8.

Angelakeris M. Magnetic nanoparticles: a multifunctional vehicle for modern theranostics. Biochimica et Biophysica Acta (BBA)—General Subjects. 2017 Jun;1861(6):1642–51.

Arora P, Sindhu A, Dilbaghi N, Chaudhury A. Biosensors as innovative tools for the detection of food borne pathogens. Biosens Bioelectron. 2011;28(1):1–12.

Aslam M, Li S, Dravid VP. Controlled synthesis and stability of Co@SiO_2 aqueous colloids. J Am Ceram Soc. 2007;90(3):950–6.

Attia ABE, Balasundaram G, Moothanchery M, Dinish US, Bi R, Ntziachristos V, et al. A review of clinical photoacoustic imaging: current and future trends. Photo-Dermatology. 2019;16:100144.

Bañobre-López M, Teijeiro A, Rivas J. Magnetic nanoparticle-based hyperthermia for cancer treatment. Rep Pract Oncol Radiother. 2013;18(6):397–400.

Ben Djemaa S, David S, Hervé-Aubert K, Falanga A, Galdiero S, Allard-Vannier E, et al. Formulation and in vitro evaluation of a siRNA delivery nanosystem decorated with gH625 peptide for triple negative breast cancer theranosis. Eur J Pharm Biopharm. 2018;131:99–108.

Burns AA, Vider J, Ow H, Herz E, Penate-Medina O, Baumgart M, et al. Fluorescent silica nanoparticles with efficient urinary excretion for nanomedicine. Nano Lett. 2009;9(1):442–8.
Buschow KHJ, Cahn RW, Flemings MC, Ilschner B, Kramer EJ, Mahajan S, editors. Encyclopedia of materials—science and technology, vol. 1-11. New York: Elsevier; 2001. References—Scientific Research Publishing [Internet]. [cited 2024 Jun 21]. Available from: https://www.scirp.org/reference/referencespapers?referenceid=595789
Cancer Today [Internet]. [cited 2024 Jun 21]. Available from: https://gco.iarc.fr/today/en/dataviz/tables?mode=population&group_populations=0
Carmen Bautista M, Bomati-Miguel O, del Puerto MM, Serna CJ, Veintemillas-Verdaguer S. Surface characterisation of dextran-coated iron oxide nanoparticles prepared by laser pyrolysis and coprecipitation. J Magn Magn Mater. 2005 May;293(1):20–7.
Chen Q, Hu Y, Chen Y, Jiang X, Yang Y. Microstructure formation and property of chitosan-poly(acrylic acid) nanoparticles prepared by macromolecular complex. Macromol Biosci. 2005;5(10):993–1000.
Chen S, Wang L, Duce SL, Brown S, Lee S, Melzer A, et al. Engineered biocompatible nanoparticles for *in vivo* imaging applications. J Am Chem Soc. 2010;132(42):15022–9.
Cherkasov VR, Mochalova EN, Babenyshev AV, Vasilyeva AV, Nikitin PI, Nikitin MP. Nanoparticle beacons: supersensitive smart materials with on/off-switchable affinity to biomedical targets. ACS Nano. 2020;14(2):1792–803.
ClinicalTrials.gov [Internet]. [cited 2024 Jun 21]. Available from: https://clinicaltrials.gov/
Curcio A, Silva AKA, Cabana S, Espinosa A, Baptiste B, Menguy N, et al. Iron oxide nanoflowers @ CuS hybrids for cancer tri-therapy: interplay of photothermal therapy, magnetic hyperthermia and photodynamic therapy. Theranostics. 2019;9(5):1288–302.
Danhier F, Le Breton A, Préat V. RGD-based strategies to target alpha(v) beta(3) integrin in cancer therapy and diagnosis. Mol Pharm. 2012;9(11):2961–73.
De M, Chou SS, Joshi HM, Dravid VP. Hybrid magnetic nanostructures (MNS) for magnetic resonance imaging applications. Adv Drug Deliv Rev. 2011;63(14–15):1282–99.
Delavari B, Bigdeli B, Mamashli F, Gholami M, Bazri B, Khoobi M, et al. Theranostic α-lactalbumin-polymer-based nanocomposite as a drug delivery carrier for cancer therapy. ACS Biomater Sci Eng. 2019;5(10):5189–208.
DeVita VT, Chu E. A history of cancer chemotherapy. Cancer Res. 2008;68(21):8643–53.
Dickerson EB, Blackburn WH, Smith MH, Kapa LB, Lyon LA, McDonald JF. Chemosensitization of cancer cells by siRNA using targeted nanogel delivery. BMC Cancer. 2010;10(1):10.
Dobiasch S, Szanyi S, Kjaev A, Werner J, Strauss A, Weis C, et al. Synthesis and functionalization of protease-activated nanoparticles with tissue plasminogen activator peptides as targeting moiety and diagnostic tool for pancreatic cancer. J Nanobiotechnol. 2016;14(1):81.
Farkona S, Diamandis EP, Blasutig IM. Cancer immunotherapy: the beginning of the end of cancer? BMC Med. 2016;14(1):73.
Feng M, Li P. Amine-containing core-shell nanoparticles as potential drug carriers for intracellular delivery. J Biomed Mater Res A. 2007;80A(1):184–93.
Fétiveau L, Paul G, Nicolas-Boluda A, Volatron J, George R, Laurent S, et al. Tailored ultra-small Prussian blue-based nanoparticles for MRI imaging and combined photothermal/photoacoustic theranostics. Chem Commun. 2019;55(98):14844–7.
Fuertges F, Abuchowski A. The clinical efficacy of poly(ethylene glycol)-modified proteins. J Control Release. 1990;11(1–3):139–48.
Gholibegloo E, Mortezazadeh T, Salehian F, Forootanfar H, Firoozpour L, Foroumadi A, et al. Folic acid decorated magnetic nanosponge: an efficient nanosystem for targeted curcumin delivery and magnetic resonance imaging. J Colloid Interface Sci. 2019;556:128–39.
Gilmore JL, Yi X, Quan L, Kabanov AV. Novel nanomaterials for clinical neuroscience. J Neuroimmune Pharmacol. 2008;3(2):83–94.
Gobbo OL, Sjaastad K, Radomski MW, Volkov Y, Prina-Mello A. Magnetic nanoparticles in cancer theranostics. Theranostics. 2015;5(11):1249–63.

Gupta AK, Naregalkar RR, Vaidya VD, Gupta M. Recent advances on surface engineering of magnetic iron oxide nanoparticles and their biomedical applications. Nanomedicine (Lond). 2007;2(1):23–39.

Hahn PF, Stark DD, Lewis JM, Saini S, Elizondo G, Weissleder R, et al. First clinical trial of a new superparamagnetic iron oxide for use as an oral gastrointestinal contrast agent in MR imaging. Radiology. 1990;175(3):695–700.

Harada-Shiba M, Yamauchi K, Harada A, Takamisawa I, Shimokado K, Kataoka K. Polyion complex micelles as vectors in gene therapy—pharmacokinetics and in vivo gene transfer. Gene Ther. 2002;9(6):407–14.

Harris JM, Chess RB. Effect of pegylation on pharmaceuticals. Nat Rev Drug Discov. 2003 Mar;2(3):214–21.

He X, Duan J, Wang K, Tan W, Lin X, He C. A novel fluorescent label based on organic dye-doped silica nanoparticles for HepG liver cancer cell recognition. J Nanosci Nanotechnol. 2004;4(6):585–9.

He X, Ma J, Mercado AE, Xu W, Jabbari E. Cytotoxicity of paclitaxel in biodegradable self-assembled core-shell poly(Lactide-co-glycolide ethylene oxide fumarate) nanoparticles. Pharm Res. 2008;25(7):1552–62.

Herranz-Blanco B, Shahbazi M, Correia AR, Balasubramanian V, Kohout T, Hirvonen J, et al. pH-switch nanoprecipitation of polymeric nanoparticles for multimodal cancer targeting and intracellular triggered delivery of doxorubicin. Adv Healthc Mater. 2016;5(15):1904–16.

Huang J, Guo M, Ke H, Zong C, Ren B, Liu G, et al. Rational design and synthesis of γFe_2O_3 @Au magnetic gold nanoflowers for efficient cancer theranostics. Adv Mater. 2015;27(34):5049–56.

Huang Y, Zhang B, Xie S, Yang B, Xu Q, Tan J. Superparamagnetic iron oxide nanoparticles modified with Tween 80 pass through the intact blood–brain barrier in rats under magnetic field. ACS Appl Mater Interfaces. 2016;8(18):11336–41.

Huang Y, Mao K, Zhang B, Zhao Y. Superparamagnetic iron oxide nanoparticles conjugated with folic acid for dual target-specific drug delivery and MRI in cancer theranostics. Mater Sci Eng C. 2017;70:763–71.

Jaiswal MK, De M, Chou SS, Vasavada S, Bleher R, Prasad PV, et al. Thermoresponsive magnetic hydrogels as theranostic nanoconstructs. ACS Appl Mater Interfaces. 2014;6(9):6237–47.

Jang MK, Jeong YI, Nah JW. Characterization and preparation of core–shell type nanoparticle for encapsulation of anticancer drug. Colloids Surf B Biointerfaces. 2010;81(2):530–6.

Jun YW, Huh YM, Choi JS, Lee JH, Song HT, Kim S, et al. Nanoscale size effect of magnetic nanocrystals and their utilization for cancer diagnosis via magnetic resonance imaging. J Am Chem Soc. 2005;127(16):5732–3.

Kim SW, Kim S, Tracy JB, Jasanoff A, Bawendi MG. Phosphine oxide polymer for water-soluble nanoparticles. J Am Chem Soc. 2005;127(13):4556–7.

Kim D, Kim K, Kim K, Lee Y. Targeting to carcinoma cells with chitosan- and starch-coated magnetic nanoparticles for magnetic hyperthermia. J Biomed Mater Res A. 2009;88A(1):1–11.

Kim KS, Hur W, Park SJ, Hong SW, Choi JE, Goh EJ, et al. Bioimaging for targeted delivery of hyaluronic acid derivatives to the livers in cirrhotic mice using quantum dots. ACS Nano. 2010;4(6):3005–14.

Kohler N, Fryxell GE, Zhang M. A bifunctional poly(ethylene glycol) silane immobilized on metallic oxide-based nanoparticles for conjugation with cell targeting agents. J Am Chem Soc. 2004;126(23):7206–11.

Kohler N, Sun C, Fichtenholtz A, Gunn J, Fang C, Zhang M. Methotrexate-immobilized poly(ethylene glycol) magnetic nanoparticles for MR imaging and drug delivery. Small. 2006;2(6):785–92.

Kudr J, Haddad Y, Richtera L, Heger Z, Cernak M, Adam V, et al. Magnetic nanoparticles: from design and synthesis to real world applications. Nano. 2017;7(9):243.

Kumar M, Yigit M, Dai G, Moore A, Medarova Z. Image-guided breast tumor therapy using a small interfering RNA nanodrug. Cancer Res. 2010;70(19):7553–61.

Law WC, Yong KT, Roy I, Xu G, Ding H, Bergey EJ, et al. Optically and magnetically doped organically modified silica nanoparticles as efficient magnetically guided biomarkers for two-photon imaging of live cancer cells. J Phys Chem C. 2008;112(21):7972–7.

Lee J, Lee K, Moon SH, Lee Y, Park TG, Cheon J. All-in-one target-cell-specific magnetic nanoparticles for simultaneous molecular imaging and siRNA delivery. Angew Chem Int Ed. 2009;48(23):4174–9.

Lee PW, Hsu SH, Tsai JS, Chen FR, Huang PJ, Ke CJ, et al. Multifunctional core-shell polymeric nanoparticles for transdermal DNA delivery and epidermal Langerhans cells tracking. Biomaterials. 2010;31(8):2425–34.

Li W, Tutton S, Vu AT, Pierchala L, Li BSY, Lewis JM, et al. First-pass contrast-enhanced magnetic resonance angiography in humans using ferumoxytol, a novel ultrasmall superparamagnetic iron oxide (USPIO)-based blood pool agent. J Magn Reson Imaging. 2005;21(1):46–52.

Li Z, Zhang Y, Jiang S. Multicolor core/shell-structured upconversion fluorescent nanoparticles. Adv Mater. 2008;20(24):4765–9.

Liao J, Wei X, Ran B, Peng J, Qu Y, Qian Z. Polymer hybrid magnetic nanocapsules encapsulating IR820 and PTX for external magnetic field-guided tumor targeting and multifunctional theranostics. Nanoscale. 2017;9(7):2479–91.

Liong M, Lu J, Kovochich M, Xia T, Ruehm SG, Nel AE, et al. Multifunctional inorganic nanoparticles for imaging, targeting, and drug delivery. ACS Nano. 2008;2(5):889–96.

Liu Y, Yang F, Yuan C, Li M, Wang T, Chen B, et al. Magnetic nanoliposomes as *in situ* microbubble bombers for multimodality image-guided cancer theranostics. ACS Nano. 2017;11(2):1509–19.

Lu Y, Yin Y, Mayers BT, Xia Y. Modifying the surface properties of superparamagnetic iron oxide nanoparticles through a sol−gel approach. Nano Lett. 2002;2(3):183–6.

Lu S, Li X, Zhang J, Peng C, Shen M, Shi X. Dendrimer-stabilized gold nanoflowers embedded with ultrasmall iron oxide nanoparticles for multimode imaging–guided combination therapy of tumors. Adv Science. 2018;5(12)

Luque-Michel E, Sebastian V, Larrea A, Marquina C, Blanco-Prieto MJ. Co-encapsulation of superparamagnetic nanoparticles and doxorubicin in PLGA nanocarriers: development, characterization and in vitro antitumor efficacy in glioma cells. Eur J Pharm Biopharm. 2019;145:65–75.

Lutz JF, Stiller S, Hoth A, Kaufner L, Pison U, Cartier R. One-pot synthesis of pegylated ultrasmall iron-oxide nanoparticles and their in vivo evaluation as magnetic resonance imaging contrast agents. Biomacromolecules. 2006;7(11):3132–8.

Ma D, Veres T, Clime L, Normandin F, Guan J, Kingston D, et al. Superparamagnetic Fe_xO_y@SiO_2 Core−Shell nanostructures: controlled synthesis and magnetic characterization. J Phys Chem C. 2007;111(5):1999–2007.

Ma M, Chen H, Chen Y, Wang X, Chen F, Cui X, et al. Au capped magnetic core/mesoporous silica shell nanoparticles for combined photothermo-/chemo-therapy and multimodal imaging. Biomaterials. 2012;33(3):989–98.

Mahmud A, Xiong XB, Aliabadi HM, Lavasanifar A. Polymeric micelles for drug targeting. J Drug Target. 2007;15(9):553–84.

Medarova Z, Pham W, Farrar C, Petkova V, Moore A. In vivo imaging of siRNA delivery and silencing in tumors. Nat Med. 2007;13(3):372–7.

Molday RS, MacKenzie D. Immunospecific ferromagnetic iron-dextran reagents for the labeling and magnetic separation of cells. J Immunol Methods. 1982;52(3):353–67.

Mornet S, Portier J, Duguet E. A method for synthesis and functionalization of ultrasmall superparamagnetic covalent carriers based on maghemite and dextran. J Magn Magn Mater. 2005;293(1):127–34.

Nie Y, Yuan W, Gong T, Lu J, Fu Y, Zhang Z. Investigation on characterization and transfection of a novel multi-polyplex gene delivery system. J Appl Polym Sci. 2007;106(2):1028–33.

Panda JJ, Mishra A, Basu A, Chauhan VS. Stimuli responsive self-assembled hydrogel of a low molecular weight free dipeptide with potential for tunable drug delivery. Biomacromolecules. 2008;9(8):2244–50.

Pandey N, Menon JU, Takahashi M, Hsieh JT, Yang J, Nguyen KT, et al. Thermo-responsive fluorescent nanoparticles for multimodal imaging and treatment of cancers. Nanotheranostics. 2020;4(1):1–13.

Patitsa M, Karathanou K, Kanaki Z, Tzioga L, Pippa N, Demetzos C, et al. Magnetic nanoparticles coated with polyarabic acid demonstrate enhanced drug delivery and imaging properties for cancer theranostic applications. Sci Rep. 2017;7(1):775.

Patra D, Mukherjee S, Chakraborty I, Dash TK, Senapati S, Bhattacharyya R, et al. Iron(III) coordinated polymeric nanomaterial: a next-generation theranostic agent for high-resolution T_1 -weighted magnetic resonance imaging and anticancer drug delivery. ACS Biomater Sci Eng. 2018;4(5):1738–49.

Peng S, Wang C, Xie J, Sun S. Synthesis and stabilization of monodisperse Fe nanoparticles. J Am Chem Soc. 2006;128(33):10676–7.

Pinho SLC, Pereira GA, Voisin P, Kassem J, Bouchaud V, Etienne L, et al. Fine tuning of the relaxometry of γ-Fe2O3@SiO2 nanoparticles by tweaking the silica coating thickness. ACS Nano. 2010a;4(9):5339–49.

Pinho SLC, Pereira GA, Voisin P, Kassem J, Bouchaud V, Etienne L, et al. Fine tuning of the Relaxometry of γ-Fe_2 O_3 @SiO_2 nanoparticles by tweaking the silica coating thickness. ACS Nano. 2010b;4(9):5339–49.

Pradhan P, Giri J, Rieken F, Koch C, Mykhaylyk O, Döblinger M, et al. Targeted temperature sensitive magnetic liposomes for thermo-chemotherapy. J Control Release. 2010;142(1):108–21.

Reimer P, Balzer T. Ferucarbotran (Resovist): a new clinically approved RES-specific contrast agent for contrast-enhanced MRI of the liver: properties, clinical development, and applications. Eur Radiol. 2003;13(6):1266–76.

Revia RA, Zhang M. Magnetite nanoparticles for cancer diagnosis, treatment, and treatment monitoring: recent advances. Mater Today. 2016;19(3):157–68.

Rodrigues JS, Santos-Magalhães NS, Coelho LCBB, Couvreur P, Ponchel G, Gref R. Novel core(polyester)-shell(polysaccharide) nanoparticles: protein loading and surface modification with lectins. J Control Release. 2003;92(1–2):103–12.

Rost NCV, Sen K, Savliwala S, Singh I, Liu S, Unni M, et al. Magnetic particle imaging performance of liposomes encapsulating iron oxide nanoparticles. J Magn Magn Mater. 2020;504:166675.

Roy Chowdhury M, Schumann C, Bhakta-Guha D, Guha G. Cancer nanotheranostics: strategies, promises and impediments. Biomed Pharmacother. 2016;84:291–304.

Roy E, Patra S, Madhuri R, Sharma PK. Stimuli-responsive poly(N -isopropyl acrylamide)-co-tyrosine@gadolinium: iron oxide nanoparticle-based nanotheranostic for cancer diagnosis and treatment. Colloids Surf B Biointerfaces. 2016;142:248–58.

Roy E, Patra S, Madhuri R, Sharma PK. Anisotropic gold nanoparticle decorated magnetopolymersome: an advanced nanocarrier for targeted photothermal therapy and dual-mode responsive T_1 MRI imaging. ACS Biomater Sci Eng. 2017;3(9):2120–35.

Saesoo S, Sathornsumetee S, Anekwiang P, Treetidnipa C, Thuwajit P, Bunthot S, et al. Characterization of liposome-containing SPIONs conjugated with anti-CD20 developed as a novel theranostic agent for central nervous system lymphoma. Colloids Surf B Biointerfaces. 2018;161:497–507.

Sahiner N, Ilgin P. Synthesis and characterization of soft polymeric nanoparticles and composites with tunable properties. J Polym Sci A Polym Chem. 2010;48(22):5239–46.

Sahoo SK, Labhasetwar V. Nanotech approaches to drug delivery and imaging. Drug Discov Today. 2003;8(24):1112–20.

Shi J, Kantoff PW, Wooster R, Farokhzad OC. Cancer nanomedicine: progress, challenges and opportunities. Nat Rev Cancer. 2017;17(1):20–37.

Siegemund T, Paulke B-R, Schmiedel H, Bordag N, Hoffmann A, Harkany T, et al. Thioflavins released from nanoparticles target fibrillar amyloid β in the hippocampus of APP/PS1 transgenic mice. Int J Dev Neurosci. 2006;24(2–3):195–201.

Silva CO, Pinho JO, Lopes JM, Almeida AJ, Gaspar MM, Reis C. Current trends in cancer nanotheranostics: metallic, polymeric, and lipid-based systems. Pharmaceutics. 2019;11(1):22.

Singh A, Sahoo SK. Magnetic nanoparticles: a novel platform for cancer theranostics. Drug Discov Today. 2014 Apr;19(4):474–81.
Song G, Zheng X, Wang Y, Xia X, Chu S, Rao J. A magneto-optical nanoplatform for multimodality imaging of tumors in mice. ACS Nano. 2019;13(7):7750–8.
Sounderya N, Zhang Y. Use of core/shell structured nanoparticles for biomedical applications. Recent Pat Biomed Eng. 2008;1(1):34–42.
Sun C, Veiseh O, Gunn J, Fang C, Hansen S, Lee D, et al. In vivo MRI detection of gliomas by chlorotoxin-conjugated superparamagnetic nanoprobes. Small. 2008;4(3):372–9.
Thorn CF, Oshiro C, Marsh S, Hernandez-Boussard T, McLeod H, Klein TE, et al. Doxorubicin pathways. Pharmacogenet Genomics. 2011 Jul;21(7):440–6.
Vallabani NVS, Singh S. Recent advances and future prospects of iron oxide nanoparticles in biomedicine and diagnostics. 3 Biotech. 2018;8(6):279.
Van Tomme SR, Storm G, Hennink WE. In situ gelling hydrogels for pharmaceutical and biomedical applications. Int J Pharm. 2008;355(1–2):1–18.
Vijayan VM, Beeran AE, Shenoy SJ, Muthu J, Thomas V. New magneto-fluorescent hybrid polymer nanogel for theranostic applications. ACS Appl Bio Mater. 2019;2(2):757–68.
Wang YX, Hussain SM, Krestin GP. Superparamagnetic iron oxide contrast agents: physicochemical characteristics and applications in MR imaging. Eur Radiol. 2001;11(11):2319–31.
Wang L, Park HY, Lim SII, Schadt MJ, Mott D, Luo J, et al. Core@shell nanomaterials: gold-coated magnetic oxide nanoparticles. J Mater Chem. 2008;18(23):2629.
Wang H, Zhao P, Liang X, Gong X, Song T, Niu R, et al. Folate-PEG coated cationic modified chitosan—cholesterol liposomes for tumor-targeted drug delivery. Biomaterials. 2010;31(14):4129–38.
Wei J, Shuai X, Wang R, He X, Li Y, Ding M, et al. Clickable and imageable multiblock polymer micelles with magnetically guided and PEG-switched targeting and release property for precise tumor theranosis. Biomaterials. 2017;145:138–53.
Weissleder R, Pittet MJ. Imaging in the era of molecular oncology. Nature. 2008;452(7187):580–9.
Wiradharma N, Khan M, Tong YW, Wang S, Yang Y. Self-assembled cationic peptide nanoparticles capable of inducing efficient gene expression in vitro. Adv Funct Mater. 2008;18(6):943–51.
Wunderbaldinger P, Josephson L, Weissleder R. Crosslinked iron oxides (CLIO): a new platform for the development of targeted MR contrast agents. Acad Radiol. 2002;9(Suppl 2):S304–6.
Xie J, Xu C, Kohler N, Hou Y, Sun S. Controlled PEGylation of monodisperse Fe_3O_4 nanoparticles for reduced non-specific uptake by macrophage cells. Adv Mater. 2007;19(20):3163–6.
Xin F, Li Y, Fu C, Javed I, Huang X, Schaschkow A, et al. Multimodal nanoprobe for pancreatic beta cell detection and amyloidosis mitigation. Chem Mater. 2020;32(3):1080–8.
Xu C, Xu K, Gu H, Zheng R, Liu H, Zhang X, et al. Dopamine as a robust anchor to immobilize functional molecules on the iron oxide shell of magnetic nanoparticles. J Am Chem Soc. 2004;126(32):9938–9.
Yan H, Zhao L, Shang W, Liu Z, Xie W, Qiang C, et al. General synthesis of high-performing magneto-conjugated polymer core–shell nanoparticles for multifunctional theranostics. Nano Res. 2017;10(2):704–17.
Yang F, Li Y, Chen Z, Zhang Y, Wu J, Gu N. Superparamagnetic iron oxide nanoparticle-embedded encapsulated microbubbles as dual contrast agents of magnetic resonance and ultrasound imaging. Biomaterials. 2009;30(23–24):3882–90.
Yang X, Hong H, Grailer JJ, Rowland IJ, Javadi A, Hurley SA, et al. cRGD-functionalized, DOX-conjugated, and 64Cu-labeled superparamagnetic iron oxide nanoparticles for targeted anticancer drug delivery and PET/MR imaging. Biomaterials. 2011;32(17):4151–60.
Yang HK, Bao JF, Mo L, Yang RM, Xu XD, Tang WJ, et al. Bioreducible amphiphilic block copolymers based on PCL and glycopolypeptide as multifunctional theranostic nanocarriers for drug delivery and MR imaging. RSC Adv. 2017;7(34):21093–106.
Yang HY, Li Y, Lee DS. Multifunctional and stimuli-responsive magnetic nanoparticle-based delivery systems for biomedical applications. Adv Ther (Weinh). 2018;1(2)
Yang H, Miao Y, Chen L, Li Z, Yang R, Xu X, et al. Redox-responsive nanoparticles from disulfide bond-linked poly-(N-ε-carbobenzyloxy-l-lysine)-grafted hyaluronan copolymers as

theranostic nanoparticles for tumor-targeted MRI and chemotherapy. Int J Biol Macromol. 2020;148:483–92.

Yoo D, Lee JH, Shin TH, Cheon J. Theranostic magnetic nanoparticles. Acc Chem Res. 2011;44(10):863–74.

Yu JH, Quan JS, Kwon JT, Xu CX, Sun B, Jiang HL, et al. Fabrication of a novel core-shell gene delivery system based on a brush-like polycation of α, β–poly (L-aspartate-graft-PEI). Pharm Res. 2009;26(9):2152–63.

Zahraei M, Marciello M, Lazaro-Carrillo A, Villanueva A, Herranz F, Talelli M, et al. Versatile theranostics agents designed by coating ferrite nanoparticles with biocompatible polymers. Nanotechnology. 2016;27(25):255702.

Zhang K, Fang H, Wang Z, Li Z, Taylor JSA, Wooley KL. Structure-activity relationships of cationic shell-crosslinked knedel-like nanoparticles: Shell composition and transfection efficiency/cytotoxicity. Biomaterials. 2010;31(7):1805–13.

Zhang KL, Zhou J, Zhou H, Wu Y, Liu R, Wang LL, et al. Bioinspired "active" stealth magneto-nanomicelles for theranostics combining efficient MRI and enhanced drug delivery. ACS Appl Mater Interfaces. 2017;9(36):30502–9.

Zhu L, Zhou Z, Mao H, Yang L. Magnetic nanoparticles for precision oncology: theranostic magnetic iron oxide nanoparticles for image-guided and targeted cancer therapy. Nanomedicine. 2017;12(1):73–87.

Zolata H, Abbasi Davani F, Afarideh H. Synthesis, characterization and theranostic evaluation of Indium-111 labeled multifunctional superparamagnetic iron oxide nanoparticles. Nucl Med Biol. 2015;42(2):164–70.

15 Innovative Organic Core–Nonmagnetic Shell Nanoconstructs: Pioneering Precision in Cancer Theranostics

Amrita Arup Roy, Rahul Pokale, Anoushka Mukharya, Sandesh Ramchandra Jadhav, Gaurisha Alias Resha Ramnath Naik, S. P. Rachana, Viola Colaco, Paniz Hedayat, Ritu Kudarha, Srinivas Mutalik, and Namdev Dhas

Abstract

Organic core–nonmagnetic shell nanoconstructs represent a cutting-edge approach in cancer theranostics, combining targeted drug delivery with diagnostic capabilities. These nanoconstructs feature an organic core, serving as the primary drug carrier, encapsulated by a nonmagnetic shell that enhances stability and bioavailability and reduces toxicity. The core–shell structure allows for the controlled release of therapeutic agents, minimizing off-target effects and improving treatment efficacy. By incorporating targeting ligands like antibodies and peptides, the nanoconstructs exhibit high specificity for cancer cells, ensuring that the drugs accumulate at the tumor site. Additionally, stimuli-responsive elements, such as pH- or temperature-sensitive triggers, enable precision drug release within the tumor microenvironment. Their theranostic potential is further enhanced by the integration of imaging agents for real-time monitoring via techniques like MRI or fluorescence imaging, offering personalized treatment options. This chapter explores the design, functionalization, and application of these nanoconstructs, along with the challenges and future directions in personalized cancer therapy and beyond.

Keywords

Nanotechnology · Cancer theranostics · Organic core · Nonmagnetic shell · Targeted drug delivery · Stimuli-responsive nanoconstructs · Personalized medicine · Diagnostic imaging · Theranostics

A. A. Roy · R. Pokale · A. Mukharya · S. R. Jadhav · G. A. R. R. Naik · S. P. Rachana
V. Colaco · P. Hedayat · R. Kudarha · S. Mutalik · N. Dhas (✉)
Department of Pharmaceutics, Manipal College of Pharmaceutical Sciences, Manipal Academy of Higher Education, Manipal, Karnataka, India

J. K. Patel et al. (eds.), *Core-Shell Nano Constructs for Cancer Theragnostic*,
https://doi.org/10.1007/978-981-96-3025-7_15

1 Introduction

1.1 Overview of Nanotechnology in Cancer Theranostics

The Greek word nano translates to "dwarf," and the prefix is the International System of Unit that stands for 10^{-9}, or 0.000000001. Nanotechnology is the tandem of advanced fabrication science and engineering, with the objective of synthesizing or assembling matter at the nanometer scale (1–100 nm), or a billionth of a meter. The benefit of nanosized material over bulk material is that it possesses a higher surface-to-volume ratio. Nanoparticles (NPs) have numerous applications in a wide range of domains, including medicine, where nanoparticles are constantly being improved for drug delivery, disease screening, and tissue engineering, among other uses (Dhas et al. 2024). As a result, nanotechnology has started to play a crucial function in catalytic processes, energy in addition to environment, agriculture, and other disciplines (Chenthamara et al. 2019). Nanoparticles (NPs) are the key to modern-day cancer therapeutics and diagnostics. They have been widely employed to accelerate and streamline traditional cancer treatment and diagnostic procedures. Larger surface area, greater volume percentage, and improved targeting capabilities are just a few of the major functional characteristics of NPs. Moreover, they have a less harmful impact on healthy cells; therefore, they can effectively permeate tissues and the epithelium layer, thereby increasing their bioavailability and overall half-life. The antioxidant function of nanoparticles has already been demonstrated in several cases to confer intrinsic anticancer activity, hence inhibiting tumor development. Nanoparticles can also boost medication release efficiency and enable regulated drug release with minimal adverse effects. Several drugs have been developed as or coated with NPs that specifically target tumors or dysfunctional organs without endangering healthy cells or tissues. Multiple nanoparticle forms, including fullerene, graphene, metal-based, magnetic, polymerized, liposomes, metal oxides, quantum dots, carbon nanotubes, and dendrimeric forms, may find application in the diagnosis and therapy of cancer (Alrushaid et al. 2023). Nanoparticles may encompass a very extensive surface area and range in size from 10 to 100 nm, which makes them a desirable substance for biological applications. The targeted tissues may be efficiently penetrated by the nanoparticles, which are then able to migrate effortlessly between organs within the body. For diagnostic applications, the nanoparticles are capable of being coupled with therapeutic molecules to target incontinent tissues and cancer cells. Nanoparticles are about the equivalent size as that of DNA and smaller than basic red blood cells. As a result, they function more efficiently and acquire unique chemical, optical, and physical characteristics that enable the treatment and diagnosis of cancer in the medical area (Joseph et al. 2023).

1.2 Core–Shell Nanoconstructs

The nucleus of core–shell nanoparticles comprise a particular substance and is wrapped up with a different layer of substance. In contrast to simple nanoparticles, core–shell nanoparticles offer substantial benefits in biological product applications. These potential benefits include reduced cytotoxicity, enhanced ability to disperse easily, improved biocompatibility and cytocompatibility, improved association with additional biologically active molecules, elevated thermal and chemical endurance, and more (Chatterjee et al. 2013).

In some cases, the target nanoparticles are highly toxic and could severely damage the host organs and tissues. These nanoparticles are significantly less harmful and biocompatible due to the neutral or benign substance strategically coated over the core. In several instances, the shell layer serves as a nontoxic layer while simultaneously enhancing the properties of the core material. As per the research conducted by Wing-Cheung Law et al., by combining magnetite (Fe_3O_4) and quantum dots (QDs) co-encapsulated into organically modified silica (ORMOSIL) nanoparticles, multifunctional nanoprobes enabling live imaging of cancer cells were fabricated. Compared to the results obtained employing these two targeting modes separately, the robust in vitro cellular uptake observed when using optically and magnetically doped ORMOSIL nanoparticles was substantially greater. These Core-Shell nanoconstructs have proved to be nontoxic based on additional cell viability investigations (Law et al. 2008).

Their hydrophilicity is crucial for nanoparticles to be dispersed in aqueous biological systems. The enhancement of bio-dispersivity, biocompatibility, and cytocompatibility makes it an advantageous replacement for traditional drug delivery methods. Core–shells based on bioinspired polymeric covering on hydrophobic drugs degrade independently with pH, temperature, and ion changes. It can assist with releasing the medicament effectively at its intended target site (Mahmud et al. 2007). Since the surface of core–shell nanoparticles contains reactive elements, the shell component is generally responsible for surface attributes like biocompatibility and conjugation of bioactive materials; however, the degree of thickness that makes up the shell can be further additionally customized to yield both sufficient contrasting properties as the contrast agent and interactions of biomolecules for precise drug delivery, biosensing, precise binding etc. (Pinho et al. 2010). In a study conducted by Bridot et al., luminous hybrid nanoparticles comprising a paramagnetic Gd_2O_3 core and a polysiloxane shell were used as contrast agents for magnetic resonance imaging and in vivo fluorescence imaging. These Core-Shell nanoconstructs are capable of migrating in blood arteries without additionally accumulating adversely in the liver and lungs, which is usually the case of a cytotoxic effect (Bridot et al. 2007).

The functionality of inorganic NPs can be enhanced by organic shells as the bio- and cytocompatibility of organic shells become more effective simply because they are primarily composed of polymers, carbonaceous materials, long-chain acidic substances, cellulose, proteins, complex carbohydrates/polysaccharides, and so on. They can also serve as the foundation sites for molecular interactions, promote biocompatibility, or even insulate the inorganic NPs from oxidation (Liu et al. 2010). Ideal for substantial electrical energy storage systems are oxide nanoparticles with elevated dielectric constant core–shell architectures and organic shells with significant breakdown fields (Arbatti et al. 2007). The dielectric oxide nanoparticles can only be obtained at an extreme growth temperature, which impacts the core–shell development process and results in suboptimal control over size, shape, and size distribution (Kim et al. 2007; Li et al. 2009). Balasubramanian et al. successfully synthesized TiO_2-paraffin core–shell monodisperse nanoparticles with enhanced dielectric characteristics and made specific improvements to the drawbacks mentioned above. It was established that a reduction in shell thickness significantly increases the functional dielectric constant of TiO_2-paraffin core–shell nanoparticles. In the frequency range of 100 Hz – 1 MHz, the capacitors exhibit minor dielectric dispersion and relatively small dielectric losses, which are especially beneficial for possibly utilizing these core–shell nanoparticles in applications (Balasubramanian et al. 2010). Furthermore, compared to basic nanoparticles, core–shell nanoparticles have been demonstrated to be more appropriate and function superiorly in biological domains. Core–shell nanotechnology is also capable of being utilized for developing exceptionally portable, stable, responsive, specific, and sensitive medical devices (Li et al. 2012). Core–shell nanocomposites, for instance, can be successfully applied to microchips, particularly for in situ medical diagnostic devices. In these kits, the outer shell anchored with sensory molecules operates as the analyte probe, while the core exclusively functions as the electrode/transducer subsystem. Given that electron discharge, transit, and multiplicity are the fundamental requirements for such sensors, the majority of the distinctive features of NPs, including magnetic, optical, or semiconductor, find significance in these categories of biosensors (Chatterjee et al. 2013).

2 Types of Core Materials and Their Properties

2.1 Organic Cores

2.1.1 Polymeric Cores

Delivering the loaded pharmaceutical is the principal function of core–shell nanostructures, primarily because they can preserve the active ingredients and release them in a controlled pattern (Naik et al. 2024). The Food and Drug Administration (FDA)-approved polymer poly(lactic-co-glycolic acid) (PLGA) is widely acknowledged as the carrier material among various biological polymeric materials (Blasi 2019). It is important for drug carriers to break down into nontoxic chemical molecules and be in agreement with healthy live tissues. Because of their high biodegradability, polyethylene glycol (PEG), polyglycolic acid (PGA), polyvinyl alcohol (PVA), polylactic acid (PLA), poly(L-lactic acid) (PLLA), and chitosan (CS) have also been employed extensively in addition to PLGA (Ayyanaar et al. 2020). Chemotherapy, photothermal therapy, hormone therapy, immunotherapy, and other treatments are among the cancer therapies that employ PLGA microspheres (Dhas et al. 2024). Similar to other uses, the goal of achieving synergistic effects is often achieved by combining more than one cancer therapy technique. In a research conducted by Liu et al., curcumin (Cur) was utilized as the central core to develop polypyrrole (PPy) after being encapsulated inside the PLGA-shell NPs. When chemotherapy and PTT are combined, the generated core–shell PLGA@PPy NPs exhibited the ability to load and release Cur into healthy cells, demonstrating significant cytotoxicity in comparison to either photothermal therapy alone or chemotherapy alone. It is possible to employ this core–shell PLGA@PPy technology as a generic drug delivery intermediary for multimodal treatment (Liu et al. 2016). Furthermore, these core–shell microparticles are additionally employed in regenerative medicine as a substitute for autologous cell recuperation and transplantation. By incorporating growth factors alongside cells into solid scaffolds or hydrogels, the application of polymeric particles utilizing PLGA is a potential technique for a temporally and spatially regulated administration of growth factors that stimulate the development and differentiation of cells in bone remodeling and regeneration (Ortega-Oller et al. 2015). By maintaining its release profile and integrating it into core–shell polymeric (Pluronic F127/chitosan) based nanoparticles, this investigation carried out by Shafiq et al. attempts to increase the oral bioavailability of methotrexate (MTX). An X-ray diffraction study was used to establish that the active ingredient was entrapped in the polymeric core. The encapsulation effectiveness of the spherical-shaped nanoparticles (181–417 nm) produced by the self-micellization and ionotropic gelation approach ranged from 80.14% to 85.54%. The use of polymeric nanoparticles loaded with MTX is demonstrated to be a promising drug delivery approach for successful care and therapeutic management of rheumatoid arthritis (Shafiq et al. 2022).

2.1.2 Lipid-Based Cores

Lipid-based formulations can be utilized to govern the absorption of therapeutic ingredients through multiple mechanisms. They are capable of communicating with enterocyte-based transport, modifying the gastrointestinal milieu, and initiating the active components' lymphatic transport (Fouad and Mohamed 2011). Owing to the increased flexibility of the membrane, fatty acids, glycerides, triglycerides, nonionic and ionic surfactants, and other lipid-based added excipients have been identified to be permeability enhancers (Arnold et al. 2010). Lipid core–shell nanoparticles were synthesized and evaluated using a prolonged protein distribution approach. Oh et al. have synthesized lecithin loaded with proteins that make up the lipid core, while Pluronics (triblock copolymer, F-127) serves as the polymeric shell. By means of electrostatic interactions, lysozyme, along with vascular endothelial growth factor (VEGF), was incorporated into core–shell nanoparticles to operate as protein drug carriers. The polymeric shell was subsequently modified to further investigate the active ingredient release profile (Oh et al. 2006).

After receiving the first chemotherapy treatment, many prostate cancers tend to relapse again. To reduce toxicity and improve medication distribution, encapsulating pharmaceuticals in nanocarriers and combining molecular therapy and chemotherapy is a useful approach. Wang et al. have synthesized core–shell lipid–polymer hybrid nanoparticles (CSLPHNPs) featuring a poly(lactic-co-glycolic acid) (PLGA) core along with a lipid layer encapsulating docetaxel, a therapeutically employed inhibitor of sphingosine kinase 1 (SK1) and fingolimod (FTY720). CSLPHNPs demonstrated superior tumor targeting and notably fewer adverse effects in a mouse model simulating human prostate cancer when compared to free drug therapy. Most notably, this study might reverse the effects of FTY720-induced lymphopenia. Evidence indicates that encapsulating nanoparticles might potentially enhance targeting, have minimal off-target toxicity, and, most importantly, lessen the lymphopenia triggered by FTY720, indicating prospective applicability in the treatment of cancer in clinical settings (Wang et al. 2017). Over the past few decades, there has been a progressive increase in the need for therapeutic delivery systems utilizing tailored nanocarriers. The overall stability of therapeutic molecules and the rapid absorption of the active in the target site are the two primary benefits of nanocarriers. These carriers provide a multitude of potential alternatives, such as anticancer therapy, gene delivery, vaccine delivery, bioimaging, and the delivery of drugs via different dosage routes in the years to come (Sivadasan et al. 2021).

2.1.3 Carbon-Based Cores

Carbon-based core–shell materials have been widely used for conserving energy owing to their substantial surface area that plays a major role during basic reactions, porosity levels that replicate the natural environment, and abundant exposure to and dispersion of active sites. These materials also take part in swift interfacial transit at varying length scales of pore space, which helps minimize or expedite diffusion routes (Feng et al. 2018). Two core–shell nanoparticles have been developed by Lukowiak et al., one with a rigid nanodiamond core and the other with a formable dendritic polyethylene shell composed of sp3 carbon. The identical polyglycerol shell and carbon core were present within the two core–shell nanoparticles. The second type was found to have a hard nanodiamond core, while the first had a soft dendritic polyethylene core. This work demonstrates that the encapsulation and transport capabilities of nanocarriers are significantly impacted by the flexibility of their cores, which will be useful in the future development of more advanced nanocarriers (Lukowiak et al. 2015). Gold nanoshelled–drug nanoparticles (GNPs) have also been produced by Li et al. for chemo-photothermal multimodal cancer therapy. The GNPs completely eliminate tumors without triggering weight loss or tumor reappearance. With the goal of increasing the GNPs' aqueous dispersity and biological environment stability, purified 10-hydroxycamptothecin nanoparticles (HCPT NPs) were first coated with Au nanoshells (referred to as pGNPs), and then, their surfaces were further functionalized. This study demonstrates the enormous future possibilities of multifunctional core–shell drug-nanostructure-based nanocomposite materials for extremely effective cancer treatment (Li et al. 2014). The research and development conducted by Wang et al., of dual-ordered mesopores (MMPS) in targeting peptide-conjugated core–shell graphitic carbon@silica nanospheres as an antitumoral doxorubicin (DOX) delivery approach (MMPSD) for complementary localized photothermal treatment of breast cancer has been executed with superior effectiveness. The hydrophilic mesoporous silica exterior shell ensures that MMPSD has good water dispersity. The hydrophobic graphitic mesoporous carbon core offers high NIR photothermal transformation efficiency, rapid drug–photothermal hotspot interaction, and efficient hydrophobic drug loading. Thus, the MMPSD system demonstrated a superb combination of anticancer effectiveness, hypersensitive NIR/pH-responsive DOX release, high targeting ability, prolonged release, and effective drug loading capabilities (Wang et al. 2014). As a result, in recent years, core–shell nanostructured materials have become an increasingly prominent field of investigation.

2.2 Nonmagnetic Shell

The scientific community is devoting a substantial amount of research to studying core–shell nanoparticles, which are made up of a core constituent encapsulated by a distinct material called shell (Fig. 15.1). Due to their distinct structural arrangement, these kinds of nanoparticles generated lots of interest in the scientific community for a variety of medicinal applications, including the treatment and diagnosis of neurological and oncological disorders.

Core–shell nanoparticles have emerged a lot in the past decades in various domains, including biomedical engineering, optics, catalysis, radio imaging, and pharmaceuticals. These nanoparticles have unique features that come from either the shell or the core, and they are characterized by their multifunctional capabilities. Additionally, the core-to-shell ratio or other alterations to the component materials may be used to tune these features, providing an adaptive platform for a wide range of applications. The intrinsic properties of core particles such as thermal stability, reactivity, bioavailability, permeation, solubility, sustainability, and controlled release can be altered by encapsulation within a shell. By this process, the overall

Fig. 15.1 Core–shell nanoparticle

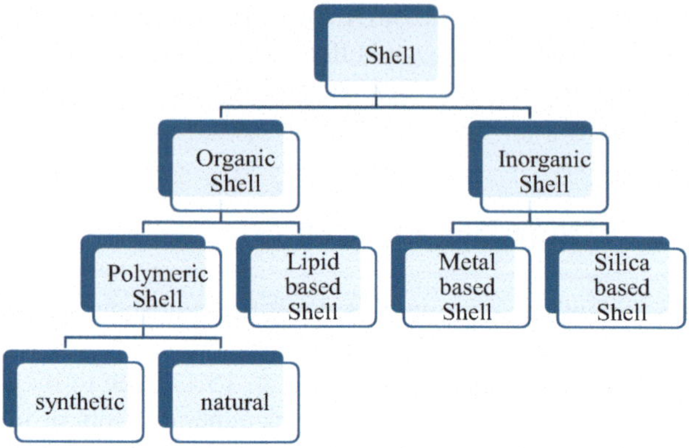

Fig. 15.2 Classification of shell materials

stability and dispersibility of the nanoparticles will greatly improve, enabling desired modifications in the core–shell nanoconstructs and, thereby, increasing their potential applications in various scientific and industrial areas (Dhas et al. 2018).

The components of the shell can be classified into the following (Fig. 15.2):

2.2.1 Organic Shell

Polymers (natural and synthetic), proteins, lipids, and complex carbohydrates are common constituents of the organic shell, which increases the biocompatibility of nanocomposites and, hence, boosts their efficacy. These organic layers act not only as a protective barrier against oxidative degradation but also in anchoring the sites for molecular attachments. Moreover, the properties like chemical and thermal robustness of the nanoparticles can be significantly enhanced by carefully controlling and designing the thickness of these shells. Additionally, this modification in shells can provide opportunities for precisely controlling the release of an encapsulated molecule from the nanoparticle's core, reduce cytotoxicity, and improve biocompatibility, cyto- compatibility, dispersibility, and chemical and thermal stability. These properties of nanoparticles have a wide range of applications in various domains, such as material science and drug delivery (Chiozzi and Rossi 2020).

2.2.2 Polymeric Shell

The coating used to enclose nanoparticles in cancer treatment mostly consists of polymeric (synthetic or natural) shells. Because of their diverse physicochemical features, synthetic polymers such as poly(methyl methacrylate) (PMMA), polyvinyl alcohol (PVA), polyethylene glycol (PEG), and polycaprolactone (PCL) are widely used as coatings in modern material sciences (Naik et al. 2024). These polymers have distinct properties such as biocompatibility, mechanical strength, chemical resistance, and the ability to sustain and control the release of the drug (Alfaifi et al. 2020; Al-Tememe et al. 2023; Vieira et al. 2022). Natural polymers like chitosan (CS), hyaluronic acid (HA), gelatine, and alginate are also employed in different applications (Khodashenas et al. 2021; Loch-Neckel et al. 2015; Rashidzadeh et al. 2023; Yu et al. 2021).

Mohammad Y. Alfaif and colleagues formulated and characterized febuxostat (FBX)-loaded PEG-coated poly lactic-co-glycolic acid (PLGA) nanoparticles to evaluate their potency in the treatment of lung cancer. The controlled release mechanism of FBX from the encapsulated PEG-PLGA nanoparticles was mainly controlled by both the erosion process as well as the diffusion of degradation product through polymeric covering pores of the particles, and this was approved in necessary experimental work (Alfaifi et al. 2020).

Jaqueline Vieira and colleagues used PCL; this research had already definitively demonstrated that nanoparticles can be synthesized with varied surface chemistries. The hydrodynamic dimensions of the PCL-coated nanoparticles were significantly altered; however, their colloidal stability was improved. Research findings on the use of these nanoparticles in targeting drug delivery, magnetic resonance imaging (MRI) diagnostics and hyperthermia applications. The nano-formulation holds much promise in the biomedical field due to its potential applications such as diagnostics and cancer treatment in times yet to come (Vieira et al. 2022).

Gecioni Loch-Neckel and team formulated polymeric nanoparticles of PCL and chitosan adsorbed on polycaprolactone (PCL) and loaded with curcumin, increasing the ability to decrease melanoma metastasis. This change of the face was a mucoadhesive system that can promote long-term local delivery for effective control of melanoma metastasis (Loch-Neckel et al. 2015).

Shaoxuan Yu and colleagues fabricated mixed nanoparticles combining curcumin and zeolitic imidazolate framework-8 (Cur-ZIF-8) in hyaluronic acid for the treatment of breast cancer therapy. The composite Cur@ZIF-8@HA has improved storage stability in various media, controlled drug release kinetics and, under neutral and acidic pH conditions, reported a higher inhibitory effect on breast cancer cells than free curcumin or only ZIF-8 loaded with curcumin (Yu et al. 2021).

B. Khodashenas and team fabricated gold (Au) nanoparticles encased in a gelatin biopolymer light to form the basis of an ideal implantable carrier system that could regulate methotrexate (MTX—antifolate therapeutic molecule used specifically for breast cancer treatment). Herein, the physicochemical characteristics of nanoparticles were altered by supplementing a gelatin coating in an intentional approach to inject pH- and temperature-responsive drug release mechanisms. The experimental findings of the current study revealed that optimal methotrexate release was only observed at 40 °C. Thus, it proved the ability of the nanocomposite system to be used in order to improve the targeted and controlled delivery of chemotherapeutic agents (Khodashenas et al. 2021).

2.2.3 Lipid-Based Shell

If complementary properties, such as enhanced biocompatibility and significant drug loading capacity, are to be overcome, the two primary nanocarrier systems for drugs must be coadministered: liposomes and polymeric nanoparticles (Colaco et al. 2024). Until now, in the nanotech domain, scientists are making hybrids from these two different types of nanoparticles, i.e., lipid and polymer (Hegde et al. 2023). These hybrids are called lipid–polymer hybrid nanoparticles (LPNs). These advanced material technologies have been used in many different types of medicinal treatments and diagnostic applications. Generally, LPNs consist of a lipidic envelope, which improves biocompatibility, and a polymeric nucleus that plays a crucial role in the controlled release of the drug (Krishnamurthy et al. 2015). There are distinct types of lipid coating according to the arrangement of the lipid layer, such as monolayer and bilayer. Supplementary coating of polymers like polyethylene glycol is occasionally added for proper coating. This coating modification gives the encapsulation a stealthy quality when circulating in the bloodstream in the body, improving its biocompatibility and its ability to elude immune monitoring (Poornima et al. 2020; Tenchov et al. 2023). Various lipids including phosphatidylcholine (PC) (Liang et al. 2020), 1,2-distearoyl-sn-glycero-3-phosphoethanolamine (DSPE) (Li et al. 2020b), cholesterol (Sharma et al. 2020), 1,2-dipalmitoyl-sn-glycero-3-phosp hocholine (DPPC) (Li et al. 2020a),14 myristic acid,15 stearic acid, and 1,2-dilauro yl-sn-glycero-3-phosphocholine (DLPC) (Krishnamurthy et al. 2015) have been employed in the formulation of lipid shells.

Jinying Liang and team developed an innovative delivery system aimed at minimizing systemic toxicity while simultaneously augmenting the antitumor efficacy of doxorubicin (DOX) against multidrug resistant (MDR) tumors that holds significant clinical implications. One of the main objectives of this work was to formulate and characterize lipid/hyaluronic acid (HA)-coated DOX-Fe_3O_4 NPs suggested for further improvements in delivery, safety, and anticancer efficiency. Ferrogel was fabricated by conjugating the anticancer agent doxorubicin (DOX) on the surface of Fe_3O_4 nanoparticles by the encapsulation of the functionalized nanoparticles with a coating of phosphatidylcholine (PC) lipid. The ability of the fabricated nanocomplexes to allow for targeted delivery and their biocompatibility were both improved by this method (Liang et al. 2020).

Saurabh Sharma and colleagues experimented on docetaxel (DTX), a therapeutic agent approved by the Food and Drug Administration (FDA) for chemotherapy, which serves as a primary therapeutic option for triple-negative breast cancer (TNBC). Nonetheless, several pharmacokinetic and pharmacodynamic challenges considerably restrict its therapeutic potential, such as its limited aqueous solubility, accelerated metabolic degradation, short plasma half-life, and suboptimal targeting efficacy toward neoplastic cells. In their study, they reported the utilization of a cholesterol-grafted, folate-targeted amphiphilic lipopolymer as a novel strategy to address the aforementioned limitations (Sharma et al. 2020).

2.2.4 Inorganic Shells

Inorganic coatings, including those composed of metals, metal oxides, silica etc. play a pivotal role in the modification of the physicochemical properties of nanostructures.

Bioactive inorganic nanomaterials, characterized by their distinctive properties and functionalities, play a pivotal role in the realm of cancer therapy. Owing to their enzyme-like and biological reactivity, these nanomaterials harness the potential for catalytic therapy, modulation of the tumor microenvironment (TME), and the facilitation of gas therapy through the emission of bioactive gaseous signaling molecules. Moreover, bioactive inorganic nanomaterials have the capability to interact with the protein and cellular structures for controlled cell death. Also, they directly induce an antitumor immune response by interacting with the immune system. Therefore, the application of these bioactive inorganic nanomaterials effectively promotes activated magnetic resonance imaging (MRI), photoacoustic imaging (PAI), and fluorescence optical systems. Synergistic method to offer the best cancer diagnostic strategy with these advancements in imaging technique (Pei et al. 2023).

2.2.5 Metallic Shell

One kind of gold core–shell structure, gold nanoshell, and gold-coated spherical silica core nanoparticles have been focused on extensively for their prospects as tools to use in cancer therapy and bioimaging modification (Patel et al. 2024). This explores the assembly and design of a metallic nanoscale shell theranostic agent (Lal et al. 2008). Therapeutic potential has been proved by Au nanoshells (AuNSs), which are presumably a spherical silica core surrounded by a gold shell at the nanoscale. Photothermal ablation and bioimaging are associated with the optical absorption of AuNSs to the near-infrared (NIR) spectrum. NIR windows are possibly modulated to a mild NIR frequency for enhancing fluorescence competently and facilitating imaging through fluorescence-based as well as photothermal ablation while tuning the resonance. In addition, AuNSs are good nanocarriers for the controlled release of antisense DNA oligonucleotides, offering a nonviral, light-triggerable vector for gene therapy (Barhoumi et al. 2009).

Kai Zhu and team worked on a core–shell nanoparticle in which gold was coated with silver (Au@Ag NP) to enhance computed tomography (CT) contrast for glioma imaging in pediatric patients. As a result, the pediatric patients were divided into experimental and control groups. The mice of the experimental group were promoted with Au@Ag nanoparticles, whereas in control groups, standard CT imaging procedures were performed. Compared with the normal group, the tumor boundary delineation (TBD), tumor contrast (TC), and tumor internal structure (TIS) scores were higher in the experimental groups. This indicated that Au@Ag nanoparticles could substantially enhance the diagnostic accuracy of CT imaging for pediatric glioma patients (Zhu and Li 2024).

2.2.6 Silica Shell

Mesoporous silica (MS) is essential both to keep the vehicle stable in biologically applicable buffers and for the delivery of a variety of drug compounds that are too large to be transported between cells using traditional diffusion or pinocytosis methods. MS efficiently prevents the rapid deposition and/or induces the clearance of core molecules from blood vessels, resulting in increased efficacy. MS serves as suitable for this purpose because it enables the accurate size and shape of silica nanoparticle production pathways and flexible surface chemistry of MS, but most importantly, MS is a food-grade material acknowledged by the FDA (Bizeau et al., 2024).

Cristina Chircov and team worked on and presented a viable therapeutic approach for the treatment of colorectal cancer since the researchers developed a novel series of magnetite-silica core–shell nanostructures and achieved selective, sustained, and controlled release of irinotecan. For the fabrication of nanoparticles, Stober method was used to deposit silica coats on the surface of the magnetite core after they were prepared using a thermal technique. By using this silica shell technique, this study was able to design a number of drug delivery systems that can increase the effectiveness of traditional chemotherapeutic medications for the treatment of cancer (Chircov et al. 2024).

The application of organic core–nonmagnetic shell nanoconstructs in cancer theranostics represents a cutting-edge approach that combines targeted therapy with diagnostic imaging, aiming to enhance treatment efficacy while minimizing side effects. The surface functionalization of these nanoconstructs plays a crucial role in optimizing their performance by improving their specificity, responsiveness to the tumor microenvironment, circulation time, and imaging capabilities. Below, we delve into the detailed strategies employed in surface functionalization, focusing on targeting ligands, stimuli-responsive functionalization, and stealth and imaging modifications.

3 Targeting Ligands

Targeting ligands are essential components that confer specificity to nanoconstructs, enabling them to selectively bind to cancer cells while sparing healthy tissues. This selectivity is achieved by decorating the surface of the nanoconstructs with molecules that recognize and bind to specific receptors or antigens overexpressed on the surface of cancer cells. This strategy not only enhances the therapeutic efficacy but also reduces off-target effects, a critical factor in cancer treatment (Fig. 15.3).

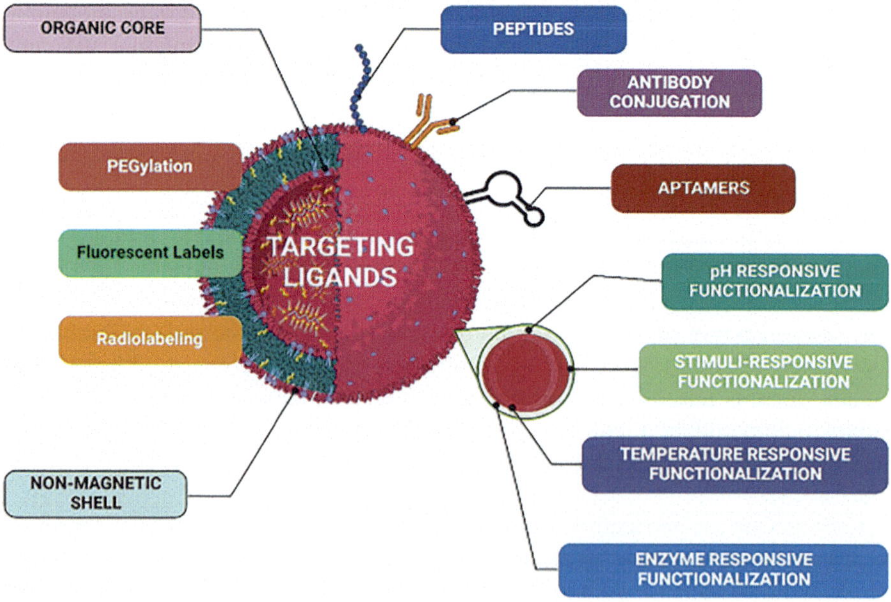

Fig. 15.3 Strategies for targeting ligands for functionalization

3.1 Antibodies

Monoclonal antibodies are a prominent class of targeting ligands due to their high specificity and affinity for antigens expressed on cancer cells. These antibodies are designed to target specific proteins such as HER2/neu, EGFR, or CD44, which are frequently overexpressed in various cancers. The process of conjugating antibodies to nanoconstructs typically involves the formation of covalent bonds between the functional groups on the antibody (e.g., amines, thiols, or carboxyls) and the surface of the nanoconstructs. This conjugation must be carefully controlled to maintain the correct orientation of the antibodies, ensuring that their antigen-binding sites remain exposed and functional. For example, trastuzumab (Herceptin), an anti-HER2 antibody, can be conjugated to the surface of nanoconstructs to target HER2-positive breast cancer cells. Once bound to the cancer cell surface, these antibody-functionalized nanoconstructs can deliver cytotoxic agents directly to the tumor, enhancing the therapeutic index while minimizing damage to normal tissues.

A study by Jack Hu and his colleagues introduced a targeted drug delivery system for pancreatic cancer using anti-carcinoembryonic antigen (CEA) half-antibody-conjugated lipid–polymer hybrid nanoparticles. These nanoparticles, designed for selective cancer cell targeting, consist of a biodegradable PLGA core, phospholipid monolayer, and PEG corona. The anti-CEA half-antibodies enhance specificity and circulation time, leading to improved cellular uptake and cytotoxicity in pancreatic cancer cell lines (BxPC-3 and XPA-3) compared to non-targeted counterparts. This approach holds promise for improving chemotherapy efficacy by delivering high drug doses directly to poorly vascularized pancreatic tumors (Hu et al. 2010).

In a recent study, a PAMAM dendrimer–antibody conjugate (G4.5/IL6) was synthesized using EDC/NHS coupling chemistry, resulting in core–shell nanoparticles with enhanced fluorescence properties. The G4.5/IL6 conjugates, approximately 70 nm in size, exhibited a significant blue shift in fluorescence due to H-aggregation and demonstrated higher cellular uptake efficiency in HeLa cells compared to unmodified G4.5. These findings suggest that the G4.5/IL6 conjugates have potential as bioimaging probes, offering a promising tool for targeted cancer therapy and diagnostics (Mekuria and Tsai 2015).

Xia and his colleagues explored the role of antibody conjugation in enhancing targeted chemotherapy against hepatocellular carcinoma (HCC) using adriamycin-loaded polymer–lipid hybrid nanoparticles (PLNP-Mal-EGFR). The anti-EGF receptor antibody was conjugated to the nanoparticles, enabling specific targeting to HCC cells overexpressing EGFR. This targeting improved cellular uptake through receptor-mediated endocytosis, leading to significantly higher cytotoxicity compared to non-targeted nanoparticles and free adriamycin. The antibody-conjugated nanoparticles also reduced the proportion of side population cells, often linked to drug resistance, and showed superior in vivo performance with increased tumor accumulation and enhanced antitumor activity. This study demonstrates that antibody conjugation enhances the specificity, cellular uptake, and therapeutic efficacy of nanoparticles, offering a promising approach for HCC chemotherapy by overcoming drug resistance and improving targeted delivery (Gao et al. 2014).

3.2 Peptides

Peptides are short chains of amino acids that can be engineered to bind specific receptors or proteins to the surface of cancer cells. Their small size and high specificity make them ideal candidates for targeting ligands in nanoconstructs. For instance, the RGD peptide, which targets integrins (e.g., αvβ3 integrins) overexpressed on tumor vasculature, can be conjugated to nanoconstructs to improve their accumulation at the tumor site. Peptides offer several advantages over larger biomolecules, including better penetration into solid tumors and reduced immunogenicity. Moreover, peptides can be chemically modified to enhance their stability, binding affinity, and resistance to enzymatic degradation, further increasing their utility in cancer targeting. The synthesis of peptide-functionalized nanoconstructs involves the attachment of the peptide to the surface of the nanoconstructs using linkers that maintain the bioactivity of the peptide. These modified nanoconstructs can then be used to deliver chemotherapeutic agents, imaging agents, or even gene therapies to the tumor, providing a multifaceted approach to cancer treatment.

Peptide conjugation in organic nanocarriers has proven to be a game-changer in drug formulation, addressing key challenges like prolonged circulation, effective tumor targeting, and efficient metabolite clearance. By incorporating coiled-coil 3-helix bundle-forming peptide–polyethylene glycol conjugates, these nanocarriers self-assemble into stable micellar nanoparticles, approximately 15 nm in diameter. This design allows for precise ligand clustering and controlled disassembly, enhancing drug loading and reducing leakage. The peptide-conjugated micelles demonstrated a 29.5-h circulation half-life and minimal off-target accumulation, making them highly effective for targeted drug delivery in oncology. This approach not only improves drug stability and delivery but also offers a promising pathway for more efficient and safer therapeutic applications (Dong et al. 2012).

In addressing the challenge of delivering antihypertensive peptides orally, this case study highlights the development of KY5-CSs, a core–shell lipid–polymeric nanoparticle designed for the peptide KY5. The core–shell structure provided critical advantages, such as enhanced stability and controlled release in the gastrointestinal tract. The lipid shell improved the nanoparticle's ability to permeate the mucus layer and be internalized by intestinal cells, significantly boosting peptide absorption. In vivo studies demonstrated that KY5-CSs had superior oral absorption and antihypertensive efficacy compared to nonconjugated nanoparticles (KY5-NPs). This peptide conjugation strategy shows great promise for improving the bioavailability and effectiveness of oral antihypertensive therapies (Yuan et al. 2023).

3.3 Aptamers

Aptamers are synthetic oligonucleotides (DNA or RNA) that fold into unique three-dimensional structures capable of binding to specific targets with high affinity and specificity. These molecules are often referred to as "chemical antibodies" due to their ability to mimic the binding properties of antibodies. Aptamers can be generated against a wide range of targets, including proteins, small molecules, and even whole cells. For example, aptamers targeting the nucleolin protein, which is overexpressed on the surface of certain cancer cells, can be conjugated to nanoconstructs to direct them specifically to tumor cells. The process of aptamer functionalization typically involves the attachment of the aptamer to the nanoconstructs through covalent bonds or other chemical linkages that do not disrupt the aptamer's binding capabilities. Aptamers offer several advantages over traditional antibodies, including lower immunogenicity, ease of synthesis and modification, and the ability to generate aptamers for virtually any target. Additionally, aptamers can be modified with chemical groups to enhance their stability in biological environments, making them highly effective targeting ligands for cancer theranostics.

In a recent study, researchers developed core–shell nanoparticles with a poly(lactic-co-glycolic acid) (PLGA) core and an exosomal membrane (EM) shell, functionalized with AS1411 aptamers (AS-EP), to enhance tumor targeting and extend circulation time. Utilizing microfluidic sonication, the team integrated the PLGA core with the EM shell and tethered cholesterol-modified AS1411 aptamers through hydrophobic interactions. This method resulted in nanoparticles with improved immune evasion and targeted delivery capabilities. The AS-EP nanoparticles exhibited significantly higher tumor accumulation compared to nonconjugated EM-coated PLGA nanoparticles and those with random-sequence oligonucleotides, underscoring the advantages of aptamer conjugation for enhanced targeting and prolonged circulation (Han et al. 2020).

3.4 Stimuli-Responsive Functionalization

Stimuli-responsive functionalization allows nanoconstructs to respond dynamically to specific triggers within the tumor microenvironment or to external stimuli. This capability enables the controlled and site-specific release of therapeutic agents, enhancing the efficacy of treatment while minimizing systemic toxicity. The design of stimuli-responsive nanoconstructs involves the incorporation of materials or functional groups that undergo structural or chemical changes in response to particular stimuli, such as pH, temperature, or enzyme activity.

3.5 pH-Responsive Functionalization

The acidic microenvironment of tumors, with pH levels typically ranging from 6.5 to 7.0, contrasts with the neutral pH of normal tissues (approximately 7.4). This pH difference can be exploited to design nanoconstructs that release their therapeutic payloads specifically in the acidic tumor environment. pH-responsive nanoconstructs are often functionalized with polymers or linkers that remain stable at physiological pH but degrade or change conformation in acidic conditions. For instance, polyhistidine, a pH-sensitive polymer, can be used to coat the surface of nanoconstructs. At the slightly acidic pH of the tumor microenvironment, the polyhistidine undergoes protonation, leading to destabilization of the nanoconstruct and subsequent release of the encapsulated drug. Another approach involves the use of acetal linkers, which are stable at neutral pH but hydrolyze in acidic environments, triggering drug release. This strategy ensures that the drug is delivered directly to the tumor site, reducing the likelihood of off-target effects and enhancing therapeutic efficacy.

Researchers tackled the challenge of balancing surface charge requirements for nanoparticles used in controlled drug delivery by developing vancomycin-loaded composite nanoparticles featuring a novel chitosan core and a poly(lactide-co-glycolide) (PLGA) shell. These spherical nanoparticles, created using a modified emulsion–gelation method, demonstrated tunable properties with sizes ranging from 316 to 573 nm, surface charges between −27.6 and 31.75 mV, and an encapsulation efficiency of up to 70.8%. The innovative core–shell design allows the particles to start with a negative charge, which switches to positive as the biodegradable PLGA shell erodes, revealing the positively charged chitosan core. This transition addresses the conflicting needs for extended circulation and effective cell affinity. Additionally, the chitosan-based hydrogel exhibited multilayer structures influenced by gelation conditions, and the composite nanoparticles showed pH-responsive drug release profiles, with distinct fast- and slow-release stages. This dual functionality makes the chitosan-core and PLGA-shell nanoparticles a promising system for controlled drug delivery, effectively overcoming previous limitations in nanoparticle design (Huang et al. 2017).

3.6 Temperature-Responsive Functionalization

Hyperthermia, which involves elevating the temperature of tumor tissues using external sources such as focused ultrasound or infrared radiation, is a technique used to enhance the effectiveness of cancer therapies. Temperature-responsive nanoconstructs are designed to exploit this by incorporating polymers that undergo a phase transition at a specific temperature. For example, poly(N-isopropylacrylamide) (PNIPAM) is a thermo-responsive polymer that exhibits a sharp phase transition at temperatures above 32 °C. When used to functionalize nanoconstructs, PNIPAM remains hydrated and stable at body temperature (37 °C) but undergoes a reversible collapse or aggregation when the temperature is raised, such as during hyperthermia treatment. This collapse can trigger the release of drugs encapsulated within the nanoconstruct, providing a targeted release mechanism that coincides with the hyperthermia-induced sensitization of tumor cells. This approach not only increases the local concentration of the drug at the tumor site but also enhances its cytotoxic effects by exploiting the hyperthermic conditions.

In a recent study, Zhang and his colleagues developed a novel temperature-responsive shell-cross-linked (SCL) nanocontainer using a specialized amphiphilic diblock copolymer: poly(2-methacryloyloxyethyl ferrocene carboxylate)-(5-propargylether-2-nitrobenzyl bromoisobutyrate)-poly(dimethylamino ethyl methacrylate) (PMAEFc-ONB-PDMAEMA). This copolymer was synthesized through atom transfer radical polymerization (ATRP) and click chemistry. The self-assembled spherical micelles were then cross-linked with N, N′-bis(bromoacetyl) cystamine (BBAC), resulting in SCL micelles that exhibited significant temperature-responsive behavior. Specifically, these SCL micelles shrank in response to higher temperatures, a property that was attributed to the protective cross-linked network. This temperature sensitivity allows the micelles to regulate the release of encapsulated drugs more precisely under varying thermal conditions. In contrast to non-cross-linked (NCL) micelles, which did not show this degree of thermal responsiveness, the SCL micelles demonstrated enhanced functionality, making them promising candidates for applications requiring controlled release triggered by temperature changes. This study underscores the benefits of incorporating a temperature-sensitive organic core and nonmagnetic shell in the design of advanced nanocontainers for precise drug delivery (Zhang et al. 2018b).

3.7 Enzyme-Responsive Functionalization

The tumor microenvironment is characterized by the overexpression of specific enzymes, such as matrix metalloproteinases (MMPs), cathepsins, and phosphatases, which play crucial roles in tumor progression, invasion, and metastasis. Enzyme-responsive nanoconstructs are functionalized with substrates or linkers that are specifically cleaved by these enzymes, leading to the controlled release of the therapeutic payload. For example, MMP-sensitive peptides can be used to link the therapeutic agent to the nanoconstruct. In the presence of MMPs, which are highly expressed in the tumor stroma, the peptide is cleaved, releasing the drug directly at the site of the tumor. This strategy ensures that the drug is activated only in the tumor microenvironment, reducing systemic exposure and minimizing side effects. Additionally, enzyme-responsive nanoconstructs can be designed to release multiple drugs or therapeutic agents in response to different enzymes, enabling combination therapies that target various aspects of tumor biology simultaneously.

Researchers developed a temperature-responsive nanocarrier using a specialized triblock copolymer. The alkynyl-terminated double hydrophilic ABC triblock copolymer, poly(oligo(ethylene glycol) monomethyl ether methacrylate)-b-poly(2-(dimethylamino) ethyl methacrylate)-b-poly(2-(diethylamino) ethyl methacrylate) (alkynyl-POEGMA-b-PDMA-b-PDEA) was synthesized via atom transfer radical polymerization (ATRP) with propargyl 2-bromoisobutyrate (PgBiB) as the initiator. This copolymer self-assembled into "onion-like" micelles with a pH-responsive PDEA core, a PDMA inner shell, and a POEGMA corona at alkaline pH. The key innovation involved cross-linking the PDMA inner shell with 2-bis(2-iodoethoxy) ethane (BIEE) to form structurally stable, surface "clickable" shell-cross-linked (SCL) micelles. To enhance temperature responsiveness, the researchers incorporated azido-terminated poly(N-isopropylacrylamide) (PNIPAM-N3) via click chemistry, creating four-layer nanoparticles (SCL-PNIPAM). These SCL-PNIPAM nanoparticles exhibited a temperature-responsive outer PNIPAM corona, which enabled controlled drug release based on temperature changes. The temperature sensitivity of the nanoparticles was demonstrated through their ability to swell or shrink in response to temperature variations, allowing precise control over drug release. This development showcases the potential of temperature-responsive nanocarriers in delivering drugs more effectively by exploiting changes in environmental temperature (Jiang et al. 2009).

4 Stealth and Imaging Modifications

Stealth and imaging modifications are critical for improving the circulation time of nanoconstructs in the bloodstream and enhancing their visualization for diagnostic and therapeutic monitoring. These modifications enable nanoconstructs to evade the immune system, prolong their half-life, and provide real-time information about their distribution and accumulation within the body.

4.1 PEGylation

PEGylation involves the attachment of polyethylene glycol (PEG) chains to the surface of nanoconstructs to create a hydrophilic shield that reduces protein adsorption and recognition by the immune system. PEGylated nanoconstructs are often referred to as "stealth" nanoparticles because they can circulate in the bloodstream for extended periods without being cleared by the mononuclear phagocyte system (MPS). The PEG chains form a hydrated corona around the nanoconstructs, preventing opsonization and subsequent clearance by macrophages. This prolonged circulation time allows the nanoconstructs to accumulate at the tumor site via the enhanced permeability and retention (EPR) effect, where the leaky vasculature of tumors facilitates the passive targeting of the nanoconstructs. PEGylation also enhances the biocompatibility and stability of the nanoconstructs, reducing the likelihood of aggregation and improving their overall therapeutic efficacy. In addition to PEGylation, other stealth strategies such as zwitterionic coatings or biomimetic surface modifications using cell membranes can also be employed to further enhance the stealth properties of nanoconstructs.

Sun and his colleagues investigated the biodistribution of shell-cross-linked (SCK) nanoparticles designed for biomedical applications, focusing on the impact of the core composition, size, and surface PEGylation. The SCK nanoparticles, which feature an amphiphilic core–shell structure, were evaluated using the positron-emitting radionuclide copper-64 for in vivo tracking. The study compared four different SCK formulations with varying core materials, diameters, and surface PEGylation. The results revealed that SCKs with a small diameter of 18 nm and a polystyrene core exhibited the most advantageous biological properties. These nanoparticles demonstrated extended blood circulation times and minimal liver accumulation, highlighting their potential for effective drug delivery. The study underscored the significance of the core composition and nanoparticle size in influencing biodistribution as the flexibility and shape adaptability of the core material were crucial factors. Conversely, surface modification with polyethylene glycol (PEG) did not significantly alter the behavior of the SCKs. This case study illustrates the advantages of tailoring the organic core and nonmagnetic shell properties to optimize nanoparticle performance in biomedical applications (Sun et al. 2005).

Researchers developed PEGylated core–shell nanoparticles to enhance the efficacy of cisplatin against GSH-overexpressing breast cancer cells. Using biodegradable carboxylic functional polycaprolactone (PCL)–block–polyethylene glycol (PEG) diblock copolymers, they created nanoparticles with a stable core size of 75 nm and variable PEG shell thickness. The PEGylation improved stability in saline and PBS, and crucially, protected the cisplatin core from detoxification by cytoplasmic thiols like GSH, maintaining less than 5% drug detoxification. In vitro tests showed that PEGylated core–shell nanoparticles achieved complete cell death in MCF-7 cells, outperforming free cisplatin. This approach demonstrates the effectiveness of PEGylation in enhancing drug delivery and reducing detoxification in cancer therapy (Surnar et al. 2015).

4.2 Fluorescent Labels

Fluorescent labeling of nanoconstructs enables their visualization and tracking in vivo and in vitro using fluorescence imaging techniques. Fluorophores such as rhodamine, fluorescein, or near-infrared (NIR) dyes can be covalently attached to the surface of nanoconstructs, allowing for the real-time monitoring of their biodistribution, cellular uptake, and clearance from the body. Fluorescently labeled nanoconstructs provide valuable insights into the pharmacokinetics and tissue distribution of the therapeutic agents they carry, helping optimize dosage and treatment regimens. Additionally, fluorescence imaging can be used to monitor the interaction of nanoconstructs with specific cellular compartments, such as endosomes or lysosomes, providing information on the intracellular trafficking and release mechanisms of the drug. The integration of fluorescent labels into nanoconstructs also facilitates the study of their interactions with the tumor microenvironment, enabling researchers to assess their penetration into tumor tissues, their localization within different regions of the tumor, and their ability to target specific cancer cell populations.

Researchers developed a novel series of water-dispersible polymer–covalent organic framework (COF) nanocomposites for advanced drug delivery applications. The nanocomposites were created by conjugating polyethylene glycol-modified curcumin derivatives (PEG-CCM) with amine-functionalized COFs (APTES-COF-1). This innovative approach enabled effective in vivo tracking through real-time fluorescence monitoring, demonstrating efficient cellular uptake and targeted drug release. The PEG-CCM@APTES-COF-1 nanocomposites exhibited superior stability, biodegradability, and high loading capacity for doxorubicin (DOX), alongside strong and stable fluorescence signals. Notably, the PEG350-CCM@APTES-COF-1 variant showed enhanced targeting capabilities for brain delivery. The study underscores the potential of fluorescent conjugation in organic core–nonmagnetic shell systems to improve drug delivery and targeting efficiency, marking a significant advancement in multifunctional drug delivery platforms for cancer treatment (Zhang et al. 2018a).

4.3 Radiolabeling

Radiolabeling nanoconstructs with radioisotopes such as ^99mTc, ^111In, or ^18F enable their detection using nuclear imaging modalities such as positron emission tomography (PET) or single-photon emission computed tomography (SPECT). These imaging techniques provide high sensitivity and spatial resolution, allowing for the noninvasive tracking of nanoconstructs in real time. Radiolabeled nanoconstructs can be used to evaluate their pharmacokinetics, biodistribution, and tumor accumulation, providing critical data for the optimization of cancer therapies. For example, ^18F-labeled nanoconstructs can be used in PET imaging to quantify their uptake in tumors and monitor the response to treatment over time. Radiolabeling can also be combined with other imaging modalities, such as magnetic resonance imaging (MRI) or fluorescence imaging, to provide complementary information on the tumor microenvironment and the therapeutic efficacy of the nanoconstructs. The integration of radiolabels into nanoconstructs also opens the possibility of using them for theranostic applications where the same nanoconstruct can be used for both imaging and therapy. For instance, ^177Lu-labeled nanoconstructs can deliver both radiotherapy and chemotherapy to tumors, providing a synergistic approach to cancer treatment.

A recent study presents a novel approach for the synthesis, radiolabeling, and in vivo evaluation of polymeric nanoparticles (NPs) functionalized with an organosilicon fluoride acceptor (SiFA). The SiFA-functionalized nanoparticles were synthesized using microemulsion polymerization, yielding particles with distinct sizes: 20 nm (NP1), 33 nm (NP2), 45 nm (NP3), and 72 nm (NP4). The SiFA served as both a polymerization initiator and a radiolabeling building block. Using simple isotopic exchange (IE) in ethanol, the nanoparticles were radiolabeled with fluorine-18 (^18F), achieving radiochemical yields of 19–26% and specific activities of 0.2–0.3 GBq/mg. In vivo evaluation in a murine mammary tumor model (EMT6) demonstrated that the 33-nm nanoparticles showed the highest tumor accumulation, with an SUVmean of 0.97 (4.4%ID/g) after 4 h, attributed to the enhanced permeability and retention (EPR) effect. This straightforward method offers a robust and efficient way to produce radiolabeled nanoparticles with desirable in vivo properties without the need for additional surface modifications (Berke et al. 2018).

This study investigates the synthesis and application of core–shell star copolymers for radiolabeled imaging. Core–shell star copolymers were created using living free radical polymerization, resulting in nanostructures with a polyethylene glycol (PEG) outer shell, a hydrophilic inner shell with reactive functional groups, and a central hydrophobic core. These nanoparticles were functionalized with a DOTA-ligand for chelating radioactive copper-64 (^64Cu), enabling positron emission tomography (PET) imaging. By varying the PEG shell thickness, researchers controlled blood circulation and organ accumulation. Findings showed that nanoparticles with thicker PEG shells had extended blood circulation times and reduced accumulation in excretory organs, demonstrating their potential as effective carriers for imaging, targeting, and therapeutic applications. This approach provides a precise method for tuning nanoparticle properties and optimizing their performance for in vivo applications (Fukukawa et al. 2008).

5 Cancer Applications

5.1 Targeted Cancer Therapy

The development of sophisticated core–shell nanoparticles intended for targeted cancer therapy is the result of recent advances in nanomedicine. These nanoparticles are designed to decrease adverse effects, increase therapeutic efficacy, and improve drug delivery.

Innovative core–shell lipid–polymer hybrid nanoparticles were created in a study for focused cancer treatment. The poly(lactic-co-glycolic acid) core of these nanoparticles is covered in a biocompatible lipid shell consisting of 1,2-dipalmitoyl-sn-glycero-3-phosphocholine. The core–shell structure allows for the co-loading of dihydroartemisinin, a hydrophobic compound, and doxorubicin, a hydrophilic drug, facilitating combination chemotherapy. Physicochemical characterization confirmed the successful dual drug loading. In vitro studies demonstrated that dihydroartemisinin provides immediate cytotoxic effects, while doxorubicin offers a delayed action. This system promotes endosomal escape, protecting the drugs from degradation and increasing nuclear accumulation. The core–shell nanoparticles show a strong potential for enhancing the efficacy of combination cancer therapies (Ma et al. 2015).

Core–shell lipid–polymer hybrid nanoparticles (CSLPHNPs), which have a lipid shell and a poly(lactic-co-glycolic acid) core, were the subject of another investigation. For targeted treatment against metastatic prostate cancer, these nanoparticles were loaded with the sphingosine kinase 1 inhibitor FTY720 (fingolimod) and docetaxel. High serum stability, a lengthy shelf life, consistent tumor cell uptake, and prolonged intracellular drug release were all displayed by the CSLPHNPs. These nanoparticles outperformed free medications in a mouse model of prostate cancer in terms of tumor targeting and side effect reduction. Interestingly, they prevented the lymphopenia caused by FTY720 and allowed for a fourfold decrease in docetaxel dosage. This approach provides increased treatment efficacy, less off-target toxicity, and better targeting (Wang et al. 2017).

Using graphene quantum dots (GQDs) loaded into hollow mesoporous silica nanoparticles (hMSNs) to create GQDs@hMSN-PEG nanoparticles was another creative method for developing a hybrid nanoplatform. After encapsulation, this device maintained the optical characteristics and effective singlet oxygen ($^1O_2$) generation of GQDs. When doxorubicin was added to GQDs@hMSN-PEG nanoparticles, tumor absorption rose dramatically and the accumulation was almost twice as high as when the nanoparticles were not loaded. This hybrid platform is a viable contender for cancer theranostics because it has the ability to combine photodynamic therapy with chemotherapy, improve tumor retention, and have a high drug loading capacity (Yang et al. 2018).

Multifunctional poly(lactic-co-glycolic acid)-lecithin polyethylene glycol core–shell nanoparticles were created for the purpose of delivering drugs targeted toward cancer. These nanoparticles, which have adjustable surface charge, size, and functional groups, combine the characteristics of polymeric and liposomal nanoparticles. The hydrophilic cisplatin and its hydrophobic prodrug were both encapsulated by the nanoparticles, resulting in great encapsulation efficiency and stability. Folic acid-based targeted delivery outperformed non-targeted nanoparticles in MCF-7 breast cancer cells expressing folic acid receptors, exhibiting increased uptake and cytotoxicity. For combination cancer treatments, these core–shell nanoparticles offer a very effective, biocompatible platform for co-delivering several therapeutic agents, such as medications and genes (Zheng et al. 2012).

5.2 Cancer Theranostics

Core–shell nanoparticles have emerged as versatile tools for integrated cancer therapy, combining radiation, chemotherapy, and advanced imaging. Also, several studies highlight their potential in this field. Core–shell nanoparticles incorporating gold nanoparticles as radiosensitizers and doxorubicin as a chemotherapeutic agent were developed. The core, consisting of vesicles encapsulating gold nanoparticles and doxorubicin, is freeze-dried in a Pluronic F-68 shell. This shell, made of polyethylene oxide, improves systemic circulation and tumor targeting. The gold nanoparticles enhance radiation sensitivity and facilitate targeted doxorubicin delivery. In vivo studies showed reduced tumor volume with combined focused radiation and doxorubicin, demonstrating the efficacy of these core–shell nanoparticles for combination cancer therapy through targeted delivery and radiosensitization (Kim et al. 2016).

Core–shell nanoparticles with a core of docetaxel-loaded poly(lactide-co-glycolide) and a gold nanoshell were synthesized. The gold nanoshell, functionalized with the tumor-targeting peptide angiopep-2, enables both active and passive targeting of glioma cells. The system allows for combined chemotherapy and photothermal therapy, with the gold nanoshell generating heat upon 808-nm laser activation to enhance anticancer effects. Additionally, these nanoparticles are suitable for X-ray imaging, combining targeted therapy with diagnostic imaging. This is one of the promising approaches for tumor-targeted chemo-photothermal therapy and diagnostic imaging.

One study reported the synthesis of hollow tantalum oxide (TaOx) nanoparticles with a core–shell structure where polypyrrole (PPy) and doxorubicin (DOX) are encapsulated in the core and a near-infrared fluorescence dye (NIRDye800) is conjugated to the shell. These PPy&DOX@TaOx-NIR Dye800-PEG nanoparticles exhibit multimodal imaging capabilities, including computed tomography (CT), photoacoustic imaging, and fluorescence imaging. They also offer pH- and thermal-sensitive drug release. The nanoparticles were tested in a subcutaneous esophageal cancer model, showing high imaging contrast, controlled photothermal therapy (PTT), and effective chemotherapy. They demonstrated comparable therapeutic efficacy via aerosol and intravenous administration, making them suitable for clinical application. Hence the PPy&DOX@TaOx-NIRDye800-PEG nanoparticles provide precise imaging and synergistic chemo-thermal therapy, with promising potential for clinical translation. Future studies should focus on tumor-targeting mechanisms, long-term biocompatibility, and larger animal models before FDA approval (Jin et al. 2017).

Using apoferritin as a template, $LuPO_4$ nanoparticles were synthesized and encapsulated within the apoferritin shell. Characterization confirmed the presence of 8-nm $LuPO_4$ nanoparticles. Biotin-modified nanoparticles were tested for pre-targeting with streptavidin-coated magnetic beads and fluorescein isothiocyanate tracers. This method offers a fast and effective way to create $LuPO_4$–apoferritin nanoparticles for potential use in radioimmunotherapy and radioimmunoimaging of cancer (Wu et al. 2008).

Selenium–polypyrrole core–shell nanoparticles were synthesized through in situ polymerization and functionalized with transferrin for the targeted imaging of human cervical cancer cells. The shell thickness and composition were controlled by adjusting the precursor concentration, enhancing the thermal stability of the selenium core. The transferrin coating significantly improved cellular uptake, making these nanoparticles promising for cancer cell targeting and imaging. Additionally, polyethylene glycol was conjugated to stabilize the nanoparticles in culture buffers. Polypyrrole was chosen for its biocompatibility and environmental stability, while transferrin, known for its role in iron transport and increased expression in malignant tissues, facilitates targeted drug delivery to cancer cells. This approach highlights the potential of these core–shell nanoparticles in cancer theranostics (Li and Liu 2008).

Tecto dendrimers with gold nanoparticles for dual-mode CT and MR imaging of tumors were synthesized. The core comprises β-cyclodextrin-modified generation 5 PAMAM dendrimers with gold nanoparticles, while the shell consists of adamantine-modified generation 3 PAMAM dendrimers, formed through host–guest interactions. The tecto dendrimers were further functionalized with Arg-Gly-Asp peptides, a Gd chelator, and 1,3-propane sultone and then loaded with Gd (III) ions. These particles, averaging 11.61 nm, demonstrated excellent stability, high X-ray attenuation, significant r1 relaxivity (9.414 mM − 1 s − 1), and good cytocompatibility. RGD-mediated targeting enabled effective CT/MR imaging of αvβ3 integrin–overexpressing cancer cells, showing promising results for diagnosing such cancers with favorable biosafety and clearance (Liu et al. 2021).

A novel approach was used to develop mesoporous core–shell up-conversion nanoparticles (mUCNPs) featuring an $NaYbF_4$:2%Er core and a mesoporous $NaGdF_4$ shell. This design enhances up-conversion luminescence and provides functionalities for drug delivery and bioimaging. The core–shell mUCNPs were further modified with polyethylenimine and folic acid to improve water dispersibility, drug delivery efficiency, and targeting of cancer cells. We verified folate receptor-mediated targeting using cell viability assays, up-conversion luminescence imaging, and flow cytometry. Additionally, the Gd and Yb elements in the nanoparticles offer strong contrast for magnetic resonance (MR) and computed tomography (CT) imaging, respectively. This multifunctional platform combines enhanced imaging with targeted drug delivery, demonstrating the potential for simultaneous multimodal imaging and cancer therapy (Zhou et al. 2014). These studies demonstrate the potential of core–shell nanoparticles in cancer theranostics, offering innovative solutions for targeted therapy and multimodal imaging.

5.3 Cancer Imaging

A study developed $NaDyF_4 - NaGdF_4$ core–shell nanoparticles (NPs) to enhance magnetic resonance imaging (MRI) of prostate cancer. These NPs feature a 21.2-nm $NaDyF_4$ core with a 0.6-nm $NaGdF_4$ shell, synthesized in a high-temperature solvent system. The nanoparticles were coated with phospholipids to improve water dispersibility and biocompatibility. The phospholipid-coated $NaDyF_4 - NaGdF_4$ NPs function as dual-modal contrast agents, enhancing both T1 and T2 relaxations in MRI at clinical (3 T) and preclinical (9.4 T) magnetic fields. In vitro studies and in vivo MRI on prostate cancer models demonstrated their effectiveness. For targeted imaging, the NPs were conjugated to anti-PSMA antibodies to specifically bind to PSMA-positive LNCaP prostate cancer cells, distinguishing them from PSMA-negative PC3 cells. The nanoparticles showed enhanced targeting efficiency (24 h) compared to passive targeting (2 h), exceeding current NP delivery methods. These $NaDyF_4 - NaGdF_4$ core–shell NPs offer significant potential for noninvasive, targeted MRI of cancerous tissues and could be adapted for diagnosing other diseases (Dash et al. 2021).

Another study introduced a novel method to produce highly efficient zinc oxide@polymer core–shell nanoparticles with green and yellow emission, suitable for biological imaging. This approach addresses the limitations of traditional zinc oxide (ZnO) quantum dots (QDs) by enhancing stability in aqueous solutions and improving photoluminescence (PL) properties. The synthesis involves preparing ZnO QDs with a polymer shell composed of polyethylene glycol methyl ether methacrylate (PEG MEMA) and zinc methacrylate. The process includes dissolving these components in ethanol, polymerizing the mixture, and transferring the resulting nanoparticles into water. The ZnO@polymer core–shell nanoparticles feature an internal hydrophobic polymethacrylate layer and an external hydrophilic PEG layer, ensuring stability and preventing aggregation. Characterization revealed excellent PL stability and high quantum yield (QY). These core–shell nanoparticles exhibit minimal toxicity to human cells at concentrations below 0.2 mg/mL and effectively label the cytoplasm of living cells. They offer notable advantages over traditional ZnO QDs, including improved water stability and tunable emission spectra. ZnO@polymer core–shell nanoparticles represent a promising, nontoxic, and cost-effective solution for cancer imaging. Their enhanced stability, tunable PL emission, and biocompatibility make them valuable for fluorescent labeling and imaging in cancer research (Xiong et al. 2008). Table 15.1 outlines the structure, drugs/agents loaded, cancer types, key features, therapeutic and imaging modalities, and outcomes for each study on core–shell nanoparticles in cancer theranostics.

6 Future Prospects

The future of organic core–nonmagnetic shell nanoconstructs in cancer theranostics lies in the continued innovation and integration of targeted therapy and diagnostic modalities. These constructs offer unparalleled versatility, and as the field advances, several key areas are poised for further exploration and development.

6.1 Enhanced Precision Through Stimuli-Responsive Functionalization

A major focus will be on refining stimuli-responsive features, enabling even more precise drug release mechanisms. While current designs utilize pH, temperature, and enzyme activity for site-specific activation, future innovations may involve multi-trigger systems that respond to a combination of internal and external stimuli. This could involve combining light-responsive materials with tumor-specific enzymes, ensuring the drug is only activated at the tumor site under highly controlled conditions. Such advancements would reduce off-target effects, a critical factor in minimizing toxicity during treatment (Sun et al. 2023).

6.2 Integration with Advanced Diagnostic Technologies

Future research will likely explore real-time diagnostics further, combining core–shell nanoconstructs with cutting-edge bioimaging technologies such as near-infrared fluorescence (NIRF) imaging, photoacoustic imaging, and magnetic resonance imaging (MRI). By incorporating imaging agents directly into the nanoconstructs, clinicians will be able to monitor treatment progress, drug release, and tumor response in real time, allowing for more adaptive therapeutic strategies. For example, nanoparticles that change their fluorescence in response to drug release could offer a direct indication of when and where the drugs are acting within the tumor (Refaat et al. 2022; Zhang et al. 2021).

6.3 Multidrug Delivery and Combination Therapies

Organic core–nonmagnetic shell nanoconstructs are ideal candidates for delivering multiple drugs simultaneously. This is particularly promising for cancer treatments, which often require combination therapies to target different pathways in tumor

Table 15.1 Core–shell nanoparticle systems, their key components, target applications, and functionalities for cancer treatment, theranostics, and imaging

Sl. No	Core material	Shell material	Loaded drugs/agents	Cancer type/ target	Key features	Application
1.	Poly(lactic-co-glycolic acid) (PLGA)	1,2-Dipalmitoyl-sn-glycero-3-phosphocholine (DPPC)	Dihydroartemisinin (hydrophobic), doxorubicin (hydrophilic)	General cancer treatment	Dual drug loading, promotes endosomal escape, increases nuclear accumulation	Combination chemotherapy
2.	Poly(lactic-co-glycolic acid) (PLGA)	Lipid	FTY720 (fingolimod), docetaxel	Metastatic prostate cancer	High serum stability, reduced side effects, increased targeting, and reduced docetaxel dose	Targeted cancer therapy
3.	Hollow mesoporous silica nanoparticles (hMSN)	PEGylated silica	Doxorubicin, graphene quantum dots (GQDs)	General cancer treatment	High drug loading, enhanced tumor retention, combination of photodynamic therapy with chemotherapy	Theranostics (PDT + chemotherapy)
4.	Poly(lactic-co-glycolic acid) (PLGA)	Lecithin-PEG	Cisplatin, hydrophobic prodrug	Breast cancer (MCF-7)	High encapsulation efficiency, folic acid targeting	Targeted cancer therapy
5.	Gold nanoparticles	Pluronic F-68 (polyethylene oxide)	Doxorubicin, gold nanoparticles	General cancer treatment	Radiosensitization, enhanced systemic circulation and tumor targeting	Combination of chemotherapy and radiosensitization
6.	Poly(lactide-co-glycolide) (PLGA)	Gold nanoshell	Docetaxel, gold nanoshell	Glioma	Photothermal therapy, active and passive targeting, suitable for X-ray imaging	Chemo-photothermal therapy, diagnostic imaging
7.	Polypyrrole (PPy), TaOx	PEG, NIR fluorescence dye	Doxorubicin	Esophageal cancer	pH- and thermal-sensitive drug release, multimodal imaging (CT, photoacoustic, fluorescence)	Chemo-thermal therapy, multimodal imaging

8.	$LuPO_4$	Apoferritin	Radiolabeled tracers	Cancer (general)	Biotin-modified for pre-targeting, high potential for radioimmunotherapy	Radioimmunotherapy and radioimmunoimaging
9.	Selenium	Polypyrrole	None	Cervical cancer	Transferrin-functionalized, targeted imaging, high cellular uptake	Cancer imaging, targeted drug delivery
10.	PAMAM dendrimers	Adamantine-modified PAMAM	Gold nanoparticles, Gd ions	Cancer (general)	Dual-mode CT/MR imaging, RGD-mediated targeting	Tumor imaging (CT/MR), biosafety
11.	$NaYbF_4$:2%Er	Mesoporous $NaGdF_4$	Folic acid, polyethylenimine (PEI)	Cancer (general)	Up-conversion luminescence, targeted drug delivery, multimodal imaging	Cancer therapy, multimodal imaging
12.	$NaDyF_4$	$NaGdF_4$	Anti-PSMA antibodies	Prostate cancer	Dual-modal MRI contrast, enhanced targeting with anti-PSMA antibodies	MRI imaging (T1/T2 relaxation)
13.	Zinc oxide (ZnO)	Polymer (PEG MEMA and zinc methacrylate)	None	Cancer (general)	Enhanced photoluminescence, high stability, low toxicity	Fluorescent labeling, cancer imaging

cells. By incorporating different drugs into the core and shell, or using layered release systems, future nanoconstructs could allow for timed release of various therapeutics. This approach could also integrate chemotherapy with immunotherapy, delivering checkpoint inhibitors or cytokines alongside traditional anticancer drugs, thereby enhancing the overall treatment efficacy and reducing tumor resistance (Gadde 2015).

6.4 Personalized and Precision Medicine

The future of nanoconstructs is closely tied to the broader movement toward personalized medicine. By tailoring the surface functionalization of these nanoparticles to target specific cancer markers unique to an individual's tumor, therapies can become more specific. As liquid biopsies and genomic profiling become more commonplace, nanoconstructs can be designed to match the molecular signature of a patient's cancer, offering truly personalized therapy. Moreover, these nanoconstructs could carry personalized combinations of drugs, imaging agents, and diagnostics tailored to an individual's specific tumor type and genetic profile (Ho et al. 2020; Johnson et al. 2021).

6.5 Improved Biocompatibility and Degradability

As the nanotechnology field progresses, improving the biocompatibility of these materials will be essential. Research into biodegradable and biocompatible organic shells that can safely break down in the body after delivering their therapeutic payload will minimize long-term toxicity. Advanced polymer chemistries that allow for programmable degradation—degrading at specific rates in response to environmental changes—could also be developed, ensuring that these nanoparticles do not accumulate in the body after their therapeutic role is complete (Saikia and Karak 2023).

6.6 Expanding Applications Beyond Oncology

While cancer remains the primary focus of nanoconstruct research, these systems have broader potential. Future applications could involve gene delivery for treating genetic disorders, vaccine development, and even tissue engineering. The controlled delivery capabilities of these nanoconstructs could be applied to other diseases that require precision medicine approaches, such as autoimmune diseases or neurodegenerative conditions. For instance, nanoparticles could be used to deliver drugs across the blood-brain barrier for treating central nervous system disorders where traditional drug delivery methods have failed (Mitchell et al. 2021).

6.7 Regulatory and Commercial Considerations

As more nanoconstructs move closer to clinical trials, regulatory pathways will need to adapt to accommodate these advanced technologies. Collaborating with regulatory bodies like the U.S. FDA and EMA will be crucial to ensure that safety, efficacy, and manufacturing standards are met. Moreover, for these technologies to become commercially viable, scalable manufacturing processes must be developed. Ensuring that the nanoconstructs are produced in a reproducible, cost-effective manner will be critical for bringing these innovations from the lab to the clinic. The focus will be on good manufacturing practices (GMP) and creating robust supply chains for the production of nanoconstructs that meet the necessary quality control requirements (Ali et al. 2023; Đorđević et al. 2022; Joyce et al. 2024).

6.8 Immunomodulation and Immunotherapy

With the rise of cancer immunotherapy, the integration of immunomodulatory agents into nanoconstructs is an exciting prospect. These nanoconstructs could be designed to deliver immune checkpoint inhibitors or stimulate immune cells directly at the tumor site, thereby enhancing the body's immune response to cancer. This synergistic approach, combining nanotechnology with the body's natural defenses, could open new doors in the treatment of cancers that are resistant to conventional therapies (Butterfield and Najjar 2024; Guo et al. 2021; Ni et al. 2023; Şimşek 2024).

7 Conclusion

In conclusion, organic core–nonmagnetic shell nanoconstructs represent an exciting and transformative technology in the field of cancer theranostics. Their ability to combine targeted therapy with advanced diagnostics offers a powerful platform for improving cancer treatment outcomes. Through sophisticated surface functionalization strategies and the integration of stimuli-responsive features, these nanoconstructs provide a precise and controlled approach to drug delivery, minimizing damage to healthy tissues and maximizing therapeutic efficacy.

Moving forward, the refinement of multidrug delivery systems, the exploration of new biocompatible materials, and the fusion of diagnostic imaging with therapeutic interventions will propel this technology toward widespread clinical application. The ability of these nanoconstructs to deliver combination therapies and adapt to individual patient profiles will play a key role in advancing personalized medicine. Moreover, as regulatory frameworks evolve, these technologies will begin to transition from experimental phases into mainstream clinical use.

The potential applications of these nanoconstructs extend far beyond oncology, with promising implications for gene therapy, vaccine delivery, and the treatment of chronic and genetic diseases. By continuing to improve their targeting accuracy, biocompatibility, and safety profiles, researchers can unlock new possibilities for treating a wide range of diseases, revolutionizing medicine as we know it.

Ultimately, the success of organic core–nonmagnetic shell nanoconstructs will depend on interdisciplinary collaboration between material scientists, biologists, clinicians, and regulatory experts. As this technology matures, it holds the potential to not only transform cancer care but to set new standards in drug delivery systems and diagnostics, offering hope for more effective, targeted, and less toxic therapies across a spectrum of medical conditions.

References

Alfaifi MY, Shati AA, Elbehairi SEI, Fahmy UA, Alhakamy NA, Md S. RETRACTED ARTICLE: Anti-tumor effect of PEG-coated PLGA nanoparticles of febuxostat on A549 non-small cell lung cancer cells. 3 Biotech. 2020;10:133. https://doi.org/10.1007/s13205-020-2077-x.

Ali F, Neha K, Parveen S. Current regulatory landscape of nanomaterials and nanomedicines: a global perspective. Journal of Drug Delivery Science and Technology. 2023;80:104118. https://doi.org/10.1016/j.jddst.2022.104118.

Alrushaid N, Khan FA, Al-Suhaimi EA, Elaissari A. Nanotechnology in cancer diagnosis and treatment. Pharmaceutics. 2023;15:1025. https://doi.org/10.3390/pharmaceutics15031025.

Al-Tememe EH, Jasim Al Hasan NH, Kareem AS, Hazim Salem K, Khmees EJ, Zabibah RS, Mohammed KA. Designing PMMA-PVA-TiO2 as new hybrid nanocomposite for anticancer applications. Int J Nanosci. 2023;22:2350029. https://doi.org/10.1142/S0219581X23500291.

Arbatti M, Shan X, Cheng Z-Y. Ceramic–polymer composites with high dielectric constant. Adv Mater. 2007;19:1369–72. https://doi.org/10.1002/adma.200601996.

Arnold Y, Bravo Gonzalez R, Versace H, Kuentz M. Comparison of different *in vitro* tests to assess oral lipid-based formulations using a poorly soluble acidic drug. Journal of Drug Delivery Science and Technology. 2010;20:143–8. https://doi.org/10.1016/S1773-2247(10)50019-4.

Ayyanaar S, Kesavan MP, Balachandran C, Rasala S, Rameshkumar P, Aoki S, Rajesh J, Webster TJ, Rajagopal G. Iron oxide nanoparticle core-shell magnetic microspheres: applications toward targeted drug delivery. Nanomedicine. 2020;24:102134. https://doi.org/10.1016/j.nano.2019.102134.

Balasubramanian B, Kraemer KL, Reding NA, Skomski R, Ducharme S, Sellmyer DJ. Synthesis of monodisperse TiO2−paraffin Core−Shell nanoparticles for improved dielectric properties. ACS Nano. 2010;4:1893–900. https://doi.org/10.1021/nn9016422.

Barhoumi A, Huschka R, Bardhan R, Knight MW, Halas NJ. Light-induced release of DNA from plasmon-resonant nanoparticles: towards light-controlled gene therapy. Chem Phys Lett. 2009;482:171–9. https://doi.org/10.1016/j.cplett.2009.09.076.

Berke S, Kampmann A-L, Wuest M, Bailey JJ, Glowacki B, Wuest F, Jurkschat K, Weberskirch R, Schirrmacher R. 18F-radiolabeling and in vivo analysis of SiFA-derivatized polymeric core-shell nanoparticles. Bioconjug Chem. 2018;29:89–95. https://doi.org/10.1021/acs.bioconjchem.7b00630.

Bizeau J, Journaux-Duclos J, Kiefer C, Freis B, Ihiawakrim D, Ramirez M. de los A, Lucante T, Parkhomenko K, Vichery C, Carrey J, Sandre O, Bertagnolli C, Ersen O, Bégin-Colin S, Gigoux V, Mertz D. Tailoring the pore structure of iron oxide core@stellate mesoporous silica shell nanocomposites: effects on MRI and magnetic hyperthermia properties and applicability to anti-cancer therapies. Nanoscale 2024;16:15585–614. https://doi.org/10.1039/D4NR01388C

Blasi P. Poly(lactic acid)/poly(lactic-co-glycolic acid)-based microparticles: an overview. J Pharm Investig. 2019;49:337–46. https://doi.org/10.1007/s40005-019-00453-z.

Bridot J-L, Faure A-C, Laurent S, Rivière C, Billotey C, Hiba B, Janier M, Josserand V, Coll J-L, Vander Elst L, Muller R, Roux S, Perriat P, Tillement O. Hybrid gadolinium oxide nanoparticles: multimodal contrast agents for in vivo imaging. J Am Chem Soc. 2007;129:5076–84. https://doi.org/10.1021/ja068356j.

Butterfield LH, Najjar YG. Immunotherapy combination approaches: mechanisms, biomarkers and clinical observations. Nat Rev Immunol. 2024;24:399–416. https://doi.org/10.1038/s41577-023-00973-8.

Chatterjee K, Sarkar S, Korukonda JR, Paria S. Core/shell nanoparticles in biomedical applications. Adv Colloid Interf Sci. 2013;209:8. https://doi.org/10.1016/j.cis.2013.12.008.

Chenthamara D, Subramaniam S, Ramakrishnan SG, Krishnaswamy S, Essa MM, Lin F-H, Qoronfleh MW. Therapeutic efficacy of nanoparticles and routes of administration. Biomater Res. 2019;23:20. https://doi.org/10.1186/s40824-019-0166-x.

Chiozzi V, Rossi F. Inorganic–organic core/shell nanoparticles: progress and applications. Nanoscale Adv. 2020;2:5090–105. https://doi.org/10.1039/D0NA00411A.

Chircov C, Petcu M-C, Vasile BS, Purcăreanu B, Nicoară AI, Oprea OC, Popescu RC. Irinotecan-loaded magnetite-silica core-shell systems for colorectal cancer treatment. Int J Pharm. 2024;661:124420. https://doi.org/10.1016/j.ijpharm.2024.124420.

Colaco V, Roy AA, Naik GARR, Mondal A, Mutalik S, Dhas N. Advancement in lipid-based nanocomposites for theranostic applications in lung carcinoma treatment. OpenNano. 2024;15:100199. https://doi.org/10.1016/j.onano.2023.100199.

Dash A, Blasiak B, Tomanek B, Latta P, van Veggel FCJM. Target-Specific Magnetic Resonance Imaging of Human Prostate Adenocarcinoma Using NaDyF4–NaGdF4 Core–Shell Nanoparticles. ACS Appl Mater Interfaces. 2021;13:24345–55. https://doi.org/10.1021/acsami.0c19273.

Dhas NL, Raval NJ, Kudarha RR, Acharya NS, Acharya SR. Core–shell nanoparticles as a drug delivery platform for tumor targeting. In: Inorganic frameworks as smart nanomedicines. Elsevier; 2018. p. 387–448. https://doi.org/10.1016/B978-0-12-813661-4.00009-2.

Dhas N, Kudarha R, Kulkarni S, Soman S, Navti PD, Kulkarni J, Roy AA, Colaco V, Raychaudhuri R, Gupta A, Pardeshi C, Bari D, Tiwari R, Patel J, Moorkoth S, Mutalik S. Nanoengineered platform-based microenvironment-triggered immunotherapy in cancer treatment. FBL. 2024;29:349. https://doi.org/10.31083/j.fbl2910349.

Dong H, Dube N, Shu JY, Seo JW, Mahakian LM, Ferrara KW, Xu T. Long-circulating 15 nm micelles based on amphiphilic 3-helix peptide-PEG conjugates. ACS Nano. 2012;6:5320–9. https://doi.org/10.1021/nn301142r.

Đorđević S, Gonzalez MM, Conejos-Sánchez I, Carreira B, Pozzi S, Acúrcio RC, Satchi-Fainaro R, Florindo HF, Vicent MJ. Current hurdles to the translation of nanomedicines from bench to the clinic. Drug Deliv and Transl Res. 2022;12:500–25. https://doi.org/10.1007/s13346-021-01024-2.

Feng H, Tang L, Zeng G, Tang J, Deng Y, Yan M, Liu Y, Zhou Y, Ren X, Chen S. Carbon-based core–shell nanostructured materials for electrochemical energy storage. J Mater Chem A. 2018;6:7310–37. https://doi.org/10.1039/C8TA01257A.

Fouad DM, Mohamed MB. Studies on the photo-catalytic activity of semiconductor nanostructures and their gold core–shell on the photodegradation of malathion. Nanotechnology. 2011;22:455705. https://doi.org/10.1088/0957-4484/22/45/455705.

Fukukawa K, Rossin R, Hagooly A, Pressly ED, Hunt JN, Messmore BW, Wooley KL, Welch MJ, Hawker CJ. Synthesis and characterization of core–shell star copolymers for in vivo PET imaging applications. Biomacromolecules. 2008;9:1329–39. https://doi.org/10.1021/bm7014152.

Gadde S. Multi-drug delivery nanocarriers for combination therapy. Med Chem Commun. 2015;6:1916–29. https://doi.org/10.1039/C5MD00365B.

Gao J, Xia Y, Chen H, Yu Y, Song J, Li W, Qian W, Wang H, Dai J, Guo Y. Polymer–lipid hybrid nanoparticles conjugated with anti-EGF receptor antibody for targeted drug delivery to hepatocellular carcinoma. Nanomedicine. 2014;9:279–93. https://doi.org/10.2217/nnm.13.20.

Guo C-X, Huang X, Xu J, Zhang X-Z, Shen Y-N, Liang T-B, Bai X-L. Combined targeted therapy and immunotherapy for cancer treatment. World J Clin Cases. 2021;9:7643–52. https://doi.org/10.12998/wjcc.v9.i26.7643.

Han Z, Lv W, Li Y, Chang J, Zhang W, Liu C, Sun J. Improving tumor targeting of exosomal membrane-coated polymeric nanoparticles by conjugation with aptamers. ACS Appl Bio Mater. 2020;3:2666–73. https://doi.org/10.1021/acsabm.0c00181.

Hao Y, Zhang B, Zheng C, Ji R, Ren X, Guo F, Sun S, Shi J, Zhang H, Zhang Z, Wang L, Zhang Y. The tumor-targeting core–shell structured DTX-loaded PLGA@Au nanoparticles for chemo-photothermal therapy and X-ray imaging. J Control Release. 2015;220:545–55. https://doi.org/10.1016/j.jconrel.2015.11.016.

Hegde AR, Paul M, Kumbham S, Roy AA, Ahmad SF, Parekh H, Biswas S, Mutalik S. Ameliorative anticancer effect of dendrimeric peptide modified liposomes of letrozole: *in vitro* and *in vivo* performance evaluations. Int J Pharm. 2023;648:123582. https://doi.org/10.1016/j.ijpharm.2023.123582.

Ho D, Quake SR, McCabe ERB, Chng WJ, Chow EK, Ding X, Gelb BD, Ginsburg GS, Hassenstab J, Ho C-M, Mobley WC, Nolan GP, Rosen ST, Tan P, Yen Y, Zarrinpar A. Enabling technologies for personalized and precision medicine. Trends Biotechnol. 2020;38:497–518. https://doi.org/10.1016/j.tibtech.2019.12.021.

Hu C-MJ, Kaushal S, Cao HST, Aryal S, Sartor M, Esener S, Bouvet M, Zhang L. Half-antibody functionalized lipid–polymer hybrid nanoparticles for targeted drug delivery to carcinoembryonic antigen presenting pancreatic cancer cells [WWW document]. ACS Publications; 2010. https://doi.org/10.1021/mp900316a.

Huang WF, Tsui CP, Tang CY, Yang M, Gu L. Surface charge switchable and pH-responsive chitosan/polymer core-shell composite nanoparticles for drug delivery application. Compos Part B: Eng. 2017;121:83–91. https://doi.org/10.1016/j.compositesb.2017.03.028.

Jiang X, Zhang G, Narain R, Liu S. Covalently stabilized temperature and pH responsive four-layer nanoparticles fabricated from surface 'clickable' shell cross-linked micelles. Soft Matter. 2009;5:1530–8. https://doi.org/10.1039/B819396G.

Jin Y, Ma X, Zhang S, Meng H, Xu M, Yang X, Xu W, Tian J. A tantalum oxide-based core/shell nanoparticle for triple-modality image-guided chemo-thermal synergetic therapy of esophageal carcinoma. Cancer Lett. 2017;397:61–71. https://doi.org/10.1016/j.canlet.2017.03.030.

Johnson KB, Wei W, Weeraratne D, Frisse ME, Misulis K, Rhee K, Zhao J, Snowdon JL. Precision medicine, AI, and the future of personalized health care. Clin Transl Sci. 2021;14:86–93. https://doi.org/10.1111/cts.12884.

Joseph TM, Kar Mahapatra D, Esmaeili A, Piszczyk Ł, Hasanin MS, Kattali M, Haponiuk J, Thomas S. Nanoparticles: taking a unique position in medicine. Nanomaterials (Basel). 2023;13:574. https://doi.org/10.3390/nano13030574.

Joyce P, Allen CJ, Alonso MJ, Ashford M, Bradbury MS, Germain M, Kavallaris M, Langer R, Lammers T, Peracchia MT, Popat A, Prestidge CA, Rijcken CJF, Sarmento B, Schmid RB, Schroeder A, Subramaniam S, Thorn CR, Whitehead KA, Zhao C-X, Santos HA. A translational framework to DELIVER nanomedicines to the clinic. Nat Nanotechnol. 2024;9:1597. https://doi.org/10.1038/s41565-024-01754-7.

Khodashenas B, Ardjmand M, Rad AS, Esfahani MR. Gelatin-coated gold nanoparticles as an effective pH-sensitive methotrexate drug delivery system for breast cancer treatment. Materials Today Chemistry. 2021;20:100474. https://doi.org/10.1016/j.mtchem.2021.100474.

Kim P, Jones SC, Hotchkiss PJ, Haddock JN, Kippelen B, Marder SR, Perry JW. Phosphonic acid-modified barium titanate polymer nanocomposites with high permittivity and dielectric strength. Adv Mater. 2007;19:1001–5. https://doi.org/10.1002/adma.200602422.

Kim K, Oh KS, Park DY, Lee JY, Lee BS, Kim IS, Kim K, Kwon IC, Sang YK, Yuk SH. Doxorubicin/gold-loaded core/shell nanoparticles for combination therapy to treat cancer through the enhanced tumor targeting. J Control Release. 2016;228:141–9. https://doi.org/10.1016/j.jconrel.2016.03.009.

Krishnamurthy S, Vaiyapuri R, Zhang L, Chan JM. Lipid-coated polymeric nanoparticles for cancer drug delivery. Biomater Sci. 2015;3:923–36. https://doi.org/10.1039/C4BM00427B.

Lal SK, Grady N, Kundu JS, Levin C, Britt Lassiter JJ, Halas N. Tailoring plasmonic substrates for surface enhanced spectroscopies. Chemical Society Reviews 2008;37:898–911. https://doi.org/10.1039/B705969H

Law W-C, Yong K-T, Roy I, Xu G, Ding H, Bergey EJ, Zeng H, Prasad PN. Optically and magnetically doped organically modified silica nanoparticles as efficient magnetically guided biomarkers for two-photon imaging of live cancer cells. J Phys Chem C. 2008;112:7972–7. https://doi.org/10.1021/jp712090y.

Li J-L, Liu X-Y. Fabrication and biofunctionalization of selenium-polypyrrole core–shell nanoparticles for targeting and imaging of cancer cells. J Nanosci Nanotechnol. 2008;8:2488–91. https://doi.org/10.1166/jnn.2008.502.

Li J, Seok SI, Chu B, Dogan F, Zhang Q, Wang Q. Nanocomposites of ferroelectric polymers with TiO2 nanoparticles exhibiting significantly enhanced electrical energy density. Adv Mater. 2009;21:217–21. https://doi.org/10.1002/adma.200801106.

Li C, Su Y, Lv X, Zuo Y, Yang X, Wang Y. Au@Pd core–shell nanoparticles: a highly active electrocatalyst for amperometric gaseous ethanol sensors. Sensors Actuators B Chem. 2012;171–172:1192–8. https://doi.org/10.1016/j.snb.2012.06.073.

Li W, Zhang X, Zhou M, Tian B, Yu C, Jie J, Hao X, Zhang X. Functional core/shell drug nanoparticles for highly effective synergistic cancer therapy. Adv Healthc Mater. 2014;3:1475–85. https://doi.org/10.1002/adhm.201300577.

Li Z, Qiao W, Wang C, Wang H, Ma M, Han X, Tang J. DPPC-coated lipid nanoparticles as an inhalable carrier for accumulation of resveratrol in the pulmonary vasculature, a new strategy for pulmonary arterial hypertension treatment. Drug Deliv. 2020a;27:736–44. https://doi.org/10.1080/10717544.2020.1760962.

Li X, Salzano G, Qiu J, Menard M, Berg K, Theodossiou T, Ladavière C, Gref R. Drug-loaded lipid-coated hybrid organic-inorganic "stealth" nanoparticles for cancer therapy. Front Bioeng Biotechnol. 2020b;8:1027. https://doi.org/10.3389/fbioe.2020.01027.

Liang J, Yang X, Liu D, Cong M, Song Y, Bai S. Lipid/hyaluronic acid–coated doxorubicin-Fe3O4 as a dual-targeting nanoparticle for enhanced cancer therapy. AAPS PharmSciTech. 2020;21:235. https://doi.org/10.1208/s12249-020-01764-3.

Liu H, Hou P, Zhang W, Wu J. Synthesis of monosized core–shell Fe3O4/Au multifunctional nanoparticles by PVP-assisted nanoemulsion process. Colloids Surf A Physicochem Eng Asp. 2010;356:21–7. https://doi.org/10.1016/j.colsurfa.2009.12.023.

Liu M, Xu N, Liu W, Xie Z. Polypyrrole coated PLGA core–shell nanoparticles for drug delivery and photothermal therapy. RSC Adv. 2016;6:84269–75. https://doi.org/10.1039/C6RA18261E.

Liu R, Guo H, Ouyang Z, Fan Y, Cao X, Xia J, Shi X, Guo R. Multifunctional Core–Shell Tecto Dendrimers Incorporated with Gold Nanoparticles for Targeted Dual Mode CT/MR Imaging of Tumors. ACS Appl Bio Mater. 2021;4:1803–12. https://doi.org/10.1021/acsabm.0c01525.

Loch-Neckel G, Santos-Bubniak L, Mazzarino L, Jacques AV, Moccelin B, Santos-Silva MC, Lemos-Senna E. Orally administered chitosan-coated polycaprolactone nanoparticles containing curcumin attenuate metastatic melanoma in the lungs. J Pharm Sci. 2015;104:3524–34. https://doi.org/10.1002/jps.24548.

Lukowiak MC, Ziem B, Achazi K, Gunkel-Grabole G, Popeney CS, Thota BNS, Böttcher C, Krueger A, Guan Z, Haag R. Carbon-based cores with polyglycerol shells—the importance of core flexibility for encapsulation of hydrophobic guests. J Mater Chem B. 2015;3:719–22. https://doi.org/10.1039/C4TB01858C.

Ma W, Xu A, Ying J, Li B, Jin Y. Biodegradable core–shell copolymer-phospholipid nanoparticles for combination chemotherapy: An in vitro study. J Biomed Nanotechnol. 2015;11:1193–200. https://doi.org/10.1166/jbn.2015.2059.

Mahmud A, Xiong X-B, Aliabadi HM, Lavasanifar A. Polymeric micelles for drug targeting. J Drug Target. 2007;15:553–84. https://doi.org/10.1080/10611860701538586.

Mekuria SL, Tsai H-C. Preparation of self-assembled core–shell nano structure of conjugated generation 4.5 poly (amidoamine) dendrimer and monoclonal anti-IL-6 antibody as bioimaging probe. Colloids Surf B: Biointerfaces. 2015;135:253–60. https://doi.org/10.1016/j.colsurfb.2015.07.052.

Mitchell MJ, Billingsley MM, Haley RM, Wechsler ME, Peppas NA, Langer R. Engineering precision nanoparticles for drug delivery. Nat Rev Drug Discov. 2021;20:101–24. https://doi.org/10.1038/s41573-020-0090-8.

Naik G a RR, Roy AA, Mutalik S, Dhas N. Unleashing the power of polymeric nanoparticles—creative triumph against antibiotic resistance: a review. Int J Biol Macromol. 2024;278:134977. https://doi.org/10.1016/j.ijbiomac.2024.134977.

Ni J, Zhang Z, Ge M, Chen J, Zhuo W. Immune-based combination therapy to convert immunologically cold tumors into hot tumors: an update and new insights. Acta Pharmacol Sin. 2023;44:288–307. https://doi.org/10.1038/s41401-022-00953-z.

Oh KS, Han SK, Lee HS, Koo HM, Kim RS, Lee KE, Han SS, Cho SH, Yuk SH. Core/Shell nanoparticles with lecithin lipid cores for protein delivery. Biomacromolecules. 2006;7:2362–7. https://doi.org/10.1021/bm060362k.

Ortega-Oller I, Padial-Molina M, Galindo-Moreno P, O'Valle F, Jódar-Reyes AB, Peula-García JM. Bone regeneration from PLGA micro-nanoparticles. Biomed Res Int. 2015;2015:415289. https://doi.org/10.1155/2015/415289.

Patel V, Arde G, Kashid G, Patel JK, Dhas N. Metal-based carriers for targeting lymphatic system. In: Dhas N, Patel JK, Pathak YV, editors. Advanced targeting of the lymphatic system. Cham: Springer Nature; 2024. p. 271–93. https://doi.org/10.1007/978-3-031-64828-1_14.

Pei Z, Lei H, Cheng L. Bioactive inorganic nanomaterials for cancer theranostics. Chem Soc Rev. 2023;52:2031–81. https://doi.org/10.1039/D2CS00352J.

Pinho SLC, Pereira GA, Voisin P, Kassem J, Bouchaud V, Etienne L, Peters JA, Carlos L, Mornet S, Geraldes CFGC, Rocha J, Delville M-H. Fine tuning of the relaxometry of γ-Fe2O3@SiO2 nanoparticles by tweaking the silica coating thickness. ACS Nano. 2010;4:5339–49. https://doi.org/10.1021/nn101129r.

Poornima K, Puri A, Gupta A. Understanding the stealth properties of PEGylated lipids: a mini-review. IJL. 2020;1:1–20. https://doi.org/10.14302/issn.2835-513X.ijl-20-3457.

Rashidzadeh H, Seidi F, Ghaffarlou M, Salehiabar M, Charmi J, Yaray K, Nosrati H, Ertas YN. Preparation of alginate coated Pt nanoparticle for radiosensitization of breast cancer tumor. Int J Biol Macromol. 2023;233:123273. https://doi.org/10.1016/j.ijbiomac.2023.123273.

Refaat A, Yap ML, Pietersz G, Walsh APG, Zeller J, del Rosal B, Wang X, Peter K. In vivo fluorescence imaging: success in preclinical imaging paves the way for clinical applications. J Nanobiotechnol. 2022;20:450. https://doi.org/10.1186/s12951-022-01648-7.

Saikia A, Karak N. Chapter 1 – Introduction to biodegradable and biocompatible polymer nanocomposites: synthesis, structure, fundamental properties, biocompatibility, and biodegradability. In: Deshmukh K, Pandey M, editors. Biodegradable and biocompatible polymer nanocomposites. Elsevier; 2023. p. 1–34. https://doi.org/10.1016/B978-0-323-91696-7.00007-6.

Shafiq A, Madni A, Khan S, Sultana H, Sumaira, Shah H, Khan S, Rehman S, Nawaz M. Core-shell Pluronic F127/chitosan based nanoparticles for effective delivery of methotrexate in the management of rheumatoid arthritis. Int J Biol Macromol. 2022;213:465–77. https://doi.org/10.1016/j.ijbiomac.2022.05.192.

Sharma S, Pukale SS, Sahel DK, Agarwal DS, Dalela M, Mohanty S, Sakhuja R, Mittal A, Chitkara D. Folate-targeted cholesterol-grafted lipo-polymeric nanoparticles for chemotherapeutic agent delivery. AAPS PharmSciTech. 2020;21:280. https://doi.org/10.1208/s12249-020-01812-y.

Şimşek GÖ. Immunotherapy in the treatment of cancer: today and tomorrow. Curr Mol Bio Rep. 2024;10:54–64. https://doi.org/10.1007/s40610-024-00161-0.

Sivadasan D, Sultan MH, Madkhali O, Almoshari Y, Thangavel N. Polymeric Lipid Hybrid Nanoparticles (PLNs) as emerging drug delivery platform—a comprehensive review of their properties, preparation methods, and therapeutic applications. Pharmaceutics. 2021;13:1291. https://doi.org/10.3390/pharmaceutics13081291.

Sun X, Rossin R, Turner JL, Becker ML, Joralemon MJ, Welch MJ, Wooley KL. An assessment of the effects of Shell cross-linked nanoparticle size, Core composition, and surface PEGylation on in vivo biodistribution. Biomacromolecules. 2005;6:2541–54. https://doi.org/10.1021/bm050260e.

Sun L, Liu H, Ye Y, Lei Y, Islam R, Tan S, Tong R, Miao Y-B, Cai L. Smart nanoparticles for cancer therapy. Sig Transduct Target Ther. 2023;8:1–28. https://doi.org/10.1038/s41392-023-01642-x.

Surnar B, Sharma K, Jayakannan M. Core–shell polymer nanoparticles for prevention of GSH drug detoxification and cisplatin delivery to breast cancer cells. Nanoscale. 2015;7:17964–79. https://doi.org/10.1039/C5NR04963F.

Tenchov R, Sasso JM, Zhou QA. PEGylated lipid nanoparticle formulations: immunological safety and efficiency perspective. Bioconjug Chem. 2023;34:941–60. https://doi.org/10.1021/acs.bioconjchem.3c00174.

Vieira J, Maurmann N, Venturini J, Pranke P, Bergmann CP. PCL-coated magnetic Fe3O4 nanoparticles: production, characterization and viability on stem cells. Materials Today Communications. 2022;31:103416. https://doi.org/10.1016/j.mtcomm.2022.103416.

Wang Y, Wang K, Zhang R, Liu X, Yan X, Wang J, Wagner E, Huang R. Synthesis of core–shell graphitic Carbon@Silica nanospheres with dual-ordered mesopores for cancer-targeted photothermochemotherapy. ACS Nano. 2014;8:7870–9. https://doi.org/10.1021/nn5027214.

Wang Q, Alshaker H, Böhler T, Srivats S, Chao Y, Cooper C, Pchejetski D. Core shell lipid-polymer hybrid nanoparticles with combined docetaxel and molecular targeted therapy for the treatment of metastatic prostate cancer. Sci Rep. 2017;7:5901. https://doi.org/10.1038/s41598-017-06142-x.

Wu H, Engelhard MH, Wang J, Fisher DR, Lin Y. Synthesis of lutetium phosphate–apoferritin core–shell nanoparticles for potential applications in radioimmunoimaging and radioimmunotherapy of cancers. J Mater Chem. 2008;18:1779–83. https://doi.org/10.1039/B717613A.

Xiong H-M, Xu Y, Ren Q-G, Xia Y-Y. Stable aqueous ZnO@Polymer Core−Shell nanoparticles with tunable photoluminescence and their application in cell imaging. J Am Chem Soc. 2008;130:7522–3. https://doi.org/10.1021/ja800999u.

Yang D, Yao X, Dong J, Wang N, Du Y, Sun S, Gao L, Zhong Y, Qian C, Hong H. Design and investigation of core/shell GQDs/hMSN nanoparticles as an enhanced drug delivery platform in triple-negative breast cancer. Bioconjug Chem. 2018;29:2776–85. https://doi.org/10.1021/acs.bioconjchem.8b00399.

Yu S, Wang S, Xie Z, Yu S, Li L, Xiao H, Song Y. Hyaluronic acid coating on the surface of curcumin-loaded ZIF-8 nanoparticles for improved breast cancer therapy: an in vitro and in vivo study. Colloids Surf B: Biointerfaces. 2021;203:111759. https://doi.org/10.1016/j.colsurfb.2021.111759.

Yuan J, Guo M, Zhao S, Li J, Wang X, Yang J, Jin Z, Song X. Core-shell lipid-polymeric nanoparticles for enhanced oral bioavailability and antihypertensive efficacy of KY5 peptide. Chin Chem Lett. 2023;34:107943. https://doi.org/10.1016/j.cclet.2022.107943.

Zhang G, Li X, Liao Q, Liu Y, Xi K, Huang W, Jia X. Water-dispersible PEG-curcumin/amine-functionalized covalent organic framework nanocomposites as smart carriers for in vivo drug delivery. Nat Commun. 2018a;9:2785. https://doi.org/10.1038/s41467-018-04910-5.

Zhang K, Liu J, Guo Y, Li Y, Ma X, Lei Z. Synthesis of temperature, pH, light and dual-redox quintuple-stimuli-responsive shell-crosslinked polymeric nanoparticles for controlled release. Mater Sci Eng C. 2018b;87:1–9. https://doi.org/10.1016/j.msec.2018.02.005.

Zhang N, Lu C, Chen M, Xu X, Shu G, Du Y, Ji J. Recent advances in near-infrared II imaging technology for biological detection. J Nanobiotechnol. 2021;19:132. https://doi.org/10.1186/s12951-021-00870-z.

Zheng M, Gong P, Jia D, Zheng C, Ma Y, Cai L. Plga–lecithin–peg core-shell nanoparticles for cancer targeted therapy. Nano LIFE. 2012;02:1250002. https://doi.org/10.1142/S1793984411000359.

Zhou L, Zheng X, Gu Z, Yin W, Zhang X, Ruan L, Yang Y, Hu Z, Zhao Y. Mesoporous NaYbF4@NaGdF4 core-shell up-conversion nanoparticles for targeted drug delivery and multimodal imaging. Biomaterials. 2014;35:7666–78. https://doi.org/10.1016/j.biomaterials.2014.05.051.

Zhu K, Li Y. Computed tomography imaging to evaluate the effects of gold-coated silver core-shell nanoparticles in pediatric glioma. Journal of Radiation Research and Applied Sciences 2024;17:101039. https://doi.org/10.1016/j.jrras.2024.101039

Part V

Inorganic/Inorganic Core Shell Nanoconstructs

Silica and Non-Silica-Based Core-Shell Nanoconstructs for Cancer Theragnostics

16

Gaurisha Alias Resha Ramnath Naik, S. P. Rachana, Viola Colaco, Paniz Hedayat, Amrita Arup Roy, Rahul Pokale, Sandesh Ramchandra Jadhav, Anoushka Mukharya, Ritu Kudarha, Srinivas Mutalik, and Namdev Dhas

Abstract

Silica core and non-silica-based shell nanoconstructs represent a promising advancement in cancer theranostics by integrating diagnostic and therapeutic functions within a single nanoparticle system. The silica core offers a robust, tunable platform with high surface area and biocompatibility, while the non-silica-based shell enhances properties such as targeted drug delivery and controlled release. These nanoconstructs can be engineered for optimal size, surface charge, and functionalization, allowing for precise targeting of tumor cells while reducing systemic toxicity. Various non-silica shell materials, such as polymers, lipids, and other inorganic compounds, enhance drug loading, improve stability, and enable multifunctional capabilities. Functionalization of the surface with targeting ligands further improves tumor specificity, making them ideal candidates for both imaging and therapy. The silica core provides excellent contrast for imaging techniques like CT and ultrasound, while the non-silica shell offers the potential for photothermal, photodynamic, or chemotherapeutic applications. These hybrid nanostructures have demonstrated their potential in transforming cancer diagnosis and treatment through innovative synthesis and design approaches, offering a versatile platform for theranostic applications.

G. A. R. R. Naik · S. R. Rachana · V. Colaco · P. Hedayat · A. A. Roy · R. Pokale · S. R. Jadhav · A. Mukharya · R. Kudarha · S. Mutalik · N. Dhas (✉)
Department of Pharmaceutics, Manipal College of Pharmaceutical Sciences, Manipal Academy of Higher Education (MAHE), Manipal, Udupi, Karnataka, India

J. K. Patel et al. (eds.), *Core-Shell Nano Constructs for Cancer Theragnostic*,
https://doi.org/10.1007/978-981-96-3025-7_16

Keywords

Cancer · Core-shell nanoparticles · Silica core · Non-silica shell · Theranostic nanoparticles

1 Introduction

1.1 Overview of Cancer Theranostics

Cancer can be explained by the definition of the most recent NCI update: "Cancer is a disease in which some of the body's cells grow uncontrollably and spread to other parts of the body." ("What Is Cancer?" 2007) Cancer can be spread to the surrounding parts of the body and invade other organs, which is called "metastasis" (Martin et al. 2013) and the majority of cancer-related fatalities are due to metastasis (Seyfried and Huysentruyt 2013).

Theranostic is derived from combining the terms therapeutic (thera) and diagnostic (gnostic) as depicted in Fig. 16.1, and it produces a dual function of diagnostic and treatment (Siddhardha and Parasuraman 2019). Theranostics can be a useful tool for diagnosing and treating various types of cancer as current improvements in cancer care are insufficient to treat the vast majority of people who have different forms of cancer, so it has been created to identify biomarkers for cancer with characteristics of "specificity" and "side effect minimization" (Kim et al. 2013). Also, combining molecular imaging (biomarker) and molecular therapy (drug) may be

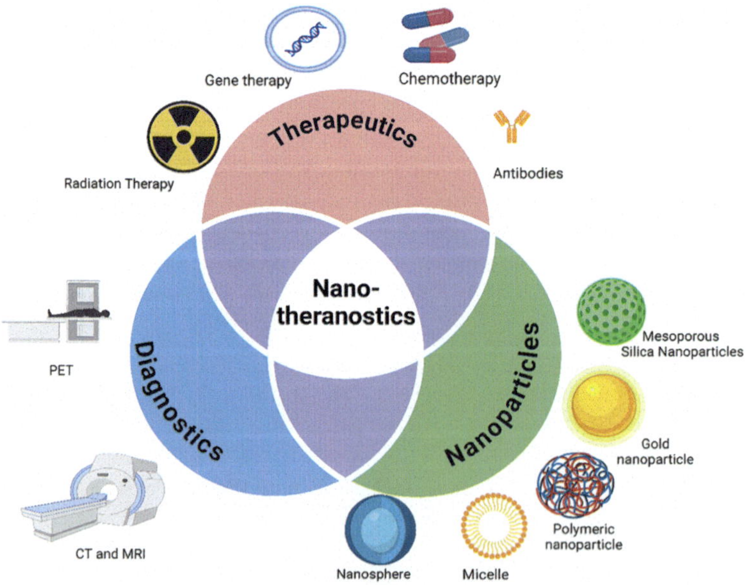

Fig. 16.1 Nanotheranostic: Integrating nanotechnology with diagnostic and therapeutics

useful in "personalized medicine" for several purposes, including early disease detection, diagnosis of various disease stages, drug selection, early drug side effect detection, and therapeutic agent continuation planning (Ryu et al. 2014).

The first step in nanotheranostic is the diagnostic phase, where the most popular molecular imaging tools including tomography (CT), fluorescence imaging (FI), magnetic resonance imaging (MRI), positron emission tomography (PET), photo-acoustic imaging (PA), and radiation imaging (Raheem et al. 2023) are used to detect cancer cells in the body. The second step involves using nanomaterials to treat various cancers, including polymeric, gold, and silica-based nanoparticles.

The history of theranostics spans more than five decades, as illustrated in Fig. 16.2, with its roots in nuclear medicine. In 1941, Saul Hertz was the first to use radioactive iodine (I^{131}) to treat thyroid disorders. Later, in 1951, sodium iodide (I^{131}) became the first radiopharmaceutical to receive FDA approval (Levine and Krenning 2017). John Funkhouser likely coined the term "theranostics" in 1998 to describe a substance that combines treatment with diagnostic imaging (Langbein et al. 2019). While this basic concept has been utilized for decades in managing thyroid conditions, the early 2000s marked a significant advancement. In Rotterdam, the first clinical trials involving LU1779 (Lutathera) featured a prospectively planned clinical protocol known as NETTER-1. In January 2018, Lutathera received FDA approval for the treatment of neuroendocrine tumors in adults (Strosberg et al. 2017). Later in March 2022, PSAM (Prostate-specific membrane antigen) obtained FDA approval for the treatment of an adult patient with metastatic castration-resistant prostate cancer who has a history of use of taxane related chemotherapy and inhibitor of the androgen receptor (Fallah et al. 2023). On April 2024, FDA approved

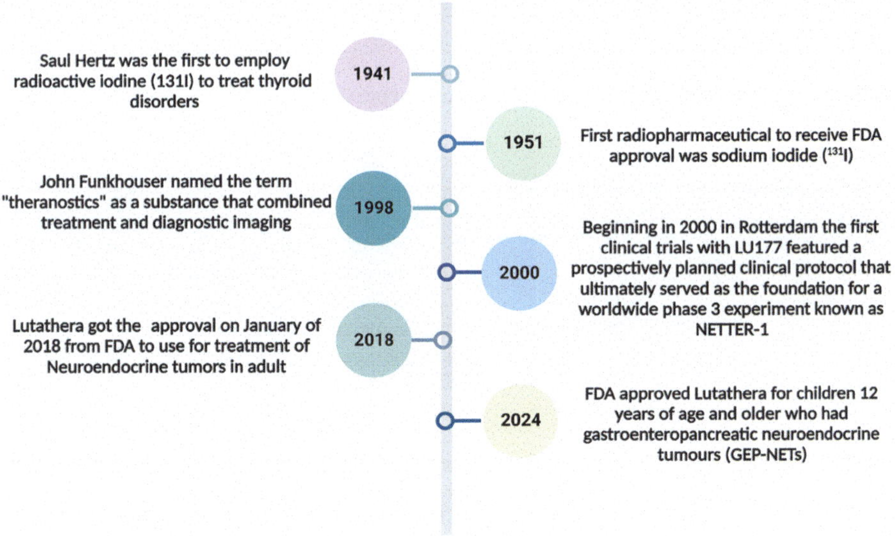

Fig. 16.2 Historical perspective on theranostic developments

Lutathera for children 12 years of age and older who had gastro-entero-pancreatic neuroendocrine tumors (GEP-NETs) (Urso et al. 2023).

1.2 Role of Nanotechnology in Cancer Theranostics

On a global basis, the incidence and death rate of cancer has increased recently, so conventional methods of managing cancer patients, such as surgery, chemotherapy, radiation, and targeted treatments, frequently result in unfavorable side effects and provide poor outcomes. Nanotechnology has been a quickly developing field of therapeutic research during the past few decades that can serve as a link between nanomedicine and the science of technology by the use of various nanoparticles (NPs). NPs with a size range of 1–100 nm are generally regarded as promising tools for cancer treatment (Fan et al. 2023).

Emerging of nanoparticles (NPs) as theranostic agents obtain several benefits such as:

- Help in the localization of disease.
- Minimization of undesirable side effects.
- Eliminate systemic toxicity.
- Longer half-life (reduce elimination).
- The high surface area-to-volume ratio of nanoparticles (NPs) enables them to carry a substantial load of drugs and imaging materials. Additionally, this increased surface area enhances the functionalization of NPs, allowing for the attachment of stabilizing agents like polyethylene glycol (PEG) (Hosseini et al. 2023) (Table 16.1).

Table 16.1 FDA-approved theranostic medicine for types of cancer

Type of cancer	Theranostic medicine	Mode of action
Neuroendocrine tumors	Lu-DOTATATE (Lutathera)	Neuroendocrine tumors have somatostatin receptors, and Lutathera binds to these receptors. By delivering radiation, it leads to the death of cancerous cells (Ichikawa et al. 2022)
Prostate cancer	Gallium-68 (^{68}Ga) PSMA	In prostate cancer, excessive expression of prostate-specific membrane antigen (PSMA) takes place that can be targeted with Ga 68 for PET (Palot Manzil et al. 2022)
Carcinogenic pheochromocytoma or paraganglioma	Iodine-131 iobenguane (Azedra)	Works by acting like norepinephrine, which makes it possible for tumors to produce a lot of norepinephrine transporters and adrenergic nerve terminals to absorb it. When given, radioactive iodine-131 releases radiation that kills and damages the targeted tumor cells while protecting the surrounding healthy tissue (Iobenguane 2024)

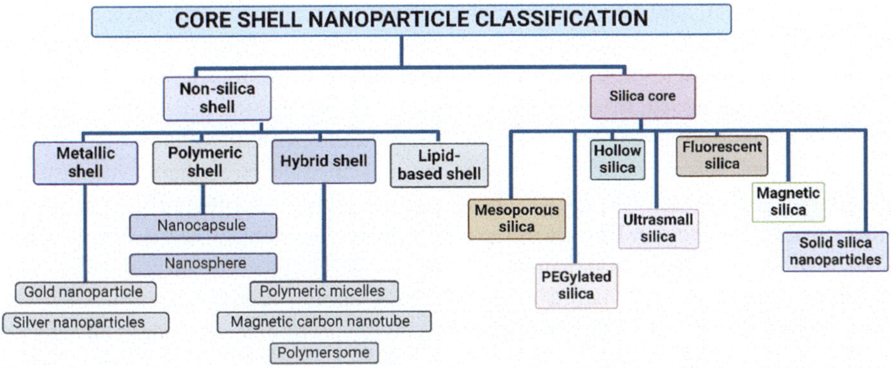

Fig. 16.3 Classification of silica core—non-silica shell nanoparticles

Nanoconstructs have two layers of a hard-core part (inner layer) and a soft shell (outer layer) with dual function of cancer detection and treatment (Dam et al. 2014). Nanoparticles can be categorized as "organic" or "inorganic." Inorganic ones can be metallic-based, silica-based, carbonated, or quantum dots with a high range of stability and specified electronic and optic characteristics that lead to use for cancer diagnostic and theranostic purposes. However, their rigid structure may lead to the degradation of attached medical compounds. Organic nanoparticles, like lipoids and large-size molecular assemblies, possess accepted bioavailability as well as multiple drug functionalization possibilities either on the surface or inner layer. However, the inorganic type has higher stability, and the organic one provides many benefits. A combination of both organic and inorganic NPs, hybrid NPs, generate better bioavailability and stability to make them incredible cancer treatment options (Jain and Thareja 2019). The various types of silica core and non-silica shell nanoparticles depicted in Fig. 16.3 focus on their structural composition, functional modifications, and specific roles in cancer theranostics (Kulkarni et al. 2023).

2 Core-Shell Nanoconstructs: A Focus on Silica Core

Core-shell nanoparticles are advanced materials characterized by a core enveloped by an exterior shell, which modifies their properties and enhances functionality. The core can be composed of various materials, such as metals, semiconductors, or organic polymers, while the shell is typically made of a different organic or inorganic material (Dhas et al. 2018; Kumar et al. 2020). This structure reduces the core's reactivity, increases dispersibility in various solvents, and allows for tailored functionalities based on the choice of materials and shell thickness. Core-shell nanoparticles can be synthesized in various shapes and sizes, influencing their behavior and applications (Chatterjee et al. 2014). They have diverse uses across multiple fields, including biomedical applications for targeted drug delivery and imaging, catalysis for improved chemical reactions, environmental remediation for pollutant removal, and electronics and photonics for sensors and devices (Yong et al. 2007). Their ability to combine different materials and functionalities makes core-shell nanoparticles valuable in advancing technology and science.

Core-shell nanoparticles are a promising tool for integrating diagnostic imaging and targeted therapy in cancer treatment. Their unique structure enables early cancer detection, precise drug delivery, and real-time treatment monitoring (Abousalman-Rezvani et al. 2024). These nanoparticles offer enhanced stability and biocompatibility and can be customized to improve specificity toward cancer cells. They also support multimodal imaging and can be engineered for controlled and sustained drug delivery (Mishra et al. 2023b). Ongoing research aims to develop more sophisticated core-shell nanoparticles to revolutionize personalized cancer treatment and management.

2.1 Types of Silica Core: Properties and Advantages

2.1.1 Solid Silica Nanoparticles

Solid silica nanoparticles (SiNPs) are emerging as a promising avenue in cancer theranostics, demonstrating significant potential for advancing more effective and personalized cancer treatment strategies. These nanoparticles are comprised chiefly of silicon dioxide (SiO_2) and are distinguished by their multifunctionality in drug delivery, imaging, and gene therapy applications (Shirshahi and Soltani 2015). The unique physicochemical and biological properties of SiNPs, which include their nanoscale size ranging from 1.0 to 100 nm, render them highly applicable across various domains, particularly in biomedical engineering and medicine (Cholkar et al. 2017).

The synthesis of SiNPs predominantly involves the condensation of silanes to create an amorphous network of silicon and oxygen. The Stöber process is the most prevalent method used for their production, whereby an alkyl silicate undergoes a reaction in an alcohol solution with water in the presence of ammonia acting as a catalyst (Fernandes et al. 2019). This process affords precise control over particle size by adjusting specific reaction parameters.

In cancer theranostics, SiNPs excel due to their ability to efficiently deliver chemotherapeutic agents, photosensitizers, and photothermal agents directly to tumor sites. This targeted delivery mechanism enhances drug bioavailability and reduces systemic toxicity (Wu et al. 2014). The surface of SiNPs can be further modified with targeting ligands for selective binding to cancer cells, thereby enabling accurate drug delivery. Additionally, doping SiNPs with imaging agents such as fluorescent dyes or quantum dots facilitates the real-time monitoring of drug delivery and treatment efficacy, which is crucial for personalizing therapy (Sun et al. 2017).

SiNPs also show promise in counteracting multidrug resistance in cancer cells, with mesoporous silica nanoparticles (MSNs) capable of delivering high drug payloads and enabling controlled drug release in response to specific stimuli. Moreover, SiNPs can potentially boost cancer immunotherapy by facilitating antigen presentation and modulating the tumor microenvironment to counteract immunosuppression

(Raheem et al. 2023). A notable application includes the encapsulation of platinum(IV)-phenolate conjugates in MSNs, which has significantly amplified anticancer efficacy in breast cancer cell lines relative to the free drug administration (Saha et al. n.d.). The biocompatibility, adjustable size and surface chemistry, and co-delivery capabilities of SiNPs establish them as a versatile platform for augmenting the effectiveness of cancer treatments. Ongoing research continues to broaden their applicative scope within oncology, underscoring their instrumental role in enhancing therapeutic outcomes.

2.1.2 Mesoporous Silica Nanoparticles (MSNs)

Nanotechnology has become a promising field for advancing cancer diagnosis and treatment in recent decades. Nanoparticles, materials with dimensions in the nanoscale range (~1–100 nm), possess unique physical, chemical, optical, and biological properties (Dutta et al. 2024). Mesoporous silica nanoparticles (MSNs) have gained significant attention for cancer-related applications due to their advantageous features, such as a large surface area, adjustable pore size and shape, high loading capacity, ease of surface modification, low toxicity, and biodegradability.

Over time, MSNs have proven effective carriers for various anticancer agents, including small molecules, macromolecules, genes, proteins, and radionuclides (Koohi et al. 2022). They can also be used for cancer imaging by incorporating contrast agents or fluorescent probes within their pores or on their surfaces. This enhances the sensitivity and accuracy of imaging techniques such as MRI, CT, PET, optical, and photoacoustic imaging, allowing researchers to better assess the tumor location, size, shape, and metabolic activity.

Ji Eun Lee et al. developed multifunctional mesoporous silica nanocomposite nanoparticles for simultaneous therapy and diagnosis, integrating magnetic and fluorescent dyes. These nanoparticles can be used for MR and fluorescence imaging and as a vehicle for delivering anticancer drugs (Lee et al. 2011). MSNs are a promising advancement in nanomedicine, particularly in cancer theranostics, with the potential to improve therapeutic outcomes and cancer treatments. They are designed to be taken up by target cells through endocytosis while causing minimal toxicity (Baeza and Vallet-Regí 2020). These nanoparticles have adjustable sizes, high pore volumes, regular pore sizes, and a basic mesoporous or hollow structure. Common types of MSNs include MCM-41 (hexagonal phase), MCM-48 (cubic pore phase), and MCM-50 (lamellar phase).

MSNs can effectively absorb drugs and control their release through diffusion while protecting the encapsulated drugs from enzymatic degradation, thereby improving pharmacokinetics (Freidus et al. 2021). They can be synthesized using three primary methods: the sol-gel process, surfactant-assisted templating, and modified Stöber methods. MSNPs are categorized into various types based on their characteristics, including molecular sieves (M41S), Santa Barbara amorphous (SBA), organically modified silica (ORMOSIL), hollow-type MSNPs (HMSNPs), and periodic mesoporous organosilicon (PMO). Advantages of MSNs include tumor targeting, biocompatibility, low toxicity, and controlled release capabilities, making them suitable for loading small molecule chemotherapy drugs, nucleic acids,

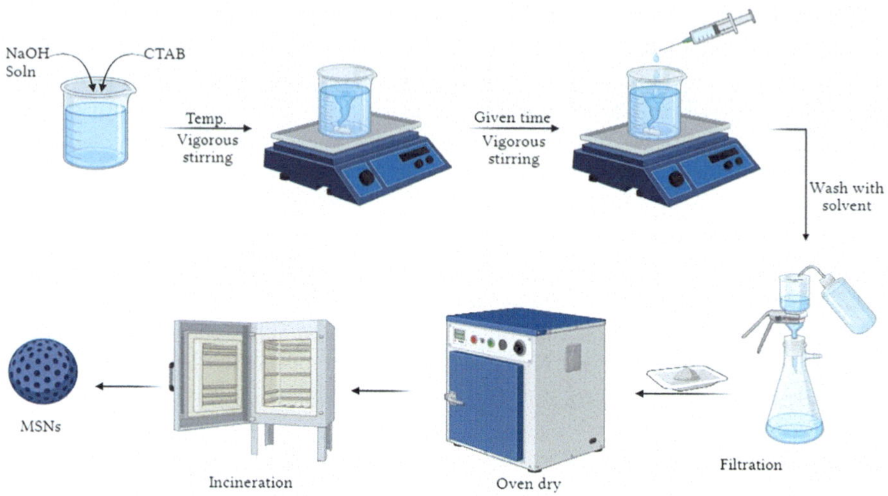

Fig. 16.4 Synthesis of MSNs by using the sol-gel method

proteins, and other biological macromolecules, with successful applications in tumor multimodal treatment research. Recent advancements in MSNs for cancer theranostics have led to innovative solutions that enhance targeted drug delivery and imaging capabilities (Živojević et al. 2021). Hybrid nanoparticle systems that combine MSNs with other nanomaterials like quantum dots and gold nanoparticles have been developed to improve targeting and enable multimodal imaging for comprehensive tumor visualization and treatment monitoring (Fig. 16.4).

Additionally, researchers are engineering stimuli-responsive MSNs that release drugs in response to specific conditions in the tumor microenvironment, such as pH changes, exemplified by the pH-responsive MSNCurNQ system demonstrating significant drug release at acidic pH levels. Biomimetic strategies are also being employed to enhance the performance of MSN-based systems by mimicking biological processes for improved cellular uptake and targeting (Abrishami et al. 2024). Furthermore, multimodal theranostic approaches are being explored, combining chemotherapy and radiotherapy within MSNs platforms to achieve targeted chemoradiotherapy, thereby enhancing therapeutic efficacy while minimizing side effects. Collectively, these advancements are paving the way for more effective and personalized cancer management through MSN-based theranostics.

2.1.3 Hollow Silica Nanoparticles

Hollow silica nanoparticles (HSNs), which have a nanosized cavity space enclosed by a porous inorganic nanoshell, have gained significant interest from both scientific and practical perspectives. These hollow nanoparticles are designed as nanoreactors, efficiently catalyzing the transformation of selectively transferred substrate molecules with minimal loss of activity and surface area of the catalysts even in harsh reaction conditions due to containing catalytic species inside the cavity enclosed by a porous nanoshell (Sharma and Polizos 2020).

Feng Chen et al. synthesized uniform hollow mesoporous silica nanoparticles (HMSN) around 150 nm in size. The process involved creating silica nanoparticles, coating them with a mesoporous silica shell, and etching with Na_2CO_3 (Chen et al. 2014). The resulting HMSN had a mesoporous shell with adjustable pore size and shell thickness. Using various methods, HSNs can be synthesized in controlled sizes, including template-based synthesis with materials like polystyrene (PS), reaction conditions, hydrothermal synthesis, etching and deposition techniques, and cross-linking and thermal treatment (Bueno and Ghoshal 2020). These methods allow for tailored sizes and properties for specific applications in drug delivery, catalysis, and other fields.

HSNs are a unique class of materials with a wide range of potential applications. They have a balloon-like structure, with an air cavity encased by a silica shell, providing several advantages. These include effective thermal insulation due to their low density and thermal conductivity, higher loading capacity for drugs compared to other materials, making them suitable for efficient drug delivery systems, and suitability for antireflection coatings due to their chemical stability, corrosion, and heat resistance, and a high degree of hardness (Zhou et al. 2014). HSNs have also been useful in enhancing energy storage capabilities in battery anodes, serving as effective catalysts in various chemical reactions, and creating functional coatings that enhance surface properties. Notable characteristics of HSNs include low density, high surface area, low thermal conductivity, and a low dielectric constant, making them suitable for diverse applications across fields such as energy storage, drug delivery, and coatings (Phan et al. 2023). Synthesis methods for HSNs include using layered double hydroxide nanoparticles as complex templates and positioning hollow silica nanoparticles as valuable materials for numerous technological and industrial applications.

Recent research on HSNs has improved synthesis methods and expanded their application. New techniques like spray drying and templated methods have developed HSNs with micro- or mesoporous shells, making them useful in thermal insulation, energy storage, drug delivery, catalysis, and functional coatings. Surface chemistry modifications have improved drug loading capacity and release profiles, making them more effective for targeted therapies (Xu et al. 2021). HSNs have shown potential in creating efficient catalysts and superhydrophobic coatings, making them versatile in industrial and biomedical applications. Overall, ongoing innovations position HSNs as promising nanomedicine, energy, and environmental science materials.

2.1.4 Fluorescent Silica Nanoparticles

Fluorescent silica nanoparticles (FSNPs) are gaining attention for their use in cancer diagnosis and treatment (theranostic). These investigations are made of silica nanoparticles containing fluorescent dyes, which make them stable, bright, and suitable for detecting cancer cells. FSNPs are initiated chemically using specific silane-based chemicals (Santra 2010). During their creation, fluorescent dyes are trapped inside through hydrolysis and condensation of the silane chemicals. Two standard methods for making FSNPs are the Stöber sol-gel method and the water-in-oil microemulsion method.

Hence, Ruth Prieto-Montero et al. devised and produced functionalized fluorescent silica nanoparticles to target cancer cells for bioimaging analysis. The report outlines in detail the process of synthesizing and characterizing these 50–60 nm nanoparticles and their internalization and subcellular localization in HeLa cells. The dye rhodamine 101 (R101) was physically embedded during the sol-gel synthesis. The team optimized dye loading by adjusting synthesis conditions and utilizing different organotriethoxysilanes as a second silica precursor. The resulting silica nanoparticles, functionalized with PEG chains, FA, and rhodamine 101, are effective fluorescent nanoplatforms for imaging cancer cells (Prieto-Montero et al. 2020).

FSNPs are versatile in cancer theranostic and capable of holding various fluorescent dyes, including organic and metal-organic types. By choosing different dyes, FSNPs can be made to emit different colors. Their silica structure is transparent, allowing light for excitation and emission to pass through easily, which is crucial for detecting their fluorescence without interference (Prieto-Montero et al. 2020). Remarkably, an FSNP can carry hundreds of dye molecules, vastly improving the detection of specific targets due to a significantly higher signal-to-noise ratio than single dyes. Their composition also enables them to mix well in water, thanks to hydrophilic (–Si-OH) and deprotonated silanol (–Si–O–) on their surface at neutral pH. FSNPs are exceptionally stable against photodegradation, with the silica protective layer guarding the dyes against damage caused by reactive oxygen species (Ambrogio et al. 2011). The silica matrix remains unaffected by changes in pH, temperature, and other environmental factors. Furthermore, FSNPs can be modified on their surface with silane chemicals to attach specific functional groups, which are useful for binding to biomolecules. The silica used in FSNPs is biocompatible, ensuring safety for biological applications (Prieto-Montero et al. 2020). The materials needed to create FSNPs, such as alkoxysilane reagents, are readily available, significantly simplifying the synthesis process.

2.1.5 Magnetic Silica Nanoparticles

Magnetic nanoparticles (MNPs) have emerged as promising theranostic agents in cancer management due to their intrinsic magnetic properties. They can act as contrast agents for MRI, deliver drugs in a targeted manner, and be used in magnetic hyperthermia. MNPs have shown potential in clinical trials for diagnostic and therapeutic applications. Challenges in optimization and long-term toxicity concerns are still being addressed through ongoing research (Li et al. 2021). Magnetic nanoparticles (MNPs) significantly enhance the efficiency of drug delivery in cancer treatment through various mechanisms. Firstly, their intrinsic magnetic properties enable the targeted delivery of therapeutic agents to tumor sites by utilizing an external magnetic field. This allows the MNPs to be precisely guided to the necessary locations, increasing drug concentration at the tumor while minimizing exposure to healthy tissues. This targeted approach is particularly beneficial in overcoming challenges such as tumor hypoxia, where traditional therapies often struggle to be effective.

Additionally, Cuilian Tao et al. have developed magnetic mesoporous silica (MMS) nanoparticles by encapsulating Fe_3O_4 nanoparticles in a mesoporous silica matrix. These MMS nanoparticles have an average particle size of 150 nm, low cytotoxicity, efficient cell uptake, and controllable magnetization to generate rapid heat when exposed to an alternating magnetic field. They promise combined chemotherapy and hyperthermia in cancer therapy (Tao and Zhu 2014). Magnetic nanoparticles (MNPs) can encapsulate therapeutic agents like chemotherapeutics and RNA to protect them and improve their stability and bioavailability in biological systems (Jiao et al. 2022). They enhance the pharmacokinetics and pharmacodynamics of conventional therapies and can be customized with targeting ligands to increase specificity for cancer cells, improving drug delivery efficiency, and reducing side effects.

Recent advancements have demonstrated the potential of MNPs as effective drug delivery systems in oncology and their role in personalized cancer therapy. MNPs can specifically target cancer cells through various mechanisms that enhance their selectivity and effectiveness in drug delivery (Zhao et al. 2020). One primary strategy involves functionalizing the surface of MNPs with specific targeting ligands, such as antibodies, peptides, or small molecules that selectively bind to receptors overexpressed on cancer cells. This functionalization enables MNPs to recognize and adhere to tumor cells while minimizing interactions with healthy tissues.

Furthermore, MNPs can be directed to tumor sites using an external magnetic field. This enables them to accumulate at the intended location and overcome barriers such as tumor hypoxia, which traditional therapies often struggle to penetrate effectively. The distinct environment of tumors, with disordered blood vessels and hypoxic regions, can be more effectively targeted with MNPs (Cai et al. 2019; Sun et al. 2008). Furthermore, these nanoparticles can be customized to release drugs in response to specific stimuli, such as temperature or pH changes commonly found in tumor tissues.

Magnetic nanoparticles (MNPs) and magnetic resonance imaging (MRI) enable real-time monitoring of their distribution and accumulation in tumors, providing valuable feedback on treatment effectiveness. This dual functionality of MNPs as therapeutic and diagnostic agents, known as theranostics, enhances their potential in personalized medicine by enabling tailored treatment strategies based on individual tumor characteristics (Gauger et al. 2020). MNPs' precise targeting of cancer cells through functionalization and external magnetic guidance significantly improves their efficiency as drug delivery systems in cancer treatment.

2.1.6 Ultrasmall Silica Nanoparticles

Ultrasmall silica nanoparticles (USNs) offer a promising platform for cancer theranostics, combining diagnostic imaging and therapeutic functions to enhance individual outcomes. These nanoparticles, typically less than 10 nm in size, possess unique properties, including excellent biocompatibility, efficient renal clearance, and the capacity for precise engineering of size and surface chemistry. Their small size enables evasion of the reticuloendothelial system, leading to prolonged circulation in the bloodstream and enhanced accumulation at tumor sites (Ferreira et al. 2021). USNs can be tailored to act as effective nanocarriers for various therapeutic agents, facilitating targeted delivery and controlled release of drugs, which are crucial in cancer therapy. Moreover, they find utility in imaging applications, providing real-time visualization of tumor dynamics and treatment responses. Recent studies underscore their potential to improve the efficacy of cancer treatments through the integration of therapeutic and diagnostic capabilities, known as theranostics (Singh et al. 2023). Nevertheless, challenges persist in the clinical translation of USNs, including the necessity for improved tumor targeting and a comprehensive understanding of their pharmacokinetics and toxicity profiles.

USNs are produced using various methods, including the sol-gel process, which allows for precise particle size and morphology control. Another approach is the template-assisted method, which creates hollow silica nanoparticles for drug delivery (Ferreira et al. 2021; Singh et al. 2023). Recent advancements in USNs synthesis include one-pot techniques and surface functionalization, making them suitable for targeted drug delivery, bioimaging, and theranostics, aiming to diagnose and treat diseases like cancer simultaneously.

Carolina A. Ferreira et al. studied ultrasmall porous silica nanoparticles (UPSN) for noninvasive imaging and internal radiotherapy. In vivo, PET imaging revealed prolonged blood circulation and excellent tumor contrast with 86Y-DOTA-UPSN. The nanoparticles significantly inhibited tumor growth and increased overall survival, demonstrating potential for clinical translation as a cancer theranostic agent (Ferreira et al. 2021). Moreover, these nanoparticles can be customized to target specific receptors on cancer cells, thereby improving drug delivery and imaging. They have shown particular promise in targeting the human epidermal growth factor receptor 2 (HER2) in certain types of breast cancer (Xu et al. 2023). This targeted approach enhances drug delivery to tumor tissues while minimizing off-target effects, making it a promising strategy for more effective cancer theranostics.

2.1.7 PEGylated Silica Nanoparticles

PEGylated silica nanoparticles (SiNPs) hold great potential as a platform for cancer theranostics. They can be tailored with polyethylene glycol (PEG) and other molecules to enhance their properties and applications. PSNPs can be synthesized in various ways, including one-pot, one-step PEGylation, post-synthesis PEGylation, silica coating on PEGylated cores, and stable PEGylation using silane linkers, providing versatile approaches for creating PSNPs customized for various biomedical applications (Akbari et al. 2015).

PEGylating silica nanoparticles enhance their effectiveness and safety in drug delivery by improving stability, preventing aggregation and hemolysis, and prolonging circulation time. Additionally, this process allows for controlled drug release, making it advantageous for targeting cancer cells and reducing side effects. Customizable surface properties further enhance targeting capabilities (Esfahani et al. 2021). PEGylation improves biocompatibility, reduces immunogenicity, and allows for customizable drug delivery systems, making PEGylated silica nanoparticles promising for enhancing drug delivery in oncology and other therapeutic areas.

PEGylated silica nanoparticles (PSNPs) have demonstrated promise in cancer theranostics, combining diagnostic imaging and therapeutic capabilities. PEGylation enhances stability and reduces nonspecific protein binding, crucial for maintaining efficacy. Despite reducing initial drug loading capacity, PEGylation improves the release profiles of therapeutic agents, facilitating targeted delivery. PSNPs have potential in photodynamic therapy (PDT), photothermal therapy (PTT), and theranostic agents (Glorani et al. 2017). Challenges include scaling up production and addressing potential long-term toxicity concerns.

PEGylation enhances the targeting efficiency of silica nanoparticles by improving stability, circulation time, and interaction with target cells. It reduces aggregation and opsonization, prolonging circulation time and facilitating accumulation at target sites, particularly in tumor tissues (Shahbaz et al. 2024). PEGylation can include specific ligands for active targeting of cancer cells, enhancing selectivity. Higher molecular weight PEGs shield nanoparticles from protein adsorption, improving biocompatibility and enabling more accurate targeting of therapeutic agents to cancer cells.

2.2 Types of Non-Silica-Based Shells

2.2.1 Metallic Shells (e.g., Gold, Silver)

Metal-based cancer theranostic nanoparticles are composed of one or more metal elements, each with unique qualities that make it beneficial in a certain kind of imaging or therapeutic approach. Gold, silver, copper, platinum, iron, lanthanide, and quantum dot (QD) nanoparticles are among them. The use of metal-based theranostic nanoparticles for either passive or active targeting in the treatment of different forms of cancer has been the subject of numerous investigations (Fernandes 2023) (Table 16.2).

2.2.2 Polymeric Shells

Polymeric shells play a vital role in cancer theranostics by integrating therapeutic and diagnostic functions into a single platform, enabling real-time monitoring and targeted treatment of tumors. These biocompatible polymer nanoconstructs facilitate the simultaneous delivery of drugs and imaging agents, allowing for precise localization of cancerous tissues and assessment of therapeutic efficacy (Naik et al. 2024). The unique properties of polymeric nanoparticles enable them to exploit the tumor microenvironment (TME), such as its acidic pH and hypoxic conditions, to enhance drug release and improve therapeutic outcomes (Hosseini et al. 2023). They can be designed to be stimuli-responsive, allowing for controlled drug release triggered by internal factors like pH, redox potential, or specific enzymes present in the TME, which minimizes side effects on healthy tissues. Additionally, polymeric shells can improve the circulation time of drugs in the bloodstream, enhancing

Table 16.2 Different types of metal-based shells for cancer theranostic

Metal nanoparticles	Properties	Imaging techniques	Treatment modalities	Advantages	References
Gold nanoparticles (AuNPs)	High surface area, excellent biocompatibility, tunable size	X-ray computed tomography (CT), photoacoustic imaging, optical imaging	Photothermal therapy (PTT), drug delivery	Strong light absorption, ease of functionalization	Zhao et al. (2016)
Silver nanoparticles (AgNPs)	Antimicrobial properties, high surface plasmon resonance (SPR)	Surface-enhanced Raman spectroscopy, optical imaging	Anticancer activity, PTT, drug delivery	Antibacterial, synergistic therapeutic effects	Zeng et al. (2020)
Copper nanoparticles (CuNPs)	High conductivity, catalytic properties	Positron emission tomography (PET), magnetic resonance imaging (MRI)	Photodynamic therapy (PDT), chemotherapy	Low cost, effective for PDT	Wang et al. (2020)
Platinum nanoparticles (PtNPs)	Catalytic, induces oxidative stress in cancer cells	CT, MRI	Chemotherapy (e.g., cisplatin delivery), radiosensitization	Enhances radiosensitivity, strong catalytic activity	Xu et al. (2018)
Iron oxide nanoparticles (IONPs)	Superparamagnetic, biocompatible	MRI, magnetic particle imaging (MPI)	Magnetic hyperthermia, drug delivery	Noninvasive imaging, responsive to magnetic fields	Kulhari et al. (2015)
Lanthanide-based nanoparticles	High luminescence, paramagnetic properties	MRI, fluorescence imaging	PDT, drug delivery	High contrast agents for imaging, tunable optical properties	Addisu et al. (2019)
Quantum dots (QDs)	Size-dependent optical properties, highly fluorescent	Fluorescence imaging, SERS	PDT, drug delivery	Bright fluorescence, tunable emission wavelengths	Soheyli et al. (2020)

tumor-selective targeting and reducing systemic toxicity (Kalita and Patowary 2023). Polymeric shells for cancer theranostics can be synthesized using various methods. Standard techniques include self-assembly, template methods, interfacial polymerization, microfluidic techniques, and functionalization. These methods facilitate the development of polymeric shells that effectively interact with the tumor microenvironment, paving the way for innovative cancer therapies and diagnostics (Luk and Zhang 2014).

Recent advancements in polymeric theranostics for cancer involve developing multifunctional nanoparticles with therapeutic and diagnostic capabilities. These nanoparticles can be designed to respond to specific conditions in the tumor environment, allowing for precise drug release at targeted sites and real-time tracking of drug distribution in tumors. They offer enhanced specificity and sensitivity in detecting cancer biomarkers, improved imaging capabilities, and the ability to deliver targeted therapies simultaneously (Chang et al. 2021). Overall, polymeric theranostic nanoparticles offer a promising alternative to traditional diagnostic methods in cancer management, combining diagnosis, imaging, and therapy in a single platform.

2.2.3 Lipid-Based Shells

Lipid-based shells are increasingly recognized for their pivotal contribution to the evolution of cancer theranostics. They provide a comprehensive platform that amalgamates precision in oncological treatment with effectively monitoring therapeutic outcomes. The continuous research and development in this field are anticipated to enhance patient prognosis in oncology significantly. The synthesis of lipid-based nanostructures for cancer theranostics can be categorized into two primary methodologies. The first is a one-step approach that concurrently synthesizes the lipid shell and core components through nanoprecipitation and self-assembly mechanisms. Alternatively, a two-step approach can be employed, involving the discrete preparation of the polymer core via techniques such as emulsion or high-pressure homogenization, followed by the formulation of the lipid shell employing methods like sonication or extrusion. Subsequently, these components are combined at precise molar ratios to fabricate stable lipid-based nanoparticles (Bukhari et al. 2021a, b). A novel, solvent-free coacervation technique, which modulates pH to induce phase separation, has also been developed for generating lipid-based nanoparticles. Furthermore, methodologies such as sonication, extrusion, high-pressure homogenization, and vortexing are utilized to modulate particle size, uniformity, and stability based on the specific requirements of the nanoparticles, including desired size, stability, and drug release dynamics. Nevertheless, the involvement of organic solvents in specific synthesis techniques may restrict their application and diminish the encapsulation efficiency for hydrophilic drugs during the incubation phase.

Lipid-based nanostructures offer a versatile platform for cancer theranostics, combining imaging and targeted treatment. These nanoparticles can encapsulate therapeutic and imaging agents, allowing for simultaneous treatment and real-time monitoring of tumor response. The outer lipid shell can be augmented with targeting ligands to facilitate active targeting of receptors predominantly overexpressed on cancer cells, which ensures targeted delivery. These systems can be further engineered to release their therapeutic payload in response to specific stimuli present within the tumor microenvironment, such as variations in pH or the existence of particular enzymes, warranting precise activation of the therapeutic agents (Waheed et al. 2024).

Lipid-based theranostic systems can integrate a variety of imaging agents catering to disparate modalities, including:

1. Magnetic resonance imaging (MRI) utilizing iron oxide nanoparticles and fluorine-containing agents.
2. Radionuclide imaging through the implementation of PET agents like fluorine-18 and SPECT agents.
3. Fluorescent imaging, facilitated by quantum dots and fluorescent silicon nanoparticles.
4. Computed tomography (CT), employing iodinated compounds.
5. Optical imaging, through the usage of photosensitizers.

Incorporating these multimodal imaging capabilities furnishes a comprehensive assessment of the tumor, facilitating the noninvasive tracking of nanoparticles in vivo. This aspect is crucial for the real-time evaluation of drug delivery, biodistribution, and therapeutic response (Luk et al. 2012).

2.2.4 Hybrid Shells

Hybrid nanoshell nanoparticles are multifunctional nanostructures that combine a nanoshell material, typically gold or silver, with another material in the core or shell, offering unique optical, electronic, and structural properties tailored for cancer theranostics (Rajana et al. 2022). These nanoparticles can be synthesized using various methods, including bioinspired approaches and one-step self-assembly, enabling applications in optical labeling, imaging, sensing, drug delivery, and targeted cancer treatment.

Their ability to absorb near-infrared (NIR) wavelengths makes them particularly suitable for biomedical applications. At the same time, their core-shell structure allows for the sequential release of therapeutic agents, enhancing their effectiveness. Despite their promise, challenges remain in improving biocompatibility, tumor targeting, and optimizing drug loading and release kinetics (Ménard 2017). Future research should focus on developing clinically viable hybrid nanoshell nanoparticles to enhance their therapeutic efficacy and improve cancer diagnosis and treatment outcomes, potentially revolutionizing the field of cancer theranostics.

Hybrid nanoshell nanoparticles can interact with cancer cells in various ways, depending on their design and surface modifications. Their unique optical properties enable photothermal therapy, while their versatile structure allows for combination therapy by co-delivering multiple therapeutic agents (Zayed et al. 2019). Overall, hybrid nanoshell nanoparticles are particularly beneficial for breast, melanoma, lung, prostate, and colon cancers because they enhance drug delivery, improve targeting, and provide multifunctional capabilities for diagnosis and therapy.

3 Functionalization of Core-Shell Nanoconstructs

The functionalization of silica-non-silica core-shell nanoconstructs, combining the unique attributes of silica with a range of core materials to form multifunctional nanoconstructs, is an extremely important aspect of research (Nayl et al. 2022). These core-shell structures are helpful for imaging, diagnosis, and drug delivery because they combine the advantages of the shell and the core. The unusual properties of silica nanoparticles—such as their high surface area, varied porosity, and exceptional biocompatibility—have attracted a lot of interest recently (Akhter et al. 2022). This subsection provides an overview of the various functionalization methods, their effects on performance, and potential applications for these nanoconstructs.

3.1 Surface Functionalization Techniques

Surface functionalization alters a nanoconstruct's surface characteristics to boost its effectiveness and performance. Surface functionalization is commonly categorized into two types: non-covalent and covalent as depicted in Fig. 16.5. Each of these is appropriate for a purpose and has distinct benefits.

3.1.1 Covalent Functionalization

Covalent functionalization is the process by which the functional group or biomolecules and the surface of nanoconstructs form robust chemical bonds. This method is appropriate for applications requiring strong ligand or medicinal agent attachment because to its long-lasting nature and specificity (Ahmad et al. 2022). Silane coupling, click chemistry, cross-linking processes, and functionalization of nucleic acids are among the methods for achieving these covalent functionalization. The mechanism of action of this functionalization method involves a few common steps. The first step of covalent functionalization is the surface activation of the shell nanoparticles, non-silica nanoparticles in this case (Hosseini et al. 2014). This step is crucial for all types of covalent functionalization. The surface of non-silica nanoparticles must be activated to introduce a reactive functional group. This can be done by either chemical reactions or physical methods (Zhang et al. 2022). The second and third steps of the functionalization are different from the other. Furthermore, characterization and stability assessment are performed for the nanoconstructs.

Silane coupling agents have a wide range of applications in surface modification, especially for silica-based core-shell nanoconstructs. They have two main kinds of functional groups: one forms bonds with organic polymers, while the other reacts with inorganic substances. The silane agents can function as a chemical

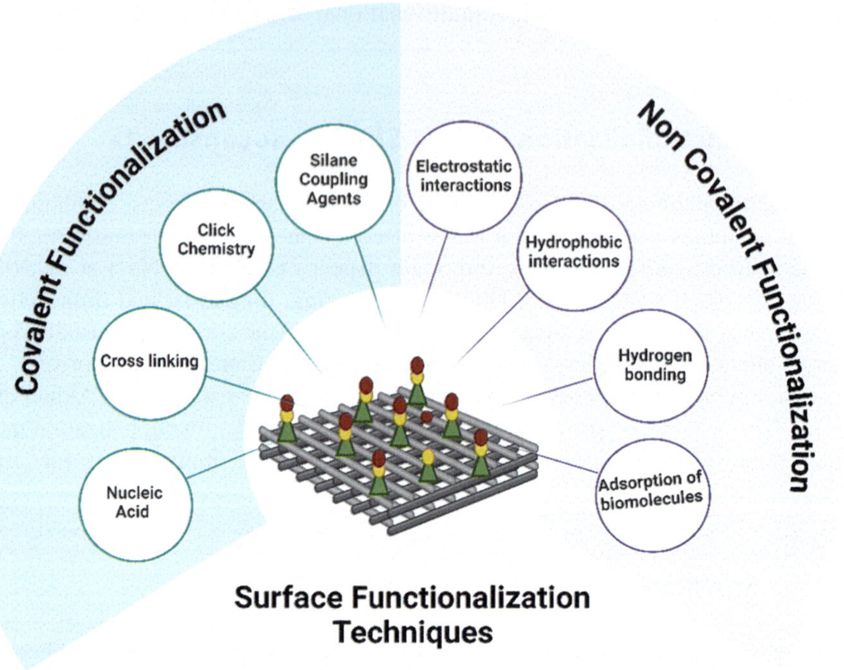

Fig. 16.5 Types of surface functionalization techniques

bridge to improve the compatibility between the inorganic core and organic shell owing to this duality (Nihei 2016). Upon activation of non-silica nanoparticles, the functional groups are introduced to the surface of the nanoparticle. This results in enhanced efficacy of functionalization by increasing the number of places where the silane coupling agent can bond (Chuang et al. 2018). Alkoxysilane groups are introduced which hydrolyze in the presence of moisture converting the alkoxy groups into silanol groups. These groups are essential owing to their ability to interact with the activated surface of non-silica nanoparticles. Siloxane linkages form as a result of these groups' further reaction with the hydroxyl groups on the surface of the non-silica nanoparticles. By forming a stable contact among silane and the surface of the nanoparticle, this bond effectively anchors the silane to the non-silica substance (Ahangaran and Navarchian 2020). Performance is increased and compatibility is enhanced by this functionalization.

Click chemistry is a robust and efficient method used for the functionalizing developed nanoconstructs, which enables the development of advanced theranostic platforms for cancer treatment. This approach enhances the specificity and efficiency of cancer therapies with real-time monitoring of the therapy outcomes (Chen et al. 2016). This strategy improves the efficacy and specificity of cancer treatments by tracking the results of those treatments in real-time. After the initial activation of the surface of non-silica nanoparticle azide or alkyne groups are introduced to the surface by using chemical modification techniques like grafting and direct functionalization (Bauer et al. 2023). This allows subsequent click reactions. In mild conditions, the non-silica shell's azide groups react with alkyne-functionalized biomolecules to form stable 1,2,3-triazole linkages (Yi et al. 2018).

Cross-linking is a method used to functionalize the nanoconstruct for cancer theranostic. It involves the formation of covalent bonds between functional groups on non-silica shell nanoparticles, which creates a stable and robust coating that will then be modified with biomolecules. The non-silica shell surface is activated to introduce reactive functional groups like amine, carboxyl, or thiol (Siddiqui et al. 2022). This is done via chemical modification or physical treatments. Further, the cross-linkers containing two different reactive functional groups on different ends are used to create a covalent bond on the shell nanoparticle/the cross-linker used includes N-hydroxysuccinimide (NHS) esters, maleimides, and carbodiimides. NHS esters react with the amine group to form a stable amide bond (Geißler et al. 2021). Similarly, maleimides react with thiol groups to form thioether bonds. Whereas carbodiimides activate carboxyl groups to form bonds with amine groups. These linker-modified nanoconstructs react with biomolecules to form stable covalent bonds.

Nucleic acid functionalization is a promising approach for developing an advanced theranostic platform for cancer treatment. This process involves surface activation of shell nanoparticles with a silane group, which introduces an amine group to the nanoparticle's surface. This is important for the ensuing nucleic acid attachment through covalent bonding. Furthermore, the cross-linkers are used to facilitate the bonding of nucleic acid, which allows a higher loading density (Siddiqui et al. 2022). Finally, the nucleic acid, often modified with the thiol group, can react with the activated surface. The maleimide group reacts with thiol groups of the nucleic acids, forming thioether bonds. This method forms a dense layer of nucleic acids on the surface of the nanoparticles which interacts effectively with the mRNA sequences of the cell.

3.1.2 Non-covalent Functionalization

Non-covalent functionalization has emerged as an important strategy for enhancing the surface properties of the core-shell nanoparticles especially in the biomedical fields like drug delivery and cancer therapy. The significance of this method is in its ability to preserve the distinct properties of nanoconstructs while imparting new functionalities (Sanità et al. 2020). Unlike covalent modifications, non-covalent functionalization maintains the stability and activity of the bonded molecules. This is an important aspect in therapeutics, where the biological activity of the biomolecules must be retained to ensure the treatment efficacy (Nell et al. 2016). This strategy exploits intermolecular forces like electrostatic interactions, hydrophobic effects, and hydrogen bonding to attach biomolecules without requiring permanent chemical bond formations.

Electrostatic interactions are fundamental mechanisms in non-covalent functionalization. These interactions occur between charged species, which is the attraction of oppositely charged molecules. When nanoparticles are functionalized with charged groups, they exhibit either a positive or negative charge, affecting their ability to attract or repel other charged molecules. For example, when the shell nanoparticles are altered to carry a net positive charge, it effectively attracts a molecule with a net negative charge like nucleic acids (Gentili and Ori 2022). This approach is widely used to improve the loading efficiency of the therapeutics on the surface of the nanoparticles, thereby ensuring the precise delivery of nanoparticles to the target cells (Vatanparast et al. 2018). Moreover, these interactions can influence the release profile of the drug from a nanoconstruct. Furthermore, studies have shown that electrostatic interactions exhibited higher adsorption than those relying on hydrogen bonding.

Hydrophobic interactions occur when non-polar molecules aggregate to reduce their exposure to water. The thermodynamic favorability of reducing the surface area of hydrophobic substances drives this phenomenon in an aqueous environment, aggregating non-polar groups to minimize their exposure to water (Petcu et al. 2017). One of the most common approaches to introduce hydrophobicity on non-silica nanoparticles is to use silane coupling agents containing alkyl or fluoroalkyl groups (Pyo and Chang 2021). These silane groups form covalent bonds with the hydroxyl group on the surface to result in the formation of hydrophobic coating to achieve the desired degree of hydrophobicity; the length and nature of the alkyl chain can be altered. The longer the alkyl chain, the more hydrophobicity is achieved.

Hydrogen bonding ensues when a hydrogen atom covalently bonds to a highly electronegative atom and interacts with another electronegative atom. This method is relatively weak but is significant in regard to biomolecules (Tang et al. 2017). When non-silica nanoparticles are functionalized with hydroxyl, amine, or carboxyl groups, they can form hydrogen bonds with biomolecules like proteins, nucleic acids, or other therapeutic agents (Herschlag and Pinney 2018). Additionally, hydrogen bonding also enhances the stability of nanoparticle dispersions in biological environments (Zhan et al. 2022). By enhancing the interactions between the nanoparticles and their surroundings, hydrogen bonding can prevent aggregation while maintaining the functionality and bioavailability of nanoparticles; it also influences the release kinetics of drug from the nanocarrier matrix.

Adsorption of biomolecules on the surface of nanoparticles ensues due to intermolecular forces mentioned above. Biomolecules like proteins, drugs, and nucleic acids can be adsorbed on the surface of the shell nanoparticles (Jena et al. 2022). Protein adsorption mostly ensues via integrating electrostatic forces, hydrophobic effects, and hydrogen bonding. Similarly, drug adsorption, especially of the hydrophobic molecules, can be done mostly through hydrophobic interaction and minutely using electrostatic attraction (Luther et al. 2020). Whereas, nucleic acid are adsorbed through electrostatic forces and hydrogen boding. The negatively charged phosphate backbone of nucleic acid interacts with positively charged surfaces.

3.2 Functionalization for Targeted Therapy

The core-shell nanoparticles have gained a lot of interest recently due to their distinctive properties and versatile approach in different aspects, especially targeted therapy for cancer treatment. In this subsection we will predominantly discuss the silica-based core nanomaterial and non-silica-based coated on the core nanoparticles (Gao et al. 2020). Silica nanoparticles are biocompatible, have low toxicity, and are easy to modify. They are often engineered to have a specific size and porosity which makes them appropriate for drug delivery (Bharti et al. 2015). These silica core nanoparticles are coated with non-silica nanomaterials as shell components, such as metals or organic compounds, to integrate additional properties to the developed nanoconstructs, such as improved stability, targeted drug delivery, or improved imaging (Şen Karaman and Kettiger 2018). Concerning targeted and precise delivery of therapeutics, core/shell nanoparticles are altered to respond to specific stimuli like pH, temperature, or redox conditions. Also, it can be developed to bind selectively to cancer cells through functionalization of targeting moiety like folic acid, hyaluronic acid transferrin, or other antibodies. This approach aims to enhance the therapeutic efficacy while minimizing the side effects.

Targeting ligands is important in the precise delivery of therapeutics to cancer cells. The commonly used ligands are folic acid, transferrin, and various antibodies. **Folic acid** is a vitamin that is known to bind to α-folate receptors which are overexpressed in cancer cells (Bax et al. 2023). The folic acid conjugated to the surface will attach to the folate receptors, leading to enhanced uptake of these carriers by cancer cells, thereby improving the delivery of therapeutics (Fernández et al. 2017). **Transferrin**, on the other hand, is a protein which transports iron in the blood and its receptors are also overexpressed in certain tumors (Daniels et al. 2012). Transferrin, used as targeting moiety, helps in the precise delivery of drugs to the cancer cells expressing transferrin receptors, which results in enhanced therapeutic efficacy (Daniels et al. 2006). Similarly, **antibody functionalization** involves altering the surface to nanoparticles to bind a monoclonal antibody. These antibodies are engineered to target specific antigens present on the surface of the cancer cells (Marques et al. 2022). This highly specific approach allows the targeted therapy to reduce damage to healthy tissues.

Stimuli-responsive functionalized nanoconstructs are developed to release their therapeutics in response to specific environmental triggers. This aids in the controlled drug release and increased therapeutic index. Most exploited environmental triggers for stimuli-responsive functionalization are pH, temperature, and redox (Zhu et al. 2023). **pH-responsive systems** use the acidic nature of cancer cells, that is lower than the normal cells. pH-responsive nanoparticles can swell and disintegrate in response to acidic conditions, resulting in drug release (Dutta et al. 2024). Many studies have shown that these nanoparticles often achieve significant drug release in the tumor microenvironment. **Temperature-responsive systems** release the therapeutic in response to variations in the temperature of the tumor tissues (Rahim et al. 2021). The nanomaterials can expand or contract in response to elevated temperature found in the tumor cells, leading to the release of drugs in tumor microenvironment (Karimi et al. 2016). Similarly, **redox-responsive systems** depend on the high levels of reducing agents present in the cancer cells such as glutathione (GSH) (Yang and Sun 2022). These nanosystems are designed in such a way that the disulfide linkages breakdown in the presence of high GSH concentrations, leading to the release of drug (Guo et al. 2018). Moreover, there are dual responsive system nanosystems that combine pH and redox responsiveness and have shown a potential for enhancing drug delivery and therapeutic outcomes (Li et al. 2020).

3.3 Functionalization for Enhanced Imaging

The development of imaging techniques has shown a significant improvement in the diagnosis and treatment of cancer treatment. Functionalization of core/shell nanoconstructs is an important aspect in the enhancement of imaging approaches. This can be accomplished by conjugating the fluorescent dyes, quantum dots, or gadolinium complexes to the surface of the nanoconstructs (Rajamanickam 2019). The fluorescent dyes and quantum dots aid in high-resolution optical images, whereas gadolinium enhances the contrast in MRI. In addition, the integration of various imaging agents in the core/shell nanoconstructs allows multimodal imaging. By integrating optical, magnetic, and radioactive agents, we can get comprehensive information of the cancer (Liu et al. 2010). This enhances the diagnostic accuracy, and real-time monitoring of treatment improves the patient outcomes.

Fluorescent dyes, quantum dots, and MRI contrast agents are of utmost importance to obtain a high-resolution image for providing functional insights into the biological process. This functionalization enhances imaging applications. This functionalization can be done by conjugating these compounds on the surface of the nanoconstructs (Serrano García et al. 2018). **Fluorescent dyes** are organic compounds widely used in fluorescence imaging. This is mainly due to their ability to emit light upon excitation. The conjugation of these dyes can significantly enhance the brightness and stability which allows a better visualization of cellular structures and processes (Walling et al. 2009). **Quantum dots** are semiconductor nanoparticles having a size-tunable fluorescence along with high photostability. Upon functionalizing with quantum dots, the nanoconstructs show enhanced imaging, which enables real-time monitoring of the treatment (Abdellatif et al. 2022). **Gadolinium (Gd) complexes** are widely used as MRI contrast agents due to their paramagnetic properties, which improve the MRI images. This functionalization can improve their relaxivity, leading to brighter MRI signals (Sun et al. 2011). Gd-doped carbon quantum dots can be used as dual-modal imaging agents for both fluorescence as well as MRI contrasts (Molaei 2022).

Multimodal imaging functionalization combines different imaging techniques to offer a more comprehensive view. It improves the diagnostic abilities of the imaging techniques by integrating their strengths. This can be done by combining different optical, magnetic, and radioactive agents (Serrano García et al. 2018). **Optical agents,** like fluorescent dyes and quantum dots, provide high-resolution optical images which gives a detailed visualization of cellular processes. When used in combination with other modalities, they can enhance the overall imaging quality with real-time insights of the treatment (Kosaka et al. 2010). **Magnetic agents** (MRI contrast agents), like gadolinium complexes, show an excellent soft tissue contrast, giving detailed anatomical information. Integration of the optical and magnetic agents showed enhanced imaging outcomes (Ibrahim et al. 2024). **Radioactive agents** are used in PET or SPECT analysis for information about the functional aspects of metabolic processes. By combining all three types of agents, we can get simultaneous insights into the structural, functional, and molecular aspects (Crişan et al. 2022). This multimodal aspect can significantly improve the understanding of the disease mechanism and treatment processes. The continued exploration of these functionalization strategies holds great promise for the future of imaging in personalized medicine (Table 16.3).

4 Biomedical Applications of Silica Core and Non-Silica-Based Shell Nanoconstructs

The silica core and non-silica-based shell nanoconstructs represent a great platform for many biomedical applications, including drug delivery and imaging. Nowadays, MSNs are most prevalent as drug delivery carriers due to their large surface area, variability in pore size, and capability to load different therapeutic agents. Adding a non-silica shell provides better targeted delivery and lowers adverse effects while increasing the potency of the antitumor drug. Functionalizing the shell with targeting ligands further enhances tumor cell specificity and drug accumulation (Chatterjee et al. 2014; Kumar et al. 2020; Rani et al. 2023).

4.1 Cancer Imaging

Designed silica nanoparticles stand out as effective contrast agents for imaging modalities such as magnetic resonance imaging (MRI) and ultrasound aiding, in the enhancement of tumor visualization for efficient diagnosis (Kesse et al. 2019; Mishra et al. 2023a). These comprised two-dimensional nanolayers that shield the nanoparticles and made them ideal in near-infrared light imaging of in vivo clinical tumors. Their predictable chemistry and biocompatibility support their use in immunodiagnostics, particularly in cancer research (Gubala et al. 2020). Tailoring their optical properties enhances their effectiveness in detecting and monitoring cancer progress. Furthermore, there is an increasing trend to create non-silica coated silica core nanoconstructs for theranostic purposes, which incorporate treatment and imaging. Such a dual functionality is beneficial in precise medicine enabling tailored treatment based on individual patient responses. These nanoparticles can be designed to target specific tumor markers, facilitating molecular imaging techniques that provide insight into tumor biology and behavior (Liu and Grodzinski 2021; Mishra et al. 2023b). Chen et al. developed NIR fluorescent core-shell silica

Table 16.3 Case studies related to surface modification

Sr. no.	Core	Shell	Drug	Targeting moiety	Type of surface modification	Cell line	Animal study	Outcome	References
1.	MSN	Cationic lipid bilayer	Paclitaxel	PSMA aptamer	Temperature-sensitive surface modification	LnCap or L929 cells used to perform MTT assay	–	The study showed that surface-modified nanoparticles displayed an enhanced cellular uptake and drug delivery efficacy compared to their uncoated counterparts. Additionally, aptamer conjugation neutralized the surface charge, leading to improved targeting and drug delivery efficacy to the cancer cells, resulting in higher cell death rates than unmodified drugs.	Saharkhiz et al. (2023)

2.	HMSN	Bovine serum albumin	Indocyanine green (ICG) and paclitaxel	Folic acid	Covalent bond	Cellular uptake efficiency, cytotoxicity, and intracellular imaging studies were performed on SGC-7901, MGC80–3, and NCI-N87	–	The study demonstrated that folic acid-modified nanoparticles showed significantly enhanced targeting efficiency and improved cellular uptake in cancer cells through receptor-mediated endocytosis resulting in induction of apoptosis in cancer cells.	Zhang and Liang (2024)
3.	HMSN	Chitosan and copper sulfide	Doxorubicin	Chitosan	CuS were capped through disulfide bonds	Cytotoxicty and confocal laser microscopy were performed on MDA-MB-231 cells	In vitro hemolysis assay, blood circulation study, in vivo biodistribution, in vivo PA imaging, in vivo antitumor efficacy, blood biochemical indices and histological studies were performed on SD female nude mice bearing MDA-MB-231 tumors	The study showed an effective targeted drug, and significant apoptosis in cancer cells. It was also observed that there was a prolonged survival in the murine breast cancer model.	Niu et al. (2021)

nanoparticles, and the surface was functionalized with different melanoma-targeting ligands. These nanoparticles coated with peptides targeting either the melanocortin-1 receptor (MC1R) or integrins were designed for dual modality imaging, combining PET/CT with interoperative fluorescence imaging for detecting sentinel lymph nodes (SLNs) in a melanoma model using miniswine. PET/CT was utilized preoperative to identify SLNs, while high-resolution optical fluorescence imaging was employed intraoperatively to achieve better spatial resolution. The combination of two spectrally distinct fluorescent particle probes enabled the detection and molecular phenotyping of metastatic nodes. Histological analysis confirmed the presence of metastatic melanoma in the resected nodes, with a string correlation between particle accumulation and tumor burden. Toxicity evaluation of the imaging agents in mice demonstrated no adverse effects, supporting the safety of this imaging strategy (Chen et al. 2019b).

Ferrauto et al., developed multifunctional nanosystems, incorporating ICG and methotrexate (MTX) into MSNs functionalized with PEG3000-COOH to prevent aggregation and stabilize the suspension. The MTX-ICG-MSNs displayed improved photophysical properties, with an enhanced quantum yield for ICG and detectable MTX photoluminescence. The nanoparticles showed stable photoluminescence and photoacoustic (PA) signals in biological fluids, with a PA peak at 700 nm for MTX and 810 nm for ICG. The release kinetics of MTX-ICG-MSNs containing MTX was found to be comparatively slow when compared to release from bladder-specific MTX-MSNs with 20% release in buffer or only modest release in human serum. When tested in vivo by PA imaging, it showed effective tumor targeting in a murine model, with distinct PA signals of MTX and ICG, confirming the localization and sustained delivery of MTX. There was a decrease in the PA700 nm/PA810 nm ratio with time which indicates the release of MTX from the nanoparticles. This dual signal approach using PA imaging offers a promising method in enhancing cancer imaging (Ferrauto et al. 2019).

4.2 Targeted Drug Delivery

Silica nanoparticles are also known to have mesoporous or hollow structures that can hold a variety of therapeutic agents. This shell can be engineered to conjugate polymered the active component or the drug to the targeted site for controlled release (Dhas et al. 2023; Rani et al. 2023). The further development of this approach is associated with the functionalization of the non-silica shell with targeting ligands (antibodies, peptides, or small molecules) which enables selective binding to cancer cells, and this approach will lead to the higher delivery of the drug at the tumor site with less exposure throughout the body, which is overcoming the challenges of drug resistance and incidence of toxicity that comes with conventional therapies (Chatterjee et al. 2014; Narayan et al. 2018; Rani et al. 2023).

Narayan et al. fabricated glycosylated cross-linked pH-sensitive capecitabine (CAP)-loaded MSNs coated with chitosan for colorectal cancer. The in vitro drug release studies of coated MSNs were conducted in simulated gastrointestinal and tumor-like pH environments. The release of CAP was higher in acidic pH 5.5, reaching 85.16% after 72 h due to chitosan swelling which opens the nanoparticle pores. At pH 6.8 release was lower (58.29%) due to chitosan collapse. Chitosan-coated MSNs exhibited control and selective drug release fitting well with the Higuchi model indicating non-Fickian transport. Cytotoxicity studies showed that CAP-loaded chitosan-glucuronic acid-coated MSNs had enhanced uptake and higher efficacy in colon cancer cells compared to pure CAP. In vivo studies in rats with induced colon cancer demonstrated that the coated MSN reduced tumor size in number, improved survival rates, and showed fewer toxic effects on major organs compared to pure CAP (Narayan et al. 2021). Ramezani-Aliakbari et al. prepared POMo-loaded MSNs coated with polydopamine and targeted with glucosamine (POMo@SBA-PDA-Glu nanoparticles) for targeting breast cancer. The POMo was synthesized under reflux conditions. The PDA coating prevented the rapid release of POMo ensuring a more sustained release, particularly under acidic conditions (pH 5). The release of POMo was more efficient in acidic environment with 42.4% of the total POMo released over 12 h compared to just 12.1% at neutral (pH 7.4). The in vivo studies preliminary focused on colloidal stability in vitro cancer activity. Colloidal stability tests demonstrated that the PDA-coated NPs maintain their size and stability in biological media, showing less than a 25% increase in hydrodynamic size in human blood serum, indicating good potential for in vivo applications. The in vitro anticancer activity showed dose-dependent cytotoxicity against breast cancer cell lines with PDA-Glu NPs exhibited enhanced efficacy. The POMo@SBA-PDA-Glu NPs showed higher cellular uptake and apoptosis levels compared to free POMo (Ramezani-Aliakbari et al. 2022).

In another study by Yin et al., prepared HMSNs were modified with folic acid and polymer p(NIPAAm-co-MAA), co-loaded with 5-fluorouracil (5-FU) and paclitaxel (PTX), and encapsulated within lipid membranes ((5-FU + PTX)@FHPL) to enable synergistic treatment for breast cancer. Folic acid-modified HMSNS achieved drug loading (DL) of 28.49% and encapsulation efficiency (EE) of 20.52% for 5-FU. For PTX in the dual load system DL was 14.29% with nearly 100% EE in the liquid membrane. At 37°C and pH 5, 5-FU and PTX released at rates of 11.34% and 15.15%, respectively, with higher release at 41°C. The release was sustained through diffusion and erosion. The anticancer activity showed that both free 5-FU and PTX inhibited MCF-7 and HepG2 cell proliferation, with greater efficacy on MCF-7 cells. Combined 5-FU and PTX demonstrated synergistic effects, enhancing apoptosis and inhibiting proliferation, particularly in MCF-7 cells. Drug-loaded NPs were more effective against MCF-7 cells than free drugs or single-drug formulations, selectively inducing apoptosis, cell cycle arrest, and inhibiting wound healing and migration. Blank nanoparticles showed no significant cytotoxicity, but (5-FU + PTX)@FHPL selectively targeted MCF-7 cells, showing targeted cancer therapy with minimal impact on normal tissues. The nanoparticles entered MCF-7 cells through clathrin-mediated, folate receptor-dependent endocytosis, while HepG2 cells entry involved multiple endocytic pathways. This nanosystem was selectively internalized by folate receptor-overexpressing MCF-7 breast cancer cells and released both drugs in a simulated tumor environment (Yin et al. 2022).

4.3 Phototherapy

MSNs have emerged as promising vectors for a wide range of biological applications, particularly in the realm of phototherapy (Garkal et al. 2023). MSNs have been engineered as nanocarriers for photothermal therapy (PTT) and photodynamic therapy (PDT) or combination of both. PDT and PTT, two prominent mobility of phototherapy, have gained significant interest in cancer treatment. In PDT, photosensitizers within MSNs produce cytotoxic reactive oxygen species (ROS) or singlet oxygen ($^1O_2$) upon light activation, leading to targeted destruction of cancer cells. In contrast, PTT relies on light-activated photothermal agents to induce localized hypothermia effectively ablating cancer cells through heat (Li et al. 2019).

In the study conducted by Agabeigi et al., they investigated the combined effects of folate and methotrexate (MTX) loaded silica-coated gold nanoparticle (Au@ SiO_2 NPs) and low-level laser therapy (LLLT) on breast cancer cells. Folate enhanced the uptake of nanoparticles via receptor-mediated endocytosis in cancer cells overexpressing folate receptors. The conjugation of MTX and FA was confirmed by shift in zeta potential. The in vitro studies demonstrated increased in cellular uptake and cytotoxicity of MTX-FA@Au-SiO_2 NPs, especially in MDA-MB-321 cells due to higher folate receptor expression. LLLT combined with MTX-FA@Au-SiO_2 NPs showed significant anticancer efficacy. Laser irradiation at 75 J/cm^2 increased cell mortality to 39% in MCF-7 cells and 45.5% in MDA-MB-231 cells. At 105 J/cm^2, mortality rates rose to 60–75%. This combination therapy significantly enhanced cell death compared to laser or nanoparticles alone, demonstrating a synergistic effect. Additionally, it induced notable apoptosis and cell cycle disruptions, particularly increasing apoptotic cells, including an increase in sub-G0/G1 population, indicative of enhanced apoptotic activity. MDA-MB-231 cells showed a higher apoptotic response compared to MCF-7 cells, and showing improved therapeutic efficacy against breast cancer cells (Agabeigi et al. 2020).

Chen et al., synthesized MSNs through a base-catalyzed sol-gel process containing doxorubicin and then coated with polydopamine and hyaluronic acid (DOX-MSNs-PDA-HA). MSNs-PDA-HA exhibited significant photothermal effects under near-infrared (NIR) laser irradiation, showing a concentration- and power-dependent temperature rise, and demonstrated high photostability. DOX-MSNs-PDA-HA showed pH-responsive and NIR-laser-triggered release behavior. DOX release was significantly higher under acidic pH 5 and further accelerated by NIR irradiation, demonstrating the potential for controlled drug release and enhanced chemo-photothermal therapy. MSNs-PDA-HA nanoparticles demonstrated good biocompatibility, low cytotoxicity, and targeted uptake in CD44-overexpressing cancer cells. The combination of DOX and these nanoparticles showed higher cytotoxicity, especially under NIR irradiation, indicating enhanced therapeutic efficacy (Chen et al. 2019a).

Shi et al. developed a DOX-loaded MSNs coated with tannic acid/Fe^{3+} (pTA) complex for combined chemo-photothermal therapy. The pTA coating effectively blocked DOX release, preventing significant drug loss. MSN-pTA exhibited broad UV-NIR absorption, with stronger absorption due to the MSN core. The photothermal effect was concentration-dependent and influenced by the TA/Fe^{3+} ratio with optimal conditions leading to a temperature rise of 33°C under 808 nm laser irradiation. MSN-pTA showed good photostability over multiple laser cycles. DOX release from MSN-pTA was pH-dependent with minimal release at neutral pH due to pTA blocking. At pH 4, 20% of DOX was released as pTA disassembly under acid conditions. The in vitro cytotoxicity studies showed negligible toxicity while DOX@MSN-pTA exhibit dose-dependent cytotoxicity. Combined chemo-photothermal therapy using DOX@MSN-pTA with laser irradiation significantly enhanced cytotoxicity, reducing cell viability from 52% to 10%. The in vivo studies with DOX in a 4 T1 tumor model showed a tumor temperature increase to 53.4°C under laser irradiation, effectively ablating tumor cells. DOX@MSN-pTA combined with laser therapy demonstrated the strongest tumor separation compared with single treatment, confirming the enhanced efficacy of combinational chemo-photothermal therapy (Shi et al. 2021) (Fig. 16.6).

Kuang et al. fabricated a PEGylated MSNs system containing curcumin (MSN-PEG@Cur) for enhanced photodynamic therapy in cancer treatment. PEGylation improved Cur solubility and bioavailability, allowing efficient loading and sustained release. The release studies of MSN-PEG@Cur showed that Cur had a pH-dependent release, with a slower and more controlled release at physiological pH (7.4) and a faster release in acidic conditions (pH 5). In in vitro cell line studies using Hela cells, MSN-PEG@Cur demonstrated enhanced cellular uptake and significantly higher production of reactive oxygen species (ROS) irradiation compared to free

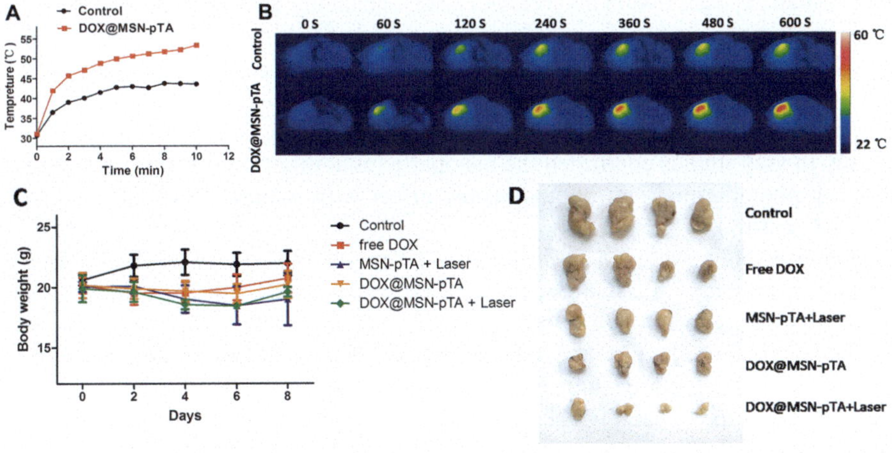

Fig. 16.6 (**a**) Temperature variation profile. (**b**) Thermal imaging of the tumor region during laser irradiation. (**c**) Tumor volume change curve. (**d**) Images of tumor from the various treatment groups. Adapted with permission (Shi et al. 2021)

Cur. This led to increased cytotoxicity, with a much lower IC50 value for MSN-PEG@Cur under irradiation, indicating its superior efficacy in eliminating cancer cells through PDT (Kuang et al. 2020).

4.4 Combination Therapies

Silica core and non-silica nanosystems are emerging as versatile platforms for theranostic applications in cancer management. The porous silica core enables efficient drug loading and controlled release, while the non-silica shell can be functionalized with targeting ligands and controlled tumor specificity. These constructs combine imaging agents within the core or shell, allowing real-time monitoring of drug delivery and treatment response. The targeted approach improves therapeutic efficacy and reduces systemic toxicity, offering a personalized strategy to overcome challenges like drug resistance (Koohi et al. 2022; Mishra et al. 2023b; Ravindran Girija and Balasubramanian 2019). Lu et al., constructed a core-shell nanocomplex with chorin e6-embeded MSNs as the core and coated with PDA and manganese ions to form protective shells [MSN-Ce6@PDA (Mn)]. The [MSN-Ce6@PDA (Mn)] nanoparticles demonstrated strong photothermal performance, rapidly heating upon 808-nm laser irradiation. The photothermal performance of [MSN-Ce6@PDA (Mn)] was evaluated under 808 nm irradiation. The [MSN-Ce6@PDA (Mn)] NPs demonstrated rapid heating in a concentration-dependent manner, reaching temperature of 36.2–57.3 °C within 10 min, compared to much lower temperature rise in phosphate buffer. Additionally, the NPs showed a significant temperature increase when exposed to varying laser intensities (0.25–1.5 W/cm^2), indicating efficient conversion of laser energy to heat. They maintained stability through multiple heating cycles. The nanoparticles effectively generated singlet oxygen, particularly when exposed to 808 and 660 nm laser, enhancing their potential for PDT. In vitro studies showed no cytotoxicity and confirmed the efficient uptake of [MSN-Ce6@PDA (Mn)] NPs by MDA-MB-231 cancer cells. The nanoparticles significantly enhanced T1-weighted MRI signals due to high Mn absorption. ROS generation was evident under laser treatment, contributing to effective cancer cell killing. In vivo studies confirmed the nanoparticles ability to accumulate in tumors, showing photothermal and photodynamic effect with minimal toxicity (Lu et al. 2021) (Fig. 16.7).

Huang et al., developed multifunctional nanoplatform for dual-modal tumor imaging and therapy. Perfluorohexane (PFH) loaded MSNs, with an indocyanine green (ICG)-polydopamine (PDA) layer and poly(ethylene glycol)-folic acid coating. It exhibits excellent monodispersity, high ICG loading, enhanced photostability, and improved cellular uptake. Laser irradiation (808 nm) demonstrated that increasing laser power and ICG concentration elevated the temperature, confirming the photothermal efficacy of nanocarriers. The nanocarriers generated ROS, indicating effective PDT activity. In vitro studies confirmed strong cellular uptake facilitated by folic acid coating, leading to enhanced PTT and PDT effects upon NIR irradiation. Ultrasound imaging demonstrated that PFH encapsulation within MSNs

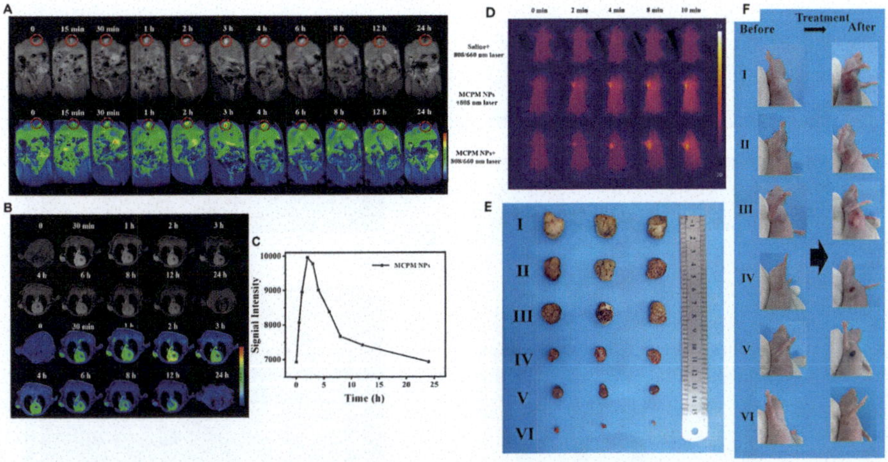

Fig. 16.7 (**a**) In vivo T1-weighted MR and pseudo-color images of mice with subcutaneous MDA-MB-231 tumor after i.v. injection of [MSN-Ce6@PDA (Mn)] NPs at various time points, with the gallbladder indicated by red circle. (**b**) Corresponding signal intensity changes in tumor regions. (**c**) In vivo FL images of mice with subcutaneous MDA-MB-231 tumor after i.v. injection of [MSN-Ce6@PDA (Mn)] NPs at different time points. (**d**) Thermal images of MDA-MB-231 tumor-bearing mice under different treatments. (**e**) Tumor images from various treatment groups. (**f**) Images of mice with subcutaneous MDA-MB-231 tumor following different treatment regimens. Adapted with permission (Lu et al. 2021)

enhanced imaging signals due to temperature-induced phase change. In vivo studies showed significant tumor accumulation and long retention time, enabling effective PTT and PDT, which led to substantial tumor growth inhibition and complete tumor elimination in some cases. The nanocarriers exhibited excellent biocompatibility and no toxicity, highlighting its potential as a powerful tool for multimodal cancer therapy (Huang et al. 2019).

Sun et al. developed MSNs into theranostic nanocomposite by doping them with gadolinium (Gd) for MRI contrast and loading them with DOX and ICG. To address the drug leakage and enhance cellular uptake, folic acid modified thermosensitive liposomes (TSLs) were coated onto the MSNs. This NIR-triggered, triple-modal imaging-guided platform (DOX@GdMSNs-ICG-TSLs) combines photodynamic and photothermal therapy with NIR fluorescence imaging, PA imaging, and MRI. The study evaluated the in vitro efficacy of DOX@GdMSNs-ICG-TSLs for photothermal and photodynamic therapy. Under 808 nm laser irradiation, the nanoparticles effectively increased the temperature, demonstrating its PTT capability, with a photothermal conversion efficacy of 11.72%. The release of DOX was enhanced by NIR-induced hyperthermia. Cellular uptake studies showed that the nanoparticles had strong internalization and ROS generation under NIR, confirming its PDT effect. The in vivo studies revealed that the nanosystems effectively targeted tumors, suppressed tumor growth and exhibited favorable safety with minimal toxicity to major organs (Sun et al. 2018) (Fig. 16.8).

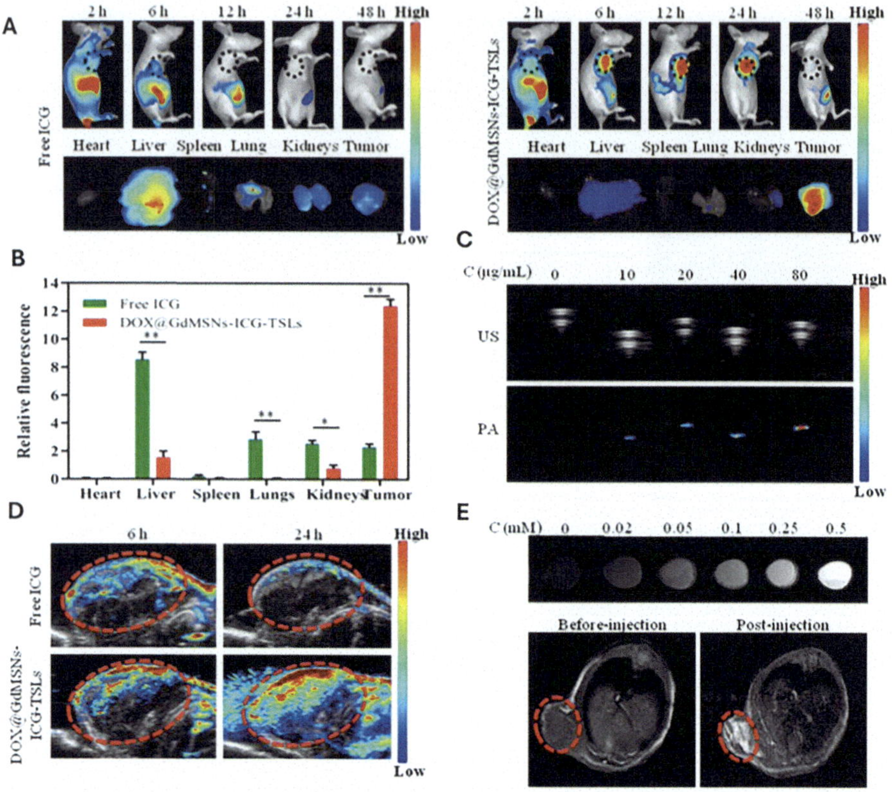

Fig. 16.8 (**a**) NIR fluorescence images of tumor bearing mice post-i.v. injection of free ICG and DOX@GdMSNs-ICG-TSLs, along with fluorescence images of various tissues at 24 h. (**b**) Relative fluorescence intensity of ICG in major organs induced by 808 nm laser irradiation at 24 h post-injection. (**c**) Ultrasound and PA images of DOX@GdMSNs-ICG-TSLs at varying ICG concentrations. (**d**) PA images of tumor site at 6 and 24 h post-injection of free ICG and DOX@GdMSNs-ICG-TSLs. (**e**) T1-weighted MR images of DOX@GdMSNs-ICG-TSLs at different Gd concentrations, and corresponding MR images of tumor-bearing mice before and after 24 h post injection. Adapted with permission (Sun et al. 2018)

5 Conclusion and Future Prospective

Silica nanoparticles are ideal drug delivery carriers due to their large surface area to volume ratio, tunable pore size in the silica, and the ability to incorporate targeting ligands to the non-silica shell to improve selectively toward cancer cells. It enhances the concentration of the drug at the tumor site, hence minimizing adverse effects, drug resistance, and toxicity. Additionally, these nanoconstructs can be designed with imaging agents to track drug delivery and therapeutic responses using fluorescence, MRI, and PA imaging for tumor targeting and site-specific therapy. The multifunctional nature enables targeted drug delivery, enhanced imaging capabilities, and effective phototherapy, positioning them as valuable assets in the development of innovative cancer therapies. These nanoconstructs demonstrated potential in preclinical research, but challenges remain for clinical translation such as engineering and fabrication of nanoparticles, characterization and addressing issues related to biodistribution, toxicity, and regulatory approval. Ongoing research aims to optimize synthesis and functionalization to enhance stability, biocompatibility, and therapeutic efficacy. However, significant improvements in nanotheranostic platforms are still required for clinical applications, with a focus on ensuring safety, stability, and managing the complexity of nanoparticles. Comprehensive toxicity assessments, including cytotoxicity, immunotoxicity, and genotoxicity, alongside evaluations of pharmacokinetics and biodistribution, are crucial. Levering AI and machine learning could accelerate design and optimization, while scalable fabrication methods and meeting regulatory requirements are important for clinical adoption. Future directions include combinatorial screening and establishing standardized fabrication and characterization protocol. Addressing these challenges will be key to realizing the potential of silica core non-silica shell nanoconstructs in advancing theranostics and improving outcomes.

References

Abdellatif AA, Younis MA, Alsharidah M, et al. Biomedical applications of quantum dots: overview, challenges, and clinical Potential. IJN. 2022;17:1951–70. https://doi.org/10.2147/IJN.S357980.

Abousalman-Rezvani Z, Refaat A, Dehghankelishadi P, et al. Insights into targeted and stimulus-responsive nanocarriers for brain cancer treatment. Adv Healthc Mater. 2024;13:2302902. https://doi.org/10.1002/adhm.202302902.

Abrishami A, Bahrami AR, Nekooei S, et al. Hybridized quantum dot, silica, and gold nanoparticles for targeted chemo-radiotherapy in colorectal cancer theranostics. Commun Biol. 2024;7:1–19. https://doi.org/10.1038/s42003-024-06043-6.

Addisu KD, Hsu W-H, Hailemeskel BZ, et al. Mixed lanthanide oxide nanoparticles coated with alginate-polydopamine as multifunctional nanovehicles for dual modality: targeted imaging and chemotherapy. ACS Biomater Sci Eng. 2019;5:5453–69. https://doi.org/10.1021/acsbiomaterials.9b01226.

Agabeigi R, Rasta SH, Rahmati-Yamchi M, et al. Novel chemo-photothermal therapy in breast cancer using methotrexate-loaded folic acid conjugated Au@SiO2 nanoparticles. Nanoscale Res Lett. 2020;15:62. https://doi.org/10.1186/s11671-020-3295-1.

Ahangaran F, Navarchian AH. Recent advances in chemical surface modification of metal oxide nanoparticles with silane coupling agents: a review. Adv Colloid Interface Sci. 2020;286:102298. https://doi.org/10.1016/j.cis.2020.102298.

Ahmad F, Salem-Bekhit MM, Khan F, et al. Unique properties of surface-functionalized nanoparticles for bio-application: functionalization mechanisms and importance in application. Nanomaterials. 2022;12:1333. https://doi.org/10.3390/nano12081333.

Akbari A, Yegani R, Pourabbas B. Synthesis of poly(ethylene glycol) (PEG) grafted silica nanoparticles with a minimum adhesion of proteins via one-pot one-step method. Colloids Surf A Physicochem Eng Aspects. 2015;484:206–15. https://doi.org/10.1016/j.colsurfa.2015.07.042.

Akhter F, Rao AA, Abbasi MN, et al. A comprehensive review of synthesis, applications and future prospects for silica nanoparticles (SNPs). Silicon. 2022;14:8295–310. https://doi.org/10.1007/s12633-021-01611-5.

Ambrogio MW, Thomas CR, Zhao Y-L, et al. Mechanized silica nanoparticles: a new frontier in theranostic nanomedicine. Acc Chem Res. 2011;44:903–13. https://doi.org/10.1021/ar200018x.

Baeza A, Vallet-Regí M. Mesoporous silica nanoparticles as theranostic antitumoral nanomedicines. Pharmaceutics. 2020;12:957. https://doi.org/10.3390/pharmaceutics12100957.

Bauer D, Cornejo MA, Hoang TT, et al. Click chemistry and radiochemistry: an update. Bioconjugate Chem. 2023;34:1925–50. https://doi.org/10.1021/acs.bioconjchem.3c00286.

Bax HJ, Chauhan J, Stavraka C, et al. Folate receptor alpha in ovarian cancer tissue and patient serum is associated with disease burden and treatment outcomes. Br J Cancer. 2023;128:342–53. https://doi.org/10.1038/s41416-022-02031-x.

Bharti C, Nagaich U, Pal AK, Gulati N. Mesoporous silica nanoparticles in target drug delivery system: a review. Int J Pharm Investig. 2015;5:124–33. https://doi.org/10.4103/2230-973X.160844.

Bueno V, Ghoshal S. Self-assembled surfactant-templated synthesis of porous hollow silica nanoparticles: mechanism of formation and feasibility of post-synthesis nanoencapsulation. Langmuir. 2020;36:14633–43. https://doi.org/10.1021/acs.langmuir.0c02501.

Bukhari SI, Imam SS, Ahmad MZ, et al. Recent progress in lipid nanoparticles for cancer theranostics: opportunity and challenges. Pharmaceutics. 2021a;13:840. https://doi.org/10.3390/pharmaceutics13060840.

Bukhari SZ, Zeth K, Iftikhar M, et al. Supramolecular lipid nanoparticles as delivery carriers for non-invasive cancer theranostics. Curr Res Pharmacol Drug Discov. 2021b;2:100067. https://doi.org/10.1016/j.crphar.2021.100067.

Cai X, Zhu Q, Zeng Y, et al. Manganese oxide nanoparticles as MRI contrast agents in tumor multimodal imaging and therapy. IJN. 2019;14:8321–44. https://doi.org/10.2147/IJN.S218085.

Chang D, Ma Y, Xu X, et al. Stimuli-responsive polymeric nanoplatforms for cancer therapy. Front Bioeng Biotechnol. 2021;9:707319. https://doi.org/10.3389/fbioe.2021.707319.

Chatterjee K, Sarkar S, Jagajjanani Rao K, Paria S. Core/shell nanoparticles in biomedical applications. Adv Colloid Interface Sci. 2014;209:8–39. https://doi.org/10.1016/j.cis.2013.12.008.

Chen F, Hong H, Shi S, et al. Engineering of hollow mesoporous silica nanoparticles for remarkably enhanced tumor active targeting efficacy. Sci Rep. 2014;4:5080. https://doi.org/10.1038/srep05080.

Chen Y, Xianyu Y, Wu J, et al. Click chemistry-mediated nanosensors for biochemical assays. Theranostics. 2016;6:969–85. https://doi.org/10.7150/thno.14856.

Chen C, Tang W, Jiang D, et al. Hyaluronic acid conjugated polydopamine functionalized mesoporous silica nanoparticles for synergistic targeted chemo-photothermal therapy. Nanoscale. 2019a;11:11012–24. https://doi.org/10.1039/C9NR01385G.

Chen F, Madajewski B, Ma K, et al. Molecular phenotyping and image-guided surgical treatment of melanoma using spectrally distinct ultrasmall core-shell silica nanoparticles. Sci Adv. 2019b;5:eaax5208. https://doi.org/10.1126/sciadv.aax5208.

Cholkar K, Hirani ND, Natarajan C. Chapter 14. Nanotechnology-based medical and biomedical imaging for diagnostics. In: Mitra AK, Cholkar K, Mandal A, editors. Emerging nanotechnologies for diagnostics, drug delivery and medical devices. Boston: Elsevier; 2017. p. 355–74.

Chuang W, Geng-sheng J, Lei P, et al. Influences of surface modification of nano-silica by silane coupling agents on the thermal and frictional properties of cyanate ester resin. Results Phys. 2018;9:886–96. https://doi.org/10.1016/j.rinp.2018.03.056.

Crişan G, Moldovean-Cioroianu NS, Timaru D-G, et al. Radiopharmaceuticals for PET and SPECT imaging: a literature review over the last decade. Int J Mol Sci. 2022;23:5023. https://doi.org/10.3390/ijms23095023.

Dam DHM, Lee RC, Odom TW. Improved in vitro efficacy of gold nanoconstructs by increased loading of G-quadruplex aptamer. Nano Lett. 2014;14:2843–8. https://doi.org/10.1021/nl500844m.

Daniels TR, Delgado T, Helguera G, Penichet ML. The transferrin receptor part II: targeted delivery of therapeutic agents into cancer cells. Clin Immunol. 2006;121:159–76. https://doi.org/10.1016/j.clim.2006.06.006.

Daniels TR, Bernabeu E, Rodríguez JA, et al. Transferrin receptors and the targeted delivery of therapeutic agents against cancer. Biochim Biophys Acta. 2012;1820:291–317. https://doi.org/10.1016/j.bbagen.2011.07.016.

Dhas NL, Raval NJ, Kudarha RR, et al. Core–shell nanoparticles as a drug delivery platform for tumor targeting. In: Inorganic frameworks as smart nanomedicines. Elsevier; 2018. p. 387–448.

Dhas N, Neyyar S, Garkal A, et al. Biomedical applications of nanocarriers in nasal delivery. In: Pathak YV, Yadav HKS, editors. Nasal drug delivery: formulations, developments, challenges, and solutions. Cham: Springer; 2023. p. 101–26.

Dutta Gupta Y, Mackeyev Y, Krishnan S, Bhandary S. Mesoporous silica nanotechnology: promising advances in augmenting cancer theranostics. Cancer Nanotechnol. 2024;15:9. https://doi.org/10.1186/s12645-024-00250-w.

Esfahani MKM, Alavi SE, Cabot PJ, et al. PEGylated mesoporous silica nanoparticles (MCM-41): a promising carrier for the targeted delivery of fenbendazole into prostrate cancer cells. Pharmaceutics. 2021;13:1605. https://doi.org/10.3390/pharmaceutics13101605.

Fallah J, Agrawal S, Gittleman H, et al. FDA approval summary: lutetium Lu 177 vipivotide tetraxetan for patients with metastatic castration-resistant prostate cancer. Clin Cancer Res. 2023;29:1651–7. https://doi.org/10.1158/1078-0432.CCR-22-2875.

Fan D, Cao Y, Cao M, et al. Nanomedicine in cancer therapy. Sig Transduct Target Ther. 2023;8:1–34. https://doi.org/10.1038/s41392-023-01536-y.

Fernandes DA. Review on metal-based theranostic nanoparticles for cancer therapy and imaging. Technol Cancer Res Treat. 2023;22:15330338231191493. https://doi.org/10.1177/15330338231191493.

Fernandes RS, Raimundo IM, Pimentel MF. Revising the synthesis of Stöber silica nanoparticles: a multivariate assessment study on the effects of reaction parameters on the particle size. Colloids Surf A Physicochem Eng Aspects. 2019;577:1–7. https://doi.org/10.1016/j.colsurfa.2019.05.053.

Fernández M, Javaid F, Chudasama V. Advances in targeting the folate receptor in the treatment/imaging of cancers. Chem Sci. 2017;9:790–810. https://doi.org/10.1039/c7sc04004k.

Ferrauto G, Carniato F, Gregorio ED, et al. Photoacoustic ratiometric assessment of mitoxantrone release from theranostic ICG-conjugated mesoporous silica nanoparticles. Nanoscale. 2019;11:18031–6. https://doi.org/10.1039/C9NR06524E.

Ferreira CA, Goel S, Ehlerding EB, et al. Ultrasmall porous silica nanoparticles with enhanced pharmacokinetics for cancer theranostics. Nano Lett. 2021;21:4692–9. https://doi.org/10.1021/acs.nanolett.1c00895.

Freidus LG, Kumar P, Marimuthu T, et al. Theranostic mesoporous silica nanoparticles loaded with a curcumin-naphthoquinone conjugate for potential cancer intervention. Front Mol Biosci. 2021;8:670792. https://doi.org/10.3389/fmolb.2021.670792.

Gao Y, Gao D, Shen J, Wang Q. A review of mesoporous silica nanoparticle delivery systems in chemo-based combination cancer therapies. Front Chem. 2020;8:598722. https://doi.org/10.3389/fchem.2020.598722.

Garkal A, Sami A, Patel J, et al. Light-responsive delivery nanoplatforms in cancer theranostics. In: Sitespecific cancer nanotheranostics. CRC Press; 2023.

Gauger AJ, Hershberger KK, Bronstein LM. Theranostics based on magnetic nanoparticles and polymers: intelligent design for efficient diagnostics and therapy. Front Chem. 2020;8:561. https://doi.org/10.3389/fchem.2020.00561.

Geißler D, Nirmalananthan-Budau N, Scholtz L, et al. Analyzing the surface of functional nanomaterials—how to quantify the total and derivatizable number of functional groups and ligands. Microchim Acta. 2021;188:321. https://doi.org/10.1007/s00604-021-04960-5.

Gentili D, Ori G. Reversible assembly of nanoparticles: theory, strategies and computational simulations. Nanoscale. 2022;14:14385–432. https://doi.org/10.1039/D2NR02640F.

Glorani G, Marin R, Canton P, et al. Pegylated silica nanoparticles: cytotoxicity and macrophage uptake. J Nanopart Res. 2017;19:294. https://doi.org/10.1007/s11051-017-3964-x.

Gubala V, Giovannini G, Kunc F, et al. Dye-doped silica nanoparticles: synthesis, surface chemistry and bioapplications. Cancer Nanotechnol. 2020;11:1. https://doi.org/10.1186/s12645-019-0056-x.

Guo X, Cheng Y, Zhao X, et al. Advances in redox-responsive drug delivery systems of tumor microenvironment. J Nanobiotechnol. 2018;16:74. https://doi.org/10.1186/s12951-018-0398-2.

Herschlag D, Pinney MM. Hydrogen bonds: simple after all? Biochemistry. 2018;57:3338–52. https://doi.org/10.1021/acs.biochem.8b00217.

Hosseini S, Ibrahim F, Djordjevic I, Koole LH. Recent advances in surface functionalization techniques on polymethacrylate materials for optical biosensor applications. Analyst. 2014;139:2933–43. https://doi.org/10.1039/C3AN01789C.

Hosseini SM, Mohammadnejad J, Salamat S, et al. Theranostic polymeric nanoparticles as a new approach in cancer therapy and diagnosis: a review. Mater Today Chem. 2023;29:101400. https://doi.org/10.1016/j.mtchem.2023.101400.

Huang C, Zhang Z, Guo Q, et al. A dual-model imaging theragnostic system based on mesoporous silica nanoparticles for enhanced cancer phototherapy. Adv Healthc Mater. 2019;8:1900840. https://doi.org/10.1002/adhm.201900840.

Ibrahim MA, Hazhirkarzar B, Dublin AB. Gadolinium magnetic resonance imaging. In: StatPearls. Treasure Island (FL): StatPearls Publishing; 2024.

Ichikawa Y, Kobayashi N, Takano S, et al. Neuroendocrine tumor theranostics. Cancer Sci. 2022;113:1930–8. https://doi.org/10.1111/cas.15327.

Jain AK, Thareja S. In vitro and in vivo characterization of pharmaceutical nanocarriers used for drug delivery. Artif Cells Nanomed Biotechnol. 2019;47:524–39. https://doi.org/10.1080/21691401.2018.1561457.

Jena S, Dutta J, Tulsiyan KD, et al. Noncovalent interactions in proteins and nucleic acids: beyond hydrogen bonding and π-stacking. Chem Soc Rev. 2022;51:4261–86. https://doi.org/10.1039/D2CS00133K.

Jiao W, Zhang T, Peng M, et al. Design of magnetic nanoplatforms for cancer theranostics. Biosensors. 2022;12:38. https://doi.org/10.3390/bios12010038.

Kalita H, Patowary M. Biocompatible polymer nano-constructs: a potent platform for cancer theranostics. Technol Cancer Res Treat. 2023;22:15330338231160391. https://doi.org/10.1177/15330338231160391.

Karimi M, Zangabad PS, Ghasemi A, et al. Temperature-responsive smart nanocarriers for delivery of therapeutic agents: applications and recent advances. ACS Appl Mater Interfaces. 2016;8:21107–33. https://doi.org/10.1021/acsami.6b00371.

Kesse S, Boakye-Yiadom KO, Ochete BO, et al. Mesoporous silica nanomaterials: versatile nanocarriers for cancer theranostics and drug and gene delivery. Pharmaceutics. 2019;11:77. https://doi.org/10.3390/pharmaceutics11020077.

Kim H, Kwak M. Structures and applications of nucleic acid-based micelles for cancer therapy. Int J Mol Sci. 2023;24:1592. https://doi.org/10.3390/ijms24021592.

Kim TH, Lee S, Chen X. Nanotheranostics for personalized medicine. Expert Rev Mol Diagn. 2013;13:257–69. https://doi.org/10.1586/erm.13.15.

Koohi Moftakhari Esfahani M, Alavi SE, Cabot PJ, et al. Application of mesoporous silica nanoparticles in cancer therapy and delivery of repurposed anthelmintics for cancer therapy. Pharmaceutics. 2022;14:1579. https://doi.org/10.3390/pharmaceutics14081579.

Kosaka N, McCann TE, Mitsunaga M, et al. Real-time optical imaging using quantum dot and related nanocrystals. Nanomedicine (Lond). 2010;5:765–76. https://doi.org/10.2217/nnm.10.49.

Kuang G, Zhang Q, He S, Liu Y. Curcumin-loaded PEGylated mesoporous silica nanoparticles for effective photodynamic therapy. RSC Adv. 2020;10:24624–30. https://doi.org/10.1039/D0RA04778C.

Kulhari H, Pooja D, Rompicharla SVK, et al. Biomedical applications of trastuzumab: as a therapeutic agent and a targeting ligand. Med Res Rev. 2015;35:849–76. https://doi.org/10.1002/med.21345.

Kulkarni D, Gadade D, Kapare H, et al. Characterization techniques for stimuli-responsive delivery nanoplatforms in cancer treatment. In: Site-specific cancer nanotheranostics. CRC Press; 2023.

Kumar R, Mondal K, Panda PK, et al. Core–shell nanostructures: perspectives towards drug delivery applications. J Mater Chem B. 2020;8:8992–9027. https://doi.org/10.1039/D0TB01559H.

Langbein T, Weber WA, Eiber M. Future of theranostics: an outlook on precision oncology in nuclear medicine. J Nucl Med. 2019;60:13S–9S. https://doi.org/10.2967/jnumed.118.220566.

Lee JE, Lee N, Kim T, et al. Multifunctional mesoporous silica nanocomposite nanoparticles for theranostic applications. Acc Chem Res. 2011;44:893–902. https://doi.org/10.1021/ar2000259.

Levine R, Krenning EP. Clinical history of the theranostic radionuclide approach to neuroendocrine tumors and other types of cancer: historical review based on an interview of Eric P. Krenning by Rachel Levine. J Nucl Med. 2017;58:3S–9S. https://doi.org/10.2967/jnumed.116.186502.

Li T, Shi S, Goel S, et al. Recent advancements in mesoporous silica nanoparticles towards therapeutic applications for cancer. Acta Biomater. 2019;89:1–13. https://doi.org/10.1016/j.actbio.2019.02.031.

Li R, Peng F, Cai J, et al. Redox dual-stimuli responsive drug delivery systems for improving tumor-targeting ability and reducing adverse side effects. Asian J Pharm Sci. 2020;15:311–25. https://doi.org/10.1016/j.ajps.2019.06.003.

Li X, Li W, Wang M, Liao Z. Magnetic nanoparticles for cancer theranostics: advances and prospects. J Control Release. 2021;335:437–48. https://doi.org/10.1016/j.jconrel.2021.05.042.

Liu CH, Grodzinski P. Nanotechnology for cancer imaging: advances, challenges, and clinical opportunities. Radiol Imaging Cancer. 2021;3:e200052. https://doi.org/10.1148/rycan.2021200052.

Liu Z, Kiessling F, Gätjens J. Advanced nanomaterials in multimodal imaging: design, functionalization, and biomedical applications. J Nanomater. 2010;2010:894303. https://doi.org/10.1155/2010/894303.

Lu J, Ni C, Huang J, et al. Biocompatible mesoporous silica–polydopamine nanocomplexes as MR/fluorescence imaging agent for light-activated photothermal–photodynamic cancer therapy in vivo. Front Bioeng Biotechnol. 2021;9:752982.

Luk BT, Zhang L. Current advances in polymer-based nanotheranostics for cancer treatment and diagnosis. ACS Appl Mater Interfaces. 2014;6:21859–73. https://doi.org/10.1021/am5036225.

Luk BT, Fang RH, Zhang L. Lipid- and polymer-based nanostructures for cancer theranostics. Theranostics. 2012;2:1117–26. https://doi.org/10.7150/thno.4381.

Luther DC, Huang R, Jeon T, et al. Delivery of drugs, proteins, and nucleic acids using inorganic nanoparticles. Adv Drug Deliv Rev. 2020;156:188–213. https://doi.org/10.1016/j.addr.2020.06.020.

Marques AC, da Costa PC, Velho S, Amaral MH. Antibody-functionalized nanoparticles for targeted drug delivery in cancer therapy. In: Rezaei N, editor. Handbook of cancer and immunology. Cham: Springer; 2022. p. 1–43.

Martin TA, Ye L, Sanders AJ, et al. Cancer invasion and metastasis: molecular and cellular perspective. In: Madame Curie bioscience database [Internet]. Landes Bioscience; 2013.

Ménard M. Synthesis of hybrid core-shell nanoparticles for theranostic applications. Doctoral dissertation, Université de Strasbourg; 2017.

Mishra P, Tripathi YB, Dhas NL, Garg N. Hypoxia-responsive delivery nanoplatforms in cancer theranostics. In: Site-specific cancer nanotheranostics. CRC Press; 2023a.

Mishra S, Bhatt T, Kumar H, et al. Nanoconstructs for theranostic application in cancer: Challenges and strategies to enhance the delivery. Front Pharmacol. 2023b;14:1101320. https://doi.org/10.3389/fphar.2023.1101320.

Molaei MJ. Gadolinium-doped fluorescent carbon quantum dots as MRI contrast agents and fluorescent probes. Sci Rep. 2022;12:17681. https://doi.org/10.1038/s41598-022-22518-0.

Naik GARR, Roy AA, Mutalik S, Dhas N. Unleashing the power of polymeric nanoparticles—creative triumph against antibiotic resistance: a review. Int J Biol Macromol. 2024;278:134977. https://doi.org/10.1016/j.ijbiomac.2024.134977.

Narayan R, Nayak UY, Raichur AM, Garg S. Mesoporous silica nanoparticles: a comprehensive review on synthesis and recent advances. Pharmaceutics. 2018;10:118. https://doi.org/10.3390/pharmaceutics10030118.

Narayan R, Gadag S, Cheruku SP, et al. Chitosan-glucuronic acid conjugate coated mesoporous silica nanoparticles: a smart pH-responsive and receptor-targeted system for colorectal cancer therapy. Carbohydr Polym. 2021;261:117893. https://doi.org/10.1016/j.carbpol.2021.117893.

Nayl AA, Abd-Elhamid AI, Aly AA, Bräse S. Recent progress in the applications of silica-based nanoparticles. RSC Adv. 2022;12:13706–26. https://doi.org/10.1039/D2RA01587K.

Nell KM, Fontenot SA, Carter TG, et al. Non-covalent functionalization of high-surface area nanomaterials: a new class of sorbent materials. Environ Sci: Nano. 2016;3:138–45. https://doi.org/10.1039/C5EN00170F.

Nihei T. Dental applications for silane coupling agents. J Oral Sci. 2016;58:151–5. https://doi.org/10.2334/josnusd.16-0035.

Niu S, Zhang X, Williams GR, et al. Hollow mesoporous silica nanoparticles gated by chitosan-copper sulfide composites as theranostic agents for the treatment of breast cancer. Acta Biomater. 2021;126:408–20. https://doi.org/10.1016/j.actbio.2021.03.024.

Palot Manzil FF, Kaur H, Szabados L. Gallium-68 prostate-specific membrane antigen positron emission tomography: a practical guide for radiologists and clinicians. Cureus. 2022;14:e22917. https://doi.org/10.7759/cureus.22917.

Petcu C, Purcar V, Spătaru C-I, et al. The influence of new hydrophobic silica nanoparticles on the surface properties of the films obtained from bilayer hybrids. Nanomaterials (Basel). 2017;7:47. https://doi.org/10.3390/nano7020047.

Phan MV, Tran TKT, Pham QN, et al. Controllable synthesis of hollow silica nanoparticles using layered double hydroxide templates and application for thermal insulation coating. ACS Omega. 2023;8:31399–409. https://doi.org/10.1021/acsomega.3c03917.

Prieto-Montero R, Katsumiti A, Cajaraville MP, et al. Functionalized fluorescent silica nanoparticles for bioimaging of cancer cells. Sensors. 2020;20:5590. https://doi.org/10.3390/s20195590.

Pyo CE, Chang JH. Hydrophobic mesoporous silica particles modified with nonfluorinated alkyl silanes. ACS Omega. 2021;6:16100–9. https://doi.org/10.1021/acsomega.1c01981.

Raheem MA, Rahim MA, Gul I, et al. Advances in nanoparticles-based approaches in cancer theranostics. OpenNano. 2023;12:100152. https://doi.org/10.1016/j.onano.2023.100152.

Rahim MA, Jan N, Khan S, et al. Recent advancements in stimuli responsive drug delivery platforms for active and passive cancer targeting. Cancers (Basel). 2021;13:670. https://doi.org/10.3390/cancers13040670.

Rajamanickam K. Multimodal molecular imaging strategies using functionalized nano probes. J Nanotechnol Res. 2019;1:119–35. https://doi.org/10.26502/jnr.2688-85210010.

Rajana N, Mounika A, Chary PS, et al. Multifunctional hybrid nanoparticles in diagnosis and therapy of breast cancer. J Control Release. 2022;352:1024–47. https://doi.org/10.1016/j.jconrel.2022.11.009.

Ramezani-Aliakbari M, Varshosaz J, Mirian M, et al. pH-responsive glucosamine anchored polydopamine coated mesoporous silica nanoparticles for delivery of Anderson-type polyoxomolybdate in breast cancer. J Microencapsul. 2022;39:433–51. https://doi.org/10.1080/02652048.2022.2096139.

Rani R, Malik P, Dhania S, Mukherjee TK. Recent advances in mesoporous silica nanoparticle-mediated drug delivery for breast cancer treatment. Pharmaceutics. 2023;15:227. https://doi.org/10.3390/pharmaceutics15010227.

Ravindran Girija A, Balasubramanian S. Theragnostic potentials of core/shell mesoporous silica nanostructures. Nanotheranostics. 2019;3:1–40. https://doi.org/10.7150/ntno.27877.

Ryu JH, Lee S, Son S, et al. Theranostic nanoparticles for future personalized medicine. J Control Release. 2014;190:477–84. https://doi.org/10.1016/j.jconrel.2014.04.027.

Saha A, Mishra P, Biswas G, Bhakta S. Greening the pathways: a comprehensive review of sustainable synthesis strategies for silica nanoparticles and their diverse applications. RSC Adv. 2023;14:11197–216. https://doi.org/10.1039/d4ra01047g.

Saharkhiz S, Nasri N, Dini G, Yousefnia S. Development of a new smart theranostic anti-PSMA-aptamer conjugated cationic-lipid coated mesoporous silica platform for targeted delivery of paclitaxel and CdSe/ZnS quantum dots to LNCaP cell line. J Drug Deliv Sci Technol. 2023;88:104964. https://doi.org/10.1016/j.jddst.2023.104964.

Sanità G, Carrese B, Lamberti A. Nanoparticle surface functionalization: how to improve biocompatibility and cellular internalization. Front Mol Biosci. 2020;7:587012. https://doi.org/10.3389/fmolb.2020.587012.

Santra S. Fluorescent silica nanoparticles for cancer imaging. Methods Mol Biol. 2010;624:151–62. https://doi.org/10.1007/978-1-60761-609-2_10.

Şen Karaman D, Kettiger H. Chapter 1. Silica-based nanoparticles as drug delivery systems: chances and challenges. In: Grumezescu AM, editor. Inorganic frameworks as smart nanomedicines. William Andrew Publishing; 2018. p. 1–40.

Serrano García R, Stafford S, Gun'ko YK. Recent progress in synthesis and functionalization of multimodal fluorescent-magnetic nanoparticles for biological applications. Appl Sci. 2018;8:172. https://doi.org/10.3390/app8020172.

Seyfried TN, Huysentruyt LC. On the origin of cancer metastasis. Crit Rev Oncog. 2013;18:43–73.

Shahbaz S, Esmaeili M, Fathian Nasab MH, et al. PEGylated mesoporous silica core–shell redox-responsive nanoparticles for delivering paclitaxel to breast cancer cells. Int J Pharm. 2024;655:124024. https://doi.org/10.1016/j.ijpharm.2024.124024.

Sharma J, Polizos G. Hollow silica particles: recent progress and future perspectives. Nanomaterials (Basel). 2020;10:1599. https://doi.org/10.3390/nano10081599.

Shi Q, Wu K, Huang X, et al. Tannic acid/Fe3+ complex coated mesoporous silica nanoparticles for controlled drug release and combined chemo-photothermal therapy. Colloids Surf A Physicochem Eng Asp. 2021;618:126475. https://doi.org/10.1016/j.colsurfa.2021.126475.

Shirshahi V, Soltani M. Solid silica nanoparticles: applications in molecular imaging. Contrast Media Mol Imaging. 2015;10:1–17. https://doi.org/10.1002/cmmi.1611.

Siddhardha B, Parasuraman P. Chapter 3. Theranostics application of nanomedicine in cancer detection and treatment. In: Grumezescu AM, editor. Nanomaterials for drug delivery and therapy. William Andrew Publishing; 2019. p. 59–89.

Siddiqui B, Rehman AU, Haq I, et al. Exploiting recent trends for the synthesis and surface functionalization of mesoporous silica nanoparticles towards biomedical applications. Int J Pharm X. 2022;4:100116. https://doi.org/10.1016/j.ijpx.2022.100116.

Singh N, Shi S, Goel S. Ultrasmall silica nanoparticles in translational biomedical research: overview and outlook. Adv Drug Deliv Rev. 2023;192:114638. https://doi.org/10.1016/j.addr.2022.114638.

Soheyli E, Ghaemi B, Sahraei R, et al. Colloidal synthesis of tunably luminescent AgInS-based/ZnS core/shell quantum dots as biocompatible nano-probe for high-contrast fluorescence bioimaging. Mater Sci Eng C Mater Biol Appl. 2020;111:110807. https://doi.org/10.1016/j.msec.2020.110807.

Strosberg J, El-Haddad G, Wolin E, et al. Phase 3 trial of 177Lu-Dotatate for midgut neuroendocrine tumors. N Engl J Med. 2017;376:125–35. https://doi.org/10.1056/NEJMoa1607427.

Sun C, Lee JSH, Zhang M. Magnetic nanoparticles in MR imaging and drug delivery. Adv Drug Deliv Rev. 2008;60:1252–65. https://doi.org/10.1016/j.addr.2008.03.018.

Sun H, Yuan Q, Zhang B, et al. GdIII functionalized gold nanorods for multimodal imaging applications. Nanoscale. 2011;3:1990–6. https://doi.org/10.1039/C0NR00929F.

Sun J, Liu Y, Ge M, et al. A distinct endocytic mechanism of functionalized-silica nanoparticles in breast cancer stem cells. Sci Rep. 2017;7:16236. https://doi.org/10.1038/s41598-017-16591-z.

Sun Q, You Q, Wang J, et al. Theranostic nanoplatform: triple-modal imaging-guided synergistic cancer therapy based on liposome-conjugated mesoporous silica nanoparticles. ACS Appl Mater Interfaces. 2018;10:1963–75. https://doi.org/10.1021/acsami.7b13651.

Tang C, Li X, Li Z, Hao J. Interfacial hydrogen bonds and their influence mechanism on increasing the thermal stability of nano-SiO2-modified meta-aramid fibres. Polymers. 2017;9:504. https://doi.org/10.3390/polym9100504.

Tao C, Zhu Y. Magnetic mesoporous silica nanoparticles for potential delivery of chemotherapeutic drugs and hyperthermia. Dalton Trans. 2014;43:15482–90. https://doi.org/10.1039/C4DT01984A.

Urso L, Nieri A, Uccelli L, et al. Lutathera® orphans: state of the art and future application of radioligand therapy with 177Lu-DOTATATE. Pharmaceutics. 2023;15:1110. https://doi.org/10.3390/pharmaceutics15041110.

Vatanparast H, Shahabi F, Bahramian A, et al. The role of electrostatic repulsion on increasing surface activity of anionic surfactants in the presence of hydrophilic silica nanoparticles. Sci Rep. 2018;8:7251. https://doi.org/10.1038/s41598-018-25493-7.

Waheed I, Ali A, Tabassum H, et al. Lipid-based nanoparticles as drug delivery carriers for cancer therapy. Front Oncol. 2024;14:1296091. https://doi.org/10.3389/fonc.2024.1296091.

Walling MA, Novak JA, Shepard JRE. Quantum dots for live cell and in vivo imaging. Int J Mol Sci. 2009;10:441–91. https://doi.org/10.3390/ijms10020441.

Wang Y, An L, Lin J, et al. A hollow Cu9S8 theranostic nanoplatform based on a combination of increased active sites and photothermal performance in enhanced chemodynamic therapy. Chem Eng J. 2020;385:123925. https://doi.org/10.1016/j.cej.2019.123925.

What Is Cancer? NCI. Iobenguane. Wikipedia. 2007. https://www.cancer.gov/about-cancer/understanding/what-is-cancer. Accessed 12 Sep 2024.

Wu X, Wu M, Xiaojun Zhao J. Recent development of silica nanoparticles as delivery vectors for cancer imaging and therapy. Nanomedicine. 2014;10:297–312. https://doi.org/10.1016/j.nano.2013.08.008.

Xu S, Zhu X, Zhang C, et al. Oxygen and Pt(II) self-generating conjugate for synergistic photo-chemotherapy of hypoxic tumor. Nat Commun. 2018;9:2053. https://doi.org/10.1038/s41467-018-04318-1.

Xu D, Song X, Zhou J, et al. Virus-like hollow mesoporous silica nanoparticles for cancer combination therapy. Colloids Surf B Biointerfaces. 2021;197:111452. https://doi.org/10.1016/j.colsurfb.2020.111452.

Xu B, Li S, Shi R, Liu H. Multifunctional mesoporous silica nanoparticles for biomedical applications. Signal Transduct Target Ther. 2023;8:435. https://doi.org/10.1038/s41392-023-01654-7.

Yang Y, Sun W. Recent advances in redox-responsive nanoparticles for combined cancer therapy. Nanoscale Adv. 2022;4:3504–16. https://doi.org/10.1039/D2NA00222A.

Yi G, Son J, Yoo J, et al. Application of click chemistry in nanoparticle modification and its targeted delivery. Biomater Res. 2018;22:13. https://doi.org/10.1186/s40824-018-0123-0.

Yin H, Yan Q, Liu Y, et al. Co-encapsulation of paclitaxel and 5-fluorouracil in folic acid-modified, lipid-encapsulated hollow mesoporous silica nanoparticles for synergistic breast cancer treatment. RSC Adv. 2022;12:32534–51. https://doi.org/10.1039/D2RA03718A.

Yong K-T, Qian J, Roy I, et al. Quantum rod bioconjugates as targeted probes for confocal and two-photon fluorescence imaging of cancer cells. Nano Lett. 2007;7:761–5. https://doi.org/10.1021/nl063031m.

Zayed DG, AbdElhamid AS, Freag MS, Elzoghby AO. Hybrid quantum dot-based theranostic nanomedicines for tumor-targeted drug delivery and cancer imaging. Nanomedicine. 2019;14:225–8. https://doi.org/10.2217/nnm-2018-0414.

Zeng X, Yan S, Di C, et al. "All-in-one" silver nanoprism platform for targeted tumor theranostics. ACS Appl Mater Interfaces. 2020;12:11329–40. https://doi.org/10.1021/acsami.9b21166.

Zhan H, Wang Y, Tang Z, et al. Exploring the influence of intermolecular hydrogen bonding on the fluorescence properties of HQCT and HQPH fluorescent chemosensors. Spectrochim Acta A Mol Biomol Spectrosc. 2022;280:121537. https://doi.org/10.1016/j.saa.2022.121537.

Zhang Y, Liang Y. Fabrication of folic acid-modified bovine serum albumin cloaked dual-drug loaded hollow mesoporous silica nanoparticles for pH-responsive and targeted delivery of gastric cancer therapy. Heliyon. 2024;10:e29274. https://doi.org/10.1016/j.heliyon.2024.e29274.

Zhang S, Malik S, Ali N, et al. Covalent and non-covalent functionalized nanomaterials for environmental restoration. Top Curr Chem (Z). 2022;380:44. https://doi.org/10.1007/s41061-022-00397-3.

Zhao L, Kim T-H, Kim H-W, et al. Enhanced cellular uptake and phototoxicity of Verteporfin-conjugated gold nanoparticles as theranostic nanocarriers for targeted photodynamic therapy and imaging of cancers. Mater Sci Eng C Mater Biol Appl. 2016;67:611–22. https://doi.org/10.1016/j.msec.2016.05.047.

Zhao S, Yu X, Qian Y, et al. Multifunctional magnetic iron oxide nanoparticles: an advanced platform for cancer theranostics. Theranostics. 2020;10:6278–309. https://doi.org/10.7150/thno.42564.

Zhou X, Cheng X, Feng W, et al. Synthesis of hollow mesoporous silica nanoparticles with tunable shell thickness and pore size using amphiphilic block copolymers as core templates. Dalton Trans. 2014;43:11834–42. https://doi.org/10.1039/C4DT01138D.

Zhu Y, Bai Y, He J, Qiu X. Advances in the stimuli-responsive mesoporous silica nanoparticles as drug delivery system nanotechnology for controlled release and cancer therapy. 3 Biotech. 2023;13:274. https://doi.org/10.1007/s13205-023-03651-7.

Živojević K, Mladenović M, Djisalov M, et al. Advanced mesoporous silica nanocarriers in cancer theranostics and gene editing applications. J Control Release. 2021;337:193–211. https://doi.org/10.1016/j.jconrel.2021.07.029.

Semiconductor/Non-semiconductor-Based Core-Shell Nanoconstructs for Cancer Theragnostics

17

Sunita Chaudhary, Mehak Bhat, Nilam Patel,
Ankit Chaudhary, and Jayvadan K. Patel

Abstract

A recently established area of study called nanotechnology aims to create instruments and tools that have precise functions at the cellular, molecular, and atomic capacities. An all-inclusionary size of these materials ranges between 01 to 100 nm. Nanomedicine is the term used to describe the employment of nanotechnology in biomedical research and clinical practice. Theragnostic platforms for cancer have been extensively researched and developed using nanoconstructs made of natural and synthetic polymers. These nanoconstructs come in various structural conformations, including vesicles, hydrogel, micelles, nanoparticles, and hyperbranched polymers. Nanomedicine offers methods for more effective diagnosis by combining developments in adjacent domains including biotechnology, pharmaceutics, and nanotechnology. This results in the identification and characterization of extremely early (pre-evidential) non-invasive diseases and the provision of assessment. The use of paramagnetic iron oxide particles was one of the first uses of nanotechnology in magnetic resonance imaging. When these particles are absorbed by healthy hepatocytes, they may aid in the differentiation of healthy from malignant liver cells. Effectiveness of drug can be enhanced by nanotechnology approaches like increase in surface area, acceler-

S. Chaudhary (✉) · M. Bhat
Arihant School of Pharmacy and Bio Research Institute, Adalaj, Gandhinagar, Gujarat, India

N. Patel
SAL Institute of Pharmacy, Ahmedabad, Gujarat, India

A. Chaudhary
Saraswati Institute of Pharmaceutical Sciences, Chiloda, Gandhinagar, Gujarat, India

J. K. Patel
Viesain Pharma LLC, GA, USA

Faculty of Pharmacy, Sankalchand Patel University, Visnagar, Gujarat, India

J. K. Patel et al. (eds.), *Core-Shell Nano Constructs for Cancer Theragnostic*,
https://doi.org/10.1007/978-981-96-3025-7_17

ates the rate of disintegration and dissolution. New formulations of nanoparticles will have longer shelf lives and better stability; developing nanoparticle formulations that enhance the absorption of insoluble substances and macromolecules allows for enhanced bioavailability and release rates, which may result in a decrease in the dosage needed and, consequently, fewer adverse effects. Sustaining release characteristics in nanoparticle formulations can enhance patient adherence to medication and dosing schedules. Targeted delivery of drugs using ligand-conjugated nanoparticles.

Keywords

Theragnostics · Semiconductor · Non-semiconductors · Cancer treatment

1 Introduction

1.1 Cancer Theragnostics

Theragnostic platforms, which merge therapeutic and investigative skills to create a single player, provide accurate ailment region location recognition, drug distribution and retention monitoring, and treatment outcome visualization (Yadav et al. 2022; Suresh et al. 2023).

In essence, the foundation of targeted medication delivery is in the capacity of a therapeutic agent to concentrate inside the intended tissue or organ in a selective manner, regardless of the administration site or method. To avoid the development of the therapeutic agent's adverse side effects in such circumstances, the local accumulation of therapeutic agents at the sick location needs to be raised, while the concentration in the other non-target organs and tissues should be low. The pharmaceutical carrier, the targeting moiety, and the therapeutic substance make up the three primary parts of a targeted drug delivery systematization (Yadav et al. 2022; Aftab et al. 2018; Box et al. 2010).

Theragnostic nanomaterials are extremely useful in the simultaneous diagnosis and treatment of all diseases. Additionally, they offer feedback methods to specify the release, location, or effectiveness of the therapeutic drug that has been included. Furthermore, the multimodal study of agent biodistribution is made possible by the coupling of medicinal moieties with imaging (Allen and Cullis 2004; Bae and Park 2011).

Several diagnostic modalities, including magnetic resonance imaging (MRI), computed tomography (CT), single-photon emission computed tomography (SPECT), near-infrared fluorescence (NIR), positron emission tomography (PET), and photoacoustic imaging, have been used to detect cancer up to this point in time. Conversely, treatment approaches include radiation, gene therapy, photodynamic therapy (PDT), immunotherapy, photothermal therapy (PTT), chemotherapy, and hyperthermia, among others. Therefore, theragnostics for cancer combines two or more of these modalities onto a single platform (Fig. 17.1) (Kiessling et al. 2014).

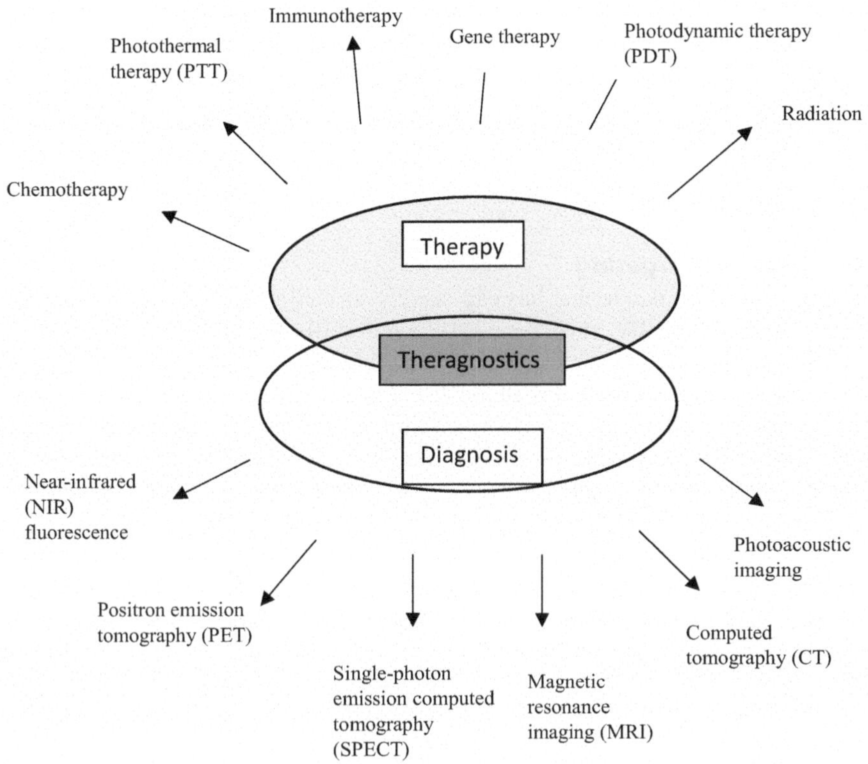

Fig. 17.1 Modalities of diagnosis and therapy in cancer theragnostics

Theragnostic platforms for cancer include metals, polymers, dendrimers, liposomes, proteins, carbon nanotubes, and quantum dots, among other nano-formulations (Murray et al. 2000). Polymers are one type of nano-formulation that have garnered a lot of attention because of their extended blood circulation time, high drug loading capacity, outstanding biocompatibility, and good biodegradability (Aftab et al. 2018).

IV and oral routes are examples of traditional medication delivery techniques. These approaches have a number of drawbacks. For exemplification, oral administration of the tablets and or either of the capsules may cause disordered pharmacokinetics since these compounds are exposed to the body's metabolic pathways. Oftentimes, the conventional IV approaches present even more challenges. Certain common intravenous medications have low specificity, which might lead to adverse effects on healthy tissues. To address some of these issues, biodegradable polymers are used in nanoparticle drug delivery, which offers a more effective and safe approach (Chai et al. 2017; Danhier 2016; Davies et al. 2013).

One way to modify the qualities of a polymer-drug conjugate model is to alter the polymer's chemical and physical characteristics. For instance, by adding solubilizing elements to the polymer, an insoluble medication can be transformed into a water-soluble one, enhancing its bioavailability and biodegradability (Moti et al. 2021) There are two main approaches to get the medication into the target tissue.

2 Drug Delivery Systems

2.1 Non-autonomous

2.1.1 Passive Targeting

A leaky vasculature is one that has excessively formed new blood vessels that are unstable, hyperpermeable, and have poor blood flow, further promoting hypoxia and the spread of disease. Vasculature is made up of vessels with varying morphologies and functions, and it distributes blood throughout all tissues while maintaining homeostasis of physiologic tissue. Solid tumors have specific characteristics in addition to their impaired lymphatic outflow (Natfji et al. 2017). The variability of the enhanced permeability and retention i.e. EPR affects the therapeutic potency of passively targeted NC both inside and between different tumors (Fabian et al. 2008; Li et al. 2016; Perry et al. 2017).

Due to this particular feature of cancer, passive targeting aids in the buildup of nanoparticles or nanocarriers within the tumor. Moreover, drug delivery to the tumor bed using a variety of invasive techniques is included in passive targeting. Certain medications can be given as prodrugs or inactive medications that, when they come into contact with tumor environs, will turn on and become intensely functional (Danhier 2016; Davies et al. 2013).

2.1.1.1 Leaky Vasculature

Many polymer nanoparticles exhibit increased permeability and retention because of the tumor site's leaky vasculature. Two elements underlie the phenomenon of improved permeability and retention: (a) Compared to the capillary-endothelium inside normal tissues, the endothelium of malignant tissue is highly invasive and hence higher penetrable to macro-molecules. This permits circulating polymeric nanoparticles to extravasate into the tumor interstitium. (b) Drug accumulation occurs due to the tumor bed's lack of tumor lymphatic outflow (Florence 2012; Garbayo et al. 2020).

If the chemotherapeutic agent is linked to a suitable polymer or another molecular carrier by a degradable linker, these carriers can increase the concentration of therapeutic substances in the tumor tissue. These characteristics allow polymer-drug conjugates to be found in tumor tissues in amounts 10–100 times higher than when the mediation is administered free.

2.1.1.2 Tumour Microenvironment

The tumor environment is utilized in this type of passive medication targeting. Prodrug therapy activated by the tumor cells refers to the administration of a drug that has been coupled with a molecule specific to the tumor and is delivered in an active state. Upon reaching its target, the tumor environment can transform the drug into a volatile and active substance (Golombek et al. 2018; Gould et al. 2015).

2.1.1.3 Local Drug Application

By applying the medication locally, tumor tissue can receive it directly without going through systemic circulation. These methods necessitate exposure to greater anticancer agent concentrations, which is not always practical. Intrathecal administration of drugs for localized distribution is a viable strategy that has been proven to work (Nichols et al. 2012). When medications are directly injected into tumor tissue, they are not able to circulate throughout the body and become accessible for metabolism by different systems. Direct drug inoculation into tumors has several drawbacks, including the potential for high levels of invasiveness, the impossibility of localizing tumors, and potential accessibility issues with specific malignancies, such as lung cancer (Gould et al. 2015; Heldin et al. 2004).

2.1.2 Active Targeting

To prioritize minimizing harm to adjacent healthy cells and maximizing bioavailability by reducing accumulation at non-target tissue, targeting moiety addition to nanoconstructs typically involves an active targeting strategy (Heldin et al. 2004; Jain and Stylianopoulos 2010) The following are the most difficult aspects of active targeting for cancer aid:

(a) Determining which receptors or surface molecules are overexpressed by cancer cells.
(b) Creating a targeted moiety whose dissociation constant is low.
(c) By employing a ligand that, upon the creation of a ligand-receptor complex, has a strong affinity for the target receptor and the potential to internalize the receptor.

The therapeutic drug or probes that are loaded onto the developed ligands that meet the aforementioned conditions are then delivered to the tumor microenvironment. The development, multiplication, together with survival of cancer cells are largely determined by the overexpressed receptors on the cell (Kohn et al. 1992).

By combining the targeting moiety with the therapeutic agent-containing nanoconstruct, one can improve the latter's tumor-specificity and decrease nonspecific interactions, which lowers toxicity, improves accumulation in tumor locations, and boosts anticancer activity.

Conjugates that have saturated surface receptors are less likely to localize to the intended site, which enhances nonspecific binding. For example, significant conjugate accumulation in the interstitial space might result from a lesser surface density of receptors caused by endocytosis of receptors, which can result in toxicity or nonspecific attachment to adjacent good tissue (Kohn et al. 1992; Lammers et al. 2012).

The efficacy and longevity of the cargo to be delivered upon internalization present another challenge for the actively targeted conjugates. Linkers come in two varieties: Cleavable and non-cleavable. The release of the cargo is dependent on the kind of linker. A linker needs to remain stable throughout the plasma for a long duration of time and cleavable upon internalization in order to be targeted (Lammers et al. 2012; Lazarovits et al. 2015).

2.1.2.1 Carbohydrate-Directed Targeting

Carbohydrates on the cell surface have an impact on how tumor cells interact with the extracellular matrix of regular cells during the growth and propagation of metastases. Tumor cell carbohydrates and the lectin proteins that attach to them can mediate these interactions. Some lectins are involved in both innate and adaptive immunity because they can identify malignant cell's "foreign" cell-surface carbohydrate patterns (Longmire et al. 2008). Research has demonstrated that lectins impact various processes essential for the growth and spread of metastases, including tumor vascularization, adherence to the extracellular matrix or endothelium, and tumor cell survival. Because they target entire organs, the medication delivery systems that have been created thus are based on this unique interaction between carbs and lectins, which may be detrimental to healthy cells. Notwithstanding these challenges, lectins are currently being studied with the objective of developing "smart carrier" molecules that will transport medications, and their unique affinity for sugar moieties on the surface of tumor tissue makes them an appealing option for further enhancing nano-drug delivery (Maeda 2015; Maeda and Matsumura 1989).

2.1.2.2 Receptor- and Antigen-Directed Targeting

Receptor-mediated endocytosis facilitates effective absorption because human malignancies overexpress receptors or antigens. This procedure allows extracellular particles to enter the environment inside cells. Generally speaking, ligand-receptor interactions allow the medicine attached to a polymer carrier to enter the target cell. After internalization, the selected drug-polymer carrier complexes can operate cytosolically along the plasma membrane or at the cell surface. The free drug is released into the cytosol when the drug dissociates from its polymer into the extracellular region, on the cell surface, or—more crucially—in lysosomes via the action of lysosomal enzymes (Suresh et al. 2023; Maeda and Matsumura 1989; Natfji et al. 2017; Nichols and Bae 2012).

Three key molecules are involved in this type of drug delivery:

(a) polymers, with which the medication can be linked,
(b) antibodies or ligands attached to the polymers, which bind highly affinity for the surface of tumour cells;
(c) receptors or antigens.

2.2 Autonomous Drug Delivery System

Targeted delivery of drugs aims to minimize nonspecific drug dispersion while optimizing drug administration to the intended areas. A perfect drug delivery system also makes sure the medication is delivered and accessible at the disease site for the right amount of time throughout the therapeutic window (Perry et al. 2017).

Although targeted nanoparticles, both passive and active, have made substantial progress in these areas, their effectiveness is still limited by a number of issues, including their inability to effectively cross several biological barriers and their inability to be cleared from the bloodstream.

2.2.1 Classification

2.2.1.1 DNA Nanotechnology

Biological applications involve designing and producing three dimensional nuclic-acid based nanotechnology platforms with tailored size, shape, and spatial address-ability by utilizing the DNA molecule's natural capacity to sense and respond toward environmental stimuli. Yin et al. have built the first DNA walker that is autonomous (Pickup et al. 2014). A DNA walker that can move in different directions on a substrate track that is complimentary was created by Lund et al. Gu et al. have presented a third component, a cargo-giving device that can be programmed, which may hold or donate the nanoparticle payloads to the passing walker, along with the origami track and the DNA walker.

These DNA nanomachine systems represent a significant turning point in the development and use of DNA nanotechnology as they were some of the first attempts at autonomous mobility and cargo delivery. The first instance of in-vivo targeted medication delivery for cancer treatment using self-assembled DNA nanotubes was described by Li et al. After being evaluated, it was found that the DNA nanotubes were able to concentrate in the tumor vascular endothelium area and release thrombin to produce thrombosis. Additionally, after 72 h of treatment, no histological anomalies were seen in the healthy organs of the tumor-bearing animals. It has been effectively shown to be a sophisticated DNA-based delivery system that can target tumors and identify their signals. It also has the ability to self-transform for the delivery of payloads (Rajora et al. 2014).

When compared to alternative delivery technologies, DNA origami has a number of benefits overall. These consist of their widely used self-assembly properties, homogeneity in size and form, and simplicity in functionalization with proteins or other biomolecules that have medicinal promise. One example is the introduction of a pH-responsive i-motif switch, which, when introduced into environments with varying pH levels, triggers a DNA structural alteration that ultimately results in drug release. The DNA nanostructures encounter various hurdles, despite the fact that their beneficial properties are crucial for the efficient and secure delivery of cancer-fighting drugs. First, there is still much to learn about DNA nanostructures' immunogenicity. According to reports, the innate immune system may quickly eliminate the DNA nanostructures if they cause an unfavorable immunological response (Palmerston et al. 2017).

Second, nuclease digestion following injection could pose a danger to the stability of DNA nanostructures. Third, there are challenges with producing DNA origami structures due to their high cost, low yield, and intricate purification procedures, particularly when large-scale production is needed for therapeutic applications. Finally, more in vivo research on the biodistribution, pharmacokinetics, and pharmacodynamics of DNA nanostructures is required to enhance their design and strike the ideal mix of effective administration and enough in vivo retention duration (Sarin 2010; Saw et al. 2021).

2.2.1.2 Bacterial Sensing

In addition to having the capacity to propel themselves forward for improved penetration into deep tumor locations, bacterial cells also exhibit the remarkable ability to perceive environmental stimuli and react appropriately. Using bacteria's molecular machinery to convey drug-loaded nanoparticles and capitalize on their skills could be a viable approach to creating an intelligent drug delivery system. Bacterial flagella, which are powered by the chemical energy within their cells, allow them to swim effectively and travel in the direction of favorable conditions. Bacterial-based nanorobots have significant prospects as an intelligent, active carrier system that can precisely locate tumors and provide therapeutic dosages of drugs. Tumor-targeting bacteria have been evaluated as treatment agents in various clinical trials (Saw et al. 2021).

2.2.1.3 Biomimetic Nanoconstructs

Through well-established top-down approaches, diverse types of cells, inclusive of malignancies, leukocytes, erythrocytes, and stem cells, have been created as drug carriers. This technique directly mimics the physiological responses of real cells in vivo by using cell membranes as a nanoparticle covering. Red blood cell (RBC) membranes, for instance, offered advantages of biomimetic nanoparticles like extended blood circulation times, biocompatibility, and large cargo volumes for various drug types. Similarly, macrophage membrane-coated nanoparticles demonstrated extended blood circulation times, but they were also better able to penetrate deep tumor regions and cross vascular barriers. Macrophage surface proteins have the ability to actively target the tumor endothelium. Additionally, these nanoparticles were protected from immunological detection by macrophage membranes, which also decreased the rate of their removal by liver metabolism and renal excretion (Pickup et al. 2014; Saw et al. 2021).

Cancer cell membranes' innate capacity for proliferation, capacity to evade the immune system, and capacity for homologous adhesion have made them desirable options for oncological applications. Coated nanoparticles with cancer cell membranes have shown multitherapeutic transport of encapsulating payloads with complementary therapeutic effects. In order to overcome hypoxia-induced chemoresistance, Tian et al. used MCF-7 cell membranes to encase PLGA nanoparticles loaded with hemoglobin and doxorubicin (DOX). Sun et al. surface-coated gold nano-cages containing 4 T1 cell membranes loaded with doxorubicin (DOX) for photothermal and chemothermal therapy.

These examples demonstrate that the application of biomimetics in drug delivery nanoconstructs offers a number of benefits, such as immune evasion capabilities, extended blood circulation times, and improved ability to get past biological barriers (Saw et al. 2021).

2.2.1.4 External Stimuli

Based on the widely recognized classification scheme, these nanocarriers can be broadly classified into two groups: (a) catalytic type nanocarriers and (b) nanocarriers driven by biocompatible external energy sources, primarily magnetic and ultrasound energy. The majority of catalytic carriers continuously create and release bubbles into the fuel as a result of converting chemical energy into kinetic energy. Numerous investigations have been carried out to determine the variables, including temperature, viscosity, and concentration. Enhancing the temperature or concentration of the fuel, while, lowering its viscosity can help increase the catalytic carriers' effectiveness and impulsion speed (Saw et al. 2021; Wheate 2012).

Furthermore, the geometric configurations of nanocarriers were investigated in order to enhance their velocity via pressure distribution and flow field modification. With the realization that toxicity concerns and fuel sources would restrict the uses of catalytic nanoparticles, research efforts have gradually shifted to the creation of natural fuel-driven nanocarriers.

Metallic nanoparticles and nanoalloys are the two main categories of magnetic nanoparticles used in medication delivery. To create nanoalloys, the magnetic components—which are often composed of ferric oxide, cobalt, or nickel—are mixed with elements like copper, zinc, strontium, or barium (Mura et al. 2013). Most magnetic-based drug delivery methods are guided to their intended site by external magnetic field sources, such as permanent magnets, electromagnetic coils, and clinical magnetic resonance imaging, after injection into the circulation. Research has indicated that drug-loaded magnetic nanoparticles in blood arteries can be drawn into tumor tissues by the external magnetic field (Saw et al. 2021; Wilhelm et al. 2016a).

All things considered, the benefits of "fuel-free" and biocompatible nanocarriers driven by magnetic or acoustic energy are comparable. Owing to their different modes of action, magnetized nanocarriers have been widely used for drug administration in extracellular and blood vessel environments, whilst acoustically assisted nanocarriers have been more frequently utilized in intracellular activities.

3 Nanoconstructs

Engineered magnetic nanoparticles (MNPs), one of the nanostructures that has garnered a lot of attention lately, are a cutting-edge tool in medicine because of their ability to be guided and functionalized by a magnetic field simultaneously. This allows for integrative theranostic applications, which include both therapeutic and diagnostic ones. MNPs are useful in tissue engineering, cell tracking, hyperthermia-based cancer therapy, guided medication and gene delivery, enhanced magnetic resonance imaging, and bioseparation. It is an ambitious goal to develop medicine delivery systems that are selective to diseased locations. As a cancer treatment, the concepts of magnetic guiding of MNP-conjugated pharmaceuticals have been used in experiments and are now in clinical studies. After being injected intravenously, MNPs are concentrated at a particular target spot using an external magnetic field—a technique that cancer patients find to be well-tolerated (Frickenstein et al. 2021; Yong et al. 2020) Delivery through the blood-brain barrier, an insurmountable barrier to the treatment of neurological illnesses, may be overcome via medication and gene delivery systems based on nanoparticles. However, there are still problems with possible embolization with the MNP aggregates in capillaries and the requirement for great distances between the diseased site and an external magnetic field. Recent developments in materials science techniques have accelerated the creation of core-shell nanostructures with customized surface properties, such as core-shell nanoparticles and core-shell nano-fibers, opening up new avenues for their application in a variety of fields (Yu et al. 2018).

When therapeutic agents or physiologically active molecules are encapsulated in core-shell nanostructures, it can help increase the bioavailability of medications with low water solubility. Additionally, this can encapsulate unstable bioactive molecules to protect their biological activity; avoid ruptured release, which might result in toxicological effects; create a sustained and extended-release; load multiple drugs concurrently within the core as well as shell region of the nanostructure to generate a distinct multidrug release; and create multifunctional nanostructures thus producing components for "theragnostic." The convergence of the domains within toxicology, bioengineering, and biomedicine will keep encouraging the creation of pertinent techniques for designing cutting-edge nanodevices with biocompatible interfaces (Zhang et al. 2008; Cho et al. 2023).

It is possible to formulate or create core-shell nanoparticles in a variety of forms, including spheres, tubes, nano-wires, rods, rings, and stars. NP systems, however, are primarily spherical, such as hollow spheres (where the core material has been entirely removed) and nanoparticle systems with spherical core-shells that have a thick, porous, or rattling core. On the other hand, the shell can also be particulate (fragmented), continuous and dense, or continuous and porous (Moti et al. 2021; Youn and Bae 2018) Additionally, intricate multiple-layered NPs have been created.

Numerous techniques have been developed for the production of core-shell nanoparticles, including metal reduction, layer-by-layer (LBL) processes, coprecipitation, sol-gel, etc. Each of these strategies offers pros and cons in addition to application-dependent details. The main issue is preserving the management of size distribution, composition, morphology, shell thickness, and core size (Kobayashi et al. 2014; Sykes et al. 2014).

3.1 Classification of Core-Shell Materials and their Surface Functionalization

3.1.1 Mesoporous Silica Nanocarriers (MSN)

In the human body, silicon is a significant (naturally occurring) essential element that is primarily acquired through diet. By synthesizing and/or stabilizing collagen, silicon contributes to the development of bones. When it comes to the structure and durability of connective tissues and bone, collagen plays a crucial role in animals. As the most prevalent protein in the matrix of bones, it provides flexibility and, in combination with elastin, is a key constituent of connective tissues, which are present in tendons, arteries, cartilage, and skin (Huo et al. 2013; Cabral et al. 2011; Li et al. 2016).

Mesoporous silica nanoparticles (MSNs) are a specialized form of silica nanoparticles characterized by their large surface area and highly organized porous structure. To improve localized concentration and enable synergistic cancer therapy and diagnostics (theranostics), their porous features can be utilized to load a variety of therapeutics (Chen et al. 2019; Mekaru et al. 2015). These include nucleotides, targeted molecular drugs, water-soluble or insoluble drugs, imaging agents, and various combinations thereof, all of which can be simultaneously released within cancer tissues.

Mesoporous silica nanoparticles (MSN), in particular, are a type of silica nanoparticle with a large surface area and well-structured porosity (Martínez-Carmona et al. 2020). In order to improve local concentration and accomplish integrated cancer therapy and diagnostics (theranostics), a number of therapeutics, such as nucleotides, targeting molecular drugs, water-insoluble or soluble drugs, imaging agents, and their various combinations, can be loaded into the porosity features and released concurrently within the cancer tissues (Black et al. 2014; Shah et al. 2011; Smith et al. 2012). MSN size is critical for biodistribution, penetration into tissues, internalization of cells, and blood circulation. Namely, the reticuloendothelial system (RES) and renal clearance are the two primary routes by which injected MSNs are eliminated.

Additionally, MSNs can be made to release loaded cargo under controlled conditions in response to external stimuli focused on cancer tissue, such as light, ultrasound (US), magnetic fields, or other biomolecules (Chauhan et al. 2011; Huang et al. 2010; Palmerston Mendes et al. 2017) They can also be made to release cargo when exposed to the extracellular environment of cancer tissues, which is weakly acidic, or when enzymes or other biomolecules overexpressed in cancer are present (Yingchoncharoen et al. 2016; Liu et al. 2018a). Functionalization of the exterior MSN surface can be used as moieties to improve the biocompatibility of nanomaterials, to actively target cancer, or to go beyond the immune system to achieve extended retention of nanocarriers in the bloodstream (Yun et al. 2015; He et al. 2010).

The achievement of efficient passive targeting is also influenced by the surface characteristics of MSNs. Specifically, the mononuclear phagocyte system (MPS) has the ability to swiftly opsonize systems that are excessively hydrophobic or charged. The duration (prolongation) of these NPs' circulation in the bloodstream and the reduction of renal clearance can both be controlled by altering the surface characteristics of MSNs (Mura et al. 2013; Erthal et al. 2021).

3.1.1.1 Different Functional Groups Can Be Functionalized on the MSN Surface (Fig. 17.2)

Peptides Integrins are heterodimeric cell surface receptors that mediate adherence to molecules of the immunoglobulin superfamily and extracellular matrix (ECM). Because of their potent and precise interaction with α vβ3 integrins, peptides find

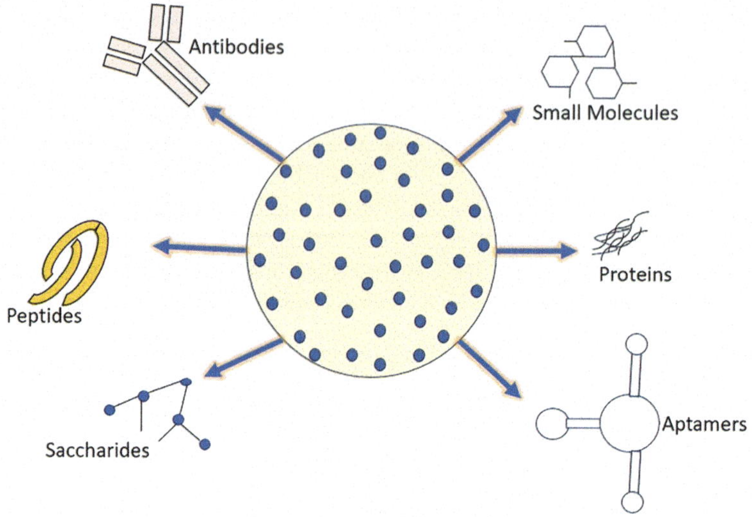

Fig. 17.2 Specific surface functionalization strategies

extensive application. It is commonly known that they play a crucial role in the development, progression, and metastasis of solid tumors by regulating a wide range of cellular activities (Anselmo and Mitragotri 2016; Tang et al. 2012; Sweeney et al. 2017) The integrins α vβ3, a4β1, and α α vβ5, a5H1, a6β4, and vβ6, whose overexpression is correlated with disease development in a variety of tumor forms, are the most studied in cancer research.

V7 is a low pH insertion peptide (pHLIP) that can also be used to target cancer. It is not preferred for pHLIP-functionalized NPs to be endocytosed unless the peptide's Asp residues are protonated in the tumor microenvironment's mildly acidic pH (Anselmo and Mitragotri 2016; Allen et al. 2021; Zhang et al. 2015).

Proteins Tumor cells can be targeted by proteins such as transferrin, which aids in the uptake of iron into cells, and epidermal growth factor (EGF), which promotes cell division and proliferation. Transferrin is a plasma iron-binding protein that attaches to TFR1 and TFR2, also known as transferrin receptors (TFRs). Whereas TFR2 is mostly expressed in the liver and has its levels controlled by transferrin saturation, TFR1 is widely expressed and its levels are determined by cellular iron levels. The literature refers to TFR1 as a receptor that is overexpressed in a variety of cancer types (Allen et al. 2021; Li et al. 2020; Chen et al. 2016).

Antibodies IgG is the primary antibody utilized for cancer therapy and diagnosis among the serum's circulating antibodies (IgA, IgM, IgE, and IgG). Its structure is Y-shaped, with effector actions being brought about by the stem structure and antigen recognition sites located in both arms. A complex MSN-based nanoconstruct containing TRC105 antibody for targeting tumor neovasculature (64 Cu-CuS@ MSN-TRC105), chelated 64Cu for PET imaging, and PEGylated core-shell CuS@ MSN for photothermal therapy (PTT) was shown to exhibit in vivo long term biological compatibility at an elevated dose of 90 mg kg^{-1} on tumor-bearing mice (Senapati et al. 2018; Raza et al. 2018). Two months following the intravenous injection, tissues from the kidney, lung, spleen, liver, and heart were analyzed for potential toxicity markers; however, no histological differences between the control and the treatment group were observed. Since the enzyme carbonic anhydrase (CA) IX is expressed on the surface of cancer cells and is not abundant in healthy tissues, an antibody against it can also be used to target cancer (Liao et al. 2020; Allen et al. 2021; Chen et al. 2020).

Aptamers Since aptamers work similarly to normal antibodies, they are often referred to as "chemical antibodies." Aptamers, on the other hand, are much smaller, can be made much more quickly and cheaply, can be chemically changed in a variety of ways, have a flexible structure, and have a low immunogenicity. Because aptamers internalize when they connect to ligands on the cell surface, they are an effective tool for delivering specific drugs to the right cells. Li et al. created anti-miR-155-loaded MSNs that were specifically designed to treat colorectal cancer. They were modified with polydopamine (PDA) and AS1411 aptamer (Wilhelm et al. 2016b; Liu et al. 2018b; Allen et al. 2021).

Tang et al. have developed a system for photo-reactive delivery of DOX based on the non-covalent binding of cancer-targeting Ci5.5-AS1411 aptamer onto the graphene oxide (GO), which is also used as the light-responsive pore-blocking constituent for the retention and administration of DOX (Liu et al. 2021a; Manzano and Vallet-Regí 2020).

Polysaccharides They are being used more and more to adorn MSNs for active cancer targeting as hydrophilic, biodegradable agents. The most typically utilized polysaccharides for this purpose are hyaluronic acid (HA), dextran, and chitosan; the latter three are frequently cited in the literature because of their unique interactions with CD44 and CD168 receptors, which are involved in a variety of cancer types (Liu et al. 2021b; Jugdaohsingh 2007; Knežević et al. 2013).

There have also been reports of core–/shell magnetic MSNs with black iron oxide-Platinum (IV) prodrug containing DOX (Wei-Hai Chen et al. 2016) and Iron oxide-gold bifunctional cores, photosensitizer chlorine e6 (Ce6), and DOX loaded into the pores. P-gp shRNA, a polyelectrolyte polylayer composed of alternating alginate and chitosan, was attached electrostatically to combat multidrug resistance while acting as pH-sensitive gatekeepers (Martínez-Carmona et al. 2020; Chen et al. 2019). Chemo-photodynamic coupled gene therapy exhibited synergistic antitumor effects in-vivo, while Fe_3 2+ O 4-Au in pairing with the core of these NPs allows for dual-modal imaging, including CT and MR imaging. FA-functionalized magnetic core/shell MSNs were developed using zinc-doped super-paramagnetic ferrous oxide (SPION) cores for concurrent MRI and tumor-targeted, pH-responsive

DOX administration (Knežević et al. 2019; Knežević and Kaluđerović 2017). In T2-weighed MRI, these improved saturated mass magnetization values offered a significant contrast capability.

A Quantum Dot (QD) yolk/shell-structured hollowed mesoporous silica nanosystem was also employed for pH-sensitive drug administration, dual-modal PET/optical imaging, and tumor vascular targeting using a chelated 64 Cu radioisotope and the cancer-targeting TRC105 antibody (García-Fernández et al. 2020; Llopis-Lorente et al. 2017).

Additionally, for SPECT imaging, a nanocomposite based on MSNs was enhanced with 99 m Tc radioisotope, packed with DOX, and functionalized with aminopropyl groups, enclosed with negatively charged MUC1 aptamer for drug delivery regulated by breast cancer, was reported (Anselmo and Mitragotri 2016). Using DNAse I activity, the NPs were able to deliver DOX to tumor cells preferentially. SPECT imaging was used to assess the NPs' in vivo targeting capacity and biodistribution (Barui and Cauda 2020; Croissant et al. 2020; Alyassin et al. 2020).

In order to facilitate successful multimodal cancer imaging and therapy, The work of Goel et al. described hollow mesoporous silica (HMSN) radiolabelled with Zr-89 and loaded with meso-tetrakis(4 carboxyphenyl) porphyrin (TCPP) within the pores and CuS NPs ornamented the outside. While TCPP is a hydrophobic molecule with inherent fluorescence (FL) and the ability to pair with radionuclides with the aid of the Cherenkov radiation energy transfer (CRET) phenomenon, CuS nanocrystals exhibit excellent PTT capabilities. The resulting HMSNs employ in vivo image-guided therapy integrating on-demand PET/FL/CRET/CL imaging, while the loaded DOX allows for improved ROS generation and hyperthermia, making this nanoconstruct appropriate for chemotherapy in conjunction with synergistic PTT and PDT (Croissant et al. 2018; Lu et al. 2010; Zhang et al. 2015).

3.1.2 Gold-Based Hybrid NPs

Because of their superior optoelectronic and improved physicochemical qualities, drug therapy, diagnosis, biosensing, bioimaging, theragnostics, and hazardous chemical sensing are merely a few of the medicinal and environmental evaluation applications involving gold nanoparticles (Au-NPs). The formulated Au-NPs are modified with polymers, surfactants, ligands, medications, proteins, or oligonucleotides with the aim of getting their intended specificity, selectivity, and sensitivity for their multiple applications in diagnostics, prognostics, and therapeutics (Abalde-Cela et al. 2018). Because of their chemical inertness, environmental friendliness, and biocompatibility when functionalized with the right ligand or combination of ligands, AuNPs are regarded as a highly significant possibility.

Its enormous spectrum of clinical and biological applications can be attributed to its diverse tuneable optical, fluorescence, SPR, and magnetic characteristics. The other elements that have contributed to AuNPs' dependable commercial use include their increased stability, biocompatibility, selectivity, and decreased toxicity in biological environments. The increased qualities of AuNPs for various biomedical applications are mostly dependent on surface changes (Bhatnagar et al. 2018). To date, a variety of chemical and biological molecules have been employed to produce such functionalized AuNPs by adjusting the AuNPs' physicochemical characteristics, such as their surface charges and capacity to bind molecules. Many biomolecules can eventually be conjugated with the aid of these modifications of AuNPs, which are frequently carried out utilizing thiols (Huo et al. 2013), amines, phosphines, silica, carboxy-terminated groups, etc.

Most frequently, one or more plasmonic particles—which can themselves be made of various composites (silica/gold, gold/silver, etc.)—are the basis for gold-based hybrid nanoparticles consisting of the components that allow diagnostic and therapeutic capabilities. A dielectric shell or coating, such as silica or biopolymeric, envelops these particles. A variety of molecules of the compounds being administered (drugs, peptides, siRNA, etc.) and/or reporter molecules (fluorescent, near-infrared (NIR), Raman, Pd, etc) can be doped into the outer coating. One significant kind of hybrid nanoparticles (HNPs) that enables the utilization of plasmonic and magnetic functions for therapeutic applications is a composite of gold and magnetic nanoparticles.

3.1.2.1 Surface Modification Strategies

Thiol-Protected AuNPs Formulated by Direct Conjugation

The thiol-based coupling has been widely used in the binding of linker molecules onto the surface of AuNPs to modify their surface. These modifications give the NPs control over their reactivity as well as the induction of hydrophobicity and hydrophilicity.

Secondary Alteration and Substitution for the Formation of Mixed Monolayer Au-NPs

By combining pre-synthesized AuNPs with end-group functionalized organothiols (ω-functionalized organothiols) in the presence of surfactants, this stabilization was brought about.

Thiol groups bond strongly to gold by covalent interaction since gold and sulfur are soft materials, which results in the formation of NPs.

The most popular techniques include chemical coupling and polymerization, where functionalization of AuNPs with carboxylic acid termination was mostly employed. Thiol ligands that create amide connections with linker molecules through N-Ethyl-N-(3 dimethylaminopropyl)-carbo di-imide coupling have also been employed in one of these instances. In addition to this, it has also been observed that NPs containing hydroxyl groups on their surface provide a variety of functions when an esterification reaction occurs in the presence of acyl moieties.

Polymer-Based Surface-Modifications

To increase stability, the polymers have also been utilized to cap the metallic NPs. Helcher made the first effort at polymer-stabilized AuNPs in 1718 in the presence of a polysaccharide; nevertheless, because of the restricted availability of characterization, these were not thoroughly and scientifically examined. Polymer-stabilized

Au-NPs offer numerous advantages, including increased stability, an elevated surface density, and the ability to adjust solubility, to name a few.

Surface Modification Using Amine

The Brust approach is frequently used to alter the surface of NPs, replacing current stabilizing thiol moieties with amine ligands. The primary benefit of amine-capped AuNPs is the ease with which a range of spectroscopic investigations can examine the surface chemistry of NPs. Hydroxylamine was first employed in the synthesis of amine-capped AuNPs as a reducing agent (Klochkov et al. 2021).

3.1.3 Carbon Nanotube-Based Hybrid NPs

Since carbon nanotubes (CNTs) exhibit special electrical, mechanical, structural, and adsorptive properties, they have garnered a great deal of attention for their many possible applications. Nevertheless, the creation of efficient CNT dispersions significantly impedes the use, growth, and implementation of CNTs. Proper surface modification of carbon nanotubes (CNTs) has garnered a lot of fascination in recent decades in an effort to overcome these restrictions and broaden the area of their application. This has contributed to the identification of several unique hybrid materials with desired applications. Many advanced improvements are made possible by the variable surface chemistry of carbon nanotubes (CNTs), which also improves their performance, sorption, selectivity, and excellent surface-to-volume ratios (Bianco et al. 2005).

CNTs can be seen as cylinders made by encircling a central hollow core structure with nanoscale graphene flakes rolled up.

The cylindrical nanotubes were usually capped with a fullerene hemisphere at least on one end. CNTs come in two varieties depending on how they were made (Fabian et al. 2008). The first type is called single layer single-walled carbon nanotubes (SWNTs), and they have a diameter of between 0.8 and 2 nm. They are made of carbon atoms arranged into seamless cylinders with one or more layers and either closed or open ends.

A two-dimensional honey-comb lattice constructed from tightly organized carbon atoms is termed graphene. It may be rolled into various shapes, including 1D carbon nanotubes (CNTs). Hydrophobic nano-structures known as carbon nanotubes (CNTs) are composed of either an individual graphite sheet (SWNT) or an axially stacked array of SWNTs, similar to tree trunk rings (MWNTs). CNTs are nanomaterials with excellent mechanical and optical characteristics. CNTs have several applications, inclusive of Raman and photoacoustic imaging, drug delivery, and photo-ablation treatment.

Carbon nanotubes are not suitable for application in biological systems owing to their limited solubility in both polar and non-polar organic compounds, short half-life of 3–3.5 h, coalescence into tightly packed bundles, bio-incompatibility, and immunogenicity. To be employed in biological systems, water solubility is an important factor for carbon nano-tubes. This allows for proper absorption, distribution, and removal, making them biocompatible and less harmful.

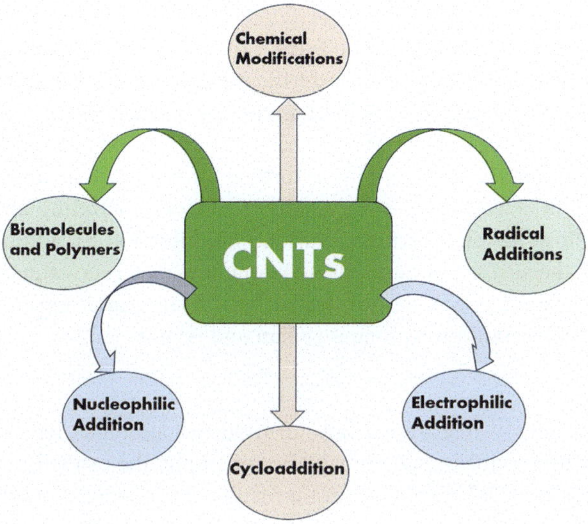

Fig. 17.3 Various surface modification strategies

CNTs are gaining popularity in medical and biological fields due to their unique topologies, wide surface areas, chemical functional areas, and nanoscale size continuity (Hutchison et al. 2001) (Fig. 17.3).

CNTs, in particular, are promising transporters for medication delivery and non-invasive therapeutic agents. CNTs can be employed as nanocarriers to transport anticancer drugs such as CP, DOX, CDDP, CPT, PTX, and DTX.

3.1.3.1 Chemical Modifications

Covalent bonds between the tiny molecules and the layers of nanotubes allow for the alteration of nanotubes. In most cases, pre-modification of nanotubes with small molecules is necessary for chemical modification using polymer molecules. So that covalent bonding can be used to graft polymer chains onto the anchor sites created during the pre-modification process. Furthermore, the interfacial interactions and, consequently, the compatibility of the fillers and matrix have an impact on the electrical, thermal, biological, and mechanical properties of manufactured composite materials.

An extremely well-known method for producing numerous oxygen-containing functional groups and then utilized for various covalent functionalizations is CNT modification by oxidation reaction. On the surface of nanotubes, carboxyl, epoxy, and hydroxyl groups are often created by various oxidation processes using oxygen, aqueous hydrogen peroxide, strong nitric acid, and sulfuric acid (Singh and Deshmukh 2022; Khan and Alamry 2022).

Amidation

The oxidized carbon nanotubes can be further modified by conducting other chemical reactions, such as amidation and esterification, to create water-soluble organic molecules based on conjugates, which include polymers, proteins, and nucleic acids. The COOH groups that connect the functional groups to the CNT surface provide active sites that can be modified further for multidimensional applications. Amidation of CNT is regulated by a variety of factors, including amine types and catalysts that allow for high levels of surface modification by adding active functional groups to the surface of nanotubes. Amine nitrogen is covalently grafted onto the nanotube surfaces, it was demonstrated by the structural analysis of amidated CNTs carried out by oxidative process (Khan and Alamry 2022).

Esterification

It has been extensively researched how to modify nanotubes by esterifying their COOH group with modifying chemicals. For example, the esterification of the porphyrin units and the CNT's COOH group results in a covalent bond that connects the end walls of the SWNT and produces molecules that are soluble in polar organic solvents. The process of esterifying oxidized carbon nanotubes (CNTs) in order to graft furfuryl alcohol groups onto their convex surface and end caps, which is then utilized to create novolac phenolic resin composites. The carbon yield value and thermal characteristics of the created composites are enhanced by the esterification-based grafting of furfuryl alcohol onto the oxidized CNT. Following a sol-gel reaction, TiO2-decorated MWNT and Pd particles are produced when an esterification reaction is triggered when CNT is exposed to the titanium isopropoxide, which encourages CooH and OH groups (Khan and Alamry 2022).

3.1.3.2 Cycloaddition

The process by which nitrenes ingest various alkylazido formates as precursors alters pure CNTs. Alkoxy carbonyl nitrene was created via the thermal breakdown of an organic azide. Alkoxy carbonyl aziridine-functionalized carbon nanotubes (CNTs) were created when the produced nitrene was affixed to the sidewalls of the nanotubes.

The attachment of pyridyl isoxazoline or 2,5-diaryl pyrazoline rings to the side walls of functionalized nanotubes via 1,3-dipolar cycloaddition of nitrile imines and nitrile oxide, respectively, to oxidized carbon nanotubes. By releasing the strain created by the functional sites, the cycloaddition folds the surface of CNTs.

A 1,3 dipolar cycloaddition process can covalently bond polymers and small molecules to the surface of carbon nanotubes. However, these changes require more time—many days—and are labor-intensive (Khan and Alamry 2022).

3.1.3.3 Radical Additions

It is possible to modify the surface of CNTs by introducing various radicals onto the sidewalls of the nanotubes. Using the radical addition technique, one can reduce and modify nanotubes using the Grignard reagent, the Bingel reaction, and reduction via alkali metals, such as organo-lithium reagents and perfluoropolyether radicals. The covalent connection of the aryl groups of aryl diazonium salts across the surface of CNT is provided by the radical addition method. The extent of functionalization and aniline-substituted derivatization determine the modified nanotubes' solubility. Polar solvents like methanol, tetrahydrofuran, and chloroform readily solubilize the aryl-functionalized CNTs thanks to their polar functional groups, carboxylic acid, and oligoethylene branches.

3.1.3.4 Biomolecules and Polymers

Biomolecules can be used to functionalize nanotubes in order to produce dispersive CNT with a biocompatible character. According to reports, the hydrophilicity and biocompatibility of the nanotubes are enhanced by the biomolecules. Amino acids, proteins, cytochrome c, polypeptides, enzymes, deoxyribonucleic acid (DNA), glucose dehydrogenase, horseradish peroxidase, and glucose oxidase can all be connected to carbon nanotubes (CNTs). By activating the COOH groups of carboxylated nanotubes, intensely oxidized carbon nanotubes can be associated with biomolecules via a carbodiimide chemical, allowing for direct interaction with the primary amines in biomolecules (Bianco et al. 2005; Khan and Alamry 2022).

The carboxylated carbon nanotubes' dispersibility in water and biocompatibility are improved by functionalizing them with polysaccharides. Polysaccharide modification of nanotubes has formed a stable dispersion of electrode materials for sensing applications.

Intense, a polyaniline-CNTs-starch nanocomposite has been designed to create a dispersible, biocompatible electrode material for the accurate measurement of hydrogen peroxide and glucose. Numerous synthetic methods, including ring-opening polymerization, atom transfer radical polymerization, and surfactant-assisted polymer wrapping, for the consistent integration of nanotubes in the polymer chain, two strategies have been found: polymerization of monomers in the presence of nanotubes and reversible addition-fragmentation chain transfer (Singh and Deshmukh 2022; Khan and Alamry 2022).

3.1.3.5 Nucleophilic Addition and Electrophilic Addition

It has also been documented that CNTs can be modified by electrophilic and nucleophilic carbene addition. A carbene derivative is added to CNTs to improve their dispersibility and manage their electrical characteristics. Pure nanotubes reacted with sec-butyllithium to address strong nucleophilic addition, and CO2 was then added to covalently functionalize both carboxylic and alkyl nanotube groups. Zeta potentiometric titration was used to confirm that the resulting nanotubes have properties similar to those of a zwitterionic polyelectrolyte. The modified carbon nanotubes (CNTs) show better dispersibility than the virgin ones and have potential applications in composite materials and molecular electronics. With the use of clay and aliphatic acid hydrothermally, side alkylated CNT-clay composites were produced.

Fractional exfoliation of the inorganic elements was demonstrated by the alkylated nanotubes intercalating within the clay layers by increasing the gap between the clay layers by approximately 2.5 nm (Singh and Deshmukh 2022; Khan and Alamry 2022).

Electrophilic substitution processes utilizing phenylated carbon nanotubes were employed to alter the sidewalls of carbon nanotubes. Four different forms of functionalized nanotubes, including thiophenol, benzyl chloride, benzene sulfonyl chloride, and diphenyl ketone-modified carbon nanotubes, were created via electrophilic substitution processes.Modified nanotubes based on benzyl and benzene sulfonyl chlorides can function as special initiators for surface-initiated atom transfer radical polymerization. The Pd metal and the thiophenol functionalized nanotubes demonstrate strong catalytic competence in the electrochemical oxidation of methanol. A sequence of alcohols and SWNTs through an electrophilic addition reaction causes various hydroxyl and alkyl groups to connect with one another on the SWNT surface. The functionalized nanotubes have better dispersibility than the unfunctionalized nanotubes in common organic solvents like ethanol and N, N-dimethyl formamide, which is important for enhancing their processability (Khan and Alamry 2022).

3.1.4 Quantum Dots

For the objective of biolabeling as well as bio-imaging, the medicinal drug's micelles can be anchored to quantum dots, organic dyes, or organic fluorophores.

Fluorescent semiconductor nanocrystals, or "quantum dots," are also known as Fluorophores that are inorganic. Organic fluorophores and quantum dots both absorb light at one wavelength and reemit it at a different one. Radioactive isotopes were a common option for biolabeling until quantum dots became available. While many applications still use radioactive isotopes, quantum dots have taken their place in many bioimaging applications. In actuality, problems in the body are treated as well as detected by radiation (Saw et al. 2021; Hola et al. 2014).

Carbon quantum dots are a potentially beneficial nanomaterial for cancer bioimaging and drug delivery by virtue of their excellent attributes, including good biocompatibility, fluorescence emission, minimal toxicity, compact size, high water solubility, and convenience of surface functionalization.

In 2004, Xu et al. discovered tiny (>10 nm) "carbogenic" fluorescent nanoparticles, which sparked a new field of nanoparticle research with many uses. CDs are abundant in elements like carbon, oxygen, and hydrogen. Many more elements, such as nitrogen, sulfur, and boron, may be doped onto CDs using suitable procedures for modifying their characteristics. Furthermore, CDs have high water solubility and customizable fluorescence properties (Hola et al. 2014). Typically, the presence of carboxyl activity on CDs' surfaces leads to their higher aqueous solubility.

CD's surface passivation and functionalization with several chemical groups can vary fluorescence behavior, reduce toxicity, and change physical attributes.

Due to their size-dependent optical characteristics, quantum dots (QDs) are useful for therapeutics, traceable delivery, and bioimaging. The employment of group IIVI QDs for these reasons is a subject of much interest and investigation (Hola et al. 2014; Jaleel and Pramod 2018).

Group II-VI quantum dots, however, must be replaced with a more biocompatible QD, such as silicon, because they contain hazardous heavy metals. By using laser pyrolysis, etching to generate luminescence, and grafting with organic molecules to build an oxidation-resistant quantum dot surface, silicon QDs are produced in an aerosol reactor. These QDs are co-encapsulated in folate and PEG-terminated phospholipid micelles in an aqueous medium together with doxorubicin, a cancer treatment agent. The photoluminescence, targeting, and cytotoxicity of these water dispersions of micelles are next investigated.

The band gap of silicon-germanium alloy nanoparticles (Si1xGex) may be tuned from the spectrum's visible to infrared regions, which could make them beneficial for a variety of electrical and optoelectronic applications (Jaleel and Pramod 2018).

4 Applications

There are many advantageous applications of nanoparticles in disparate areas and related biomedical fields, and cancer theragnostics is not an exception owing to their exciting performance in bioimaging, targeted drug and gene delivery, and sensors, among various others (Prasad 2008). Core/shell nanoparticles have been identified as an up-and-coming class for several biomedical applications due to several benefits over simple nanoparticles. The majority of difficulties allied with simple nanoparticles can be overcome by using core/shell nanoparticles. The utilization of nanotechnology in the medical field has advanced in various specialized areas, including bioimaging, genetic manipulation, drug targeting, and biodiagnostics areas (Sudimack and Lee 2000; Cheng et al. 2013).

Core-shell nanoparticles offer significant benefits over basic nanoparticles in biological uses for nanoparticles, improving characteristics like:

(a) **Reduced cytotoxicity:** The recipient tissues and organs can be significantly damaged when the desired nanoparticles are hazardous. The nanoparticles' toxicity and biocompatibility are decreased by adding a harmless material to the center. In certain cases, the shell layer improves the qualities of the core material while also acting as a protective layer. The shell of a different material improves optical characteristics and photostability in semiconductor core/shell nanoparticles (Cheng Zhou and Xing DaYuan 2019).

(b) **Increase in dispersibility, bio- and cytocompatibility:** For nanoparticles to disperse in biological systems (aqueous), they must be hydrophilic. It is a good substitute for conventional drug administration methods due to its enhanced biodispersity, biocompatibility, and cytocompatibility. Another important feature in attracting the attention of researchers to this class of materials is the ease of synthesis. Dispersibility issues, as well as bio- and cytocompatibility issues, can be resolved when the core material is hydrophobic by covering the core surface with a hydrophilic material in the form of core/shell nanoparticles (Chernyak et al. 2020; Prasad 2008).

(c) **Better conjugation with other bioactive molecules:** Biomolecules must conjugate onto particle surfaces in order to be used in numerous bio applications. It can be challenging to conjugate the material of interest with a particular type of biomolecule in many situations; in these situations, coating with a suitable biocompatible substance might help resolve the issue (Ghosh Chaudhuri and Paria 2012).

(d) **Enhanced thermal and chemical stability:** Applying an inert substance to core materials increases their stability when they are susceptible to thermal or chemical changes brought on by exposure to the environment. When it comes to biological applications, core/shell nanoparticles show greater promise than individual nanoparticles (Cho et al. 2008; Kelly et al. 2003).

4.1 Core/Shell Nanoparticle Applications

4.1.1 Bioimaging

The use of "imaging" techniques to improve disease diagnosis, treatment, and prevention has been a significant contribution of bioimaging to human health. These days, a range of imaging methods, such as optical imaging, positron emission tomography (PET), magnetic resonance imaging (MRI), and ultrasound, are crucial for identifying diseases, comprehending the fundamental molecular characteristics of living things, and evaluating medical interventions (Liu et al. 2010).

4.1.2 Biosensor

An analytical tool known as a "biosensor" converts a biological reaction into an electrical signal in order to analyze or identify biological materials. It is basically a biocompatible diagnostic tool that can convert a signal into an electronic mode in response to a biochemical reaction (such as an enzyme-substrate reaction) or a bimolecular interaction (antigen-antibody, receptor-ligand, nucleic acid-protein, nucleic acid-nucleic acid, or metal-macromolecule).

4.1.3 Targeted Drug Delivery

Following the 1950s, with advances in genetics and biotechnology, as well as a general increase in hygiene consciousness, life expectancy in many affluent countries unexpectedly rose to 77–90 years, and the majority of known harmful diseases were thought treatable. Nonetheless, non-pathogenic diseases and novel virus-induced biological syndrome have grown statistically important enough to warrant medical attention (Sharma et al. 2006).

4.1.4 Nanoparticles' Interactions with RNA and DNA

Researchers are interested in how nanoparticles interact with human cells because it may be the secret to future advances in biodiagnosis and treatment, among other fields. It has been discovered that the most efficient nanoparticles for cell absorption are those with sizes between 50 and 200 nm, which opens up new application avenues. The most typical applications for gold nanoparticles are DNA detection, spectroscopy, and electrophoresis (Arora et al. 2011; Murray et al. 2000).

4.1.5 Targeted Delivery of Genes

Genes are essentially functional components of DNA, and genetic engineering frequently entails introducing one DNA sequence into another using plasmids, viral vectors, and other naturally infectious cells or their constituent parts both in vitro and in vivo. The practice known as "transduction" was the first known method of transferring genes from one living cell to another. Eventually, "transformation" a term used to describe non-viral gene transfer techniques in prokaryotic cells formed (Mader et al. 2010; Cui et al. 2009).

4.1.6 Synthesis of Novel Nanoparticles

The surface chemistry of the core/shell nanoparticles increases their ability to bind with ligands, receptors, medications, and other substances, making them primarily suited for biological applications (Buzea et al. 2007; Sahoo and Labhasetwar 2003). As a result, rather than using bulk materials, special nanoparticles that cooperate with the biological system have been produced. Its cytocompatibility and biocompatibility enhance its therapeutic value and create a new pathway for the creation of innovative drug carriers with better qualities like longer residence time, higher bioavailability, lower frequency and quantity of doses, and higher specificity. For instance, the bio-inspired polymeric coating on hydrophobic drugs can enable the proper release of the drug at its intended location due to the polymer's breakdown specific to ions, temperature, and pH (Selvan et al. 2010; Hao et al. 2010). Being more biocompatible than basic nanoparticles, core/shell nanoparticles are also frequently used in bioimaging (Gilmore et al. 2008). The overall contrasting ability of core/shell nanoparticles is caused by the core substance. In core/shell nanoparticles, the shell material is typically in charge of surface characteristics including biocompatibility and bioactive material conjugation due to the reactive moieties present on the surface. For targeted drug delivery, selective binding, biosensing, and other uses, the shell thickness can be changed to provide appropriate contrasting properties as a contrast agent and biomolecule binding (Haidar 2010; Lee et al. 2010).

5 Conclusion and Future Perspectives

Nanoconstructs developed for cancer diagnostics allow improved detection, precise drug delivery to tumors, and fewer adverse effects on healthy organs. Furthermore, imaging approaches can use bespoke probes to evaluate medication efficacy in real-time. A variety of nanotechnology-based systems for theragnostics applications have been studied, including organic and inorganic nanomaterials. Exogenous and endogenous cues play significant roles in modifying medication release and improving the independent and dependent mechanisms of target ability. DNA origami, nanorobots, artificial intelligence, and machine learning have yet to be thoroughly researched for the purpose of developing nanoconstructs for cancer therapy. Although significant progress has been made in the production of nanoconstructs for theragnostics applications, only a few have advanced to the clinical stage of investigation.

However, the solution is always found in advanced nanomaterial, which will pave the way for the creation of multimodal nanoparticles with improved diagnostic and therapeutic capacities. Multidisciplinary studies on a large patient cohort are necessary to obtain the expected benefits of nanotechnology use. The versatility of nanoparticles has been exploited in a range of research applications, including disease diagnostics, early detection investigations, and enhanced contrast agents for imaging techniques. The invention of novel drug delivery vehicles has not only minimized the drug payload but additionally boosted medicine efficacy in the system through improved biocompatibility and cytocompatibility, as well as improved circulation time. Thus, nanoparticles have had a wide-ranging impact on biotechnology for medicine and medical engineering by expanding and improving existing procedures while also testing new and improved medication delivery and monitoring strategies. This chapter covers two broad areas of application: diagnosis (analytical-biosensor/nucleotide interactions or visual-bioimaging) and transportation.

It's also worth noting that nanoparticles' biomedical applications would not be possible without the core/shell structure. Single-layered/multi-layered/layered layer template designs have been used in bioimaging, biosensors, and DNA interactions, but matrix-embedded, hollow core/shell templates are more suited for medication and gene delivery. Because pure nanoparticles are dangerous and unstable, various shell materials are utilized to coat the core surface, resulting in a core/shell arrangement. Most nanoparticles feature a shell layer composed of organic polymers or biomolecules such as complex polysaccharides, cholesterol, amino acids, and their derivatives, with the exception of a few common materials like gold, silver, and mesoporous silica. In addition, a variety of stabilizers, mediators, trans locators, targeting ligands, oligonucleotides, probe molecules, and processors or reactants are immobilized on the shell as required. Therefore, the most complex and unique features of engineered nanoparticles are surface modifications and interactions.

References

Abalde-Cela S, Taladriz-Blanco P, de Oliveira MG, Abell C. Droplet microfluidics for the highly controlled synthesis of branched gold nanoparticles. Sci Rep. 2018;8(1):2440.

Aftab S, et al. Nanomedicine: an effective tool in cancer therapy. Int J Pharm. 2018;540(1–2):132–49.

Agrawal B, Chandra P, Goyal RN, Shim YB. Detection of norfloxacin and monitoring its effect on caffeine catabolism in urine samples. Biosens Bioelectron. 2013;47:307–12.

Akhtar MH, Hussain KK, Gurudatt NG, Chandra P, Shim YB. Ultrasensitive dual probe immunosensor for the monitoring of nicotine induced-brain derived neurotrophic factor released from cancer cells. Biosens Bioelectron. 2018;116:108–15.

Akhter S, Ahmad MZ, Ahmad FJ, Storm G, Kok RJ. Gold nanoparticles in theranostic oncology: current state-of-the-art. Expert Opin Drug Deliv. 2012;9(10):1225–43.

Alex S, Tiwari A. Functionalized gold nanoparticles: synthesis, properties and applications—a review. J Nanosci Nanotechnol. 2015;15(3):1869–94.

Allen TM, Cullis PR. Drug delivery systems: entering the mainstream. Science. 2004;303(5665):1818–22.

Allen SD, Liu X, Jiang J, Liao YP, Chang CH, Nel AE, Meng H. Immune checkpoint inhibition in syngeneic mouse cancer models by a silicasome nanocarrier delivering a GSK3 inhibitor. Biomaterials. 2021;269:120635.

Alyassin Y, Sayed EG, Mehta P, Ruparelia K, Arshad MS, Rasekh M, Shepherd J, Kucuk I, Wilson PB, Singh N, Chang MW. Application of mesoporous silica nanoparticles as drug delivery carriers for chemotherapeutic agents. Drug Discov Today. 2020;25(8):1513–20.

Anselmo AC, Mitragotri S. Nanoparticles in the clinic. Bioengin Transl Med. 2016;1(1):10–29.

Anselmo AC, Mitragotri S. Nanoparticles in the clinic: an update. Bioeng Transl Med. 2019;4(3):e10143.

Aqil A, Qiu H, Greisch JF, Jérôme R, De Pauw E, Jérôme C. Coating of gold nanoparticles by thermosensitive poly (N-isopropylacrylamide) end-capped by biotin. Polymer. 2008;49(5):1145–53.

Arora P, Sindhu A, Dilbaghi N, Chaudhury A. Biosensors as innovative tools for the detection of food borne pathogens. Biosens Bioelectron. 2011;28(1):1–12.

Ascencio JA, Pérez M, José-Yacamán M. A truncated icosahedral structure observed in gold nanoparticles. Surf Sci. 2000;447(1–3):73–80.

Aslam M, Fu L, Su M, Vijayamohanan K, Dravid VP. Novel one-step synthesis of amine-stabilized aqueous colloidal gold nanoparticles. J Mater Chem. 2004;14(12):1795–7.

Bae YH, Park K. Targeted drug delivery to tumors: myths, reality and possibility. Elsevier; 2011. p. 198–205.

Balasubramanian R, Kim B, Tripp SL, Wang X, Lieberman M, Wei A. Dispersion and stability studies of resorcinarene-encapsulated gold nanoparticles. Langmuir. 2002;18(9):3676–81.

Baranwal A, Chandra P. Clinical implications and electrochemical biosensing of monoamine neurotransmitters in body fluids, in vitro, in vivo, and ex vivo models. Biosens Bioelectron. 2018;121:137–52.

Baranwal A, Mahato K, Srivastava A, Maurya PK, Chandra P. Phytofabricated metallic nanoparticles and their clinical applications. RSC Adv. 2016;6(107):105996–6010.

Baranwal A, Kumar A, Priyadharshini A, Oggu GS, Bhatnagar I, Srivastava A, Chandra P. Chitosan: an undisputed bio-fabrication material for tissue engineering and bio-sensing applications. Int J Biol Macromol. 2018a;110:110–23.

Baranwal A, Srivastava A, Kumar P, Bajpai VK, Maurya PK, Chandra P. Prospects of nanostructure materials and their composites as antimicrobial agents. Front Microbiol. 2018b;9:309424.

Barui S, Cauda V. Multimodal decorations of mesoporous silica nanoparticles for improved cancer therapy. Pharmaceutics. 2020;12(6):527.

Bellamy TC, Garthwaite J. Sub-second kinetics of the nitric oxide receptor, soluble guanylyl cyclase, in intact cerebellar cells. J Biol Chem. 2001;276(6):4287–92.

Bhatnagar I, Mahato K, Ealla KKR, Asthana A, Chandra P. Chitosan stabilized gold nanoparticle mediated self-assembled gliP nanobiosensor for diagnosis of invasive aspergillosis. Int J Biol Macromol. 2018;110:449–56.

Bianco A, Kostarelos K, Prato M. Applications of carbon nanotubes in drug delivery. Curr Opin Chem Biol. 2005;9(6):674–9.

Black KC, Wang Y, Luehmann HP, Cai X, Xing W, Pang B, Zhao Y, Cutler CS, Wang LV, Liu Y, Xia Y. Radioactive 198Au-doped nanostructures with different shapes for in vivo analyses of their biodistribution, tumor uptake, and intratumoral distribution. ACS Nano. 2014;8(5):4385–94.

Box, C., et al. Tumour-microenvironmental interactions: paths to progression and targets for treatment. in Seminars in cancer biology. 2010. Elsevier, 20, 128.

Buzea C, Pacheco II, Robbie K. Nanomaterials and nanoparticles: sources and toxicity. Biointerphases. 2007;2(4):MR17-MR71.

Cabral H, Matsumoto Y, Mizuno K, Chen Q, Murakami M, Kimura M, Terada Y, Kano MR, Miyazono K, Uesaka MJNN, Nishiyama N. Accumulation of sub-100 nm polymeric micelles in poorly permeable tumours depends on size. Nat Nanotechnol. 2011;6(12):815–23.

Chai H-J, et al. Renal targeting potential of a polymeric drug carrier, poly-L-glutamic acid, in normal and diabetic rats. Int J Nanomedicine. 2017;12:577–91.

Chauhan VP, Popović Z, Chen O, Cui J, Fukumura D, Bawendi MG, Jain RK. Fluorescent nanorods and nanospheres for real-time in vivo probing of nanoparticle shape-dependent tumor penetration. Angew Chem Int Ed. 2011;50(48):11417–20.
Chen WH, Luo GF, Lei Q, Cao FY, Fan JX, Qiu WX, Jia HZ, Hong S, Fang F, Zeng X, Zhuo RX. Rational design of multifunctional magnetic mesoporous silica nanoparticle for tumor-targeted magnetic resonance imaging and precise therapy. Biomaterials. 2016;76:87–101.
Chen W, Glackin CA, Horwitz MA, Zink JI. Nanomachines and other caps on mesoporous silica nanoparticles for drug delivery. Acc Chem Res. 2019;52(6):1531–42.
Chen C, Ma T, Tang W, Wang X, Wang Y, Zhuang J, Zhu Y, Wang P. Reversibly-regulated drug release using poly (tannic acid) fabricated nanocarriers for reduced secondary side effects in tumor therapy. Nanoscale Horiz. 2020;5(6):986–98.
Cheng R, Meng F, Deng C, Klok HA, Zhong Z. Dual and multi-stimuli responsive polymeric nanoparticles for programmed site-specific drug delivery. Biomaterials. 2013;34(14):3647–57.
Chernyak S, Podgornova A, Dorofeev S, Maksimov S, Maslakov K, Savilov S, Lunin V. Synthesis and modification of pristine and nitrogen-doped carbon dots by combining template pyrolysis and oxidation. Appl Surf Sci. 2020;507:145027.
Cho K, Wang XU, Nie S, Chen Z, Shin DM. Therapeutic nanoparticles for drug delivery in cancer. Clin Cancer Res. 2008;14(5):1310–6.
Cho E, et al. Tumor-targeted liposomes with platycodin D2 promotes apoptosis in colorectal cancer. Mater Today Bio. 2023;22:100745.
Croissant JG, Fatieiev Y, Almalik A, Khashab NM. Mesoporous silica and organosilica nanoparticles: physical chemistry, biosafety, delivery strategies, and biomedical applications. Adv Healthc Mater. 2018;7(4):1700831.
Croissant JG, Butler KS, Zink JI, Brinker CJ. Synthetic amorphous silica nanoparticles: toxicity, biomedical and environmental implications. Nat Rev Mater. 2020;5(12):886–909.
Cui H, Feng Y, Ren W, Zeng T, Lv H, Pan Y. Strategies of large scale synthesis of monodisperse nanoparticles. Recent Pat Nanotechnol. 2009;3(1):32–41.
Danhier F. To exploit the tumor microenvironment: since the EPR effect fails in the clinic, what is the future of nanomedicine? J Control Release. 2016;244:108–21.
Davies LC, et al. Tissue-resident macrophages. Nat Immunol. 2013;14(10):986–95.
Erthal LC, Gobbo OL, Ruiz-Hernandez E. Biocompatible copolymer formulations to treat glioblastoma multiforme. Acta Biomater. 2021;121:89–102.
Fabian E, et al. Tissue distribution and toxicity of intravenously administered titanium dioxide nanoparticles in rats. Arch Toxicol. 2008;82:151–7.
Florence AT. "Targeting" nanoparticles: the constraints of physical laws and physical barriers. J Control Release. 2012;164(2):115–24.
Frickenstein AN, Hagood JM, Britten CN, Abbott BS, McNally MW, Vopat CA, Patterson EG, MacCuaig WM, Jain A, Walters KB, McNally LR. Mesoporous silica nanoparticles: properties and strategies for enhancing clinical effect. Pharmaceutics. 2021;13(4):570.
Garbayo E, et al. Nanomedicine and drug delivery systems in cancer and regenerative medicine. Wiley Interdiscip Rev Nanomed Nanobiotechnol. 2020;12(5):e1637.
García-Fernández A, Aznar E, Martínez-Máñez R, Sancenon F. New advances in in vivo applications of gated mesoporous silica as drug delivery nanocarriers. Small. 2020;16(3):1902242.
Ghosh Chaudhuri R, Paria S. Core/shell nanoparticles: classes, properties, synthesis mechanisms, characterization, and applications. Chem Rev. 2012;112(4):2373–433.
Gilmore JL, Yi X, Quan L, Kabanov AV. Novel nanomaterials for clinical neuroscience. J Neuroimmune Pharmacol. 2008;3:83–94.
Golombek SK, et al. Tumor targeting via EPR: strategies to enhance patient responses. Adv Drug Deliv Rev. 2018;130:17–38.
Gould SE, Junttila MR, de Sauvage FJ. Translational value of mouse models in oncology drug development. Nat Med. 2015;21(5):431–9.
Haidar ZS. Bio-inspired/-functional colloidal core-shell polymeric-based nanosystems: technology promise in tissue engineering, bioimaging and nanomedicine. Polymers. 2010;2(3):323–52.

Hao R, Xing R, Xu Z, Hou Y, Gao S, Sun S. Synthesis, functionalization, and biomedical applications of multifunctional magnetic nanoparticles. Adv Mater. 2010;22(25):2729–42.
He Q, Zhang J, Shi J, Zhu Z, Zhang L, Bu W, Guo L, Chen Y. The effect of PEGylation of mesoporous silica nanoparticles on nonspecific binding of serum proteins and cellular responses. Biomaterials. 2010;31(6):1085–92.
Heldin C-H, et al. High interstitial fluid pressure—an obstacle in cancer therapy. Nat Rev Cancer. 2004;4(10):806–13.
Hola K, Zhang Y, Wang Y, Giannelis EP, Zboril R, Rogach AL. Carbon dots—emerging light emitters for bioimaging, cancer therapy and optoelectronics. Nano Today. 2014;9(5):590–603.
Huang X, Teng X, Chen D, Tang F, He J. The effect of the shape of mesoporous silica nanoparticles on cellular uptake and cell function. Biomaterials. 2010;31(3):438–48.
Huo S, Ma H, Huang K, Liu J, Wei T, Jin S, Zhang J, He S, Liang XJ. Superior penetration and retention behavior of 50 nm gold nanoparticles in tumors. Cancer Res. 2013;73(1):319–30.
Hutchison JL, Kiselev NA, Krinichnaya EP, Krestinin AV, Loutfy RO, Morawsky AP, Muradyan VE, Obraztsova ED, Sloan J, Terekhov SV, Zakharov DN. Double-walled carbon nanotubes fabricated by a hydrogen arc discharge method. Carbon. 2001;39(5):761–70.
Jain RK, Stylianopoulos T. Delivering nanomedicine to solid tumors. Nat Rev Clin Oncol. 2010;7(11):653–64.
Jaleel JA, Pramod K. Artful and multifaceted applications of carbon dot in biomedicine. J Control Release. 2018;269:302–21.
Jugdaohsingh R. Silicon and bone health. J Nutr Health Aging. 2007;11(2):99.
Kelly KL, Coronado E, Zhao LL, Schatz GC. The optical properties of metal nanoparticles: the influence of size, shape, and dielectric environment. J Phys Chem B. 2003;107(3):668–77.
Khan A, Alamry KA. Surface modified carbon nanotubes: an introduction. In Surface modified carbon nanotubes volume 1: fundamentals, synthesis and recent trends. American Chemical Society, 2022 pp. 1–25.
Kiessling F, et al. Nanoparticles for imaging: top or flop? Radiology. 2014;273(1):10–28.
Klochkov SG, Neganova ME, Nikolenko VN, Chen K, Somasundaram SG, Kirkland CE, Aliev G. Implications of nanotechnology for the treatment of cancer: recent advances. Semin Cancer Biol. 2021;69:190–9. Academic Press
Knežević NŽ, Kaluđerović GN. Silicon-based nanotheranostics. Nanoscale. 2017;9(35):12821–9.
Knežević NŽ, Ruiz-Hernández E, Hennink WE, Vallet-Regí M. Magnetic mesoporous silica-based core/shell nanoparticles for biomedical applications. RSC Adv. 2013;3(25):9584–93.
Knežević NŽ, Gadjanski I, Durand JO. Magnetic nanoarchitectures for cancer sensing, imaging and therapy. J Mater Chem B. 2019;7(1):9–23.
Kobayashi H, Watanabe R, Choyke PL. Improving conventional enhanced permeability and retention (EPR) effects; what is the appropriate target? Theranostics. 2014;4(1):81.
Kohn S, et al. Pathways of macromolecular tracer transport across venules and small veins. Structural basis for the hyperpermeability of tumor blood vessels. Lab Invest. 1992;67(5):596–607.
Lammers T, et al. Drug targeting to tumors: principles, pitfalls and (pre-) clinical progress. J Control Release. 2012;161(2):175–87.
Lazarovits J, et al. Nanoparticle–blood interactions: the implications on solid tumour targeting. Chem Commun. 2015;51(14):2756–67.
Lee PW, Hsu SH, Wang JJ, Tsai JS, Lin KJ, Wey SP, Chen FR, Lai CH, Yen TC, Sung HW. The characteristics, biodistribution, magnetic resonance imaging and biodegradability of superparamagnetic core–shell nanoparticles. Biomaterials. 2010;31(6):1316–24.
Li HJ, Du JZ, Du XJ, Xu CF, Sun CY, Wang HX, Cao ZT, Yang XZ, Zhu YH, Nie S, Wang J. Stimuli-responsive clustered nanoparticles for improved tumor penetration and therapeutic efficacy. Proc Natl Acad Sci. 2016;113(15):4164–9.
Li Y, Miao Y, Chen M, Chen X, Li F, Zhang X, Gan Y. Stepwise targeting and responsive lipid-coated nanoparticles for enhanced tumor cell sensitivity and hepatocellular carcinoma therapy. Theranostics. 2020;10(8):3722.
Liao Z, Wong SW, Yeo HL, Zhao Y. Smart nanocarriers for cancer treatment: clinical impact and safety. NanoImpact. 2020;20:100253.

Liu G, Swierczewska M, Lee S, Chen X. Functional nanoparticles for molecular imaging guided gene delivery. Nano Today. 2010;5(6):524–39.
Liu J, Dong J, Zhang T, Peng Q. Graphene-based nanomaterials and their potentials in advanced drug delivery and cancer therapy. J Control Release. 2018a;286:64–73.
Liu X, Jiang J, Chan R, Ji Y, Lu J, Liao YP, Okene M, Lin J, Lin P, Chang CH, Wang X. Improved efficacy and reduced toxicity using a custom-designed irinotecan-delivering silicasome for orthotopic colon cancer. ACS Nano. 2018b;13(1):38–53.
Liu X, Jiang J, Chang CH, Liao YP, Lodico JJ, Tang I, Zheng E, Qiu W, Lin M, Wang X, Ji Y. Development of facile and versatile platinum drug delivering silicasome nanocarriers for efficient pancreatic cancer chemo-immunotherapy. Small. 2021a;17(14):2005993.
Liu X, Jiang J, Liao YP, Tang I, Zheng E, Qiu W, Lin M, Wang X, Ji Y, Mei KC, Liu Q. Combination chemo-immunotherapy for pancreatic cancer using the immunogenic effects of an irinotecan silicasome nanocarrier plus anti-PD-1. Adv Sci. 2021b;8(6):2002147.
Llopis-Lorente A, Lozano-Torres B, Bernardos A, Martínez-Máñez R, Sancenón F. Mesoporous silica materials for controlled delivery based on enzymes. J Mater Chem B. 2017;5(17):3069–83.
Longmire M, Choyke PL, Kobayashi H. Clearance properties of nano-sized particles and molecules as imaging agents: considerations and caveats. Nanomedicine (Lond). 2008;3(5):703–17.
Lu J, Liong M, Li Z, Zink JI, Tamanoi F. Biocompatibility, biodistribution, and drug-delivery efficiency of mesoporous silica nanoparticles for cancer therapy in animals. Small. 2010;6(16):1794–805.
Mader HS, Kele P, Saleh SM, Wolfbeis OS. Upconverting luminescent nanoparticles for use in bioconjugation and bioimaging. Curr Opin Chem Biol. 2010;14(5):582–96.
Maeda H. Toward a full understanding of the EPR effect in primary and metastatic tumors as well as issues related to its heterogeneity. Adv Drug Deliv Rev. 2015;91:3–6.
Maeda H, Matsumura Y. Tumoritropic and lymphotropic principles of macromolecular drugs. Crit Rev Ther Drug Carrier Syst. 1989;6(3):193–210.
Manzano M, Vallet-Regí M. Mesoporous silica nanoparticles for drug delivery. Adv Funct Mater. 2020;30(2):1902634.
Martínez-Carmona M, Ho QP, Morand J, García A, Ortega E, Erthal LC, Ruiz-Hernandez E, Santana MD, Ruiz J, Vallet-Regí M, Gun'ko, Y.K. Amino-functionalized mesoporous silica nanoparticle-encapsulated octahedral organoruthenium complex as an efficient platform for combatting cancer. Inorg Chem. 2020;59(14):10275–84.
Mekaru H, Lu J, Tamanoi F. Development of mesoporous silica-based nanoparticles with controlled release capability for cancer therapy. Adv Drug Deliv Rev. 2015;95:40–9.
Moti LAA, et al. Multi-functionalization, a promising adaptation to overcome challenges to clinical translation of nanomedicines as nano-diagnostics and nano-therapeutics for breast cancer. Curr Pharm Des. 2021;27(43):4356–75.
Mura S, Nicolas J, Couvreur P. Stimuli-responsive nanocarriers for drug delivery. Nat Mater. 2013;12(11):991–1003.
Murray CB, Kagan CR, Bawendi MG. Synthesis and characterization of monodisperse nanocrystals and close-packed nanocrystal assemblies. Annu Rev Mater Sci. 2000;30(1):545–610.
Natfji AA, et al. Parameters affecting the enhanced permeability and retention effect: the need for patient selection. J Pharm Sci. 2017;106(11):3179–87.
Nichols JW, Bae YH. Odyssey of a cancer nanoparticle: from injection site to site of action. Nano Today. 2012;7(6):606–18.
Palmerston Mendes L, Pan J, Torchilin VP. Dendrimers as nanocarriers for nucleic acid and drug delivery in cancer therapy. Molecules. 2017;22(9):1401.
Perry JL, et al. Mediating passive tumor accumulation through particle size, tumor type, and location. Nano Lett. 2017;17(5):2879–86.
Pickup MW, Mouw JK, Weaver VM. The extracellular matrix modulates the hallmarks of cancer. EMBO Rep. 2014;15(12):1243–53.
Prasad GL. Biomedical applications of nanoparticles. *In* Safety of nanoparticles: from manufacturing to medical applications (pp. 89–109). New York, NY: Springer New York; 2008.

Rajora AK, et al. Impact of the enhanced permeability and retention (EPR) effect and cathepsins levels on the activity of polymer-drug conjugates. Polymers. 2014;6(8):2186–220.
Raza A, Hayat U, Rasheed T, Bilal M, Iqbal HM. Redox-responsive nano-carriers as tumor-targeted drug delivery systems. Eur J Med Chem. 2018;157:705–15.
Rizwanullah M, Ahmad MZ, Garg A, Ahmad J. Advancement in design of nanostructured lipid carriers for cancer targeting and theranostic application. Biochimica et Biophysica Acta (BBA)-General Subjects. 2021;1865(9):129936.
Sahoo SK, Labhasetwar V. Nanotech approaches to drug delivery and imaging. Drug Discov Today. 2003;8(24):1112–20.
Sarin H. Physiologic upper limits of pore size of different blood capillary types and another perspective on the dual pore theory of microvascular permeability. J Angiogenes Res. 2010,11;2:14.
Sarin H, Kanevsky AS, Wu H, Sousa AA, Wilson CM, Aronova MA, Griffiths GL, Leapman RD, Vo HQ. J Transl Med. 2009;7:1–13.
Saw WS, Anasamy T, Foo YY, Kwa YC, Kue CS, Yeong CH, Kiew LV, Lee HB, Chung LY. Delivery of nanoconstructs in cancer therapy: challenges and therapeutic opportunities. Adv Therap. 2021;4(3):2000206.
Selvan ST, Tan TTY, Yi DK, Jana NR. Functional and multifunctional nanoparticles for bioimaging and biosensing. Langmuir. 2010;26(14):11631–41.
Senapati S, Mahanta AK, Kumar S, Maiti P. Controlled drug delivery vehicles for cancer treatment and their performance. Signal Transduct Target Ther. 2018;3(1):7.
Shah S, Liu Y, Hu W, Gao J. Modeling particle shape-dependent dynamics in nanomedicine. J Nanosci Nanotechnol. 2011;11(2):919–28.
Sharma P, Brown S, Walter G, Santra S, Moudgil B. Nanoparticles for bioimaging. Adv Colloid Interf Sci. 2006;123:471–85.
Singh R, Deshmukh R. Carbon nanotube as an emerging theranostic tool for oncology. J Drug Deliv Sci Technol. 2022;74:103586.
Smith BR, Kempen P, Bouley D, Xu A, Liu Z, Melosh N, Dai H, Sinclair R, Gambhir SS. Shape matters: intravital microscopy reveals surprising geometrical dependence for nanoparticles in tumor models of extravasation. Nano Lett. 2012;12(7):3369–77.
Soo Choi H, et al. Renal clearance of quantum dots. Nat Biotechnol. 2007;25(10):1165–70.
Sudimack J, Lee RJ. Targeted drug delivery via the folate receptor. Adv Drug Deliv Rev. 2000;41(2):147–62.
Suresh PK, Banerjee A, Singh P. Promises and challenges for targeting the immunological players in the tumor micro-environment–critical determinants for NP-based therapy. OpenNano. 2023;10:100134.
Swartz MA, Lund AW. Lymphatic and interstitial flow in the tumour microenvironment: linking mechanobiology with immunity. Nat Rev Cancer. 2012;12(3):210–9.
Sweeney S, Adamcakova-Dodd A, Thorne PS, Assouline JG. Biocompatibility of multi-imaging engineered mesoporous silica nanoparticles: in vitro and adult and fetal in vivo studies. J Biomed Nanotechnol. 2017;13(5):544–58.
Sykes EA, Chen J, Zheng G, Chan WC. Investigating the impact of nanoparticle size on active and passive tumor targeting efficiency. ACS Nano. 2014;8(6):5696–706.
Tang F, Li L, Chen D. Mesoporous silica nanoparticles: synthesis, biocompatibility and drug delivery. Adv Mater. 2012;24(12):1504–34.
Wheate NJ. Nanoparticles: the future for platinum drugs or a research red herring? Nanomedicine. 2012;7(9):1285–7.
Wicki A, et al. Nanomedicine in cancer therapy: challenges, opportunities, and clinical applications. J Control Release. 2015;200:138–57.
Wilhelm A, Tavares QD, Ohta S, Audet J, Dvorak HF, Chan WCW. Nat Rev Mater. 2016a;1:16014.
Wilhelm S, Tavares AJ, Dai Q, Ohta S, Audet J, Dvorak HF, Chan WC. Analysis of nanoparticle delivery to tumours. Nat Rev Mater. 2016b;1(5):1–12.
Yadav K, Singh D, Singh MR, Pradhan M. Nano-constructs targeting the primary cellular energy source of cancer cells for modulating tumor progression. OpenNano. 2022;8:100107.

Yingchoncharoen P, Kalinowski DS, Richardson DR. Lipid-based drug delivery systems in cancer therapy: what is available and what is yet to come. Pharmacol Rev. 2016;68(3):701–87.

Yong T, et al. Extracellular vesicles for tumor targeting delivery based on five features principle. J Control Release. 2020;322:555–65.

Youn YS, Bae YH. Perspectives on the past, present, and future of cancer nanomedicine. Adv Drug Deliv Rev. 2018;130:3–11.

Yu L-X, Ling Y, Wang H-Y. Role of nonresolving inflammation in hepatocellular carcinoma development and progression. NPJ precision oncology. 2018;2(1):6.

Yun YH, Lee BK, Park K. Controlled drug delivery: historical perspective for the next generation. J Control Release. 2015;219:2–7.

Zhang C, et al. Pharmacokinetics, biodistribution, efficacy and safety of N-octyl-O-sulfate chitosan micelles loaded with paclitaxel. Biomaterials. 2008;29(9):1233–41.

Zhang Y, Hou Z, Ge Y, Deng K, Liu B, Li X, Li Q, Cheng Z, Ma PA, Li C, Lin J. DNA-hybrid-gated photothermal mesoporous silica nanoparticles for NIR-responsive and aptamer-targeted drug delivery. ACS Appl Mater Interfaces. 2015;7(37):20696–706.

Lanthanide-Based Core-Shell Nanoconstructs for Cancer Theragnostics

18

Dipak Bari, Vivek Rajkule, Shradhha Tiwari, and Chandrakantsing Pardeshi

Abstract

The distinct luminescent, magnetic characteristics and unique optical properties such as narrow emission bands, emission tunability, multiple emission wavelengths, long fluorescence lifetime, and large Stokes shift of lanthanide ions (Ln^{3+}) make multifaceted lanthanide-based nanoparticles (NPs) predicted to have a significant influence in nanomedicine as well as growing in prominence for usage in a multitude of theranostic applications. Numerous imaging techniques have been extensively used in conjunction with probes that can increase the contrast between normally healthy and malignant tissues to identify malignancy and its pathological activities. Because of the compartmentalised framework, nanoassemblies present an intriguing prospect for cancer multidimensional imaging and theranostics. This allows for the combination of diagnosis and therapy on a single platform, facilitating biological system investigations, molecular imaging, and therapeutic monitoring.

Keywords

Core-shell nanoassemblies · Lanthanide-based · Cancer · Theranostics

D. Bari · V. Rajkule · C. Pardeshi (✉)
Department of Pharmaceutics, R. C. Patel Institute of Pharmaceutical Education and Research, Shirpur, Maharashtra, India

S. Tiwari
School of Pharmacy, MIT Vishwaprayag University, Solapur-Pune highway, Kegaon, Solapur, India

J. K. Patel et al. (eds.), *Core-Shell Nano Constructs for Cancer Theragnostic*,
https://doi.org/10.1007/978-981-96-3025-7_18

1 Introduction

Combining therapies and diagnostics, theranostics is used to evaluate early response to therapy or stratify individuals for treatment response. Co-development is possible for entities like antibody-drug conjugates for treatment and nanoplatforms for imaging. Cancer theranostics is crucial for individualised cancer therapy, lowering delays, and enhancing patient care (Chen and Wong 2014). Cancer theranostics, including chemotherapeutics, radiotherapy, immunotherapy, and phototherapy, hold great potential for cancer care (Tang et al. 2022).

Nanoscale nanomedicine provides a large specific surface area, customisable physiochemical characteristics, versatile imaging/therapeutic activities, and enhanced biocompatibility (Malik et al. 2023). It has been developed to be more effective than conventional cancer treatments. Nanostructures with imaging capabilities can include disease-specific biomarker-targeting ligands to aid in early cancer diagnosis and precise tumour profiling (Alrushaid et al. 2023). This might transform the entire cancer treatment process. Furthermore, the ability to modify and manipulate nanostructures opens up new possibilities for cancer therapy (Chehelgerdi et al. 2023).

Theranostic nanomedicine has the potential to deliver an enormous drug load to tumour locations via increased permeability and retention or customisation with active targeting molecules (Chehelgerdi et al. 2023). Stimuli-activatable strategies for developing and building theranostic nanotechnology have been presented, with a focus on the specific properties of the tumour microenvironment (TME) and mechanical forces produced by light, radiation, magnetic fields, and ultrasound. These bioactivatable nanomedicinal materials are frequently constructed with an activatable structure prior to reaching tumour locations, and their structures are dismantled, reassembled, or activated in response to TME properties or external physical pressures. Once these exogenous/endogenous stimuli change the structure of nanotechnology products, imaging agents/drugs may be launched from the nanostructures to directly exert imaging/therapeutic effects, or they may be aggregated to improve their imaging and/or therapeutic potency. Novel ways for stimuli-triggered signal switching in nanomedicine have evolved, such as adjusting the distance between a signal emitter/enhancer and a quencher in a hollow nanostructure. These stimuli-activatable strategies for producing theranostic nanomedicine might have a high promise for treating advanced cancer. This review emphasises the critical significance of theranostics in the long-term clinical requirement for cancer care (Li et al. 2023a, b).

Radiopharmaceuticals differ in their molecular targeting mechanism and ionising radiation output, which affects their advantages and limits. Radionuclide treatment employs an unsteady nuclide and a targeting vehicle to provide therapeutic radiation to cancer cells (Salih et al. 2022; Cherry et al. 2012). The term 'theranostics' is associated with radionuclide treatment, it particularly refers to the application of two radiopharmaceuticals including radionuclides for diagnostic imaging along with therapeutic purposes (Burkett et al. 2023). Radionuclides bind to their target when they attach themselves to a ligand, such as the outer receptor of a cell for specificity (Lepareur et al. 2023; Bavelaar et al. 2018). The quantity in each radiopharmaceutical region may be measured using standardised uptake values, which can be obtained by imaging the area with a single-photon emission computed tomography (Spect) and positron emission tomography (PET) (Crişan et al. 2022). For diagnosis and treatment, most radionuclides require a specific vector to reach the lesion site, such as folate, peptides, antibodies, or aptamers (Fig. 18.1) (Sun et al. 2022). At small levels of circulation in the blood, the ideal radioactive material is either mostly attached to the target or has already left the body (Lau et al. 2020). These bio-images help with tumour diagnosis and staging, target expression patient selection, delivery of treatment confirmation, effective radiation dosage calculation, and therapy monitoring (Fass 2008). Nuclear medicine has employed radionuclide imaging, a routine molecular imaging technique, from the initial investigations employing ^{131}I sodium iodide for thyroid cancer diagnosis and treatment (Yavuz and Puckett 2024). Targeted radionuclide therapy includes the internal radiotherapy of cancer cells with ligands combined with cytotoxic alpha- or beta-emitting radionuclides. These therapeutic radionuclides cause DNA damage, resulting in cell death. Theranostics uses the same ligand for both imaging and treatment (Lepareur et al. 2023).

Phototherapy for tumour treatment has obstacles such as low specificity, restricted absorption of stimulated light, hypoxia, and cytotoxicity (Correia et al. 2021). Phototherapy drugs need large doses, which might have negative effects (Rathod et al. 2024). Nanomaterials including micelles (Kumar et al. 2012), liposomes (Seleci et al. 2017), dendrimers (Cruz et al. 2023), metal oxide nanoparticles (Fernandes 2023), quantum dots (Tripathi et al. 2015), silica nanoparticles (Baeza and Vallet-Regí 2020), and metal-organic frameworks (Sun et al. 2019) have been produced to enhance antitumor activity. These nanomaterials improve therapeutic effectiveness via their adaptable size, high therapeutic drug loading,

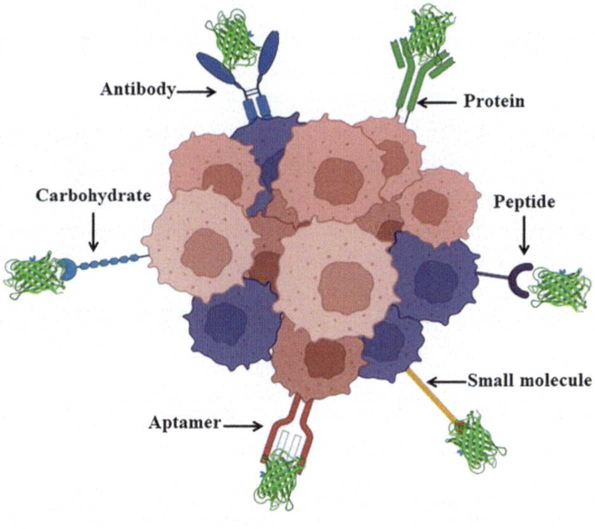

Fig. 18.1 Schematic representation of various radionuclides vector

tumour-targeting capability, biocompatibility, and versatile surfaces (Yetisgin et al. 2020; Chehelgerdi et al. 2023). However, these limitations continue to restrict phototherapy's practical applicability (Luo et al. 2024).

Lanthanide-based nanoparticles (LNPs) show potential for increased phototherapy due to their distinct optical features, which include narrow-band emission (Luo et al. 2023), super photostability (Casar et al. 2021), high effectiveness, and extended emission lifetime (Fan et al. 2023). Lanthanide-based nanoparticles (LNPs) can emit upconversion by six different mechanisms: excited state absorption, energy transfer upconversion, energy migration upconversion, cooperative sensitised upconversion, cross-relaxation, and photon avalanche (Qin et al. 2019). LNPs may also downconvert under near-infrared (NIR) irradiation, a process termed the Stokes shift, to produce second near-infrared luminescence, which is perfect for deep-tissue bioimaging (Li et al. 2023a, b). LNPs' high atomic number radiosensitisation capabilities make them useful radiosensitisers for cancer radiation and X-ray-induced photodynamic therapy (X-PDT) (Cline et al. 2019). Surface modification can enhance tumour targeting and enable combination oncological therapy (Ferro-Flores et al. 2024).

2 A Concept of Cancer Nanotheranostic

Nanoparticles are employed in medicine to serve both therapeutic and diagnostic objectives. Their capacity to penetrate deep into tumours enables them to administer therapeutic drugs and diagnostics using non-invasive procedures such as CT, MRI, and PET. Nanoparticulate agents can be utilised to silence genes, deliver anticancer drugs, target tumours, and provide phototherapy (Chavda et al. 2023).

Cancer theranostics consists of a diagnostic element, a treatment element, and a targeted mechanism (Song et al. 2024). Common methods include nanoparticle-based systems, radionuclide theranostics, and photodynamic theranostics (Xue et al. 2021). Cancer theranostics offers personalised therapy, fewer side effects, greater effectiveness, and early identification and treatment (Anitha et al. 2024). The challenges include design complexity, regulatory barriers, and high prices (Shi et al. 2017). Prostate cancer is treated with prostate-specific membrane antigen (PSMA) ligands labelled with gallium-68 for PET imaging and lutetium-177 for treatment, whereas neuroendocrine tumours are treated with somatostatin receptor imaging and peptide receptor radionuclide therapy (Plichta et al. 2021). Cancer theranostics in clinical practice include imaging-guided treatment and radiotheranostics with diagnostic/therapeutic radionuclide pairs (Jadvar et al. 2018). Real-time imaging may advise therapies including ultrasound-guided puncture and resection, tumour removal using stained fluorescent images, CT/MRI-guided tumour separation for accurate radiation therapy, and gastroscopy/proctoscopy-guided surgery (Mondal et al. 2014; Grégoire et al. 2020). Nanotechnology in cancer theranostics has led to the development of theranostic nanoagents that combine imaging and therapy properties or multiple diagnostic/therapeutic agents through a single nanocarrier. This allows for better imaging-guided therapy, quantitative drug management, and extended therapeutic regimes (Gupta et al. 2024).

3 Nanoparticle-Based Tumour Targeting

The tumour microenvironment (TME) is a critical component of malignant tissues that influences cancer growth. It consists of distinct cellular and noncellular entities that modify the extracellular matrix (Anderson and Simon 2020). The TME's hypoxic, hypoglycemia, and acidic circumstances cause medication release, enabling researchers to develop TME-responsive delivery devices (Dekker et al. 2022). Nanobiotechnological methods can release active ingredients in response to a variety of stimuli, including temperature, pH, redox potential, and enzymes (Liu et al. 2020). Tumour cells that overexpress certain cell surface receptors can target cancer cells using antibodies or smaller compounds, resulting in more targeted antitumour effects while minimising negative effects on normal tissues (Hong et al. 2023).

4 Lanthanide Elements: A Prime Summary

Nuclides having short half-lives, a large yield of γ-rays, and the ability to prevent excessive tissue irradiation are necessary for radiopharmaceuticals (Holland et al. 2010). They have a high specific activity and are ready for in vivo use (Zhang et al. 2022). The physical and biological decay constants of the radionuclide determine the effective radiopharmaceutical decay constant (Morris et al. 2021). Radionuclides, such as technetium-99 m, which is frequently employed in nuclear medicine, must be physiologically site-specific (Kane et al. 2024). For certain uses, such as acid lability, time to maximal chelation, and stability in acidic or serum environments, other radioisotopes should be thoroughly assessed (Holik et al. 2022). Lanthanide radionuclides have important properties for imaging and therapeutic applications (Cutler et al. 2000). Lanthanum and lutetium are different elements on the periodic table, with a primarily trivalent chemistry when compared to silver and yttrium. However, when seen as transition metals, they have structural, electrical, and energetic properties. These distinctions have just been completely grasped in the last decade. Early attempts at characterisation were hampered by difficulty in isolating components and creating pure compounds. NBS Circular 500 summarises the results of measurements taken before 1940 (Morrs 1976). Radiotherapy uses lanthanide isotopes such as low-energy ^{177}Lu, medium-energy ^{149}Pm and ^{153}Sm, and high-energy ^{166}Ho and ^{90}Y. To select the best isotope for radiotherapy, evaluate tumour absorption, blood clearance, radiation delivery rate, half-life, specific activity, and cost-effective large-scale manufacture. The objective is to administer a tumoricidal dosage without causing severe adverse effects (Liu and Edwards 2001).

The half-life of a medicinal radionuclide should correspond to the biological half-life of the radioactive substance at the tumour location. A half-life that is too short might result in insufficient integrated tumour dosage for sterilisation and unwanted radiation exposure to normal tissues. Ideally, the radionuclide should have a long half-life to produce a low dose rate while also providing enough time for manufacture, release, and transportation (Gudkov et al. 2015). Additional concerns include availability, quality, target receptor locations, specific activity, trace

Table 18.1 Radionuclides useful for radiotherapy (Liu and Edwards 2001)

Nuclide	Half-life (days)	Energy (MeV)	Maximum range (mm)	Gamma (keV)	Source	Specific activity
^{67}Cu	2.58	0.575	1.8	185 (40%)	Accelerator	Low
^{90}Y	2.66	2.27	12.0		Reactor/ generator	High
^{109}Pd	0.56	1.03	5.0	88 (5%)	Reactor	Low
^{111}Ag	7.47	1.05	4.8		Reactor	Low
^{149}Pm	2.21	1.07	5.0	286 (3%)	Reactor	Low/ medium
^{153}Sm	1.95	0.80	3.0	103 (28%)	Reactor	Low/ medium
^{166}Ho	1.1	1.6	8.0	81 (6.3%)	Reactor/ generator	High
^{177}Lu	6.7	0.497	1.5	208 (28%)	Reactor	Medium/ high
^{186}Re	3.7	1.02	5.0	137 (9%)	Reactor/ accelerator	Low/ medium
^{188}Re	0.71	2.12	11.0	155 (15%)	Reactor/ generator	High

metal contamination, and both α- and β-emitters α-emitters are effective cytotoxic agents, whereas β-emitters have a comparatively large penetration range. The choice of α- and β-emitter is determined by the tumour size and location. When adopting a radionuclide for radiation treatment planning and staging, it is important to prevent γ-emission. Dosimetry can be determined using a nearly comparable diagnostic surrogate (Kleynhans et al. 2021) (Table 18.1).

5 Synthesis and Classification of Core-Shell Nanomaterials

Lanthanide-doped rare-earth nanocrystals may be manufactured utilising a variety of processes, including aqueous thermal, co-precipitation, and thermal decomposition. Aqueous thermal and co-precipitation are particularly prevalent for rare-earth core nanocrystals (Zhao et al. 2024).

5.1 Thermal Synthesis in Water

There are several ways to classify core-shell nanoparticles. Core-shell nanomaterials are classed based on their structure and number of cores and shells. A single core-shell construction consists of a core made of one material and a shell made of another. The core-shell nanostructure's physico-chemical characteristics vary based on the materials used. Multicore shell structures, which consist of earth metal cores encased by another material's shell, are ubiquitous in polymer, silica, and carbon-designed lanthanide nanocomposite materials. The research focuses on developing core-multishell systems with single cores and multiple shells. Self-catalytic growth, the Stober technique, and the hydrothermal approach are utilised to create spindle-shaped NPs encapsulated with lanthanide nanoparticles (Singh and Bhateria 2021).

Rare-earth nanocrystals are synthesised via hydrothermal and solvent-thermal techniques, with varying sizes, morphologies, and optical characteristics. The solvothermal approach is chosen because it provides perfect control over the synthesis process. The liquid-solid solution (LSS) phase transfer and separation technology has become a popular method for producing several types of nanocrystals (Wang et al. 2011).

Yinyan and Shiqing developed lanthanide hydroxide micro/nanorods as a rapid and simple method for producing functional lanthanide materials with high uniformity by using the hydrothermal synthesis method. The process combines $Ln(OH)_3$ nano/microcrystals with $Ln(NO_3)_3{\cdot}6H_2O$ and tetrabutylammonium hydroxide (TBAH) in an autoclave at 160 °C. The XRD patterns of the synthesised micro/nanocrystals exhibit substantial line widening, which indicates the presence of tiny crystallites. The hexagonal $Gd(OH)_3{:}Eu^{3+}$ was converted into cubic Gd2O3:Eu^{3+}, showing that the hosts had well-doped ions (Yinyan and Shiqing 2016).

5.2 Co-Precipitation Method

The co-precipitation technique is a popular synthesis method since it is simple, inexpensive, and does not require specialised apparatus. It includes adding a precipitant to a solution containing various cations, resulting in a new substance with a consistent composition. The approach was initially described for downconversion nanocrystals doped with lanthanide ions (Zheng et al. 2019). The procedure includes creating a core and layer-by-layer encapsulation of the shell layer. The co-precipitation approach is useful for precisely controlling stoichiometry, synthesising ultrasmall lanthanide-doped nanoparticles with tight size distributions, and integrating with other synthesis methods for rare-earth core-shell nanoparticles (Zhao et al. 2024).

Tinwala et al., developed $La_2Ce_2O_7$ utilising cerium nitrate hexahydrate $Ce(NO_3)_3{\cdot}6H_2O)$ and lanthanum nitrate hexahydrate $La(NO_3)_3{\cdot}6H_2O)$ as precursors. Triethylamine $(C2H_5)_3N)$ and ethanol were utilised as precipitant and solvent, respectively, along with analytical grade reagents. Separate solutions were generated by dissolving $Ce(NO_3)_3{\cdot}6H2O)$, $La((NO_3)_3{\cdot}6H2O)$, and $(C2H_5)_3N)$ into the solvent. The precipitation was carried out at room temperature, and the optimum pH state recorded during the procedure was 9.0. The as-prepared materials were then calcined at various temperatures for 3 h. The materials were characterised using analytical procedures such as thermogravimetric analysis (TGA), X-ray diffraction, transmission electron microscopy, electron dispersive spectroscopy, and Fourier-transformed infrared spectroscopy (FTIR). $La_2Ce_2O_7$ nanoparticles were prepared using a coprecipitation method using triethylamine-formed particles of a cubic fluorite structure after heating at 600–900 °C for 3 h. The particles were 55–80 nm in size and contained both La and Ce. The co-precipitation method was simple and cost-effective (Tinwala et al. 2014).

5.3 Thermal Injection Method

Thermal injection is a common approach for producing rare-earth core-shell nanocrystals, as it is more efficient than co-precipitation and solvothermal procedures. It carefully controls the thickness of the shell layer, allowing for upconversion nanoparticles of various thicknesses. This approach streamlines synthesis by adding the shell layer precursor solution in a single step (Vykoukal et al. 2019).

Ren et al., synthesised ScF_3:Yb/Er (18/2%)@ScF_3 core-shell nanoparticles by injecting rare-earth trifluoroacetate into a solvent mixture of oleic acid, oleylamine, and 1-octadecene at 330 °C. The undoped ScF_3 shell was designed to shield the dopants from surface quenchers, enhancing luminescence intensity. TEM images (Fig. 18.2a, b) show well-arranged pure cubic ScF_3 core and core-shell particles, with d-spacing in lattice fringes in good agreement with the cubic phase ScF_3 (Fig. 18.2c). X-ray diffraction patterns (Fig. 18.2d) show characteristic peaks with high crystallinity (Ren et al. 2021).

6 Characterisations of Lanthanide-Based Core-Shell Nanomaterials

Rare-earth core-shell nanostructures provide luminous nanocrystals with unique functional features such as morphological control and chemical and physical property modulation. Structural synthesis characterisation methods are essential for properly realising these features (Yan et al. 2011). Researchers frequently utilise a variety of characterisation methods, including transmission electron microscopy (TEM) (Malatesta 2021), electron energy loss spectroscopy (EELS) (Aronova and Leapman 2012), and energy dispersive X-ray spectroscopy (EDS) (Scimeca et al. 2018). TEM pictures may disclose the size and shape of produced nanoparticles, whereas EELS quantifies the energy shift in incoming energetic electrons following inelastic scattering. EELS has been verified as an efficient method for investigating the structural characteristics of lanthanide-doped upconversion nanoparticles and revealing the inherent relationship between their structure and properties. Scanning transmission electron microscopy combined with EDS (STEM-EDS) is commonly used to determine ion distribution and movement inside individual nanocrystals. Time-resolved and steady-state luminescence curves may confirm the synthesis of the core-shell structure, with the relationship between elements mostly occurring at the core-shell interface (Zhao et al. 2024).

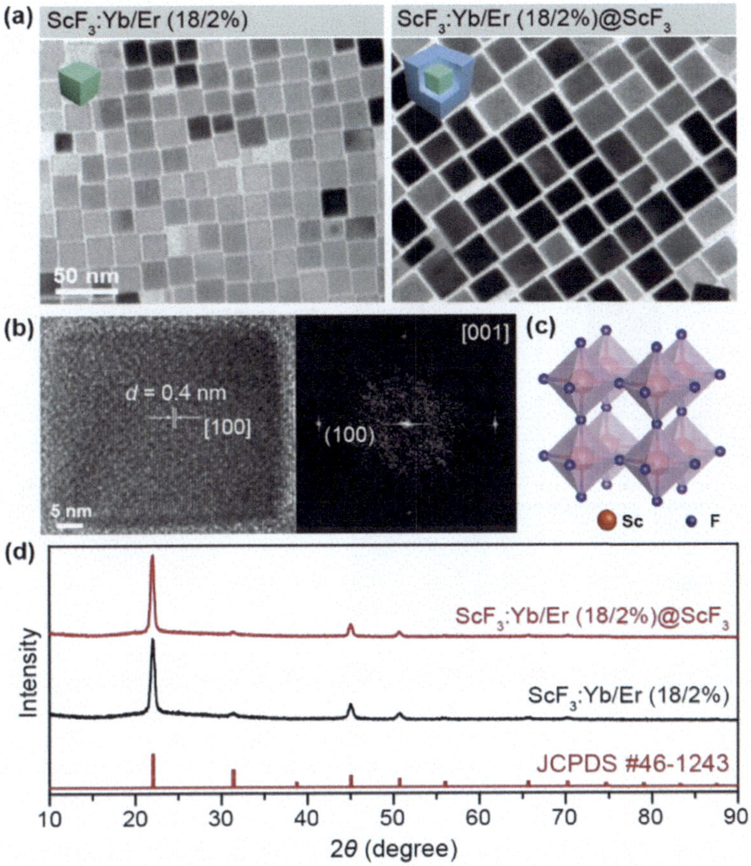

Fig. 18.2 Characterisation of the ScF_3 nanoparticles. (**a**) Typical TEM images of the as-synthesised ScF_3:Yb/Er core and ScF_3:Yb/Er@ ScF_3 core–shell nanoparticles. (**b**) HR-TEM of a randomly selected core–shell nanoparticle and the corresponding FFT pattern. (**c**) Schematic illustration of the ScF_3 structure composed of cornersharing ScF_6 octahedra. (**d**) XRD patterns of the core and core–shell nanoparticles and the literature reference of cubic ScF3 (JCPDS #46–1243) (Adapted from Ren et al. 2021 with kind permission of the copyright holder)

7 Recent Studies and Latest Advancements on Core-Shell Nanoassemblies of Lanthanides

By loading adamantane-modified doxorubicin (Dox) into polyethylenimine-crosslinked-γ-cyclodextrin (PC) through supramolecular assembly, researchers created (a single core multiple shell NPs) interior Dox-loaded PC (PCD). Next, PCD and siRNA self-assembled electrostatically to form PCD/siRNA nanocomplexes. In this study, Jin et al. fabricated cationic supramolecular polyethylenimine-crosslinked-γ-cyclodextrin (PC) conjugates and formed PC films by dropping a solution onto a silicon plate. Neodymium (Nd) ion implantation was carried out using a metal ion implanter to obtain Nd-PC. Four

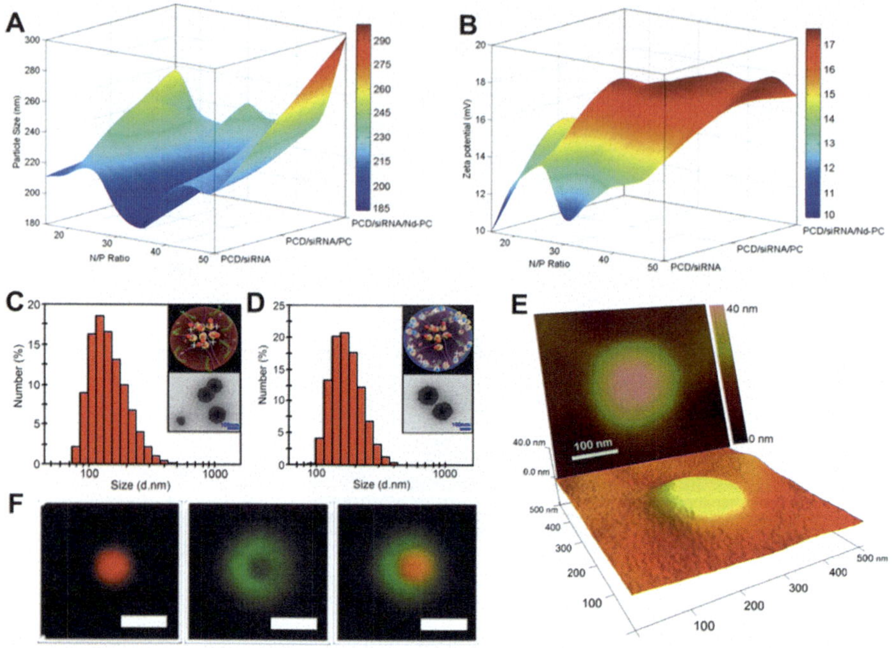

Fig. 18.3 Characterisation of the PCD/siRNA, PCD/siRNA/PC, and PCD/siRNA/Nd-PC nanoassemblies: (**a**) Particle size and (**b**) zeta potential analysis of the PCD/siRNA, PCD/siRNA/PC, and PCD/siRNA/Nd-PC nanoassemblies with various N/P ratios by DLS. Particle size distribution inserted schematic structure and TEM images of (**c**) PCD/siRNA and (**d**) PCD/siRNA/Nd-PC with a N/P ratio of 30. (**e**) AFM amplitude image and height image of the PCD/siRNA/Nd-PC nanoassembly with an N/P ratio of 30. (**f**) Super-resolution microscopy images of the PCD/siRNA/Nd-PC nanoassembly with an N/P ratio of 30. The FAM-siRNA is shown in green and the drug Dox is shown in red. The scale bar is 200 nm. (Adapted from Jin et al. 2017 with kind permission of the copyright holder)

more metal components were inserted into the PC for comparison. The elemental depth profiles and chemical states of untreated PC, Nd-PC, and other metal-doped PC were characterised by X-ray photoelectron spectroscopy. The Nd-PC was created by electrostatically mixing the solution with SiO_2 particles. A transmission electron microscope was used to perform energy-dispersive X-ray spectroscopy. The doxorubicin-loaded PC (PCD) PCD/siRNA nanoassembly was created by combining PCD and siRNA and incubating for 20 min. The PC or Nd-PC solution was added to the complexes, with the final N/P ratio kept at 30/1. The particle size and zeta potential of the nanoassemblies were measured using dynamic light scattering. The morphology of the complexes was investigated by TEM, atomic force microscopy, and super-resolution stochastic optical reconstruction microscopy (Fig. 18.3).

The structure and properties of the PCD/siRNA/Nd-PC nanoassembly were investigated using dynamic light scattering (DLS), TEM, atomic force microscopy (AFM), and super-resolution microscopy. The DLS results revealed a limited size distribution, a N/P ratio of 20–45, and a positive charge of 16–18 mV. The tertiary nanoassembly has a greater diameter than the PCD/siRNA complex. The AFM pictures revealed a spherical structure with a diameter of 200 nm and N/P ratio of 30. The nanoassembly's N/P ratio of 30 achieved an appropriate balance of particle size and surface charge (Jin et al. 2017).

8 Theranostics Applications of Lanthanide-Based Nanoassemblies

Applications for lanthanide-coated nanocomposites (LCNPs) are many and include cytotoxicity, tumour-targeted imaging, and cellular internalisation research. The influence of surface chemistry and doped elements on the prospective uses of LCNPs in drug delivery, cancer treatment, and bioimaging has been brought to light by recent investigations. Different coating groups and structures, such as biomolecules, polymers, inorganic nanoparticles, and anticancer medicines, have been used to create LCNPs. High stability, dispersibility, ease of functionalisation, low nonspecific adsorption, little disturbance of native molecules, and effective luminescence imaging are among their benefits. It has been discovered that LCNPs are the best probes for tracking target distribution in tumour cells in real-time. The location and size of tumours can only be estimated with the use of in vivo fluorescence imaging. These LCNPs have been employed in a number of ways, such as drug carriers, photothermal therapy, and PDT, to aid in the detection of cancer (Rafique et al. 2019) (Table 18.2).

A 10-step procedure was used by Zang et al. to synthesise lanthanide–cyclen complexes (CycLn)—disulfide bond (ss)—camptothecin (CPT) (CycLn-ss-CPT) (CycLn-ss-CPT), an amphiphile suitable for MRI, hybrid near-infrared imaging (NIR), and luminescence investigation. It was purified, characterised, and

Table 18.2 Different lanthanides used in cancer therapy

Drug	Type of cancer	Element of Lanthanide	Reference
–	Pulmonary and hepatic tumor metastasis	Nd^{3+}	Li et al. (2016)
–	Epithelial cancer	Terbium	Zeng et al. (2024)
Doxorubicin	Breast cancer	Eu_{60} and Tb_{60} (collectively Ln_{60})	Wang et al. (2024)
–	Breast (MCF7) and colon (HT29) cancer	La,Sm	Mejía-Méndez et al. (2024)
Camptothecin	–	Lanthanum-doped carbon dots	Williams et al. (2024)
Cisplatin	Triple-negative breast cancer (TNBC)	$LaCl_3$ and $CeCl_3$	Huang et al. (2024)
Doxorubicin	Targeted liver cancer cell	Sodium alginate-lanthanum (SA-La) cross-linked microspheres	Yin et al. (2024)
Flumequine		La(III), Sm(III), and Tb (III)	El-Habeeb et al. (2024)

confirmed by UV-vis and emission spectral analysis. The luminescence of CPT was quenched in cycLn-ss-CPT, likely due to the metal quenching effect. The hydration numbers of the amphiphiles were calculated, showing one H_2O molecule was directly coordinated to the Ln(III) core, indicating its use as an MRI contrast agent. CycLn-ss-CPT nanoparticles can self-assemble in an aqueous solution, retaining their amphiphilic nature. They were prepared using a simple dialysis technique, resulting in Gd/YbNPs. The critical aggregation concentrations (CAC) value, which reflects self-assembling ability and stability, was measured to be around 0.62 μM. DLS measurements showed well-defined nanoparticles with hydrophobic diameters of 128–152 nm. TEM studies showed (Fig. 18.4a–f) Gd/YbNPs were spherical with uniform sizes of 59–87 nm. The size of Gd/YbNPs measured by DLS was twice the size observed in TEM. The study evaluated the release of CPT by Gd/YbNPs using DLS, TEM, and luminescence analyses. The results showed a significant size increase for every LnNP in the presence of glutathione (GSH), including Gd/YbNPs. The GdNPs maintained their hydrodynamic size under these conditions, consistent with colloidal stability (Fig. 18.4g). The ζ-potential of Gd/YbNPs (Fig. 18.4h) also increased with GSH. The TEM images showed the collapse of the nanoparticle structures upon GSH treatment, resulting in the rearrangement of hydrophobic fragments. The addition of GSH caused an increase in CPT luminescence intensity at 430 nm, and an increase in Eu(III) and Yb(III) emissions. The study also examined the stability of Gd/YbNPs in solution (PBS) and ex vivo (mice blood serum). The high stability and monodispersity index make Gd/YbNPs highly attractive as a safe and controllable CPT-delivering platform for cancer therapy (Zang et al. 2021).

Confocal imaging and co-staining experiments were conducted to investigate the cellular uptake, localisation profiles, and in vitro CPT release of Gd/YbNPs. Results showed that Gd/YbNPs exhibited more intense blue CPT emission in HeLa cells than MRC-5 cells, with luminescence intensity increasing with prolonged incubation times. GSH treatment also increased NIR emission in HeLa cells, suggesting a more efficient release of CPT in cancer cells (Fig. 18.5a). The CPT release mechanism in Gd/YbNPs has better selectivity in cancer cells compared to free CPT (Fig. 18.5b). Co-staining experiments confirmed the effective and selective CPT release in cancer cell lines, with merged purple emission in HeLa cells (Fig. 18.5c) indicating the release was located in the mitochondria (Zang et al. 2021).

The study also investigated the potential of Gd/YbNPs as MRI contrast agents. The contrast-enhanced performance was evaluated in water and in vivo MR imaging of mice bearing HeLa tumours (Fig. 18.6a). GdNPs efficiently enhanced the T_1 signal in mice (Fig. 18.6b). However, Gd was detected in most organs, notably the liver and kidney, and a significant amount was found in the tumour site but not in the brain (Fig. 18.6c). The nanoparticles were further investigated for tumour inhibition therapy in a HeLa tumour-bearing xenograft model and compared to chemotherapy-induced tumour growth (CPT) (Fig. 18.6d). The tumour inhibition efficacy of CPT was maintained by Gd/YbNPs after release in vivo (Fig. 18.6e) (Zang et al. 2021).

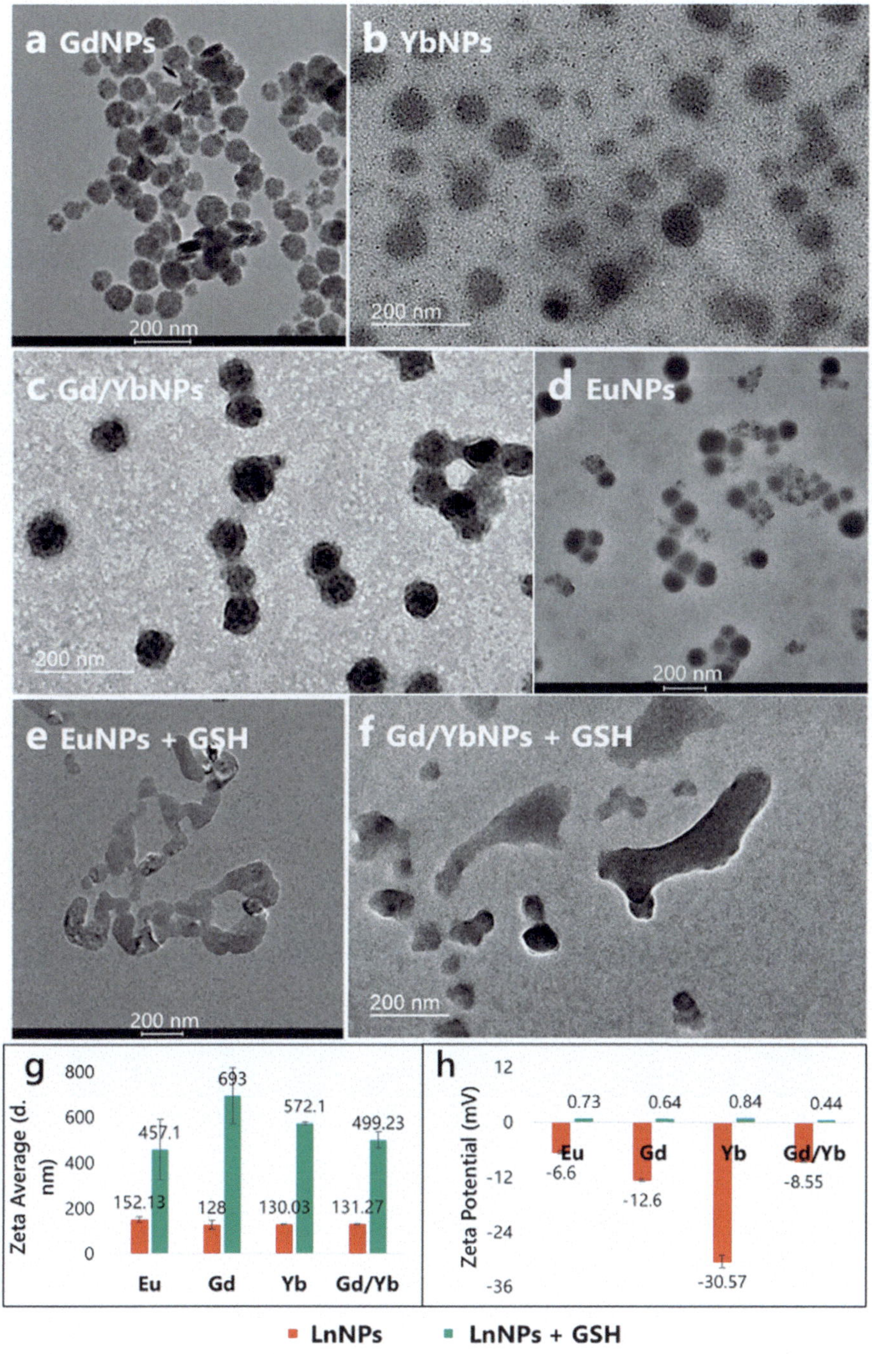

Fig. 18.4 TEM images of (**a**) GdNPs, (**b**) YbNPs, (**c**) Gd/YbNPs, (**d**) EuNPs, (**e**) EuNPs + GSH, and (**f**) Gd/YbNPs + GSH. (**g**) Size distribution and (**h**) ζ-potential of EuNPs, GdNPs, YbNPs, and Gd/YbNPs with GSH (in green bars) and without GSH (in red bars) determined by DLS. (Adapted from Zang et al. 2021, with kind permission of the copyright holder)

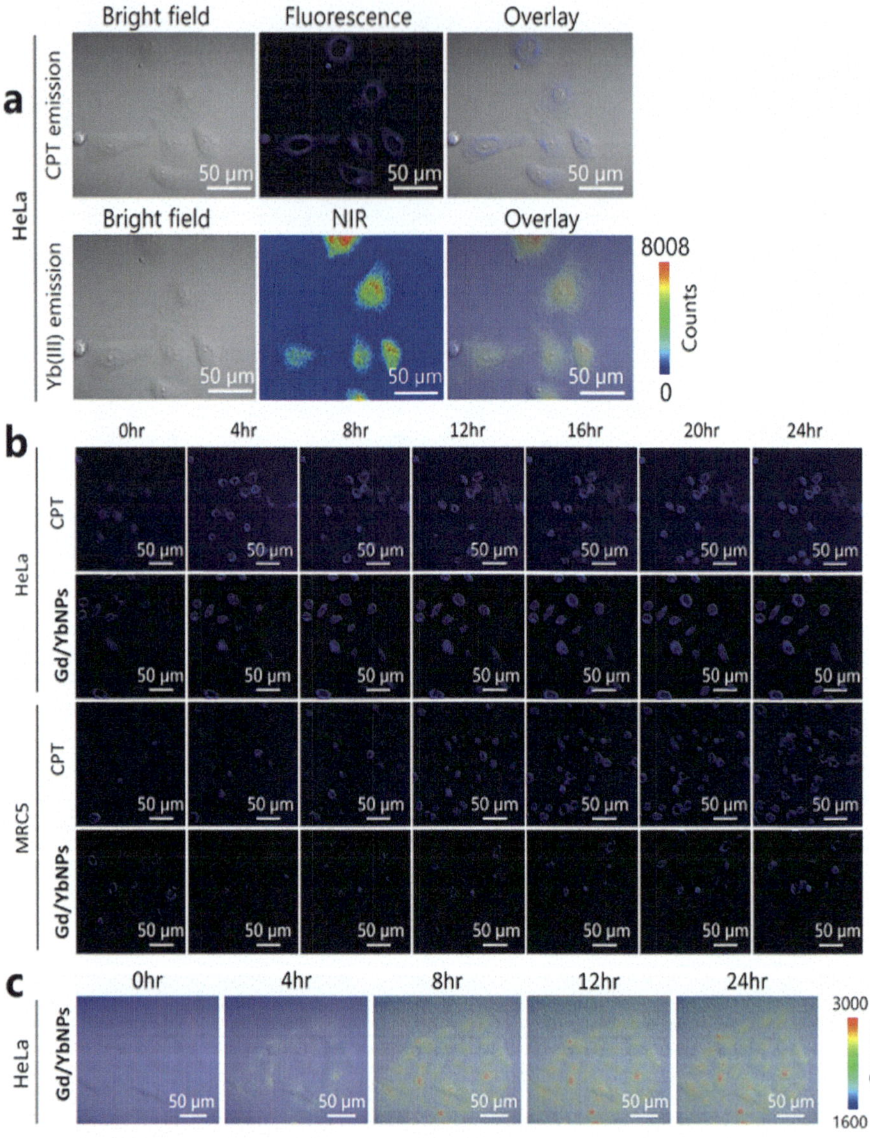

Fig. 18.5 (**a**) In vitro NIR imaging of 5 μM YbNPs in HeLa cells (incubation time: 24 h; λex: 370 nm for CPT emission, detection wavelength range: 400–450 nm; λex: 380 nm for Yb(III) emission, detection wavelength range: 900–1700 nm; scale bar: 50 μm). Time-dependent in vitro emission changes of (**b**) live imaging signal of the CPT moiety from free CPT only and from Gd/YbNPs in HeLa and MRC-5 cells (λex: 370 nm, detection wavelength range: 400–450 nm; scale bar: 50 μm) and (**c**) Yb(III) (λex: 380 nm) of 5 μM Gd/YbNPs by the NIR camera (λex: 380 nm, detection wavelength range: 900–1700 nm; scale bar: 50 μm). (Adapted from Zang et al. 2021 with kind permission of the copyright holder)

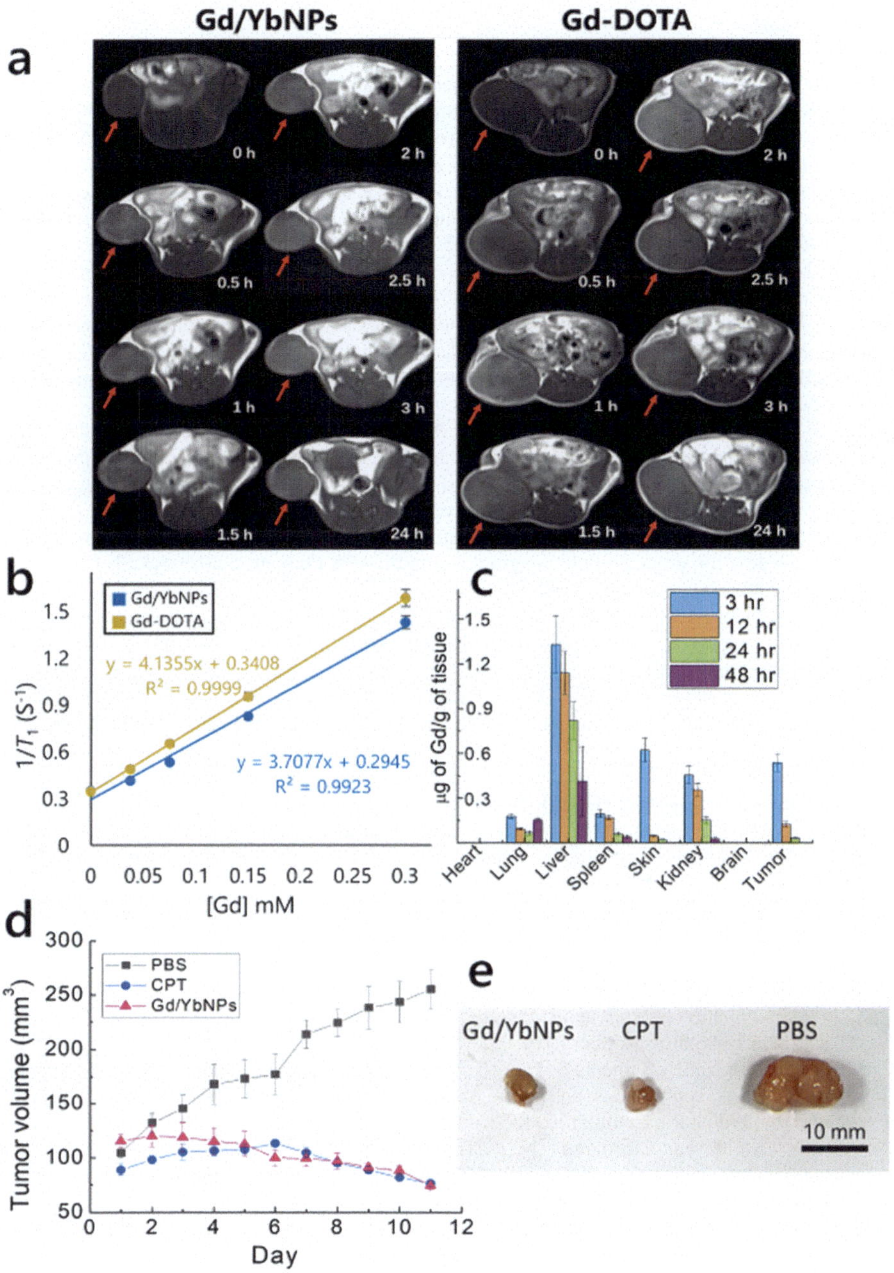

Fig. 18.6 (**a**) In vivo T1-weighted MR images of the mice bearing HeLa xenograft from 0 to 24 h after the injection of Gd/YbNPs and Gd-DOTA (control). (**b**) Plots of $1/T_1$ vs [Gd] for the determination of r1 of Gd-DOTA and Gd/YbNPs hybrid nanoparticles. (**c**) In vivo biodistribution of Gd/YbNPs in the mice bearing HeLa xenograft via ICP-MS. (**d**) In vivo antitumor activity of Gd/YbNPs, three mice per group, data are expressed as mean ± SEM. (**e**) Representative photo of tumours treated by Gd/YbNPs, CPT, and PBS control, scale bar = 10 mm. (Adapted from Zang et al. 2021 with kind permission of the copyright holder)

References

Alrushaid N, Khan FA, Al-Suhaimi EA, Elaissari A. Nanotechnology in cancer diagnosis and treatment. Pharmaceutics. 2023;15(3):1025. https://doi.org/10.3390/pharmaceutics15031025.

Anderson NM, Simon MC. The tumor microenvironment. Curr Biol. 2020;30(16):R921–5. https://doi.org/10.1016/j.cub.2020.06.081.

Anitha K, Chenchula S, Surendran V, Shvetank B, Ravula P, Milan R, Chikatipalli RRP. Advancing cancer theranostics through biomimetics: a comprehensive review. Heliyon. 2024;10(6):e27692. https://doi.org/10.1016/j.heliyon.2024.e27692.

Aronova MA, Leapman RD. Development of electron energy loss spectroscopy in the biological sciences. MRS Bull. 2012;37(1):53–62. https://doi.org/10.1557/mrs.2011.329.

Baeza A, Vallet-Regí M. Mesoporous silica nanoparticles as theranostic antitumoral nanomedicines. Pharmaceutics. 2020;12(10):957. https://doi.org/10.3390/pharmaceutics12100957.

Bavelaar BM, Lee BQ, Gill MR, Falzone N, Vallis KA. Subcellular targeting of Theranostic radionuclides. Front Pharmacol. 2018;9:996. https://doi.org/10.3389/fphar.2018.00996.

Burkett BJ, Bartlett DJ, McGarrah PW, Lewis AR, Johnson DR, Berberoğlu K, Pandey MK, Packard AT, Halfdanarson TR, Hruska CB, Johnson GB, Kendi AT. A review of theranostics: perspectives on emerging approaches and clinical advancements. Radiol Imaging Cancer. 2023;5(4):e220157. https://doi.org/10.1148/rycan.220157.

Casar JR, McLellan CA, Siefe C, Dionne JA. Lanthanide-based nanosensors: refining nanoparticle responsiveness for single particle imaging of stimuli. ACS Photonics. 2021;8(1):3–17. https://doi.org/10.1021/acsphotonics.0c00894.

Chavda VP, Khadela A, Shah Y, Postwala H, Balar P, Vora L. Current status of cancer nanotheranostics: emerging strategies for cancer management. Nano. 2023;7(4):368–79. https://doi.org/10.7150/ntno.82263.

Chehelgerdi M, Chehelgerdi M, Allela OQB, Pecho RDC, Jayasankar N, Rao DP, Thamaraikani T, Vasanthan M, Viktor P, Lakshmaiya N, Saadh MJ, Amajd A, Abo-Zaid MA, Castillo-Acobo RY, Ismail AH, Amin AH, Akhavan-Sigari R. Progressing nanotechnology to improve targeted cancer treatment: overcoming hurdles in its clinical implementation. Mol Cancer. 2023;22(1):169. https://doi.org/10.1186/s12943-023-01865-0.

Chen X, Wong STC. Chapter 1—cancer theranostics: an introduction. In: Chen X, Wong STC, editors. Cancer theranostics. Academic Press; 2014. p. 3–8. https://doi.org/10.1016/B978-0-12-407722-5.00001-3.

Cherry SR, Sorenson JA, Phelps ME. Chapter 5—radionuclide and radiopharmaceutical production. In: Cherry SR, Sorenson JA, Phelps ME, editors. Physics in nuclear medicine, vol. 4; 2012. p. 43–61. https://doi.org/10.1016/B978-1-4160-5198-5.00005-8.

Cline B, Delahunty I, Xie J. Nanoparticles to mediate X-ray-induced photodynamic therapy and Cherenkov radiation photodynamic therapy. Wiley Interdiscip Rev Nanomed Nanobiotechnol. 2019;11(2):e1541. https://doi.org/10.1002/wnan.1541.

Correia JH, Rodrigues JA, Pimenta S, Dong T, Yang Z. Photodynamic therapy review: principles, photosensitizers, applications, and future directions. Pharmaceutics. 2021;13(9):1332. https://doi.org/10.3390/pharmaceutics13091332.

Crişan G, Moldovean-Cioroianu NS, Timaru DG, Andrieş G, Căinap C, Chiş V. Radiopharmaceuticals for PET and SPECT imaging: a literature review over the last decade. Int J Mol Sci. 2022;23(9):5023. https://doi.org/10.3390/ijms23095023.

Cruz A, Barbosa J, Antunes P, Bonifácio VDB, Pinto SN. A glimpse into dendrimers integration in cancer imaging and theranostics. Int J Mol Sci. 2023;24(6):5430. https://doi.org/10.3390/ijms24065430.

Cutler CS, Smith CJ, Ehrhardt GJ, Tyler TT, Jurisson SS, Deutsch E. Current and potential therapeutic uses of lanthanide radioisotopes. Cancer Biother Radiopharm. 2000;15(6):531–45. https://doi.org/10.1089/cbr.2000.15.531.

Dekker Y, Le Dévédec SE, Danen EHJ, Liu Q. Crosstalk between hypoxia and extracellular matrix in the tumor microenvironment in breast cancer. Genes (Basel). 2022;13(9):1585. https://doi.org/10.3390/genes13091585.

El-Habeeb AA, El-Sayed MY, Alatawy IM, Refat MS. Preparation, spectroscopic and anticancer investigations of metal-drug complexes associated between flumequine antibiotic drug with lanthanum (III), samarium (III) and terbium (III) chloride. Bull Chem Soc Ethiop. 2024;38(3):671–84. https://doi.org/10.4314/bcse.v38i3.10.

Fan Q, Sun C, Hu B, Wang Q. Recent advances of lanthanide nanomaterials in tumor NIR fluorescence detection and treatment. Mater Today Bio. 2023;20:100646. https://doi.org/10.1016/j.mtbio.2023.100646.

Fass L. Imaging and cancer: a review. Mol Oncol. 2008;2(2):115–52. https://doi.org/10.1016/j.molonc.2008.04.001. Epub 2008 May 10. Erratum in: Mol Oncol. 2022;16(15):2896. https://doi.org/10.1002/1878-0261.13283

Fernandes DA. Review on metal-based theranostic nanoparticles for cancer therapy and imaging. Technol Cancer Res Treat. 2023;22:15330338231191493. https://doi.org/10.1177/15330338231191493.

Ferro-Flores G, Ancira-Cortez A, Ocampo-García B, Meléndez-Alafort L. Molecularly targeted lanthanide nanoparticles for cancer theranostic applications. Nanomaterials (Basel). 2024;14(3):296. https://doi.org/10.3390/nano14030296.

Grégoire V, Guckenberger M, Haustermans K, Lagendijk JJW, Ménard C, Pötter R, Slotman BJ, Tanderup K, Thorwarth D, van Herk M, Zips D. Image guidance in radiation therapy for better cure of cancer. Mol Oncol. 2020;14(7):1470–91. https://doi.org/10.1002/1878-0261.12751.

Gudkov SV, Shilyagina NY, Vodeneev VA, Zvyagin AV. Targeted radionuclide therapy of human tumors. Int J Mol Sci. 2015;17(1):33. https://doi.org/10.3390/ijms17010033.

Gupta D, Roy P, Sharma R, Kasana R, Rathore P, Gupta TK. Recent nanotheranostic approaches in cancer research. Clin Exp Med. 2024;24(1):8. https://doi.org/10.1007/s10238-023-01262-3.

Holik HA, Ibrahim FM, Elaine AA, Putra BD, Achmad A, Kartamihardja AHS. The chemical scaffold of theranostic radiopharmaceuticals: radionuclide, bifunctional chelator, and pharmacokinetics modifying linker. Molecules. 2022;27(10):3062. https://doi.org/10.3390/molecules27103062.

Holland JP, Williamson MJ, Lewis JS. Unconventional nuclides for radiopharmaceuticals. Mol Imaging. 2010;9(1):1–20.

Hong L, Li W, Li Y, Yin S. Nanoparticle-based drug delivery systems targeting cancer cell surfaces. RSC Adv. 2023;13(31):21365–82. https://doi.org/10.1039/d3ra02969g.

Huang YM, Hsu TY, Liu CY, Hsieh YC, Lai KY, Yang YW, Lo KY. Exploring the multifaceted impact of lanthanides on physiological pathways in human breast cancer cells. Toxicology. 2024;502:153731. https://doi.org/10.1016/j.tox.2024.153731.

Jadvar H, Chen X, Cai W, Mahmood U. Radiotheranostics in cancer diagnosis and management. Radiology. 2018;286(2):388–400. https://doi.org/10.1148/radiol.2017170346.

Jin W, Wang Q, Wu M, Li Y, Tang G, Ping Y, Chu PK. Lanthanide-integrated supramolecular polymeric nanoassembly with multiple regulation characteristics for multidrug-resistant cancer therapy. Biomaterials. 2017;129:83–97. https://doi.org/10.1016/j.biomaterials.2017.03.020.

Kane SM, Padda IS, Patel P, et al. Technetium-99m. [updated 2024 Feb 29]. In: StatPearls [internet]. Treasure Island (FL): StatPearls Publishing; 2024. https://www.ncbi.nlm.nih.gov/books/NBK559013/.

Kleynhans J, Sathekge M, Ebenhan T. Obstacles and recommendations for clinical translation of nanoparticle system-based targeted alpha-particle therapy. Materials (Basel). 2021;14(17):4784. https://doi.org/10.3390/ma14174784.

Kumar R, Kulkarni A, Nagesha DK, Sridhar S. *In vitro* evaluation of theranostic polymeric micelles for imaging and drug delivery in cancer. Theranostics. 2012;2(7):714–22. https://doi.org/10.7150/thno.3927.

Lau J, Rousseau E, Kwon D, Lin KS, Bénard F, Chen X. Insight into the development of PET radiopharmaceuticals for oncology. Cancers (Basel). 2020;12(5):1312. https://doi.org/10.3390/cancers12051312.

Lepareur N, Ramée B, Mougin-Degraef M, Bourgeois M. Clinical advances and perspectives in targeted radionuclide therapy. Pharmaceutics. 2023;15(6):1733. https://doi.org/10.3390/pharmaceutics15061733.

Li Y, Tang J, Pan DX, Sun LD, Chen C, Liu Y, Wang YF, Shi S, Yan CH. A versatile imaging and therapeutic platform based on dual-band luminescent lanthanide nanoparticles toward tumor metastasis inhibition. ACS Nano. 2016;10(2):2766–73. https://doi.org/10.1021/acsnano.5b07873.

Li H, Bai G, Lian Y, Li Y, Chen L, Zhang J, Xu S. Advances in near-infrared-activated lanthanide-doped optical nanomaterials: imaging, sensing, and therapy. Mater Des. 2023a;231 https://doi.org/10.1016/j.matdes.2023.112036.

Li H, Feng Y, Luo Q, Li Z, Li X, Gan H, Gu Z, Gong Q, Luo K. Stimuli-activatable nanomedicine meets cancer theranostics. Theranostics. 2023b;13(15):5386–417. https://doi.org/10.7150/thno.87854.

Liu S, Edwards DS. Bifunctional chelators for therapeutic lanthanide radiopharmaceuticals. Bioconjug Chem. 2001;12(1):7–34. https://doi.org/10.1021/bc000070v.

Liu G, Lovell JF, Zhang L, Zhang Y. Stimulus-responsive nanomedicines for disease diagnosis and treatment. Int J Mol Sci. 2020;21(17):6380. https://doi.org/10.3390/ijms21176380.

Luo X, Zhang C, Yu Z, Wen S, Xian Y. Recent advances in responsive lanthanide-doped luminescence nanoprobes in the near-infrared-II window. TrAC Trends Anal Chem. 2023;169 https://doi.org/10.1016/j.trac.2023.117368.

Luo Z, Mao D, Li X, Luo J, Gong C, Liu X. Lanthanide-based nanoparticles for cancer phototherapy. Coord Chem Rev. 2024:508. https://doi.org/10.1016/j.ccr.2024.215773.

Malatesta M. Transmission electron microscopy as a powerful tool to investigate the interaction of nanoparticles with subcellular structures. Int J Mol Sci. 2021;22(23):12789. https://doi.org/10.3390/ijms222312789.

Malik S, Muhammad K, Waheed Y. Emerging applications of nanotechnology in healthcare and medicine. Molecules. 2023;28(18):6624. https://doi.org/10.3390/molecules28186624.

Mejía-Méndez JL, Navarro-López DE, Sanchez-Martinez A, Ceballos-Sanchez O, Garcia-Amezquita LE, Tiwari N, Juarez-Moreno K, Sanchez-Ante G, López-Mena ER. Lanthanide-doped ZnO nanoparticles: unraveling their role in cytotoxicity, antioxidant capacity, and Nanotoxicology. Antioxidants (Basel). 2024;13(2):213. https://doi.org/10.3390/antiox13020213.

Mondal SB, Gao S, Zhu N, Liang R, Gruev V, Achilefu S. Real-time fluorescence image-guided oncologic surgery. Adv Cancer Res. 2014;124:171–211. https://doi.org/10.1016/B978-0-12-411638-2.00005-7.

Morris ZS, Wang AZ, Knox SJ. The radiobiology of radiopharmaceuticals. Semin Radiat Oncol. 2021;31(1):20–7. https://doi.org/10.1016/j.semradonc.2020.07.002.

Morrs LR. Thermochemical properties of yttrium, lanthanum, and the lanthanide elements and ions. Chem Rev. 1976;76(6):827–41. https://doi.org/10.1021/cr60304a007.

Plichta KA, Graves SA, Buatti JM. Prostate-specific membrane antigen (PSMA) theranostics for treatment of oligometastatic prostate cancer. Int J Mol Sci. 2021;22(22):12095. https://doi.org/10.3390/ijms222212095.

Qin X, Xu J, Wu Y, Liu X. Energy-transfer editing in lanthanide-activated upconversion nanocrystals: a toolbox for emerging applications. ACS Cent Sci. 2019;5(1):29–42. https://doi.org/10.1021/acscentsci.8b00827.

Rafique R, Kailasa SK, Park TJ. Recent advances of upconversion nanoparticles in theranostics and bioimaging applications. TrAC Trends Anal Chem. 2019;120:115646. https://doi.org/10.1016/j.trac.2019.115646.

Rathod DG, Muneer H, Masood S. Phototherapy. [updated 2023 Feb 16]. In: StatPearls [internet]. Treasure Island (FL): StatPearls Publishing; 2024. https://www.ncbi.nlm.nih.gov/books/NBK563140/.

Ren B, Chen B, Zhao J, Guo Y, ZhangX CX, Du Y, Deng Z, Zhu G, Wang F. Synthesis of Core–Shell ScF3 nanoparticles for thermal enhancement of upconversion. Chem Mater. 2021;33(1):158–63. https://doi.org/10.1021/acs.chemmater.0c03124.

Salih S, Alkatheeri A, Alomaim W, Elliyanti A. Radiopharmaceutical treatments for cancer therapy, radionuclides characteristics, applications, and challenges. Molecules. 2022;27(16):5231. https://doi.org/10.3390/molecules27165231.

Scimeca M, Bischetti S, Lamsira HK, Bonfiglio R, Bonanno E. Energy dispersive X-ray (EDX) microanalysis: a powerful tool in biomedical research and diagnosis. Eur J Histochem. 2018;62(1):2841. https://doi.org/10.4081/ejh.2018.2841.

Seleci M, Ag Seleci D, Scheper T, Stahl F. Theranostic liposome-nanoparticle hybrids for drug delivery and bioimaging. Int J Mol Sci. 2017;18(7):1415. https://doi.org/10.3390/ijms18071415.

Shi J, Kantoff PW, Wooster R, Farokhzad OC. Cancer nanomedicine: progress, challenges and opportunities. Nat Rev Cancer. 2017;17(1):20–37. https://doi.org/10.1038/nrc.2016.108.

Singh R, Bhateria R. Core-shell nanostructures: a simplest two-component system with enhanced properties and multiple applications. Environ Geochem Health. 2021;43(7):2459–82. https://doi.org/10.1007/s10653-020-00766-1.

Song Y, Zou J, Castellanos EA, Matsuura N, Ronald JA, Shuhendler A, Weber WA, Gilad AA, Müller C, Witney TH, Chen X. Theranostics—a sure cure for cancer after 100 years? Theranostics. 2024;14(6):2464–88. https://doi.org/10.7150/thno.96675.

Sun W, Li S, Tang G, Luo Y, Ma S, Sun S, Ren J, Gong Y, Xie C. Recent progress of nanoscale metal-organic frameworks in cancer theranostics and the challenges of their clinical application. Int J Nanomedicine. 2019;14:10195–207. https://doi.org/10.2147/IJN.S230524.

Sun J, Huangfu Z, Yang J, Wang G, Hu K, Gao M, Zhong Z. Imaging-guided targeted radionuclide tumor therapy: from concept to clinical translation. Adv Drug Deliv Rev. 2022;190:114538. https://doi.org/10.1016/j.addr.2022.114538.

Tang L, Li J, Pan T, Yin Y, Mei Y, Xiao Q, Wang R, Yan Z, Wang W. Versatile carbon nanoplatforms for cancer treatment and diagnosis: strategies, applications and future perspectives. Theranostics. 2022;12(5):2290–321. https://doi.org/10.7150/thno.69628.

Tinwala H, Shah DV, Menghani J, Pati R. Synthesis of La2Ce2O7 nanoparticles by co-precipitation method and its characterization. J Nanosci Nanotechnol. 2014;14(8):6072–6. https://doi.org/10.1166/jnn.2014.8834.

Tripathi SK, Kaur G, Khurana RK, Kapoor S, Singh B. Quantum dots and their potential role in cancer theranostics. Crit Rev Ther Drug Carrier Syst. 2015;32(6):461–502. https://doi.org/10.1615/critrevtherdrugcarriersyst.2015012360.

Vykoukal V, Bursik J, Roupcova P, Cullen DA, Pinkas J. Solvothermal hot injection synthesis of core-shell AgNi nanoparticles. J Alloys Compd. 2019;770:377–85. https://doi.org/10.1016/j.jallcom.2018.08.082.

Wang M, Abbineni G, Clevenger A, Mao C, Xu S. Upconversion nanoparticles: synthesis, surface modification and biological applications. Nanomedicine. 2011;7(6):710–29. https://doi.org/10.1016/j.nano.2011.02.013.

Wang SQ, Wang Y, Yang X, Liu Y, Li H, Yang Z, Sun WY, Sessler JL. High-nuclearity luminescent lanthanide nanocages for tumor drug delivery. Angew Chem Int Ed Engl. 2024;63(12):e202317775. https://doi.org/10.1002/anie.202317775.

Williams MJ, Grace SR, Manikantan V, Alexander A, Varalakshmi GS, Ramasamy S, Pillai AS, Enoch IV, Núñez D. Lanthanum-doped carbon dots for anticancer drug transport. Mater Lett. 2024;355:135540. https://doi.org/10.1016/j.matlet.2023.135540.

Xue Y, Gao Y, Meng F, Luo L. Recent progress of nanotechnology-based theranostic systems in cancer treatments. Cancer Biol Med. 2021;18(2):336–51. https://doi.org/10.20892/j.issn.2095-3941.2020.0510.

Yan CH, Yan ZG, Du YP, Shen J, Zhang C, Feng W. Controlled synthesis and properties of rare earth nanomaterials. In: Gschneidner KA, Bünzli JCG, Pecharsky VK, editors. Handbook on the physics and chemistry of rare earths, vol. 41. Elsevier; 2011. p. 275–472. https://doi.org/10.1016/B978-0-444-53590-0.00004-2.

Yavuz S, Puckett Y. Iodine-131 uptake study. [updated 2023 Oct 29]. In: StatPearls [internet]. Treasure Island (FL): StatPearls Publishing; 2024.: https://www.ncbi.nlm.nih.gov/books/NBK559283/.

Yetisgin AA, Cetinel S, Zuvin M, Kosar A, Kutlu O. Therapeutic nanoparticles and their targeted delivery applications. Molecules. 2020;25(9):2193. https://doi.org/10.3390/molecules25092193.

Yin M, Yuan K, Lv Z, Yuan Y, Feng X, Wu Q. Novel lanthanum-based fluorescent drug-loaded microspheres for tracing and inhibition of the HepG2 cells line. J Macromol Sci, Part B. 2024:1–2. https://doi.org/10.1080/00222348.2024.2316494.

Yinyan LI, Shiqing XU. Hydrothermal synthesis of lanthanide (hydr)oxide micro/nanorods in presence of tetrabutylammonium hydroxide. J Rare Earths. 2016;34:6. https://doi.org/10.1016/S1002-0721(16)60070-4.

Zang Y, Ma X, Chau HF, Thor W, Jiyang L, Zha S, Fok WY, Mak HM, Zhang J, Cai J, Ng CF, Li H, Parker D, Li L, Law GL, Wong KL. Lanthanide–Cyclen–Camptothecin nanocomposites for cancer theranostics guided by near-infrared and magnetic resonance maging. ACS Appl Nano Mater. 2021;4(1):271–8. https://doi.org/10.1021/acsanm.0c02597.

Zeng Y, Wang X, Zhu N, Yu Y, Wang X, Kang K, Wu Y, Yi Q. Magnetic lanthanide sensor with self-ratiometric time-resolved luminescence for accurate detection of epithelial cancerous exosomes. J Mater Chem B. 2024;12(29):7203–14. https://doi.org/10.1039/D4TB00497C.

Zhang Q, O'Brien S, Grimm J. Biomedical applications of lanthanide nanomaterials, for imaging, sensing and therapy. Nano. 2022;6(2):184–94. https://doi.org/10.7150/ntno.65530.

Zhao Q, Tian X, Ren L, Su Y, Su Q. Understanding of lanthanide-doped core–shell structure at the nanoscale level. Nano. 2024;14(12):1063. https://doi.org/10.3390/nano14121063.

Zheng K, Loh KY, Wang Y, Chen Q, Fan J, Jung T, Nam SH, Suh YD, Liu X. Recent advances in upconversion nanocrystals: expanding the kaleidoscopic toolbox for emerging applications. Nano Today. 2019:29. https://doi.org/10.1016/j.nantod.2019.100797.

Upconversion Core-Shell Nanoconstructs for Cancer Theragnostics

19

Aditya Singh, Shubhrat Maheshwari,
Vishal Kumar Vishwakarma, and Bhupendra Prajapati

Abstract

Cancer theragnostics, the integration of diagnostic and therapeutic modalities, has emerged as a promising strategy for personalized medicine. In this study, we present the design and characterization of upconversion core-shell nanoconstructs tailored for advanced cancer theragnostics. The nanoconstructs consist of a core of upconversion nanoparticles, which efficiently convert near-infrared (NIR) light into higher energy emission, surrounded by a versatile shell with multifunctional components. The upconversion core facilitates deep tissue penetration and minimizes background signal, making it an ideal candidate for imaging in the NIR window. Additionally, the core-shell architecture allows for the incorporation of various functional components, such as targeting ligands for specific cancer cell recognition, therapeutic agents for localized drug delivery, and imaging probes for real-time monitoring. The synthesis and optimization of these nanoconstructs were carried out systematically, ensuring their biocompatibility and stability under physiological conditions. In vitro and in vivo studies

A. Singh
Department of Pharmacy, Integral University, Lucknow, Uttar Pradesh, India

S. Maheshwari
Faculty of Pharmaceutical Sciences, Rama University, Kanpur, India

Bioorganic and Medicinal Chemistry Research Laboratory, Department of Pharmaceutical Sciences, am Higginbottom University of Agriculture Technology and Sciences, Prayagraj, India

V. K. Vishwakarma
Institute of Pharmacy, Deen Dayal Upadhyay Gorakhpur University, Gorakhpur, Uttar Pradesh, India

B. Prajapati (✉)
Shree S.K.Patel College of Pharmaceutical Education and Research, Ganpat University, Mehsana, Gujarat, India

J. K. Patel et al. (eds.), *Core-Shell Nano Constructs for Cancer Theragnostic*,
https://doi.org/10.1007/978-981-96-3025-7_19

demonstrate the efficient targeting of cancer cells, triggered drug release, and simultaneous imaging, providing a comprehensive platform for theragnostics. The versatility of the nanoconstructs enables customization for different cancer types, optimizing the therapeutic efficacy while minimizing off-target effects. Furthermore, the presented upconversion core-shell nanoconstructs exhibit enhanced photothermal and photodynamic therapy capabilities, offering a synergistic approach to cancer treatment. The combination of imaging, targeted drug delivery, and therapeutic interventions within a single nanoconstruct underscores the potential of these multifunctional platforms in advancing precision medicine for cancer patients. Overall, our study showcases the potential of upconversion core-shell nanoconstructs as a versatile and effective tool for cancer theragnostics, paving the way for the development of tailored and personalized approaches in the diagnosis and treatment of various malignancies.

Keywords

Cancer theragnostics · Upconversion · Core-shell · Triggered drug release · Off-target effects · Nanoconstruct · Core-shell nanoconstructs · Versatility

1 Introduction

Cancer, a formidable adversary in global health, continues to pose significant challenges, with breast tumors being a particular focal point for researchers (Abbasi et al. 2024). The uncontrolled proliferation and abnormal biological changes in cells define cancer, presenting a widespread threat that jeopardizes the well-being of millions worldwide (Adhalrao et al. 2024). The intricate development of malignant tumors involves gene mutations in epigenetic codes, intensifying the risk of harm to adjacent tissues (Alam et al. 2024). The conventional arsenal of chemical anti-cancer drugs, while effective, comes with severe long-term side effects. In this landscape, nanotechnology emerges as a beacon of hope, offering a promising avenue for more efficient and targeted cancer therapy (Asfiya et al. 2024). Nanoparticles (NPs), specifically those with dimensions below 100 nm, prove to be pivotal in designing drug delivery systems that overcome the limitations of conventional treatments (Ateş et al. 2024). The unique characteristics of NPs, such as their size and surface charge, play a crucial role in tumor penetration and systemic toxicity prevention during in vivo studies. Surface modifications, including various coatings like polymers, further enhance the capabilities of NPs, ensuring protection against rapid clearance in the bloodstream (Barba-Rosado et al. 2024). This adaptability in building and adjusting the surface properties of NPs positions them as sophisticated materials, ushering in a new era for tumor targeting and drug delivery.

The landscape of cancer treatment methods has witnessed a transformative wave with the harnessing of nanoparticles. From chemotherapy to innovative approaches like photothermal therapy, sonodynamic therapy, chemodynamic therapy, and radiotherapy, NPs have become instrumental in advancing diverse facets of cancer care. This chapter explores a particularly intriguing facet of nanotechnology, focusing on upconversion nanoparticles (UCNPs) designed for cancer therapy. UCNPs, with their core-shell structures incorporating lanthanide elements, present a paradigm shift in cancer therapy. The integration of photoinitiators and dyes within the core shell facilitates the generation of reactive oxygen species (ROS) in the tumor site through photodynamic therapy (PDT). This activation, in turn, induces caspase for apoptosis within the targeted cells. The use of dyes and photoinitiators as essential components in theranostic applications highlights the versatility of UCNPs in the evolving landscape of cancer research (Caro et al. 2021). The term ROS, crucial in the mechanism of UCNPs, encompasses short-lived diffusible radicals and medium-lifetime species generated by inflammatory cells. While conventional fluorophores face drawbacks like high background noise and photobleaching, UCNPs present a viable solution for in vitro and in vivo fluorescence imaging. The incorporation of Yb3+ and Tm3+ in UCNPs, with an excitation wavelength of 980 nm, yields NIR photoluminescence in the range of 750–850 nm (Chakravarty et al. 2017). This unique property allows UCNPs to navigate through tissues with minimal autofluorescence and light scattering, enhancing their imaging contrast. Moreover, UCNPs prove beneficial for designing multimodal imaging probes, serving as contrast agents in magnetic resonance imaging (MRI), computed tomography (CT), and positron emission tomography (PET).

The pivotal components of UCNPs—host matrix, sensitizer (absorber), and activator—set the stage for a detailed exploration of their design principles. The host matrix, in fulfilling specific requirements like enduring the presence of bright centers, maintaining modest phonon energy, ensuring significant transparencies, and exhibiting excellent chemical and thermal stability, forms the foundation for the successful integration of UCNPs in cancer therapy. Lanthanide dopants, utilized as activators, play a crucial role, and their concentration is carefully calibrated to prevent undesirable interactions. Ytterbium (Yb) stands out as a widely employed sensitizer, aligning with the energy transitions of lanthanides used as activators (Chen et al. 2024). The narrative extends to the synthesis and applications of UCNPs, with a particular focus on their luminescence capabilities, penetration depth, and biocompatibility. Noteworthy examples include NaYF4-related UCNPs and the progress made with CaF2 NPs in therapeutic and diagnostic applications. The ongoing efforts in the laboratory, centered on CaF2 host matrix with lanthanide doping, demonstrate a commitment to advancing UCNPs for bioimaging and NIR-regulating drug release. The incorporation of heterogeneous shells aims to enhance the observed therapeutic efficacy of core-shell UCNPs, moving closer to the targeted levels in nanomedicine. UCNPs require thorough characterization (Table 19.1), including physicochemical properties, biological interactions, and potential toxicity in various environments, ensuring clinical applicability and safety (Chen et al. 2020).

Five distinct mechanisms govern upconversion: excited state absorption (ESA), photon avalanche (PA), energy transfer upconversion (ETU), cooperative upconversion (CUC), and energy migration-mediated upconversion (EMU) (Girija and Balasubramanian 2019). Among these, ETU emerges as particularly popular due to its high upconversion efficiency, requiring two nearby ions to pump a photon of the

Table 19.1 Methods for assessing the physicochemical characteristics of UCNPs

S. no	Characterization	Result
1	Analytical ultracentrifugation (AUC)	Hydrodynamic diameter (Chen et al. 2020)
2	Agarose gel electrophoresis	hydrodynamic diameter and surface charge (Colaco et al. 2023)
3	Scanning electron microscopy (SEM	Size and surface shape (Dhaliwal et al. 2024)
4	Isothermal calorimetry	Binding constants and thermody-namic parameters (Dilnawaz et al. 2017)
5	Atomic force microscopy (AFM)	Adhesion forces and surface free energy (Dongsar et al. 2024)
6	Quartz crystal microbalance (QCM)	Interactions result in mass changes (Du et al. 2022)
7	X-ray diffraction (XRD)	Crystal size and structure (Falagan-Lotsch et al. 2017)

same energy. This mechanism, along with others, forms the basis for the high-throughput multiplex molecule detection capabilities of UCNPs, characterized by good photostability, tunable multicolor emission, and non-overlapping emission bands (Girija and Balasubramanian 2019). Despite the remarkable capabilities of UCNPs, challenges persist, limiting their transition from the laboratory bench to clinical practice. Quantum yield limitations and issues related to excitation (Goel et al. 2017).

2 Mechanism of UCNPs

Upconversion core-shell nanoconstructs play a pivotal role in cancer theragnostics by absorbing multiple low-energy photons and converting them into high-energy photons within host materials, often referred to as activators (Sahu et al. 2023). The efficacy of these nanoparticles relies heavily on the phonon energies of the host materials, aiming to minimize nonradiative energy losses. Additionally, the accessibility of cut-off phonon energy, particularly concerning the dominant peak of Raman spectroscopy, is a crucial determinant (Bhattacharya et al. 2022). Materials characterized by low phonon energy, such as NaYF4:Yb/Er (~350 cm^{-1}), stand out as promising candidates for UCNPs. Notably, NaYF4:Yb, Tm@NaF4 UCNP demonstrates the capability to transform 980 nm wavelength light into emission at 365 nm (Gupta et al. 2024). The use of near-infrared (NIR) excitation light provides deeper penetration compared to ultraviolet (UV) light, coupled with reduced photodamage, autofluorescence, light scattering, and phototoxicity. The mechanism of UCNPs involves five essential steps: excited-state absorption (ESA), energy transfer upconversion (ETU), cooperative sensitization upconversion (CSU), cross-relaxation (CR), and photon avalanche. Understanding these processes is paramount for unlocking the full potential of upconversion core-shell nanoconstructs in biomedical applications (Guryev et al. 2018).

3 Current Research on Cancer Theranostic Nanoparticles Based on MSN

Mesoporous silica nanoparticles (MSN) have become versatile platforms in current cancer theranostic research, seamlessly integrating diagnostic and therapeutic functionalities (Hu et al. 2019). Their unique properties, including a high surface area, tunable pore size, and excellent biocompatibility, make them ideal candidates for simultaneous drug delivery and imaging (Huang et al. 2020). In drug delivery, MSN efficiently carry various chemotherapeutic agents, utilizing tunable pores for controlled release kinetics, ensuring optimal drug concentrations at the target site while minimizing off-target effects (Jeong et al. 2021). Surface functionalization enhances specificity, enabling targeted delivery to cancer cells. MSN-based drug delivery systems also facilitate combination therapy by co-loading multiple therapeutic agents, offering a synergistic approach to enhance treatment efficacy (Jia et al. 2017). The multifunctionality of MSN extends beyond drug delivery to imaging, serving as contrast agents for various modalities such as magnetic resonance imaging (MRI), computed tomography (CT), and fluorescence imaging. Integrating diagnostic capabilities with therapeutic functions allows real-time monitoring of treatment responses and personalized medicine approaches (Jin et al. 2020). Ongoing research reflects the promising potential of MSN in advancing cancer treatment strategies. Efforts are directed toward improving biocompatibility, stability, and mitigating potential toxicity concerns. Surface modifications, including biocompatible coatings or stimuli-responsive moieties, overcome limitations, while integration with other nanomaterials explores hybrid systems with synergistic properties (Kostiv et al. 2021). In imaging, MSN is explored for multimodal applications, enhancing comprehensive cancer diagnostics (Kowalik et al. 2017). Theranostic MSN formulations responsive to the tumor microenvironment aim for precise and targeted therapeutic interventions. Exploration beyond traditional drug delivery and imaging includes the integration of MSN with emerging technologies like nanomedicine and immunotherapy. MSN's role as carriers for immunomodulators and their ability to modulate genes through nucleic acid loading contribute to advancements in cancer immunotherapy and precision medicine (Kumar et al. 2023). In regenerative medicine, MSN's potential for tissue regeneration and repair broadens its therapeutic scope. Efforts are underway to optimize systemic circulation, biodistribution, and clearance challenges in MSN-based theranostics (Kumar et al. 2024). Strategies for responsive MSN designs, such as stimuli-responsive behavior in the tumor microenvironment, improve targeted drug release and minimize systemic side effects. Diagnostic innovation focuses on novel imaging modalities and contrast agents, while collaborative interdisciplinary approaches involving material scientists, chemists, biologists, and clinicians contribute to holistic understanding and translation from bench to bedside. As the field evolves, cutting-edge technologies such as artificial intelligence (AI) and machine learning enhance data analysis from MSN-based diagnostics and therapeutics (Kumar et al. 2019). These computational approaches contribute to refining treatment strategies, predicting patient responses, and optimizing future MSN-based theranostic systems (Kuthati et al. 2013).

4 Synthesis of UCNPs and Surface Modification of UCNPs

The synthesis and surface modification of upconversion nanoparticles (UCNPs) are integral steps in optimizing their utility for diverse biomedical applications (Fig. 19.1). During synthesis, methods like thermal decomposition, coprecipitation, and hydrothermal techniques are employed to control the nanocrystal properties such as size, shape, and crystal structure. However, the inherent hydrophobicity and limited surface functional groups of UCNPs pose challenges to their effective use in biological environments. To overcome these limitations, strategic surface modifications are essential (Le et al. 2024). Silica coating forms core-shell structures, enhancing hydrophilicity without compromising luminescent properties. Ligand exchange introduces hydrophilic ligands, improving water solubility and enabling bioconjugation. The hydrophobic-hydrophilic interaction involves introducing amphiphilic molecules, facilitating water solubility, and providing active sites for bioconjugation and drug loading. These modifications collectively enhance the biocompatibility, water solubility, and functionality of UCNPs, unlocking their potential for advanced theranostic applications in medicine. Upconversion nanostructures (UCNs) show promise in theranostic applications, particularly in cancer diagnosis (Table 19.2). By conjugating ligands or aptamers onto UCN-based carriers, they can recognize and attach to specific biomarkers on cancer cells. Upon NIR light activation, UCL allows for targeted site visualization, distinguishing cancer from normal cells. The pH-responsive UCN carriers enhance tumor-specific accumulation, delivering therapeutic agents like DOX. Advanced designs, such as pH-controlled dual drug delivery systems combining UCNs and MOFs, showcase synergistic effects for improved cytotoxicity and bioimaging potential (Li et al. 2017).

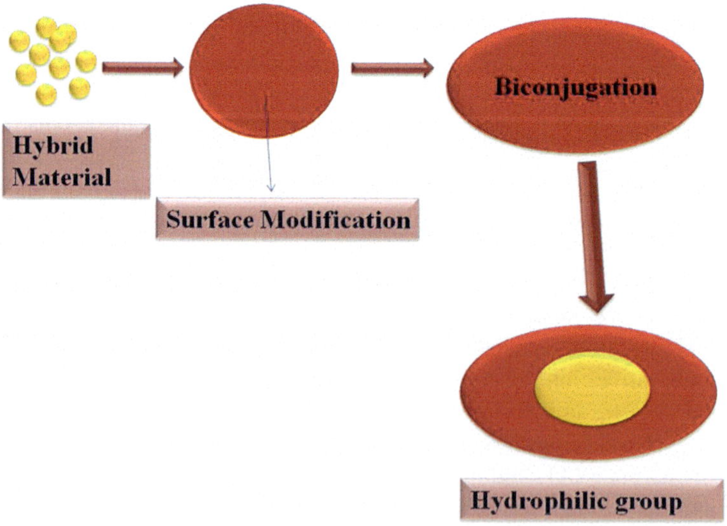

Fig. 19.1 The surface modification of UCNPs to regulate their characteristics

Table 19.2 UCN composition and surface modifications optimize theranostic applications in biomedicine

Surface modification strategies	Composition of upconversion nanostructures	Therapy	In-vivo	In-vitro
Silica and manganese dioxide coating, PEGylated surface	NaGdF4:20%Yb,2%Er@ NaGdF4:20%Yb NPs	Loading DOX for chemotherapy. Loading chlorin e6 for PDT.	U14 tumor bearing mice (UCL, T1-weighted MR and CT imaging)	HeLa cells (Li et al. 2019)
Silica and cerium oxide coating, PEGylated surface	NaGdF4:20%Yb,1%Tm@ NaGdF4	Cerium oxide for PDT. DOX for chemotherapy.	U14 tumor-bearing mice. (UCL, T1-weighted MR and CT imaging)	L929 cells and HeLa cells (Li et al. 2020)
PEGylated	NaYF4:18%Yb,0.6%Tm@ NaYF	Resonant excitation 2F5/2 → 2F7/2 of Yb3+ for PTT	A375 Male Balb/c nu/nu mice	A375 and HEK 293 cells (Liang et al. 2017)
PEG and folic acid-modified	NaYF4:25%Yb,0.3%Tm NPs	DOX for chemotherapy. MoS2 for PDT	–	HeLa and HepG2 cells (Liao et al. 2017)
Stabilization by polyetherimide, followed by chitosan wrapped surface	NaYF4:Yb/Er	Pyropheophorbide a for PDT. Conjugate with RGD peptide c for targeting.	–	U87-MG cells (Lin et al. 2018)
Gelatin-modified	BaGdF5:20%Yb3+,2%Tm3 + @ BaGdF5:x%Yb3+ Ultra-small NPs	DOX for chemotherapy. pH triggered drug releasing.	Anesthetized white Kunming mice (UCL)	HeLa cells (Liu et al. 2024)
Encapsulation into liposomes	NaYF4:60%Yb,2%Er	DOX and methylene blue for chemotherapy and PDT, respectively. Conjugated to the anti-HER2 peptide to target breast cancer cells.	–	SKBR-3 breast cancer cell lines (Lu et al. 2022)
Mesoporous silica coating	NaYF4:25%Yb,2%Er,0.5%Tm NPs	Rose Bengal and zinc(II) phthalocyanine for PDT	–	HeLa cells (Lv et al. 2018)

4.1 Thermal Decomposition

The thermal decomposition synthesis of upconversion nanoparticles (UCNPs) involves the high-temperature breakdown of organometallic precursors, such as metal trifluoroacetate, in the presence of high-boiling organic solvents like 1-octadecene at temperatures around 300 °C. Long-chain hydrocarbons and functional groups, including oleic acid (COOH), oleyl amine (NH2), and trioctyl phosphine oxide (PO3H), are often incorporated into the synthesis to prevent nanoparticle aggregation. Yan and colleagues initially applied this technique to generate LaF3 triangular nanoplates and later improved it for producing high-quality cubic (α-phase) and hexagonal (β-phase) UCNPs in the Na (RE) F4 host crystal. Further utilized trifluoroacetate precursors to synthesize UCNPs in the cubic NaYF4 host crystal. Octadecene and oleic acid serve as a solvent and passivation ligand, ensuring the separation of nucleation and growth processes. Notably, the addition of precursors needs to be slow to achieve highly monodispersed nanoparticles (Mahata et al. 2021).

While thermal decomposition yields high-quality UCNPs with pure crystal phases, strong upconversion emission, and large production volumes, it comes with challenges such as the need for air-sensitive precursors, anaerobic and water-free reaction environments, and potential toxicity concerns. Alternative solvents, like paraffin liquid, have been proposed to mitigate some of these challenges, offering a more stable and cost-effective option, especially in biological applications. Thermal decomposition, hydrothermal decomposition, and coprecipitation are widely employed methods in UCNP synthesis, with each method offering unique advantages. Thermal decomposition provides precise control over crystallinity, while hydrothermal decomposition and coprecipitation yield stable UCNPs with varying morphologies. The sol-gel method and combustion approach contribute to the versatility of UCNP synthesis, with scalability and simplicity, respectively. However, challenges may arise in achieving uniform size distribution and stability. The synthesis of UCNPs is a critical aspect that significantly influences their performance characteristics. Switching focus, natural fibers derived from plant sources, with constituents like cellulose, hemicellulose, lignin, pectin, and waxes, pose challenges such as poor thermal properties and degradation (Mishra and Ahmed 2024).

The temperature-dependent nature of thermal stability and moisture absorption properties makes these fibers susceptible to degradation and the release of volatile products during elevated temperatures. Strategies to address flammability challenges include the use of nanoparticles, fire retardant coatings, impregnation of fibers with fire retardants, non-flammable binders, resins, polymer matrices, and composite insulation. Commonly used fire retardants include ammonium, halogens, boron, phosphorous, bromine, aluminum and magnesium-based compounds, zinc borate, silica, graphite, and alkaline earth metal compounds. Phosphorous-based fire retardants exhibit auto-extinguishing behavior, while bromine-based compounds terminate chemical reactions during combustion. Despite their effectiveness, phosphorous-based fire retardants face environmental and health concerns.

This well-established technique involves the controlled decomposition of lanthanide-based precursors under elevated temperatures, providing precise control over nanoparticle characteristics (Moghaddam et al. 2024).

4.2 Coprecipitation

Coprecipitation is a widely employed synthesis method in the production of upconversion nanoparticles (UCNPs), offering a relatively straightforward and cost-effective approach. This method involves the simultaneous precipitation of multiple ions or compounds from a solution, leading to the formation of solid nanocrystals. In the context of UCNPs, coprecipitation is particularly valuable for obtaining spherical or irregularly shaped nanoparticles. The process typically starts with precursor salts or compounds containing the necessary ions for the desired UCNP composition. These precursors are dissolved in a solvent, forming a solution. The addition of a precipitating agent triggers the simultaneous precipitation of the various components, resulting in the formation of nanocrystals. In the case of UCNPs, the precipitating agent initiates the conversion of precursor ions into solid nanocrystals with specific optical properties. One advantage of coprecipitation is its simplicity and scalability. It is relatively easy to perform and can be adapted for large-scale production. The straightforward nature of coprecipitation makes it an attractive choice for synthesizing UCNPs with diverse applications, ranging from bioimaging to drug delivery. However, coprecipitation does come with certain challenges. Achieving precise control over the size and uniformity of the resulting UCNPs can be more challenging compared to other synthesis methods. Variations in reaction conditions, such as temperature, pH, and concentration, can influence the morphology and properties of the nanoparticles. Despite its limitations, coprecipitation remains a valuable tool in the toolkit of UCNP synthesis methods. Its simplicity and cost-effectiveness make it suitable for various applications where precise control over nanocrystal properties may not be the primary concern. Researchers continue to explore and optimize coprecipitation techniques, aiming to enhance the reproducibility and versatility of UCNPs for diverse biomedical uses (Nady et al. 2023).

4.3 Sol-Gel Method

The sol-gel method is a versatile and widely used synthesis technique for producing upconversion nanoparticles (UCNPs) with controlled properties, including size, composition, and morphology. This method is particularly valuable in tailoring UCNPs for diverse applications in biomedical fields, such as bioimaging and drug delivery. In the sol-gel method, the process begins with a sol, which is a colloidal suspension of nanoparticles in a liquid phase (Mohite et al. 2023). This sol undergoes gelation, transforming into a three-dimensional network of interconnected nanoparticles. The gel is then subjected to further processing steps, such as drying and annealing, to obtain the final UCNP product. Several key advantages make the sol-gel method attractive for UCNP synthesis. Firstly, it offers precise control over the composition of the nanoparticles, allowing for the incorporation of various dopants, including lanthanide ions, to achieve specific optical properties. Secondly, the method allows for the formation of intricate core-shell structures and surface modifications by introducing additional materials during the gelation process. This flexibility enhances the versatility of UCNPs for various applications. The sol-gel method also facilitates the production of UCNPs with well-defined crystallinity, contributing to enhanced optical performance. The controlled conditions of the sol-gel process allow for the optimization of particle size and morphology, which are critical factors in determining the properties of UCNPs. Moreover, the sol-gel method is scalable, making it suitable for both laboratory-scale research and large-scale production. Its adaptability and versatility have led to the development of diverse sol-gel approaches, such as reverse micelle, microemulsion, and template-assisted methods, each offering unique advantages for UCNP synthesis (Osuchowski et al. 2021).

4.4 Combustion Approach

The combustion approach is an effective and straightforward synthesis method for producing upconversion nanoparticles (UCNPs) with unique optical properties. This method is known for its simplicity, cost-effectiveness, and ability to generate UCNPs in a single step. The combustion approach is particularly valuable in tailoring UCNPs for various applications, ranging from bioimaging to drug delivery. In the combustion approach, precursor materials containing the necessary components for UCNPs are mixed and ignited. The combustion of the mixture generates high temperatures, leading to the rapid formation of UCNPs. This process typically involves the use of fuel-rich environments, such as glycine, which serves both as a fuel and a capping agent. One of the notable advantages of the combustion approach is its simplicity and efficiency. The synthesis occurs in a short time frame, making it suitable for high-throughput production. Additionally, the method does not require complex equipment or elaborate reaction setups, contributing to its cost-effectiveness. The combustion approach allows for the production of UCNPs with well-defined crystalline structures, influencing their optical properties and upconversion efficiency (Park et al. 2024).

The rapid reaction kinetics and high temperatures contribute to the formation of nanocrystals with unique optical transitions, crucial for applications in bioimaging and sensing. Furthermore, the combustion approach enables the incorporation of various dopants, such as lanthanide ions, to achieve specific luminescent properties. This flexibility in composition, combined with the simplicity of the method, makes it attractive for researchers seeking to customize UCNPs for specific biomedical applications. While the combustion approach offers advantages in terms of simplicity and efficiency, researchers need to carefully optimize reaction conditions to control particle size, morphology, and other properties. Despite these challenges, the combustion approach remains a valuable tool for the rapid and cost-effective synthesis of UCNPs with tailored optical characteristics for diverse biomedical uses (Parry and Pandey 2017).

4.5 Luminescent Properties

The luminescent properties of upconversion nanoparticles (UCNPs) are central to their functionality in various biomedical applications, particularly in bioimaging and sensing. Understanding and manipulating these properties is crucial for optimizing the performance of UCNPs in diverse settings. UCNPs exhibit unique luminescent behavior, wherein they absorb lower-energy photons, typically in the near-infrared (NIR) region, and emit higher-energy photons through an upconversion process. This phenomenon allows UCNPs to overcome the limitations of traditional fluorophores, such as autofluorescence and tissue penetration issues, making them valuable for deep-tissue imaging. The key factors influencing the luminescent properties of UCNPs include the choice of core materials, dopants, crystal structures, and surface modifications. Lanthanide-doped cores, incorporating ions like erbium (Er), ytterbium (Yb), and thulium (Tm), play a pivotal role in the upconversion process. The specific transitions of these ions contribute to the emission of photons at shorter wavelengths, resulting in a variety of colors in the visible and ultraviolet spectra (Patel et al. 2024a).

Crystal structures, such as hexagonal-phase NaYF4 and cubic-phase NaGdF4, also impact luminescent properties. The well-defined phases contribute to efficient upconversion, influencing the intensity and color of emitted light. Additionally, strategic doping of UCNPs with different lanthanide ions allows for precise control over emission wavelengths, facilitating customization for specific applications. Surface modifications, including silica coatings and ligand exchanges, can influence the luminescent properties of UCNPs. Silica-coated UCNPs, for instance, provide a protective layer that enhances stability without compromising luminescence. Ligand exchanges with molecules like polyethylene glycol (PEG) can improve water solubility and biocompatibility, ensuring optimal performance in biological environments. The luminescent properties of UCNPs are not only crucial for bioimaging but also for applications such as biosensing and therapeutics. The ability to customize emission wavelengths, intensity, and overall luminescent behavior provides researchers with a versatile toolkit to design UCNPs tailored to specific biomedical needs. Continued advancements in understanding and manipulating the luminescent properties of UCNPs pave the way for innovative applications in diagnostics, imaging, and targeted therapies. As the field evolves, researchers explore novel core materials, surface modifications, and design strategies to further enhance the luminescent capabilities of UCNPs for a wide range of biomedical applications (Pillarisetti et al. 2019).

4.6 Size Distribution

Size distribution is a crucial parameter in the synthesis of upconversion nanoparticles (UCNPs) as it directly impacts their optical properties, stability, and performance in various biomedical applications. Achieving a narrow and uniform size distribution is essential for ensuring consistency and reliability in the behavior of UCNPs. The size distribution of UCNPs refers to the range of particle sizes present within a sample and is typically characterized by metrics such as the mean size, standard deviation, and polydispersity index (Singh et al. 2023). Controlling size distribution is essential for tailoring the optical characteristics of UCNPs, as their upconversion efficiency and emission spectra are often size-dependent. Several synthesis methods influence the size distribution of UCNPs. For instance, thermal decomposition allows for precise control over crystallinity and, consequently, size. Hydrothermal decomposition and coprecipitation methods may lead to a broader size distribution, requiring careful optimization of reaction conditions to achieve uniformity. Size distribution also plays a crucial role in the stability of UCNPs (Prasad and Selvaraj 2024). Monodisperse UCNPs are less prone to aggregation, ensuring better colloidal stability and preventing potential issues in biological applications. Silica coating is often employed to improve the stability of UCNPs by providing an additional layer that hinders particle aggregation. In biomedical applications such as bioimaging and drug delivery, a narrow size distribution is desirable for consistent and controlled behavior within biological systems. The ability to precisely control the size distribution of UCNPs allows researchers to tailor their properties for specific applications, optimizing their performance in diagnostics, imaging, and therapeutic interventions. Ongoing research focuses on refining synthesis techniques to achieve narrower and more uniform size distributions of UCNPs. Advances in controlling size variability contribute to the continual improvement of UCNPs, enhancing their versatility and reliability in the ever-expanding field of nanomedicine (Purkait and Sontakke 2023).

4.7 Stability

Stability is a critical aspect of upconversion nanoparticles (UCNPs) that significantly influences their performance and applicability, especially in biomedical and theranostic applications. The stability of UCNPs refers to their ability to maintain their physical and chemical properties over time and under various environmental conditions. The stability of UCNPs is influenced by factors such as the core material, surface modifications, and the surrounding medium. Crystal structures with well-defined phases, such as hexagonal or cubic, contribute to the stability of UCNPs. Additionally, the choice of core material, such as lanthanide-doped cores or silica-coated UCNPs, can impact stability under different conditions. Surface modifications, including coatings like silica or polymeric hybrids, play a crucial role in enhancing the stability of UCNPs. Silica-coated UCNPs, for instance, provide an additional layer that acts as a protective shell, preventing aggregation and improving colloidal stability (Qin et al. 2023). These modifications not only enhance stability but also contribute to biocompatibility, making UCNPs suitable for applications in biological systems. The stability of UCNPs is particularly important for their performance in bioimaging, drug delivery, and theranostic applications. In biological environments, where UCNPs may encounter varying pH levels, ionic strength, and other factors, maintaining stability ensures consistent and reliable behavior. Furthermore, stability is essential for the long-term storage and transport of UCNPs, allowing for their widespread distribution and use in diverse applications. Researchers continuously explore strategies to improve the stability of UCNPs, considering factors such as surface modifications, core material selection, and the incorporation of stabilizing agents. As UCNPs continue to advance in nanomedicine and related fields, achieving and maintaining stability is a key focus of research. The development of stable UCNPs contributes to their versatility and effectiveness in addressing challenges and realizing the full potential of these nanoparticles in various biomedical applications (Qin et al. 2023).

4.8 Morphology

Morphology plays a pivotal role in determining the physical structure and characteristics of upconversion nanoparticles (UCNPs), influencing their interactions with biological systems and performance in various applications. The synthesis methods employed significantly impact the morphology of UCNPs, and different shapes offer distinct advantages for specific biomedical applications. Thermal decomposition, a common synthesis technique, yields UCNPs with well-defined shapes. This method provides precise control over the crystallinity of the nanoparticles, resulting in uniform and reproducible morphologies. Hydrothermal decomposition, an alternative approach, leads to various morphologies, such as nanorods and nanowires. The morphology of UCNPs can be tailored based on reaction conditions, offering flexibility in design. Coprecipitation is another method that can yield spherical or irregularly shaped UCNPs (Rezende et al. 2022). The sol-gel method and combustion approach contribute to the versatility of UCNP synthesis, allowing for diverse morphologies. While thermal decomposition offers control over crystallinity, hydrothermal decomposition, and coprecipitation provide stable UCNPs with varying morphologies. The sol-gel method and combustion approach offer scalability and simplicity, with potential challenges in achieving uniform size distribution and stability. The choice of morphology is crucial for optimizing UCNPs for specific applications. For instance, mesoporous silica-coated UCNPs with a hollow core encapsulated by a shell (yolk-shell structure) allow for efficient loading of therapeutic agents within the core, making them promising candidates for drug delivery applications. Nanorods and nanowires, obtained through hydrothermal decomposition, may exhibit enhanced properties for certain imaging or therapeutic applications. Understanding and controlling the morphology of UCNPs are essential for tailoring their properties to meet the demands of various biomedical scenarios. Ongoing advancements in synthesis techniques and a comprehensive understanding of the relationship between morphology and performance contribute to the continued optimization of UCNPs for applications ranging from bioimaging to therapeutic interventions. Interdisciplinary collaborations and innovative research in this field are vital for driving advancements in the design and application of UCNPs with diverse morphologies (Rodrigues and Correia 2023).

5 Types of Core Materials and their Properties

The choice of core materials plays a pivotal role in determining the properties and functionalities of upconversion nanoparticles (UCNPs) for diverse applications. Lanthanide-doped cores, featuring ions like erbium (Er), ytterbium (Yb), and thulium (Tm), contribute to the upconversion process, enabling unique optical transitions and making them widely employed in bioimaging and sensing applications. Crystal structures such as hexagonal-phase NaYF4 and cubic-phase β-NaGdF4 enhance stability and upconversion efficiency, making them suitable for robust performance in various environmental conditions. Silica-coated UCNPs improve biocompatibility, reduce potential cytotoxicity, and offer a versatile platform for surface functionalization, enabling targeted drug delivery (Shanwar et al. 2020).

Magnetic core materials like iron oxide (Fe3O4) impart additional functionalities, allowing controlled movement and localization for applications such as magnetic targeting in drug delivery and magnetic resonance imaging (MRI). Gold nanoparticles serve as core materials, introducing unique optical properties and enhanced photothermal effects, making them promising candidates for combined photothermal therapy and imaging applications. Perovskite-based UCNPs, emerging as novel core materials, offer tunable bandgaps and excellent photoluminescence properties, showing promise in applications requiring precise control over emission wavelengths. Quantum dots (QDs) provide exceptional tunability of emission wavelengths, facilitating diverse imaging and sensing applications. UCNPs with a yolk-shell structure, featuring a hollow core encapsulated by a shell, offer efficient loading of therapeutic agents, making them promising candidates for drug delivery applications. Strategic doping of core materials with different lanthanide ions or other dopants allows for fine-tuning of optical properties, providing opportunities for customizing UCNPs for various applications as shown in Fig. 19.2. Overall, the selection of core materials in UCNPs is a critical determinant of their optical, magnetic, and structural properties, influencing their performance in diagnostics, imaging, drug delivery, and therapeutic interventions (Shapoval et al. 2023).

6 Properties of Core Materials

The properties of core materials in upconversion nanoparticles (UCNPs) play a pivotal role in shaping their overall characteristics, influencing their behavior in various applications ranging from bioimaging to drug delivery. Lanthanide-doped cores, incorporating ions such as erbium (Er), ytterbium (Yb), and thulium (Tm), serve as the foundation for UCNPs' unique optical properties. These dopants

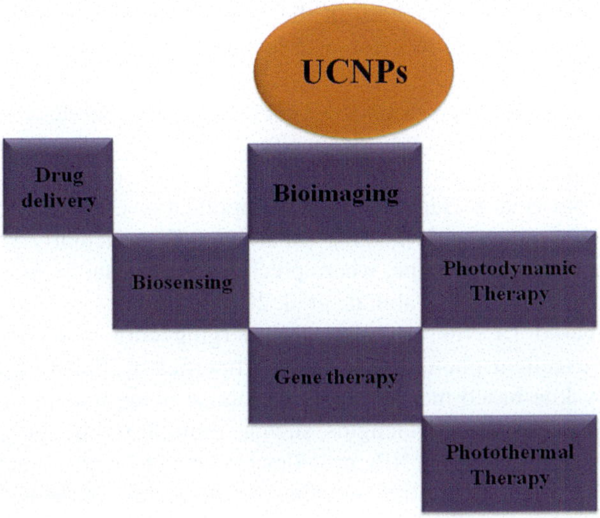

Fig. 19.2 The possible biomedical uses of UCNPs

facilitate efficient upconversion processes, contributing to the generation of higher-energy photons from lower-energy excitation sources. This intrinsic property is particularly advantageous in bioimaging and sensing applications, where the ability to harness specific optical transitions is crucial for signal detection and analysis. The crystal structure of UCNPs is another fundamental aspect that directly impacts their properties. Hexagonal-phase NaYF4 and cubic-phase β-NaGdF4 are examples of crystal structures that enhance the stability and upconversion efficiency of UCNPs. The choice of crystal structure not only ensures robust performance under various environmental conditions but also contributes to maximizing the conversion of incident light into higher energy emissions. This efficiency is paramount in applications where signal strength and precision are critical, such as in medical imaging (Sikora-Dobrowolska et al. 2024).

Biocompatibility is a key consideration for UCNPs intended for use in biological systems. Silica-coated UCNPs and polymeric hybrids represent core materials that enhance biocompatibility and reduce potential cytotoxicity. The addition of biocompatible coatings ensures that UCNPs can be safely utilized in bioimaging and drug delivery applications without adverse effects on living organisms. Furthermore, the core material's influence on surface functionalization is a crucial factor. Silica-coated UCNPs, for instance, provide a versatile platform for additional functionalization, enabling researchers to tailor these nanoparticles for specific applications, including targeted drug delivery and enhanced specificity in bioimaging (Sneider et al. 2017).

Magnetic core materials, such as iron oxide (Fe_3O_4), introduce additional functionalities to UCNPs. These magnetic UCNPs can be manipulated and guided to specific locations using external magnetic fields, enabling targeted drug delivery with spatial precision. The integration of magnetic properties broadens the scope of applications for UCNPs, especially in the field of magnetic resonance imaging (MRI) where controlled movement and localization are essential for accurate imaging.

Gold nanoparticles, when utilized as core materials, introduce unique optical properties and surface plasmon resonance to UCNPs. This characteristic leads to enhanced photothermal effects, making gold-core UCNPs promising candidates for combined photothermal therapy and imaging applications. The ability to integrate therapeutic interventions with imaging capabilities is particularly valuable in the context of personalized medicine, where precise targeting and monitoring of treatment responses are essential (Tahiliani et al. 2024).

Perovskite-based UCNPs represent an emerging core material with tunable bandgaps and excellent photoluminescence properties. CsPbX3-based UCNPs, a subset of perovskite-based materials, show promise in applications requiring precise control over emission wavelengths, such as multiplexed imaging. The tunable nature of perovskite-based UCNPs provides researchers with a versatile tool for tailoring nanoparticles to specific imaging requirements, allowing for advancements in diagnostic capabilities.

Quantum dots (QDs) serve as core materials that provide exceptional tunability of emission wavelengths. Quantum dot-based UCNPs allow for a broad range of colors in upconversion luminescence, facilitating diverse imaging and sensing applications. The ability to manipulate emission colors enhances the versatility of UCNPs in multiplexed imaging scenarios, where the simultaneous detection of multiple targets is desirable (Tang et al. 2019).

The yolk-shell structure in UCNPs introduces a hollow core encapsulated by a shell, offering unique properties for drug delivery applications. This design allows for efficient loading of therapeutic agents within the core, making yolk-shell UCNPs promising candidates for drug delivery applications. The high loading capacity within the hollow core enhances the efficiency of drug delivery systems, providing a potential avenue for targeted therapeutic interventions.

Fine-tuning the properties of UCNPs involves strategic doping of the core material with different lanthanide ions or other dopants. This approach enables precise control over emission wavelengths, quantum efficiency, and overall performance. By carefully selecting and manipulating the dopants within the core, researchers can tailor UCNPs to exhibit specific optical characteristics, ensuring their suitability for a wide array of applications.

Magnetic core materials, in addition to introducing magnetization properties to UCNPs, enable controlled movement and localization. This capability is particularly valuable in applications like magnetic targeting in drug delivery, where the ability to guide nanoparticles to specific locations enhances the precision and effectiveness of therapeutic interventions (Tang et al. 2018).

Gold-core UCNPs leverage surface plasmon resonance effects, enhancing their photothermal properties. This feature is advantageous in applications requiring both imaging and therapeutic interventions. The integration of gold as a core material showcases the multifunctionality of UCNPs, where a single nanoparticle can serve dual roles in imaging and therapy.

Certain core materials, such as perovskite and quantum dots, offer tunable properties. Researchers can tailor UCNPs for specific applications by adjusting bandgaps and emission wavelengths. This level of customization ensures that UCNPs can be optimized for diverse imaging and therapeutic scenarios, accommodating the unique requirements of each application (Tao et al. 2020).

Yolk-shell UCNPs, characterized by a high loading capacity within the hollow core, emerge as efficient carriers for therapeutic agents in drug delivery applications. The design of the yolk-shell structure allows for precise control over the release of therapeutic payloads, ensuring targeted and controlled drug delivery.

Doping the core material represents a sophisticated approach for achieving precise control over the optical properties of UCNPs. By introducing specific lanthanide ions or other dopants into the core, researchers can fine-tune the emission wavelengths, quantum efficiency, and overall performance of UCNPs. This level of control offers opportunities for customizing the performance of UCNPs to suit the requirements of different applications, showcasing the versatility of these nanoparticles in the realm of nanomedicine (Patel et al. 2024b).

7 Types of Shell Materials and Their Properties

The properties of shell materials in upconversion nanoparticles (UCNPs) play a crucial role in determining their overall behavior, stability, and functionality in diverse biomedical applications. The choice of shell material contributes to the nanoparticles' biocompatibility, surface functionality, and protection of the core, influencing their performance in areas such as bioimaging, drug delivery, and theranostics (Targhotra and Chauhan 2024). One common shell material used in UCNPs is silica (SiO_2). Silica-coated UCNPs form core-shell structures where the core, typically composed of lanthanide-doped materials, is encapsulated by a silica shell. This silica coating serves multiple purposes. Firstly, it enhances the hydrophilicity and biocompatibility of UCNPs, making them suitable for biological applications. Secondly, the silica shell provides a protective barrier, preventing the leaching of toxic elements from the core and improving the stability of UCNPs under physiological conditions. Importantly, the silica coating does not compromise the photoluminescent properties of UCNPs, allowing for efficient upconversion while maintaining biocompatibility.

Another approach involves the use of polymer-based shell materials in hybrid structures. The combination of UCNPs with polymers offers flexibility, allowing for the tailoring of properties based on specific application requirements. Polymer shells enhance the biocompatibility of UCNPs and can be engineered for targeted drug delivery. Polyethylene glycol (PEG), for example, is a commonly used polymer for surface modification due to its hydrophilic nature and ability to reduce nonspecific interactions with biological components, thus extending the circulation time of UCNPs in the bloodstream (Thanasekaran et al. 2018). Metal-organic frameworks (MOFs) represent another category of shell materials for UCNPs. MOFs are porous structures formed by metal ions or clusters coordinated with organic ligands. The unique porous nature of MOFs allows for the loading and release of therapeutic agents within the shell. This property is particularly advantageous for drug delivery applications, providing a controlled release mechanism. Additionally, MOFs can contribute to the overall stability of UCNPs.

The ligand exchange method is widely employed for surface modification and shell formation on UCNPs. Ligands such as citric acid, poly(acrylic acid), polyethyleneimine (PEI), hyaluronic acid, 3-mercaptopropionic acid, and 11-mercaptoundecanoic acid have been utilized to stabilize UCNPs and increase their water solubility. The choice of ligands can influence the surface charge, stability, and interaction with biological entities (Tong et al. 2020). Amphiphilic molecules, including surfactants (e.g., TWEEN), synthetic polymers (e.g., PEG), natural polymers (e.g., polysaccharide polymers), and proteins (e.g., transferrin), are employed for hydrophilic-hydrophobic interactions. These interactions contribute to improved water solubility and provide active sites for bioconjugation on the hydrophilic segment, while the hydrophobic area facilitates the loading of hydrophobic drugs. This versatility allows for the customization of UCNPs for specific applications, combining imaging with therapeutic functionalities. The properties of shell materials in UCNPs are dynamic and can be tailored based on the desired application. The continuous exploration of new shell materials and optimization of existing ones contribute to the versatility and expanding utility of UCNPs in the rapidly evolving field of nanomedicine (Tripathi et al. 2023).

8 Upconversion Nanoparticles with Porous Core Structures

The incorporation of mesoporous silica as a core material in upconversion nanoparticles (UCNPs) introduces a porous structure, revolutionizing drug delivery systems. This design enhances drug-loading capacities, allowing for more efficient delivery with controlled release kinetics. The mesoporous silica core provides an ideal platform for loading therapeutic agents, contributing to the development of advanced drug delivery systems that hold significant promise in precision medicine. UCNPs designed for theranostic applications showcase the integration of core materials that enable simultaneous imaging and therapeutic functions. This innovative approach allows for real-time monitoring of treatment responses, marking a significant advancement in the field of personalized medicine (Tsai et al. 2018). The integration of imaging and therapeutic agents within a single nanoparticle enhances the efficiency of diagnostics and treatment strategies. In the pursuit of enhancing the applicability of UCNPs in biomedical settings, designing UCNPs with core materials exhibiting near-infrared (NIR) absorption becomes crucial. Such core materials extend the penetration depth of UCNPs in biological tissues, enabling applications in deep-tissue imaging and therapy. This advancement addresses a critical challenge in biomedical research, paving the way for more effective imaging and therapeutic interventions in deeper tissues. Exploring radiosensitive core materials, characterized by elements with high atomic numbers, reveals their potential to improve the radiosensitivity of cancer cells (Veeranarayanan and Maekawa 2019). This opens new avenues for combining radiation therapy with the unique properties of UCNPs, offering the prospect of enhanced therapeutic outcomes in cancer treatment. The integration of radiosensitive core materials represents a synergistic approach that aligns nanotechnology with radiation therapy for more targeted and effective cancer treatments. Temperature-sensitive cores contribute to the development of UCNPs that respond to specific environmental cues, introducing stimuli-responsive capabilities. This innovation provides opportunities for controlled drug release and optical modulation, allowing for precise and on-demand therapeutic interventions. The integration of temperature-sensitive core materials adds a dynamic dimension to the field of UCNPs, enabling tailored responses to varying physiological conditions. Multimodal UCNPs, integrating diverse core materials, offer versatility in imaging applications. This design allows for the collection of complementary information from different imaging modalities, enhancing the overall diagnostic capabilities of UCNPs. The integration of multiple core materials in a single nanoparticle provides a multifaceted approach to imaging, enabling more comprehensive insights into biological systems. UCNPs with bio-orthogonal chemistry-friendly cores provide a precise functionalization platform, supporting the development of targeted and specific bioimaging probes (Verma et al. 2024). This capability facilitates the creation of imaging agents that can be tailored for specific biomolecules or cellular targets, enhancing the specificity and accuracy of bioimaging applications. Mesoporous silica cores continue to prove effective in the realm of drug delivery for UCNPs. Their porous structure offers an efficient platform for drug loading, resulting in improved loading capacities and controlled release kinetics. This feature is instrumental in advancing drug delivery systems, contributing to the development of more effective and patient-friendly therapeutic interventions. Tailoring core materials specifically for theranostic UCNPs facilitates the seamless integration of diagnostic and therapeutic functions. This approach represents a significant leap towards personalized medicine, where real-time monitoring and simultaneous treatment can be customized for individual patients. The development of theranostic UCNPs holds immense potential for advancing patient care and treatment strategies. UCNPs designed with near-infrared (NIR)-absorbing core materials demonstrate enhanced applicability in biomedical settings. This is particularly advantageous for applications in deep-tissue imaging and therapies requiring increased tissue penetration. The utilization of NIR-absorbing core materials addresses a critical challenge in biomedical research, enhancing the potential of UCNPs in diverse clinical applications (Vyas and Patel 2024).

9 Specific Surface Functionalization Strategies

Surface functionalization strategies play a pivotal role in optimizing the properties and applications of mesoporous silica nanoparticles (MSN) in the context of cancer theranostics. One prominent approach involves amine functionalization, wherein amino groups are introduced onto the MSN surface, enhancing reactivity for controlled drug loading and allowing for further modification. Polyethylene glycol (PEG) coating is another widely employed strategy, imparting stealth properties to MSN surfaces, reducing nonspecific interactions, and improving circulation time in the bloodstream, making them suitable for drug delivery applications (Wang et al. 2019). Ligand conjugation, such as with folic acid or antibodies, enables targeted delivery to cancer cells, enhancing the selectivity of MSN-based therapeutics. A pH-responsive functionalization involves incorporating pH-sensitive moieties onto MSN surfaces, enabling selective drug release in the acidic tumor microenvironment (Wang et al. 2020). Stimuli-responsive coatings, like temperature-sensitive polymers, provide controlled drug release in response to environmental cues. Incorporating targeting peptides, such as arginine-glycine-aspartic acid (RGD) peptides, enhances MSN's recognition and binding to cancer cells. Lastly, enzyme-responsive coatings enable selective drug release triggered by specific enzymes in the tumor microenvironment. These specific surface functionalization strategies collectively contribute to the development of multifunctional MSN platforms, offering enhanced targeting specificity, controlled drug release, and improved overall biocompatibility for advanced cancer theranostics. Continued advancements in surface functionalization strategies for mesoporous silica nanoparticles (MSN) aim to address specific challenges and expand their applications in cancer theranostics. Researchers are exploring innovative modifications to enhance the precision and effectiveness of MSN-based platforms. Strategies such as dual-functionalization, combining two or more surface modifications, offer synergistic effects, providing enhanced targeting capabilities and controlled drug release. Moreover, efforts are directed toward developing "smart" MSN systems with responsive coatings that can dynamically adapt to the tumor microenvironment. The pH-responsive and enzyme-responsive coatings enable tailored drug release, ensuring therapeutic agents are deployed precisely when and where needed. This contributes to minimizing off-target effects and optimizing treatment outcomes. In addition to functionalization for drug delivery, MSN surfaces are being tailored for advanced imaging capabilities. Integration with contrast agents for various imaging modalities, including magnetic resonance imaging (MRI), positron emission tomography (PET), and fluorescence imaging, enhances MSN's diagnostic potential. These multifunctional MSN systems enable real-time monitoring of treatment responses, supporting personalized medicine approaches. Collaborative efforts involving materials scientists, chemists, biologists, and clinicians are crucial for the translation of these innovative surface functionalization strategies into clinical applications. Preclinical studies, involving animal models, provide valuable insights into safety, efficacy, and feasibility (Wang et al. 2023). Clinical trials are underway to evaluate the performance of MSN-based theranostic platforms in human subjects, with a focus on assessing tolerability, pharmacokinetics, and therapeutic outcomes. The integration of artificial intelligence (AI) and machine learning in the analysis of complex data generated by MSN-based diagnostics and therapeutics is becoming increasingly prevalent. These computational approaches contribute to refining treatment strategies, predicting patient responses, and optimizing the design of future MSN-based theranostic systems. As research in surface functionalization strategies for MSN progresses, the goal is to develop versatile and customizable platforms that can revolutionize cancer theranostics. The synergy between innovative surface modifications, advanced imaging capabilities, and personalized treatment approaches holds promise for more effective, targeted, and integrated cancer management. Recent innovations in the realm of surface functionalization for mesoporous silica nanoparticles (MSN) within cancer theranostics are characterized by a focus on precision, multifunctionality, and safety. Advances include the integration of dual-functionalization strategies that combine different surface modifications synergistically. For instance, combining amine functionalization with pH-responsive or stimuli-responsive coatings provides a versatile platform with enhanced targeting precision and controlled drug release, offering a dynamic response to the intricacies of the tumor microenvironment. Biocompatibility has become a central theme in recent developments, with researchers exploring biomimetic coatings and smart polymers to mimic biological interfaces. These approaches aim to reduce immune responses, improve compatibility, and extend circulation time. Concurrently, the integration of specific targeting ligands, such as aptamers or peptides, enriches the surface of MSN with a heightened level of specificity, ensuring tailored interactions with cancer cells (Xie et al. 2023). The evolving landscape of theranostics is witnessing a growing emphasis on simultaneous imaging and therapy. Dual-functionalized MSN, equipped with both imaging contrast agents and therapeutic payloads, enables real-time monitoring of drug delivery and treatment response. This integration contributes to a more comprehensive understanding of the therapeutic process, facilitating personalized treatment strategies. Safety considerations are paramount in the design of next-generation MSN, leading to innovations that enhance biodegradability. Surface modifications, including enzyme-triggered degradation pathways and the use of biocompatible coatings, address concerns related to potential toxicity, paving the way for safer clinical applications. The horizon of surface functionalization is expanding into the realm of gene therapies. Researchers are exploring the incorporation of nucleic acids onto MSN surfaces, unlocking possibilities for precision medicine. Tailoring MSNs to individual genetic profiles holds promise for personalized and targeted cancer treatment strategies. In the ongoing evolution of surface functionalization for mesoporous silica nanoparticles (MSNs) in cancer theranostics, researchers are exploring novel approaches to enhance precision, versatility, and therapeutic outcomes. Advanced strategies now involve the integration of dual-functionalization, combining multiple surface modifications for synergistic effects. For instance, the combination of amine functionalization with pH-responsive or stimuli-responsive coatings offers a dynamic platform that not only enhances targeting capabilities but also provides controlled drug release tailored to the complex conditions of the tumor microenvironment. Biocompatibility remains a key focus, with biomimetic coatings and smart polymers emerging as innovative solutions to mimic biological interfaces (Xu et al. 2019). These advancements aim to minimize immune responses, improve compatibility, and extend circulation time. Simultaneously, the incorporation of specific targeting ligands, such as aptamers or peptides, adds an extra layer of precision to MSN, ensuring tailored interactions with cancer cells. Recent developments in theranostics underscore the importance of simultaneous imaging and therapy. Dual-functionalized MSN, equipped with imaging contrast agents and therapeutic payloads, enables real-time monitoring of drug delivery and treatment response. This integration contributes to a more comprehensive understanding of the therapeutic process, facilitating personalized treatment strategies. Safety considerations are guiding innovations, with a focus on enhancing biodegradability. Surface modifications, including enzyme-triggered degradation pathways and the use of biocompatible coatings, address concerns related to potential toxicity, paving the way for safer clinical applications. The horizon of surface functionalization extends into the realm of gene therapies, with researchers exploring the incorporation of nucleic acids onto MSN surfaces. This opens up avenues for precision medicine, where MSN can be tailored to individual genetic profiles, promising personalized and targeted cancer treatment strategies (Xu et al. 2018).

10 Future Perspectives and Challenges

Despite the remarkable capabilities of UCNPs, challenges persist, limiting their transition from the laboratory bench to clinical practice. Quantum yield limitations and issues related to the excitation mechanism demand focused research efforts. Additionally, the optimization of synthesis techniques and scalability for large-scale production is crucial for the practical implementation of UCNPs in cancer therapy.

Continued advancements in the laboratory, particularly those centered around host matrix modifications and lanthanide doping, demonstrate a commitment to advancing UCNPs for bioimaging and NIR-regulating drug release. The incorporation of heterogeneous shells aims to enhance the observed therapeutic efficacy of core-shell UCNPs, moving closer to the targeted levels in nanomedicine (Xu et al. 2020).

In the quest for the seamless integration of UCNPs into mainstream cancer treatments, a multifaceted approach is required. Collaborative efforts between materials scientists, chemists, biologists, and clinicians are crucial for translating these innovative nanotechnological advancements into clinical applications. Preclinical studies, involving animal models, provide valuable insights into safety, efficacy, and feasibility. Clinical trials are underway to evaluate the performance of UCNP-based theranostic platforms in human subjects, with a focus on assessing tolerability, pharmacokinetics, and therapeutic outcomes.

Furthermore, the integration of artificial intelligence (AI) and machine learning in the analysis of complex data generated by UCNP-based diagnostics and therapeutics is becoming increasingly prevalent. These computational approaches contribute to refining treatment strategies, predicting patient responses, and optimizing the design of future UCNP-based theranostic systems. The synergy between innovative surface modifications, advanced imaging capabilities, and personalized treatment approaches holds promise for more effective, targeted, and integrated cancer management (Yang et al. 2019).

As research in UCNPs progresses, the goal is to develop versatile and customizable platforms that can revolutionize cancer theranostics. The convergence of nanotechnology, surface functionalization, and computational approaches stands as a testament to the relentless pursuit of innovative solutions for the global health challenge posed by cancer. The intricate landscape of cancer research and therapy is witnessing a transformative shift fueled by the innovative integration of nanotechnology, particularly the application of UCNPs and MSN. These nanoparticles, with their unique properties and multifunctional capabilities, offer a beacon of hope in overcoming the formidable challenges posed by cancer, especially in the case of breast tumors. UCNPs, characterized by their core-shell structures and integration of lanthanide elements, have emerged as promising candidates for efficient and targeted cancer therapy. Their ability to induce apoptosis through photodynamic therapy (PDT) and serve as contrast agents in various imaging modalities presents a paradigm shift in cancer care. Concurrently, MSN, with advanced surface functionalization strategies, showcases precision, versatility, and safety in theranostics (Yao et al. 2020). The evolution towards dual-functionalization, biomimetic coatings, and smart polymers signifies a focus on enhancing targeting specificity, controlled drug release, and overall biocompatibility. Looking ahead, future perspectives center on addressing current challenges associated with UCNPs, including quantum yield limitations, and further optimizing their therapeutic efficacy. The exploration of multimodal imaging and therapeutics, integrating UCNPs into multifunctional platforms for simultaneous imaging and therapy, promises a more comprehensive understanding of treatment processes. Surface functionalization strategies for MSN are progressing towards advanced precision, multifunctionality, and safety. The incorporation of nucleic acids onto MSN surfaces opens new avenues for personalized medicine, tailoring cancer treatments to individual genetic profiles. The collaborative integration of artificial intelligence and machine learning in data analysis adds another layer of sophistication, refining treatment strategies and predicting patient responses. In this evolving landscape, the convergence of nanotechnology, surface functionalization, and computational approaches holds the key to reshaping the future of cancer theranostics (Zhang et al. 2023, 2020; Zhao et al. 2022, 2020, 2018; Živojević et al. 2021).

11 Conclusions and Future Perspectives

The successful integration of UCNPs in cancer therapy heavily relies on the careful calibration of the concentration of lanthanide dopants, with ytterbium (Yb) being a widely employed sensitizer. Understanding the mechanism of UCNPs is crucial for unlocking their full potential in biomedical applications. Excited state absorption (ESA), energy transfer upconversion (ETU), cooperative sensitization upconversion (CSU), cross-relaxation (CR), and photon avalanche collectively define the steps involved in upconversion. To elaborate on the intricacies of the UCNPs' mechanism, it's essential to highlight the specific surface functionalization strategies employed. Surface functionalization plays a pivotal role in optimizing the properties and applications of UCNPs in the context of cancer theranostics. Various strategies enhance the versatility of UCNPs, making them effective tools in the fight against cancer. One key strategy involves tailoring the surface of UCNPs with specific ligands or coatings, which not only improves their stability but also facilitates targeted drug delivery. For instance, the incorporation of polymers as coatings enhances the biocompatibility of UCNPs and prevents their rapid clearance in the bloodstream. Additionally, the modification of UCNPs with ligands such as folic acid or antibodies enables targeted delivery to cancer cells, thereby enhancing the selectivity of the therapeutic approach. The adaptability of UCNPs extends beyond drug delivery to include advanced imaging capabilities. The unique property of UCNPs to emit NIR photoluminescence allows for highly efficient fluorescence imaging with minimal interference from background signals, autofluorescence, or light scattering. This makes UCNPs well-suited for both in vitro and in vivo fluorescence imaging applications. Furthermore, UCNPs are instrumental in the design of multimodal imaging probes, serving as contrast agents for various imaging modalities such as MRI, CT, and PET. This capability allows for a comprehensive and detailed assessment of the tumor microenvironment, aiding in precise diagnostics and monitoring treatment responses. In the quest for improved therapeutic outcomes, researchers are exploring innovative modifications to UCNPs, such as the incorporation of heterogeneous shells. These modifications aim to enhance the therapeutic efficacy of core-shell UCNPs, bringing them closer to achieving targeted levels in the field of nanomedicine. The journey towards the clinical application of UCNPs involves addressing challenges related to their quantum yield limitations and excitation mechanisms. Ongoing research efforts focus on refining synthesis techniques and optimizing the scalability of UCNPs for large-scale production. Bridging the gap between laboratory experimentation and clinical practice requires interdisciplinary collaboration and a comprehensive understanding of UCNPs' behavior in complex biological systems.

References

Abbasi K, Siddiqui K, Bano S, Iqbal S, Shaikh SA. Future trends and innovation in nano drug delivery for cancer therapy, application of siRNA (nanoparticle-based RNA) therapy, ultrasound linked nano-cancer therapeutics, and application of exosomes-based cancer therapy. In Nano drug delivery for cancer therapy: principles and practices. Singapore: Springer Nature Singapore; 2024. pp. 197–251.

Adhalrao SB, Jadhav KR, Patil PL, Kadam VJ. Engineering Platelet Membrane Imitating Nanoparticles for Targeted Therapeutic Delivery. Current Pharmaceutical Biotechnology. 2024;25(10):1230–44.

Alam MS, Garg M, Bhalla V, Kumari P, Kadyan R, Anjum MM, Yadav T, Yadav M, Kumar M, Waziri A, Munawar SM. Trends in theranostic applications of metallic nanoparticles. In Metallic nanoparticles for health and the environment. CRC Press; 2024. pp. 121–159.

Asfiya R, Xu L, Paramanantham A, Kabytaev K, Chernatynskaya A, McCully G, Yang H, Srivastava A. Physio-chemical Modifications to Re-engineer Small Extracellular Vesicles for Targeted Anticancer Therapeutics Delivery and Imaging. ACS Biomaterials Science & Engineering. 2024;10(2):697–722.

Ateş B, Ulu A, Asiltürk M, Noma SA, Topel SD, Dik G, Özhan O, Bakar B, Yıldız A, Vardı N, Parlakpınar H. Enhancement of enzyme activity by laser-induced energy propulsion of upconverting nanoparticles under near-infrared light: a comprehensive methodology for in vitro and in vivo applications. Int J Biol Macromol. 2024;17:129343.

Barba-Rosado LV, Carrascal-Hernández DC, Insuasty D, Grande-Tovar CD. Graphene oxide (GO) for the treatment of bone cancer: a systematic review and bibliometric analysis. Nano. 2024;14(2):186.

Bhattacharya S, Saindane D, Prajapati BG. Liposomal drug delivery and its potential impact on cancer research. Anticancer Agents Med Chem. (Formerly Current Medicinal Chemistry-Anti-Cancer Agents). 2022;22(15):2671–83.

Caro C, Gámez F, Quaresma P, Páez-Muñoz JM, Domínguez A, Pearson JR, Pernía Leal M, Beltrán AM, Fernandez-Afonso Y, De la Fuente JM, Franco R. Fe3O4-Au core-shell nanoparticles as a multimodal platform for in vivo imaging and focused photothermal therapy. Pharmaceutics. 2021;13(3):416.

Chakravarty R, Shreya GOEL, Ashutosh DASH, Weibo CAI. Radiolabeled inorganic nanoparticles for positron emission tomography imaging of cancer: an overview. The quarterly journal of nuclear medicine and molecular imaging: official publication of the Italian Association of Nuclear Medicine (AIMN)[and] the International Association of Radiopharmacology (IAR),[and] Section of the Society of..., 2017 61(2), p. 181.

Chen X, Zhu X, Hu Y, Yuan W, Qiu X, Jiang T, Xia C, Xiong L, Li F, Gao Y. EDTA-modified 17β-estradiol-laden upconversion nanocomposite for bone-targeted hormone replacement therapy for osteoporosis. Theranostics. 2020;10(7):3281.

Chen GS, Wan KC, Gao M, Wei RP, Flournoy TH. Transition from pitting to fatigue crack growth—modeling of corrosion fatigue crack nucleation in a 2024-T3 aluminum alloy. Materials Science and Engineering: A. 1996;219(1-2):126–32.

Colaco V, Roy AA, Naik GA, Mondal A, Mutalik S, Dhas N. Advancement in lipid-based nanocomposites for theranostic applications in lung carcinoma treatment. OpenNano. 2023;25:100199.

Dhaliwal A, Ma J, Zheng M, Lyu Q, Rajora MA, Ma S, Oliva L, Ku A, Valic M, Wang B, Zheng G. Deep learning for automatic organ and tumor segmentation in nanomedicine pharmacokinetics. Theranostics. 2024;14(3):973.

Dilnawaz F, Iqbal Z, Kumar D. Nanoconstructs for cancer therapy and imaging. J Mol Nanot Nanom. 2017;1(1):103.

Dongsar TS, Dongsar TT, Gupta G, Alsayari A, Wahab S, Kesharwani P. PLGA nanomedical consignation: a novel approach for the management of prostate cancer. Int J Pharm. 2024;14:123808.

Du K, Feng J, Gao X, Zhang H. Nanocomposites based on lanthanide-doped upconversion nanoparticles: diverse designs and applications. Light Sci Appl. 2022;11(1):222.

Falagan-Lotsch P, Grzincic EM, Murphy CJ. New advances in nanotechnology-based diagnosis and therapeutics for breast cancer: an assessment of active-targeting inorganic nanoplatforms. Bioconjug Chem. 2017;28(1):135–52.

Girija AR, Balasubramanian S. Theragnostic potentials of core/shell mesoporous silica nanostructures. Nano. 2019;3(1):1.

Goel S, England CG, Chen F, Cai W. Positron emission tomography and nanotechnology: a dynamic duo for cancer theranostics. Adv Drug Deliv Rev. 2017;113:157–76.

Gupta D, Roy P, Sharma R, Kasana R, Rathore P, Gupta TK. Recent nanotheranostic approaches in cancer research. Clin Exp Med. 2024;24(1):1–30.

Guryev EL, Volodina NO, Shilyagina NY, Gudkov SV, Balalaeva IV, Volovetskiy AB, Lyubeshkin AV, Sen AV, Ermilov SA, Vodeneev VA, Petrov RV. Radioactive (90Y) upconversion nanoparticles conjugated with recombinant targeted toxin for synergistic nanotheranostics of cancer. Proc Natl Acad Sci. 2018;115(39):9690–5.

Hu F, Liu B, Chu H, Liu C, Li Z, Chen D, Li L. Real-time monitoring of pH-responsive drug release using a metal-phenolic network-functionalized upconversion nanoconstruct. Nanoscale. 2019;11(18):9201–6.

Huang WT, Chan MH, Chen X, Hsiao M, Liu RS. Theranostic nanobubble encapsulating a plasmon-enhanced upconversion hybrid nanosystem for cancer therapy. Theranostics. 2020;10(2):782.

Jeong K, Kim D, Kim HJ, Lee YD, Yoo J, Jang D, Lee S, Park H, Kim Y, Singh A, Ahn DJ. Photoechogenic inflatable nanohybrids for upconversion-mediated sonotheranostics. ACS Nano. 2021;15(11):18394–402.

Jia HR, Jiang YW, Zhu YX, Li YH, Wang HY, Han X, Yu ZW, Gu N, Liu P, Chen Z, Wu FG. Plasma membrane activatable polymeric nanotheranostics with self-enhanced light-triggered photosensitizer cellular influx for photodynamic cancer therapy. J Control Release. 2017;255:231–41.

Jin Y, Wang H, Li X, Zhu H, Sun D, Sun X, Liu H, Zhang Z, Cao L, Gao C, Wang H. Multifunctional DNA polymer-assisted upconversion therapeutic nanoplatform for enhanced photodynamic therapy. ACS Appl Mater Interfaces. 2020;12(24):26832–41.

Kostiv U, Natile MM, Jirák D, Pulpanova D, Jiráková K, Vosmanská M, Horák D. PEG-neridronate-modified NaYF4: Gd3+, Yb3+, Tm3+/NaGdF4 core–shell upconverting nanoparticles for bimodal magnetic resonance/optical luminescence imaging. ACS Omega. 2021;6(22):14420–9.

Kowalik P, Elbaum D, Mikulski J, Fronc K, Kamińska I, Morais PC, De Souza PE, Nunes RB, Veiga-Souza FH, Gruzeł G, Minikayev R. Upconversion fluorescence imaging of HeLa cells using ROS generating SiO 2-coated lanthanide-doped NaYF 4 nanoconstructs. RSC Adv. 2017;7(48):30262–73.

Kumar B, Murali A, Mattan I, Giri S. Near-infrared-triggered photodynamic, photothermal, and on demand chemotherapy by multifunctional upconversion nanocomposite. J Phys Chem B. 2019;123(17):3738–55.

Kumar A, Kulkarni S, Pandey A, Mutalik S, Subramanian S. Nano-tracers for sentinel lymph node detection: current trends in technique and application. Nanomedicine. 2024;19(1):59–77.

Kumar A, Parmar A, Singh R, Dhiman S. Nanoscience: an overview about nanotheranostics for cancer treatment. Egypt J Basic Appl Sci. 2024;11(1):55–68.

Kuthati Y, Sung PJ, Weng CF, Mou CY, Lee CH. Functionalization of mesoporous silica nanoparticles for targeting, biocompatibility, combined cancer therapies and theragnosis. J Nanosci Nanotechnol. 2013;13(4):2399–430.

Le JQ, Song XH, Tong LW, Lin YQ, Feng KK, Tu YF, Hu YS, Shao JW. Dual-drug controllable co-assembly nanosystem for targeted and synergistic treatment of hepatocellular carcinoma. J Colloid Interface Sci. 2024;656:177–88.

Li S, Cui S, Yin D, Zhu Q, Ma Y, Qian Z, Gu Y. Dual antibacterial activities of a chitosan-modified upconversion photodynamic therapy system against drug-resistant bacteria in deep tissue. Nanoscale. 2017;9(11):3912–24.

Li X, Jiang M, Zeng S, Liu H. Polydopamine coated multifunctional lanthanide theranostic agent for vascular malformation and tumor vessel imaging beyond 1500 nm and imaging-guided photothermal therapy. Theranostics. 2019;9(13):3866.

Li Y, Zhang X, Zhang Y, Zhang Y, He Y, Liu Y, Ju H. Activatable photodynamic therapy with therapeutic effect prediction based on a self-correction upconversion nanoprobe. ACS Appl Mater Interfaces. 2020;12(17):19313–23.

Liang L, Lu Y, Zhang R, Care A, Ortega TA, Deyev SM, Qian Y, Zvyagin AV. Deep-penetrating photodynamic therapy with KillerRed mediated by upconversion nanoparticles. Acta Biomater. 2017;51:461–70.

Liao J, Wei X, Ran B, Peng J, Qu Y, Qian Z. Polymer hybrid magnetic nanocapsules encapsulating IR820 and PTX for external magnetic field-guided tumor targeting and multifunctional theranostics. Nanoscale. 2017;9(7):2479–91.

Lin SL, Chen ZR, Chang CA. Nd3+ sensitized core-shell-shell nanocomposites loaded with IR806 dye for photothermal therapy and up-conversion luminescence imaging by a single wavelength NIR light irradiation. Nano. 2018;2(3):243.

Liu S, Sun Y, Ye J, Li C, Wang Q, Liu M, Cui Y, Wang C, Jin G, Fu Y, Xu J. Targeted Delivery of Active Sites by Oxygen Vacancy-Engineered Bimetal Silicate Nanozymes for Intratumoral Aggregation-Potentiated Catalytic Therapy. ACS nano. 2024;18(2):1516–30.

Lu C, Joulin E, Tang H, Pouri H, Zhang J. Upconversion nanostructures applied in theranostic systems. Int J Mol Sci. 2022;23(16):9003.

Lv R, Feng M, Xiao L, Damasco JA, Tian J, Prasad PN. Multilevel nanoarchitecture exhibiting biosensing for cancer diagnostics by dual-modal switching of optical and magnetic resonance signals. ACS Appl Bio Mater. 2018;1(5):1505–11.

Mahata MK, De R, Lee KT. Near-infrared-triggered upconverting nanoparticles for biomedicine applications. Biomedicines. 2021;9(7):756.

Mishra VK, Ahmed M. Polymeric gold nanoparticles for gene therapy. In Gold nanoparticles for drug delivery. Academic Press. 2024. pp. 393–417.

Moghaddam FD, Zare EN, Hassanpour M, Bertani FR, Serajian A, Ziaei SF, Paiva-Santos AC, Neisiany RE, Makvandi P, Iravani S, Xu Y. Chitosan-based nanosystems for cancer diagnosis and therapy: stimuli-responsive, immune response, and clinical studies. Carbohydr Polym. 2024;20:121839.

Mohite P, Rajput T, Pandhare R, Sangale A, Singh S, Prajapati BG. Nanoemulsion in Management of Colorectal Cancer: challenges and future prospects. Nano. 2023;3(2):139–66.

Nady DS, Hassan A, Amin MU, Bakowsky U, Fahmy SA. Recent innovations of mesoporous silica nanoparticles combined with photodynamic therapy for improving cancer treatment. Pharmaceutics. 2023;16(1):14.

Osuchowski M, Osuchowski F, Latos W, Kawczyk-Krupka A. The use of upconversion nanoparticles in prostate cancer photodynamic therapy. Life. 2021;11(4):360.

Park D, Lee SJ, Park JW. Aptamer-based smart targeting and spatial trigger–response drug-delivery systems for anticancer therapy. Biomedicines. 2024;12(1):187.

Parry ZA, Pandey R. Dendrimers in medical science. CRC Press; 2017.

Patel V, Khopade AJ, Patel JK. Subcellular Targeting of Nanoparticles for Cancer Theranostics. InSite-specific Cancer Nanotheranostics 2024 (pp. 296–321). CRC Press.

Patel P, Raval M, Airao V, Ali N, Shazly GA, Khan R, Prajapati B. Formulation of folate receptor-targeted Silibinin-loaded inhalable chitosan nanoparticles by the QbD approach for lung cancer targeted delivery. ACS Omega. 2024b;

Pillarisetti S, Uthaman S, Huh KM, Koh YS, Lee S, Park IK. Multimodal composite iron oxide nanoparticles for biomedical applications. Tissue Eng Regen Med. 2019;16:451–65.

Prasad R, Selvaraj K. Choice of nanoparticles for theranostics engineering: surface coating to nanovalves approach. Nano. 2024;8(1):12.

Purkait MK, Sontakke AD. Carbon-based nanocarriers for drug delivery. CRC Press; 2023.

Qin J, Guo N, Yang J, Chen Y. Recent advances of metal–polyphenol coordination polymers for biomedical applications. Biosensors. 2023;13(8):776.

Rezende TK, Barbosa HP, Dos Santos LF, Lima KDO, Alves De Matos P, Tsubone TM, Gonçalves RR, Ferrari JL. Upconversion rare earths nanomaterials applied to photodynamic therapy and bioimaging. Front Chem. 2022;10:1035449.

Rodrigues JA, Correia JH. Photodynamic therapy for colorectal cancer: an update and a look to the future. Int J Mol Sci. 2023;24(15):12204.

Sahu R, Shah K, Malviya R, Paliwal D, Sagar S, Singh S, Prajapati BG, Bhattacharya S. E-cigarettes and associated health risks: an update on cancer potential. Adv Respir Med. 2023;91(6):516–31.

Shanwar S, Liang L, Nechaev AV, Balalaeva IV, Vodeneev VA, Zvyagin AV, Guryev EL. The assembly of a photoluminescent nanocomplex based on upconversion nanoparticles. Opera Medica et Physiologica. 2020;7(4):42–8.

Shapoval O, Větvička D, Patsula V, Engstová H, Kočková O, Konefał M, Kabešová M, Horák D. Temoporfin-conjugated upconversion nanoparticles for NIR-induced photodynamic therapy: studies with pancreatic adenocarcinoma cells in vitro and in vivo. Pharmaceutics. 2023;15(12):2694.

Sikora-Dobrowolska B, Borodziuk A, Kulpa-Greszta M, Pazik R, Wojciechowski T, Sobczak K, Rybusinski J, Szczytko J, Klopotowski L. Opto-magnetic nanoparticles with upconverting properties for optical imaging and photothermal therapies. J Magn Magn Mater. 2024;4:171714.

Singh AK, Malviya R, Prajapati B, Singh S, Goyal P. Utilization of stimuli-responsive biomaterials in the formulation of cancer vaccines. J Funct Biomater. 2023;14(5):247.

Sneider A, VanDyke D, Paliwal S, Rai P. Remotely triggered nano-theranostics for cancer applications. Nano. 2017;1(1):1.

Tahiliani S, Lukhmana N, Aravamudhan S. Theranostic applications of functional nanomaterials using microscopic and spectroscopic techniques. In Functional smart nanomaterials and their theranostics approaches. Singapore: Springer Nature Singapore, 2024, pp. 87–118.
Tang XL, Wu J, Lin BL, Cui S, Liu HM, Yu RT, Shen XD, Wang TW, Xia W. Near-infrared light-activated red-emitting upconverting nanoplatform for T1-weighted magnetic resonance imaging and photodynamic therapy. Acta Biomater. 2018;74:360–73.
Tang M, Zhu X, Zhang Y, Zhang Z, Zhang Z, Mei Q, Zhang J, Wu M, Liu J, Zhang Y. Near-infrared excited orthogonal emissive upconversion nanoparticles for imaging-guided on-demand therapy. ACS Nano. 2019;13(9):10405–18.
Tao K, Sun K, Choi SK. Upconversion nanocrystals for near-infrared-controlled drug delivery. In Photonanotechnology for therapeutics and imaging 2020, pp. 345–371. Elsevier.
Targhotra M, Chauhan MK. Multifunctional nanocomposites for blood cancer. In Multifunctional nanocomposites for targeted drug delivery in cancer therapy 2024, pp. 323–339. Academic Press.
Thanasekaran P, Chu CH, Wang SB, Chen KY, Gao HD, Lee MM, Sun SS, Li JP, Chen JY, Chen JK, Chang YH. Lipid-wrapped upconversion nanoconstruct/photosensitizer complex for near-infrared light-mediated photodynamic therapy. ACS Appl Mater Interfaces. 2018;11(1):84–95.
Tong Q, Qiu N, Ji J, Ye L, Zhai G. Research progress in bioinspired drug delivery systems. Expert Opin Drug Deliv. 2020;17(9):1269–88.
Tripathi D, Hajra K, Maity D. Recent advancement of bio-inspired nanoparticles in cancer theragnostic. J Nanotheranostics. 2023;4(3):299–322.
Tsai YC, Vijayaraghavan P, Chiang WH, Chen HH, Liu TI, Shen MY, Omoto A, Kamimura M, Soga K, Chiu HC. Targeted delivery of functionalized upconversion nanoparticles for externally triggered photothermal/photodynamic therapies of brain glioblastoma. Theranostics. 2018;8(5):1435.
Veeranarayanan S, Maekawa T. External stimulus responsive inorganic nanomaterials for cancer theranostics. Adv Drug Deliv Rev. 2019;138:18–40.
Verma R, Akter R, Kumar M, Bhatt S, Tiwari A, Tiwari V, Tagde P, Pandey P, Mittal V, Purohit D, Redhu R. Exploring the role of nanotherapeutics for diagnosis and treatment of solid tumor. Curr Nanosci. 2024;20(1):109–29.
Vyas K, Patel MM. Dual/multi-responsive delivery nanoplatforms in cancer theranostics. In Site-specific cancer nanotheranostics. CRC Press, 2024, pp. 120–141.
Wang H, Li Y, Yang M, Wang P, Gu Y. FRET-based upconversion nanoprobe sensitized by Nd^{3+} for the ratiometric detection of hydrogen peroxide in vivo. ACS Appl Mater Interfaces. 2019;11(7):7441–9.
Wang Y, Lü WD, Lin B, Yang F, Feng M, Lv R. An optimized lanthanide-chlorophyll nanocomposite for dual-modal imaging-guided surgery navigation and anti-cancer theranostics. Biomater Sci. 2020;8(5):1270–8.
Wang Y, Ren L, Peng H, Guo L, Wang L. DNA-programmed biomolecular spatial pattern recognition. Chemosensors. 2023;11(7):362.
Xie X, Yue T, Gu W, Cheng W, He L, Ren W, Li F, Piao JG. Recent advances in mesoporous silica nanoparticles delivering siRNA for cancer treatment. Pharmaceutics. 2023;15(10):2483.
Xu J, Han W, Yang P, Jia T, Dong S, Bi H, Gulzar A, Yang D, Gai S, He F, Lin J. Tumor microenvironment-responsive mesoporous MnO_2-coated upconversion nanoplatform for self-enhanced tumor theranostics. Adv Funct Mater. 2018;28(36):1803804.
Xu J, Gulzar A, Yang D, Gai S, He F, Yang P. Tumor self-responsive upconversion nanomedicines for theranostic applications. Nanoscale. 2019;11(38):17535–56.
Xu J, Shi R, Chen G, Dong S, Yang P, Zhang Z, Niu N, Gai S, He F, Fu Y, Lin J. All-in-one theranostic nanomedicine with ultrabright second near-infrared emission for tumor-modulated bioimaging and chemodynamic/photodynamic therapy. ACS Nano. 2020;14(8):9613–25.
Yang M, Wang H, Wang Z, Han Z, Gu Y. A Nd^{3+} sensitized upconversion nanosystem with dual photosensitizers for improving photodynamic therapy efficacy. Biomater Sci. 2019;7(4):1686–95.
Yao J, Huang C, Liu C, Yang M. Upconversion luminescence nanomaterials: a versatile platform for imaging, sensing, and therapy. Talanta. 2020;208:120157.

Zhang Z, Jayakumar MK, Shikha S, Zhang Y, Zheng X, Zhang Y. Modularly assembled upconversion nanoparticles for orthogonally controlled cell imaging and drug delivery. ACS Appl Mater Interfaces. 2020;12(11):12549–56.

Zhang J, Hu M, Wen C, Liu J, Yu F, Long J, Lin XC. CeO2@ CuS@ PDA-FA as targeted near-infrared PTT/CDT therapeutic agents for cancer cells. Biomed Mater. 2023;18(6):065006.

Zhao L, Ge X, Zhao H, Shi L, Capobianco JA, Jin D, Sun L. Double-sensitive drug release system based on MnO2 assembled upconversion nanoconstruct for double-model guided chemotherapy. ACS Appl Nano Mater. 2018;1(4):1648–56.

Zhao L, Choi J, Lu Y, Kim SY. NIR photoregulated theranostic system based on hexagonal-phase upconverting nanoparticles for tumor-targeted photodynamic therapy and fluorescence imaging. Nano. 2020;10(12):2332.

Zhao H, Li Y, Zhang X, Wu K, Lv J, Chen C, Liu H, Shi Z, Ju H, Liu Y. Orthogonal excitations of lanthanide nanoparticle up/down conversion emissions via switching NIR lights for in-vivo theranostics. Biomaterials. 2022;291:121873.

Živojević K, Mladenović M, Djisalov M, Mundzic M, Ruiz-Hernandez E, Gadjanski I, Knežević NŽ. Advanced mesoporous silica nanocarriers in cancer theranostics and gene editing applications. J Control Release. 2021;337:193–211.

Part VI

Toxicological and Regulatory Aspects

Toxicological Aspects of Core–Shell Nanoconstructs

20

Deshmukh Aaishwaryadevi, Jayvadan K. Patel, and Bharat Mishra

Abstract

The field of nanotechnology has experienced significant growth, with nanoparticles (NPs) and nanoconstructs finding diverse applications, especially in the medical field. These materials, due to their unique physical and chemical properties, have considerable potential in improving both diagnostic and therapeutic approaches, particularly in oncology. However, as engineered nanomaterials (ENMs) are increasingly used, concerns about their safety and potential toxicity have emerged, emphasizing the need for comprehensive safety assessments. Core-shell nanoconstructs, which are composed of nanoparticles and surface-modifying ligands, are designed to improve the precision and efficiency of drug delivery. These constructs are capable of specifically targeting disease sites, overcoming issues often encountered in cancer treatments, such as off-target effects, uncontrolled drug release, and high toxicity. While these advantages are promising, the use of nanoconstructs also presents challenges, particularly in ensuring accurate delivery and refining their design through advanced computational modeling. Additionally, the behavior of these nanomaterials in biological systems can be unpredictable and highly dependent on factors such as size, surface chemistry, and the method of synthesis. In vitro toxicity studies often fail to simulate the actual effects seen in vivo, which underlines the necessity for standardized testing protocols to assess their safety more reliably. A collaborative, multidisciplinary approach involving toxicologists, biologists, and engineers is crucial to overcoming these challenges. This will help to deepen the understanding of nanoconstructs' interactions with biological systems and ensure their safe integration into medical practice.

D. Aaishwaryadevi (✉)
School of Pharmaceutical Sciences, JSPM University, Pune, Maharashtra, India

J. K. Patel
Viesain Pharma LLC, GA, USA

Faculty of Pharmacy, Sankalchand Patel University, Visnagar, Gujarat, India

B. Mishra
Dr. Shakuntla Misra National Rehabilitation University, Lucknow, Uttar Pradesh, India

J. K. Patel et al. (eds.), *Core-Shell Nano Constructs for Cancer Theragnostic*,
https://doi.org/10.1007/978-981-96-3025-7_20

Keywords

Core shell nanoconstructs · Nanotoxicology · Cytotoxicity · Nanomaterials safety

1 Introduction

Engineered nanoparticles (NPs) are defined as materials with a minimum of one dimension having a measurement of fewer than 100 nm. They have become inherent elements across several modern industries. The advent of nanotechnology has led to transformative developments in numerous fields, employing the exclusive physicochemical as well as electrical properties of nanoscale materials for applications in various engineering fields like electronics, biotechnology, and aerospace. Specifically, NPs have expanded projection in medicine as novel systems for delivering not only drugs but also proteins, DNA, and monoclonal antibodies (Nowrouzi et al. 2010). These NPs can be produced from distinct substances, including metals, nonmetals, polymers, bio-ceramics, etc. Especially, liposomes, polyethylene glycol, and dendrimers are normally underlined for their medical applications (Dreher 2004). Nonetheless, the extensive usage of nanoscale materials has construed apprehensions about potential health risks as humans are exposed to these particles from an early age.

Because of their remarkably small size, NPs can penetrate the human body intersecting biological barriers to access vital organs. The NPs that are smaller than 10 nm are assumed to act similarly to gases, allowing them to infiltrate tissues and thereby interrupt cellular biochemical environments (Vishwakarma et al. 2010). Research involving both animal as well as human models has shown that NPs can disperse to multiple organs, such as the liver, heart, spleen, brain, lungs, and gastrointestinal system, following inhalation or ingestion (Hagens et al. 2007). The immune system is crucial in eliminating these particles though their half-life in the lungs is estimated to be around 700 days, indicating long-term risks to the respiratory system. Moreover, during metabolic processes, NPs may accumulate in the liver, potentially aggravating their lethal influences (Garnett and Kallinteri 2006). Likewise, the toxicity of NPs habitually goes beyond that of larger particles of the same chemical composition, with smaller particles characteristically showing greater toxicity (Mostafalou et al. 2013). The distinct physicochemical traits of NPs make their health effects erratic, accenting the requirement to focus on these uncertainties via innovative safety strategies. Thus, closing the knowledge gaps regarding NP toxicity is necessary for their secure application.

Nanoconstructs, classically consisting of a "hard" nanoparticle core bounded by a "soft" shell of biomolecular ligands and allowing precise therapeutic delivery, characterize a substantial advancement in targeted drug delivery systems (Dam et al. 2014). Acting as carriers for active agents, these systems are able to transport and protect drugs as they navigate biological barriers, releasing them at projected sites. Nanoparticle-based chemotherapy poses numerous advantages, including improved treatment accuracy, enhanced drug concentration at target sites, and reduced systemic toxicity. Furthermore, nanoconstructs can be tailored to integrate numerous agents on their surface or within their core, enabling combination therapy approaches. Once administered, the behavior of nanoconstructs is persuaded by their surface chemistry as well as vascular dynamics, including factors like pressure, velocity, and tissue variability. Computational modeling, using 4S parameters—size, shape, surface, and stiffness—has been employed to analyze the distribution and retention of nanoconstructs within the vasculature, extending significant perceptions into their pharmacokinetic behavior (Coclite et al. 2017). These computational tools are involved in recognizing the complex interactions of nanoconstructs, enabling the rational design of more effective formulations for biomedical purposes (Cervadoro et al. 2018). By optimizing design parameters through computational approaches, researchers can enhance therapeutic outcomes while minimizing unintended effects.

2 Toxicological Profiling in Cellular, Animal, and Human Models

Nanomaterials are exemplified by a remarkably high surface-area-to-volume ratio in comparison to bulk materials of the same mass. This occurs due to the exponential increase in the ratio of surface atoms or molecules to total atoms as particle size reduces. Accordingly, these materials demonstrate increased surface reactivity, which substantially influences their strength and physical properties. While the nanoscale dimensions and larger surface area of nanoparticles (NPs) deliberate unique and beneficial traits, these same traits can also increase biological activity, possibly leading to unexpected interactions with biological systems. In addition, the decreased size of NPs modifies their biokinetic behavior, allowing them to access remote areas of the body more efficiently (Oberdorster et al. 2005).

Over the past two decades, the application of nanoparticles in experimental research and clinical practices has grown substantially, driven by their versatility in biomedical fields such as drug delivery, imaging, and cell tracking (Thanh and Raton 2011). Despite these advancements, it is imperative to not only recognize the benefits of NPs but also address the potential adverse and unpredictable effects of human exposure. Nanoparticle toxicity pertains to their ability to adversely affect normal physiological processes and compromise the structural integrity of organs and tissues in both humans and animals. Toxicity is influenced by various physicochemical factors, including particle size, shape, surface charge, surface chemistry, composition, and nanoparticle stability. Although the precise mechanisms remain uncertain, emerging research indicates that cytotoxic effects may be associated with oxidative stress and the activation of pro-inflammatory gene pathways (Hussain et al. 2005). Beyond particle-specific factors, parameters such as dosage, route of administration, and the extent of tissue distribution are critical determinants of the cytotoxicity of nanoparticles (NPs). Typically, cellular toxicity studies employ escalating concentrations of nanoparticles to assess dose-dependent effects on cells or tissues. However, translating these in vitro findings to in vivo conditions presents significant challenges. These challenges include the efficient delivery of nanoparticles to target tissues and the identification of biochemical alterations induced by nanoparticles in living organisms. To address the potential risks posed by unpredictable, non-dose-dependent effects of NPs in vivo, innovative predictive approaches are required.

The route of nanoparticle administration is another underexplored yet influential factor in nanotoxicity. Nanoparticles can enter the body via various pathways, including oral ingestion, inhalation, dermal penetration, or intravascular injection, each leading to distinct bio-distribution patterns. These routes have unique advantages and limitations that affect nanoparticle accumulation, metabolism, excretion, and potential toxicity. For instance, pulmonary drug delivery has shown promise, but concerns remain regarding local and systemic toxicity, which may arise from nanoparticle aggregation or inflammatory responses in the lung tissues (Liu et al. 2009). In the case of topical applications, such as sunscreen or cosmetic formulations, questions about dermal or systemic toxicity and their potential for transdermal drug delivery have been raised. Intravenous and oral administration routes facilitate rapid systemic effects but expose nanoparticles to first-pass metabolism in the liver, followed by distribution to various organs, including the brain. Despite the blood-brain barrier's (BBB) role in protecting the brain, some nanoparticles can cross tight junctions, raising concerns about neurotoxicity (Mutlu et al. 2010; Muller et al. 2005). Given these risks, obtaining reliable data on nanoparticle toxicity is essential to minimize adverse outcomes. While nanoparticles have the potential to revolutionize fields such as medical imaging, diagnostics, and therapeutics, their unique properties may also contribute to toxicity. Furthermore, the severity of toxic effects can vary significantly depending on the administration route and the specific sites of nanoparticle deposition.

3 Nanomaterials and Their Toxicities

3.1 NPs of Metallic Substances

3.1.1 Aluminum Oxide Nanoparticles

Nanoparticles (NPs) based on aluminum represent nearly 20% of all nanoscale chemicals. As per the report titled *The Global Market for Aluminum Oxide NPs*, these nanoparticles are utilized across numerous industries, including fuel cells, polymers, paints, coatings, textiles, and biomaterials. Regarding their toxicity, Chen et al. (Chen et al. 2008) documented that aluminum oxide NPs can compromise cell viability, impair mitochondrial functions, increase oxidative stress levels, and alter the expression of tight junction proteins within the blood-brain barrier (BBB). In contrast, Radziun et al. (Radziun et al. 2011) reported no notable toxic effects on mammalian cell viability at aluminum oxide NP concentrations ranging between 10 and 400 μg/mL when assessed using the EZ4U assay, as opposed to the MTT assay. Further, another study exploring the dose-dependent effects of aluminum oxide NPs (160 nm) on human mesenchymal stem cells revealed cytotoxicity at concentrations between 25 and 40 μg/mL, evaluated through the MTT assay (Alshatwi et al. 2012). Aluminum oxide NPs have also been studied for their potential genotoxic effects. Balasubramanyam et al. (Balasubramanyam et al. 2009) found that aluminum oxide NPs (30–40 nm) exhibited dose-dependent genotoxicity, as evidenced by the comet assay and micronucleus test on rat blood cells. Similarly, research involving a mouse lymphoma cell line demonstrated that aluminum oxide NPs smaller than 50 nm could induce DNA damage without showing mutagenic properties (Kim et al. 2009). Although in vivo studies on aluminum oxide NP toxicity remain limited, current research predominantly focuses on their cytotoxic and genotoxic potential. Given their widespread applications and inevitable human exposure, it is crucial to thoroughly assess aluminum-based NPs for potential health risks using standardized and reliable evaluation protocols.

3.1.2 Gold Nanoparticles

Gold nanoparticles (AuNPs) exhibit distinctive physicochemical properties, such as ease of functionalization and ability to bind with amine and thiol groups. These features allow for surface modification, making them highly suitable for diverse applications, including cancer drug delivery, thermal therapy, and as contrast agents (Jain et al. 2012). AuNPs are largely deemed safe due to the inert and nontoxic nature of their core structure. An experimental study evaluated gold nanoparticles of various sizes (4, 12, and 18 nm) with different capping agents for their cytotoxicity against leukemia cell lines. The findings revealed that spherical AuNPs could penetrate cells without impairing cellular functions, as demonstrated using the MTT assay (Connor et al. 2005). However, other studies suggest that the cytotoxic effects of AuNPs are influenced by factors such as dosage, the nature of the side chains (e.g., cationic), and the type of stabilizer used (Boisselier and Astruc 2009). Additionally, cytotoxicity varies depending on the assay type, cell line, and the nanoparticles' physical and chemical characteristics. For instance, differences in toxicity have been noted across cell lines, including the human lung and liver cancer cell lines (Patra et al. 2007).

Given their straightforward synthesis process and prospective for bio-functionalization, gold nanoparticles have been actively studied in clinical contexts, e.g., dermal drug delivery applications. Research by Sonavane and colleagues exhibited that the skin penetration of AuNPs varies depending on their size when applied topically on excised rat skin. The study compared particles sized 15, 102, and 198 nm and found that smaller nanoparticles penetrated deeper layers of the skin, while larger ones remained confined to the superficial epidermis and dermis. These findings suggest promising applications for nanoparticle-based dermal drug delivery systems (Sonavane et al. 2008). Although gold compounds have been clinically utilized for years, such as in the treatment of rheumatoid arthritis, their nanoscale reduction can alter their biochemical behavior, necessitating further exploration of their cytotoxic profiles. Despite extensive research on AuNP toxicity, conflicting outcomes continue to pose challenges for their clinical translation. Studies indicate that the cellular uptake of AuNPs depends on factors such as particle size, concentration, and exposure time. For example, Mironava et al. observed that larger particles (45 nm) caused significant cytotoxicity in human dermal fibroblasts at lower concentrations, while smaller particles (13 nm) were less toxic even at higher concentrations (Mironava et al. 2010). On the other hand, Pan et al. reported the highest toxicity for particles sized 1.4 nm (Pan et al. 2007). These inconsistencies could result from variations in the intracellular distribution of nanoparticles, underscoring the need for more in-depth investigations.

Gold nanoparticles (AuNPs) exhibit remarkable properties that make them highly promising for biomedical applications, including not just imaging applications but also drug delivery as well as gene therapy. A study conducted by Chen and their group investigated the size-dependent toxicity and destructive aspects of AuNPs (Chen et al. 2009). Mice administered intraperitoneal (IP) injections of bare AuNPs sized between 3 and 100 nanometers, at a dose of 8 mg/kg/week for 4 weeks, displayed varying toxicity levels. While nanoparticles sized 3, 5, 50, and 100 nm caused no notable adverse effects, those varying from 8 to 37 nm caused chronic systemic toxicity. Test subjects experienced symptoms such as instance exhaustion, diminished hunger, and weight loss, with changes in fur color, accompanied by high mortality rates. Histological analysis revealed gold accumulation in organs, leading to liver Kupffer cell activation, diffusion of white pulp in the spleen, and essential abnormalities in lung parenchymal cells. Notably, modifying particle surfaces with a highly immunogenic peptide enhanced cytocompatibility by triggering a host antibody response. Notwithstanding this, liver toxicity, driven by acute inflammation and subsequent apoptosis, remained a concern even with biologically functionalized coatings (Cho et al. 2009). Furthermore, AuNP toxicity is influenced by both surface functionalization and the method of administration. Intraperitoneal delivery has been linked to significantly higher toxicity compared to intravenous routes (Zhang et al. 2010). However, caution is necessary when comparing adverse effects across different delivery methods as results may vary significantly between studies.

AuNPs have also demonstrated potential as imaging agents for both in vitro as well as in vivo tracking of cells, with research proposing their applicability for central nervous system (CNS) imaging (Lu et al. 2010). However, the interaction of AuNPs with CNS cells remains poorly understood. Primary brain microvascular endothelial cells (BMECs) are commonly exploited to study the transport characteristics and molecular interactions of nanoparticles at the blood-brain barrier (BBB) (Franke et al. 2000). Evidence suggests that exposure to silver nanoparticles (AgNPs) in BMECs may result in the liberation of pro-inflammatory cytokines, leading to increased permeability and toxicity in a time- as well as size-dependent manner (Trickler et al. 2010). On the contrary, studies using AuNPs of similar magnitude observed neither elevated secretion of inflammatory markers nor superior cellular permeability. For instance, macrophage cell line RAW264.7 did not show pro-inflammatory cytokine release upon exposure to AuNPs, though smaller particles (3 nm) exhibited slightly higher toxicity than larger ones (5 nm). Size-dependent toxicity was perceived across numerous cell lines and in vivo systems. Ultra-small AuNPs are capable of widespread distribution across tissues, comprising the tissue of the brain, while larger particles are often restricted from penetrating cerebral tissues, likely due to the protective role of the BBB or their rapid clearance from circulation. Consequently, smaller concentrations of large particles reach and interact with the BBB. This was further validated in findings where continual administration of reasonably larger AuNPs (12.5 nm) resulted in significantly lower brain accumulation compared to other organs like the liver, spleen, and kidneys (Lasagna-Reeves et al. 2010).

3.1.3 Copper Oxide Nanoparticles

Copper oxide nanoparticles (CuO NPs) have gained considerable relevance owing to their varied applications, embracing usage in semiconductors and heat transfer fluids, antimicrobial agents, intrauterine contraceptive devices, and many more. However, experimental studies have highlighted their toxicological impact. CuO NPs have been observed to initiate critical injury to organs like the liver, kidneys, and spleen in experimental animals (Lei et al. 2008). When administered orally, these nanoparticles interact with gastric juice, producing highly reactive ionic copper, which tends to accumulate in the kidneys of exposed animals (Chen et al. 2006). In vitro studies further support their toxic potential as CuO NPs (50 nm) have demonstrated genotoxic and cytotoxic effects, including cell membrane disruption and oxidative stress (Ahamed et al. 2010).

3.1.4 Silver Nanoparticles (AgNP)

Silver nanoparticles (AgNPs), widely recognized for their antibacterial properties, are extensively used in products such as wound dressings, surgical instrument coatings, and prosthetic devices. These nanoparticles have access to the human body via numerous routes ranging from inhalation and ingestion to injection, subsequently accumulating in organs like the lungs, kidneys, spleen, liver, and even the brain as they can cross the blood-brain barrier (BBB) (Tang et al. 2009). AgNPs exhibit greater toxicity compared to other nanoparticles. For example, their cytotoxic effects, such as increased reactive oxygen species (ROS) production and leakage of lactate dehydrogenase (LDH), have been well-documented (Hussain et al. 2005). Studies have shown that AgNP toxicity varies depending on their coating materials. For instance, polyvinyl-pyrrolidone-coated silver nanoparticles (6–20 nm) have been found to induce dose-dependent cytotoxicity and prompt damage to the DNA in human lung cancer cells (Foldbjerg et al. 2011). Similarly, peptide-coated silver nanoparticles demonstrate higher toxicity compared to citrate-coated ones of the same size, as observed in leukemia cell lines (Haase et al. 2011).

The primary route for the exposure of silver nanoparticles (AgNPs) to the pulmonary tissue is typically through inhalation, especially through occupational activities such as manufacturing processes (Maynard et al. 2004). A study by Oberdorster et al. in a rat model demonstrated that inhaled nanoparticles could migrate from the lungs, the initial sites of deposition, to the rest of the organs throughout the body (Oberdörster et al. 2002). The American Conference of Governmental Industrial Hygienists (ACGIH) at present establishes the occupational exposure threshold for silver dust at 100 $\mu g/m^3$. Sung et al. conducted a series of inhalation studies in rats to evaluate different toxicity tests viz. acute, subacute (28-day), and sub-chronic (90-day) exposure of AgNP (Sung et al. 2009). In their acute exposure study, rats were exposed to various concentrations of AgNP in the setting of a whole-body inhalation chamber for 4 h/day, followed by 2 weeks of monitoring. At the greatest concentration tested (750 $\mu g/m^3$, 7.5 times the permissible limit), unsubstantial variations were stated in the body weight or clinical parameters. Tests related to assessing lung function also showed no substantial distinctions concerning the exposed and control groups. Similarly, frequent exposure over 4 weeks resulted in no meaningful changes. However, sub-chronic exposure over a period of 13 weeks at a concentration of 515 μg/m^3 (five times the limit) caused dose- and time-dependent alveolar inflammation, granulomatous changes, and impaired lung function (Sung et al. 2008). These findings suggest that while prolonged exposure to high concentrations of AgNP may be harmful, current exposure limits provide adequate protection against severe inhalation risks.

In contrast to gold nanoparticles (AuNP), AgNP demonstrates neurodegenerative effects. Current studies investigating the impact of AgNP on the blood-brain barrier (BBB) have demonstrated disruptions in BBB function, which led to the formation of brain edema (Sharma et al. 2010). Supporting these results, Tang et al. observed BBB damage along with inflammation in astrocyte cells, and neuronal degeneration in rats exposed to AgNP (Tang et al. 2010). The toxicity and distribution patterns observed were dependent on the concentration and size of the nanoparticles. Research data by Costa et al. identified mitochondrial dysfunction and decreased energy production as potential mechanisms contributing to neurodegeneration (Costa et al. 2010). Studies involving repeated exposure revealed substantial accumulation of AgNP in various tissues. AgNPs having a size of 100 nm primarily accumulated in the spleen, likely due to the scope of the capillary fenestrae (~100 nm). Moreover, rats exposed to repeated oral doses of AgNP exhibited a dose-dependent accumulation of nanoparticles in their tissues. In humans, ingestion of AgNP has been linked to a grayish discoloration of the skin, indicating gastrointestinal absorption and subsequent distribution of nanoparticles throughout the body. These findings underscore the need for further research into how nanoparticle size influences their bio-distribution and toxic effects. Powers et al. also exhibited the neurotoxic potential of AgNP, revealing developmental toxicity in zebrafish embryos that led to long-term synaptic changes (Powers et al. 2011). Exposure to ionized silver (Ag^+) leads to a considerable boost in serotonin (5-HT) and dopamine (DA) turnover, both of which are critical neurotransmitters for regulating functions related to sensorimotor nerves, anxiety, and reward. This Ag^+-tempted hyperactivity of neurotransmitters led to behavioral transformations steady with a lowered anxiety threshold.

3.1.5 Zinc Oxide Nanoparticles

Zinc oxide nanoparticles (ZnO NPs) are extensively utilized in numerous industries, including in paints, UV protectants, gas sensors, sunscreens, and personal care products. This wide usage results in inevitable human exposure. Numerous reports have assessed the potentially lethal effects of ZnO NPs on both bacterial and mammalian cells, revealing harmful outcomes such as cytotoxicity, oxidative stress, and damage to cell membranes (Huang et al. 2010). For example, Brunner et al. conducted experiments on human mesothelioma and rodent fibroblast cells, exposing them to high concentrations of ZnO NPs (49 mg/mL), which caused nearly complete cell death (Brunner et al. 2006). Additionally, another in vitro study noted significant mitochondrial dysfunction, DNA damage, and changes in cell morphology in hepatocytes, as well as embryonic kidney cells exposed to ZnO NPs, with the comet and MTT assays employed to evaluate DNA damage and cell viability, respectively (Guan et al. 2012). Meyer et al. further supported these results, observing a marked decrease in cell viability in human dermal fibroblasts after exposure to 20 nm of ZnO NPs, as measured by the MTT assay (Meyer et al. 2011). In addition to cytotoxicity, the genotoxic effects of ZnO NPs are detailed in both in vitro and in vivo studies. An in vitro study with HEp-2 cells revealed DNA damage due to ZnO NPs, which was assessed through comet assays and cytokinesis-blocked micronucleus assays (Osman et al. 2010). Chronic exposure to ZnO NPs at a dose of 300 mg/kg led to oxidative DNA damage and altered liver enzyme activities, with comet assays used to assess DNA damage (Sharma et al. 2012a). These findings emphasize the significance of continual research into the mechanistic aspects of ZnO NP toxicity to understand the prospective threats to human health and the environment, as well as to establish safer usage protocols.

3.1.6 Iron Oxide Nanoparticles

Iron oxide nanoparticles (NPs) are commonly applied in medical fields like drug delivery and diagnostics and are known to accumulate in organs linked to the reticuloendothelial system, particularly the liver. Upon internalization, iron oxide NPs localize in organelles such as lysosomes and endosomes, where they release iron into the cytoplasm, affecting cellular iron levels. Magnetic iron oxide NPs, in particular, are found in organs like the spleen, lungs, liver, and even the brain after inhalation, demonstrating their capability of crossing the blood-brain barrier (Liu et al. 2013). These particles are associated with several toxic effects, including damage to cell membranes, inflammatory responses, and interference with blood coagulation. In vitro studies generally report a decrease in cell viability due to iron oxide NPs, with toxicity levels varying depending on the type of coating. For example, Naqvi et al. found that Tween-coated superparamagnetic iron oxide NPs were more toxic at lower concentrations in murine macrophage cells (Naqvi et al. 2010). In contrast, dextran-coated iron oxide NPs reduced cell viability in human macrophages after prolonged exposure (Pawelczyk et al. 2008). Another study on mouse neuroblastoma cells reported minimal toxicity from iron oxide NPs (Jeng and Swanson 2006). However, chitosan-coated iron oxide NPs led to lower viability in human hepatocellular carcinoma cells, while 1-hydroxy-ethylidene-1,1-bisphospho nic acid–coated iron oxide NPs showed better viability in rat mesenchymal stem cells (Ge et al. 2009). The lethal impacts of iron oxide NPs are thought to stem primarily from the overproduction of ROS, ensuing oxidative damage to cellular lipids and DNA molecules. A comprehensive understanding of the routes implicated in the toxicity of iron oxide NPs is essential for improving their safety profile and maximizing their potential in biomedical applications.

3.1.7 Titanium Oxide and Titanium Dioxide (TiO_2) Nanoparticles

Although titanium dioxide (TiO_2) nanoparticles (NPs) are considered chemically inert, they have been associated with lethal consequences in investigational animals, involving DNA damage and genotoxicity, with lung inflammation (Trouiller et al. 2009). Studies suggest that TiO_2 NPs, especially those smaller than 100 nm, can prompt oxidative stress and contribute to the formation of DNA adducts (Bhattacharya et al. 2009). Additionally, these nanoparticles have been shown to affect immune function and disrupt the function of organs like the liver, kidneys, spleen, and heart muscles, as well as impair glucose and lipid metabolism in animal models (Liu et al. 2010). TiO_2 NPs are valued in sunscreens and cosmetics for their ability to block UV light while enhancing the transparency and appearance of products. However, in vitro studies have highlighted their toxicity, which varies depending on the cell type and can influence processes such as cell proliferation, differentiation, movement, and apoptosis (Pan et al. 2009). Interestingly, the adverse effects observed in vitro are not always replicated in in vivo settings.

To assess whether TiO_2 NPs can penetrate the skin, dermal absorption studies on human volunteers using different methods have been conducted. Lademan et al. examined the effects of frequent application of TiO_2-containing sunscreen, finding that the nanoparticles predominantly accessed the open areas of hair follicles rather than the viable epidermis or dermis. The titanium detected in the follicles was minimal, accounting for less than 1% of the total sunscreen applied (Lademann et al. 1999). Similarly, studies by Bennat and Muller-Goymann showed that surface penetration occurred mainly through hair follicles or pores, particularly in areas of skin with more hair (Bennat and Müller-Goymann 2000). Mavon et al. demonstrated that almost all the sunscreen applied could be recovered through tape stripping, with no significant deposition of TiO_2 found in the skin layers or hair follicles (Mavon et al. 2007). Factors such as surface coatings, functionalization, and the density of follicular pores influence the degree of nanoparticle penetration and uptake. Additional research is needed to fully understand the potential risks associated with TiO_2 NPs in consumer products.

3.1.8 Quantum Dots

Semiconductor nanocrystals, commonly known as quantum dots (QDs), hold considerable promise for biomedical applications owing to their exceptional organization, which includes an inorganic core–shell design and an organic coating. This structure facilitates the conjugation of QDs with biomolecules for targeted delivery within the body. However, their interaction with biological systems necessitates a thorough toxicological evaluation. Studies investigating the toxic effects of QDs in cellular models have found that their toxicity is primarily attributed to the release of metal ions from the heavy metal core. This toxicity is exacerbated under oxidative conditions that promote the degradation of the core and the leaching of metal ions. The liver, as a central organ involved in first-pass metabolism, is especially susceptible to the toxic effects induced by quantum dots (QDs), as exhibited in findings by Yang et al. (Yang et al. 2007). The size of QDs plays a significant role in their distribution across organs, with smaller QDs (<20 nm) being able to pass through liver capillaries, which are approximately 100 nm in diameter. Given the liver's susceptibility to heavy metal ion exposure, the accumulation of QDs poses a substantial risk. To mitigate this risk, researchers have explored surface coatings designed to protect the QD core from degradation. Traditional coatings like zinc sulfide (ZnS) and mercaptoacetic acid have shown some effectiveness, but persistent cellular toxicity highlights the need for more advanced or comprehensive coating strategies.

In a study by Clift et al., QDs with a CdTe/CdSe core and a ZnS shell were evaluated with various surface modifications, including organic, carboxylated (COOH), amino (NH_2), and polyethylene glycol (PEG) coatings. Despite these modifications, considerable cytotoxicity was observed, with the coating materials and the release of cadmium ions being the main contributors to the toxicity (Yildirimer et al. 2011). Recent work by Seifalian and colleagues has introduced polyhedral oligomeric silsesquioxane (POSS), a novel synthetic nanomaterial, as an alternative coating material. When applied to CdTe core QDs, POSS significantly enhanced cytocompatibility, even without the ZnS shell, as shown in unpublished data. The use of POSS effectively prevented ion escape from the center, demonstrating the significance of selecting the right coating materials to minimize QD toxicity. These studies emphasize the significant function that surface coatings portray in controlling the release of heavy metal ions from QD cores, which remains a key factor in determining their toxicity.

Quantum dots (QDs) have emerged as highly promising tools for functionalization, enabling their application in both diagnostic and therapeutic settings. Currently, QDs are widely used in experimental studies, such as in vivo imaging of neural stem and progenitor cells in developing mouse embryos, targeted drug delivery to the brain, and the identification of brain tumors to assist in precise surgical removal. However, concerns have been raised about the potential for nanoparticles (NPs), including QDs, to cross the blood-brain barrier (BBB) and enter the brain through pathways such as the olfactory epithelium or trigeminal nerve, possibly leading to harmful effects. To explore the potential for QDs to reach the central nervous system (CNS) structures, Kato et al. conducted a study in which they administered CdSe/ZnS core–shell QDs (functionalized with captopril and measuring 13.5 nm in diameter) intraperitoneally to mice (Kato et al. 2010). Six hours after the injection, organ samples were collected, and the brain was dissected into regions such as the hippocampus, thalamus olfactory bulb, cerebral cortex, and brainstem. The study found surprisingly excessive levels of cadmium in neuronal tissues but did not report any signs of inflammation, neuronal damage, or functional damage. Numerous potential mechanisms for QD entry into the brain have been proposed. One possibility is that small QDs could pass through gaps (<20 nm) between the astrocytic foot processes that form the BBB, thereby moving from the capillaries into the brain tissue. Another proposed mechanism involves the uptake of QDs through processes viz. phagocytosis or pinocytosis, which could be mediated by the interactions between the QD surface and BBB receptors (Zhang and Monteiro-Riviere 2009). However, receptor-mediated uptake is considered unlikely in this case as no evidence of captopril receptor expression at the BBB has been found. A third theory suggests that retrograde axonal transport may play a role, where QDs could travel from neurons in the gut through peripheral nerves to the CNS structures.

3.1.9 Superparamagnetic Nanoparticles

Superparamagnetic iron oxide nanoparticles (SPIONs) and ultra-small superparamagnetic iron oxide nanoparticles (USPIONs) have garnered significant attention in theranostic applications owing to their distinctive magnetic properties. These nanoparticles contain an iron oxide core, often coated with a carbohydrate layer, which plays a crucial role in modulating their cellular uptake, bio-distribution, and biological half-life. The degree of surface coating can influence the nanoparticles' behavior: incomplete coating can lead to opsonization and rapid endocytosis by the reticuloendothelial system, while fully coated nanoparticles tend to avoid opsonization, thus extending their plasma half-life. Despite this, recent studies suggest that particle size is the primary determinant of macrophage uptake rates rather than the extent of coating (Raynal et al. 2004). SPIONs are one of the few nanoparticles that have been FDA-approved for medical use, primarily for magnetic resonance imaging (MRI). They are employed for imaging purposes, particularly for visualizing vasculature, lymph nodes, and intracerebral tumors, as well as neurological lesions within the central nervous system (CNS) (Baumjohann et al. 2006). However, while SPIONs have shown promise in various applications, their long-term impacts and impending neurotoxicity are insufficiently studied. The distinctive physicochemical properties of SPIONs—small size and large surface area—make them ideal candidates for a variety of diagnostic and therapeutic applications. However, decreasing the size of these particles can lead to cytotoxic effects, disrupting normal cellular functions. In vitro studies confirmed that SPIONs can induce the production of ROS, weaken mitochondrial function, and increase the leak of lactate dehydrogenase (LDH), leading to neurotoxicity and exacerbating existing neuronal damage. Additionally, studies have shown that the toxicity of SPIONs is influenced by several factors, including particle size, coating type, degradation products, concentration, and opsonization degree (Mahmoudi et al. 2010).

For instance, Berry et al. found that the toxicity of SPIONs could be modulated by surface coatings in fibroblast cultures. They compared the toxicities of uncoated magnetic iron oxide nanoparticles as well as those coated with dextran or albumin. Their findings revealed that both uncoated and dextran-coated nanoparticles had similar toxicities, while albumin-coated nanoparticles promoted cell proliferation (Berry et al. 2003). Yet another study by Muldoon et al. evaluated the distribution, cellular uptake, and toxicity of various FDA-approved SPIONs of different sizes and surface coatings. They found that ferumoxtran-10 (USPION: 20–50 nm, fully coated with native dextran) was effectively taken up by the tumor tissue and remained in the lesion for up to five days, while ferumoxide (SPION: 60–185 nm, incompletely coated with dextran) did not enhance tumor targeting (Muldoon et al. 2005). To evaluate the capability of SPIONs to cross the BBB and assess their neurotoxicity, researchers disrupted the BBB osmotically. The results indicated that, despite the presence of SPIONs, no significant pathological changes were observed, suggesting that the clinical application of ultra-small SPIONs (USPIONs) via intracerebral injection may be feasible without causing major neurotoxic effects. Overall, SPIONs hold significant potential for use in diagnostics and therapy, particularly in imaging and targeted drug delivery. Nonetheless, additional investigation is required to understand their long-term safety, the impact of their physicochemical properties, and their potential neurotoxic effects when used in the human body.

3.2 NPs of Nonmetallic Substances

3.2.1 Carbon-Based Nanomaterials

Carbon-based nanomaterials, including carbon nanotubes (CNTs), fullerenes, and both single- and multi-walled variants, are extensively studied for their diverse applications. Despite their potential, several studies have identified them as cytotoxic agents. Magrez et al. (Magrez et al. 2006) investigated the cytotoxicity of carbon-based nanomaterials and found that their toxicity was size-dependent in lung cancer cells, as assessed by the MTT assay. Herzog et al. (Herzog et al. 2007) observed similar cytotoxic effects using clonogenic assays to examine the toxicity, cell proliferation, and death in various human cell lines such as those from alveolar carcinoma, bronchial epithelial tissue, and keratinocytes. The toxicity of CNTs is also dependent on their size, with multi-walled CNTs showing carcinogenic properties akin to asbestos when injected into the peritoneal cavity of animals. Conversely, single-walled CNTs are more easily internalized by macrophages, and their accumulation in the liver over time has been associated with biochemical alterations and organ changes in animal models. Factors such as the method of preparation, the presence of trace metals, and the size of the nanoparticles are crucial in determining the extent of their toxicity. Fullerenes, another class of carbon-based nanomaterials, are commonly present in the environment due to fuel combustion. Non-functionalized C60 fullerenes accumulate in the kidney, spleen, liver, and bones over time. In vitro studies have shown that fullerenes can induce genotoxic effects, including DNA strand breaks, chromosomal damage, and micronucleus formation, particularly following extended exposure to various cell lines (Dhawan et al. 2006; Niwa and Iwai 2006). However, some studies present conflicting results on the genotoxicity of fullerenes, likely due to differences in experimental conditions.

3.2.2 Carbon Nanotubes (CNTs)

CNTs (single-walled carbon nanotubes (SWCNTs) and multi-walled carbon nanotubes (MWCNTs)) have gained significant attention in in vivo inhalation models. It also holds promise for biomedical applications viz. drug delivery, photodynamic therapy (PDT), and tissue engineering scaffolds. Nonetheless, their potential for toxicity remains a concern. SWCNTs are generally considered to have higher cytotoxicity than MWCNTs, often due to their larger surface area. Research has shown that when CNTs are instilled into the trachea of mice, it results in the formation of pulmonary granulomas, interstitial inflammation, and fibrosis (Takagi et al. 2008). This toxicity is believed to be linked to the upregulation of TNF-α, an inflammatory cytokine. Though current research by Mutlu et al. suggested that the observed toxicity in mice after the instillation of SWCNTs may result from the aggregation of nanotubes rather than the substantial aspect ratio of individual nanotubes (Mutlu et al. 2010). As a result, various strategies have been extended to enhance the dispersal of CNTs to reduce their toxicity. In comparison, the effects of MWCNTs have been more inconsistent, with some studies indicating toxic effects similar to asbestos exposure, while others report that MWCNTs are biocompatible and noncytotoxic (Takagi et al. 2008). These varying outcomes may be attributed to differences in the composition of the nanotube and the methods of administration (e.g., intratracheal instillation vs. whole-body inhalation), highlighting the need for further studies to better understand their biological effects.

3.2.3 Silica (SiO_2) Nanoparticles

Silica (SiO_2) nanoparticles (NPs) are widely appreciated for their potential in drug delivery systems due to their ease of functionalization. They are also commonly present in the ambient air, making up roughly 8% of all airborne nanoparticles.

While SiO_2 NPs have been regarded as biocompatible in drug delivery applications, recent studies indicate that they can cause an upsurge of ROS-induced oxidative stress. Lin et al. (Lin et al. 2006) observed increased ROS production, lactate dehydrogenase (LDH) release, and malondialdehyde accumulation in human bronchoalveolar carcinoma cells exposed to SiO_2 NPs (15–46 nm) at concentrations ranging from 10 to 100 μg/mL. Various inflammatory biomarkers, e.g., IL-1, IL-6, IL-8, and TNF-α, along with mitochondrial damage, were also linked to SiO_2 NPs. Additional in vitro studies on liver cells showed that silica-based NPs (70 nm) at a dose of 30 mg/kg led to hepatotoxicity and altered biochemical markers (Nishimori et al. 2009). Beyond medical applications, SiO_2 NPs are found in chemical polishing, cosmetics, varnishes, and food products. They are increasingly used in biomedical domains, viz. cancer therapeutics and drug delivery, and as biomarkers. At moderate doses (<20 g/mL), SiO_2 NPs are considered safe, in contrast to their crystalline counterparts, which are classified as class 1 carcinogens (Lu et al. 2009). Conversely, at higher concentrations (≥25 g/mL), SiO_2 NPs exhibit a tendency to agglomerate, leading to dose-dependent cytotoxicity. At concentrations above 50 g/mL, these nanoparticles cause cell death through ROS generation and lipid peroxidation (Lison et al. 2008). Smaller SiO_2 NPs are found to be more cytotoxic than larger ones, showing toxicity that is size-dependent. In animal models of lung exposure, SiO_2 NPs triggered acute pulmonary inflammation, neutrophil infiltration, and chronic granulomatous changes after 14 weeks in a dose-dependent manner. Prolonged exposure resulted in the induction of anti-inflammatory mediators, and the inflammatory and fibrotic changes were partially reversed, approaching near-control levels. IL-4-, IL-10-, and IL-13-like fibrogenic mediators were upregulated, contributing to fibrosis, which was countered by the overexpression of matrix metalloproteinases (MMP-2) and interferon-gamma (INF-γ) (Barbarin et al. 2005). The recovery of lung tissue is likely due to the gradual reduction in NP concentration within the alveoli, through diffusion and translocation to extrapulmonary organs.

In the context of drug delivery and cosmetics, SiO_2 NPs serve as nano-vehicles though limited research is available on their skin toxicity. Studies suggest that smaller particle sizes increase cytotoxicity, as observed when murine epidermal Langerhans cells were exposed to SiO_2 particles ranging from 70 to 1000 nm (Nabeshi et al. 2010). Smaller particles (<100 nm) are more efficiently absorbed by the cells, leading to greater toxicity. A study by Park and colleagues assessed the toxic effects of SiO_2 NPs of varying sizes on human keratinocytes and evaluated the results with a human skin equivalent model (HSEM) and an in vivo rabbit model (Park et al. 2010). SiO_2 NPs caused considerable, dose-dependent toxicity in keratinocytes, with a marked decline in cell viability at 50 g/ml (Jiang et al. 2008). However, Jiang et al.'s proposed size-dependent increase in toxicity could not be confirmed. In contrast, exposure of NPs to HSEM did not show any inflammatory effects even at 500 g/mL, and in vivo rabbit skin model showed low acute toxicity. These findings underscore the importance of using HSEM over cell culture models for evaluating dermal toxicity and highlight the inconsistencies concerning in vitro and in vivo results when considering the skin effects of NPs.

3.2.4 NPs of Polymeric Materials

Polymeric nanoparticles (NPs) have garnered significant interest due to their biocompatibility, modifiable surfaces, and ability to provide sustained drug release, making them suitable for treating a variety of pulmonary and extrapulmonary diseases, for instance, tuberculosis, bronchial asthma, chronic obstructive pulmonary disease, lung cancer, and diabetes. Natural organic polymers like gelatin, collagen, chitosan, alginate, and bovine serum albumin (BSA) are commonly used in nanoparticle formulations. In recent decades, synthetic polymers like poly(lactic-co-glycolic acid) (PLGA) have gained recognition as biocompatible and biodegradable carriers for drug delivery (Romero et al. 2010). These polymeric NPs have demonstrated the potential to improve therapeutic outcomes while minimizing side effects through sustained and targeted delivery. A key advantage of polymeric NPs is their ability to incorporate biomaterials such as chitosan or BSA, which enhance their biocompatibility and biodegradability. These biological capping agents can further decrease cytotoxicity by mirroring the natural biological environment, preventing the immune system from recognizing the nanoparticles as foreign. For instance, Romero et al. found that PLGA NPs stabilized with BSA showed lower cytotoxicity than those with synthetic coatings in lung cancer cell cultures. BSA, a highly abundant protein in serum, is known for its superior biocompatibility, making it an ideal stabilizer for drug delivery systems. Similarly, chitosan-stabilized NPs demonstrated improved cell preservation and better adhesion to the pulmonary mucosa in an in vivo lung cancer model (Tahara et al. 2009).

Biodegradable and polymeric NPs hold great promise for targeted drug delivery, especially in cancer therapy. They are capable of encapsulating a variety of therapeutic agents, such as peptides, nucleic acids, and proteins, while offering sustained release and good tissue compatibility. These NPs are generally considered nontoxic, non-immunogenic, and noninflammatory, without triggering neutrophil activation. PLGA, in particular, has shown high efficacy in targeted drug delivery systems due to its ability to undergo hydrolysis, yielding biologically compatible metabolites like lactic acid and glycolic acid. Nonetheless, recent data has elevated fear about the potentially toxic effects of surface coatings on polymeric NPs, particularly toward human-like macrophages (Grabowski et al. 2015). This highlights the need for further research to evaluate the impact of surface modifications on the biocompatibility and toxicity of these nanoparticles to ensure their cautious use in biomedical applications.

4 Evaluating the Biological Responses to Nanoparticulate Exposure

The growing concerns about the health risks associated with nanoparticles have led to an increase in research focused on their toxicological assessment. This section outlines the primary biological effects of various nanomaterials and the methods employed to assess these impacts.

5 Nanoparticle-Induced DNA Damage

Genotoxicity plays a crucial role in assessing the potential risks posed by substances commonly encountered by humans, such as those found in food and personal care products. Understanding the genotoxic effects of nanoparticles is particularly important due to their ability to damage DNA, potentially leading to cancer, infertility, and genetic issues in future generations if reproductive cells are affected. Several studies have highlighted the genotoxic properties of nanoparticles, with several factors influencing this potential.

6 Direct Interactions with the DNA

Nanoparticles are capable of entering the nucleus and directly interacting with the DNA (Shukla et al. 2011). Although nanoparticles typically cannot pass through the nuclear membrane, they can gather in the cytoplasm and enter the nucleus during cell division at which time the membrane is temporarily compromised. This interaction can physically damage the genetic material and interfere with DNA repair mechanisms within the nucleus, which in turn further contributes to this impairment.

7 Oxidative DNA Damage

Nanomaterials, because of their large area relative to their volume, stimulate the levels of ROS. Previous studies have identified a relationship between the perinuclear accumulation of TiO_2 nanoparticles in BEAS-2B cells and the ROS production in that area, observed using fluorescence dyes. ROS have the potential to cause single- and double-strand breaks, base modifications, and DNA cross-links (Toyokuni 1998). Furthermore, ROS can be produced through interactions with cellular organelles, including the disruption of mitochondrial electron transport chain activity. During inflammatory responses, activated phagocytes release ROS in an attempt to combat pathogens; this can lead to DNA damage of both the infected cells and adjoining host cells. The comet assay, or single-cell gel electrophoresis, is a general technique for quantitatively and qualitatively evaluating DNA damage in discrete cells. It requires only a modest number of cells/sample (~10,000) and can identify specific types of DNA damage by employing lesion-specific antibodies or DNA repair enzymes. More than 60 studies examining nanoparticle-induced genotoxicity using the comet assay have yielded positive findings, confirming the DNA-damaging effects of engineered nanoparticles. The micronucleus assay, which reveals chromosomal damage, is another method employed to assess the clastogenic or aneugenic potential of substances. However, the presence of nanoparticles on prepared slides can deter the visibility of micronuclei, making the counting process more challenging (Gonzalez et al. 2011).

The genotoxic effects of nanoparticles affect humans as well as other organisms, such as bacteria. The Ames test (Salmonella reverse mutation assay) has been employed to gauge the mutagenic ability of nanoparticles. Nanoparticles like TiO_2 and ZnO have shown mild mutagenic effects in certain bacterial strains although some studies have reported no mutagenic activity. Variations in results may be due to differences in the sources, physical and chemical properties, and preparation techniques of nanoparticle suspensions (e.g., concentration and sonication). Since the conventional Ames test was designed to assess chemical mutagenesis, it may not fully capture the physical interactions between nanoparticles and DNA. Despite these limitations, the Ames test continues to be an important method for evaluating the mutagenicity of nanoparticles. It identifies point mutations in single genes in prokaryotic cells, both with and without the S9 metabolic activation fraction. Research using the Salmonella reverse mutation assay to investigate the TiO_2 and ZnO nanoparticle ecotoxicity in *Salmonella typhimurium* and *Escherichia coli* has found weak mutagenic prospective in TA98, TA1537, and *E. coli* strains (Kumar et al. 2011a, b). However, discrepancies in the results underscore the importance of considering nanoparticle-specific characteristics and methodological differences. While the Ames test is a useful tool for detecting chemical mutagens, it may not fully account for mutations caused by nanoparticles. The differences between bacterial and mammalian cell membranes, along with the role of endocytosis in nanoparticle uptake, complicate the interactions between nanoparticles and DNA. Therefore, although the Ames test is crucial for mutagenicity testing, it might not completely capture the complex mechanisms through which nanoparticles interact within biological systems.

8 Cytotoxicity of Nanoparticles

Nanoparticles of various types have been shown to reduce cell viability, leading to their widespread use in in vitro assays for toxicity evaluation. These methods are popular due to their simplicity, cost-effectiveness, and efficiency. Common assays used to assess nanoparticle toxicity include mitochondrial dehydrogenase activity, neutral red uptake, and lactate dehydrogenase (LDH) release. Mitochondrial dehydrogenase activity is measured using a tetrazolium dye, 3-[4,5-dimethylthiazol-2-yl]-2,5 diphenyl tetrazolium bromide (MTT). Viable cells with active mitochondrial enzymes reduce the tetrazolium ring, producing purple insoluble crystals. These crystals are dissolved in acidified isopropanol or dimethyl sulfoxide (DMSO), and absorbance is measured spectrophotometrically (Mosmann 1983). The neutral red uptake assay assesses lysosomal membrane integrity by measuring the active transport of neutral red dye into lysosomes in viable cells. Nonviable cells, however, cannot retain the dye, and it is swept away. The presence of LDH in the extracellular medium can be detected by the LDH release assay. Elevated LDH levels indicate irreversible cell death resulting from membrane damage. A study by Zaquat et al. (Zaqout et al. 2012) found that TiO_2 nanoparticles bind to LDH, which could lead to an underestimation of their toxicity in LDH release assays. This underscores the importance of considering nanoparticle interactions with assay components when evaluating cytotoxicity.

9 Nanoparticle Internalization in Biological Systems

The size of nanostructures is a key factor influencing their cellular uptake, underscoring the importance of understanding the processes that regulate nanoparticle (NP) entry and exit from cells. This knowledge is vital for assessing both NP toxicity and their potential biomedical uses. Research has shown that NP uptake is inhibited at 4 °C or under conditions of ATP depletion, suggesting that endocytosis, rather than passive diffusion, is the main route for cellular internalization (Chithrani and Chan 2007). Endocytic pathways can include clathrin-mediated endocytosis, caveolae-mediated endocytosis, or mechanisms independent of both clathrin and caveolae. Importantly, nanoparticles that enter the cells via caveolae-mediated endocytosis or those that accumulate near the nucleus may interact physically with nuclear pore complexes, potentially facilitating their entry into the nucleus. Phagocytosis, which is primarily associated with professional phagocytic cells, is not a major contributor to NP uptake as even non-phagocytic cells can efficiently internalize nanoparticles. Additionally, phagocytosis typically occurs with particles larger than 500 nm, which excludes smaller nanoparticles from this process (Rupper and Cardelli 2001).

To gain a deeper understanding of nanoparticle (NP) internalization and its correlation with biological effects, advanced tracking methods are crucial. Transmission electron microscopy (TEM) provides high-resolution imaging of NP interactions with cellular components, capturing events such as membrane invaginations, vesicle formation, and interactions with organelles. However, TEM is primarily a qualitative technique, and its use is constrained by labor-intensive sample preparation, potential artifacts from staining, and challenges in analyzing the resulting images.

For quantitative analysis, inductively coupled plasma mass spectrometry (ICP-MS) delivers superior sensitivity, making it possible to discover trace amounts of nanoparticles in biological samples. This method is particularly useful in in vivo studies, as it can map the distribution and accumulation of nanoparticles in specific organs of the body. Flow cytometry, a sensitive and versatile technique, is also employed to probe into NP uptake mechanisms in both bacterial and mammalian cells. It allows for the discovery of both fluorescent and nonfluorescent nanoparticles within cells, offering valuable insights into the mechanisms of cellular internalization (Sharma et al. 2012b).

10 In Vivo Toxicity of Nanoparticles

Evaluating the toxic effects of new entities in vivo is essential as in vitro systems cannot fully replicate the complex interactions between cells, the extracellular matrix, hormonal influences, and the variety of cell types present in a living organism. In vivo studies also enable the investigation of long-term chronic effects, which are difficult to assess in vitro. The significance of in vivo research in nanomaterial toxicology is well-documented (Fischer and Chan 2007). Currently, standardized protocols or methodologies are not available for conducting comprehensive risk assessments of nanomaterials. However, advisory organizations such as the UK's "Committees on Toxicity (COT), Mutagenicity (COM), and Carcinogenicity (COC) of Chemicals in Food, Consumer Products, and the Environment" have recommended using in vitro nanotoxicity findings as an initial step, followed by in vivo validation. These committees also stress the importance of selecting the appropriate exposure route in in vivo studies to ensure a precise evaluation.

Numerous studies have investigated the in vivo effects of nanoparticles, revealing outcomes such as organ accumulation (particularly in the liver), DNA damage, immune system disruptions, and changes in neurobehavioral patterns. Despite these findings, conducting in vivo research with nanomaterials presents several challenges. Measuring nanoparticle concentrations in samples like that of water, air, soil, and/or consumer products is technically difficult owing to their minuscule volume and low concentrations. Even when nanoparticle doses are determined, exceeding certain limits in experimental settings is not recommended as higher doses can cause nanoparticle agglomeration, potentially leading to misleading results. Additionally, assessing the bio-distribution or biodosimetry of nanoparticles that are nonfluorescent, nonradioactive, or nonmagnetic remains a major challenge with existing technologies. These limitations highlight the need for the development of advanced methods and tools to enable accurate in vivo toxicity evaluation of nanomaterials (Sharma et al. 2012b).

11 Conclusion

The rapid expansion of nanomaterial production has outpaced the development of effective safety and toxicity assessment methods. Bridging this gap requires a collaborative approach that involves material scientists, molecular biologists, toxicologists, and physicists to fully comprehend the cellular responses and mechanisms associated with nanomaterial exposure. This integrated effort is crucial for assessing potential risks to human health and the environment and for fostering the development of environmentally friendly and biologically safe nanoparticles.

Nanoparticles (NPs) possess unique properties that are both advantageous and potentially harmful, especially in biomedical applications. The relationship between particle characteristics, dosage, and host immune response plays a key role in NP-induced biotoxicity. This toxicity can manifest as inflammatory responses, cellular necrosis, or apoptosis, often triggered by ROS. However, current toxicity studies face significant challenges. Many of these studies use exposure concentrations much higher than those typically encountered in clinical or therapeutic settings, leading to results that may be unreliable or unnecessarily alarming. Additionally, differences in particle formulations across studies and the chronic nature of NP exposure contribute to inconsistent findings, which compromise the reliability of conclusions. To address these challenges, future research must focus on determining relevant NP doses for specific routes of administration, considering factors such as organ-specific accumulation, bio-distribution, metabolism, and excretion. The establishment of standardized toxicological profiling protocols is crucial for generating consistent and actionable data. These protocols should reflect realistic exposure scenarios to ensure that the findings are meaningful for human applications of nanoparticles. In conclusion, overcoming the current challenges in NP toxicity evaluation requires a comprehensive approach that emphasizes standardized study designs, accurate dosing, and careful consideration of administration routes. Such an approach will ensure the safe application of nanotechnology in clinical practice and protect both human health and the environment, enabling nanotechnology to reach its full potential.

References

Ahamed M, Siddiqui MA, Akhtar MJ, Ahmad I, Pant AB, Alhadlaq HA. Genotoxic potential of copper oxide nanoparticles in human lung epithelial cells. Biochem Biophys Res Commun. 2010;396(2):578–83.

Alshatwi AA, Vaiyapuri Subbarayan P, Ramesh E, ALHazzani AA, Alsaif MA, Alwarthan AA. Al2O3 nanoparticles induce mitochondria-mediated cell death and upregulate the expression of signaling genes in human mesenchymal stem cells. J Biochem Mol Toxicol. 2012;26(11):469–76.

Balasubramanyam A, Sailaja N, Mahboob M, Rahman MF, Hussain SM, Grover P. In vivo genotoxicity assessment of aluminium oxide nanomaterials in rat peripheral blood cells using the comet assay and micronucleus test. Mutagenesis. 2009;24(3):245–51.

Barbarin V, Nihoul A, Misson P, Arras M, Delos M, Leclercq I, Lison D, Huaux F. The role of pro- and anti-inflammatory responses in silica-induced lung fibrosis. Respir Res. 2005;6:112.

Baumjohann D, Hess A, Budinsky L, Brune K, Schuler G, Lutz MB. In vivo magnetic resonance imaging of dendritic cell migration into the draining lymph nodes of mice. Eur J Immunol. 2006;36(9):2544–55.

Bennat C, Müller-Goymann CC. Skin penetration and stabilization of formulations containing microfine titanium dioxide as physical UV filter. Int J Cosmet Sci. 2000;22(4):271–83.

Berry CC, Wells S, Charles S, Curtis AS. Dextran and albumin derivatised iron oxide nanoparticles: influence on fibroblasts in vitro. Biomaterials. 2003;24(25):4551–7.

Bhattacharya K, Davoren M, Boertz J, Schins R, Hoffmann E, Dopp E. Titanium dioxide nanoparticles induce oxidative stress and DNA-adduct formation but not DNA-breakage in human lung cells. Part Fibre Toxicol. 2009;6:17.

Boisselier E, Astruc D. Gold nanoparticles in nanomedicine: preparations, imaging, diagnostics, therapies and toxicity. Chem Soc Rev. 2009;38(6):1759–82.

Brunner TJ, Wick P, Manser P, Spohn P, Grass RN, Limbach LK, Bruinink A, Stark WJ. In vitro cytotoxicity of oxide nanoparticles: comparison to asbestos, silica, and the effect of particle solubility. Environ Sci Technol. 2006;40(14):4374–81.

Cervadoro A, Coclite A, Di Mascolo D, Ferreira M, Palange AL, Palomba R, et al. Smart nanoconstructs for theranostics in cancer and cardiovascular diseases. Core-shell nanostructures drug delivery and theranostics challenges, strategies prospects for novel carrier systems 2018:297–321. doi:https://doi.org/10.1016/C2016-0-03458-7.

Chen Z, Meng H, Xing G, Chen C, Zhao Y, Jia G, Wang T, Yuan H, Ye C, Zhao F, Chai Z, Zhu C, Fang X, Ma B, Wan L. Acute toxicological effects of copper nanoparticles in vivo. Toxicol Lett. 2006;163(2):109–20.

Chen L, Yokel RA, Hennig B, Toborek M. Manufactured aluminum oxide nanoparticles decrease expression of tight junction proteins in brain vasculature. J Neuroimmune Pharmacol. 2008;3(4):286–95.

Chen YS, Hung YC, Liau I, Huang GS. Assessment of the in vivo toxicity of gold nanoparticles. Nanoscale Res Lett. 2009;4(8):858–64.

Chithrani BD, Chan WC. Elucidating the mechanism of cellular uptake and removal of protein-coated gold nanoparticles of different sizes and shapes. Nano Lett. 2007;7:1542–50.

Cho WS, Cho M, Jeong J, Choi M, Cho HY, Han BS, Kim SH, Kim HO, Lim YT, Chung BH, Jeong J. Acute toxicity and pharmacokinetics of 13 nm-sized PEG-coated gold nanoparticles. Toxicol Appl Pharmacol. 2009;236(1):16–24.

Coclite A, Mollica H, Ranaldo S, Pascazio G, de Tullio MD, Decuzzi P. Predicting different adhesive regimens of circulating particles at blood capillary walls. Microfluid Nanofluid. 2017;21:168. https://doi.org/10.1007/s10404-017-2003-7.

Connor EE, Mwamuka J, Gole A, Murphy CJ, Wyatt MD. Gold nanoparticles are taken up by human cells but do not cause acute cytotoxicity. Small. 2005;1(3):325–7.

Costa CS, Ronconi JV, Daufenbach JF, Goncalves CL, Rezin GT, Streck EL, Paula MM. In vitro effects of silver nanoparticles on the mitochondrial respiratory chain. Mol Cell Biochem. 2010;342:51–6.

Dam DHM, Lee RC, Odom TW. Improved in vitro efficacy of gold nanoconstructs by increased loading of G-quadruplex aptamer. Nano Lett. 2014;14:2843–8. https://doi.org/10.1021/nl500844.

Dhawan A, Taurozzi JS, Pandey AK, Shan W, Miller SM, Hashsham SA, Tarabara VV. Stable colloidal dispersions of C60 fullerenes in water: evidence for genotoxicity. Environ Sci Technol. 2006;40(23):7394–401.

Dreher KL. Health and environmental impact of nanotechnology: toxicological assessment of manufactured nanoparticles. Toxicol Sci. 2004;77(1):3–5.

Fischer HC, Chan WC. Nanotoxicity: the growing need for in vivo study. Curr Opin Biotechnol. 2007;18:565–71.

Foldbjerg R, Dang DA, Autrup H. Cytotoxicity and genotoxicity of silver nanoparticles in the human lung cancer cell line, A549. Arch Toxicol. 2011;85(7):743–50.

Franke H, Galla H, Beuckmann CT. Primary cultures of brain microvessel endothelial cells: a valid and flexible model to study drug transport through the blood-brain barrier in vitro. Brain Res Brain Res Protoc. 2000;5(3):248–56.

Garnett MC, Kallinteri P. Nanomedicines and nanotoxicology: some physiological principles. Occup Med (Lond). 2006;56(5):307–11.

Ge Y, Zhang Y, He S, Nie F, Teng G, Gu N. Fluorescence modified chitosan-coated magnetic nanoparticles for high-efficient cellular imaging. Nanoscale Res Lett. 2009;4(4):287–95.

Gonzalez L, Sanderson BJ, Kirsch-Volders M. Adaptations of the in vitro MN assay for the genotoxicity assessment of nanomaterials. Mutagenesis. 2011;26:185–91.

Grabowski N, Hillaireau H, Vergnaud J, Tsapis N, Pallardy M, Kerdine-Röme S, Fattal E. Surface coating mediates the toxicity of polymeric nanoparticles towards human-like macrophages. Int J Pharm. 2015;482(1–2):75–83.

Guan R, Kang T, Lu F, Zhang Z, Shen H, Liu M. Cytotoxicity, oxidative stress, and genotoxicity in human hepatocyte and embryonic kidney cells exposed to ZnO nanoparticles. Nanoscale Res Lett. 2012;7(1):602–7.

Haase A, Tentschert J, Jungnickel H, Graf P, Mantion A, Draude F, Plendl J, Goetz ME, Galla S, Masic A. Toxicity of silver nanoparticles in human macrophages: uptake, intracellular distribution and cellular responses. J Phys Conf Ser. 2011;304(1):012030.

Hagens WI, Oomen AG, de Jong WH, Casse FR, Sips AJ. What do we need to know about the kinetic properties of nanoparticles in the body? Regul Toxicol Pharmacol. 2007;49(3):217–29.

Herzog E, Casey A, Lyng FM, Chambers G, Byrne HJ, Davoren M. A new approach to the toxicity testing of carbon-based nanomaterials the clonogenic assay. Toxicol Lett. 2007;174(1–3):49–60.

Huang CC, Aronstam RS, Chen DR, Huang YW. Oxidative stress, calcium homeostasis, and altered gene expression in human lung epithelial cells exposed to ZnO nanoparticles. Toxicol In Vitro. 2010;24(1):45–55.

Hussain SM, Hess KL, Gearhart JM, Geiss KT, Schlager JJ. In vitro toxicity of nanoparticles in BRL 3A rat liver cells. Toxicol In Vitro. 2005;19(7):975–83.

Jain S, Hirst DG, O'sullivan JM. Gold nanoparticles as novel agents for cancer therapy. Br J Radiol. 2012;85(1010):101–13.

Jeng HA, Swanson J. Toxicity of metal oxide nanoparticles in mammalian cells. J Environ Sci Health A Tox Hazard Subst Environ Eng. 2006;41(12):2699–711.

Jiang W, Kim BY, Rutka JT, Chan WC. Nanoparticle-mediated cellular response is size-dependent. Nat Nanotechnol. 2008;3:145–50.

Kato S, Itoh K, Yaoi T, Tozawa T, Yoshikawa Y, Yasui H, Kanamura N, Hoshino A, Manabe N, Yamamoto K. Organ distribution of quantum dots after intraperitoneal administration, with special reference to area-specific distribution in the brain. Nanotechnology. 2010;21:335103.

Kim YJ, Choi HS, Song MK, Youk DY, Kim JH, Ryu JC. Genotoxicity of aluminum oxide (Al2O3) nanoparticle in mammalian cell lines. Mol Cell Toxicol. 2009;5:172–8.

Kumar A, Pandey AK, Singh SS, Shanker R, Dhawan A. Cellular response to metal oxide nanoparticles in bacteria. J Biomed Nanotechnol. 2011a;7:102–3.

Kumar A, Dhawan A, Shanker R. The need for novel approaches in ecotoxicity of engineered nanomaterials. J Biomed Nanotechnol. 2011b;7:79–80.

Lademann J, Weigmann H, Rickmeyer C, Barthelmes H, Schaefer H, Mueller G, Sterry W. Penetration of titanium dioxide microparticles in a sunscreen formulation into the horny layer and the follicular orifice. Skin Pharmacol Appl Ski Physiol. 1999;12(5):247–56.

Lasagna-Reeves C, Gonzalez-Romero D, Barria MA, Olmedo I, Clos A, Sadagopa Ramanujam VM, Urayama A, Vergara L, Kogan MJ, Soto C. Bioaccumulation and toxicity of gold nanoparticles after repeated administration in mice. Biochem Biophys Res Commun. 2010;393(4):649–55.

Lei R, Wu C, Yang B, Ma H, Shi C, Wang Q, Yuan Y, Liao M. Integrated metabolomic analysis of the nanosized copper particle-induced hepatotoxicity and nephrotoxicity in rats: a rapid in vivo screening method for nanotoxicity. Toxicol Appl Pharmacol. 2008;232(2):292–301.

Lin W, Huang YW, Zhou XD, Ma Y. In vitro toxicity of silica nanoparticles in human lung cancer cells. Toxicol Appl Pharmacol. 2006;217(3):252–9.

Lison D, Thomassen LC, Rabolli V, Gonzalez L, Napierska D, Seo JW, et al. Nominal and effective dosimetry of silica nanoparticles in cytotoxicity assays. Toxicol Sci. 2008;104:155–16.

Liu Z, Tabakman S, Welsher K, Dai H. Carbon nanotubes in biology and medicine: in vitro and in vivo detection, imaging and drug delivery. Nano Res. 2009;2(2):85–120.

Liu R, Zhang X, Pu Y, Yin L, Li Y, Zhang X, Liang G, Li X, Zhang J. Small-sized titanium dioxide nanoparticles mediate immune toxicity in rat pulmonary alveolar macrophages in vivo. J Nanosci Nanotechnol. 2010;10(8):5161–9.

Liu G, Gao J, Ai H, Chen X. Applications and potential toxicity of magnetic iron oxide nanoparticles. Small. 2013;9(9–10):1533–45.

Lu S, Duffin R, Poland C, Daly P, Murphy F, Drost E, MacNee W, Stone V, Donaldson K. Efficacy of simple short-term in vitro assays for predicting the potential of metal oxide nanoparticles to cause pulmonary inflammation. Environ Health Perspect. 2009;117:241–7.

Lu W, Huang Q, Ku G, Wen X, Zhou M, Guzatov D, Brecht P, Su R, Oraevsky A, Wang LV, Li C. Photoacoustic imaging of living mouse brain vasculature using hollow gold nanospheres. Biomaterials. 2010;31(9):2617–26.

Magrez A, Kasas S, Salicio V, Pasquier N, Seo JW, Celio M, Catsicas S, Schwaller B, Forro L. Cellular toxicity of carbon-based nanomaterials. Nano Lett. 2006;6(6):1121–5.

Mahmoudi M, Simchi A, Imani M, Shokrgozar MA, Milani AS, Häfeli UO, Stroeve P. A new approach for the in vitro identification of the cytotoxicity of superparamagnetic iron oxide nanoparticles. Colloids Surf B Biointerfaces. 2010;75(1):300–9.

Mavon A, Miquel C, Lejeune O, Payre B, Moretto P. In vitro percutaneous absorption and in vivo stratum corneum distribution of an organic and a mineral sunscreen. Skin Pharmacol Physiol. 2007;20(1):10–20.

Maynard AD, Baron PA, Foley M, Shvedova AA, Kisin ER, Castranova V. Exposure to carbon nanotube material: aerosol release during the handling of unrefined single-walled carbon nanotube material. J Toxicol Environ Health A. 2004;67(1):87–107.

Meyer K, Rajanahalli P, Ahamed M, Rowe JJ, Hong Y. ZnO nanoparticles induce apoptosis in human dermal fibroblasts via p53 and p38 pathways. Toxicol In Vitro. 2011;25(8):1721–6.

Mironava T, Hadjiargyrou M, Simon M, Jurukovski V, Rafailovich MH. Gold nanoparticles cellular toxicity and recovery: effect of size, concentration and exposure time. Nanotoxicology. 2010;4(1):120–37.

Mosmann T. Rapid colorimetric assay for cellular growth and survival: application to proliferation and cytotoxicity assays. J Immunol Methods. 1983;65:55–63.

Mostafalou S, Mohammadi H, Ramazani A, Abdollahi M. Different biokinetics of nanomedicines linking to their toxicity; an overview. Daru. 2013;21(1):14.

Muldoon LL, Sàndor M, Pinkston KE, Neuwelt EA. Imaging, distribution, and toxicity of superparamagnetic iron oxide magnetic resonance nanoparticles in the rat brain and intracerebral tumor. Neurosurgery. 2005;57(4):785–96.

Muller J, Huaux F, Moreau N, Misson P, Heilier JF, Delos M, Arras M, Fonseca A, Nagy JB, Lison D. Respiratory toxicity of multi-wall carbon nanotubes. Toxicol Appl Pharmacol. 2005;207(3):221–31.

Mutlu GM, Budinger GR, Green AA, Urich D, Soberanes S, Chiarella SE, Alheid GF, McCrimmon DR, Szleifer I, Hersam MC. Biocompatible nanoscale dispersion of single-walled carbon nanotubes minimizes in vivo pulmonary toxicity. Nano Lett. 2010;10(5):1664–70.

Nabeshi H, Yoshikawa T, Matsuyama K, Nakazato Y, Arimori A, Isobe M, et al. Size-dependent cytotoxic effects of amorphous silica nanoparticles on Langerhans cells. Pharmazie. 2010;65(3):199–201.

Naqvi S, Samim M, Abdin M, Ahmed FJ, Maitra A, Prashant C, Dinda AK. Concentration-dependent toxicity of iron oxide nanoparticles mediated by increased oxidative stress. Int J Nanomedicine. 2010;5:983–9.

Nishimori H, Kondoh M, Isoda K, Tsunoda S, Tsutsumi Y, Yagi K. Silica nanoparticles as hepatotoxicants. Eur J Pharm Biopharm. 2009;72(3):496–501.

Niwa Y, Iwai N. Genotoxicity in cell lines induced by chronic exposure to water-soluble fullerenes using micronucleus test. Environ Health Prev Med. 2006;11(6):292–7.

Nowrouzi A, Meghrazi K, Golmohammadi T, Golestani A, Ahmadian S, Shafiezadeh M, Shajary Z, Khaghani S, Amiri AN. Cytotoxicity of subtoxic AgNP in human hepatoma cell line (HepG2) after long-term exposure. Iran Biomed J. 2010;14(1–2):23–32.

Oberdörster G, Sharp Z, Atudorei V, Elder A, Gelein R, Lunts A, Kreyling W, Cox C. Extrapulmonary translocation of ultrafine carbon particles following whole-body inhalation exposure of rats. J Toxicol Environ Health A. 2002;65(20):1531–43.

Oberdorster G, Oberdorster E, Oberdorster J. Nanotoxicology: an emerging discipline evolving from studies of ultrafine particles. Environ Health Perspect. 2005;113:823–39.

Osman IF, Baumgartner A, Cemeli E, Fletcher JN, Anderson D. Genotoxicity and cytotoxicity of zinc oxide and titanium dioxide in HEp-2 cells. Nanomedicine (Lond). 2010;5(8):1193–203.

Pan Y, Neuss S, Leifert A, Fischler M, Wen F, Simon U, Schmid G, Brandau W, Jahnen-Dechent W. Size-dependent cytotoxicity of gold nanoparticles. Small. 2007;3(11):1941–9.

Pan Z, Lee W, Slutsky L, Clark RA, Pernodet N, Rafailovich MH. Adverse effects of titanium dioxide nanoparticles on human dermal fibroblasts and how to protect cells. Small. 2009;5(4):511–20.

Park YH, Kim JN, Jeong SH, Choi JE, Lee SH, Choi BH, Lee JP, Sohn KH, et al. Assessment of dermal toxicity of nanosilica using cultured keratinocytes, a human skin equivalent model and an in vivo model. Toxicology. 2010;267(1–3):178–81.

Patra HK, Banerjee S, Chaudhuri U, Lahiri P, Dasgupta AK. Cell selective response to gold nanoparticles. Nanomedicine. 2007;3(2):111–9.

Pawelczyk E, Arbab AS, Chaudhry A, Balakumaran A, Robey PG, Frank JA. In vitro model of bromodeoxyuridine or iron oxide nanoparticle uptake by activated macrophages from labeled stem cells: implications for cellular therapy. Stem Cells. 2008;26(5):1366–75.

Powers CM, Levin ED, Seidler FJ, Slotkin TA. Silver exposure in developing zebrafish produces persistent synaptic and behavioral changes. Neurotoxicol Teratol. 2011;33(2):329–32.

Radziun E, Dudkiewicz Wilczyńska J, Książek I, Nowak K, Anuszewska EL, Kunicki A, Olszyna A, Zabkowski T. Assessment of the cytotoxicity of aluminum oxide nanoparticles on selected mammalian cells. Toxicol In Vitro. 2011;25(8):1694–700.

Raynal I, Prigent P, Peyramaure S, Najid A, Rebuzzi C, Corot C. Macrophage endocytosis of superparamagnetic iron oxide nanoparticles: mechanisms and comparison of ferumoxides and ferumoxtran-10. Investig Radiol. 2004;39(1):56–63.

Romero G, Estrela-Lopis I, Zhou J, Rojas E, Franco A, Espinel CS, et al. Surface engineered poly(lactide-co-glycolide) nanoparticles for intracellular delivery: uptake and cytotoxicity—a confocal Raman microscopic study. Biomacromolecules. 2010;11:2993–9.

Rupper A, Cardelli J. Regulation of phagocytosis and endo-phagosomal trafficking pathways in Dictyostelium discoideum. Biochim Biophys Acta. 2001;1525:205–16.
Sharma HS, Hussain S, Schlager J, Ali SF, Sharma A. Influence of nanoparticles on blood-brain barrier permeability and brain edema formation in rats. Acta Neurochir Suppl. 2010;106:359–64.
Sharma V, Singh P, Pandey AK, Dhawan A. Induction of oxidative stress, DNA damage and apoptosis in mouse liver after sub-acute oral exposure to zinc oxide nanoparticles. Mutat Res Genet Toxicol Environ Mutagen. 2012a;745(1–2):84–91.
Sharma V, Kumar A, Dhawan A. Nanomaterials: exposure, effects and toxicity assessment. Proc Natl Acad Sci India Sect B Biol Sci. 2012b;82(S1):3–11.
Shukla RK, Sharma V, Pandey AK, Singh S, Sultana S, Dhawan A. ROS-mediated genotoxicity induced by titanium dioxide nanoparticles in human epidermal cells. Toxicol In Vitro. 2011;25:231–41.
Sonavane G, Tomoda K, Sano A, Ohshima H, Terada H, Makino K. In vitro permeation of gold nanoparticles through rat skin and rat intestine: effect of particle size. Colloids Surf B: Biointerfaces. 2008;65(1):1–10.
Sung JH, Ji JH, Yoon JU, Kim DS, Song MY, Jeong J, Han BS, Han JH, Chung YH, Kim J, Kim TS, Chang HK, Lee EJ, Lee JH, Yu IJ. Lung function changes in Sprague-Dawley rats after prolonged inhalation exposure to silver nanoparticles. Inhal Toxicol. 2008;20:567–74.
Sung JH, Ji JH, Park JD, Yoon JU, Kim DS, Jeon KS, Song MY, Jeong J, Han BS, Han JH, Chung YH, Chang HK, Lee JH, Cho MH, Kelman BJ, Yu IJ. Subchronic inhalation toxicity of silver nanoparticles. Toxicol Sci. 2009;108(2):452–61.
Tahara K, Sakai T, Yamamoto H, Takeuchi H, Hirashima N, Kawashima Y. Improved cellular uptake of chitosan-modified PLGA nanospheres by A549 cells. Int J Pharm. 2009;382(1–2):198–204.
Takagi A, Hirose A, Nishimura T, Fukumori N, Ogata A, Ohashi N, Kitajima S, Kanno J. Induction of mesothelioma in p53+/− mouse by intraperitoneal application of multi-wall carbon nanotube. J Toxicol Sci. 2008;33(1):105–16.
Tang J, Xiong L, Wang S, Wang J, Liu L, Li J, Yuan F, Xi T. Distribution, translocation and accumulation of silver nanoparticles in rats. J Nanosci Nanotechnol. 2009;9(8):4924–32.
Tang J, Xiong L, Zhou G, Wang S, Wang J, Liu L, Li J, Yuan F, Lu S, Wan Z. Silver nanoparticles crossing through and distribution in the blood-brain barrier in vitro. J Nanosci Nanotechnol. 2010;10:6313–7.
Thanh NTK, Raton B, editors. Magnetic nanoparticles: from fabrication to clinical applications. CRC Press/Taylor and Francis; 2011.
Toyokuni S. Oxidative stress and cancer: the role of redox regulation. Biotherapy. 1998;11:147–54.
Trickler WJ, Lantz SM, Murdock RC, Schrand AM, Robinson BL, Newport GD, Schlager JJ, Oldenburg SJ, Paule MG, Slikker W Jr, Hussain SM, Ali SF. Silver nanoparticle induced blood-brain barrier inflammation and increased permeability in primary rat brain microvessel endothelial cells. Toxicol Sci. 2010;118(1):160–70.
Trouiller B, Reliene R, Westbrook A, Solaimani P, Schiestl RH. Titanium dioxide nanoparticles induce DNA damage and genetic instability in vivo in mice. Cancer Res. 2009;69(22):8784–9.
Vishwakarma V, Samal SS, Manoharan N. Safety and risk associated with nanoparticles-a review. J Miner Mater Charact Eng. 2010;9(5):455.
Yang RS, Chang LW, Wu JP, Tsai MH, Wang HJ, Kuo YC, Yeh TK, Yang CS, Lin P. Persistent tissue kinetics and redistribution of nanoparticles, quantum dot 705, in mice: ICP-MS quantitative assessment. Environ Health Perspect. 2007;115(9):1339–43.
Yildirimer L, Thanh NT, Loizidou M, Seifalian AM. Toxicology and clinical potential of nanoparticles. Nano Today. 2011;6(6):585–607.
Zaqout MS, Sumizawa T, Igisu H, Wilson D, Myojo T, Ueno S. Binding of titanium dioxide nanoparticles to lactate dehydrogenase. Environ Health Prev Med. 2012;17:341–5.
Zhang LW, Monteiro-Riviere NA. Mechanisms of quantum dot nanoparticle cellular uptake. Toxicol Sci. 2009;110(1):138–55.
Zhang XD, Wu HY, Wu D, Wang YY, Chang JH, Zhai ZB, Meng AM, Liu PX, Zhang LA, Fan FY. Toxicologic effects of gold nanoparticles in vivo by different administration routes. Int J Nanomedicine. 2010;5:771–81.

Clinical Applications and Commercialization Challenges of Core–Shell Nanoconstructs

21

Snigdha Das Mandal, Surjyanarayan Mandal, and Amitkumar K. Patel

Abstract

Designing of core–shell nanoconstructs has become a priority research goal nowadays with its very specific applications in biomedical fields. Basically, the optimization of the developed nanoconstructs is still in the developmental stage due to lack of in-depth understandings on the design and production of nanomaterials. Particle size distribution, shape, surface charge, and microstructures along with the stability of ligand-gated structures are important factors for their in vitro and in vivo performance. Particularly in cancer treatment, the involvement of nanotechnologies for the purpose of diagnosis followed by therapeutic (healing) purposes is considered as a most innovative and effective approach to compensate the limitations of conventional or existing chemotherapy such as non-efficacious (toxicity), undesirable drug distribution along with their unpredictable release. Core–shell nanoconstructs are made up of nanoparticles with specific ligands and are engineered with anticancer drugs to cross the main biological barriers and simultaneously to target the cancerous cells through intravenous route in order to obtain better therapeutic effect with reduced side effect. In-depth study of stability of these developed nanoconstructs in complex media, protein, and cell interaction will be required for their successful transformation to the preclinical and clinical prospects. Tailor-made nanoconstructs designed with explicit and accurate ligands help to invade cancer cells in a foreseeable

S. D. Mandal (✉)
Department of Pharmacology, Parul Institute of Pharmacy and Research, Parul University, Vadodara, Gujarat, India

S. Mandal
Themis Medicare Ltd, Haridwar, Uttarakhand, India

A. K. Patel
Saffron Health LLC, East Brunswick, NJ, USA

J. K. Patel et al. (eds.), *Core-Shell Nano Constructs for Cancer Theragnostic*,
https://doi.org/10.1007/978-981-96-3025-7_21

manner by effectively delivering the drugs. So, nanoconstructs offer numerous benefits in cancer treatment; however, they possess numerous challenges, including the issues of nanotoxicity. To deal such formidable tasks, methods derived from artificial intelligence (AI) are being explored. Regulators also struggle with nanotoxicology and biological response. Regulatory agencies like US-FDA, MHRA, and the European Medicines Agency (EMA) also do not provide more specific direction. In this chapter, an overview of the development with commercialization, clinical applications, and the regulatory status as well as challenges to market authorization of nanoconstructs especially for cancer treatment is presented. This chapter emphasizes the chemotherapeutic linked to core–shell nanoconstructs.

Keywords

Nanoconstructs · Theranostic · Regulatory framework · Nanotoxicity · Chemotherapy

1 Introduction

Cancer is representing itself as a most deadly disease that resulted in high mortality globally (with an expected 70% of all deaths occurring within the next 20 years); hence, development of an effective pharmaceutical treatment modality for anticancer becomes a hot topic of discussion nowadays (Dai et al. 2017). Several scientific journals on oncology research have revealed that almost 10 million death cases were observed due to cancer in 2020 (Sung et al. 2021; Chaturvedi et al. 2019) and the situation will be further deteriorated in the coming time. Although a degree of technological development happened in the recent times to detect the cancer even in its early stage followed by its prevention or treatment to a larger extent, still adequate therapy remains indefinable due to late-stage diagnosis of the disease and even lack of clinical procedures for eliminating disseminated cancer, inadequate approaches for addressing multidrug resistance (MDR), and proper understanding of cancer cell physiology and its plasticity (Tsuruo et al. 2003; Donnenberg and Donnenberg 2005). Current treatment technique that deals with cancer is either chemotherapy, surgery, and/or radiotherapy. Both radiotherapy and chemotherapy are already associated with significant side effects as non-cancerous cells adjacent to cancerous part are also targeted non-specifically (Naidu et al. 2004). Due to poor aqueous solubility and their relative instability in the biological environment, the biodistribution of anticancer drugs becomes inadequate leaving the drug in-efficacious. Thus, extensive research in the targeting technologies should be done in order to overcome the above huddles of conventional therapy, which in turn increase the therapeutic index of the anticancer molecules (Palma et al. 2003; Navya et al. 2019; Bajaj et al. 2009).

Scientists have documented the application and importance of nanotechnologies in the field of cancer treatment and the documents reveal the most extensive development in nano-drug delivery field with very reassuring consequences. But, so far the predictable delivery of anti-cancer drugs is cosidered, this targeting technology witnesses very few biomedical applicability due to their lack of their proper optimisation as well as the characteristic of these entities (Boisseau and Loubaton 2011). The anticancer agents are entrapped or encapsulated within the matrices of the polymers used in the nanoformulations, providing a range of not only architectures but also functionalities. So, these nanoformulations utilize the targeting phenomenon and help to solve the solubility and permeability issues of the anticancer drugs so that bioavailability can increase, which in turn facilitates the adequate drug accumulation in the cancerous sites (Wang and Thanou 2010; Mattheolabakis et al. 2012). Nano-systems are capable of overcoming the biological barriers and also due to their targeting ability, malignant tissues/cells are hammered with the required drug, resulting in enhanced therapeutic index (Peer et al. 2007; Rotello 2009). However, bare nanoparticles are toxic and are reported to cause damage to host tissues (Sharma et al. 2024). The therapeutic benefits of anti-cancer drugs can be expected to be increaseed to manifolds if their intracellular delivery inside the cancer cells achieved. Therefore, a well-designed approach is established for the development and optimization of nanocarriers to obtain the efficacious drug delivery systems with lower undesirable effects. Furthermore, these nanoconstructs are also capable of overcoming several issues associated with conventional drug delivery systems. So, nanoformulations can be utilized for drug targetability through utilizing their intrinsic properties, i.e., penetration through bio-barriers and not allowing the multidrug resistance (MDR) to develop. On the other hand, a ligand-gated nanocarriers possess many potential benefits starting from diagnosis of metastatic cancer to its treatment due to their ability of targeting intricate anti-cancer molecules to lungs and other vascular organs like liver and lymph nodes as well as intra and inter organ targeting (Schroeder et al. 2012).

As demonstrated in Fig. 21.1, several nano-compositions have been fabricated using organic and inorganic compounds including lipoproteins. These compositions

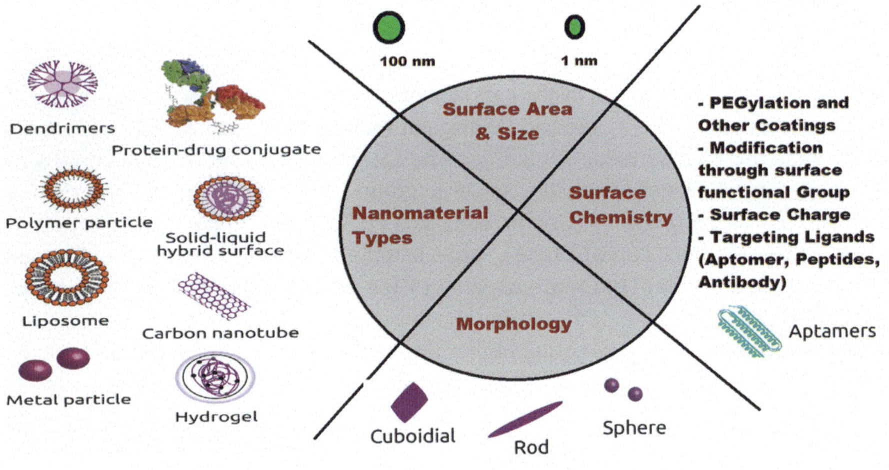

Fig. 21.1 Schematic display of engineered nanomaterials used in cancer therapy

with 10–100 nm particle size are observed to deliver various drugs to the metastasis region by optimizing in terms of the formulation compositions, size, and size distribution or even modifying the morphology by definite ligands. Nano-drug carriers provide several advantages: (1) solubility and stability enhancement of anticancer drugs; (2) prevent or minimize the enzymatic drug degradation and increase its circulation half-life; (3) drug distribution and targeting efficiency; (4) provide controlled drug release at the site of action, and (5) facilitate multiple drug delivery for reducing drug resistance (Navya and Daima 2016). Thus, nanotechnology is generating an innovative prospect that can reform or upgrade the approaches for treatment of cancer (Navya et al. 2019; Wicki et al. 2015).

2 Nanoconstruct Development and Scale-Up Challenges

Core–shell nanomaterials or nanoconstructs have numerous hands-on applications, especially in the field of cancer treatment. The term "nanoconstruct" refers to a dual-component system and is made of nanoparticle core (hard core) with suitable ligands as soft envelope. These ligands help the formulation to target the affected requisite sites, and as a result, the encapsulated drug moieties are being auto-targeted (Dam et al. 2014). Nanoconstructs when engineered adequately are referred as drug carriers, which protect the entrapped drugs when transported through the biological barriers, to target the drugs to the affected area only (Allen and Cullis 2004). Out of several proven benefits, nanoconstruct-based chemotherapy provides adequate deposition at the target that enhances the specificity and nullifies the systemic toxicity (Saw et al. 2021). Furthermore, for combination therapy, these nanoconstructs should be engineered to retain both hydrophilic and lipophilic drug molecules inside their core and adhere to the polymeric surface and hence can be referred as an ideal tool for chemotherapy. Several factors combinedly influence the effectiveness of the nanoconstructs and these factors are vasculature differences, tissue heterogeneities, and chemistry of modified surface of the polymeric layer. Hence, a 4S model like shape, size, surface, and stiffness is being explored nowadays to determine the localization or accumulation of the nanoconstructs after their in vivo administration (Başagaoglu et al. 2013). Therefore, computational methods utilize these 4S model to optimize the nanoconstructs which can be categorized based on the overall size and shape, surface layer chemistry, and core composition as well as better targeting ability to the cancerous sites (Coclite et al. 2017). Several polymers such as poly-l-lysine (PLL), poly(ε-caprolactone), poly(lactic-co-glycolic acid) (PLGA), and poly-(propylene oxide) (PPO) are capable of increasing the drug solubility, better control over drug release and distribution that help to enhance the bioavailability which in turn favors their targeting ability and hence minimizing systemic toxicity, making them suitable candidates to be used as building blocks for nanoconstructs which will finally be utilized for anticancer therapy. Polymers or these building blocks are selected for nanoconstruct design based on their compatibility with the drug, toxicity status, and more importantly targeting ability to the tumor microenvironment (TME). Several lipidic nanoformulations are also being discovered or developed in order to be utilized in the treatment of cancer (Alven and Aderibigbe 2020). Representatives of nanoconstructs used in cancer therapy are shown in Fig. 21.1.

The development of nanoparticles is a complex process and hence there is a wide range of techniques available for the development and production of different kinds of nanoparticles. However, broadly all these techniques are essentially of three types: (1) condensation from vapor, (2) development through chemical reaction, and (3) solid-state recovery processes, such as milling. Through these above-mentioned technologies, hybrid or pure nanoparticles as well as coated nanoparticles or nanoconstructs with either hydrophilic or hydrophobic materials can be synthesized depending on their suitability of applications. All the techniques described above can be categorized as bottom-up technique (i.e., starting with first solubilizing the drug and further to agglomerate through precipitation) and top-down processes (i.e., micronizing the drug powder to nano size range). The top-down approach is found to be very useful in pharmaceutical applications, while bottom-up process is considered as relatively less effective due to the involvement of some organic solvents during synthesis which cannot be removed completely, leaving some traces which many times generate some undesirable reaction inside the body (Prasad 2008; Cui et al. 2009). Despite their distinct advantages, ligand-gated nanoformulations or nanoconstructs are having better penetrability through the biomembranes and hence sometimes regarded as toxic to human body system if adequate targeting is not achieved, which in turn interferes with metabolic reactions. Also, these nanoconstructs based on their size and surface charge are easily distributed throughout the body via the circulatory system or even the neural system. Body's normal detoxification system and even the macrophage clearance in the liver and spleen are not always effective enough to control the nanoparticle agglomeration at the site of application and hence more often some undesirable side effect can be expected, which can be manifested as incurable diseases such as Alzheimer's and other CNS disorders like Parkinson disease (Buzea et al. 2007).

Several established technologies were used to develop nanoconstructs and are categorized into three main types: polymeric nanoparticles, lipid-based nanoformulations, nanoemulsions, dendrimers, and the third one is the magnetic nanoconstructs. Once the development is done, the scale-up of these nanoconstructs is influenced by several components before they reach the market. These components include the physiochemical properties like size and shape with their GRAS (generally regarded as safe) status along with the chemical stability at elevated temperature and finally the in-vivo biodegradability of nanoparticle (Ryu et al. 2012). So, the careful selection of materials and solvent to be used and the procedure of manufacturing is important as they influence the quality of these finished products. Most of the time when the batch size of the R&D lab batches is increased, it is observed that quality attributes of the nanoconstructs changed and hence, proper optimisation in the overall quantity and the process parameters like temperature, stirring, size reduction, extent of nitrogen application should be taken care. Therefore, the scale-up procedure of the nanoconstructs has to pass through lots of quality attributes in order to generate a robust and stable formulation. After optimization of research and development followed by scale-up feasibilities, the nanoconstructs are subjected for preclinical study trials, clinical study trials, and finally if regulatory compliances meet, then only a nanoformulation gets approval for clinical use and reaches patients. So, it could be clearly understood that both time and money is involved for the commercialisation of these nanoconstructs. However, the therapeutic index is very high for these nanoconstructs as compared to the existing marketed delivery systems, which lead to their high applicability and adaptability, especially for cancer treatment. Cancer therapy through nanoconstructs can be referred as innovative intervention which minimizes the drug-related toxicity and offers noticeable benefits to patients. This system as already discussed possesses diagnostic capabilities and most importantly has absolute targeting ability to deliver the encapsulated anticancer active moieties to cancer-affected cells (Ryu et al. 2012). Diagnosis of cancer followed by its treatment is the basis of the cancer treatment which depends on the description of different phenotypes to maximum vicinity of tumor cellular level distinguished by diagnostic agents followed by adequate cancer therapy and observation of these anticancer drugs as compiled in Table 21.1. Nanotechnology as discussed by various scientists offers a unique and very possible alternative to the existing treatment for the diagnosis cancer and hence its remedy. The developed non-aqueous based nanoformulations like polymer or lipid based or liposomal or dendrimer types, etc., as well as other nanoconstruct like metallic nanoparticles are reported to serve the purposes like imaging and treatment through their targeting abilities. Based on the outcome of the substantial research, the application areas of such well-designed nanoconstructs like chemotherapy and siRNA/miRNA or other anticancer drug delivery seem to be a matter of discussion nowadays.

Out of several nanoconstructs, important nanoconstructs are discussed in this chapter as follows.

Table 21.1 Nanoconstructs and their applications in cancer treatment

Formulations	Important building block and drug(s)	Ligand	Indication	References
Polymeric nanoparticles	PCL-PEG-PCL (PCEP)–doxorubicin	Folic acid	Breast cancer therapy	Hu et al. (2018)
	PLGA–DOX	AS1411 aptamer	Colon carcinoma	Mosafer et al. (2017)
	PLGA–I131	Anti-EpCAM	Breast cancer, osteosarcoma	Marshall et al. (2022)
Lipid nanoparticles	DMKE–siRNA	Anti-EGFR aptamer	Gene delivery and bioimaging	Kim et al. (2017)
Metallic nanoparticles	Gold–triptorelin	–	Triple-negative breast cancer treatment	Uzonwanne et al. (2022)
Nanoemulsion	Gadolinium DTX-docetaxel	Folic acid	Ovarian cancer therapy	Patel et al. (2018)

2.1 Inorganic Material Based Nanoconstructs

A very common nanomaterial like black phosphorus gets huge attention due to its adaptable properties and structures. Heterogeneous doping is used to stabilize nanoconstructs and also enhance their shelf life. Black phosphorus when engineered with different metals along with folic acid, fragments of albumin, etc. adequate targetability can be obtained and hence their diagnostic properties in chemotherapy and other biomedical applications can be utilized. Stimuli-sensitive nanoconstructs made up of black phosphorus for anticancer therapy can be developed by exploring the pH gradient activation with NIR irradiation (Pandey et al. 2021). An additional example reveals the potential use of such nanoconstructs in the gene delivery for the treatment of cancer and CNS disorders. Out of the several nanoconstructs activated by some stimulus, nanoconstructs those are activated by pH of internal bioenvironment shows potential advantage for cancer treatment over other nanoconstructs. A novel formulation like mesoporous silica-based nanoparticles (MSNs) was designed with pH-responsive surface modification for finding a new way to treat cancer. For example, tannic acid loaded mesoporous silica-coated nanoconstruct is prepared with surface modification using cupric ions and metal having phenolic functional groups. With these types of modifications, predictable release of encapsulated drug from these nanoconstructs can be observed at the target tissues. And when these delivery systems are engineered with anticancer drugs, then an alternative to conventional therapy but efficient cancer therapy can be expected (Hu et al. 2018). Moreover, for better imaging of cancer cells and to obtain therapy, copper sulfide (CuS) with porphyrin loaded with porous silica-based nanoconstruct labeled with zirconium [Zr^{89}] can be developed (Goel et al. 2018). These hybrid nanoconstructs were further reported to increase its circulation time and be compatible with biological system and also capable of targeting the tumor cells efficiently and hence can be referred to have combined benefits like imaging and therapeutic. Hyaluronic acid can target CD44 receptor of cancer cells and this property is exploited to target doxorubicin through nanoconstruct triggered by a photothermal agent such as CuS. The novel formulations can provide predictable drug release at the tumor site which helps in the higher drug retention and hence tumor growth inhibited causing apoptosis leading to establish them as better alternative in cancer treatment (Poudel et al. 2020). Although these designed nanoconstructs were precisely cytotoxic, but their stability sometimes limits their usefulness.

2.2 Polymer-Based Nanoconstructs

Since the inception, polymer-based nanoformulation are used for cancer treatment which are capable of accommodate both types of drug molecules and when their surface modification is done, better targeting to the affected cancer cells is experienced. But when they are designed as stimuli-responsive formulations, then these nanoconstructs are very useful not only as diagnostics but also as new pharmaceutical formulation capable of delivering these anticancer molecules. The polymerization of nanoconstructs using PEG and PLGA polymer matrices helps to include both hydrophobic and hydrophilic agents and is found effective for various imaging and therapeutic action. The surface modification of polymeric nanoparticles with PEG is known as PEGylation, which escapes the mononuclear phagocytic system to obtain enhanced blood circulation time (Perry et al. 2017). Furthermore, these polymeric networks can easily be modulated with contrasting agents, lipidic drugs, and drug-loaded polymeric formulation, leading to an effective nanoconstructs that are considered as theranostic moieties (Palange et al. 2017). Currently, fluorescence resonance energy transfer (FRET) emerges as an effective optical imaging technique that is used to monitor the localization and drug release from the developed nanoconstructs around the tumor cell (Caldorera-Moore et al. 2011). In connection to this technique, a PEGylated doxorubicin (PEG-DOX)-loaded coupled with peptide (FFKY)-block-tetraphenylethylene nanoformulations were developed and investigated the effect of FRET magnitudes in cancer A549 cells. The observation revealed that Doxorubicin and the peptide loaded self-assembled formulations display first-order release of drug (Wang et al. 2020). Nanoformulations made up of PLGA were also used for anticancer therapy. This formulation utilized the hypoxia condition of the tumor site and generates reactive oxygen (ROS). So, the example on this regard is the verteporfin loaded PLGA based nanoconstructs showed to increase the reactive oxygen species (ROS) when comes in contact with especially the hypoxic environment. These treatment modalities are found to be ~60% efficient in pancreatic cancer in human being within a month time of treatment. These nanoconstructs based on radiodynamic treatment modality can be referred as a better and non-operative option for intense hypoxic cancer (Clement et al. 2021). Polymer-based nanoformulations made up of polyurethanes (PU) can be used for treatment applications by exploring the biodegradable nature of the PU and more importantly stimuli-responsive and hence referred as alternative drug carriers for cancer. These types of nanoconstructs exhibit several important properties improving solubility of entrapped hydrophobic drugs, modulate their release, improving efficiency and targetability with stimuli-responsive. The newer formulations can actively target the cancerous cells when engineered with suitable ligands. Several stimuli like physiological pH, temperature, and other bio-stimuli can be utilized to release the encapsulated drugs from these nanocarriers and hence targeting the encapsulated drugs to the desired sites through these nanocarriers (Gajbhiye et al. 2020). Further, from the list of the lipid-based nanoconstructs, a special type, i.e., PEGylated squalene (SQPEG)-based nanoformulations was already found beneficial for cancer treatment. When they are engineered with water insoluble pyropheophorbide-a (Ppa), they are observed to skip the RES system with precise targeting, if especially their average particle size is 200 nm or less. These novel formulations are triggered by fluorescence leading to provide higher bioavailability when they are evaluated in in-vivo condition.

2.3 Dendrimer-Based Nanoconstructs

Apart from the lipid-based and polymer-based nanoconstructs, a novel dendrimer also provides several advantages so far, the therapeutic as well as diagnostic property concerned. Poly(amidoamine) (PAMAM) is a synthetic molecule with several functional groups and with well-defined molecular structure, when used to develop dendrimers, a novel nanomedicine is evolved to treat the tumor. However, the size of these dendrimers limits their applicability due to inadequate drug loading and hence not considered appropriate for passive tumor targeting. PAMAM dendrimers are adequately branched, which are found to stabilize nanoparticles, like gold NPs. When the curcumin encapsulated dendrimer with its branching structure gets PEGylated and was then attached to the MUC-1 aptamer, then the modified dendrimer is capable of targeting in HT29 and C26 cells efficiently and hence can be explored for both for tumor imaging and therapy (Alibolandi et al. 2018). Another study demonstrated that micelles of PAMAM dendrimers conjugated with suitable peptide to target cellular nucleolin which manipulated maximum in MDA-MB-231 cells. For positron emission tomography (PET) imaging, the ^{64}Cu was coupled with the dendrimers to demonstrate their suitable distribution inside the body. Serial of PET imaging revealed that this docetaxel loaded dendrimer when conjugated with F3 peptide, MDA-MB-231 tumor cells are targeted so insistently compared to non-peptide based ^{64}Cu-PAMAM-DOX. Furthermore, their distribution in other non-targetable organs inside the bodies was noticeably comparable (Alibolandi et al. 2018).

2.4 Miscellaneous Nanoconstructs

Since only the most potent drug molecule cannot completely cure the diseases like cancer, several novel drug carriers are developed in order to obtain therapeutic index. Hence, several novel formulations capable of encapsulating both hydrophilic and lipophilic active ingredients are under development process and are tried with other suitable treatment approaches in order to find out a very effective and safe formulation to treat deadly diseases like cancer (Gane et al. 2017). Hence, synergistically active agents are administered through the new method in order to experience more intracellular penetration so as to get rapid action (Gee et al. 2005). Protein-engineered nanoparticles loaded with docetaxel (DTX) and gold dendrimers modified with poly(sodium-4-styrenesulfonate) and entrapped Doxorubicin were designed for chemotherapy combined with photothermal therapy (PPTT). The release of encapsulated doxorubicin through diffusion was triggered through NIR radiation. The synergistically administered two anticancer drugs activated by NIR irradiation show toxic to MDA-MB-231 cells (Villar-Alvarez et al. 2019). To increase the solubility of 3,4-difluorobenzylidene-curcumin (CDF) and evaluate its therapeutic potential, FAPAMAM dendrimer was developed.

3 Cancer Targeting by Nanoconstructs

The engineered drug delivery system with relatively higher encapsulating capacity to target tumors with almost no harm to healthy cells is very much important for effective cancer therapy. It is imperative to understand the targetability behavior for the cancer cells. Nanoformulations with adequate targeting ability and activated by suitable bio-stimuli can be considered as alternatives to the conventional formulations and are able to localize themselves to the cancer cells.

3.1 Passive-Based Targeting

The targeting by diffusion-controlled delivery system is also known as passive targeting. During this passive targeting, the drug formulation is capable of targeting the drug to the site of action through the circulation system. Nanoconstructs in order to obtain passive targeting utilize the unique leaky vasculature of solid tumors as well as the lesser lymphatic clearance that allow nanoformulations to outflow from the leaky capillaries of tumor and experience ERP (enhances retention and permeability) to accumulate at the target, i.e., tumor (Black et al. 2014). Parameters like molecular weight, the surface's nature, observed surface charge, solubility in biofluid, and size distribution of the active ingredients play critical role in the passive way of drug targeting to be experienced. So categorically, stealth or PEGylated liposomes will delay the disposition, while the surface charge determines their flow duration in the blood. PLGA-based nanoparticles with high drug loading are reported to provide higher physical stability, controlled drug release, and more importantly the specificity toward the targeting site (Rezvantalab et al. 2018). Polyvinyl alcohol (PVA) is another polymer which also can be used for passive targeting and also reported to help in encapsulation of lipophilic molecules and their predicted release at specific tumor sites (Wei et al. 2020). The passive targeting is the most used technique to deliver the anticancer drugs to the tumor site, as displayed in Fig. 21.2.

3.2 Active Targeting

Active targeting of drug moieties can be accomplished through ligands or certain molecules which are having specific affinity for the cell receptors. So, this mode of targeting depends on establishing the binding of the ligands and the surface receptors which are normally overexpressed on the target cell surface (Goldberg et al. 2013). This method of targeting drug molecules to the affected site is also known as ligand-mediated targeting (Ko et al. 2013). The engineered nanoparticles with specific ligands having higher affinity for the target receptor and their retention should be in such a way that the ligands are directed toward the target's proximity in order to generate maximum affinity for the target cells. In active targeting through nanoconstructs, the drug is directly delivered into the microenvironment of tumor (TME),

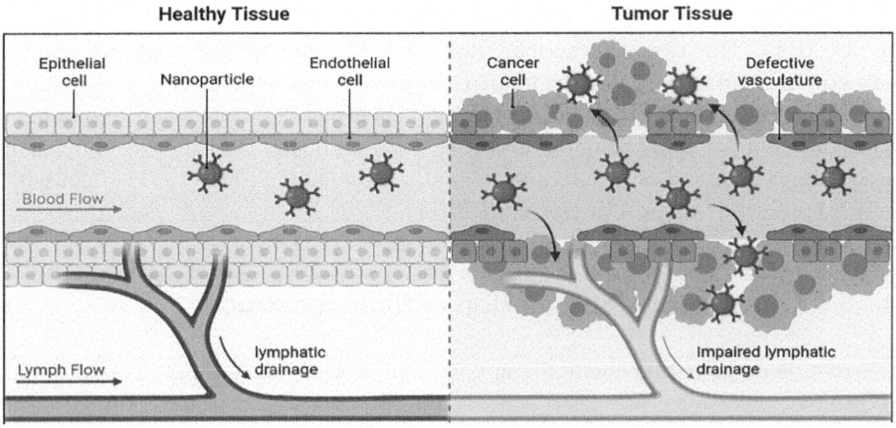

Fig. 21.2 Passive targeting of anticancer drugs through nanoparticles

which in turn result in minimal damage to normal cells around the tumor. In this approach, adequate binding of ligands with the target receptors is initiated and hence ligand–receptor complex is formed (Kue et al. 2016). Targeting the surface receptors like folate and transferrin followed by receptor-mediated endocytosis can be experienced when the nanoconstructs are engineered with ligands such as proteins, peptides, and sugars. This approach rather helps the nanoconstructs in penetration the encapsulated drug molecules to the affected cancer cells and hence exhibit higher therapeutic index (Chaturvedi et al. 2019).

3.3 Stimuli-Responsive Targeting

The deadly disease, i.e., cancer, is evaluated based on their complex pathophysiology, exclusive TME, and uncontrolled cell proliferation. As already discussed, conventional treatment modalities for cancer have several restrictions due to their non-specific way of targeting causing equal damage to the normal cells. Several external stimuli like magnetic field, light intensity as well as internal biological stimuli like pH gradient at the affected site to that of normal site, enzymes activation status, oxidative stress in the microenvironment are observed to alter the performance of nanoconstructs to a larger extent. When developing nanoconstructs, calcium carbonate and glutathione are used to achieve optimum pH-responsiveness and targeting. In addition to the responsiveness to pH or oxygen gradient (H_2O_2) towards the human serum albumin (HSA), the studies revealed that manganese dioxide (MnO_2) coated nanoconstructs are explored for mineralization of Mn^{2+} inside the simulated biosystem. In general, transducers are used to convert external stimuli like temperature, photoenergy, or magnetic fields, which further modulate into the heat for effective cancerous cell killing (Fang et al. 2021). A separately published literature revealed that the temperature sensitive nanoconstructs utilises the magnetic field in order to target the delivery of the anticancer drug (Alexiou et al. 2011). The results described that such chemotherapeutic agents loaded nanoparticles are applied into the human bodies under the influence of considerably high magnetic field in order to invade the cancerous tissue and hence adequate localization in the intra and inter-cancerous cells can be obtainined. This innovative procedure is signified as an alternative approach due to their lesser side effect with higher therapeutic index (Alexiou et al. 2011).

4 Ligands for Manipulation of Nanoconstructs

Conventional or existing chemotherapy without precise targetability is observed to show noticeable side effects, and hence, reduced effectiveness of the formulation is experienced. Hence, it becomes very essential to develop newer drug delivery alternatives which will warrant the predictable drug release from the newer formulations like nanoformulations which in turn resulted in adequate intracellular drug accumulation in these affected sites of the patient. In this regard, selection of proper surface modification through the ligands plays a vital role which will selectively get stick to particular receptors in cancerous cells. This form can be referred as a technique that promotes tumor targeting. To design a precise and effective tumor-targeting system, ligands like hyaluronic acid, peptides, even some antibodies, are used because these ligands are not only providing the targetability but also facilitate the intracellular accumulation of anticancer drugs so as to manipulate the tight biological junctions which in turn resulted into higher selectivity and cellular uptake of these encapsulated anticancer drugs (Nguyen et al. 2021). However, either single or ligands in hybrid form can serve this purpose of cancer treatment.

4.1 Ligand Optimization Process

As discussed, specific ligand/s (in combination) when used to engineer the formulation, the therapeutic moiety can be delivered to the pathological site which lead to enhance the effectiveness of this anticancer formulation. Moreover, ligands with smaller molecules or single-stranded nuclear part (aptamers) and antibody fragments may be the good choice for this intended purpose. These ligand-gated forms are not only used for the treatment purpose but also have extended applications in diagnosis purposes. During the selection of ligands, several factors including the ligand size, its chemical form etc are evaluated in order to modify the surface of the formulations in order to obtain adequate adherence to the receptor. The inherent ability of the ligand to get adhered to the specific receptors is heavily depending on its size as well as the changes in the pharmacokinetics of the drug in the in-vivo condition. The extent of adherence of the ligand to the targeted site will directly relate to the amount of the accumulated drug at that position of the targeted site. When an antibody is carefully chosen as ligand in order to obtain adequate targetability, then immunogenicity has to be considered as an important factor which may influence the fate of the drug like its accumulation, distribution and finally its excretion. In addition to the above factors so discussed, the production cost and time required to optimize the ligand-gated nanoformulation also an important parameter to be evaluated for before its finalization for the surface modification. Moreover, as the in-vitro as well as in-vivo stability and in-vivo circulation time get modified through the PEGylation of the modified nanoconstructs, so an extensive evaluation is required to optimise a receptor-targeted drug delivery system (Srinivasarao and Low 2017).

5 Challenges During Scale-Up of Nanoconstructs: Methods and Their Challenges

The scaling-up of nanoconstructs includes the complete development and optimization of the manufacturing methods along with transfer of the optimized technology to the GMP site for their large-scale production. However, the limitations of both CMC data, analytical profiling and manufacturing process of these nanoconstructs should be evaluated extensively otherwise the technology transfer may not be fruitful during commercial scaling-up owing to its reproducibility and stability (Rishi et al. 2014). Generally, the manufacturing of nanomedicines is a thought-provoking job when the reproducibility followed by the quality of the product of interest is concerned. Moreover, a complete validated scale-up procedure like lab batch to R&D scale-up and from exhibit batch to validation batch scale-up will guarantee the superiority of these nanomedicine, its cost-effectiveness and timely market entry. Market report suggested that scale-up of only few polymer-based nanoconstructs and nanomedicines developed by lipid emulsion technique and nanoprecipitation methods was gained partial success and showed limited commercialization (Galindo-Rodriguez et al. 2005). The limitations involved in the preparative methods of nanoconstructs followed by their commercial manufacturing are referred as the formidable tasks during their commercialization. First, scale-up has mainly been encountered by the characteristics of nanomedicines, such as particle size, size distribution, stability in bioenvironment, drug encapsulation, and surface morphology. Furthermore, some publications indicated that the drug encapsulation capacity is comparable if the scale-up of a commercial nanoconstruct from the R&D lab batch is by emulsion-based method (Saraf 2009). But during the commercial batch manufacturing, the same drug to polymer ratio (as obtained from the R&D batch optimization) may not work sometimes and an increase in agitation and its duration during the scale-up may result further into drug leaking leading to comparatively less entrapment efficiency. As already described, during the scale-up of nanoformulations, the noticeable point is the reduction in drug encapsulation efficiency due to alteration in composition ration and process parameters pertaining to the equipment size. It is also reflected from the outcomes of many scientific studies that nanoconstructs are able to target cancer cells by using the patient's physiological situation as a triggering factor for the drug release at the site of action, i.e., cancer. However, designing efficient nanoconstructs for treating cancer still remains as a huge task and only a few of them stood up to the pharmacodynamic expectation after clinical study (Mahmood et al. 2022). Furthermore, the unfavorable properties of nanocarriers concluded the demand of toxicity assessment by the regulators when the innovator applies for regulatory approval for their developed nanoconstructs for cancer treatment.

Referring to the above practicalities, a significant obstacle in commercialization of these nanoconstructs is nothing but the inadequate clinically useful data generation and hence, many of the formulation fails in this study due to some unseen interaction between the nanoconstructs and the biological barriers. Furthermore, the difference in the in vitro pharmacokinetic to that of in vivo observed results leads to noncompliance in the in vitro-in vivo correlation (IVIVC) (Jain et al. 2017). It is also true that very few nanomedicines or nanoconstructs are reaching from the stage of preclinical to their clinical trial stage especially in the cancer treatment.

The commercial-scale manufacturing of nanoconstructs becomes a real task for the formulators because of the generation of inconvenient properties from batch to batch. Moreover, the complex multi-stage manufacturing processes of nanomedicines along with high raw material cost contribute them as costly alternatives. After properly designed R&D process, R&D scale-up followed by validation manufacturing process is the normal way of product development, so considerably high manufacturing cost is anticipated due to the involvement of sophisticated equipment and novel material, but if the clinical benefit is observed to be as per expectation, then the manufacturing cost can be justified (Betker et al. 2013). Few formulators have also suggested the scale-up possibility of nanoconstructs by the nanoprecipitation method (Catherine and Hatem 2005). The nanoprecipitation is also observed for not supporting the commercial scale-up of nanoconstructs and another concern is the stability of formulation component as well as the cumulative effect of the toxic organic solvent (like dichloromethane). Hence, most of the manufacturer is not in the favor of this method and are trying to adapt the manufacturing process where the organic solvent is completely replaced with aqueous system.

Another crucial point of discussion is the regulatory sanction of nanomedicines, although no specific FDA strategies for the products manufactured with nanomaterials or nanoconstructs are laid down. The regulatory strategies on the nanoconstructs are complete evaluations that cause delay in commercialization. This is also apparent that the approval formalities will enforce the time-consuming developmental feature of nanoformulations. Thus, to nullify the issues involved with nanomedicines for treating cancer, a brief optimization on the developmental strategy needs to be merged before the validation batch execution or scaling-up. So, the pharmacokinetic protocol of nanoconstructs in animal model should be designed to minimize the complications involved in cancer treatment (Dawidczyk et al. 2014). Furthermore, case-to-case thorough examination is also essential. Furthermore, FDA with consultation with the concerned scientists should provide a full set of guidelines for approval so that the scale-up of these nanoconstructs can be more predictable (Ando et al. 2012; Chan 2006).

6 Regulatory Challenges of Nanoconstructs

With the advancement of the nanotechnology and the particular outcome as nanomedicines, several areas of pharmaceutical as well as medical science were upgraded to a greater extent. However, numbers of disagreements for the basic concepts related to these nanoformulations raised; hence, their scale-up and commercialization were questioned. Because of the challenges associated with the nanomedicines, respective regulatory issues specifically in the pharmaceutical development of nanoconstructs are discussed and are presented in details. With the new development, a newer platform is coming up and is termed as theranostics which combines the imaging and diagnostics front to that of the therapy part especially for cancer and also in CNS disorders. Further, these nanoformulations are expected to meet the unmet medical needs in cancer treatment, either through adequate drug targeting or site-specific controlled release of the anticancer agents. This is expected to be achieved by understanding the exact mechanism to improve the anticancer drug transport across bio-barriers so as to maximize their efficacy and reduction in systemic toxicity through perfect distribution, particularly in the cancerous areas (Moghimi et al. 2001). Owning to the nano size and higher surface energy, these nanoconstructs are observed to be extensively reactive. Nanomaterials after administered into bio systems, an adsorption layer of biomolecules like proteins, etc., is observed to cover-up their surface once they are exposed to the bio-fluid (Zhao and Chen 2016). However, the composition of the adsorption layer influences the release of entrapped drugs followed by their transference from one biological compartment to another (Pino et al. 2014).

Since last 2 decades, nanomedicines or nanoconstructs are entered into the clinical practice. In Europe, the nanomedicines are understood to be novel formulations like selective drug loaded liposomes, nanoparticles, nanoemulsions, polymeric formulations, etc. (Agrahari and Hiremath 2017). In the process of regulatory approval, these nanoconstructs were initially evaluated based on the therapeutic index, more specifically the risk associated with the formulation. Another major obstacle is the pharmaceutical equivalence as well as pharmacological equivalence of the developed nanoconstructs to that of the referred reference-listed drug (RLD) (Bleeker et al. 2013). Nanomedicine are developed from various formulation ingredients either sourced from either biologicals, or semibiologicals or even from the synthetic front. These synthetica-based nanomedicines are developed using synthetically manufactured active ingredients, hence are categorized as non-biological (synthetic) but complex drug formulations (NBCD) (Tinkle et al. 2014; Hussaarts et al. 2017). In order to reach to regulatory, semi-regulatory market or even Rest of World (ROW) market, a complete formulation optimization is required in order to be considered as pharmaceutical equivalent drug product and hence parameters like qualitative and quantitative equivalency to that of the RLD and other the quality parameters like impurity profile and drug release behavior, etc., should be compared and established. Whatever the type of the nanoconstructs may be, an in-vitro in-vivo correlation to be established in order to evaluate the safety and efficacy of the generic product which eventually represented as a bioequivalent product (Astier et al. 2017).

The bioequivalence should be comparable with the respective RLD. However, especially connected to the biological nanoconstructs, European Medicines Agency (EMA) has established a pathway for the introduction of generic products to the regulatory market which is nothing but the set of studies required for the developed product to be proven as comparable quality product assisted by the nonclinical and finally data from the clinical studies (Muhlebach et al. 2015). Regulatory checklist for non-biological complex drug products (NBCDs) approval is under review. Moreover, the biological complex drug products (BCD) and NBCD have some similar or common characteristics, and hence, the NBCD framework provided by EMA can be useful to be exercised for that of the BCD. Since quality control outputs of BCD is not observed to be reproducible due to the difference in characterization from batch to batch because of the clinical outcome are heavily influenced by the manufacturing process and altering parameters itself. Based on these data, a draft of the regulatory for specially NBCDs shall be made and evaluated through the guideline provided by the EMA for marketing permission (Schellekens et al. 2014).

Published literatures on the therapeutically non-equivalent nanoformulation give a hint on the gap between the regulatory demand and their actual quality attributes and after this gap analysis, in order to obtain a market approval, time-consuming process optimization is performed. This gap analysis should be done effectively and the required studies are carried out for complying so that the therapeutic outcome of the nanoconstructs can be well accepted and the complex formulations can be conferred to be as safe and used as therapeutic alternative. This is a straight pathway for obtaining market approval of generic formulations. However, all quality attributes like safety, efficacy, side effect (unpredictable biological interactions) are always emerging as challenges before the acceptance of nanoconstructs as an efficient treatment modality, especially for cancer (Ehmann et al. 2013). The EMA has provided certain regulatory pathway for the approval of lipid formulation like liposome, etc. (EMA 2013) and iron nanomedicines (EMA CHMP/SWP 2013) for intravenous use. In addition to the above parameters discussed earlier, biosimilars cannot be developed without the immunogenic profiling of the nanoconstructs, especially when intravenous use is inevitable (FDA 2013). The FDA also additionally drafted the requirement on the nanoparticle characterization, including the bioequivalence study of several colloidal injections for intravenous use (Alibolandi et al. 2018). US-FDA demands the evaluation of the quality of the optimal nanoconstructs upon nanosizing through clinical bioequivalence; however, the European Medicines Agency urges for clinical bioequivalence to assess safety and efficacy for nanoconstructs (Table 21.2). The US-FDA demands all sorts of evidences for the biosimilar comparison of nanomedicine products.

7 Clinical Challenges of Nanomedicines

Nanomedicines must be proven to be clinically effective and safe formulations for human use, a process that is both costly and laborious. The technologies required for manufacturing nanoconstructs are usually sophisticated and hence deep

Table 21.2 Variances in the prerequisites for EMA and the US-FDA for the evaluation of intravenous preparations

Evaluation requirement	EMA	US-FDA
Quality	Qualitative and quantitative composition Close/identical to the reference product	Qualitative and quantitative composition Close/identical to the reference product
	Physicochemical characterization Particle size distribution Morphology Complex stability Labile iron release in plasma	Physicochemical characterization Particle size distribution morphology
Nonclinical	In a relevant animal model (distribution, metabolism excretion). Emphasis on targeting, accumulation and retention at least in: • Plasma • Reticular endothelial system (RES) • Target tissues/organs	No specific recommendations
Clinical bioequivalence	Single-dose parallel or crossover study Analytes (serum): • Total iron • Transferrin-bound iron	Single-dose, randomized, parallel trial analytes (serum): • Total iron • Transferrin-bound iron
Clinical safety/ efficacy	Therapeutic equivalence study: • Not necessary when totality of data (quality, nonclinical data, human PK study) demonstrates similarity	No specific recommendations

understanding is required for their interpretation against the reality (FDA 2013; Teli et al. 2010; Sainz et al. 2015). Starting from R&D development to scale-up, biological challenges like immunogenicity and biocompatibility and lastly intellectual property (IP) and approval regulations are the rate limiting factors for a nanomedicine to be referred as clinically proven formulations. These factors are limiting the commercialization probability of nanomedicines (Allen and Cullis 2004; Zhang et al. 2008). The parameters like accumulation, distribution, retention, and efficacy of the nanomedicines which are more likely to be impacted by the disease pathophysiology and their interactions in vivo condition in animals and humans will determine their fate as alternative treatment modalities. Therefore, optimizing the nanoconstructs in the similar diseased condition further strengthens the clinical translation (Hare et al. 2017). So, in order to reduce investment menace for nanomedicines development, initially the preclinical data required to be examined extensively before to be exercised in human volunteers.

Furthermore, the PK-PD study of nanoconstructs should be carried out in an animal model very similar to the human physiology before establishing their suitability for the clinical interpretation. Additionally, pathologically relevant animal models that follow the same route of administration and provide a narrow range of intended disease manifestations can be considered more suitable candidates for the study, such as cancer (Zhang et al. 2008). Since certain manifestations are not published in all recent literatures, hence it emerge as another limitation for their clinical interpretation. The specific protocol for the preclinical animal model studies for the developed nanoformulations is required in order to obtain better in-vivo performance, should be designed and approved. Furthermore, dosing schedules as well as treatment schedules are also important parameters for the efficacy of the nanomedicine or nanoconstructs performance specially to combat cancer which further can be modified based on the severity of the disease (Park 2017). Despite of the availability of several publications on the development of nanoformulations, but the translation of the published data to preclinical or clinical usefulness is observed to be very substandard and unsatisfactory (Puri et al. 2009). Generally, cancer shows inter-patient and intra-patient variability with time as the disease progresses. That is why the approach for engineering the nanoconstructs to treat the cancerous cells is not fruitful to interpret the expected clinically beneficial outcomes. However, the enhanced permeation and retention (EPR) effect is helping these nanoconstructs for adequate accumulation of drug at the site of cancerous tissue. Since the properties of nanoconstructs substantially affect the drug pharmacokinetics and residence duration and distribution at the tumor site, hence outcomes of these preclinical studies are not predictable as per expectations. Furthermore, differences in effective dose based on the tumor size and its severity along with the above discussed inconveniences related to the PK-PD data is likely to impact to a greater extent so far, the treatment of cancer is concerned. On the other hand, several important factors like physiochemical properties of nanoconstructs, their surface modification through PEGylation and targetability along with their stability in the biological or pathological conditions also play a notable role on their efficacy in the field of cancer treatment (Hare et al. 2017; Yang et al. 2018). Furthermore, the expected benefits from the engineered nanoconstructs is observed to not matching with the actual clinical outcomes to a larger extent. Possible reasons for this variation may be due to the factors include nanoconstructs behavior in the biological fluid, biological barriers and finally their targetability for the effected site. It is also quite important to investigate the nanoformulation accumulation at cancerous site followed by their internalization and functionality with degradation inside the cancerous cell and the clinical outcomes. Therefore, a newer approach to be followed for the nanoconstructs development and further the collection of complete but product-wise preclinical data and their overall study is required to establish a new path to treat the cancer.

8 Applications of Core–Shell Nanostructures and Future Prospectives

Core–shell nanostructures show several unique applications specially in medial front like diagnosis through imaging and precise targeting the drug molecules to the site of action or the site of absorption for the treatment of the cancer. Furthermore, these novel drug formulations have shown a very promising ability in the treatment of cancer (LaVan et al. 2003). This review highlights the development of core/shell nanoconstructs during over the past two decades. To categorize the applications of core-shell nanoconstructs, literature was reviewed and classified into five major types: targeted drug delivery, bioimaging, biosensor, nuclear material interaction, and targeted gene delivery.

Despite significant developments in healthcare, life expectancy around the globe has decreased to a considerable level. Hence, the efforts of biomedical science have been focused on treating the manifestations of diseases such as cancer and AIDS. According to the American Cancer Society, a significant decline in mortality due to cancer is being observed in the USA and other developed countries when the treatment of cancer through the novel nanoformulations is done. Due to advancements in these novel techniques, cancer is now detected at an early stage and as a result of their combined safety and targetability, more promising results are being observed. The limitations of conventional cancer treatments, such as chemotherapy, radiation therapy, and immunotherapy, may be due to their inability to selectively target cancer cells. But sometimes, the multimodal cancer therapy through conventional ways is shown to have higher survival chances of cancer patients. Categorically, nanoformulations or nanoconstructs such as dendrimers, nanoparticles, liposomes, micelles, and emulsions are now utilized in cancer treatment. Based on particle size and size range, these nanoconstructs may occasionally generate side effects such as blushing, rashes, and inflammation. However, in other cases, immune responses or toxicity reduction can be referred as their better side and should be watched out (De Jong and Borm 2008). For an efficacious drug-carrier system, nanoparticle-based drug delivery system (nanoconstructs) should have unique characteristics like capable of adsorb or encapsulate all drug moieties, in-vivo biodegradability, activate through bio stimulus like temperature, pH as well as other in-vitro stimulus like light, sound, and magnetism and with flexible to the surface modification by ligands in order to obtain absolute targetability.

References

Agrahari V, Hiremath P. Challenges associated and approaches for successful translation of nanomedicines into commercial products. Nanomedicine. 2017;12:819–23.

Alexiou C, Tietze R, Schreiber E, et al. Cancer therapy with drug loaded magnetic nanoparticles magnetic drug targeting. J Magn Magn Mater. 2011;323:1404–7.

Alibolandi M, Hoseini F, Mohammadi M, et al. Curcumin-entrapped MUC-1 aptamer targeted dendrimer-gold hybrid nanostructure as a theranostic system for colon adenocarcinoma. Int J Pharm. 2018;549:67–75.

Allen TM, Cullis PR. Drug delivery systems: entering the mainstream. Science. 2004;303:1818–22.

Alven S, Aderibigbe BA. Efficacy of polymer-based nanocarriers for codelivery of curcumin and selected anticancer drugs. Nanomaterials. 2020;10:1556–628.

Ando M, Yonemori K, Katsumata N, et al. Phase I and pharmacokinetic study of nab-paclitaxel, nanoparticle albumin-bound paclitaxel, administered weekly to Japanese patients with solid tumors and metastatic breast cancer. Cancer Chemother Pharmacol. 2012;69:457–65.

Astier A, Barton Pai A, Bissig M, et al. How to select a nanosimilar. Ann N Y Acad Sci. 2017;1407:50–62.
Bajaj A, Miranda OR, Kim IB, et al. Detection and differentiation of normal, cancerous, and metastatic cells using nanoparticle-polymer sensor arrays. Proc Natl Acad Sci. 2009;106(27):10912–6.
Başagaoglu H, Allwein S, Succi S, et al. Two- and three-dimensional lattice Boltzmann simulations of particle migration in microchannels. Microfluid Nanofluid. 2013;15:785–96.
Betker JL, Gomez J, Anchordoquy TJ. The effects of lipoplex formulation variables on the protein corona and comparisons with in vitro transfection efficiency. J Control Release. 2013;171(3):261–8.
Black KCL, Wang Y, Luehmann HP, et al. Radioactive ^{198}Au-doped nanostructures with different shapes for in vivo analyses of their biodistribution, tumor uptake, and intratumoral distribution. ACS Nano. 2014;8:4385–94.
Bleeker EA, de Jong WH, Geertsma RE, et al. Considerations on the EU definition of a nanomaterial: science to support policy making. Regul Toxicol Pharmacol. 2013;65:119–25.
Boisseau P, Loubaton B. Nanomedicine, nanotechnology in medicine. Comptes Rendus Phys. 2011;12:620–36.
Buzea C, Pacheco II, Robbie K. Nanomaterials and nanoparticles: sources and toxicity. Biointerphases. 2007;2:17–71.
Caldorera-Moore ME, Liechty WB, Peppas NA. Responsive theranostic systems: integration of diagnostic imaging agents and responsive controlled release drug delivery carriers. Acc Chem Res. 2011;44:1061–70.
Catherine C, Hatem F. Preparation of nanoparticles with a membrane contactor. JMSR. 2005;266(1):115–20.
Chan VS. Nanomedicine: an unresolved regulatory issue. Regul Toxicol Pharmacol. 2006;46:218–24.
Chaturvedi VK, Singh A, Singh VK, et al. Cancer nanotechnology: a new revolution for cancer diagnosis and therapy. Curr Drug Metab. 2019;20:416–29.
Clement S, Guller A, Mahbub SB, et al. Oxygen-carrying polymer nanoconstructs for radiodynamic therapy of deep hypoxic malignant tumors. Biomedicine. 2021;9:322–8.
Coclite A, Mollica H, Ranaldo S, et al. Predicting different adhesive regimens of circulating particles at blood capillary walls. Microfluid Nanofluid. 2017;21:168–71.
Cui H, Feng Y, Ren W, et al. Strategies of large-scale synthesis of monodisperse nanoparticles. Recent Pat Nanotechnol. 2009;3:32–41.
Dai W, Wang X, Song G, et al. Combination antitumor therapy with targeted dual-nanomedicines. Adv Drug Deliv Rev. 2017;115:23–45.
Dam DH, Lee M, Odom TW, et al. Improved in vitro efficacy of gold nanoconstructs by increased loading of G-quadruplex aptamer. Nano Lett. 2014;14:2843–8.
Dawidczyk CM, Russell LM, Searson PC. Nanomedicines for cancer therapy: state-of-the-art and limitations to pre-clinical studies that hinder future developments. Front Chem. 2014;69(2):1–13.
De Jong WH, Borm PJA. Drug delivery and nanoparticles: applications and hazards. Int J Nanomedicine. 2008;3:133–49.
Donnenberg VS, Donnenberg AD. Multiple drug resistance in cancer revisited: the cancer stem cell hypothesis. J Clin Pharmacol. 2005;45:872–7.
Ehmann F, Sakai-Kato K, Duncan R, et al. Next-generation nanomedicines and nanosimilars: EU regulators' initiatives relating to the development and evaluation of nanomedicines. Nanomedicine. 2013;8:849–56.
EMA. Reflection paper on the data requirements for intravenous liposomal products developed with reference to an innovator liposomal product. 2013. www.ema.europa.eu.
EMA CHMP/SWP. Draft reflection paper on the data requirements for intravenous iron-based nano-colloidal products developed with reference to an innovator medicinal product. 2013. www.ema.europa.eu.

Fang Z, Shen Y, Gao D. Stimulus-responsive nanocarriers for targeted drug delivery. New J Chem. 2021;45:4534–44.
FDA. Draft guidance for industry on bioequivalence recommendations for iron sucrose injection. 2013. http://google2.fda.gov/.
Gajbhiye KR, Chaudhari BP, Pokharkar VB, et al. Stimuli-responsive biodegradable polyurethane nano-constructs as a potential triggered drug delivery vehicle for cancer therapy. Int J Pharm. 2020;588:119–26.
Galindo-Rodriguez SA, Puel F, Briançon S, et al. Comparative scale-up of three methods for producing ibuprofen-loaded nanoparticles. Eur J Pharm Sci. 2005;25(4):357–67.
Gane EJ, Hyland RH, Yang Y, et al. Efficacy of ledipasvir plus sofosbuvir for 8 or 12 weeks in patients with hepatitis C virus genotype 2 infection. Gastroenterology. 2017;152:1366–71.
Gee JMW, Howell A, Gullick WJ, et al. Workshop on therapeutic resistance in breast cancer: impact of growth factor signalling pathways and implications for future treatment. Endocr Relat Cancer. 2005;12:S1–7.
Goel S, Ferreira CA, Chen F, et al. Activatable hybrid nanotheranostics for tetramodal imaging and synergistic photothermal/photodynamic therapy. Adv Mater. 2018;30:170–7.
Goldberg MS, Hook SS, Wang AZ, et al. Biotargeted nanomedicines for cancer: six tenets before you begin. Nanomedicine. 2013;8:299–308.
Hare JI, Lammers T, Ashford MB, et al. Challenges and strategies in anti-cancer nanomedicine development: an industry perspective. Adv Drug Deliv Rev. 2017;108:25–38.
Hu C, Fan F, Qin Y, et al. Redox-sensitive folate-conjugated polymeric nanoparticles for combined chemotherapy and photothermal therapy against breast cancer. J Biomed Nanotechnol. 2018;14:2018–30.
Hussaarts L, Muhlebach S, Shah VP, et al. Equivalence of complex drug products: advances in and challenges for current regulatory frameworks. Ann N Y Acad Sci. 2017;1407:39–49.
Jain P, Pawar RS, Pandey RS, et al. In-vitro in-vivo correlation (IVIVC) in nanomedicine: is protein corona the missing link? Biotechnol Adv. 2017;35:889–904.
Kim MW, Jeong HY, Kang SJ, et al. Cancer-targeted nucleic acid delivery and quantum dot imaging using EGF receptor aptamer-conjugated lipid nanoparticles. Sci Rep. 2017;7:94–9.
Ko AH, Tempero MA, Shan YS, et al. A multinational phase 2 study of nanoliposomal irinotecan sucrosofate (PEP02, MM-398) for patients with gemcitabine-refractory metastatic pancreatic cancer. Br J Cancer. 2013;109:920–5.
Kue CS, Kamkaew A, Burgess K, et al. Small molecules for active targeting in cancer. Med Res Rev. 2016;36:494–575.
LaVan DA, McGuire T, Langer R. Small-scale systems for in vivo drug delivery. Nat Biotechnol. 2003;21:1184–91.
Mahmood R, Malik A, Waquar S, et al. New challenges in the use of nanomedicine in cancer therapy. Bioengineered. 2022;13(1):759–73.
Marshall S, Panrak Y, Makchuchit N, et al. Anti-EpCAM and anti-EGFR targeted theranostic functionalized I-131 labeled poly(lactic-co-glycolic acid) (PLGA) radio-nanotherapeutic for cancer treatment. J Med Imaging Radiat Sci. 2022;53:S33–8.
Mattheolabakis G, Rigas B, Constantinides PP. Nanodelivery strategies in cancer chemotherapy: biological rationale and pharmaceutical perspectives. Nanomedicine. 2012;7:1577–90.
Moghimi SM, Hunter A, Murray J. Long-circulating and target specific nanoparticles: theory to practice. Pharmacol Rev. 2001;53:283–318.
Mosafer J, Abnous K, Tafaghodi M, et al. In vitro and in vivo evaluation of anti-nucleolin-targeted magnetic PLGA nanoparticles loaded with doxorubicin as a theranostic agent for enhanced targeted cancer imaging and therapy. Eur J Pharm Biopharm. 2017;113:60–74.
Muhlebach S, Borchard G, Yildiz S. Regulatory challenges and approaches to characterize nanomedicines and their follow-on similars. Nanomedicine. 2015;10:659–74.
Naidu MUR, Gogula VR, Pingali UR, Iyyapu, et al. Chemotherapy-induced and/or radiation therapy induced oral mucositis-complicating the treatment of cancer. Neoplasia. 2004;6(5):423–31.
Navya PN, Daima HK. Rational engineering of physicochemical properties of nanomaterials for biomedical applications with nanotoxicological perspectives. Nano Converg. 2016;3(1):16–22.

Navya PN, Kaphle A, Daima HK, et al. Nanomedicine in sensing, delivery, imaging and tissue engineering: advances, opportunities and challenges. Nanoscience. 2019;5:30–56.
Nguyen PV, Herve-Aubert K, Chourpa I, et al. Active targeting strategy in nanomedicines using anti-EGFR ligands - a promising approach for cancer therapy and diagnosis. Int J Pharm. 2021;609:121–34.
Palange AL, Palomba R, Rizzuti IF, et al. Deformable discoidal polymeric nanoconstructs for the precise delivery of therapeutic and imaging agents. Mol Ther. 2017;25:1514–21.
Palma MD, Venneri MA, Roca C, et al. Targeting exogenous genes to tumor angiogenesis by transplantation of genetically modified hematopoietic stem cells. Nat Med. 2003;9(6):789–95.
Pandey A, Nikam AN, Padya BS, et al. Surface architectured black phosphorous nanoconstructs based smart and versatile platform for cancer theranostics. Coord Chem Rev. 2021;435:21–8.
Park K. The drug delivery field at the inflection point: time to fight its way out of the egg. J Control Release. 2017;267:2–14.
Patel NR, Piroyan A, Ganta S, et al. In vitro and in vivo evaluation of a novel folate-targeted theranostic nanoemulsion of docetaxel for imaging and improved anticancer activity against ovarian cancers. Cancer Biol Ther. 2018;19:554–64.
Peer D, Karp JM, Hong S, et al. Nanocarriers as an emerging platform for cancer therapy. Nat Nanotechnol. 2007;2:751–60.
Perry JL, Reuter KG, Luft JC, et al. Mediating passive tumor accumulation through particle size, tumor type, and location. Nano Lett. 2017;17:2879–86.
Pino P, Pelaz B, Zhang Q, et al. Protein corona formation around nanoparticles—from the past to the future. Mater Horiz. 2014;1:301–13.
Poudel K, Banstola A, Tran TH, et al. Hyaluronic acid wreathed, trio-stimuli receptive and on-demand triggerable nanoconstruct for anchored combinatorial cancer therapy. Carbohydr Polym. 2020;249:116–21.
Prasad G. Biomedical applications of nanoparticles. In: Safety of nanoparticles: nanostructure science and technology. Cham: Springer; 2008. p. 89–109.
Puri A, Loomis K, Smith B, et al. Lipid-based nanoparticles as pharmaceutical drug carriers: from concepts to clinic. Crit Rev Ther Drug Carrier Syst. 2009;26:523–80.
Rezvantalab S, Drude NI, Moraveji MK, et al. PLGA-based nanoparticles in cancer treatment. Front Pharmacol. 2018;9:1260–7.
Rishi P, Jayachandra Babu R, Srinath P. Nanomedicine scale-up technologies: feasibilities and challenges. AAPS PharmSci Tech. 2014;15(6):1527–34.
Rotello VM. Sniffing out cancer using "chemical nose" sensors. Cell Cycle. 2009;8(22):3615–6.
Ryu JH, Koo H, Sun IC, et al. Tumor targeting multi-functional nanoparticles for theragnosis: new paradigm for cancer therapy. Adv Drug Deliv Rev. 2012;64:1447–58.
Sainz V, Conniot J, Matos AI, et al. Regulatory aspects on nanomedicines. Biochem Biophys Res Commun. 2015;468:504–10.
Saraf S. Process optimization for the production of nanoparticles for drug delivery applications. Expert Opin Drug Deliv. 2009;6:187–96.
Saw WS, Anasamy T, Foo YY, et al. Delivery of nanoconstructs in cancer therapy: challenges and therapeutic opportunities. Adv Ther. 2021;4:200–6.
Schellekens H, Stegemann S, Weinstein V, et al. How to regulate nonbiological complex drugs (NBCD) and their follow-on versions: points to consider. AAPSJ. 2014;16:15–21.
Schroeder A, Heller DA, Winslow MM, et al. Treating metastatic cancer with nanotechnology. Nat Rev Cancer. 2012;12:39–50.
Sharma N, Das Kurmi B, Singh D, et al. Nanoparticles toxicity: an overview of its mechanism and plausible mitigation strategies. J Drug Target. 2024;32(5):457–69.
Srinivasarao M, Low PS. Ligand-targeted drug delivery. Chem Rev. 2017;117:12133–64.
Sung H, Ferlay J, Siegel RL, et al. Global cancer statistics 2020: GLOBOCAN estimates of incidence and mortality worldwide for 36 cancers in 185 countries. CA Cancer J Clin. 2021;71:209–49.
Teli MK, Mutalik S, Rajanikant GK. Nanotechnology and nanomedicine: going small means aiming big. Curr Pharm Des. 2010;16:1882–92.

Tinkle S, McNeil SE, Mühlebach S, et al. Nanomedicines: addressing the scientific and regulatory gap. Ann N Y Acad Sci. 2014;1313:35–56.
Tsuruo T, Naito M, Tomida A, et al. Molecular targeting therapy of cancer: drug resistance, apoptosis and survival signal. Cancer Sci. 2003;94:15–21.
Uzonwanne VO, Navabi A, Obayemi JD, et al. Triptorelin-functionalized PEG-coated biosynthesized gold nanoparticles: effects of receptor-ligand interactions on adhesion to triple negative breast cancer cells. Biomater Adv. 2022;136:21–8.
Villar-Alvarez E, Cambón A, Pardo A, et al. Combination of light-driven co-delivery of chemodrugs and plasmonic-induced heat for cancer therapeutics using hybrid protein nanocapsules. J Nanobiotechnol. 2019;17:106–15.
Wang M, Thanou M. Targeting nanoparticles to cancer. Pharmacol Res. 2010;62:90–9.
Wang TT, Wei QC, Zhang ZT, et al. AIE/FRET-based versatile PEG-Pep-TPE/DOX nanoparticles for cancer therapy and real-time drug release monitoring. Biomater Sci. 2020;8:118–24.
Wei QY, Xu YM, Lau ATY. Recent progress of nanocarrier-based therapy for solid malignancies. Cancer. 2020;12:2783–837.
Wicki A, Witzigmann D, Balasubramanian V, et al. Nanomedicine in cancer therapy: challenges, opportunities, and clinical applications. J Control Release. 2015;200:138–57.
Yang J, Lu W, Xiao J, et al. A positron emission tomography image-guidable unimolecular micelle nanoplatform for cancer theranostic applications. Acta Biomater. 2018;79:306–16.
Zhang L, Gu FX, Chan JM, et al. Nanoparticles in medicine: therapeutic applications and developments. Clin Pharmacol Ther. 2008;83:761–9.
Zhao Y, Chen C. Nano on reflection. Nat Nanotechnol. 2016;11:828–34.

The manufacturer's authorised representative in the EU is Springer
Nature Customer Service Centre GmbH, Europaplatz 3, 69115 Heidelberg,
Germany. If you have any concerns regarding our products, please
contact ProductSafety@springernature.com

Printed and bound by CPI Group (UK) Ltd, Croydon, CR0 4YY
01/10/2026
02225937-0018